Student's Solutions Manual

to accompany

Precalculus

Second Edition

John W. Coburn
St. Louis Community College at Florissant Valley

Written by
Rosemary M. Karr, Ph.D.
Collin County Community College

 **Higher Education**

Boston   Burr Ridge, IL   Dubuque, IA   New York   San Francisco   St. Louis
Bangkok   Bogotá   Caracas   Kuala Lumpur   Lisbon   London   Madrid   Mexico City
Milan   Montreal   New Delhi   Santiago   Seoul   Singapore   Sydney   Taipei   Toronto

 **Higher Education**

Student's Solutions Manual to accompany
PRECALCULUS, SECOND EDITION
JOHN COBURN

Published by McGraw-Hill Higher Education, an imprint of The McGraw-Hill Companies, Inc., 1221 Avenue of the Americas, New York, NY 10020. Copyright © 2010 and 2007 by The McGraw-Hill Companies, Inc. All rights reserved.

♲ This book is printed on recycled, acid-free paper containing 10% post consumer waste.

2 3 4 5 6 7 8 9 0 QPD/QPD 0 9

ISBN: 978-0-07-336087-4
MHID: 0-07-336087-2

www.mhhe.com

Table of Contents

Chapter 1: Equations and Inequalities

1.1 Technology Highlight

1. 12 pounds of premium ground beef;
   40 pounds of peanuts

1.1 Exercises

1. Identity, unknown

3. Literal, two

5. Answers will vary.

7. $4x + 3(x-2) = 18 - x$
   $4x + 3x - 6 = 18 - x$
   $7x - 6 = 18 - x$
   $8x - 6 = 18$
   $8x = 24$
   $x = 3$;
   Check:
   $4(3) + 3(3-2) = 18 - 3$
   $12 + 3(1) = 15$
   $12 + 3 = 15$
   $15 = 15$

9. $21 - (2v + 17) = -7 - 3v$
   $21 - 2v - 17 = -7 - 3v$
   $-2v + 4 = -7 - 3v$
   $v + 4 = -7$
   $v = -11$;
   Check:
   $21 - (2(-11) + 17) = -7 - 3(-11)$
   $21 - (-22 + 17) = -7 + 33$
   $21 - (-5) = 26$
   $21 + 5 = 26$
   $26 = 26$

11. $8 - (3b + 5) = -5 + 2(b+1)$
    $8 - 3b - 5 = -5 + 2b + 2$
    $3 - 3b = -3 + 2b$
    $-5b = -6$
    $b = \dfrac{6}{5}$;
    Check:
    $8 - \left(3\left(\dfrac{6}{5}\right) + 5\right) = -5 + 2\left(\dfrac{6}{5} + 1\right)$
    $8 - \left(\dfrac{18}{5} + \dfrac{25}{5}\right) = -5 + 2\left(\dfrac{6}{5} + \dfrac{5}{5}\right)$
    $8 - \left(\dfrac{43}{5}\right) = -5 + 2\left(\dfrac{11}{5}\right)$
    $\dfrac{40}{5} - \dfrac{43}{5} = \dfrac{-25}{5} + \dfrac{22}{5}$
    $\dfrac{-3}{5} = \dfrac{-3}{5}$

13. $\dfrac{1}{5}(b + 10) - 7 = \dfrac{1}{3}(b - 9)$
    $\dfrac{1}{5}b + 2 - 7 = \dfrac{1}{3}b - 3$
    $\dfrac{1}{5}b - 5 = \dfrac{1}{3}b - 3$
    $15\left(\dfrac{1}{5}b - 5\right) = 15\left(\dfrac{1}{3}b - 3\right)$
    $3b - 75 = 5b - 45$
    $-2b - 75 = -45$
    $-2b = 30$
    $b = -15$

15. $\dfrac{2}{3}(m + 6) = \dfrac{-1}{2}$
    $\dfrac{2}{3}m + 4 = \dfrac{-1}{2}$
    $6\left(\dfrac{2}{3}m + 4\right) = 6\left(\dfrac{-1}{2}\right)$
    $4m + 24 = -3$
    $4m = -27$
    $m = -\dfrac{27}{4}$

1.1 Exercises

17. $\dfrac{1}{2}x + 5 = \dfrac{1}{3}x + 7$

$6\left(\dfrac{1}{2}x + 5\right) = 6\left(\dfrac{1}{3}x + 7\right)$

$3x + 30 = 2x + 42$

$x + 30 = 42$

$x = 12$

19. $\dfrac{x+3}{5} + \dfrac{x}{3} = 7$

$15\left(\dfrac{x+3}{5} + \dfrac{x}{3}\right) = 15(7)$

$3(x+3) + 5x = 105$

$3x + 9 + 5x = 105$

$8x + 9 = 105$

$8x = 96$

$x = 12$

21. $15 = -6 - \dfrac{3p}{8}$

$21 = -\dfrac{3p}{8}$

$\left(\dfrac{-8}{3}\right)(21) = \dfrac{-8}{3}\left(-\dfrac{3p}{8}\right)$

$-56 = p$

23. $0.2(24 - 7.5a) - 6.1 = 4.1$

$4.8 - 1.5a - 6.1 = 4.1$

$-1.5a - 1.3 = 4.1$

$-1.5a = 5.4$

$a = -3.6$

25. $6.2v - (2.1v - 5) = 1.1 - 3.7v$

$6.2v - 2.1v + 5 = 1.1 - 3.7v$

$4.1v + 5 = 1.1 - 3.7v$

$7.8v + 5 = 1.1$

$7.8v = -3.9$

$v = -0.5$

27. $\dfrac{n}{2} + \dfrac{n}{5} = \dfrac{2}{3}$

$30\left(\dfrac{n}{2} + \dfrac{n}{5}\right) = 30\left(\dfrac{2}{3}\right)$

$15n + 6n = 20$

$21n = 20$

$n = \dfrac{20}{21}$

29. $3p - \dfrac{p}{4} - 5 = \dfrac{p}{6} - 2p + 6$

$12\left(3p - \dfrac{p}{4} - 5\right) = 12\left(\dfrac{p}{6} - 2p + 6\right)$

$36p - 3p - 60 = 2p - 24p + 72$

$33p - 60 = -22p + 72$

$55p - 60 = 72$

$55p = 132$

$p = \dfrac{12}{5}$

31. $-3(4z + 5) = -15z - 20 + 3z$

$-12z - 15 = -15z - 20 + 3z$

$-12z - 15 = -12z - 20$

$-15 \neq -20$

Contradiction; $\{\ \}$

33. $8 - 8(3n + 5) = -5 + 6(1 + n)$

$8 - 24n - 40 = -5 + 6 + 6n$

$-24n - 32 = 1 + 6n$

$-30n = 33$

$n = -\dfrac{11}{10}$

Conditional; $n = -\dfrac{11}{10}$

35. $-4(4x + 5) = -6 - 2(8x + 7)$

$-16x - 20 = -6 - 16x - 14$

$-16x - 20 = -20 - 16x$

$0 = 0$

Identity; $\{x \mid x \in ¡\ \}$

Chapter 1: Equations and Inequalities

37.
$$P = C + CM$$
$$P = C(1+M)$$
$$\frac{P}{1+M} = C$$
$$C = \frac{P}{1+M}$$

39.
$$C = 2\pi r$$
$$\frac{C}{2\pi} = r$$
$$r = \frac{C}{2\pi}$$

41.
$$\frac{P_1 V_1}{T_1} = \frac{P_2 V_2}{T_2}$$
$$T_1 T_2 \left(\frac{P_1 V_1}{T_1}\right) = T_1 T_2 \left(\frac{P_2 V_2}{T_2}\right)$$
$$P_1 V_1 T_2 = T_1 P_2 V_2$$
$$T_2 = \frac{T_1 P_2 V_2}{P_1 V_1}$$

43.
$$V = \frac{4}{3}\pi r^2 h$$
$$\frac{3}{4}V = \pi r^2 h$$
$$\frac{3V}{4\pi r^2} = h$$
$$h = \frac{3V}{4\pi r^2}$$

45.
$$S_n = n\left(\frac{a_1 + a_n}{2}\right)$$
$$\left(\frac{2}{a_1 + a_n}\right)\cdot S_n = \left(\frac{2}{a_1 + a_n}\right)\cdot n\left(\frac{a_1 + a_n}{2}\right)$$
$$\frac{2S_n}{a_1 + a_n} = n$$
$$n = \frac{2S_n}{a_1 + a_n}$$

47.
$$S = B + \frac{1}{2}PS$$
$$S - B = \frac{1}{2}PS$$
$$2(S-B) = PS$$
$$\frac{2(S-B)}{S} = P$$

49.
$$Ax + By = C$$
$$By = -Ax + C$$
$$y = \frac{-A}{B}x + \frac{C}{B}$$

51.
$$\frac{5}{6}x + \frac{3}{8}y = 2$$
$$\frac{3}{8}y = -\frac{5}{6}x + 2$$
$$\left(\frac{8}{3}\right)\left(\frac{3}{8}y\right) = \left(\frac{8}{3}\right)\left(-\frac{5}{6}x + 2\right)$$
$$y = -\frac{20}{9}x + \frac{16}{3}$$

53.
$$y - 3 = \frac{-4}{5}(x+10)$$
$$y - 3 = \frac{-4}{5}x - 8$$
$$y = \frac{-4}{5}x - 5$$

55.
$$3x + 2 = -19$$
$$a = 3, b = 2, c = -19$$
$$x = \frac{-19-2}{3}$$
$$x = -7$$

57.
$$-6x + 1 = 33$$
$$a = -6, b = 1, c = 33$$
$$x = \frac{33-1}{-6}$$
$$x = -\frac{16}{3}$$

1.1 Exercises

59. $7x - 13 = -27$
$a = 7, b = -13, c = -27$
$$x = \frac{-27 - (-13)}{7}$$
$x = -2$

61. $SA = 2\pi r^2 + 2\pi rh$
$1256 = 2(3.14)(8)^2 + 2(3.14)(8)h$
$1256 = 401.92 + 50.24h$
$854.08 = 50.24h$
$17 = h$
$h = 17 \text{cm}$

63. Let x represent the length of the second descent.
$2x + 198 = 1218$
$2x = 1218 - 198$
$2x = 1020$
$x = 510$
The second spelunker descended 510 feet.

65. Let L represent the length of the package.
$2(14 + 12) + L = 108$
$2(26) + L = 108$
$52 + L = 108$
$L = 56$
The package can be up to 56 inches long.

67. Let L represent the length of the Shimotsui bridge.
$364 + 2L = 6532$
$2L = 6168$
$L = 3084$
The Shimotsui bridge is 3084 feet long.

69. Let x represent the first consecutive even integer.
Let $x + 2$ represent the second consecutive even integer.
$2x + x + 2 = 146$
$3x + 2 = 146$
$3x = 144$
$x = 48$
The first integer is 48.
The second integer is 50.

71. Let x represent the first consecutive odd integer.
Let $x + 2$ represent the second consecutive odd integer.
$7x = 5(x + 2)$
$7x = 5x + 10$
$2x = 10$
$x = 5$
The first integer is 5.
The second integer is 7.

73. Let t represent the number of hours when Bruce overtakes Linda.
$D_{Linda} = D_{Bruce}$
$60(t + 0.5) = 75t$
$60t + 30 = 75t$
$30 = 15t$
$2 = t$
2 hours after 9:30am is 11:30am.

75. Let t represent the number of hours Jeff was driving in the construction zone.
$D_1 + D_2 = 72$
$30t + 60(1.5 - t) = 72$
$30t + 90 - 60t = 72$
$-30t + 90 = 72$
$-30t = -18$
$t = 0.6$ hour
$0.6(60) = 36$ minutes.

77. 2 quarts + 2 quarts = 4 quarts;
$2(1.00) + 2(0.00) = 2$
$\frac{2}{4} = 50\%$ juice
4 quart mixture of 50% orange juice

79. 8 lbs + 8 lbs = 16 lbs;
$8(2.50) + 8(1.10) = 28.8$
$\frac{\$28.80}{16} = \$1.80$ per pound
Sixteen pound mixture at a cost of $1.80 per pound

81. Let x represent the number of pounds of premium ground beef.
$3.10(x) + 2.05(8) = 2.68(x + 8)$
$3.10x + 16.4 = 2.68x + 21.44$
$0.42x + 16.4 = 21.44$
$0.42x = 5.04$
$x = 12$
12 pounds of premium ground beef

83. Let x represent the pounds of walnuts.
$0.84x + 1.20(20) = 1.04(x + 20)$
$0.84x + 24 = 1.04x + 20.8$
$-0.2x = -3.2$
$x = 16$
16 pounds of walnuts

85. Answers will vary.

Chapter 1: Equations and Inequalities

87. $P + Q + S = 40$
$P + R + U = 34$
$S + T + U = 30$
$Q + R = 26$
$Q + T = 23$
$R + T = 19$;

$$Q + R = 26$$
$$\underline{-Q - T = -23}$$
$$R - T = 3;$$

$R - T = 3$
$\underline{R + T = 19}$
$2R = 22$
$R = 11$;

$Q + R = 26$
$Q + 11 = 26$
$Q = 15$;

$Q + T = 23$
$15 + T = 23$
$T = 8$;

$P + R + U = 34$
$P + 11 + U = 34$
$P + U = 23$;

$P + Q + S = 40$
$P + 15 + S = 40$
$P + S = 25$;

$$P + U = 23$$
$$\underline{-P - S = -25}$$
$$U - S = -2;$$

$S + T + U = 30$

$S + 8 + U = 30$

$S + U = 22$;

$U - S = -2$
$\underline{S + U = 22}$
$2U = 20$
$U = 10$;

$S + U = 22$
$S + 10 = 22$
$S = 12$;

$P + Q + S = 40$
$P + 15 + 12 = 40$
$P = 13$;

$P + Q + R + S + T + U$

$= 13 + 15 + 11 + 12 + 8 + 10 = 69$

89. $-2 - 6^2 \div 4 + 8$
$= -2 - 36 \div 4 + 8$
$= -2 - 9 + 8$
$= -11 + 8$
$= -3$

91. a. $4x^2 - 9$
$= (2x + 3)(2x - 3)$

b. $x^3 - 27$
$= (x - 3)(x^2 + 3x + 9)$

1.2 Exercises

1. Set, interval

3. Intersection, union

5. Answers will vary.

7. $w \geq 45$

9. $250 < T < 450$

11. $y < 3$

13. $m \leq 5$

15. $x \neq 1$

17. $5 > x > 2$

19. $\{x | x \geq -2\}$; $[-2, \infty)$

21. $\{x | -2 \leq x \leq 1\}$; $[-2, 1]$

23. $5a - 11 \geq 2a - 5$
$3a \geq 6$
$a \geq 2$
$\{a | a \geq 2\}$, Interval notation: $a \in [2, \infty)$
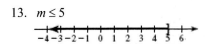

25. $2(n + 3) - 4 \leq 5n - 1$
$2n + 6 - 4 \leq 5n - 1$
$2n + 2 \leq 5n - 1$
$-3n \leq -3$
$n \geq 1$
$\{n | n \geq 1\}$, Interval notation: $n \in [1, \infty)$
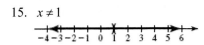

27. $\dfrac{3x}{8}+\dfrac{x}{4}<-4$

$8\left(\dfrac{3x}{8}+\dfrac{x}{4}\right)<8(-4)$

$3x+2x<-32$

$5x<-32$

$x<-\dfrac{32}{5}$

$\left\{x\Big|x<\dfrac{-32}{5}\right\},$

Interval notation: $x\in\left(-\infty,\dfrac{-32}{5}\right)$

29. $7-2(x+3)\ge 4x-6(x-3)$

$7-2x-6\ge 4x-6x+18$

$-2x+1\ge -2x+18$

$1\ge 18$ false

$\{\,\}$

31. $4(3x-5)+18<2(5x+1)+2x$

$12x-20+18<10x+2+2x$

$12x-2<12x+2$

$-2<2$ true

$\{x|x\in \mathbb{i}\,\}$

33. $-6(p-1)+2p\le -2(2p-3)$

$-6p+6+2p\le -4p+6$

$-4p+6\le -4p+6$

$6\le 6$ true

$\{p|p\in \mathbb{i}\,\}$

35. $A\cap B=\{2\}$

$A\cup B=\{-3,-2,-1,0,1,2,3,4,6,8\}$

37. $A\cap D=\{\,\}$

$A\cup D=\{-3,-2,-1,0,1,2,3,4,5,6,7\}$

39. $B\cap D=\{4,6\}$

$B\cup D=\{2,4,5,6,7,8\}$

41. $x<-2$ or $x>1$

$(-\infty,-2)\cup(1,\infty)$

43. $x<5$ and $x\ge -2$

$[-2,5)$

45. $x\ge 3$ and $x\le 1$

no solution

47. $4(x-1)\le 20$ or $x+6>9$

$4x-4\le 20$ or $x>3$

$4x\le 24$

$x\le 6$ or $x>3$

$x\in(-\infty,\infty)$

49. $-2x-7\le 3$ and $2x\le 0$

$-2x\le 10$ and $x\le 0$

$x\ge -5$ and $x\le 0$

$x\in[-5,0]$

51. $\dfrac{3}{5}x+\dfrac{1}{2}>\dfrac{3}{10}$ and $-4x>1$

$10\left(\dfrac{3}{5}x+\dfrac{1}{2}\right)>\left(\dfrac{3}{10}\right)10$ and $-4x>1$

$6x+5>3$ and $x<-\dfrac{1}{4}$

$6x>-2$ and $x<-\dfrac{1}{4}$

$x>-\dfrac{1}{3}$ and $x<-\dfrac{1}{4}$

$x\in\left(\dfrac{-1}{3},\dfrac{-1}{4}\right)$

53. $\dfrac{3x}{8}+\dfrac{x}{4}<-3 \qquad \text{or } x+1>-5$

$8\left(\dfrac{3x}{8}+\dfrac{x}{4}\right)<8(-3) \ \text{or } x>-6$

$3x+2x<-24 \qquad \text{or } x>-6$

$5x<-24 \qquad\qquad \text{or } x>-6$

$x<-\dfrac{24}{5} \qquad\qquad \text{or } x>-6$

$x\in(-\infty,\infty)$

55. $-3\le 2x+5<7$

$-8\le 2x<2$

$-4\le x<1$

$x\in[-4,1)$

57. $-0.5\le 0.3-x\le 1.7$

$-0.8\le -x\le 1.4$

$0.8\ge x\ge -1.4$

$x\in[-1.4,0.8]$

59. $-7<-\dfrac{3}{4}x-1\le 11$

$-6<-\dfrac{3}{4}x\le 12$

$\left(-\dfrac{4}{3}\right)(-6)>\left(-\dfrac{4}{3}\right)\left(-\dfrac{3}{4}x\right)\ge\left(-\dfrac{4}{3}\right)12$

$8>x\ge -16$

$x\in[-16,8)$

61. $\dfrac{12}{m}$

$m\ne 0$

$m\in(-\infty,0)\cup(0,\infty)$

63. $\dfrac{5}{y+7}$

$y+7\ne 0$

$y\ne -7$

$y\in(-\infty,-7)\cup(-7,\infty)$

65. $\dfrac{a+5}{6a-3}$

$6a-3\ne 0$

$6a\ne 3$

$a\ne\dfrac{1}{2}$

$a\in\left(-\infty,\dfrac{1}{2}\right)\cup\left(\dfrac{1}{2},\infty\right)$

67. $\dfrac{15}{3x-12}$

$3x-12\ne 0$

$3x\ne 12$

$x\ne 4$

$x\in(-\infty,4)\cup(4,\infty)$

69. $\sqrt{x-2}$

$x-2\ge 0$

$x\ge 2$

$x\in[2,\infty)$

71. $\sqrt{3n-12}$

$3n-12\ge 0$

$3n\ge 12$

$n\ge 4$

$n\in[4,\infty)$

73. $\sqrt{b-\dfrac{4}{3}}$

$b-\dfrac{4}{3}\ge 0$

$b\ge\dfrac{4}{3}$

$b\in\left[\dfrac{4}{3},\infty\right)$

75. $\sqrt{8-4y}$

$8-4y\ge 0$

$-4y\ge -8$

$y\le 2$

$y\in(-\infty,2]$

1.2 Exercises

77. a) $B = \dfrac{704W}{H^2}$

$$BH^2 = 704W$$

$$\dfrac{BH^2}{704} = W$$

$$W = \dfrac{BH^2}{704}$$

b) $W < \dfrac{BH^2}{704}$

$$W < \dfrac{(27)(68)^2}{704}$$

$$W < 177.34$$

Weight could be 177.34 pounds or less.

79. $\dfrac{82+76+65+71+x}{5} \ge 75$

$$82+76+65+71+x \ge 375$$

$$294+x \ge 375$$

$$x \ge 81$$

81. $\dfrac{1125+850+625+400+b}{5} \ge 1000$

$$1125+850+625+400+b \ge 5000$$

$$3000+b \ge 5000$$

$$b \ge \$2000$$

83. $0 < 20W < 150$

$$0 < W < 7.5m$$

85. $45 < \dfrac{9}{5}C+32 < 85$

$$13 < \dfrac{9}{5}C < 53$$

$$7.2° < C < 29.4°$$

87. $20+4.50h < 11+6.00h$

$$20-1.50h < 11$$

$$-1.50h < -9$$

$$h > 6$$

89. Answers may vary.

91. $<$

93. $<$

95. $<$

97. $>$

99. $2n-8$

101. $2\left(\dfrac{5}{9}x-1\right)-\left(\dfrac{1}{6}x+3\right)$

$$= \dfrac{10}{9}x-2-\dfrac{1}{6}x-3$$

$$= \dfrac{20}{18}x-2-\dfrac{3}{18}x-3$$

$$= \dfrac{17}{18}x-5$$

1.3 Technology Highlight

1. Algebraic verification:

$$3|x+1|-2 \ge 7$$

$$3|x+1| \ge 9$$

$$|x+1| \ge 3$$

$$x+1 \ge 3 \text{ or } x+1 \le -3$$

$$x \ge 2 \text{ or } x \le -4$$

$$x \in (-\infty, -4] \cup [2, \infty)$$

3. Algebraic verification:

$$-1 \le 4|x-3|-1$$

$$0 \le 4|x-3|$$

$$0 \le |x-3|$$

$$|x-3| \ge 0$$

Absolute value is always greater than or equal to zero.

$$x \in ¡$$

Chapter 1: Equations and Inequalities

1.3 Exercises

1. Reverse

3. -7; 7

5. No solution; answers will vary.

7. $2|m-1|-7=3$
$2|m-1|=10$
$|m-1|=5$
$m-1=5$ or $m-1=-5$
$m=6$ or $m=-4$
$\{-4,6\}$

9. $-3|x+5|+6=-15$
$-3|x+5|=-21$
$|x+5|=7$
$x+5=7$ or $x+5=-7$
$x=2$ or $x=-12$
$\{-12,2\}$

11. $2|4v+5|-6.5=10.3$
$2|4v+5|=16.8$
$|4v+5|=8.4$
$4v+5=8.4$ or $4v+5=-8.4$
$4v=3.4$ or $4v=-13.4$
$v=0.85$ or $v=-3.35$
$\{-3.35,0.85\}$

13. $-|7p-3|+6=-5$
$-|7p-3|=-11$
$|7p-3|=11$
$7p-3=11$ or $7p-3=-11$
$7p=14$ or $7p=-8$
$p=2$ or $p=\dfrac{-8}{7}$
$\left\{\dfrac{-8}{7},2\right\}$

15. $-2|b|-3=-4$
$-2|b|=-1$
$|b|=\dfrac{1}{2}$
$b=\dfrac{1}{2}$ or $b=-\dfrac{1}{2}$
$\left\{-\dfrac{1}{2},\dfrac{1}{2}\right\}$

17. $-2|3x|-17=-5$
$-2|3x|=12$
$|3x|=-6$
$\{\ \}$

19. $-3\left|\dfrac{w}{2}+4\right|-1=-4$
$-3\left|\dfrac{w}{2}+4\right|=-3$
$\left|\dfrac{w}{2}+4\right|=1$
$\dfrac{w}{2}+4=1$ or $\dfrac{w}{2}+4=-1$
$\dfrac{w}{2}=-3$ or $\dfrac{w}{2}=-5$
$w=-6$ or $w=-10$
$\{-6,-10\}$

21. $8.7|p-7.5|-26.6=8.2$
$8.7|p-7.5|=34.8$
$|p-7.5|=4$
$p-7.5=4$ or $p-7.5=-4$
$p=11.5$ or $p=3.5$
$\{3.5,11.5\}$

23. $8.7|-2.5x|-26.6=8.2$
$8.7|-2.5x|=34.8$
$|-2.5x|=4$
$-2.5x=4$ or $-2.5x=-4$
$x=-1.6$ or $x=1.6$
$\{-1.6,1.6\}$

25. $|x-2| \le 7$

$-7 \le x-2 \le 7$

$-5 \le x \le 9$

$x \in [-5, 9]$

27. $-3|m| - 2 > 4$

$-3|m| > 6$

$|m| < -2$

Absolute value is never less than a negative number.

$\varnothing$

29. $\dfrac{|5v+1|}{4} + 8 < 9$

$\dfrac{|5v+1|}{4} < 1$

$|5v+1| < 4$

$-4 < 5v+1 < 4$

$-5 < 5v < 3$

$-1 < v < \dfrac{3}{5}$

$v \in \left(-1, \dfrac{3}{5}\right)$

31. $3|p+4| + 5 < 8$

$3|p+4| < 3$

$|p+4| < 1$

$-1 < p+4 < 1$

$-5 < p < -3$

$p \in (-5, -3)$

33. $|3b-11| + 6 \le 9$

$|3b-11| < 3$

$-3 \le 3b-11 \le 3$

$8 \le 3b \le 14$

$\dfrac{8}{3} \le b \le \dfrac{14}{3}$

$b \in \left[\dfrac{8}{3}, \dfrac{14}{3}\right]$

35. $|4-3z| + 12 < 7$

$|4-3z| < -5$

No solution

37. $\left|\dfrac{4x+5}{3} - \dfrac{1}{2}\right| \le \dfrac{7}{6}$

$-\dfrac{7}{6} \le \dfrac{4x+5}{3} - \dfrac{1}{2} \le \dfrac{7}{6}$

$6\left(-\dfrac{7}{6}\right) \le 6\left(\dfrac{4x+5}{3} - \dfrac{1}{2}\right) \le 6\left(\dfrac{7}{6}\right)$

$-7 \le 8x+10-3 \le 7$

$-7 \le 8x+7 \le 7$

$-14 \le 8x \le 0$

$-\dfrac{14}{8} \le x \le 0$

$-\dfrac{7}{4} \le x \le 0$

$\left[-\dfrac{7}{4}, 0\right]$

39. $|n+3| > 7$

$n+3 > 7$ or $n+3 < -7$

$n > 4$ or $n < -10$

$n \in (-\infty, -10) \cup (4, \infty)$

41. $-2|w| - 5 \le -11$

$-2|w| \le -6$

$|w| \ge 3$

$w \ge 3$ or $w \le -3$

$w \in (-\infty, -3] \cup [3, \infty)$

43. $\dfrac{|q|}{2} - \dfrac{5}{6} \ge \dfrac{1}{3}$

$6\left(\dfrac{|q|}{2} - \dfrac{5}{6}\right) \ge 6\left(\dfrac{1}{3}\right)$

$3|q| - 5 \ge 2$

$3|q| \ge 7$

$|q| \ge \dfrac{7}{3}$

$q \ge \dfrac{7}{3}$ or $q \le -\dfrac{7}{3}$

$q \in \left(-\infty, -\dfrac{7}{3}\right] \cup \left[\dfrac{7}{3}, \infty\right)$

45. $3|5-7d|+9 \geq 15$

    $3|5-7d| \geq 6$

    $|5-7d| \geq 2$

    $5-7x \geq 2$ or $5-7x \leq -2$

    $-7d \geq -3$

    $\qquad -7d \leq -7$

    $d \leq \dfrac{3}{7}$ $\qquad d \geq 1$

    $d \in \left(-\infty, \dfrac{3}{7}\right] \cup [1, \infty)$

47. $|4z-9|+6 \geq 4$

    $|4z-9| \geq -2$

    $z \in (-\infty, \infty)$

49. $4|5-2h|-9 > 11$

    $4|5-2h| > 20$

    $|5-2h| > 5$

    $5-2h > 5$ or $5-2h < -5$

    $-2h > 0$ $\qquad -2h < -10$

    $h < 0$ $\qquad\quad h > 5$

    $h \in (-\infty, 0) \cup (5, \infty)$

51. $-3.9|4q-5|+8.7 \leq -22.5$

    $-3.9|4q-5| \leq -31.2$

    $|4q-5| \geq 8$

    $4q-5 \geq 8$ or $4q-5 \leq -8$

    $4q \geq 13$ $\qquad 4q \leq -3$

    $q \geq \dfrac{13}{4}$ $\qquad q \leq -\dfrac{3}{4}$

    $q \geq 3.25$ $\qquad q \leq -0.75$

    $q \in (-\infty, -0.75] \cup [3.25, \infty)$

53. $2 < \left| -3m + \dfrac{4}{5} \right| - \dfrac{1}{5}$

    $\dfrac{11}{5} < \left| -3m + \dfrac{4}{5} \right|$

    $\left| -3m + \dfrac{4}{5} \right| > \dfrac{11}{5}$

    $-3m + \dfrac{4}{5} > \dfrac{11}{5}$ or $-3m + \dfrac{4}{5} < -\dfrac{11}{5}$

    $-3m > \dfrac{7}{5}$ $\qquad -3m < -3$

    $m < -\dfrac{7}{15}$ $\qquad m > 1$

    $m \in \left(-\infty, \dfrac{-7}{15}\right) \cup (1, \infty)$

55. $|d-x| \leq L$

    4 ft = 48 in.

    $|d-48| \leq 3$

    $d-48 \leq 3$ and $d-48 \geq -3$

    $d \leq 51$ and $\qquad d \geq 45$

    $45 \leq d \leq 51$ in.

57. $|h-35,050| \leq 2,550$

    $h-35050 \leq 2550$ and $h-35050 \geq -2550$

    $h \leq 37600$ $\qquad h \geq 32500$

    $32,500 \leq h \leq 37,600$; yes, if between 32,500 feet and 37,600 feet, inclusive.

59. $|d-394|-20 > 164$

    $|d-394| > 184$

    $d-394 > 184$ or $d-394 < -184$

    $d > 578$ or $\qquad d < 210$;

    $d < 210$ or $d > 578$

    Less than 210 feet to go over the net, more than 578 feet to go under the net.

61. (a) $|s-37.58| \leq 3.35$

    (b) $s-37.58 \leq 3.35$ and $s-37.58 \geq -3.35$

    $s \leq 40.93$ and $s \geq 34.23$

    $34.23 \leq s \leq 40.93$

    $[34.23, 40.93]$

63. (a) $|s - 125| \leq 23$

    (b) $-23 \leq s - 125 \leq 23$

        $102 \leq s \leq 148$

        $[102, 148]$

65. a. $|d - 42.7| < 0.03$

    b. $|d - 73.78| < 1.01$

    c. $|d - 57.150| < 0.127$

    d. $|d - 2171.05| < 12.05$

    e. golf: $t = \dfrac{2(0.03)}{42.7} \approx 0.0014$;

       baseball: $t = \dfrac{2(1.01)}{73.78} \approx 0.0274$;

       billiard: $t = \dfrac{2(0.127)}{57.150} \approx 0.0044$;

       bowling: $t = \dfrac{2(12.05)}{2171.05} \approx 0.0111$;

       Golf balls.

67. a. $x = 4$

    b. $\left[\dfrac{4}{3}, 4\right]$

    c. $x = 0$

    d. $\left(-\infty, \dfrac{3}{5}\right]$

    e. $\{\ \}$

69. $18x^3 + 21x^2 - 60x$

    $3x(6x^2 + 7x - 20)$

    $3x(2x + 5)(3x - 4)$

71. $\dfrac{-1}{3 + \sqrt{3}}$

    $\dfrac{-1}{3 + \sqrt{3}} \cdot \dfrac{3 - \sqrt{3}}{3 - \sqrt{3}}$

    $\dfrac{-3 + \sqrt{3}}{9 - 3\sqrt{3} + 3\sqrt{3} - 3}$

    $\dfrac{-3 + \sqrt{3}}{6} \approx -0.21$

Chapter 1 Mid-Chapter Check

1. a. $\dfrac{r}{3} + 5 = 2$

    $\dfrac{r}{3} = -3$

    $r = -9$

  b. $5(2x - 1) + 4 = 9x - 7$

    $10x - 5 + 4 = 9x - 7$

    $10x - 1 = 9x - 7$

    $x = -6$

  c. $m - 2(m + 3) = 1 - (m + 7)$

    $m - 2m - 6 = 1 - m - 7$

    $-m - 6 = -m - 6$

    $0 = 0$

    Identity; $x \in ¡$

  d. $\dfrac{1}{5}y + 3 = \dfrac{3}{2}y - 2$

    $10\left(\dfrac{1}{5}y + 3\right) = 10\left(\dfrac{3}{2}y - 2\right)$

    $2y + 30 = 15y - 20$

    $-13y + 30 = -20$

    $-13y = -50$

    $y = \dfrac{50}{13}$

  e. $\dfrac{1}{2}(5j - 2) = \dfrac{3}{2}(j - 4) + j$

    $\dfrac{5}{2}j - 1 = \dfrac{3}{2}j - 6 + j$

    $\dfrac{5}{2}j - 1 = \dfrac{5}{2}j - 6$

    $-1 = -6$

    Contradiction; $\{\ \}$

  f. $0.6(x - 3) + 0.3 = 1.8$

    $0.6x - 1.8 + 0.3 = 1.8$

    $0.6x - 1.5 = 1.8$

    $0.6x = 3.3$

    $x = 5.5$

3. $S = 2\pi x^2 + \pi x^2 y$

   $S = x^2(2\pi + \pi y)$

   $\dfrac{S}{2\pi + \pi y} = x^2$

   $\dfrac{S}{\pi(2 + y)} = x^2$

   $\sqrt{\dfrac{S}{\pi(2 + y)}} = x$

5. a. $\dfrac{3x+1}{2x-5}$

   $2x - 5 \neq 0$

   $2x \neq 5$

   $x \neq \dfrac{5}{2}$

   $x \in \left(-\infty, \dfrac{5}{2}\right) \cup \left(\dfrac{5}{2}, 0\right)$

   b. $\sqrt{17 - 6x}$

   $17 - 6x \geq 0$

   $-6x \geq -17$

   $x \leq \dfrac{17}{6}$

   $x \in \left(-\infty, \dfrac{17}{6}\right]$

7. a. $3|q+4| - 2 < 10$

   $3|q+4| < 12$

   $|q+4| < 4$

   $-4 < q + 4 < 4$

   $-8 < q < 0$

   $x \in (-8, 0)$

b. $\left|\dfrac{x}{3} + 2\right| + 5 \leq 5$

   $\left|\dfrac{x}{3} + 2\right| \leq 0$

   $\dfrac{x}{3} + 2 = 0$

   $\dfrac{x}{3} = -2$

   $x = -6$

   $\{-6\}$

9. $\dfrac{x}{30} + \dfrac{115 - x}{50} = 2 + \dfrac{50}{60}$

   $\dfrac{x}{30} + \dfrac{115 - x}{50} = \dfrac{17}{6}$

   $150\left(\dfrac{x}{30} + \dfrac{115 - x}{50}\right) = 150\left(\dfrac{17}{6}\right)$

   $5x + 3(115 - x) = 425$

   $5x + 345 - 3x = 425$

   $2x = 80$

   $x = 40$ miles;

   $\dfrac{40}{30} = 1\dfrac{1}{3}$ hours or 1 hour 20 minutes

Reinforcing Basic Concepts

1. $|x - 2| = 5$

   $x - 2 = -5$ or $x - 2 = 5$

   $x = -3$ or $x = 7$

3. $|2x - 3| \geq 5$

   $2x - 3 \geq 5$ or $2x - 3 \leq -5$

   $2x \geq 8$ or $2x \leq -2$

   $x \geq 4$ or $x \leq -1$

   $x \in (-\infty, -1] \cup [4, \infty)$

1.4 Exercises

1. $3 - 2i$

3. $2, 3\sqrt{2}$

5. b is correct.

7. a. $\sqrt{-16} = 4i$

   b. $\sqrt{-49} = 7i$

   c. $\sqrt{27} = \sqrt{9(3)} = 3\sqrt{3}$

   d. $\sqrt{72} = \sqrt{36(2)} = 6\sqrt{2}$

9. a. $-\sqrt{-18} = -\sqrt{-1(9)(2)} = -3i\sqrt{2}$

   b. $-\sqrt{-50} = -\sqrt{-1(25)(2)} = -5i\sqrt{2}$

   c. $3\sqrt{-25} = 3(5i) = 15i$

   d. $2\sqrt{-9} = 2(3i) = 6i$

11. a. $\sqrt{-19} = i\sqrt{19}$

   b. $\sqrt{-31} = i\sqrt{31}$

   c. $\sqrt{\dfrac{-12}{25}} = \dfrac{\sqrt{-1(4)3}}{\sqrt{25}} = \dfrac{2\sqrt{3}}{5}i$

   d. $\sqrt{\dfrac{-9}{32}} = \dfrac{\sqrt{-9}}{\sqrt{32}} \cdot \dfrac{\sqrt{2}}{\sqrt{2}} = \dfrac{3\sqrt{2}}{\sqrt{64}}i = \dfrac{3\sqrt{2}}{8}i$

13. a. $\dfrac{2 + \sqrt{-4}}{2} = \dfrac{2 + 2i}{2} = 1 + i$

   $a = 1, b = 1$

   b. $\dfrac{6 + \sqrt{-27}}{3} = \dfrac{6 + 3i\sqrt{3}}{3} = 2 + \sqrt{3}\,i$

   $a = 2, b = \sqrt{3}$

15. a. $\dfrac{8 + \sqrt{-16}}{2} = \dfrac{8 + 4i}{2} = 4 + 2i$

   $a = 4, b = 2$

   b. $\dfrac{10 - \sqrt{-50}}{5} = \dfrac{10 - 5i\sqrt{2}}{5} = 2 - \sqrt{2}\,i$

   $a = 2, b = -\sqrt{2}$

17. a. $5 = 5 + 0i$

   $a = 5, b = 0$

   b. $3i = 0 + 3i$

   $a = 0, b = 3$

19. a. $2\sqrt{-81} = 2(9i) = 0 + 18i = 18i$

   $a = 0, b = 18$

   b. $\dfrac{\sqrt{-32}}{8} = \dfrac{4\sqrt{2}}{8}i = 0 + \dfrac{\sqrt{2}}{2}i = \dfrac{\sqrt{2}}{2}i$

   $a = 0, b = \dfrac{\sqrt{2}}{2}$

21. a. $4 + \sqrt{-50} = 4 + 5\sqrt{2}\,i$

   $a = 4, b = 5\sqrt{2}$

   b. $-5 + \sqrt{-27} = -5 + 3\sqrt{3}\,i$

   $a = -5, b = 3\sqrt{3}$

23. a. $\dfrac{14 + \sqrt{-98}}{8} = \dfrac{14 + 7i\sqrt{2}}{8} = \dfrac{7}{4} + \dfrac{7\sqrt{2}}{8}i$

   $a = \dfrac{7}{4}, b = \dfrac{7\sqrt{2}}{8}$

   b. $\dfrac{5 + \sqrt{-250}}{10} = \dfrac{5 + 5i\sqrt{10}}{10} = \dfrac{1}{2} + \dfrac{\sqrt{10}}{2}i$

   $a = \dfrac{1}{2}, b = \dfrac{\sqrt{10}}{2}$

25. a. $\left(12 - \sqrt{-4}\right) + \left(7 + \sqrt{-9}\right)$

   $= (12 - 2i) + (7 + 3i)$

   $= 19 + i$

   b. $\left(3 + \sqrt{-25}\right) + \left(-1 - \sqrt{-81}\right)$

   $= (3 + 5i) + (-1 - 9i)$

   $= 2 - 4i$

   c. $\left(11 + \sqrt{-108}\right) - \left(2 - \sqrt{-48}\right)$

   $= 11 + \sqrt{-108} - 2 + \sqrt{-48}$

   $= 11 + \sqrt{-1(36)(3)} - 2 + \sqrt{-1(16)(3)}$

   $= 9 + 6\sqrt{3}\,i + 4\sqrt{3}\,i$

   $= 9 + 10\sqrt{3}\,i$

27. a. $(2 + 3i) + (-5 - i)$

   $= 2 + 3i - 5 - i$

   $= -3 + 2i$

   b. $(5 - 2i) + (3 + 2i)$

   $= 5 - 2i + 3 + 2i$

   $= 8$

   c. $(6 - 5i) - (4 + 3i)$

   $= 6 - 5i - 4 - 3i$

   $= 2 - 8i$

29. a. $(3.7 + 6.1i) - (1 + 5.9i)$
$= 3.7 + 6.1i - 1 - 5.9i$
$= 2.7 + 0.2i$

 b. $\left(8 + \dfrac{3}{4}i\right) - \left(-7 + \dfrac{2}{3}i\right)$

 $= 8 + \dfrac{3}{4}i + 7 - \dfrac{2}{3}i$

 $= 15 + \dfrac{1}{12}i$

 c. $\left(-6 - \dfrac{5}{8}i\right) + \left(4 + \dfrac{1}{2}i\right)$

 $= -6 - \dfrac{5}{8}i + 4 + \dfrac{1}{2}i$

 $= -2 - \dfrac{1}{8}i$

31. a. $5i \cdot (-3i)$

 $= -15i^2$
 $= 15$

 b. $4i \cdot (-4i)$

 $= -16i^2$
 $= 16$

33. a. $-7i(5 - 3i)$

 $= -35i + 21i^2$
 $= -21 - 35i$

 b. $6i(-3 + 7i)$

 $= -18i + 42i^2$
 $= -42 - 18i$

35. a. $(-3 + 2i)(2 + 3i)$

 $= -6 - 9i + 4i + 6i^2$
 $= -12 - 5i$

 b. $(3 + 2i)(1 + i)$

 $= 3 + 3i + 2i + 2i^2$
 $= 1 + 5i$

37. a. conjugate $4 - 5i$

 $= (4 + 5i)(4 - 5i)$
 $= 16 - 20i + 20i - 25i^2$
 $= 16 + 25$
 $= 41$

 b. conjugate $3 + i\sqrt{2}$

 $= (3 + i\sqrt{2})(3 - i\sqrt{2})$
 $= 9 - 3\sqrt{2}i + 3\sqrt{2}i - 2i^2$
 $= 9 + 2$
 $= 11$

39. a. conjugate $-7i$

 $(7i)(-7i)$
 $= -49i^2$
 $= 49$

 b. conjugate $\dfrac{1}{2} + \dfrac{2}{3}i$

 $\left(\dfrac{1}{2} + \dfrac{2}{3}i\right)\left(\dfrac{1}{2} - \dfrac{2}{3}i\right)$

 $= \dfrac{1}{4} - \dfrac{1}{3}i + \dfrac{1}{3}i - \dfrac{4}{9}i^2$

 $= \dfrac{25}{36}$

41. a. $(4 - 5i)(4 + 5i)$

 $= 16 + 20i - 20i - 25i^2$
 $= 41$

 b. $(7 - 5i)(7 + 5i)$

 $= 49 + 35i - 35i - 25i^2$
 $= 74$

43. a. $(3 - i\sqrt{2})(3 + i\sqrt{2})$

 $= 9 + 3i\sqrt{2} - 3i\sqrt{2} - 2i^2$
 $= 11$

 b. $\left(\dfrac{1}{6} + \dfrac{2}{3}i\right)\left(\dfrac{1}{6} - \dfrac{2}{3}i\right)$

 $= \dfrac{1}{36} - \dfrac{1}{9}i + \dfrac{1}{9}i - \dfrac{4}{9}i^2$

 $= \dfrac{17}{36}$

45. a. $(2 + 3i)^2$

 $= (2 + 3i)(2 + 3i)$
 $= 4 + 6i + 6i + 9i^2$
 $= -5 + 12i$

 b. $(3 - 4i)^2$

 $= (3 - 4i)(3 - 4i)$
 $= 9 - 12i - 12i + 16i^2$
 $= -7 - 24i$

47. a. $(-2 + 5i)^2$

 $= (-2 + 5i)(-2 + 5i)$
 $= 4 - 10i - 10i + 25i^2$
 $= -21 - 20i$

 b. $(3 + i\sqrt{2})^2$

 $= (3 + i\sqrt{2})(3 + i\sqrt{2})$
 $= 9 + 3i\sqrt{2} + 3i\sqrt{2} + 2i^2$
 $= 7 + 6\sqrt{2}\ i$

1.4 Exercises

49. $x^2 + 36 = 0, x = -6;$

$$(-6)^2 + 36 = 0$$
$$36 + 36 = 0$$
$$72 \neq 0 \text{ no}$$

51. $x^2 + 49 = 0, x = -7i;$

$$(-7i)^2 + 49 = 0$$
$$49i^2 + 49 = 0$$
$$-49 + 49 = 0$$
$$0 = 0 \text{ yes}$$

53. $(x-3)^2 = -9, x = 3 - 3i;$

$$(3 - 3i - 3)^2 = -9$$
$$(-3i)^2 = -9$$
$$9i^2 = -9$$
$$-9 = -9 \text{ yes}$$

55. $x^2 - 2x + 5 = 0, x = 1 - 2i;$

$$(1 - 2i)^2 - 2(1 - 2i) + 5 = 0$$
$$1 - 4i + 4i^2 - 2 + 4i + 5 = 0$$
$$4 + 4i^2 = 0$$
$$4 - 4 = 0$$
$$0 = 0 \text{ yes}$$

57. $x^2 - 4x + 9 = 0, x = 2 + i\sqrt{5};$

$$(2 + i\sqrt{5})^2 - 4(2 + i\sqrt{5}) + 9 = 0$$
$$4 + 4i\sqrt{5} + 5i^2 - 8 - 4i\sqrt{5} + 9 = 0$$
$$5 + 5i^2 = 0$$
$$5 - 5 = 0$$
$$0 = 0 \text{ yes}$$

59. $x^2 - 2x + 17 = 0, x = 1 + 4i, 1 - 4i;$

$$(1 + 4i)^2 - 2(1 + 4i) + 17 = 0$$
$$1 + 8i + 16i^2 - 2 - 8i + 17 = 0$$
$$16 + 16i^2 = 0$$
$$16 - 16 = 0$$
$$0 = 0$$
$1 + 4i$ is a solution.
$$(1 - 4i)^2 - 2(1 - 4i) + 17 = 0$$
$$1 - 8i + 16i^2 - 2 + 8i + 17 = 0$$
$$16 + 16i^2 = 0$$
$$16 - 16 = 0$$
$$0 = 0$$
$1 - 4i$ is a solution.

61. a. $i^{48} = (i^4)^{12} = (1)^{12} = 1$

b. $i^{26} = (i^4)^6 i^2 = (1)^6(-1) = -1$

c. $i^{39} = (i^4)^9 i^3 = (1)^9(-i) = -i$

d. $i^{53} = (i^4)^{13} i^1 = (1)^{13}(i) = i$

63. a. $\dfrac{-2}{\sqrt{-49}} = \dfrac{-2}{7i} \cdot \dfrac{i}{i} = \dfrac{-2i}{7i^2} = \dfrac{2}{7}i$

b. $\dfrac{4}{\sqrt{-25}} = \dfrac{4}{5i} \cdot \dfrac{i}{i} = \dfrac{4i}{5i^2} = \dfrac{-4}{5}i$

65. a. $\dfrac{7}{3 + 2i} \cdot \dfrac{3 - 2i}{3 - 2i} = \dfrac{21 - 14i}{9 - 4i^2} = \dfrac{21 - 14i}{13}$

$$= \dfrac{21}{13} - \dfrac{14}{13}i$$

b. $\dfrac{-5}{2 - 3i} \cdot \dfrac{2 + 3i}{2 + 3i} = \dfrac{-10 - 15i}{4 - 9i^2} = \dfrac{-10 - 15i}{13}$

$$= \dfrac{-10}{13} - \dfrac{15}{13}i$$

67. a. $\dfrac{3 + 4i}{4i} \cdot \dfrac{i}{i} = \dfrac{3i + 4i^2}{4i^2} = \dfrac{-4 + 3i}{-4} = 1 - \dfrac{3}{4}i$

b. $\dfrac{2 - 3i}{3i} \cdot \dfrac{i}{i} = \dfrac{2i - 3i^2}{3i^2} = \dfrac{3 + 2i}{-3} = -1 - \dfrac{2}{3}i$

69. $|a + bi| = \sqrt{a^2 + b^2}$

a. $|2 + 3i| = \sqrt{(2)^2 + (3)^2} = \sqrt{13}$

b. $|4 - 3i| = \sqrt{(4)^2 + (-3)^2} = 5$

c. $|3 + \sqrt{2}\, i| = \sqrt{(3)^2 + (\sqrt{2})^2} = \sqrt{11}$

71. $5 + \sqrt{15}\, i + 5 - \sqrt{15}\, i = 10$
$10 = 10$
verified;
$$(5 + \sqrt{15}\, i)(5 - \sqrt{15}\, i) = 40$$
$$25 - 5\sqrt{15}\, i + 5\sqrt{15}\, i - 15i^2 = 40$$
$40 = 40$
verified

73. $Z = R + iX_L - iX_C$

$Z = 7 + i(6) - i(11) = 7 - 5i$ Ω

75. $V = IZ$

$V = (3 - 2i)(5 + 5i)$

$V = 15 + 15i - 10i - 10i^2$

$V = 25 + 5i$ volts

77. $Z = \dfrac{Z_1 Z_2}{Z_1 + Z_2}$

$Z = \dfrac{(1 + 2i)(3 - 2i)}{1 + 2i + 3 - 2i}$

$Z = \dfrac{3 - 2i + 6i - 4i^2}{4}$

$Z = \dfrac{7 + 4i}{4}$

$Z = \dfrac{7}{4} + i$ Ω

79. a. $x^2 + 36$

$(x + 6i)(x - 6i)$

b. $m^2 + 3$

$(m + i\sqrt{3})(m - i\sqrt{3})$

c. $n^2 + 12$

$(n + 2i\sqrt{3})(n - 2i\sqrt{3})$

d. $4x^2 + 49$

$(2x + 7i)(2x - 7i)$

81. $i^{17}(3 - 4i) - 3i^3(1 + 2i)^2$

$i(3 - 4i) + 3i(1 + 2i)^2$

$3i - 4i^2 + 3i(1 + 4i + 4i^2)$

$3i + 4 + 3i(1 + 4i - 4)$

$3i + 4 + 3i(4i - 3)$

$3i + 4 + 12i^2 - 9i$

$-6i + 4 - 12$

$-8 - 6i$

83. a. $P = 4s, A = s^2$

b. $P = 2L + 2W, A = LW$

c. $P = a + b + c, A = \dfrac{bh}{2}$

d. $C = \pi d, A = \pi r^2$

85. John takes $\dfrac{200}{10} = 20$ seconds.

Rick takes $\dfrac{200}{9} = 22.\overline{2}$ seconds.

Even with a 2 second head start, John will finish first.

1.5 Exercises

1. Descending, 0

3. Quadratic, 1

5. GCF factoring;

$4x^2 - 5x = 0$

$x(4x - 5) = 0$

$x = 0$ or $4x - 5 = 0$

$4x = 5$

$x = \dfrac{5}{4}$

7. $2x - 15 - x^2 = 0$

$-x^2 + 2x - 15 = 0$

Quadratic; $a = -1, b = 2, c = -15$

9. $\dfrac{2}{3}x - 7 = 0$

not quadratic

11. $\dfrac{1}{4}x^2 = 6x$

$\dfrac{1}{4}x^2 - 6x = 0$

Quadratic; $a = \dfrac{1}{4}, b = -6, c = 0$

13. $2x^2 + 7 = 0$

Quadratic; $a = 2, b = 0, c = 7$

15. $-3x^2 + 9x - 5 + 2x^3 = 0$
Not quadratic

17. $(x-1)^2 + (x-1) + 4 = 9$
$x^2 - 2x + 1 + x - 1 - 5 = 0$
$x^2 - x - 5 = 0$
Quadratic; $a = 1, b = -1, c = -5$

19. $x^2 - 15 = 2x$
$x^2 - 2x - 15 = 0$
$(x-5)(x+3) = 0$
$x - 5 = 0$ or $x + 3 = 0$
$x = 5$ or $x = -3$

21. $m^2 = 8m - 16$
$m^2 - 8m + 16 = 0$
$(m-4)(m-4) = 0$
$m - 4 = 0$
$m = 4$

23. $5p^2 - 10p = 0$
$5p(p-2) = 0$
$5p = 0$ or $p - 2 = 0$
$p = 0$ or $p = 2$

25. $-14h^2 = 7h$
$-14h^2 - 7h = 0$
$-7h(2h+1) = 0$
$-7h = 0$ or $2h + 1 = 0$
$h = 0$ or $2h = -1$
$h = 0$ or $h = \dfrac{-1}{2}$

27. $a^2 - 17 = -8$
$a^2 - 9 = 0$
$(a+3)(a-3) = 0$
$a + 3 = 0$ or $a - 3 = 0$
$a = -3$ or $a = 3$

29. $g^2 + 18g + 70 = -11$
$g^2 + 18g + 81 = 0$
$(g+9)(g+9) = 0$
$g + 9 = 0$
$g = -9$

31. $m^3 + 5m^2 - 9m - 45 = 0$
$m^2(m+5) - 9(m+5) = 0$
$(m+5)(m^2 - 9) = 0$
$(m+5)(m+3)(m-3) = 0$
$m + 5 = 0$ or $m + 3 = 0$ or $m - 3 = 0$
$m = -5$ or $m = -3$ or $m = 3$

33. $(c-12)c - 15 = 30$
$c^2 - 12c - 15 = 30$
$c^2 - 12c - 45 = 0$
$(c-15)(c+3) = 0$
$c - 15 = 0$ or $c + 3 = 0$
$c = 15$ or $c = -3$

35. $9 + (r-5)r = 33$
$9 + r^2 - 5r = 33$
$r^2 - 5r - 24 = 0$
$(r-8)(r+3) = 0$
$r - 8 = 0$ or $r + 3 = 0$
$r = 8$ or $r = -3$

37. $(t+4)(t+7) = 54$
$t^2 + 11t + 28 = 54$
$t^2 + 11t - 26 = 0$
$(t+13)(t-2) = 0$
$t + 13 = 0$ or $t - 2 = 0$
$t = -13$ or $t = 2$

39. $2x^2 - 4x - 30 = 0$
$2(x^2 - 2x - 15) = 0$
$2(x-5)(x+3) = 0$
$x - 5 = 0$ or $x + 3 = 0$
$x = 5$ or $x = -3$

41. $2w^2 - 5w = 3$
$2w^2 - 5w - 3 = 0$
$(2w+1)(w-3) = 0$
$2w + 1 = 0$ or $w - 3 = 0$
$2w = -1$ or $w = 3$
$w = -\dfrac{1}{2}$ or $w = 3$

43. $m^2 = 16$
$m = \pm 4$

45. $y^2 - 28 = 0$

$y^2 = 28$
$y = \pm\sqrt{28}$
$y = \pm 2\sqrt{7} \approx \pm 5.29$

47. $p^2 + 36 = 0$

$p^2 = -36$
$p = \pm\sqrt{-36}$
No real solutions

49. $x^2 = \dfrac{21}{16}$

$x = \pm\dfrac{\sqrt{21}}{4} \approx \pm 1.15$

51. $(n-3)^2 = 36$

$n - 3 = \pm 6$
$n = 6 + 3$ or $n = -6 + 3$
$n = 9$ or $n = -3$

53. $(w+5)^2 = 3$

$w + 5 = \pm\sqrt{3}$
$w = -5 \pm \sqrt{3}$
$w \approx -3.27$ or $w \approx -6.73$

55. $(x-3)^2 + 7 = 2$

$(x-3)^2 = -5$
$x - 3 = \pm\sqrt{-5}$
No real solutions

57. $(m-2)^2 = \dfrac{18}{49}$

$m - 2 = \pm\dfrac{\sqrt{18}}{7}$
$m = 2 \pm \dfrac{3\sqrt{2}}{7}$
$m \approx 2.61$ or $m \approx 1.39$

59. $x^2 + 6x + \underline{9}$

$(x+3)^2$

61. $n^2 + 3n + \dfrac{9}{4}$

$\left(n + \dfrac{3}{2}\right)^2$

63. $p^2 + \dfrac{2}{3}p + \dfrac{1}{\underline{9}}$

$\left(p + \dfrac{1}{3}\right)^2$

65. $x^2 + 6x = -5$

$x^2 + 6x + 9 = -5 + 9$
$(x+3)^2 = 4$
$x + 3 = \pm 2$
$x = -3 \pm 2$
$x = -1$ or $x = -5$

67. $p^2 - 6p + 3 = 0$

$p^2 - 6p = -3$
$p^2 - 6p + 9 = -3 + 9$
$(p-3)^2 = 6$
$p - 3 = \pm\sqrt{6}$
$p = 3 \pm \sqrt{6}$
$p \approx 5.45$ or $p \approx 0.55$

69. $p^2 + 6p = -4$

$p^2 + 6p + 9 = -4 + 9$
$(p+3)^2 = 5$
$p + 3 = \pm\sqrt{5}$
$p = -3 \pm \sqrt{5}$
$p \approx -0.76$ or $p \approx -5.24$

71. $m^2 + 3m = 1$

$m^2 + 3m + \dfrac{9}{4} = 1 + \dfrac{9}{4}$
$\left(m + \dfrac{3}{2}\right)^2 = \dfrac{13}{4}$
$m + \dfrac{3}{2} = \pm\dfrac{\sqrt{13}}{2}$
$m = -\dfrac{3}{2} \pm \dfrac{\sqrt{13}}{2}$
$m \approx 0.30$ or $m \approx -3.30$

73. $n^2 = 5n + 5$

$n^2 - 5n = 5$

$n^2 - 5n + \dfrac{25}{4} = 5 + \dfrac{25}{4}$

$\left(n - \dfrac{5}{2}\right)^2 = \dfrac{45}{4}$

$n - \dfrac{5}{2} = \pm\dfrac{\sqrt{45}}{2}$

$n = \dfrac{5}{2} \pm \dfrac{3\sqrt{5}}{2}$

$n \approx 5.85$ or $n \approx -0.85$

75. $2x^2 = -7x + 4$

$2x^2 + 7x = 4$

$x^2 + \dfrac{7}{2}x = 2$

$x^2 + \dfrac{7}{2}x + \dfrac{49}{16} = 2 + \dfrac{49}{16}$

$\left(x + \dfrac{7}{4}\right)^2 = \dfrac{81}{16}$

$x + \dfrac{7}{4} = \pm\dfrac{9}{4}$

$x = -\dfrac{7}{4} \pm \dfrac{9}{4}$

$x = \dfrac{1}{2}$ or $x = -4$

77. $2n^2 - 3n - 9 = 0$

$2n^2 - 3n = 9$

$n^2 - \dfrac{3}{2}n = \dfrac{9}{2}$

$n^2 - \dfrac{3}{2}n + \dfrac{9}{16} = \dfrac{9}{2} + \dfrac{9}{16}$

$\left(n - \dfrac{3}{4}\right)^2 = \dfrac{81}{16}$

$n - \dfrac{3}{4} = \pm\dfrac{9}{4}$

$n = \dfrac{3}{4} \pm \dfrac{9}{4}$

$n = 3$ or $n = -\dfrac{3}{2}$

79. $4p^2 - 3p - 2 = 0$

$4p^2 - 3p = 2$

$p^2 - \dfrac{3}{4}p = \dfrac{1}{2}$

$p^2 - \dfrac{3}{4}p + \dfrac{9}{64} = \dfrac{1}{2} + \dfrac{9}{64}$

$\left(p - \dfrac{3}{8}\right)^2 = \dfrac{41}{64}$

$p - \dfrac{3}{8} = \pm\dfrac{\sqrt{41}}{8}$

$p = \dfrac{3}{8} \pm \dfrac{\sqrt{41}}{8}$

$p \approx 1.18$ or $p \approx -0.43$

81. $m^2 = 7m - 4$

$m^2 - 7m = -4$

$m^2 - 7m + \dfrac{49}{4} = \dfrac{49}{4} - 4$

$\left(m - \dfrac{7}{2}\right)^2 = \dfrac{33}{4}$

$m - \dfrac{7}{2} = \dfrac{\pm\sqrt{33}}{2}$

$m = \dfrac{7}{2} \pm \dfrac{\sqrt{33}}{2}$

$m \approx 6.37$ or $m \approx 0.63$

83. $x^2 - 3x = 18$

$x^2 - 3x - 18 = 0$

$(x - 6)(x + 3) = 0$

$x - 6 = 0$ or $x + 3 = 0$

$x = 6$ or $x = -3$

85. $4m^2 - 25 = 0$

$4m^2 = 25$

$m^2 = \dfrac{25}{4}$

$m = \pm\dfrac{5}{2}$

Chapter 1: Equations and Inequalities

87. $4n^2 - 8n - 1 = 0$

$a = 4, b = -8, c = -1$

$$n = \frac{-(-8) \pm \sqrt{(-8)^2 - 4(4)(-1)}}{2(4)}$$

$$n = \frac{8 \pm \sqrt{80}}{8}$$

$$n = \frac{8 \pm 4\sqrt{5}}{8}$$

$$n = \frac{2 \pm \sqrt{5}}{2}$$

$n \approx 2.12$ or $n \approx -0.12$

89. $6w^2 - w = 2$

$6w^2 - w - 2 = 0$

$(2w + 1)(3w - 2) = 0$

$2w + 1 = 0$ or $3w - 2 = 0$

$2w = -1$ or $3w = 2$

$w = -\dfrac{1}{2}$ or $w = \dfrac{2}{3}$

91. $4m^2 = 12m - 15$

$4m^2 - 12m + 15 = 0$

$a = 4, b = -12, c = 15$

$$m = \frac{-(-12) \pm \sqrt{(-12)^2 - 4(4)(15)}}{2(4)}$$

$$m = \frac{12 \pm \sqrt{-96}}{8}$$

$$m = \frac{12 \pm 4\sqrt{6}\, i}{8}$$

$$m = \frac{3 \pm \sqrt{6}\, i}{2}$$

$$m = \frac{3}{2} \pm \frac{\sqrt{6}}{2}\, i$$

$m \approx 1.5 \pm 1.22i$

93. $4n^2 - 9 = 0$

$4n^2 = 9$

$n^2 = \dfrac{9}{4}$

$n = \pm\dfrac{3}{2}$

95. $5w^2 = 6w + 8$

$5w^2 - 6w - 8 = 0$

$(5w + 4)(w - 2) = 0$

$5w + 4 = 0$ or $w - 2 = 0$

$5w = -4$ or $w = 2$

$w = -\dfrac{4}{5}$ or $w = 2$

97. $3a^2 - a + 2 = 0$

$a = 3, b = -1, c = 2$

$$a = \frac{-(-1) \pm \sqrt{(-1)^2 - 4(3)(2)}}{2(3)}$$

$$a = \frac{1 \pm \sqrt{-23}}{6}$$

$$a = \frac{1 \pm \sqrt{23}\, i}{6}$$

$$a = \frac{1}{6} \pm \frac{\sqrt{23}}{6}\, i$$

$a \approx 0.1\overline{6} \pm 0.80i$

99. $5p^2 = 6p + 3$

$5p^2 - 6p - 3 = 0$

$a = 5, b = -6, c = -3$

$$p = \frac{-(-6) \pm \sqrt{(-6)^2 - 4(5)(-3)}}{2(5)}$$

$$p = \frac{6 \pm \sqrt{96}}{10}$$

$$p = \frac{6 \pm 4\sqrt{6}}{10}$$

$$p = \frac{3 \pm 2\sqrt{6}}{5}$$

$$p = \frac{3}{5} \pm \frac{2\sqrt{6}}{5}$$

$p \approx 1.58$ or $p \approx -0.38$

1.5 Exercises

101. $5w^2 - w = 1$

$5w^2 - w - 1 = 0$
$a = 5, b = -1, c = -1$

$w = \dfrac{-(-1) \pm \sqrt{(-1)^2 - 4(5)(-1)}}{2(5)}$

$w = \dfrac{1 \pm \sqrt{21}}{10}$

$w = \dfrac{1}{10} \pm \dfrac{\sqrt{21}}{10}$

$w \approx 0.56$ or $w \approx -0.36$

103. $2a^2 + 5 = 3a$

$2a^2 - 3a + 5 = 0$
$a = 2, b = -3, c = 5$

$a = \dfrac{-(-3) \pm \sqrt{(-3)^2 - 4(2)(5)}}{2(2)}$

$a = \dfrac{3 \pm \sqrt{-31}}{4}$

$a = \dfrac{3 \pm \sqrt{31}\,i}{4}$

$a = \dfrac{3}{4} \pm \dfrac{\sqrt{31}}{4}\,i$

$a \approx 0.75 \pm 1.39i$

105. $2p^2 - 4p + 11 = 0$

$a = 2, b = -4, c = 11$

$p = \dfrac{-(-4) \pm \sqrt{(-4)^2 - 4(2)(11)}}{2(2)}$

$p = \dfrac{4 \pm \sqrt{-72}}{4}$

$p = \dfrac{4 \pm 6\sqrt{2}\,i}{4}$

$p = 1 \pm \dfrac{3\sqrt{2}}{2}\,i$

$p \approx 1 \pm 2.12i$

107. $w^2 + \dfrac{2}{3}w = \dfrac{1}{9}$

$9\left[w^2 + \dfrac{2}{3}w = \dfrac{1}{9} \right]$

$9w^2 + 6w = 1$
$9w^2 + 6w - 1 = 0$
$a = 9, b = 6, c = -1$

$w = \dfrac{-(6) \pm \sqrt{(6)^2 - 4(9)(-1)}}{2(9)}$

$w = \dfrac{-6 \pm \sqrt{72}}{18}$

$w = \dfrac{-6 \pm 6\sqrt{2}}{18}$

$w = \dfrac{-1 \pm \sqrt{2}}{3}$

$w = \dfrac{-1}{3} \pm \dfrac{\sqrt{2}}{3}$

$w \approx 0.14$ or $w \approx -0.80$

109. $0.2a^2 + 1.2a + 0.9 = 0$

$a = 0.2, b = 1.2, c = 0.9$

$a = \dfrac{-(1.2) \pm \sqrt{(1.2)^2 - 4(0.2)(0.9)}}{2(0.2)}$

$a = \dfrac{-1.2 \pm \sqrt{0.72}}{0.4}$

$a = \dfrac{-1.2 \pm 0.6\sqrt{2}}{0.4}$

$a = \dfrac{-6 \pm 3\sqrt{2}}{2}$

$a = \dfrac{-6}{2} \pm \dfrac{3\sqrt{2}}{2}$

$a = -3 \pm \dfrac{3\sqrt{2}}{2}$

$a \approx -0.88$ or $a \approx -5.12$

111. $\dfrac{2}{7}p^2 - 3 = \dfrac{8}{21}p$

$21\left[\dfrac{2}{7}p^2 - 3 = \dfrac{8}{21}p\right]$

$6p^2 - 63 = 8p$
$6p^2 - 8p - 63 = 0$
$a = 6, b = -8, c = -63$

$p = \dfrac{-(-8) \pm \sqrt{(-8)^2 - 4(6)(-63)}}{2(6)}$

$p = \dfrac{8 \pm \sqrt{1576}}{12}$

$p = \dfrac{8 \pm 2\sqrt{394}}{12}$

$p = \dfrac{4 \pm \sqrt{394}}{6}$

$p = \dfrac{4}{6} \pm \dfrac{\sqrt{394}}{6}$

$p = \dfrac{2}{3} \pm \dfrac{\sqrt{394}}{6}$

$p \approx 3.97$ or $p \approx -2.64$

113. $-3x^2 + 2x + 1 = 0$

$a = -3, b = 2, c = 1$

$(2)^2 - 4(-3)(1) = 16$
two rational solutions

115. $-4x + x^2 + 13 = 0$

$x^2 - 4x + 13 = 0$
$a = 1, b = -4, c = 13$

$(-4)^2 - 4(1)(13) = -36$
two complex solutions

117. $15x^2 - x - 6 = 0$

$a = 15, b = -1, c = -6$

$(-1)^2 - 4(15)(-6) = 361$
two rational solutions

119. $-4x^2 + 6x - 5 = 0$

$a = -4, b = 6, c = -5$

$(6)^2 - 4(-4)(-5) = -44$
two complex solutions

121. $2x^2 + 8 = -9x$

$2x^2 + 9x + 8 = 0$
$a = 2, b = 9, c = 8$

$(9)^2 - 4(2)(8) = 17$
two irrational solutions

123. $4x^2 + 12x = -9$

$4x^2 + 12x + 9 = 0$
$a = 4, b = 12, c = 9$

$(12)^2 - 4(4)(9) = 0$
one repeated solution

125. $-6x + 2x^2 + 5 = 0$

$2x^2 - 6x + 5 = 0$
$a = 2, b = -6, c = 5$

$x = \dfrac{-(-6) \pm \sqrt{(-6)^2 - 4(2)(5)}}{2(2)}$

$x = \dfrac{6 \pm \sqrt{-4}}{4}$

$x = \dfrac{6 \pm 2i}{4}$

$x = \dfrac{3}{2} \pm \dfrac{1}{2}i$

1.5 Exercises

127. $5x^2 + 5 = -5x$

$5x^2 + 5x + 5 = 0$
$a = 5, b = 5, c = 5$

$x = \dfrac{-(5) \pm \sqrt{(5)^2 - 4(5)(5)}}{2(5)}$

$x = \dfrac{-5 \pm \sqrt{-75}}{10}$

$x = \dfrac{-5 \pm 5\sqrt{3}\,i}{10}$

$x = -\dfrac{1}{2} \pm \dfrac{\sqrt{3}}{2}\,i$

129. $-2x^2 = -5x + 11$

$0 = 2x^2 - 5x + 11$
$a = 2, b = -5, c = 11$

$x = \dfrac{-(-5) \pm \sqrt{(-5)^2 - 4(2)(11)}}{2(2)}$

$x = \dfrac{5 \pm \sqrt{-63}}{4}$

$x = \dfrac{5 \pm 3\sqrt{7}\,i}{4}$

$x = \dfrac{5}{4} \pm \dfrac{3\sqrt{7}}{4}\,i$

131. $h = -16t^2 + vt$

$16t^2 - vt + h = 0$
$a = 16, b = -v, c = h$

$t = \dfrac{-(-v) \pm \sqrt{(-v)^2 - 4(16)(h)}}{2(16)}$

$t = \dfrac{v \pm \sqrt{v^2 - 64h}}{32}$

133. $-16t^2 + 96t + 408 = 0$

$16t^2 - 96t - 408 = 0$
$8(2t^2 - 12t - 51) = 0$
$a = 2, b = -12, c = -51$

$t = \dfrac{-(-12) \pm \sqrt{(-12)^2 - 4(2)(-51)}}{2(2)}$

$t = \dfrac{12 \pm \sqrt{552}}{4}$

$t = \dfrac{12 \pm 2\sqrt{138}}{4}$

$t = \dfrac{6 \pm \sqrt{138}}{2}$

$t \approx 8.87$ seconds

135. $R = x\left(40 - \dfrac{1}{3}x\right)$

$900 = 40x - \dfrac{1}{3}x^2$

$\dfrac{1}{3}x^2 - 40x + 900 = 0$

$x^2 - 120x + 2700 = 0$
$a = 1, b = -120, c = 2700$

$x = \dfrac{-(-120) \pm \sqrt{(-120)^2 - 4(1)(2700)}}{2(1)}$

$x = \dfrac{120 \pm \sqrt{3600}}{2}$

$x = \dfrac{120 \pm 60}{2}$

$x = 90$ or $x = 30$
30 thousand ovens

Chapter 1: Equations and Inequalities

137.a. $P = -x^2 + 122x - 1965 - (2x + 35)$

$P = -x^2 + 120x - 2000$

b. $x^2 - 120x + 2000 = 0$
$(x - 100)(x - 20) = 0$
$x - 100 = 0$ or $x - 20 = 0$
$x = 100$ or $x = 20$
10,000 toys

139. $260 = -16t^2 + 144t$

$16t^2 - 144t + 260 = 0$

$4t^2 - 36t + 65 = 0$
$(2t - 13)(2t - 5) = 0$
$2t - 13 = 0$ or $2t - 5 = 0$
$2t = 13$ or $2t = 5$
$t = \dfrac{13}{2}$ seconds or $t = \dfrac{5}{2}$ seconds
$t = 6.5$ seconds or $t = 2.5$ seconds

141. $3750 = 17.4x^2 + 36.1x + 83.3$
$0 = 17.4x^2 + 36.1x - 3666.7$
$a = 17.4, b = 36.1, c = -3666.7$

$x = \dfrac{-36.1 + \sqrt{(36.1)^2 - 4(17.4)(-3666.7)}}{2(17.4)}$

$x = \dfrac{-36.1 + \sqrt{256505.53}}{34.8}$

$x = \dfrac{-36.1 + 506.4637499}{34.8}$

$x = \dfrac{470.3637499}{34.8}$

$x \approx 13.5$

$1995 + 13.5 = 2008.5$

In the year 2008.

143. Let w represent the width of the doubles court.
Let $2w + 6$ represent the length of the doubles court.
$w(2w + 6) = 2808$
$2w^2 + 6w - 2808 = 0$
$w^2 + 3w - 1404 = 0$
$(w + 39)(w - 36) = 0$
$w + 39 = 0$ or $w - 36 = 0$
$w = -39$ or $w = 36$
The width is 36 feet.
The length is 78 feet.

145.a. $x^2 + 6x - 16 = 0$
$b = 6, c = -16$
$b^2 - 4ac$
$6^2 - 4a(-16)$
$= 36 + 64a = $ perfect square
If $a = 7$,
$= 36 + 64(7) = 484$ (perfect square);
$7x^2 + 6x - 16 = 0$

b. $x^2 + 5x - 14 = 0$
$b = 5, c = -14$
$b^2 - 4ac$
$5^2 - 4a(-14)$
$= 25 + 56a = $ perfect square
If $a = 6$,
$= 25 + 56(6) = 361$ (perfect square);
$6x^2 + 5x - 14 = 0$

c. $x^2 - x - 6 = 0$
$b = -1, c = -6$
$b^2 - 4ac$
$(-1)^2 - 4a(-6)$
$= 1 + 24a = $ perfect square
If $a = 5$,
$= 1 + 24(5) = 121$ (perfect square);
$5x^2 + x - 6 = 0$

147. $z^2 - 3iz = -10$

   $z^2 - 3iz + 10 = 0$
   $a = 1, b = -3i, c = 10$

   $$z = \frac{-(-3i) \pm \sqrt{(-3i)^2 - 4(1)(10)}}{2(1)}$$

   $$z = \frac{3i \pm \sqrt{9i^2 - 40}}{2}$$

   $$z = \frac{3i \pm \sqrt{-49}}{2}$$

   $$z = \frac{3i \pm 7i}{2}$$

   $z = 5i$ or $z = -2i$

149. $4iz^2 + 5z + 6i = 0$

   $a = 4i, b = 5, c = 6i$

   $$z = \frac{-(5) \pm \sqrt{(5)^2 - 4(4i)(6i)}}{2(4i)}$$

   $$z = \frac{-5 \pm \sqrt{25 - 96i^2}}{8i}$$

   $$z = \frac{-5 \pm \sqrt{121}}{8i}$$

   $$z = \frac{-5 \pm 11}{8i}$$

   $z = \frac{6}{8i}$ or $z = \frac{-16}{8i}$ $\left(\text{Recall } \frac{1}{i} = -i \right)$

   $z = -\frac{3}{4}i$ or $z = 2i$

151. $0.5z^2 + (7 + i)z + (6 + 7i) = 0$

   $a = 0.5, b = 7 + i, c = 6 + 7i$

   $$z = \frac{-(7 + i) \pm \sqrt{(7 + i)^2 - 4(0.5)(6 + 7i)}}{2(0.5)}$$

   $$z = \frac{-7 - i \pm \sqrt{49 + 14i + i^2 - 12 - 14i}}{1}$$

   $z = -7 - i \pm \sqrt{36}$
   $z = -7 - i \pm 6$
   $z = -1 - i$ or $z = -13 - i$

153. a. $P = 2L + 2W, A = LW$

   b. $P = 2\pi r, A = \pi r^2$

   c. $A = \frac{1}{2}h(b_1 + b_2), P = c + h + b_1 + b_2$

   d. $A = \frac{1}{2}bh, P = a + b + c$

155. Let x represent the number of good seats sold.
   Let $900 - x$ represent the number of cheap seats sold.

   $30x + 20(900 - x) = 25,000$

   $30x + 18,000 - 20x = 25,000$

   $10x = 7,000$

   $x = 700$
   700, \$30 tickets
   200, \$20 tickets

1.6 Exercises

1. Excluded

3. Extraneous

5. Answers will vary.

7. $22x = x^3 - 9x^2$
$0 = x^3 - 9x^2 - 22x$
$0 = x(x^2 - 9x - 22)$
$0 = x(x - 11)(x + 2)$
$x = 0$ or $x - 11 = 0$ or $x + 2 = 0$
$x = 0, x = -2, x = 11$

9. $3x^3 = -7x^2 + 6x$
$3x^3 + 7x^2 - 6x = 0$
$x(3x^2 + 7x - 6) = 0$
$x(3x - 2)(x + 3) = 0$
$x = 0$ or $3x - 2 = 0$ or $x + 3 = 0$
$\qquad 3x = 2 \qquad$ or $x = -3$
$\qquad x = \dfrac{2}{3}$
$x = 0, x = -3, x = \dfrac{2}{3}$

11. $2x^4 - 3x^3 = 9x^2$
$2x^4 - 3x^3 - 9x^2 = 0$
$x^2(2x^2 - 3x - 9) = 0$
$x^2(2x + 3)(x - 3) = 0$
$x^2 = 0$ or $2x + 3 = 0$ or $x - 3 = 0$
$x = 0 \qquad 2x = -3 \qquad x = 3$
$\qquad\qquad x = -\dfrac{3}{2}$
$x = 0, x = -\dfrac{3}{2}, x = 3$

13. $2x^4 - 16x = 0$
$2x(x^3 - 8) = 0$
$2x(x - 2)(x^2 + 2x + 4) = 0$
$2x = 0$ or $x - 2 = 0$ or $x^2 + 2x + 4 = 0$
$x = 0$ or $x = 2$ or $x = \dfrac{-2 \pm \sqrt{(2)^2 - 4(1)(4)}}{2(1)}$
$\qquad x = \dfrac{-2 \pm \sqrt{-12}}{2}$
$\qquad x = \dfrac{-2 \pm 2\sqrt{3}i}{2}$
$\qquad x = -1 \pm i\sqrt{3}$
$x = 0, x = 2, x = -1 \pm i\sqrt{3}$

15. $x^3 - 4x = 5x^2 - 20$
$x^3 - 5x^2 - 4x + 20 = 0$
$x^2(x - 5) - 4(x - 5) = 0$
$(x - 5)(x^2 - 4) = 0$
$(x - 5)(x + 2)(x - 2) = 0$
$x - 5 = 0$ or $x + 2 = 0$ or $x - 2 = 0$
$\quad x = 5$ or $\quad x = -2$ or $x = 2$
$x = 5, x = 2, x = -2$

17. $4x - 12 = 3x^2 - x^3$
$x^3 - 3x^2 + 4x - 12 = 0$
$x^2(x - 3) + 4(x - 3) = 0$
$(x - 3)(x^2 + 4) = 0$
$x - 3 = 0$ or $x^2 + 4 = 0$
$\quad x = 3$ or $x^2 = -4$
$x = 3, x = \pm 2i$

19. $2x^3 - 12x^2 = 10x - 60$
$2x^3 - 12x^2 - 10x + 60 = 0$
$2(x^3 - 6x^2 - 5x + 30) = 0$
$2[x^2(x - 6) - 5(x - 6)] = 0$
$2(x - 6)(x^2 - 5) = 0$
$x - 6 = 0$ or $x^2 - 5 = 0$
$x = 6$ or $x^2 = 5$
$x = 6, x = \pm\sqrt{5}$

1.6 Exercises

21. $x^4 - 7x^3 + 4x^2 = 28x$

$x^4 - 7x^3 + 4x^2 - 28x = 0$

$x(x^3 - 7x^2 + 4x - 28) = 0$

$x[x^2(x-7) + 4(x-7)] = 0$

$x(x-7)(x^2+4) = 0$

$x = 0$ or $x - 7 = 0$ or $x^2 + 4 = 0$

$\qquad x = 7 \quad$ or $\quad x^2 = -4$

$x = 0, x = 7, x = \pm 2i$

23. $x^4 - 81 = 0$

$(x^2 + 9)(x^2 - 9) = 0$

$x^2 + 9 = 0$ or $x^2 - 9 = 0$

$x^2 = -9$ or $\quad x^2 = 9$

$x = \pm 3i, x = \pm 3$

25. $x^4 - 256 = 0$

$(x^2 + 16)(x^2 - 16) = 0$

$x^2 + 16 = 0$ or $x^2 - 16 = 0$

$x^2 = -16$ or $\quad x^2 = 16$

$x = \pm 4i, x = \pm 4$

27. $x^6 - 2x^4 - x^2 + 2 = 0$

$x^4(x^2 - 2) - 1(x^2 - 2) = 0$

$(x^2 - 2)(x^4 - 1) = 0$

$(x^2 - 2)(x^2 + 1)(x^2 - 1) = 0$

$(x^2 - 2)(x^2 + 1)(x + 1)(x - 1) = 0$

$x^2 - 2 = 0$ or $x^2 + 1 = 0$

$\qquad$ or $x + 1 = 0$ or $x - 1 = 0$

$x^2 = 2$ or $x^2 = -1$

$\qquad$ or $x = -1$ or $x = 1$

$x = \pm\sqrt{2}, x = \pm i, x = -1, x = 1$

29. $x^5 - x^3 - 8x^2 + 8 = 0$

$x^3(x^2 - 1) - 8(x^2 - 1) = 0$

$(x^2 - 1)(x^3 - 8) = 0$

$(x + 1)(x - 1)(x - 2)(x^2 + 2x + 4) = 0$

$x + 1 = 0$ or $x - 1 = 0$

$\qquad$ or $x - 2 = 0$ or $x^2 + 2x + 4 = 0$

$x = -1$ or $x = 1$ or $x = 2$

$\qquad$ or $x = \dfrac{-2 \pm \sqrt{(2)^2 - 4(1)(4)}}{2(1)}$

$\qquad x = \dfrac{-2 \pm \sqrt{-12}}{2}$

$\qquad x = \dfrac{-2 \pm 2i\sqrt{3}}{2}$

$x = \pm 1, x = 2, x = -1 \pm i\sqrt{3}$

31. $x^6 - 1 = 0$

$(x^3 + 1)(x^3 - 1) = 0$

$(x + 1)(x^2 - x + 1)(x - 1)(x^2 + x + 1) = 0$

$x + 1 = 0$ or $x^2 - x + 1 = 0$

or $x - 1 = 0$ or $x^2 - x + 1 = 0$;

$x = -1$

or $x = \dfrac{-(-1) \pm \sqrt{(-1)^2 - 4(1)(1)}}{2(1)} = \dfrac{1 \pm \sqrt{-3}}{2} = \dfrac{1 \pm i\sqrt{3}}{2}$

or $x = 1$

or $x = \dfrac{-(1) \pm \sqrt{1^2 - 4(1)(1)}}{2(1)} = \dfrac{-1 \pm \sqrt{-3}}{2} = \dfrac{-1 \pm i\sqrt{3}}{2}$;

$x = \pm 1, x = \dfrac{1}{2} \pm \dfrac{i\sqrt{3}}{2}, x = -\dfrac{1}{2} \pm \dfrac{i\sqrt{3}}{2}$

33. $\dfrac{2}{x} + \dfrac{1}{x+1} = \dfrac{5}{x^2 + x}$

$\dfrac{2}{x} + \dfrac{1}{x+1} = \dfrac{5}{x(x+1)}$

$x(x+1)\left[\dfrac{2}{x} + \dfrac{1}{x+1} = \dfrac{5}{x(x+1)}\right]$

$2x + 2 + x = 5$

$3x = 3$

$x = 1$

35. $\dfrac{21}{a+2} = \dfrac{3}{a-1}$

$(a+2)(a-1)\left[\dfrac{21}{a+2} = \dfrac{3}{a-1}\right]$

$21a - 21 = 3a + 6$

$18a = 27$

$a = \dfrac{27}{18} = \dfrac{3}{2}$

37. $\dfrac{1}{3y} - \dfrac{1}{4y} = \dfrac{1}{y^2}$

$12y^2\left[\dfrac{1}{3y} - \dfrac{1}{4y} = \dfrac{1}{y^2}\right]$

$4y - 3y = 12$

$y = 12$

39. $x + \dfrac{14}{x-7} = 1 + \dfrac{2x}{x-7}$

$(x-7)\left[x + \dfrac{14}{x-7} = 1 + \dfrac{2x}{x-7}\right]$

$x(x-7) + 14 = 1(x-7) + 2x$

$x^2 - 7x + 14 = x - 7 + 2x$

$x^2 - 7x + 14 = 3x - 7$

$x^2 - 10x + 21 = 0$

$(x-7)(x-3) = 0$

$x - 7 = 0$ or $x - 3 = 0$

$x = 7$ or $x = 3$

$x = 3;\ x = 7$ is extraneous

41. $\dfrac{6}{n+3} + \dfrac{20}{n^2+n-6} = \dfrac{5}{n-2}$

$\dfrac{6}{n+3} + \dfrac{20}{(n+3)(n-2)} = \dfrac{5}{n-2}$

$(n+3)(n-2)\left[\dfrac{6}{n+3} + \dfrac{20}{(n+3)(n-2)} = \dfrac{5}{n-2}\right]$

$6(n-2) + 20 = 5(n+3)$

$6n - 12 + 20 = 5n + 15$

$6n + 8 = 5n + 15$

$n = 7$

43. $\dfrac{a}{2a+1} - \dfrac{2a^2+5}{2a^2-5a-3} = \dfrac{3}{a-3}$

$\dfrac{a}{2a+1} - \dfrac{2a^2+5}{(2a+1)(a-3)} = \dfrac{3}{a-3}$

$(2a+1)(a-3)\left[\dfrac{a}{2a+1} - \dfrac{2a^2+5}{(2a+1)(a-3)} = \dfrac{3}{a-3}\right]$

$a(a-3) - (2a^2+5) = 3(2a+1)$

$a^2 - 3a - 2a^2 - 5 = 6a + 3$

$-a^2 - 3a - 5 = 6a + 3$

$0 = a^2 + 9a + 8$

$(a+8)(a+1) = 0$

$a + 8 = 0$ or $a + 1 = 0$

$a = -8$ or $a = -1$

45. $\dfrac{1}{f} = \dfrac{1}{f_1} + \dfrac{1}{f_2}$

$f f_1 f_2\left[\dfrac{1}{f} = \dfrac{1}{f_1} + \dfrac{1}{f_2}\right]$

$f_1 f_2 = f f_2 + f f_1$

$f_1 f_2 = f(f_2 + f_1)$

$\dfrac{f_1 f_2}{f_1 + f_2} = f$

47. $I = \dfrac{E}{R+r}$

$(R+r)\left[I = \dfrac{E}{R+r}\right]$

$IR + Ir = E$

$Ir = E - IR$

$r = \dfrac{E - IR}{I}$ or $r = \dfrac{E}{I} - R$

1.6 Exercises

49. $V = \dfrac{1}{3}\pi r^2 h$

$3V = \pi r^2 h$

$\dfrac{3V}{\pi r^2} = h$

51. $V = \dfrac{4}{3}\pi r^3$

$3V = 4\pi r^3$

$\dfrac{3V}{4\pi} = r^3$

53. a. $-3\sqrt{3x-5} = -9$

$\sqrt{3x-5} = 3$

$3x-5 = 9$

$3x = 14$

$x = \dfrac{14}{3}$

b. $x = \sqrt{3x+1} + 3$

$x-3 = \sqrt{3x+1}$

$(x-3)^2 = \left(\sqrt{3x+1}\right)^2$

$x^2 - 6x + 9 = 3x + 1$

$x^2 - 9x + 8 = 0$

$(x-8)(x-1) = 0$

$x-8 = 0 \ \text{ or } \ x-1 = 0$

$x = 8 \quad \text{ or } \ x = 1$

$x = 8; \ x = 1 \text{ is extraneous.}$

55. a. $2 = \sqrt[3]{3m-1}$

$8 = 3m-1$

$9 = 3m$

$3 = m$

b. $2\sqrt[3]{7-3x} - 3 = -7$

$2\sqrt[3]{7-3x} = -4$

$\sqrt[3]{7-3x} = -2$

$7 - 3x = -8$

$-3x = -15$

$x = 5$

c. $\dfrac{\sqrt[3]{2m+3}}{-5} + 2 = 3$

$\dfrac{\sqrt[3]{2m+3}}{-5} = 1$

$\sqrt[3]{2m+3} = -5$

$2m+3 = -125$

$2m = -128$

$m = -64$

d. $\sqrt[3]{2x-9} = \sqrt[3]{3x+7}$

$2x-9 = 3x+7$

$-x = 16$

$x = -16$

57. a. $\sqrt{x-9} + \sqrt{x} = 9$

$\sqrt{x-9} = 9 - \sqrt{x}$

$\left(\sqrt{x-9}\right)^2 = \left(9 - \sqrt{x}\right)^2$

$x-9 = 81 - 18\sqrt{x} + x$

$-90 = -18\sqrt{x}$

$5 = \sqrt{x}$

$25 = x$

b. $\sqrt{x+9} + \sqrt{x-7} = 8$

$\sqrt{x+9} = 8 - \sqrt{x-7}$

$\left(\sqrt{x+9}\right)^2 = \left(8 - \sqrt{x-7}\right)^2$

$x+9 = 64 - 16\sqrt{x-7} + x - 7$

$x+9 = 57 - 16\sqrt{x-7} + x$

$-48 = -16\sqrt{x-7}$

$3 = \sqrt{x-7}$

$9 = x-7$

$16 = x$

c. $\sqrt{x-2} - \sqrt{2x} = -2$

$\sqrt{x-2} = \sqrt{2x} - 2$

$\left(\sqrt{x-2}\right)^2 = \left(\sqrt{2x} - 2\right)^2$

$x - 2 = 2x - 4\sqrt{2x} + 4$

$-x - 6 = -4\sqrt{2x}$

$x + 6 = 4\sqrt{2x}$

$\left(x+6\right)^2 = \left(4\sqrt{2x}\right)^2$

$x^2 + 12x + 36 = 16(2x)$

$x^2 + 12x + 36 = 32x$

$x^2 - 20x + 36 = 0$

$(x-2)(x-18) = 0$

$x - 2 = 0$ or $x - 18 = 0$

$x = 2$ or $x = 18$

d. $\sqrt{12x+9} - \sqrt{24x} = -3$

$\sqrt{12x+9} = \sqrt{24x} - 3$

$\left(\sqrt{12x+9}\right)^2 = \left(\sqrt{24x} - 3\right)^2$

$12x + 9 = 24x - 6\sqrt{24x} + 9$

$-12x = -6\sqrt{24x}$

$2x = \sqrt{24x}$

$\left(2x\right)^2 = \left(\sqrt{24x}\right)^2$

$4x^2 = 24x$

$4x^2 - 24x = 0$

$4x(x-6) = 0$

$4x = 0$ or $x - 6 = 0$

$x = 0$ or $x = 6$

$x = 6; x = 0$ is extraneous

59. $x^{\frac{3}{5}} + 17 = 9$

$x^{\frac{3}{5}} = -8$

$\left(x^{\frac{3}{5}}\right)^{\frac{5}{3}} = \left(-8\right)^{\frac{5}{3}}$

$x = -32$

61. $0.\overline{3}x^{\frac{5}{2}} - 39 = 42$

$\frac{1}{3}x^{\frac{5}{2}} = 81$

$x^{\frac{5}{2}} = 243$

$\left(x^{\frac{5}{2}}\right)^{\frac{2}{5}} = \left(243\right)^{\frac{2}{5}}$

$x = 9$

63. $2(x+5)^{\frac{2}{3}} - 11 = 7$

$2(x+5)^{\frac{2}{3}} = 18$

$(x+5)^{\frac{2}{3}} = 9$

$\left((x+5)^{\frac{2}{3}}\right)^{\frac{3}{2}} = 9^{\frac{3}{2}}$

$x + 5 = 27$ or $x + 5 = -27$

$x = 22$ $x = -32$

65. $x^{\frac{2}{3}} - 2x^{\frac{1}{3}} - 15 = 0$

Let $u = x^{\frac{1}{3}}$

$u^2 = x^{\frac{2}{3}}$;

$u^2 - 2u - 15 = 0$

$\left(u-5\right)\left(u+3\right) = 0$

$u - 5 = 0$ or $u + 3 = 0$

$u = 5$ or $u = -3$

$x^{\frac{1}{3}} = 5$ or $x^{\frac{1}{3}} = -3$

$\left(x^{\frac{1}{3}}\right)^3 = (5)^3$ or $\left(x^{\frac{1}{3}}\right)^3 = (-3)^3$

$x = 125$ or $x = -27$

67. $x^4 - 24x^2 - 25 = 0$

Let $u = x^2$, then $u^2 = x^4$;

$u^2 - 24u - 25 = 0$

$\left(u-25\right)\left(u+1\right) = 0$

$u - 25 = 0$ or $u + 1 = 0$

$u = 25$ or $u = -1$;

$x^2 = 25$ or $x^2 = -1$

$x = \pm 5$ or $x = \pm i$

1.6 Exercises

69. $\left(x^2-3\right)^2+\left(x^2-3\right)-2=0$

Let $u = x^2-3$

$\quad u^2 = \left(x^2-3\right)^2;$

$u^2+u-2=0$
$(u-1)(u+2)=0$
$u-1=0$ or $u+2=0$
$u=1 \quad$ or $u=-2$
$x^2-3=1 \qquad$ or $x^2-3=-2$
$x^2=4 \quad$ or $x^2=1$
$x=\pm2 \quad$ or $x=\pm1$

71. $x^{-2}-3x^{-1}-4=0$

Let $u = x^{-1}$

$\quad u^2 = x^{-2};$

$u^2-3u-4=0$
$(u-4)(u+1)=0$
$u-4=0$ or $u+1=0$
$u=4 \quad$ or $u=-1$
$x^{-1}=4 \qquad$ or $x^{-1}=-1$
$\left(x^{-1}\right)^{-1}=(4)^{-1}$ or $\left(x^{-1}\right)^{-1}=(-1)^{-1}$
$x=\dfrac{1}{4} \qquad$ or $x=-1$

73. $x^{-4}-13x^{-2}+36=0$

Let $u = x^{-2}$

$\quad u^2 = x^{-4};$

$u^2-13u+36=0$
$(u-9)(u-4)=0$
$u-9=0$ or $u-4=0$
$u=9 \quad$ or $u=4$
$x^{-2}=9 \quad$ or $x^{-2}=4$
$\dfrac{1}{x^2}=9 \quad$ or $\dfrac{1}{x^2}=4$
$x^2=\dfrac{1}{9} \quad$ or $x^2=\dfrac{1}{4}$
$x=\pm\dfrac{1}{3} \quad$ or $x=\pm\dfrac{1}{2}$

75. $x+4=7\sqrt{x+4}$

Let $u = (x+4)^{\frac{1}{2}}$

$\quad u^2 = x+4;$

$u^2 = 7u$
$u^2-7u=0$
$u(u-7)=0$
$u=0$ or $u-7=0$
$u=0$ or $u=7$
$\sqrt{x+4}=0$ or $\sqrt{x+4}=7$
$x+4=0 \quad$ or $x+4=49$
$x=-4 \qquad$ or $x=45$

77. $2\sqrt{x+10}+8=3(x+10)$

Let $u = (x+10)^{\frac{1}{2}}$

$u^2 = x+10;$

$2u+8=3u^2$
$0=3u^2-2u-8$
$0=(3u+4)(u-2)$
$3u+4=0$ or $u-2=0$
$3u=-4 \quad$ or $u=2$
$u=-\dfrac{4}{3} \quad$ or $u=2$
$\sqrt{x+10}=-\dfrac{4}{3}$ or $\sqrt{x+10}=2$
$x+10=\dfrac{16}{9} \quad$ or $x+10=4$
$x=-\dfrac{74}{9} \qquad$ or $x=-6$
$x=-6; \; x=-\dfrac{74}{9}$ is extraneous

Chapter 1: Equations and Inequalities

79. a.
$$S = \pi r \sqrt{r^2 + h^2}$$
$$\frac{S}{\pi r} = \sqrt{r^2 + h^2}$$
$$\left(\frac{S}{\pi r}\right)^2 = r^2 + h^2$$
$$\left(\frac{S}{\pi r}\right)^2 - r^2 = h^2$$
$$\sqrt{\left(\frac{S}{\pi r}\right)^2 - r^2} = h$$

b.
$$S = \pi(6)\sqrt{(6)^2 + (10)^2}$$
$$S = 6\pi\sqrt{36 + 100}$$
$$S = 6\pi\sqrt{136}$$
$$S = 12\pi\sqrt{34} \text{ m}^2$$

81. Let x represent the number.
$$x^3 + 2x^2 = 18 + 9x$$
$$x^3 + 2x^2 - 9x - 18 = 0$$
$$x^2(x+2) - 9(x+2) = 0$$
$$(x+2)(x^2-9) = 0$$
$$(x+2)(x+3)(x-3) = 0$$
$$x+2 = 0 \text{ or } x+3 = 0 \text{ or } x-3 = 0$$
$$x = -2 \text{ or } x = -3 \text{ or } x = 3$$
$$x = -2, \ x = \pm 3$$

83. Let x represent the first integer.
Let $x + 2$ represent the second integer.
Let $x + 4$ represent the third integer.
$$4(x+4) + x^4 = (x+2)^2 + 24$$
$$4x + 16 + x^4 = x^2 + 4x + 4 + 24$$
$$4x + 16 + x^4 = x^2 + 4x + 28$$
$$x^4 - x^2 - 12 = 0$$
$$(x^2 - 4)(x^2 + 3) = 0$$
$$x^2 - 4 = 0 \text{ or } x^2 + 3 = 0$$
$$x^2 = 4 \quad \text{ or } \quad x^2 = -3$$
$$x = \pm 2 \quad \text{ or } \quad \text{Not Real;}$$
if $x = 2, x + 2 = 4, x + 4 = 6$;
if $x = -2, x + 2 = 0, x + 4 = 2$;
$x = 2, 4, 6$ or $x = -2, 0, 2$

85. Let w represent the width.
$$w(w+2) = 143$$
$$w^2 + 2w = 143$$
$$w^2 + 2w - 143 = 0$$
$$(w-11)(w+13) = 0$$
$$w - 11 = 0 \text{ or } w + 13 = 0$$
$$w = 11 \quad \text{ or } w = -13$$
11 inches by 13 inches

87.
$$24\pi r = \frac{2}{3}\pi r^3 + \pi r^2(6)$$
$$0 = \frac{2}{3}\pi r^3 + 6\pi r^2 - 24\pi r$$
$$3(0) = 3\left(\frac{2}{3}\pi r^3 + 6\pi r^2 - 24\pi r\right)$$
$$0 = 2\pi r^3 + 18\pi r^2 - 72\pi r$$
$$0 = 2\pi r(r^2 + 9r - 36)$$
$$0 = 2\pi r(r-3)(r+12)$$
$$2\pi r = 0 \text{ or } r - 3 = 0 \text{ or } r + 12 = 0$$
$$r = 0 \quad \text{ or } \quad r = 3 \text{ or } \quad r = -12$$
$r = 3$ m;
$r = 0$m and $r = 12$m do not fit the context.

89. Let x represent the number of decreases in price.
$$(70 - 2x)(15 + 3x) = 2250$$
$$1050 + 210x - 30x - 6x^2 = 2250$$
$$6x^2 - 180x + 1200 = 0$$
$$x^2 - 30x + 200 = 0$$
$$(x - 10)(x - 20) = 0$$
$$x - 10 = 0 \text{ or } x - 20 = 0$$
$$x = 10 \text{ or } x = 20$$
10, \$2 decreases results in a price of \$50 and a sale of 45 shoes.
20, \$2 decreases results in a price of \$30 and a sale of 75 shoes.

91. $h = -16t^2 + vt + k$
    $h = -16t^2 + 176t - 480$

    a. $h = -16(4)^2 + 176(4) - 480$
    $h = -32$ feet
    32 feet below the rim.

    b. $-480 = -16t^2 + 176t - 480$
    $0 = -16t^2 + 176t$
    $0 = -16t(t - 11)$
    $-16t = 0$ or $t - 11 = 0$
    $t = 0$ or $t = 11$
    The pebble returns after 11 seconds.

    c. $h = -16(5)^2 + 176(5) - 480 = 0;$
    $h = -16(6)^2 + 176(6) - 480 = 0;$
    The pebble is at the canyon's rim.

93. $\dfrac{1}{20} + \dfrac{1}{30} = \dfrac{1}{x}$

    $60x\left(\dfrac{1}{20} + \dfrac{1}{30}\right) = 60x\left(\dfrac{1}{x}\right)$

    $3x + 2x = 60$
    $5x = 60$
    $x = 12$ minutes

95. Let v represent the rate Tom can row in still water.
    Then $v + 4$ is the rate downstream,
    $v - 4$ is the rate upstream.
    $t_{up} + t_{down} = 3$

    $\dfrac{5}{v-4} + \dfrac{5}{v+4} = 3$

    $(v+4)(v-4)\left(\dfrac{5}{v-4} + \dfrac{5}{v+4}\right) = 3(v+4)(v-4)$

    $5(v+4) + 5(v-4) = 3(v^2 - 16)$
    $5v + 20 + 5v - 20 = 3v^2 - 48$
    $10v = 3v^2 - 48$
    $0 = 3v^2 - 10v - 48$
    $0 = (3v + 8)(v - 6)$
    $3v + 8 = 0$ or $v - 6 = 0$
    $v = -\dfrac{8}{3}$ $v = 6;$
    $v = 6$ mph

97. $C = \dfrac{92P}{100 - P}$

    $100 = \dfrac{92P}{100 - P}$

    $(100 - P)\left[100 = \dfrac{92P}{100 - P}\right]$

    $10000 - 100P = 92P$
    $10000 = 192P$
    $52.1\% = P$

99. $T = 0.407R^{\frac{3}{2}}$

    a. $88 = 0.407R^{\frac{3}{2}}$
    $216.22 \approx R^{\frac{3}{2}}$
    $(216.22)^{\frac{2}{3}} \approx \left(R^{\frac{3}{2}}\right)^{\frac{2}{3}}$
    $R \approx 36$ million miles

    b. $225 = 0.407R^{\frac{3}{2}}$
    $552.83 \approx R^{\frac{3}{2}}$
    $(552.83)^{\frac{2}{3}} \approx \left(R^{\frac{3}{2}}\right)^{\frac{2}{3}}$
    $R \approx 67$ million miles

    c. $365 = 0.407R^{\frac{3}{2}}$
    $896.81 \approx R^{\frac{3}{2}}$
    $(896.81)^{\frac{2}{3}} \approx \left(R^{\frac{3}{2}}\right)^{\frac{2}{3}}$
    $R \approx 93$ million miles

    d. $687 = 0.407R^{\frac{3}{2}}$
    $1687.96 \approx R^{\frac{3}{2}}$
    $(1687.96)^{\frac{2}{3}} \approx \left(R^{\frac{3}{2}}\right)^{\frac{2}{3}}$
    $R \approx 142$ million miles

e. $4333 = 0.407R^{\frac{3}{2}}$

$10646.19 \approx R^{\frac{3}{2}}$

$\left(10646.19\right)^{\frac{2}{3}} \approx \left(R^{\frac{3}{2}}\right)^{\frac{2}{3}}$

$R \approx 484$ million miles

f. $10759 = 0.407R^{\frac{3}{2}}$

$26434.89 \approx R^{\frac{3}{2}}$

$\left(26434.89\right)^{\frac{2}{3}} \approx \left(R^{\frac{3}{2}}\right)^{\frac{2}{3}}$

$R \approx 887$ million miles

101. The constant "3" was not multiplied by the LCD;

$3 - \dfrac{8}{x+3} = \dfrac{1}{x}$

$3x(x+3) - 8x = x+3$

$3x^2 + 9x - 8x = x+3$

$3x^2 + x = x+3$

$3x^2 = 3$

$x^2 = 1$

$x = \pm 1$

103. $\dfrac{\sqrt{x-1}}{x^2-4}$

$x - 1 \geq 0$ and $x^2 - 4 \neq 0$

$x \geq 1$ and $(x+2)(x-2) \neq 0$

$x \geq 1$ and $x + 2 \neq 0$ and $x - 2 \neq 0$

$x \geq 1$ and $x \neq -2$ and $x \neq 2$

$x \in [1,2) \cup (2,\infty)$

105. a. $\left|x^2 - 2x - 25\right| = 10$

$x^2 - 2x - 25 = 10$ or $x^2 - 2x - 25 = -10$

$x^2 - 2x - 35 = 0$ or $x^2 - 2x - 15 = 0$

$(x-7)(x+5) = 0$ or $(x-5)(x+3) = 0$

$x - 7 = 0$ or $x + 5 = 0$ or $x - 5 = 0$ or $x + 3 = 0$

$x = 7$ or $x = -5$ or $x = 5$ or $x - 3$

$x = -5, -3, 5, 7$

b. $\left|x^2 - 5x - 10\right| = 4$

$x^2 - 5x - 10 = 4$ or $x^2 - 5x - 10 = -4$

$x^2 - 5x - 14 = 0$ or $x^2 - 5x - 6 = 0$

$(x-7)(x+2) = 0$ or $(x-6)(x+1) = 0$

$x - 7 = 0$ or $x + 2 = 0$ or $x - 6 = 0$ or $x + 1 = 0$

$x = 7$ or $x = -2$ or $x = 6$ or $x = 1$

$x = -2, -1, 6, 7$

c. $\left|x^2 - 4\right| = x + 2$

$x^2 - 4 = x + 2$ or $x^2 - 4 = -(x+2)$

$x^2 - x - 6 = 0$ or $x^2 - 4 = -x - 2$

$(x-3)(x+2) = 0$ or $x^2 + x - 2 = 0$

$x - 3 = 0$ or $x + 2 = 0$ or $(x+2)(x-1) = 0$

$x = 3$ or $x = -2$ or $x + 2 = 0$ or $x - 1 = 0$

$x = -2, x = 1$

$x = -2, 1, 3$

d. $\left|x^2 - 9\right| = -x + 3$

$x^2 - 9 = -x + 3$ or $x^2 - 9 = -(-x+3)$

$x^2 + x - 12 = 0$ or $x^2 - 9 = x - 3$

$(x+4)(x-3) = 0$ or $x^2 - x - 6 = 0$

$x + 4 = 0$ or $x - 3 = 0$ or $(x-3)(x+2) = 0$

$x = -4$ $x = 3$ or $x - 3 = 0$ or $x + 2 = 0$

$x = 3, x = -2$

$x = -4, -2, 3$

e. $\left|x^2 - 7x\right| = -x + 7$

$x^2 - 7x = -x + 7$ or $x^2 - 7x = -(-x + 7)$

$x^2 - 6x - 7 = 0$ $x^2 - 7x = x - 7$

$(x - 7)(x + 1) = 0$ $x^2 - 8x + 7 = 0$

$x - 7 = 0$ or $x + 1 = 0$ $(x - 7)(x - 1) = 0$

$x = 7$ $x = -1$ $x - 7 = 0$ $x - 1 = 0$

$x = 7, x = 1$

$x = -1, 1, 7$

f. $\left|x^2 - 5x - 2\right| = x + 5$

$x^2 - 5x - 2 = x + 5$ or $x^2 - 5x - 2 = -(x + 5)$

$x^2 - 6x - 7 = 0$ $x^2 - 5x - 2 = -x - 5$

$(x - 7)(x + 1) = 0$ $x^2 - 4x + 3 = 0$

$x - 7 = 0$ or $x + 1 = 0$ $(x - 3)(x - 1) = 0$

$x = 7$ $x = -1$ $x - 3 = 0$ $x - 1 = 0$

$x = 3, x = 1$

$x = -1, 1, 3, 7$

107. $x^2 + 10^2 = 12^2$

$x^2 + 100 = 144$

$x^2 = 44$

$x = \pm\sqrt{44}$

$x = \pm 2\sqrt{11}$

$x = 2\sqrt{11}$ cm

109. $2x - 3 < 7$ and $x + 2 > 1$

$2x < 10$ and $x > -1$

$x < 5$ and $x > -1$

$-1 < x < 5$

Chapter 1 Summary and Concept Review

1. a. $6x - (2 - x) = 4(x - 5)$

$6(-6) - \left(2 - (-6)\right) = 4(-6 - 5)$

$-36 - (2 + 6) = 4(-11)$

$-36 - 8 = -44$

$-44 = -44$; yes

b. $\dfrac{3}{4}b + 2 = \dfrac{5}{2}b + 16$

$\dfrac{3}{4}(-8) + 2 = \dfrac{5}{2}(-8) + 16$

$-6 + 2 = -20 + 16$

$-4 = -4$

yes

c. $4d - 2 = -\dfrac{1}{2} + 3d$

$4\left(\dfrac{3}{2}\right) - 2 = -\dfrac{1}{2} + 3\left(\dfrac{3}{2}\right)$

$6 - 2 = -\dfrac{1}{2} + \dfrac{9}{2}$

$4 = 4$

yes

3. $3(2n - 6) + 1 = 7$

$6n - 18 + 1 = 7$

$6n - 17 = 7$

$6n = 24$

$n = 4$

5. $\dfrac{1}{2}x + \dfrac{2}{3} = \dfrac{3}{4}$

$12\left[\dfrac{1}{2}x + \dfrac{2}{3} = \dfrac{3}{4}\right]$

$6x + 8 = 9$

$6x = 1$

$x = \dfrac{1}{6}$

7. $-\dfrac{g}{6} = 3 - \dfrac{1}{2} - \dfrac{5g}{12}$

$12\left[-\dfrac{g}{6} = 3 - \dfrac{1}{2} - \dfrac{5g}{12} \right]$

$-2g = 36 - 6 - 5g$

$3g = 30$

$g = 10$

9. $P = 2L + 2W$

$P - 2W = 2L$

$\dfrac{P - 2W}{2} = L$

11. $2x - 3y = 6$

$-3y = -2x + 6$

$y = \dfrac{2}{3}x - 2$

13. $3(4) + \dfrac{1}{2}\pi(1.5)^2$

$12 + \dfrac{9}{8}\pi \ \text{ft}^2 \approx 15.5 \ \text{ft}^2$

15. $a \geq 35$

17. $s \leq 65$

19. $7x > 35$

$x > 5$

$(5, \infty)$

21. $2(3m - 2) \leq 8$

$6m - 4 \leq 8$

$6m \leq 12$

$m \leq 2$

$(-\infty, 2]$

23. $-4 < 2b + 8$ and $3b - 5 > -32$

$-12 < 2b$ and $3b > -27$

$-6 < b$ and $b > -9$

$b > -6$ and $b > -9$

$(-6, \infty)$

25. a. $\dfrac{7}{n - 3}$

$n - 3 \neq 0$

$n \neq 3$

$(-\infty, 3) \cup (3, \infty)$

b. $\dfrac{5}{2x - 3}$

$2x - 3 \neq 0$

$2x \neq 3$

$x \neq \dfrac{3}{2}$

$\left(-\infty, \dfrac{3}{2}\right) \cup \left(\dfrac{3}{2}, \infty\right)$

c. $\sqrt{x + 5}$

$x + 5 \geq 0$

$x \geq -5$

$[-5, \infty)$

d. $\sqrt{-3n + 18}$

$-3n + 18 \geq 0$

$-3n \geq -18$

$n \leq 6$

$(-\infty, 6]$

27. $7 = |x - 3|$

$x - 3 = 7$ or $x - 3 = -7$

$x = 10$ or $x = -4$

$\{-4, 10\}$

29. $|-2x + 3| = 13$

$-2x + 3 = 13$ or $-2x + 3 = -13$

$-2x = 10$ or $-2x = -16$

$x = -5$ $x = 8$

$\{-5, 8\}$

31. $-3|x + 2| - 2 < -14$

$-3|x + 2| < -12$

$|x + 2| > 4$

$x + 2 > 4$ or $x + 2 < -4$

$x > 2$ or $x < -6$

$x \in (-\infty, -6) \cup (2, \infty)$

33. $|3x + 5| = -4$

Absolute value can never be negative.

$\{\ \}$

35. $2|x + 1| > -4$

$|x + 1| > -2$

Absolute value is always greater than a negative number.

$(-\infty, \infty)$

37. $\dfrac{|3x-2|}{2}+6 \geq 10$

$\dfrac{|3x-2|}{2} \geq 4$

$|3x-2| \geq 8$

$3x-2 \geq 8$ or $3x-2 \leq -8$

$3x \geq 10$ or $3x \leq -6$

$x \geq \dfrac{10}{3}$ or $x \leq -2$

$(-\infty,-2] \cup \left[\dfrac{10}{3},\infty\right)$

39. $\sqrt{-72}=\sqrt{-1(36)(2)}=6\sqrt{2}\,i$

41. $\dfrac{-10+\sqrt{-50}}{5}=\dfrac{-10+\sqrt{-1(25)(2)}}{5}$

$=\dfrac{-10+5\sqrt{2}\,i}{5}=-2+\sqrt{2}\,i$

43. $i^{57}=\left(i^{4}\right)^{14}i=i$

45. $\dfrac{5i}{1-2i} \cdot \dfrac{1+2i}{1+2i}=\dfrac{5i+10i^{2}}{1+2i-2i-4i^{2}}$

$=\dfrac{-10+5i}{5}=-2+i$

47. $(2+3i)(2-3i)=4-6i+6i-9i^{2}=13$

49. $x^{2}-9=-34, x=5i$;

$(5i)^{2}-9=-34$

$25i^{2}-9=-34$

$-25-9=-34$

$-34=-34$ verified;

$(-5i)^{2}-9=-34$

$25i^{2}-9=-34$

$-25-9=-34$

$-34=-34$ verified

51. a. $-3=2x^{2}$

$2x^{2}+3=0$

$a=2,b=0,c=3$

b. $7=-2x+11$ is not quadratic

c. $99=x^{2}-8x$

$x^{2}-8x-99=0$

$a=1,b=-8,c=-99$

d. $20=4-x^{2}$

$x^{2}+16=0$

$a=1,b=0,c=16$

53. a. $x^{2}-9=0$

$x^{2}=9$

$x=\pm 3$

b. $2(x-2)^{2}+1=11$

$2(x-2)^{2}=10$

$(x-2)^{2}=5$

$x-2=\pm\sqrt{5}$

$x=2\pm\sqrt{5}$

c. $3x^{2}+15=0$

$3x^{2}=-15$

$x^{2}=-5$

$x=\pm\sqrt{5}\,i$

d. $-2x^{2}+4=-46$

$-2x^{2}=-50$

$x^{2}=25$

$x=\pm 5$

55. a. $x^{2}-4x=-9$

$x^{2}-4x+9=0$

$a=1,b=-4,c=9$

$x=\dfrac{-(-4)\pm\sqrt{(-4)^{2}-4(1)(9)}}{2(1)}$

$x=\dfrac{4\pm\sqrt{-20}}{2}$

$x=\dfrac{4\pm 2\sqrt{5}\,i}{2}$

$x=2\pm\sqrt{5}\,i$

$x\approx 2\pm 2.24\,i$

b. $4x^2 + 7 = 12x$

$4x^2 - 12x + 7 = 0$

$a = 4, b = -12, c = 7$

$x = \dfrac{-(-12) \pm \sqrt{(-12)^2 - 4(4)(7)}}{2(4)}$

$x = \dfrac{12 \pm \sqrt{32}}{8}$

$x = \dfrac{12 \pm 4\sqrt{2}}{8}$

$x = \dfrac{3 \pm \sqrt{2}}{2}$

$x \approx 2.21 \text{ or } x \approx 0.79$

c. $2x^2 - 6x + 5 = 0$

$a = 2, b = -6, c = 5$

$x = \dfrac{-(-6) \pm \sqrt{(-6)^2 - 4(2)(5)}}{2(2)}$

$x = \dfrac{6 \pm \sqrt{-4}}{4}$

$x = \dfrac{6 \pm 2i}{4}$

$x = \dfrac{3}{2} \pm \dfrac{1}{2}i$

57. a. $120 = -16t^2 + 64t + 80$

$16t^2 - 64t + 40 = 0$

$2t^2 - 8t + 5 = 0$

$a = 2, b = -8, c = 5$

$t = \dfrac{-(-8) \pm \sqrt{(-8)^2 - 4(2)(5)}}{2(2)}$

$t = \dfrac{8 \pm \sqrt{24}}{4}$

$t = \dfrac{8 \pm 2\sqrt{6}}{4}$

$t = \dfrac{4 \pm \sqrt{6}}{2}$

$t \approx 3.22 \text{ or } t \approx 0.78$

0.8 seconds

b. 3.2 seconds

c. $0 = -16t^2 + 64t + 80$

$0 = -16(t^2 - 4t - 5)$

$0 = (t - 5)(t + 1)$

$t - 5 = 0 \text{ or } t + 1 = 0$

$t = 5 \quad \text{ or } t = -1$

5 seconds

59. Let x represent the time for the smaller pump.

$\dfrac{1}{x-3} + \dfrac{1}{x} = \dfrac{1}{2}$

$2x(x-3)\left[\dfrac{1}{x-3} + \dfrac{1}{x} = \dfrac{1}{2}\right]$

$2x + 2(x-3) = x(x-3)$

$2x + 2x - 6 = x^2 - 3x$

$0 = x^2 - 7x + 6$

$0 = (x-6)(x-1)$

$x - 6 = 0 \text{ or } x - 1 = 0$

$x = 6 \quad \text{ or } x = 1$

It takes the smaller pump 6 hours.

61. $3x^3 + 5x^2 = 2x$

$3x^3 + 5x^2 - 2x = 0$

$x(3x^2 + 5x - 2) = 0$

$x(3x - 1)(x + 2) = 0$

$x = 0 \text{ or } 3x - 1 = 0 \text{ or } x + 2 = 0$

$3x = 1 \quad \text{ or } \quad x = -2$

$x = \dfrac{1}{3}$

$x = -2, x = 0, x = \dfrac{1}{3}$

63. $x^4 - \dfrac{1}{16} = 0$

$\left(x^2 + \dfrac{1}{4}\right)\left(x^2 - \dfrac{1}{4}\right) = 0$

$x^2 + \dfrac{1}{4} = 0 \text{ or } x^2 - \dfrac{1}{4} = 0$

$x^2 = -\dfrac{1}{4} \quad \text{ or } \quad x^2 = \dfrac{1}{4}$

$x = \pm\dfrac{1}{2}i \quad \text{ or } \quad x = \pm\dfrac{1}{2}$

$x = \pm\dfrac{1}{2}, x = \pm\dfrac{1}{2}i$

Chapter 1 Summary and Concept Review

65. $\dfrac{3h}{h+3} - \dfrac{7}{h^2+3h} = \dfrac{1}{h}$

$\dfrac{3h}{h+3} - \dfrac{7}{h(h+3)} = \dfrac{1}{h}$

$h(h+3)\left[\dfrac{3h}{h+3} - \dfrac{7}{h(h+3)} = \dfrac{1}{h}\right]$

$3h^2 - 7 = h+3$

$3h^2 - h - 10 = 0$

$(3h+5)(h-2) = 0$

$3h+5 = 0 \quad \text{or} \quad h-2 = 0$

$h = -\dfrac{5}{3} \quad \text{or} \quad h = 2$

67. $\dfrac{\sqrt{x^2+7}}{2} + 3 = 5$

$\dfrac{\sqrt{x^2+7}}{2} = 2$

$\sqrt{x^2+7} = 4$

$x^2 + 7 = 16$

$x^2 = 9$

$x = \pm 3$

69. $\sqrt{3x+4} = 2 - \sqrt{x+2}$

$\left(\sqrt{3x+4}\right)^2 = \left(2-\sqrt{x+2}\right)^2$

$3x+4 = 4 - 4\sqrt{x+2} + x + 2$

$2x - 2 = -4\sqrt{x+2}$

$2(x-1) = -4\sqrt{x+2}$

$x - 1 = -2\sqrt{x+2}$

$(x-1)^2 = \left(-2\sqrt{x+2}\right)^2$

$x^2 - 2x + 1 = 4(x+2)$

$x^2 - 2x + 1 = 4x + 8$

$x^2 - 6x - 7 = 0$

$(x-7)(x+1) = 0$

$x-7 = 0 \text{ or } x+1 = 0$

$x = 7 \quad \text{or } x = -1$

$x = -1\,; x = 7$ is extraneous.

71. $-2(5x+2)^{\frac{2}{3}} + 17 = -1$

$-2(5x+2)^{\frac{2}{3}} = -18$

$(5x+2)^{\frac{2}{3}} = 9$

$\left[(5x+2)^{\frac{2}{3}}\right]^{\frac{3}{2}} = 9^{\frac{3}{2}}$

$5x+2 = 27 \text{ or } 5x+2 = -27$

$5x = 25 \quad \text{or} \quad 5x = -29$

$x = 5 \quad \text{or} \quad x = -5.8$

73. $x^4 - 7x^2 = 18$

$x^4 - 7x^2 - 18 = 0$

$(x^2 - 9)(x^2 + 2) = 0$

$x^2 - 9 = 0 \text{ or } x^2 + 2 = 0$

$x^2 = 9 \qquad x^2 = -2$

$x = \pm 3 \qquad x = \pm i\sqrt{2}$

$x = -3, x = 3, x = -i\sqrt{2}, x = i\sqrt{2}$

75. Let w represent the width.
Let $w + 3$ represent the length.

$w(w+3) = 54$

$w^2 + 3w - 54 = 0$

$(w+9)(w-6) = 0$

$w+9 = 0 \text{ or } w-6 = 0$

$w = -9 \quad \text{or } w = 6$

6 inches by 9 inches
width, 6 in; length, 9 in

77. Let x represent the number of \$2 decreases.

$(50 - 2x)(40 + 5x) = 2520$

$2000 + 250x - 80x - 10x^2 = 2520$

$0 = 10x^2 - 170x + 520$

$0 = x^2 - 17x + 52$

$0 = (x-4)(x-13)$

$x - 4 = 0 \text{ or } x - 13 = 0$

$x = 4 \quad \text{or } x = 13$

$50 - 2(4) = \$42$ or $50 - 2(13) = \$24$

Chapter 1 Mixed Review

1. a. $\dfrac{10}{\sqrt{x-8}}$

   $x - 8 > 0$
   $x > 8$
   $x \in (8, \infty)$

   b. $\dfrac{-5}{3x+4}$

   $3x + 4 \neq 0$
   $3x \neq -4$
   $x \neq -\dfrac{4}{3}$

   $x \in \left(-\infty, \dfrac{-4}{3}\right) \cup \left(\dfrac{-4}{3}, \infty\right)$

3. a. $-2x^3 + 4x^2 = 50x - 100$

   $0 = 2x^3 - 4x^2 + 50x - 100$

   $0 = 2(x^3 - 2x^2 + 25x - 50)$

   $0 = 2\left[x^2(x-2) + 25(x-2)\right]$

   $0 = 2(x-2)(x^2+25)$

   $x - 2 = 0$ or $x^2 + 25 = 0$

   $x = 2$ or $x^2 = -25$

   or $x = \pm 5i$;

   $x = 2, \quad x = \pm 5i$

   b. $-3x^4 - 375x = 0$

   $-3x(x^3 + 125) = 0$

   $-3x(x+5)(x^2 - 5x + 25) = 0$

   $-3x = 0$ or $x+5 = 0$ or $x^2 - 5x + 25 = 0$

   $x = 0$ or $x = -5$

   or $x = \dfrac{-(-5) \pm \sqrt{(-5)^2 - 4(1)(25)}}{2(1)}$

   $x = \dfrac{5 \pm \sqrt{-75}}{2}$

   $x = \dfrac{5 \pm 5i\sqrt{3}}{2} = \dfrac{5}{2} \pm \dfrac{5i\sqrt{3}}{2}$;

   $x = 0, x = -5, x = \dfrac{5}{2} \pm \dfrac{5i\sqrt{3}}{2}$

   c. $-2|3x+1| = -12$

   $|3x+1| = 6$

   $3x + 1 = 6$ or $3x + 1 = -6$

   $3x = 5 \qquad 3x = -7$

   $x = \dfrac{5}{3} \qquad x = -\dfrac{7}{3}$;

   $x = -\dfrac{7}{3}, x = \dfrac{5}{3}$

   d. $-3\left|\dfrac{x}{3} - 5\right| \leq -12$

   $\left|\dfrac{x}{3} - 5\right| \geq 4$

   $\dfrac{x}{3} - 5 \geq 4$ or $\dfrac{x}{3} - 5 \leq -4$

   $\dfrac{x}{3} \geq 9$ or $\dfrac{x}{3} \leq 1$

   $x \geq 27$ or $x \leq 3$

   $(-\infty, 3] \cup [27, \infty)$

   e. $v^{\frac{4}{3}} = 81$

   $\left(v^{\frac{4}{3}}\right)^{\frac{3}{4}} = 81^{\frac{3}{4}}$

   $v = \pm 27$

   f. $-2(x+1)^{\frac{1}{4}} = -6$

   $(x+1)^{\frac{1}{4}} = 3$

   $\left[(x+1)^{\frac{1}{4}}\right]^4 = 3^4$

   $x + 1 = 81$

   $x = 80$

5. $3x + 4y = -12$

   $4y = -3x - 12$

   $y = -\dfrac{3}{4}x - 3$

Chapter 1 Mixed Review

7. a. $5x - (2x - 3) + 3x = -4(5 + x) + 3$

       $5x - 2x + 3 + 3x = -20 - 4x + 3$

       $3 + 6x = -17 - 4x$

       $10x = -20$

       $x = -2$

   b. $\dfrac{n}{5} - 2 = 2 - \dfrac{5}{3} - \dfrac{4}{15}n$

       $15\left[\dfrac{n}{5} - 2 = 2 - \dfrac{5}{3} - \dfrac{4}{15}n\right]$

       $3n - 30 = 30 - 25 - 4n$

       $3n - 30 = 5 - 4n$

       $7n = 35$

       $n = 5$

9. $x^2 - 18x + 77 = 0$

   $(x - 7)(x - 11) = 0$

   $x - 7 = 0$ or $x - 11 = 0$

   $x = 7$ or $x = 11$

11. $4x^2 - 5 = 19$

   $4x^2 = 24$

   $x^2 = 6$

   $x = \pm\sqrt{6}$

13. $25x^2 + 16 = 40x$

   $25x^2 - 40x + 16 = 0$

   $(5x - 4)(5x - 4) = 0$

   $5x - 4 = 0$

   $5x = 4$

   $x = \dfrac{4}{5}$

15. $2x^4 - 50 = 0$

   $2\left(x^4 - 25\right) = 0$

   $2\left(x^2 + 5\right)\left(x^2 - 5\right) = 0$

   $x^2 + 5 = 0$ or $x^2 - 5 = 0$

   $x^2 = -5$ or $x^2 = 5$

   $x = \pm\sqrt{5}\,i$ or $x = \pm\sqrt{5}$

17. a. $\sqrt{2v - 3} + 3 = v$

       $\sqrt{2v - 3} = v - 3$

       $2v - 3 = v^2 - 6v + 9$

       $0 = v^2 - 8v + 12$

       $0 = (v - 6)(v - 2)$

       $v - 6 = 0$ or $v - 2 = 0$

       $v = 6$ or $v = 2$

       $v = 6$; $v = 2$ is extraneous

   b. $\sqrt[3]{x^2 - 9} + \sqrt[3]{x - 11} = 0$

       $\sqrt[3]{x^2 - 9} = -\sqrt[3]{x - 11}$

       $x^2 - 9 = -(x - 11)$

       $x^2 + x - 20 = 0$

       $(x - 4)(x + 5) = 0$

       $x - 4 = 0$ or $x + 5 = 0$

       $x = 4$ or $x = -5$

   c. $\sqrt{x + 7} - \sqrt{2x} = 1$

       $\sqrt{x + 7} = 1 + \sqrt{2x}$

       $\left(\sqrt{x + 7}\right)^2 = \left(1 + \sqrt{2x}\right)^2$

       $x + 7 = 1 + 2\sqrt{2x} + 2x$

       $-x + 6 = 2\sqrt{2x}$

       $(-x + 6)^2 = \left(2\sqrt{2x}\right)^2$

       $x^2 - 12x + 36 = 4(2x)$

       $x^2 - 12x + 36 = 8x$

       $x^2 - 20x + 36 = 0$

       $(x - 2)(x - 18) = 0$

       $x - 2 = 0$ or $x - 18 = 0$

       $x = 2$ or $x = -18$;

       $x = 2$; $x = 18$ is extraneous

19. $\dfrac{2(75) + 2(79) + x}{5} = 78$

   $2(75) + 2(79) + x = 390$

   $150 + 158 + x = 390$

   $x = 82$ inches

   $6'10''$

Chapter 1: Equations and Inequalities

Chapter 1 Practice Test

1. a. $-\dfrac{2}{3}x-5=7-(x+3)$

   $-\dfrac{2}{3}x-5=7-x-3$

   $-\dfrac{2}{3}x-5=4-x$

   $\dfrac{1}{3}x=9$

   $x=27$

   b. $-5.7+3.1x=14.5-4(x+1.5)$

   $-5.7+3.1x=14.5-4x-6$

   $-5.7+3.1x=8.5-4x$

   $7.1x=14.2$

   $x=2$

   c. $P=C+kC$

   $P=C(1+k)$

   $\dfrac{P}{1+k}=C$

   d. $2|2x+5|-17=-11$

   $2|2x+5|=6$

   $|2x+5|=3$

   $2x+5=3$ or $2x+5=-3$

   $2x=-2$ or $2x=-8$

   $x=-1$ or $x=-4$

3. a. $-\dfrac{2}{5}x+7<19$

   $-\dfrac{2}{5}x<12$

   $\left(-\dfrac{5}{2}\right)\left(-\dfrac{2}{5}x\right)>\left(-\dfrac{5}{2}\right)(12)$

   $x>-30$

   b. $-1<3-x\le8$

   $-4<-x\le5$

   $4>x\ge-5$

   $-5\le x<4$

c. $\dfrac{1}{2}x+3<9$ or $\dfrac{2}{3}x-1\ge3$

   $\dfrac{1}{2}x<6$ or $\dfrac{2}{3}x\ge4$

   $x<12$ or $x\ge6$

   $x\in\mathbb{R}$

d. $\dfrac{1}{2}|x-3|+\dfrac{5}{4}=\dfrac{7}{4}$

   $\dfrac{1}{2}|x-3|=\dfrac{2}{4}$

   $|x-3|=1$

   $x-3=1$ or $x-3=-1$

   $x=4$ or $x=2$

e. $-\dfrac{2}{3}|x+1|-5<-7$

   $-\dfrac{2}{3}|x+1|<-2$

   $|x+1|>3$

   $x+1>3$ or $x+1<-3$

   $x>2$ or $x<-4$

5. $z^2-7z-30=0$

   $(z-10)(z+3)=0$

   $z-10=0$ or $z+3=0$

   $z=10$ or $z=-3$

7. $(x-1)^2+3=0$

   $(x-1)^2=-3$

   $x-1=\pm i\sqrt{3}$

   $x=1\pm i\sqrt{3}$

9. $3x^2-20x=-12$

   $3x^2-20x+12=0$

   $(3x-2)(x-6)=0$

   $3x-2=0$ or $x-6=0$

   $3x=2$ or $x=6$

   $x=\dfrac{2}{3}$ or $x=6$

11. $\dfrac{2}{x-3}+\dfrac{2x}{x+2}=\dfrac{x^2+16}{x^2-x-16}$

$(x-3)(x+2)\left[\dfrac{2}{x-3}+\dfrac{2x}{x+2}=\dfrac{x^2+16}{(x-3)(x+2)}\right]$

$2(x+2)+2x(x-3)=x^2+16$

$2x+4+2x^2-6x=x^2+16$

$2x^2-4x+4=x^2+16$

$x^2-4x-12=0$

$(x-6)(x+2)=0$

$x-6=0$ or $x+2=0$

$x=6$ or $x=-2$

$x=6, x=-2$ is extraneous

13. $\sqrt{x}+1=\sqrt{2x-7}$

$\left(\sqrt{x}+1\right)^2=\left(\sqrt{2x-7}\right)^2$

$x+2\sqrt{x}+1=2x-7$

$2\sqrt{x}=x-8$

$\left(2\sqrt{x}\right)^2=(x-8)^2$

$4x=x^2-16x+64$

$0=x^2-20x+64$

$0=(x-4)(x-16)$

$x-4=0$ or $x-16=0$

$x=4$ $x=16$

Check:

$\sqrt{4}+1=\sqrt{2(4)-7}$

$2+1=\sqrt{1}$

$3\neq 1$

Check:

$\sqrt{16}+1=\sqrt{2(16)-7}$

$4+1=\sqrt{32-7}$

$5=\sqrt{25}$

$5=5$

$x=16; x=4$ is extraneous

15. $P=(120-2x)(3+0.10x)$

$P=360+12x-6x-0.2x^2$

$P=-0.2x^2+6x+360$

a. $405=-0.2x^2+6x+360$

$0.2x^2-6x+45=0$

$2x^2-60x+450=0$

$x^2-30x+225=0$

$(x-15)^2=0$

$x=15$

$3+0.10(15)=\$4.50$ per tin

b. $120-2\,(15)=90$ tins

17. $\dfrac{-8+\sqrt{-20}}{6}=\dfrac{-8+\sqrt{-4(5)}}{6}=\dfrac{-8+2\sqrt{5}\,i}{6}$

$=-\dfrac{4}{3}+\dfrac{\sqrt{5}}{3}i$

19. a. $\left(\dfrac{1}{2}+\dfrac{\sqrt{3}}{2}i\right)+\left(\dfrac{1}{2}-\dfrac{\sqrt{3}}{2}i\right)=1$

b. $\left(\dfrac{1}{2}+\dfrac{\sqrt{3}}{2}i\right)-\left(\dfrac{1}{2}-\dfrac{\sqrt{3}}{2}i\right)$

$=\dfrac{1}{2}+\dfrac{\sqrt{3}}{2}i-\dfrac{1}{2}+\dfrac{\sqrt{3}}{2}i=i\sqrt{3}$

c. $\left(\dfrac{1}{2}+\dfrac{\sqrt{3}}{2}i\right)\cdot\left(\dfrac{1}{2}-\dfrac{\sqrt{3}}{2}i\right)$

$=\dfrac{1}{4}-\dfrac{\sqrt{3}}{4}i+\dfrac{\sqrt{3}}{4}i-\dfrac{3}{4}i^2=1$

21. $(3i+5)(5-3i)=15i-9i^2+25-15i=34$

23. a. $2x^2 - 20x + 49 = 0$

$2x^2 - 20x = -49$

$x^2 - 10x = -\dfrac{49}{2}$

$x^2 - 10x + 25 = -\dfrac{49}{2} + 25$

$(x-5)^2 = \dfrac{1}{2}$

$x - 5 = \pm\sqrt{\dfrac{1}{2}}$

$x - 5 = \pm\sqrt{\dfrac{1}{2}}\sqrt{\dfrac{2}{2}}$

$x = 5 \pm \dfrac{\sqrt{2}}{2}$

b. $2x^2 - 5x = -4$

$x^2 - \dfrac{5}{2}x = -2$

$x^2 - \dfrac{5}{2}x + \dfrac{25}{16} = -2 + \dfrac{25}{16}$

$\left(x - \dfrac{5}{4}\right)^2 = \dfrac{-7}{16}$

$x - \dfrac{5}{4} = \pm\dfrac{\sqrt{7}}{4}i$

$x = \dfrac{5}{4} \pm \dfrac{\sqrt{7}}{4}i$

25. $F \approx 0.3 W^{\frac{3}{4}}$

a. $F \approx 0.3(1296)^{\frac{3}{4}}$

$F \approx 64.8$ g

b. $19.2 \approx 0.3 W^{\frac{3}{4}}$

$64 \approx W^{\frac{3}{4}}$

$(64)^{\frac{4}{3}} \approx \left(W^{\frac{3}{4}}\right)^{\frac{4}{3}}$

256 g $\approx W$

Chapter 1 Calculator Exploration and Discovery

1. They differ by 0.2.

3. They differ by $\sqrt{2}$ or ≈ 1.41.

Strengthening Core Skills

1. $2x^2 - 5x - 7 = 0$

$a = 2, b = -5, c = -7$

$x_1 = \dfrac{7}{2}, x_2 = -1$

$\dfrac{7}{2} + (-1) = \dfrac{5}{2} = -\dfrac{b}{a};$

$\dfrac{7}{2} \cdot (-1) = \dfrac{-7}{2} = \dfrac{c}{a}$

3. $x^2 - 10x + 37 = 0$

$a = 1, b = -10, c = 37$

$x_1 = 5 + 2\sqrt{3}\,i, x_2 = 5 - 2\sqrt{3}\,i$

$\left(5 + 2\sqrt{3}\,i\right) + \left(5 - 2\sqrt{3}\,i\right) = 10 = \dfrac{-b}{a};$

$\left(5 + 2\sqrt{3}\,i\right)\left(5 - 2\sqrt{3}\,i\right) = 25 + 12 = 37 = \dfrac{c}{a}$

Technology Highlight

1. $y = \pm 4.8$; $y = \pm 3.6$, Answers will vary.

2.1 Exercises

1. First, second

3. Radius, center

5. Answers will vary.

7.

Domain = {1, 2, 3, 4, 5}
Range = {2.75, 3.00, 3.25, 3.50, 3.75}

9. D = {1, 3, 5, 7, 9}
   R = {2, 4, 6, 8, 10}

11. D = {4, −1, 2, −3}
    R = {0, 5, 4, 2, 3}

13. $y = -\dfrac{2}{3}x + 1$

x	y
−6	$-\dfrac{2}{3}(-6)+1 = 4+1 = 5$
−3	$-\dfrac{2}{3}(-3)+1 = 2+1 = 3$
0	$-\dfrac{2}{3}(0)+1 = 0+1 = 1$
3	$-\dfrac{2}{3}(3)+1 = -2+1 = -1$
6	$-\dfrac{2}{3}(6)+1 = -4+1 = -3$
8	$-\dfrac{2}{3}(8)+1 = -\dfrac{16}{3}+1 = -\dfrac{13}{3}$

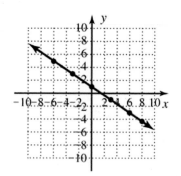

15. $x + 2 = |y|$

x	y
−2	0
0	2, -2
1	3, -3
3	5, -5
6	8, -8
7	9, -9

$$-2+2 = |y| \qquad\qquad 0+2 = |y|$$
$$0 = |y| \qquad\qquad 2 = |y|$$
$$0 = y; \qquad\qquad \pm 2 = y;$$

$$1+2 = |y| \qquad\qquad 3+2 = |y|$$
$$3 = |y| \qquad\qquad 5 = |y|$$
$$\pm 3 = y; \qquad\qquad \pm 5 = y;$$

$$6+2 = |y| \qquad\qquad 7+2 = |y|$$
$$8 = |y| \qquad\qquad 9 = |y|$$
$$\pm 8 = y; \qquad\qquad \pm 9 = y;$$

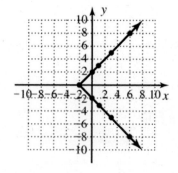

17. $y = x^2 - 1$

x	y
-3	$(-3)^2 - 1 = 9 - 1 = 8$
-2	$(-2)^2 - 1 = 4 - 1 = 3$
0	$(0)^2 - 1 = 0 - 1 = -1$
2	$(2)^2 - 1 = 4 - 1 = 3$
3	$(3)^2 - 1 = 9 - 1 = 8$
4	$(4)^2 - 1 = 16 - 1 = 15$

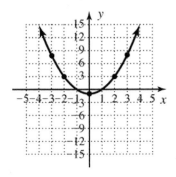

19. $y = \sqrt{25 - x^2}$

x	y
-4	$\sqrt{25 - (-4)^2} = \sqrt{25 - 16} = \sqrt{9} = 3$
-3	$\sqrt{25 - (-3)^2} = \sqrt{25 - 9} = \sqrt{16} = 4$
0	$\sqrt{25 - (0)^2} = \sqrt{25} = 5$
2	$\sqrt{25 - (2)^2} = \sqrt{25 - 4} = \sqrt{21}$
3	$\sqrt{25 - (3)^2} = \sqrt{25 - 9} = \sqrt{16} = 4$
4	$\sqrt{25 - (4)^2} = \sqrt{25 - 16} = \sqrt{9} = 3$

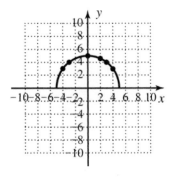

21. $x - 1 = y^2$

$y = \pm\sqrt{x - 1}$

x	y
10	$\sqrt{(10) - 1} = \sqrt{9} = \pm 3$
5	$\sqrt{(5) - 1} = \sqrt{4} = \pm 2$
4	$\sqrt{(4) - 1} = \pm\sqrt{3}$
2	$\sqrt{(2) - 1} = \sqrt{1} = \pm 1$
1.25	$\sqrt{(1.25) - 1} = \sqrt{0.25} = \pm 0.5$
1	$\sqrt{(1) - 1} = \sqrt{0} = 0$

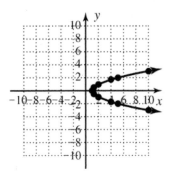

23. $y = \sqrt[3]{x + 1}$

x	y
-9	$\sqrt[3]{(-9) + 1} = \sqrt[3]{-8} = -2$
-2	$\sqrt[3]{(-2) + 1} = \sqrt[3]{-1} = -1$
-1	$\sqrt[3]{(-1) + 1} = \sqrt[3]{0} = 0$
0	$\sqrt[3]{(0) + 1} = \sqrt[3]{1} = 1$
4	$\sqrt[3]{(4) + 1} = \sqrt[3]{5}$
7	$\sqrt[3]{(7) + 1} = \sqrt[3]{8} = 2$

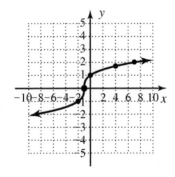

2.1 Exercises

25. $M = \left(\dfrac{x_1 + x_2}{2}, \dfrac{y_1 + y_2}{2} \right)$

$M = \left(\dfrac{1+5}{2}, \dfrac{8+(-6)}{2} \right)$

$M = \left(\dfrac{6}{2}, \dfrac{2}{2} \right)$

$M = (3,1)$

27. $M = \left(\dfrac{x_1 + x_2}{2}, \dfrac{y_1 + y_2}{2} \right)$

$M = \left(\dfrac{-4.5+3.1}{2}, \dfrac{9.2+(-9.8)}{2} \right)$

$M = \left(\dfrac{-1.4}{2}, \dfrac{-0.6}{2} \right)$

$M = (-0.7, -0.3)$

29. $M = \left(\dfrac{x_1 + x_2}{2}, \dfrac{y_1 + y_2}{2} \right)$

$M = \left(\dfrac{\dfrac{1}{5} + \left(\dfrac{-1}{10} \right)}{2}, \dfrac{\dfrac{-2}{3} + \dfrac{3}{4}}{2} \right)$

$M = \left(\dfrac{\dfrac{1}{10}}{2}, \dfrac{\dfrac{1}{12}}{2} \right)$

$M = \left(\dfrac{1}{20}, \dfrac{1}{24} \right)$

31. $(-5, -4)\,(5, 2)$

$M = \left(\dfrac{x_1 + x_2}{2}, \dfrac{y_1 + y_2}{2} \right)$

$M = \left(\dfrac{-5+5}{2}, \dfrac{-4+2}{2} \right)$

$M = \left(\dfrac{0}{2}, \dfrac{-2}{2} \right)$

$M = (0, -1)$

33. $(-4, -4)\,(2, 4)$

$M = \left(\dfrac{x_1 + x_2}{2}, \dfrac{y_1 + y_2}{2} \right)$

$M = \left(\dfrac{-4+2}{2}, \dfrac{-4+4}{2} \right)$

$M = \left(\dfrac{-2}{2}, \dfrac{0}{2} \right)$

$M = (-1, 0)$

The center of the circle is $(-1, 0)$.

35. $(-5, -4)\,(5, 2)$

$d = \sqrt{(x_2 - x_1)^2 + (y_2 - y_1)^2}$

$d = \sqrt{(5-(-5))^2 + (2-(-4))^2}$

$d = \sqrt{(10)^2 + (6)^2}$

$d = \sqrt{100 + 36}$

$d = \sqrt{136}$

$d = 2\sqrt{34}$

37. $(-4, -4)\,(2, 4)$

$d = \sqrt{(x_2 - x_1)^2 + (y_2 - y_1)^2}$

$d = \sqrt{(2-(-4))^2 + (4-(-4))^2}$

$d = \sqrt{6^2 + 8^2}$

$d = \sqrt{36 + 64}$

$d = \sqrt{100}$

$d = 10$

39. $(5, 2)\,(0, -3)$

$m = \dfrac{-3-2}{0-5} = \dfrac{-5}{-5} = 1\,;$

$(0, -3)\,(4, -4)$

$m = \dfrac{-4-(-3)}{4-0} = \dfrac{-4+3}{4} = \dfrac{-1}{4}\,;$

$(5, 2)\,(4, -4)$

$m = \dfrac{-4-2}{4-5} = \dfrac{-6}{-1} = 6$

Not a right triangle. Lines are not perpendicular. Slopes: 1; $\dfrac{-1}{4}$; 6

41. $(-4, 3)(-7, -1)$

$m = \dfrac{-1-3}{-7-(-4)} = \dfrac{-4}{-7+4} = \dfrac{-4}{-3} = \dfrac{4}{3}$;

$(-7, -1)(3, -2)$

$m = \dfrac{-2-(-1)}{3-(-7)} = \dfrac{-2+1}{3+7} = \dfrac{-1}{10}$;

$(-4, 3)(3, -2)$

$m = \dfrac{-2-3}{3-(-4)} = \dfrac{-5}{7}$

Not a right triangle. Lines are not

perpendicular. Slopes: $\dfrac{4}{3}$; $\dfrac{-1}{10}$; $\dfrac{-5}{7}$

43. $(-3, 2)(-1, 5)$

$m = \dfrac{5-2}{-1-(-3)} = \dfrac{3}{-1+3} = \dfrac{3}{2}$;

$(-3, 2)(-6, 4)$

$m = \dfrac{4-2}{-6-(-3)} = \dfrac{2}{-6+3} = -\dfrac{2}{3}$

Right triangle because these two lines are

perpendicular. Slopes: $\dfrac{3}{2}$; $\dfrac{-2}{3}$

45. Center (0,0), radius 3

$x^2 + y^2 = 9$

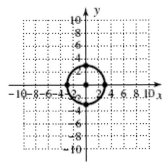

47. Center (5,0), radius $\sqrt{3}$

$(x-5)^2 + y^2 = 3$

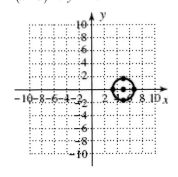

49. Center $(4, -3)$, radius 2

$(x-4)^2 + (y+3)^2 = 4$

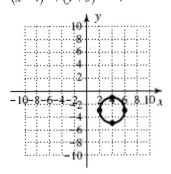

51. Center $(-7, -4)$, radius $\sqrt{7}$

$(x+7)^2 + (y+4)^2 = 7$

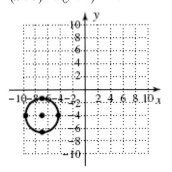

53. Center $(1, -2)$, radius $2\sqrt{3}$

$(x-1)^2 + (y+2)^2 = 12$

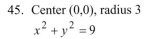

55. Center (4,5), diameter $4\sqrt{3}$

$\text{radius} = \dfrac{1}{2} \cdot \text{diameter}$

$r = \dfrac{1}{2}\left(4\sqrt{3}\right) = 2\sqrt{3}$

$(x-4)^2 + (y-5)^2 = \left(2\sqrt{3}\right)^2$

$(x-4)^2 + (y-5)^2 = 12$

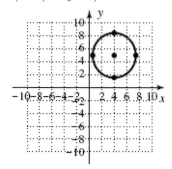

57. Center at (7,1),
    graph contains the point (1, −7)

$(x-7)^2 + (y-1)^2 = r^2;$

$(1-7)^2 + (-7-1)^2 = r^2$

$36 + 64 = r^2$

$100 = r^2;$

$(x-7)^2 + (y-1)^2 = 100$

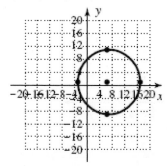

59. Center at (3,4),
    graph contains the point (7,9)

$(x-3)^2 + (y-4)^2 = r^2;$

$(7-3)^2 + (9-4)^2 = r^2$

$16 + 25 = r^2$

$41 = r^2;$

$(x-3)^2 + (y-4)^2 = 41$

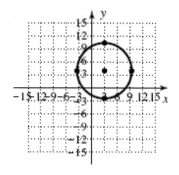

61. Diameter has endpoints (5,1) and (5,7);
    midpoint of diameter = center of circle

$\left(\dfrac{5+5}{2}, \dfrac{1+7}{2}\right) = (5,4);$

radius = distance from center to endpt

$r = \sqrt{(5-5)^2 + (1-4)^2} = 3;$

$(x-5)^2 + (y-4)^2 = 9$

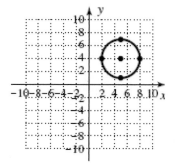

63. Center: $(2,3), r = 2$

$D : x \in [0,4]$

$R : y \in [1,5]$

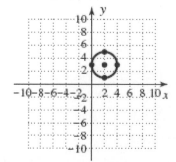

65. Center: $(-1,2), r = 2\sqrt{3}$

$D: x \in \left[-1 - 2\sqrt{3}, -1 + 2\sqrt{3}\right]$

$R: y \in \left[2 - 2\sqrt{3}, 2 + 2\sqrt{3}\right]$

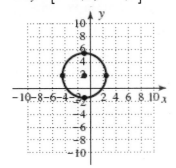

67. Center: $(-4,0), r = 9$

$D: x \in \left[-13, 5\right]$

$R: y \in \left[-9, 9\right]$

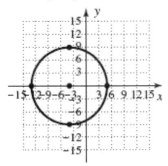

69. $x^2 + y^2 - 10x - 12y + 4 = 0$

$x^2 - 10x + y^2 - 12y = -4$

$x^2 - 10x + 25 + y^2 - 12y + 36 = -4 + 25 + 36$

$(x-5)^2 + (y-6)^2 = 57$

Center: $(5,6)$, Radius: $r = \sqrt{57}$

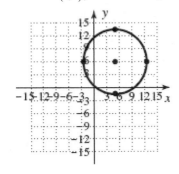

71. $x^2 + y^2 - 10x + 4y + 4 = 0$

$x^2 - 10x + y^2 + 4y = -4$

$x^2 - 10x + 25 + y^2 + 4y + 4 = -4 + 25 + 4$

$(x-5)^2 + (y+2)^2 = 25$

Center: $(5,-2)$, Radius: $r = 5$

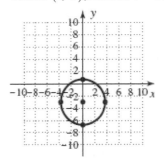

73. $x^2 + y^2 + 6y - 5 = 0$

$x^2 + y^2 + 6y = 5$

$x^2 + y^2 + 6y + 9 = 5 + 9$

$x^2 + (y+3)^2 = 14$

Center: $(0,-3)$, Radius: $r = \sqrt{14}$

75. $x^2 + y^2 + 4x + 10y + 18 = 0$

$x^2 + 4x + y^2 + 10y = -18$

$x^2 + 4x + 4 + y^2 + 10y + 25 = -18 + 4 + 25$

$(x+2)^2 + (y+5)^2 = 11$

Center: $(-2,-5)$, Radius: $r = \sqrt{11}$

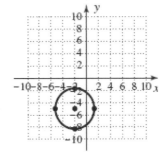

77. $x^2 + y^2 + 14x + 12 = 0$

$x^2 + 14x + y^2 = -12$

$x^2 + 14x + 49 + y^2 = -12 + 49$

$(x + 7)^2 + y^2 = 37$

Center: $(-7, 0)$, Radius: $r = \sqrt{37}$

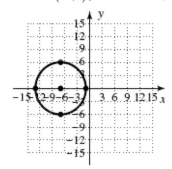

79. $2x^2 + 2y^2 - 12x + 20y + 4 = 0$

$x^2 + y^2 - 6x + 10y + 2 = 0$

$x^2 - 6x + y^2 + 10y = -2$

$x^2 - 6x + 9 + y^2 + 10y + 25 = -2 + 9 + 25$

$(x - 3)^2 + (y + 5)^2 = 32$

Center: $(3, -5)$, Radius: $r = 4\sqrt{2}$

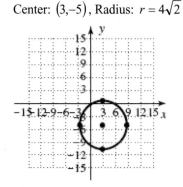

81. $s = 12.5t + 59$

a. Let $t = 1, s = 12.5(1) + 59 = 71.5$;

Let $t = 2, s = 12.5(2) + 59 = 84$;

Let $t = 3, s = 12.5(3) + 59 = 96.5$;

Let $t = 5, s = 12.5(5) + 59 = 121.5$;

Let $t = 7, s = 12.5(7) + 59 = 146.5$;

$(1, 71.5), (2, 84), (3, 96.5), (5, 121.5), (7, 146.5)$

b. Let $t = 8, s = 12.5(8) + 59 = 159$

Average amount spend in 2008 is \$159.

c. Let $s = 196$,

$196 = 12.5t + 59$

$137 = 12.5t$

$10.96 = t$

In 2011, annual spending surpasses \$196.

d.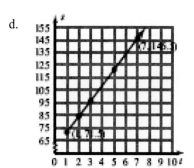

83. a. $(x - 5)^2 + (y - 12)^2 = 25^2$

$(x - 5)^2 + (y - 12)^2 = 625$

b. $d = \sqrt{(15 - 5)^2 + (36 - 12)^2}$

$d = \sqrt{10^2 + 24^2} = \sqrt{676} = 26$

No, radar cannot pick up the liner's sister ship.

85. Red: $(x - 2)^2 + (y - 2)^2 = 4$;

Center: (2,2), Radius: 2

Blue: $(x - 2)^2 + y^2 = 16$;

Center: (2,0), Radius: 4

Area of blue: $\pi(16) - \pi(4) = 12\pi$ units^2

87. $x^2 + y^2 + 8x - 6y = 0$

$x^2 + 8x + y^2 - 6y = 0$

$x^2 + 8x + 16 + y^2 - 6y + 9 = 0 + 16 + 9$

$(x+4)^2 + (y-3)^2 = 25$;

$x^2 + y^2 - 10x + 4y = 0$

$x^2 - 10x + y^2 + 4y = 0$

$x^2 - 10x + 25 + y^2 + 4y + 4 = 0 + 25 + 4$

$(x-5)^2 + (y+2)^2 = 29$;

Distance between centers: $(-4,3)$, $(5, -2)$

$d = \sqrt{(-4-5)^2 + (3-(-2))^2}$

$= \sqrt{81+25} = \sqrt{106} \approx 10.30$;

Sum of the radii: $5 + \sqrt{29} \approx 10.39$

No, Distance between the centers is less than the sum of the radii.

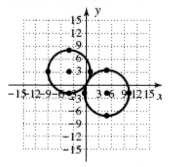

89. Answers will vary.

91. a. $x^2 + y^2 - 12x + 4y + 40 = 0$

$x^2 - 12x + y^2 + 4y = -40$

$x^2 - 12x + 36 + y^2 + 4y + 4 = -40 + 36 + 4$

$(x-6)^2 + (y+2)^2 = 0$

Center $(6, -2)$, $r = 0$, degenerate case

b. $x^2 + y^2 - 2x - 8y - 8 = 0$

$x^2 - 2x + y^2 - 8y = 8$

$x^2 - 2x + 1 + y^2 - 8y + 16 = 8 + 1 + 16$

$(x-1)^2 + (y-4)^2 = 25$

Center $(1, 4)$, $r = 5$

c. $x^2 + y^2 - 6x - 10y + 35 = 0$

$x^2 - 6x + y^2 - 10y = -35$

$x^2 - 6x + 9 + y^2 - 10y + 25 = -35 + 9 + 25$

$(x-3)^2 + (y-5)^2 = -1$

Center $(3, 5)$, $r^2 = -1$, degenerate case

93. a. 0
    b. not possible
    c. 0.3 ;many answers possible
    d. not possible
    e. not possible
    f. $\sqrt{3}$;many answers possible

95. $1 - \sqrt{n+3} = -n$

$-\sqrt{n+3} = -n - 1$

$\sqrt{n+3} = n + 1$

$\left(\sqrt{n+3}\right)^2 = (n+1)^2$

$n + 3 = n^2 + 2n + 1$

$0 = n^2 + n - 2$

$0 = (n+2)(n-1)$

$n + 2 = 0$ or $n - 1 = 0$

$n = -2$ or $n = 1$

Check: $n = -2$

$1 - \sqrt{-2+3} = -(-2)$

$1 - \sqrt{1} = 2$

$0 \neq 2$;

Check: $n = 1$

$1 - \sqrt{1+3} = -1$

$1 - \sqrt{4} = -1$

$-1 = -1$;

$n = 1$ is a solution, $n = -2$ is extraneous.

2.2 Exercises

2.2 Technology Highlight

Exercise 1: $Y_1 = \dfrac{2}{3}x + 1;$ $(-1.5, 0),$ $(0, 1)$

2.2 Exercises

1.   0; 0.

3.   negative, downward

5.   yes; slopes are not equal $m_1 \neq m_2$;
     No; $m_1 \cdot m_2 \neq -1$

7. $2x + 3y = 6$
     $3y = -2x + 6$

     $y = -\dfrac{2}{3}x + 2$

x	y
-6	$-\dfrac{2}{3}(-6) + 2 = 4 + 2 = 6$
-3	$-\dfrac{2}{3}(-3) + 2 = 2 + 2 = 4$
0	$-\dfrac{2}{3}(0) + 2 = 0 + 2 = 2$
3	$-\dfrac{2}{3}(3) + 2 = -2 + 2 = 0$

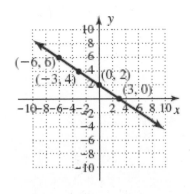

9. $y = \dfrac{3}{2}x + 4$

x	y
-2	$\dfrac{3}{2}(-2) + 4 = -3 + 4 = 1$
0	$\dfrac{3}{2}(0) + 4 = 0 + 4 = 4$
2	$\dfrac{3}{2}(2) + 4 = 3 + 4 = 7$
4	$\dfrac{3}{2}(4) + 4 = 6 + 4 = 10$

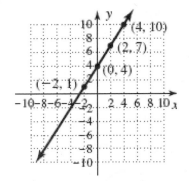

11. $y = \dfrac{3}{2}x + 4$

     $-0.5 = \dfrac{3}{2}(-3) + 4$

     $-0.5 = -\dfrac{9}{2} + 4$

     $-0.5 = -0.5$;

     $\dfrac{19}{4} = \dfrac{3}{2}\left(\dfrac{1}{2}\right) + 4$

     $\dfrac{19}{4} = \dfrac{3}{4} + 4$

     $\dfrac{19}{4} = \dfrac{19}{4}$

Chapter 2: Relations, Functions and Graphs

13. $3x + y = 6$

x-intercept: $(2, 0)$

$3x + 0 = 6$

$3x = 6$

$x = 2$

y-intercept: $(0, 6)$

$3(0) + y = 6$

$y = 6$

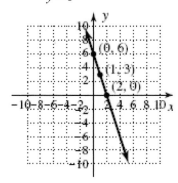

15. $5y - x = 5$

x-intercept: $(-5, 0)$

$5(0) - x = 5$

$-x = 5$

$x = -5$

y-intercept: $(0, 1)$

$5y - 0 = 5$

$5y = 5$

$y = 1$

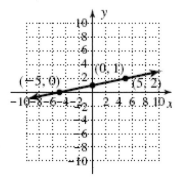

17. $-5x + 2y = 6$

x-intercept: $\left(-\dfrac{6}{5}, 0\right)$

$-5x + 2(0) = 6$

$-5x = 6$

$x = -\dfrac{6}{5}$

y-intercept: $(0, 3)$

$-5(0) + 2y = 6$

$2y = 6$

$y = 3$

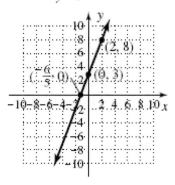

19. $2x - 5y = 4$

x-intercept: $(2, 0)$

$2x - 5(0) = 4$

$2x = 4$

$x = 2$

y-intercept: $\left(0, -\dfrac{4}{5}\right)$

$2(0) - 5y = 4$

$-5y = 4$

$y = -\dfrac{4}{5}$

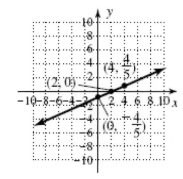

21. $2x + 3y = -12$

    x-intercept: $(-6, 0)$

    $2x + 3(0) = -12$

    $2x = -12$

    $x = -6$

    y-intercept: $(0, -4)$

    $2(0) + 3y = -12$

    $3y = -12$

    $y = -4$

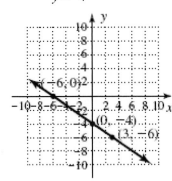

23. $y = -\dfrac{1}{2}x$

    $y = -\dfrac{1}{2}(2)$

    $y = -1$

    $(2, -1);$

    $y = -\dfrac{1}{2}x$

    $y = -\dfrac{1}{2}(4)$

    $y = -2$

    $(4, -2);$

    $y = -\dfrac{1}{2}x$

    $y = -\dfrac{1}{2}(0)$

    $y = 0$

    $(0, 0)$

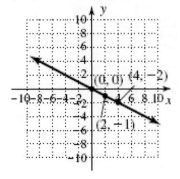

25. $y - 25 = 50x$

    $y - 25 = 50(-1)$

    $y - 25 = -50$

    $y = -25$

    $(-1, -25);$

    $y - 25 = 50x$

    $y - 25 = 50(1)$

    $y - 25 = 50$

    $y = 75$

    $(1, 75)$

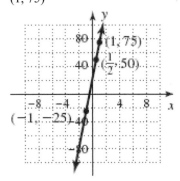

27. $y = -\dfrac{2}{5}x - 2$

    x-intercept: $(-5, 0)$

    $0 = -\dfrac{2}{5}x - 2$

    $2 = -\dfrac{2}{5}x$

    $\left(-\dfrac{5}{2}\right)(2) = \left(-\dfrac{5}{2}\right)\left(-\dfrac{2}{5}x\right)$

    $-5 = x$

    $(-5, 0);$

    y-intercept: $(0, -2)$

    $y = -\dfrac{2}{5}(0) - 2$

    $y = -2$

    $(0, -2)$

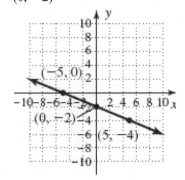

29. $2y - 3x = 0$

$2y - 3(2) = 0$

$2y - 6 = 0$

$2y = 6$

$y = 3$

$(2, 3);$

$2y - 3x = 0$

$2y - 3(4) = 0$

$2y - 12 = 0$

$2y = 12$

$y = 6$

$(4, 6);$

$2y - 3x = 0$

$2y - 3(0) = 0$

$2y = 0$

$y = 0$

$(0, 0)$

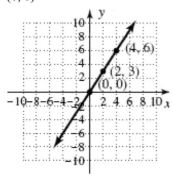

31. $3y + 4x = 12$

x-intercept: $(3, 0)$

$3(0) + 4x = 12$

$4x = 12$

$x = 3$

y-intercept: $(0, 4)$

$3y + 4(0) = 12$

$3y = 12$

$y = 4$

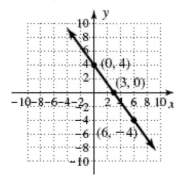

33. $m = \dfrac{6-5}{4-3} = \dfrac{1}{1} = 1$

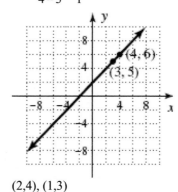

$(2,4), (1,3)$

35. $m = \dfrac{3-(-5)}{10-4} = \dfrac{8}{6} = \dfrac{4}{3}$

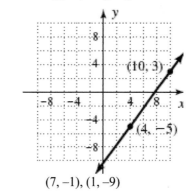

$(7,-1), (1,-9)$

37. $m = \dfrac{-8-7}{1-(-3)} = \dfrac{-15}{4} = -\dfrac{15}{4}$

$(1,-8), \left(-1,-\dfrac{1}{2}\right)$

39. $m = \dfrac{2-6}{4-(-3)} = \dfrac{-4}{7} = -\dfrac{4}{7}$

$(-10,10), (11,-2)$

41. a. $m = \dfrac{500-250}{4-2} = \dfrac{250}{2} = 125$

Cost increased \$125,000 per 1000 square feet.

b. \$375,000

43. a. $m = \dfrac{270-90}{12-4} = \dfrac{180}{8} = 22.5$

Distance increases 22.5 miles per hour.

   b.   186 miles

45. a. $m = \dfrac{165-142}{70-64} = \dfrac{23}{6}$

       A person weighs 23 pounds more for each additional 6 inches in height.

    b. $\dfrac{23}{6} \approx 3.8$ pounds

47. Convert 48 feet to inches: $48(12) = 576$;
    $(0, -6)$ represents position of the sewer line at edge of house;
    $(576, -18)$ represents position of sewer line at the main line.

$$m = \dfrac{-18-(-6)}{576-0} = \dfrac{-12}{576} = -\dfrac{1}{48}$$

    The sewer line is one inch deeper for each 48 inches in length.

49. $x = -3$
    $x + 0y = -3$
    $x + 0(4) = -3$
    $x = -3$
    $(-3, 4)$;
    $x + 0y = -3$
    $x + 0(-4) = -3$
    $x = -3$
    $(-3, -4)$

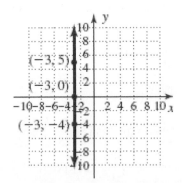

51. $x = 2$
    $x + 0y = 2$
    $x + 0(2) = 2$

$x = 2$
$(2, 0)$
$x + 0y = 2$
$x + 0(-2) = 2$
$x = 2$
$(2, 0)$

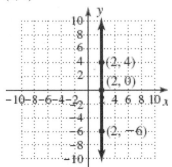

53. $L_1 : x = 2$
    $L_2 : y = 4$
    Point of intersection: $(2, 4)$

55. a. Choose any two points (t,j).
      $(0,9)$, $(10,9)$

$$m = \dfrac{9-9}{10-0} = \dfrac{0}{10} = 0$$

    Which indicates there is no increase or decrease in the number of Supreme Court justices.

    b. Choose any two points (t,n).
      $(0,0)$, $(10,1)$

$$m = \dfrac{1-0}{10-0} = \dfrac{1}{10}$$

    Which indicates that over the last 5 decades, one non-white or non-female justice has been added to the court every ten years.

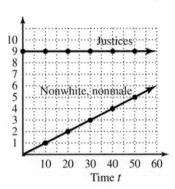

57. L_1: $m = \dfrac{6-0}{0-(-2)} = \dfrac{6}{2} = 3$

    L_2: $m = \dfrac{5-8}{0-1} = \dfrac{-3}{-1} = 3$

Parallel

59. L_1: $m = \dfrac{-4-1}{-3-0} = \dfrac{-5}{-3} = \dfrac{5}{3}$

    L_2: $m = \dfrac{4-0}{-4-0} = \dfrac{4}{-4} = -1$

    Neither

61. L_1: $m = \dfrac{7-3}{8-6} = \dfrac{4}{2} = 2$

    L_2: $m = \dfrac{2-0}{7-6} = \dfrac{2}{1} = 2$

    Parallel

63. $(5, 2)\ (0, -3)$

    $m = \dfrac{-3-2}{0-5} = \dfrac{-5}{-5} = 1$;

    $(0, -3)\ (4, -4)$

    $m = \dfrac{-4-(-3)}{4-0} = \dfrac{-4+3}{4} = \dfrac{-1}{4}$;

    $(5, 2)\ (4, -4)$

    $m = \dfrac{-4-2}{4-5} = \dfrac{-6}{-1} = 6$

    Not a right triangle. Lines are not

    perpendicular. Slopes: 1; $\dfrac{-1}{4}$; 6

65. $(-4, 3)\ (-7, -1)$

    $m = \dfrac{-1-3}{-7-(-4)} = \dfrac{-4}{-7+4} = \dfrac{-4}{-3} = \dfrac{4}{3}$;

    $(-7, -1)\ (3, -2)$

    $m = \dfrac{-2-(-1)}{3-(-7)} = \dfrac{-2+1}{3+7} = \dfrac{-1}{10}$;

    $(-4, 3)\ (3, -2)$

    $m = \dfrac{-2-3}{3-(-4)} = \dfrac{-5}{7}$

    Not a right triangle. Lines are not

    perpendicular. Slopes: $\dfrac{4}{3}$; $\dfrac{-1}{10}$; $\dfrac{-5}{7}$

67. $(-3, 2)\ (-1, 5)$

    $m = \dfrac{5-2}{-1-(-3)} = \dfrac{3}{-1+3} = \dfrac{3}{2}$;

    $(-3, 2)\ (-6, 4)$

    $m = \dfrac{4-2}{-6-(-3)} = \dfrac{2}{-6+3} = -\dfrac{2}{3}$

    Right triangle because these two lines are

    perpendicular. Slopes: $\dfrac{3}{2}$; $\dfrac{-2}{3}$

69. $L = 0.11T + 74.2$
    a. $L(20) = 0.11(20) + 74.2 = 76.4$ years

    b. $77.5 = 0.11T + 74.2$
       $3.3 = 0.11T$
       $30 = T$
       $1980 + 30 = 2010$

71. $V = 8500 - 1250y$
    a. $V = 8500 - 1250(4) = \$3500$

    b. $2250 = 8500 - 1250y$
       $-6250 = -1250y$
       $5 = y$
       5 years

73. Let h represent the water level, in inches.
    Let t represent the time, in months.
    $h = -3t + 300$
    a. $h = -3(9) + 300 = 273$ in .

    b. Convert feet to inches: $20(12) = 240$;
       $240 = -3t + 300$
       $-60 = -3t$
       $20 = t$
       20 months

75. Slope of FM 1960: $\dfrac{38}{12}$;

    Slope of FM 380: $\dfrac{30}{9.5}$;

    Since $\dfrac{38}{12} \neq \dfrac{30}{9.5}$, the roads are not parallel
    and yes, the roads will meet.

77. $y = 144x + 621$

    a. $y = 144(22) + 621$

        $y = 3789$

        $3,789

    b. $5250 = 144x + 621$

        $4629 = 144x$

        $32.15 \approx x$

        $1980 + 32 = 2012$

        Year 2012

79. $y = -\dfrac{7}{15}x + 32$

    a. $y = -\dfrac{7}{15}(20) + 32$

        $y = \dfrac{-28}{3} + 32$

        $y = 22\dfrac{2}{3}$

        23%

    b. $20 = -\dfrac{7}{15}x + 32$

        $-12 = -\dfrac{7}{15}x$

        $-180 = -7x$

        $25.7 = x$

        $1980 + 25.7 = 2005.7$

        During the year 2005

81. $4y + 2x = -5$

    $4y = -2x - 5$

    $y = -\dfrac{1}{2}x - \dfrac{5}{4}$;

    $3y + ax = -2$

    $3y = -ax - 2$

    $y = -\dfrac{a}{3}x - \dfrac{2}{3}$;

    $-\dfrac{a}{3} \cdot -\dfrac{1}{2} = -1$

    $\dfrac{a}{6} = -1$

    $a = -6$

83. $t_n = t_1 + (n-1)d$

    a. $n = 21, t_1 = 2, d = 9 - 2 = 7$

        $t_{21} = 2 + (21-1)7 = 142$

    b. $n = 31, t_1 = 7, d = 4 - 7 = -3$

        $t_{31} = 7 + (31-1)(-3) = -83$

    c. $n = 27, t_1 = 5.10, d = 5.25 - 5.10 = 0.15$

        $t_{27} = 5.10 + (27-1)(0.15) = 9$

    d. $n = 17, t_1 = \dfrac{3}{2}, d = \dfrac{9}{4} - \dfrac{3}{2} = \dfrac{3}{4}$

        $t_{17} = \dfrac{3}{2} + (17-1)\left(\dfrac{3}{4}\right) = \dfrac{27}{2}$

85. $P = 2L + 2W$

    Perimeter of a rectangle;

    $V = LWH$

    Volume of a rectangular prism;

    $V = \pi r^2 h$

    Volume of a cylinder;

    $C = 2\pi r$

    Circumference of a circle

87.

	Distance	Rate	Time
Westbound Boat	D	15	t
Eastbound Boat	$70 - D$	20	t

$\begin{cases} D = 15t \\ 70 - D = 20t \end{cases}$

$70 - 15t = 20t$

$70 = 35t$

$2 = t$

2 hours

Chapter 2: Relations, Functions and Graphs

2.3 Exercises

1. $-\dfrac{7}{4}$; $(0, 3)$

3. 2.5

5. Answers will vary.

7. $4x + 5y = 10$
$5y = -4x + 10$
$y = -\dfrac{4}{5}x + 2$

x	$y = -\dfrac{4}{5}x + 2$
-5	$y = -\dfrac{4}{5}(-5) + 2 = 4 + 2 = 6$
-2	$y = -\dfrac{4}{5}(-2) + 2 = \dfrac{8}{5} + 2 = \dfrac{18}{5}$
0	$y = -\dfrac{4}{5}(0) + 2 = 0 + 2 = 2$
1	$y = -\dfrac{4}{5}(1) + 2 = -\dfrac{4}{5} + 2 = \dfrac{6}{5}$
3	$y = -\dfrac{4}{5}(3) + 2 = -\dfrac{12}{5} + 2 = -\dfrac{2}{5}$

9. $-0.4x + 0.2y = 1.4$
$0.2y = 0.4x + 1.4$
$y = 2x + 7$

x	$y = 2x + 7$
-5	$y = 2(-5) + 7 = -10 + 7 = -3$
-2	$y = 2(-2) + 7 = -4 + 7 = 3$
0	$y = 2(0) + 7 = 0 + 7 = 7$
1	$y = 2(1) + 7 = 2 + 7 = 9$
3	$y = 2(3) + 7 = 6 + 7 = 13$

11. $\dfrac{1}{3}x + \dfrac{1}{5}y = -1$
$\dfrac{1}{5}y = -\dfrac{1}{3}x - 1$
$y = -\dfrac{5}{3}x - 5$

x	$y = -\dfrac{5}{3}x - 5$
-5	$y = -\dfrac{5}{3}(-5) - 5 = \dfrac{25}{3} - 5 = \dfrac{10}{3}$
-2	$y = -\dfrac{5}{3}(-2) - 5 = \dfrac{10}{3} - 5 = -\dfrac{5}{3}$
0	$y = -\dfrac{5}{3}(0) - 5 = 0 - 5 = -5$
1	$y = -\dfrac{5}{3}(1) - 5 = -\dfrac{5}{3} - 5 = -\dfrac{20}{3}$
3	$y = -\dfrac{5}{3}(3) - 5 = -5 - 5 = -10$

13. $6x - 3y = 9$
$-3y = -6x + 9$
$y = 2x - 3$
New Coefficient: 2
New Constant: -3

15. $-0.5x - 0.3y = 2.1$
$-0.3y = 0.5x + 2.1$
$y = \dfrac{-5}{3}x - 7$
New Coefficient: $\dfrac{-5}{3}$
New Constant: -7

17. $\dfrac{5}{6}x + \dfrac{1}{7}y = -\dfrac{4}{7}$
$\dfrac{1}{7}y = -\dfrac{5}{6}x - \dfrac{4}{7}$
$y = -\dfrac{35}{6}x - 4$
New Coefficient: $-\dfrac{35}{6}$
New Constant: -4

19. $y = -\dfrac{4}{3}x + 5$

x	$y = -\dfrac{4}{3}x + 5$
0	$y = -\dfrac{4}{3}(0) + 5 = 0 + 5 = 5$
3	$y = -\dfrac{4}{3}(3) + 5 = -4 + 5 = 1$
6	$y = -\dfrac{4}{3}(6) + 5 = -8 + 5 = -3$

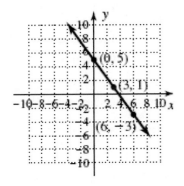

23. $y = -\dfrac{1}{6}x + 4$

x	$y = -\dfrac{1}{6}x + 4$
−6	$y = -\dfrac{1}{6}(-6) + 4 = 1 + 5 = 5$
0	$y = -\dfrac{1}{6}(0) + 4 = 0 + 4 = 4$
6	$y = -\dfrac{1}{6}(6) + 4 = -1 + 4 = 3$

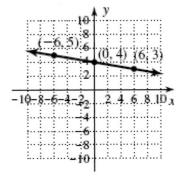

21. $y = -\dfrac{3}{2}x - 2$

x	$y = -\dfrac{3}{2}x - 2$
0	$y = -\dfrac{3}{2}(0) - 2 = 0 - 2 = -2$
2	$y = -\dfrac{3}{2}(2) - 2 = -3 - 2 = -5$
4	$y = -\dfrac{3}{2}(4) - 2 = -6 - 2 = -8$

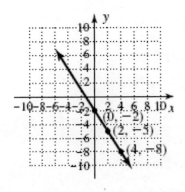

25. $3x + 4y = 12$

x-intercept: (4, 0)   y-intercept: (0, 3)

$3x + 4(0) = 12$

$3x = 12$

$x = 4$

$3(0) + 4y = 12$

$4y = 12$

$y = 3$

a. $m = \dfrac{0 - 3}{4 - 0} = -\dfrac{3}{4}$

b. $y = -\dfrac{3}{4}x + 3$

c. The coefficient of x is the slope and the constant is the y-intercept.

Chapter 2: Relations, Functions and Graphs

27. $2x - 5y = 10$

   x-intercept: $(5, 0)$ y-intercept: $(0, -2)$
   $2x - 5(0) = 10$
   $$2x = 10$$
   $$x = 5$$
   $$2(0) - 5y = 10$$
   $$-5y = 10$$
   $$y = -2$$

   a. $m = \dfrac{0 - (-2)}{5 - 0} = \dfrac{2}{5}$

   b. $y = \dfrac{2}{5}x - 2$

   c.   The coefficient of x is the slope and the constant is the y-intercept.

29. $4x - 5y = -15$

   x-intercept: $\left(-\dfrac{15}{4}, 0\right)$ y-intercept: $(0, 3)$

   $$4x - 5(0) = -15$$
   $$4x = -15$$
   $$x = -\dfrac{15}{4}$$
   $$4(0) - 5y = -15$$
   $$-5y = -15$$
   $$y = 3$$

   a. $m = \dfrac{0 - 3}{-\dfrac{15}{4} - 0} = \dfrac{-3}{-\dfrac{15}{4}} = \dfrac{12}{15} = \dfrac{4}{5}$

   b. $y = \dfrac{4}{5}x + 3$

   c.   The coefficient of x is the slope and the constant is the y-intercept.

31. $2x + 3y = 6$

   $$3y = -2x + 6$$
   $$y = -\dfrac{2}{3}x + 2$$
   $$m = -\dfrac{2}{3} \text{; } y\text{-intercept } (0, 2)$$

33. $5x + 4y = 20$

   $$4y = -5x + 20$$
   $$y = -\dfrac{5}{4}x + 5$$
   $$m = -\dfrac{5}{4} \text{; } y\text{-intercept } (0, 5)$$

35. $x = 3y$

   $$y = \dfrac{1}{3}x$$
   $$m = \dfrac{1}{3} \text{; } y\text{-intercept } (0, 0)$$

37. $3x + 4y - 12 = 0$

   $$4y = -3x + 12$$
   $$y = -\dfrac{3}{4}x + 3$$
   $$m = -\dfrac{3}{4} \text{; } y\text{-intercept } (0, 3)$$

39. $m = \dfrac{2}{3}$; y-intercept $(0, 1)$

   $$y = mx + b$$
   $$y = \dfrac{2}{3}x + 1$$

41. $m = 3$; y-intercept $(0, 3)$
   $$y = mx + b$$
   $$y = 3x + 3$$

43. $m = 3$; y-intercept $(0, 2)$
   $$y = mx + b$$
   $$y = 3x + 2$$

45. $m = 250$; $(14, 4000)$
   $$y - y_1 = m(x - x_1)$$
   $$y - 4000 = 250(x - 14)$$
   $$y - 4000 = 250x - 3500$$
   $$y = 250x + 500$$
   $$f(x) = 250x + 500$$

47. $m = \dfrac{75}{2}$; $(24, 1050)$

2.3 Exercises

$$y - y_1 = m(x - x_1)$$

$$y - 1050 = \frac{75}{2}(x - 24)$$

$$y - 1050 = \frac{75}{2}x - 900$$

$$y = \frac{75}{2}x + 150$$

$$f(x) = \frac{75}{2}x + 150$$

49. $m = 2;\ (5,\ -3)$

$$y - y_1 = m(x - x_1)$$

$$y + 3 = 2(x - 5)$$

$$y + 3 = 2x - 10$$

$$y = 2x - 13$$

51. $3x + 5y = 20$

$$5y = -3x + 20$$

$$y = -\frac{3}{5}x + 4$$

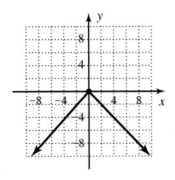

53. $2x - 3y = 15$

$$-3y = -2x + 15$$

$$y = \frac{2}{3}x - 5$$

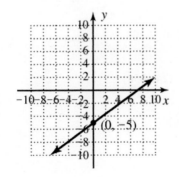

55. $y = \frac{2}{3}x + 3$

$m = \frac{2}{3}$; y-intercept $(0,\ 3)$

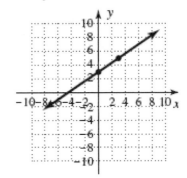

57. $y = -\frac{1}{3}x + 2$

$m = \frac{-1}{3}$; y-intercept $(0,\ 2)$

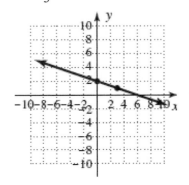

59. $y = 2x - 5$

$m = 2$; y-intercept $(0,\ -5)$

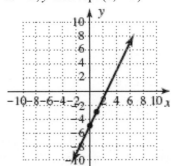

61. $f(x) = \dfrac{1}{2}x - 3$

$m = \dfrac{1}{2}$; y-intercept $(0, -3)$

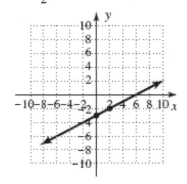

63. $2x - 5y = 10$

$-5y = -2x + 10$

$y = \dfrac{2}{5}x - 2$

$m = \dfrac{2}{5}$; $(-5, 2)$

$y - y_1 = m(x - x_1)$

$y - 2 = \dfrac{2}{5}(x - (-5))$

$y - 2 = \dfrac{2}{5}x + 2$

$y = \dfrac{2}{5}x + 4$

65. $5y - 3x = 9$

$5y = 3x + 9$

$y = \dfrac{3}{5}x + \dfrac{9}{5}$

$m = -\dfrac{5}{3}$; $(6, -3)$;

$y - y_1 = m(x - x_1)$

$y - (-3) = -\dfrac{5}{3}(x - 6)$

$y + 3 = -\dfrac{5}{3}x + 10$

$y = -\dfrac{5}{3}x + 7$

67. $12x + 5y = 65$

$5y = -12x + 65$

$y = -\dfrac{12}{5}x + 13$

$m = -\dfrac{12}{5}$; $(-2, -1)$

$y - y_1 = m(x - x_1)$

$y + 1 = -\dfrac{12}{5}(x + 2)$

$y + 1 = -\dfrac{12}{5}x - \dfrac{24}{5}$

$y = -\dfrac{12}{5}x - \dfrac{29}{5}$

69. $y = -3$ has slope of zero.

Slope of any line parallel to this line has the same slope, 0.

$y = mx + b$

$5 = 0(2) + b$

$5 = b$;

$y = 0x + 5$

$y = 5$

71. $4y - 5x = 8$

$4y = 5x + 8$

$y = \dfrac{5}{4}x + 2$;

$5y + 4x = -15$

$5y = -4x - 15$

$y = -\dfrac{4}{5}x - 3$

perpendicular

73. $2x - 5y = 20$

$-5y = -2x + 20$

$y = \dfrac{2}{5}x - 4$;

$4x - 3y = 18$

$-3y = -4x + 18$

$y = \dfrac{4}{3}x - 6$

Neither

75. $-4x + 6y = 12$

$6y = 4x + 12$

$y = \dfrac{2}{3}x + 2;$

$2x + 3y = 6$

$3y = -2x + 6$

$y = -\dfrac{2}{3}x + 2$

Neither

77. $(0,1),(4,-2)$

$m = \dfrac{-2-1}{4-0} = -\dfrac{3}{4}$

a. $y - (-4) = -\dfrac{3}{4}(x-2)$

$y + 4 = -\dfrac{3}{4}x + \dfrac{3}{2}$

$y = -\dfrac{3}{4}x - \dfrac{5}{2}$

b. $y - (-4) = \dfrac{4}{3}(x-2)$

$y + 4 = \dfrac{4}{3}x - \dfrac{8}{3}$

$y = \dfrac{4}{3}x - \dfrac{20}{3}$

79. $(-4,0),(5,4)$

$m = \dfrac{4-0}{5-(-4)} = \dfrac{4}{9}$

a. $y - 3 = \dfrac{4}{9}(x-(-1))$

$y - 3 = \dfrac{4}{9}(x+1)$

$y - 3 = \dfrac{4}{9}x + \dfrac{4}{9}$

$y = \dfrac{4}{9}x + \dfrac{31}{9}$

b. $y - 3 = \dfrac{-9}{4}(x-(-1))$

$y - 3 = \dfrac{-9}{4}(x+1)$

$y - 3 = \dfrac{-9}{4}x - \dfrac{9}{4}$

$y = \dfrac{-9}{4}x + \dfrac{3}{4}$

81. $(-2,3),(4,0)$

$m = \dfrac{0-3}{4-(-2)} = \dfrac{-3}{6} = \dfrac{-1}{2}$

a. $y - (-2) = \dfrac{-1}{2}(x-0)$

$y + 2 = \dfrac{-1}{2}x$

$y = \dfrac{-1}{2}x - 2$

b. $y - (-2) = 2(x-0)$

$y + 2 = 2x$

$y = 2x - 2$

83. $m = 2;\ P_1 = (2,-5)$

$y - y_1 = m(x - x_1)$

$y + 5 = 2(x-2)$

$y + 5 = 2x - 4$

$y = 2x - 9$

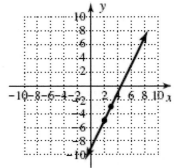

85. $P_1(3,-4), P_2(11,-1)$

$$m = \frac{-1-(-4)}{11-3} = \frac{3}{8};$$

$$y - y_1 = m(x - x_1)$$

$$y - (-4) = \frac{3}{8}(x - 3)$$

$$y + 4 = \frac{3}{8}x - \frac{9}{8}$$

$$y = \frac{3}{8}x - \frac{41}{8}$$

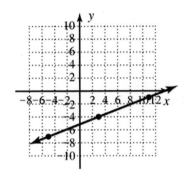

87. $m = 0.5; \ P_1 = (1.8, -3.1)$

$$y - y_1 = m(x - x_1)$$

$$y + 3.1 = 0.5(x - 1.8)$$

$$y + 3.1 = 0.5x - 0.9$$

$$y = 0.5x - 4$$

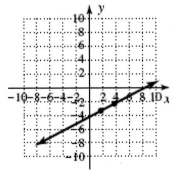

89. $m = \frac{6}{5}; (4, 2)$

$$y - y_1 = m(x - x_1)$$

$$y - 2 = \frac{6}{5}(x - 4)$$

For each 5000 additional sales, income rises $6000.

91. $m = -20; (0.5, 100)$

$$y - y_1 = m(x - x_1)$$

$$y - 100 = -20(x - 0.5)$$

For every hour of television, a student's final grade falls 20%.

93. $m = \frac{35}{2}; (0.5, 10)$

$$y - y_1 = m(x - x_1)$$

$$y - 10 = \frac{35}{2}(x - 0.5)$$

Every 2 inches of rainfall increases the number of cattle raised per acre by 35.

95. C

97. A

99. B

101. D

103. $ax + by = c$

$$by = -ax + c$$

$$y = -\frac{a}{b}x + \frac{c}{b};$$

Slope $-\frac{a}{b}$, y-intercept $\left(0, \frac{c}{b}\right)$

a. $3x + 4y = 8$

$$m = -\frac{a}{b} = -\frac{3}{4};$$

$$y\text{-int} = \frac{c}{b} = \frac{8}{4} = 2, \ (0,2)$$

2.3 Exercises

b. $2x + 5y = -15$

$$m = -\frac{a}{b} = -\frac{2}{5};$$

$$y\text{-int} = \frac{c}{b} = -\frac{15}{5} = -3, (0, -3)$$

c. $5x - 6y = -12$

$$m = -\frac{5}{-6} = \frac{5}{6};$$

$$y\text{-int} = \frac{c}{b} = \frac{-12}{-6} = 2, (0,2)$$

d. $3y - 5x = 9$

$$m = -\frac{a}{b} = -\frac{-5}{3} = \frac{5}{3};$$

$$y\text{-int} = \frac{c}{b} = \frac{9}{3} = 3, (0,3)$$

105.a. As the temperature increases 5°C, the velocity of sound waves increases 3 m/s. At a temperature of 0°C, the velocity is 331 m/s.

b. $V(20) = \frac{3}{5}(20) + 331 = 343$ m/s

c. $361 = \frac{3}{5}C + 331$

$$30 = \frac{3}{5}C$$
$$50 = C$$
$$50°C$$

107.a. $m = \frac{190 - 150}{6 - 0} = \frac{40}{6} = \frac{20}{3}$

$$V(t) = \frac{20}{3}t + 150$$

b. Every three years, the coin increased in value by $20. The initial value was $150.

109.a. $m = \frac{51 - 9}{2001 - 1995} = \frac{42}{6} = 7$

$$N(t) = 7t + 9$$

b. Every 1 year, the number of homes hooked to the internet increases by 7 million.

c. $0 = 7t + 9$

$$-9 = 7t$$

$$-\frac{9}{7} = t$$

$$-1.29 = t$$

1.29 years prior to 1995 is 1993.

111 $m = \frac{1320000 - 740000}{2000 - 1990}$

$$= \frac{580000}{10} = 58000$$

$P(t) = 58000t + 740000$

Grows 58,000 every year.

$P(17) = 58000(17) + 740000 = 1726000$

113. Answers will vary.

115.a. $ax + by = c$

Find x-intercept by letting $y = 0$.

$$ax + b(0) = c$$

$$x = \frac{c}{a}$$

$$\left(\frac{c}{a}, 0\right);$$

Find y-intercept by letting $x = 0$.

$$a(0) + by = c$$

$$y = \frac{c}{b}$$

$$\left(0, \frac{c}{b}\right)$$

The intercept method works most efficiently when a and b are factors of c.

b. Solve $ax + by = c$ for y.

$$ax + by = c$$

$$by = -ax + c$$

$$y = -\frac{a}{b}x + \frac{c}{b};$$

$$m = -\frac{a}{b}; y-\text{intercept} \left(0, \frac{c}{b}\right)$$

The slope-intercet method works most efficiently when b is a factor of c.

117. $3x^2 - 10x = 9$

$$3x^2 - 10 - 9 = 0$$

$$x = \frac{10 \pm \sqrt{(-10)^2 - 4(3)(-9)}}{2(3)}$$

$$x = \frac{10 \pm \sqrt{100 + 108}}{6}$$

$$x = \frac{10 \pm \sqrt{208}}{6}$$

$$x = \frac{10 \pm 4\sqrt{13}}{6}$$

$$x = \frac{5 \pm 2\sqrt{13}}{3}$$

$$x \approx 4.07 \text{ or } x \approx -0.74$$

119. $A = \pi r^2$

Larger circle: Smaller Circle

$$A = \pi(10)^2 \qquad A = \pi(8)^2$$

$$A = 100\pi \qquad A = 64\pi$$

$$100\pi - 64\pi = 36\pi \approx 113.10 \text{ yds}^2$$

2.4 Exercises

1. First

3. Range

5. Answers will vary.

7. Function

9. Not a function. The Shaq is paired with two heights.

11. Not a function, 4 is paired with 2 and -5.

13. Function

15. Function

17. Not a function, -2 is paired with 3 and -4.

19. Function

21. Function

23. Not a function, 0 is paired with 4 and -4.

25. Function

27. Not a function, 5 is paired with -1 and 1.

29. Function

31.

x	$y = x$
-2	$y = -2$
-1	$y = -1$
0	$y = 0$
1	$y = 1$
2	$y = 2$

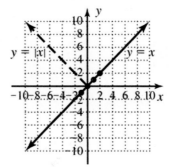

Function

2.4 Exercises

33.

x	$y = (x+2)^2$
-4	$y = (-4+2)^2 = 4$
-3	$y = (-3+2)^2 = 1$
-2	$y = (-2+2)^2 = 0$
-1	$y = (-1+2)^2 = 1$
0	$y = (0+2)^2 = 4$
1	$y = (1+2)^2 = 9$

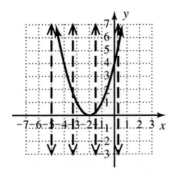

Function

35. Function; $x \in [-4,-5]$ $y \in [-2,3]$

37. Function; $x \in [-4,\infty)$ $y \in [-4,\infty)$

39. Function; $x \in [-4,4]$ $y \in [-5,-1]$

41. Function; $x \in (-\infty,\infty)$ $y \in (-\infty,\infty)$

43. Not a function; $x \in [-3,5]$ $y \in [-3,3]$

45. Not a function; $x \in (-\infty,3]$ $y \in (-\infty,\infty)$

47. $f(x) = \dfrac{3}{x-5}$

$x - 5 = 0$

$x = 5$

$x \in (-\infty,5) \cup (5,\infty)$

49. $h(a) = \sqrt{3a+5}$

$3a + 5 \geq 0$

$3a \geq -5$

$a \geq -\dfrac{5}{3}$

$a \in \left[-\dfrac{5}{3},\infty\right)$

51. $v(x) = \dfrac{x+2}{x^2 - 25}$

$x^2 - 25 = 0$

$x^2 = 25$

$x = \pm 5$

$x \in (-\infty,-5) \cup (-5,5) \cup (5,\infty)$

53. $u = \dfrac{v-5}{v^2 - 18}$

$v^2 - 18 = 0$

$v^2 = 18$

$v = \pm 3\sqrt{2}$

$v \in \left(-\infty,-3\sqrt{2}\right) \cup \left(-3\sqrt{2},3\sqrt{2}\right) \cup \left(3\sqrt{2},\infty\right)$

55. $y = \dfrac{17}{25}x + 123$

$x \in (-\infty,\infty)$

57. $m = n^2 - 3n - 10$

$n \in (-\infty,\infty)$

59. $y = 2|x| + 1$

$x \in (-\infty,\infty)$

61. $y_1 = \dfrac{x}{x^2 - 3x - 10}$

$x^2 - 3x - 10 = 0$

$(x-5)(x+2) = 0$

$x = 5$ or $x = -2$

$x \in (-\infty,-2) \cup (-2,5) \cup (5,\infty)$

63. $y = \dfrac{\sqrt{x-2}}{2x-5}$, $x \geq 2$

$2x - 5 = 0$

$2x = 5$

$x = \dfrac{5}{2}$

$x \in \left[2, \dfrac{5}{2}\right) \cup \left(\dfrac{5}{2}, \infty\right)$

65. $f(x) = \sqrt{\dfrac{5}{x-2}}$

Since the radicand must be non-negative,

solve the inequality: $\dfrac{5}{x-2} \geq 0, x \neq 2$

Use test points to each side of 2.

If $x = 0, \dfrac{5}{0-2} \geq 0$ false

If $x = 3, \dfrac{5}{3-2} \geq 0$ true

Domain: $x \in (2, \infty)$

67. $h(x) = \dfrac{-2}{\sqrt{4+x}}$

Since the radicand must be non-negative and the denominator cannot equal zero, solve the inequality: $4 + x > 0, x > -4$.

Domain: $x \in (-4, \infty)$

69. $f(x) = \dfrac{1}{2}x + 3$

$f(-6) = \dfrac{1}{2}(-6) + 3 = -3 + 3 = 0$;

$f\left(\dfrac{3}{2}\right) = \dfrac{1}{2}\left(\dfrac{3}{2}\right) + 3 = \dfrac{3}{4} + 3 = \dfrac{15}{4}$;

$f(2c) = \dfrac{1}{2}(2c) + 3 = c + 3$

71. $f(x) = 3x^2 - 4x$

$f(-6) = 3(-6)^2 - 4(-6) = 108 + 24 = 132$;

$f\left(\dfrac{3}{2}\right) = 3\left(\dfrac{3}{2}\right)^2 - 4\left(\dfrac{3}{2}\right) = 3\left(\dfrac{9}{4}\right) - 6$

$= \dfrac{27}{4} - 6 = \dfrac{3}{4}$;

$f(2c) = 3(2c)^2 - 4(2c) = 3\left(4c^2\right) - 8c$

$= 12c^2 - 8c$

73. $h(x) = \dfrac{3}{x}$

$h(3) = \dfrac{3}{(3)} = 1$;

$h\left(-\dfrac{2}{3}\right) = \dfrac{3}{\left(-\dfrac{2}{3}\right)} = -\dfrac{9}{2}$;

$h(3a) = \dfrac{3}{3a} = \dfrac{1}{a}$

75. $h(x) = \dfrac{5|x|}{x}$

$h(3) = \dfrac{5|3|}{3} = \dfrac{5(3)}{3} = 5$;

$h\left(-\dfrac{2}{3}\right) = \dfrac{5\left|-\dfrac{2}{3}\right|}{-\dfrac{2}{3}} = \dfrac{5\left(\dfrac{2}{3}\right)}{-\dfrac{2}{3}} = -5$;

$h(3a) = \dfrac{5|3a|}{3a} = \dfrac{15|a|}{3a} = \dfrac{5|a|}{a}$;

-5 if $a < 0$; 5 if $a > 0$

77. $g(r) = 2\pi r$

$g(0.4) = 2\pi(0.4) = 0.8\pi$;

$g\left(\dfrac{9}{4}\right) = 2\pi\left(\dfrac{9}{4}\right) = \dfrac{9}{2}\pi$;

$g(h) = 2\pi(h) = 2\pi h$;

79. $g(r) = \pi r^2$

$g(0.4) = \pi(0.4)^2 = 0.16\pi$;

$g\left(\dfrac{9}{4}\right) = \pi\left(\dfrac{9}{4}\right)^2 = \dfrac{81}{16}\pi$;

$g(h) = \pi(h)^2 = \pi h^2$

81. $p(x) = \sqrt{2x+3}$

$p(0.5) = \sqrt{2(0.5)+3} = \sqrt{1+3} = \sqrt{4} = 2$;

$p\left(\dfrac{9}{4}\right) = \sqrt{2\left(\dfrac{9}{4}\right)+3} = \sqrt{\dfrac{9}{2}+3} = \sqrt{\dfrac{15}{2}} = \dfrac{\sqrt{30}}{2}$;

$p(a) = \sqrt{2(a)+3} = \sqrt{2a+3}$

83. $p(x) = \dfrac{3x^2 - 5}{x^2}$

$$p(0.5) = \dfrac{3(0.5)^2 - 5}{(0.5)^2} = \dfrac{3(0.25) - 5}{0.25}$$

$$= \dfrac{0.75 - 5}{0.25} = \dfrac{-4.25}{0.25} = -17$$

$$p\left(\dfrac{9}{4}\right) = \dfrac{3\left(\dfrac{9}{4}\right)^2 - 5}{\left(\dfrac{9}{4}\right)^2} = \dfrac{3\left(\dfrac{81}{16}\right) - 5}{\dfrac{81}{16}}$$

$$= \dfrac{\dfrac{243}{16} - 5}{\dfrac{81}{16}} = \dfrac{\dfrac{163}{16}}{\dfrac{81}{16}} = \dfrac{163}{81} \ ;$$

$$p(a) = \dfrac{3(a)^2 - 5}{(a)^2} = \dfrac{3a^2 - 5}{a^2}$$

85. a. D:$\{-1, 0, 1, 2, 3, 4, 5\}$
    b. R:$\{-2, -1, 0, 1, 2, 3, 4\}$
    c. $f(2) = 1$
    d. $f(-1) = 4$

87. a. $x \in [-5, 5]$
    b. $y \in [-3, 4]$
    c. $f(2) = -2$
    d. when $y = 1$, $x = 0$ and $x = -4$.

89. a. $x \in [-3, \infty)$
    b. $y \in (-\infty, 4]$
    c. $f(2) = 2$
    d. when $y = 2$, $x = 2$ and $x = -2$

91. $W(H) = \dfrac{9}{2}H - 151$

    a. $W(75) = \dfrac{9}{2}(75) - 151 = 186.5$ lb

    b. $W(72) = \dfrac{9}{2}(72) - 151 = 173$ lb
        $210 - 173 = 37$ lb

93. $A = \dfrac{1}{2}B + I - 1$

    $\square PQR$

    $P(-3, 1), Q(3, 9), R(7, 6)$

    $$m = \dfrac{9 - 1}{3 - (-3)} = \dfrac{4}{3}$$

    $$y - 1 = \dfrac{4}{3}(x + 3)$$

    $$y - 1 = \dfrac{4}{3}x + 4$$

    $$y = \dfrac{4}{3}x + 5$$

    (0,5) lies on PQ;
    Lattice points are points that join vertical and horizontal grids in a Cartesian coordinate system.
    There are four lattice points on the boundary; three vertices and point (0,5), thus $B = 8$. There are 24 lattice points in the interior of the triangle, thus $I = 24$.

    $$A = \dfrac{1}{2}(8) + 22 - 1 = 25 \text{ units}^2$$

95. a. $N(g) = 2.5g$

    b. $g \in [0, 5]$; $N \in [0, 12.5]$

97. a. $D \in [0, \infty)$

    b. $V(7.5) = 100\pi(7.5) = 750\pi$

    c. $V\left(\dfrac{8}{\pi}\right) = 100\pi\left(\dfrac{8}{\pi}\right) = 800 \text{ cm}^3$

99. a. $c(t) = 42.50t + 50$

    b. $c(2.5) = 42.50(2.5) + 50 = \$156.25$

    c. $262.50 = 42.50t + 50$
        $212.50 = 42.50t$
        $5 \text{ hr} = t$

    d. $500 = 42.50t + 50$
        $450 = 42.50t$
        $10.6 \text{ hr} \approx t$
        $t \in [0, 10.6]$; $c \in [0, 500]$

101.a. Yes.

Each "x" is paired with exactly one "y".

b. 10 P.M.

c. 0.9 m

d. 7 P.M. and 1 A.M.

103.a. Average rate of change from 1920 to 1940, use (20,3.2) and (40,2.2).

$\dfrac{\Box fertility}{\Box time} = \dfrac{2.2 - 3.2}{40 - 20} = -\dfrac{1}{20}$; Negative;

Fertility is decreasing by one child every 20 years.

b. Average rate of change from 1940 to 1950, use (40,2.2) and (50,3.0).

$\dfrac{\Box fertility}{\Box time} = \dfrac{3.0 - 2.2}{50 - 40} = \dfrac{0.8}{10}$; Positive;

Fertility is increasing by less than one child every 10 years.

c. from 1980 to 1990, use (80,1.8) and (90,2.0).

$\dfrac{\Box fertility}{\Box time} = \dfrac{2.0 - 1.8}{90 - 80} = \dfrac{0.2}{10}$; The

fertility rate was increasing four times as fast from 1940 to 1950.

105. The y-values of the negative x integers would become positive.

All points would be in Quadrants I and III.

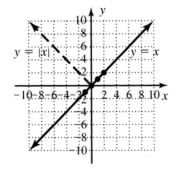

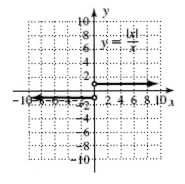

107.a. $y = \dfrac{x-3}{x+2}, x \neq -2$

Domain: $x \in (-\infty, -2) \cup (-2, \infty)$;

$(x+2)y = (x+2)\left(\dfrac{x-3}{x+2}\right)$

$xy + 2y = x - 3$

$xy - x = -2y - 3$

$x(y-1) = -2y - 3$

$x = \dfrac{-2y-3}{y-1} = \dfrac{2y+3}{1-y}, y \neq 1$

Range: $y \in (-\infty, 1) \cup (1, \infty)$

b. $y = x^2 - 3$

Domain: $x \in \Box$;

$y = x^2 - 3$

$y + 3 = x^2$

$\pm\sqrt{y+3} = x$;

$y + 3 \geq 0$

$y \geq -3$

Range: $y \in [-3, \infty)$

109.a. $\sqrt{24} + 6\sqrt{54} - \sqrt{6}$

$= 2\sqrt{6} + 6 \cdot 3\sqrt{6} - \sqrt{6}$

$= 2\sqrt{6} + 18\sqrt{6} - \sqrt{6}$

$= 19\sqrt{6}$

b. $\left(2 + \sqrt{3}\right)\left(2 - \sqrt{3}\right)$

$= 4 - 2\sqrt{3} + 2\sqrt{3} - 3$

$= 1$

111.a. $x^3 - 3x^2 - 25x + 75$

$= \left(x^3 - 3x^2\right) - \left(25x - 75\right)$

$= x^2(x-3) - 25(x-3)$

$= (x-3)\left(x^2 - 25\right)$

$= (x-3)(x-5)(x+5)$

b. $2x^2 - 13x - 24 = (2x+3)(x-8)$

c. $8x^3 - 125 = (2x-5)\left(4x^2 + 10x + 25\right)$

Chapter 2 Mid-Chapter Check

1. $4x - 3y = 12$
$$-3y = -4x + 12$$
$$y = \frac{4}{3}x - 4$$

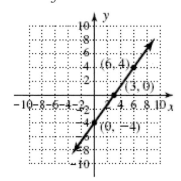

3. $m = \dfrac{-0.5 - (-2)}{2003 - 2002} = \dfrac{1.5}{1} = 1.5$;

Positive, loss is decreasing, profit is increasing.
Data.com's loss decreases by 1.5 million dollars per year.

5. $x = -3$; not a function. Input -3 is paired with more than one output.

7. a. $h(2) = 0$

   b. $x \in [-3,5]$

   c. $x = -1$ when $h(x) = -3$

   d. $y \in [-4,5]$

9. $m = \dfrac{3}{4}$; $(1, 2)$
$$y - y_1 = m(x - x_1)$$
$$y - 2 = \frac{3}{4}(x - 1)$$
$$y - 2 = \frac{3}{4}x - \frac{3}{4}$$
$$y = \frac{3}{4}x + \frac{5}{4}$$
$$F(p) = \frac{3}{4}p + \frac{5}{4}$$

For every 4000 pheasants, the fox population increases by 300.
$$F(20) = \frac{3}{4}(20) + \frac{5}{4} = 15 + 1.25 = 16.25$$

Fox population is 1625 when the pheasant population is 20,000.

Chapter 2 Reinforcing Basic Concepts

1. $P_1(0,5)$; $P_2(6,7)$

   a. $m = \dfrac{7-5}{6-0} = \dfrac{2}{6} = \dfrac{1}{3}$; increasing

   b. $y - 5 = \dfrac{1}{3}(x - 0)$

   c. $y = \dfrac{1}{3}x + 5$

   d. $y = \dfrac{1}{3}x + 5$
$$-\frac{1}{3}x + y = 5$$
$$x - 3y = -15$$

   e. x-intercept: $(-15, 0)$ y-intercept: $(0, 5)$
$$
\begin{array}{ll}
x - 3(0) = -15 & 0 - 3y = -15 \\
x = -15 & -3y = -15 \\
& y = 5
\end{array}
$$

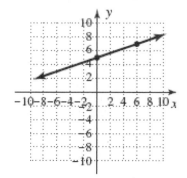

3. $P_1(3,2)$; $P_2(9,5)$

    a. $m = \dfrac{5-2}{9-3} = \dfrac{3}{6} = \dfrac{1}{2}$; increasing

    b. $y - 2 = \dfrac{1}{2}(x - 3)$

    c. $y - 2 = \dfrac{1}{2}x - \dfrac{3}{2}$

        $y = \dfrac{1}{2}x + \dfrac{1}{2}$

    d. $y = \dfrac{1}{2}x + \dfrac{1}{2}$

        $-\dfrac{1}{2}x + y = \dfrac{1}{2}$

        $x - 2y = -1$

    e. x-intercept: $(-1, 0)$ y-intercept: $\left(0, \dfrac{1}{2}\right)$

    $x - 2(0) = -1 \qquad\qquad 0 - 2y = -1$

    $\qquad x = -1 \qquad\qquad\qquad y = \dfrac{1}{2}$

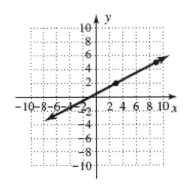

5. $P_1(-2,5)$; $P_2(6,-1)$

    a. $m = \dfrac{-1-5}{6-(-2)} = \dfrac{-6}{8} = -\dfrac{3}{4}$; decreasing

    b. $y - 5 = -\dfrac{3}{4}(x + 2)$

    c. $y - 5 = -\dfrac{3}{4}x - \dfrac{3}{2}$

        $y = -\dfrac{3}{4}x + \dfrac{7}{2}$

    d. $y = -\dfrac{3}{4}x + \dfrac{7}{2}$

        $\dfrac{3}{4}x + y = \dfrac{7}{2}$

        $3x + 4y = 14$

    e. x-intercept: $\left(\dfrac{14}{3}, 0\right)$ y-intercept: $\left(0, \dfrac{7}{2}\right)$

    $3x + 4(0) = 14$

    $\qquad 3x = 14$

    $\qquad x = \dfrac{14}{3}$

    $3(0) + 4y = 14$

    $\qquad 4y = 14$

    $\qquad y = \dfrac{7}{2}$

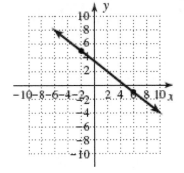

2.5 Technology Highlight

Exercise 1: $x \approx -2.87$, $x \approx 0.87$,
min: $y = -7$ at $(-1, -7)$, no max

Exercise 3: $x \approx 1.35$, $x \approx 6.65$,
min: $y = -7$ at $(4, -7)$, no max

Exercise 5: $x = -2$, $x = 0$, $x \approx 2.41$,
min: $y = -3.20$ at $(-1.47, -3.20)$,
min: $y \approx -9.51$ at $(1.67, -9.51)$,
max: $y = 0$ at $(0, 0)$

2.5 Exercises

1. Linear; bounce

3. Increasing

5. Answers will vary.

7.

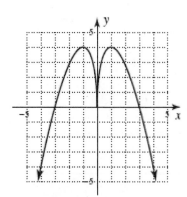

9. $f(x) = -7|x| + 3x^2 + 5$
$f(k) = -7|k| + 3(k)^2 + 5$;
$f(-k) = -7|-k| + 3(-k)^2 + 5$
$= -7|k| + 3(k)^2 + 5 = f(k)$;
Even

11. $g(x) = \frac{1}{3}x^4 - 5x^2 + 1$
$g(k) = \frac{1}{3}(k)^4 - 5(k)^2 + 1$
$= \frac{1}{3}k^4 - 5k^2 + 1$;
$g(-k) = \frac{1}{3}(-k)^4 - 5(-k)^2 + 1$
$= \frac{1}{3}k^4 - 5k^2 + 1$;
$g(k) = g(-k)$
Even

13.

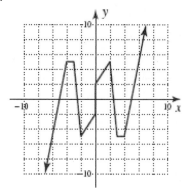

15. $f(x) = 4\sqrt[3]{x} - x$
$f(k) = 4\sqrt[3]{k} - k$
$f(-k) = 4\sqrt[3]{-k} - (-k)$
$= -4\sqrt[3]{k} + k = -\left(4\sqrt[3]{k} - k\right)$;
$f(k) = -f(k)$
Odd

17. $p(x) = 3x^3 - 5x^2 + 1$
$p(k) = 3(k)^3 - 5(k)^2 + 1$
$= 3k^3 - 5k^2 + 1$
$p(-k) = 3(-k)^3 - 5(-k)^2 + 1$
$= -3k^3 - 5k^2 + 1$
$p(k) \neq -p(k)$; Not Odd

19. $w(x) = x^3 - x^2$
$w(-x) = (-x)^3 - (-x)^2$
$= -x^3 - x^2$; neither

21. 20. $p(x) = 2\sqrt[3]{x} - \dfrac{1}{4}x^3$

$p(-x) = 2\sqrt[3]{(-x)} - \dfrac{1}{4}(-x)^3$

$= -2\sqrt[3]{x} + \dfrac{1}{4}x^3 = -\left(2\sqrt[3]{x} - \dfrac{1}{4}x^3\right)$; odd

23. $v(x) = x^3 + 3|x|$

$v(-x) = (-x)^3 + 3|-x|$

$= -x^3 + 3|x|$; neither

25. $f(x) = x^3 - 3x^2 - x + 3$

Verify Zeros: Let $f(x) = 0$

$0 = x^2(x-3) - (x-3)$

$0 = (x-3)(x^2 - 1)$

$0 = (x-3)(x+1)(x-1)$

Zeros: $(-1,0),(1,0),(3,0)$

For $f(x) \geq 0$, $x \in [-1,1], \ [3,\infty)$

27. $f(x) = x^4 - 2x^2 + 1$

Verify Zeros: Let $f(x) = 0$

$0 = x^4 - 2x^2 + 1$

$0 = (x^2 - 1)(x^2 - 1)$

$0 = (x+1)(x-1)(x+1)(x-1)$

Zeros: $(-1,0),(1,0)$

For $f(x) > 0$, $x \in (-\infty,-1) \cup (-1,1) \cup (1,\infty)$

29. $p(x) = \sqrt[3]{x-1} - 1$

$p(x) \geq 0$ for $x \in [2,\infty)$

31. $f(x) = (x-1)^3 - 1$

$f(x) \leq 0$ for $x \in (-\infty, 2]$

33. $f(x) \uparrow: (-3,1) \cup (4,6)$

$f(x) \downarrow: (-\infty,-3),(1,4)$

Constant : None

35. $f(x) \uparrow: (1,4)$

$f(x) \downarrow: (-2,1) \cup (4,\infty)$

Constant: $(-\infty,-2)$

37. $p(x) = 0.5(x+2)^3$

a. $p(x) \uparrow: x \in (-\infty,\infty)$

$p(x) \downarrow:$ None

b. down, up

39. $y = p(x)$

a. $p(x) \uparrow: x \in (-3,0) \cup (3,\infty)$

$p(x) \downarrow: x \in (-\infty,-3) \cup (0,3)$

b. up, up

41. $H(x) = -5|x-2| + 5$

a. $x \in (-\infty,\infty)$

$y \in (-\infty,5]$

b. $(1, 0), (3, 0)$

c. $H(x) \geq 0: x \in [1,3]$

$H(x) \leq 0: x \in (-\infty,1] \cup [3,\infty)$

d. $H(x) \uparrow: x \in (-\infty,2)$

$H(x) \downarrow: x \in (2,\infty)$

e. local maximum: $y = 2$ at $(2, 5)$

43. $y = g(x)$

a. $x \in (-\infty,\infty)$

$y \in (-\infty,\infty)$

b. $(-1,0), (5, 0)$

c. $g(x) \geq 0: x \in [-1,\infty)$

$g(x) \leq 0: x \in (-\infty,-1] \cup \{5\}$

d. $g(x) \uparrow: x \in (-\infty,1) \cup (5,\infty)$

$g(x) \downarrow: x \in (1,5)$

e. local maximum: $y = 6$ at $(1,6)$

local minimum: $y = 0$ at $(5, 0)$

45. $y = Y_2$

a. $x \in (-\infty,\infty)$

$y \in (-\infty,3]$

b. $(0, 0), (2, 0)$

c. $Y_2 \geq 0: x \in [0,2]$

$Y_2 \leq 0: x \in (-\infty,0] \cup [2,\infty)$

d. $Y_2 \uparrow: x \in (-\infty,1)$

$Y_2 \downarrow: x \in (1,\infty)$

e. local maximum: $y = 3$ at $(1, 3)$

47. $p(x) = (x+3)^3 + 1$

   a. $x \in ¡$, $y \in ¡$

   b. $x = -4$

   c. $p(x) \geq 0 : x \in [-4, \infty)$;

      $p(x) \leq 0 : x \in (-\infty, -4]$

   d. $p(x) \uparrow : x \in (-\infty, -3) \cup (-3, \infty)$

      $p(x) \downarrow$: never decreasing

   e. Local max: none

      Local min: none

49. $y = \dfrac{1}{3}\sqrt{4x^2 - 36}$

   a. $x \in (-\infty, -3] \cup [3, \infty)$

      $y \in [0, \infty)$

   b. $(-3, 0), (3, 0)$

   c. $f(x) \uparrow : x \in (3, \infty)$

      $f(x) \downarrow : x \in (-\infty, -3)$

   d. Even

51. a. $x \in [0, 260]$

      $y \in [0, 80]$

   b. 80 feet

   c. 120 feet

   d. Yes

   e. (0, 120)

   f. (120, 260)

53. $f(x) = x^{\frac{2}{3}} - 1$

   a. $x \in (-\infty, \infty)$

      $y \in [-1, \infty)$

   b. $(-1, 0), (1, 0)$

   c. $f(x) \geq 0 : x \in (-\infty, -1] \cup [1, \infty)$

      $f(x) < 0 : x \in (-1, 1)$

   d. $f(x) \uparrow : x \in (0, \infty)$

      $f(x) \downarrow : x \in (-\infty, 0)$

   e. Minimum: $(0, -1)$

55. a. $D : t \in [72, 96]$

      $R : I \in [7.25, 16]$

   b. $I(t) \uparrow$ for $t \in$ (72,74) $\cup$ (77,81) $\cup$ (83, 84) $\cup$ (93, 94)

      $I(t) \downarrow$ for $t \in$ (74, 75) $\cup$ (81, 83) $\cup$ (84, 86) $\cup$ (90, 93) $\cup$ (94, 95)

      $I(t)$ constant for $t \in$ (75, 77) $\cup$ (86, 90) $\cup$ (95, 96)

   c. Maximum: (74, 9.25), (81, 16) (global max), (84, 13), (94, 8.5)

      Minimum: (72,7.5), (83, 12.75), (93, 7.2)

   d. Increase: 1980 to 1981

      Decrease: 1982 to 1983 or 1985 to 1986

57.

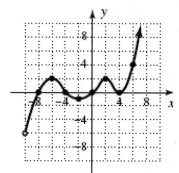

Zeroes: (–8, 0), (–4, 0), (0, 0), (4, 0)

Maximum: $(-6, 2), (2, 2)$

Minimum: $(-2, -1), (4, 0)$

59. $f(x) = x^3$

   a. $\dfrac{\Delta f}{\Delta x} = \dfrac{f(-1) - f(-2)}{-1 - (-2)} = \dfrac{-1 - (-8)}{1} = 7$

   b. $\dfrac{\Delta f}{\Delta x} = \dfrac{f(2) - f(1)}{2 - 1} = \dfrac{8 - 1}{1} = 7$

   c. They are the same.

   d. Slopes of the lines are the same.

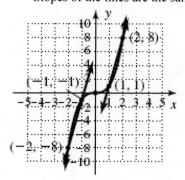

61. $h(t) = -16t^2 + 192t$

   a. $h(1) = -16(1)^2 + 192(1) = 176$ ft

   b. $h(2) = -16(2)^2 + 192(2) = 320$ ft

   c. $\dfrac{\Delta h}{\Delta t} = \dfrac{h(2) - h(1)}{2 - 1} = \dfrac{320 - 176}{1}$
   $= 144$ ft/sec

   d. $\dfrac{\Delta h}{\Delta t} = \dfrac{h(11) - h(10)}{11 - 10} = \dfrac{176 - 320}{1}$
   $= -144$ ft/sec
   The arrow is going down.

63. $v = \sqrt{2gs}$, $v = \sqrt{2(32)s} = 8\sqrt{s}$

   a. $v = \sqrt{2(32)(5)} = \sqrt{320} = 17.89$ ft/sec ;
   $v = \sqrt{2(32)(10)} = \sqrt{640} = 25.30$ ft/sec

   b. $v = \sqrt{2(32)(15)} = \sqrt{960} = 30.98$ ft/sec ;
   $v = \sqrt{2(32)(20)} = \sqrt{1280} = 35.78$ ft/sec

   c. Between $s = 5$ and $s = 10$

   d. $\dfrac{\Delta v}{\Delta s} = \dfrac{v(10) - v(5)}{10 - 5} = \dfrac{25.3 - 17.89}{5}$
   $= 1.482$ ft/sec;
   $\dfrac{\Delta v}{\Delta s} = \dfrac{v(20) - v(15)}{20 - 15} = \dfrac{35.78 - 30.98}{5}$
   $= 0.96$ ft/sec

65. $f(x) = 2x - 3$

   $\dfrac{\Delta f}{\Delta x} = \dfrac{f(x+h) - f(x)}{h}$

   $= \dfrac{[2(x+h) - 3] - (2x - 3)}{h}$

   $= \dfrac{2x + 2h - 3 - 2x + 3}{h}$

   $= \dfrac{2h}{h} = 2$

67. $h(x) = x^2 + 3$

   $\dfrac{\Delta f}{\Delta x} = \dfrac{h(x+h) - h(x)}{h} = \dfrac{[(x+h)^2 + 3] - (x^2 + 3)}{h}$

   $= \dfrac{x^2 + 2xh + h^2 + 3 - x^2 - 3}{h}$

   $= \dfrac{2xh + h^2}{h} = \dfrac{h(2x + h)}{h} = 2x + h$

69. $g(x) = x^2 + 2x - 3$

   $\dfrac{\Delta g}{\Delta x} = \dfrac{g(x+h) - g(x)}{h}$

   $= \dfrac{[(x+h)^2 + 2(x+h) - 3] - (x^2 + 2x - 3)}{h}$

   $= \dfrac{x^2 + 2xh + h^2 + 2x + 2h - 3 - x^2 - 2x + 3}{h}$

   $= \dfrac{2xh + h^2 + 2h}{h} = \dfrac{h(2x + h + 2)}{h} = 2x + 2 + h$

71. $f(x) = \dfrac{2}{x}$

   $\dfrac{\Delta f}{\Delta x} = \dfrac{f(x+h) - f(x)}{h}$

   $= \dfrac{\dfrac{2}{x+h} - \dfrac{2}{x}}{h} = \dfrac{\dfrac{2x - 2(x+h)}{x(x+h)}}{h}$

   $= \dfrac{\dfrac{2x - 2x - 2h}{x(x+h)}}{h} = \dfrac{-2h}{x(x+h)} \cdot \dfrac{1}{h} = \dfrac{-2}{x(x+h)}$

73. a. $g(x) = x^2 + 2x$

   $\dfrac{\Delta g}{\Delta x} = \dfrac{g(x+h) - g(x)}{h}$

   $= \dfrac{[(x+h)^2 + 2(x+h)] - (x^2 + 2x)}{h}$

   $= \dfrac{x^2 + 2xh + h^2 + 2x + 2h - x^2 - 2x}{h}$

   $= \dfrac{2xh + h^2 + 2h}{h} = \dfrac{h(2x + h + 2)}{h} = 2x + 2 + h$

   b. For $[-3.0, -2.9]$, $x = -3.0$ and $h = 0.1$
   Rate of change:
   $2(-3.0) + 2 + 0.1 = -3.9$

c. For [0.50, 0.51], $x = 0.50$ and $h = 0.01$
Rate of change:
$2(0.50) + 2 + 0.01 = 3.01$

d.

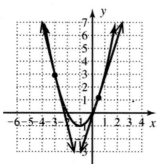

The rates of change have opposite signs, with the secant line to the left being more steep.

75. a. $g(x) = x^3 + 1$

$$\frac{\Delta g}{\Delta x} = \frac{g(x+h) - g(x)}{h}$$

$$= \frac{[(x+h)^3 + 1] - (x^3 + 1)}{h}$$

$$= \frac{x^3 + 3x^2h + 3xh^2 + h^3 + 1 - x^3 - 1}{h}$$

$$= \frac{3x^2h + 3xh^2 + h^3}{h}$$

$$= \frac{h(3x^2 + 3xh + h^2)}{h} = 3x^2 + 3xh + h^2$$

b. For [–2.1, –2], $x = -2.1$ and $h = 0.1$
Rate of change:
$3(-2.1)^2 + 3(-2.1)(0.1) + (0.1)^2 = 12.61$

c. For [0.40, 0.41], $x = 0.40$ and $h = 0.01$
Rate of change:
$3(0.40)^2 + 3(0.40)(0.01) + (0.01)^2 \approx 0.49$

d.

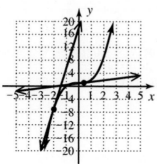

Both lines have a positive slope, but the line at $x = -2$ is much steeper.

77. $d(x) = 1.5\sqrt{x}$

$$\frac{\Delta d}{\Delta x} = \frac{d(x+h) - d(x)}{h}$$

$$= \frac{1.5\sqrt{x+h} - 1.5\sqrt{x}}{h}$$

a. For [9, 9.01], $x = 9$ and $h = 0.01$
Rate of change:
$$\frac{1.5\sqrt{9+0.01} - 1.5\sqrt{9}}{0.01} \approx 0.25$$

b. For [225, 225.01], $x = 225$ and $h = 0.01$
Rate of change:
$$\frac{1.5\sqrt{225+0.01} - 1.5\sqrt{225}}{0.01} \approx 0.05$$

c.

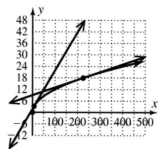

As height increases, you can see farther, the sight distance is increasing much slower.

79. No; No; Answers will vary.

81. Answers will vary.

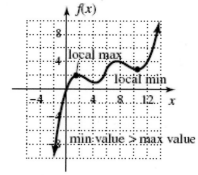

Chapter 2: Relations, Functions and Graphs

83. $x^2 - 8x - 20 = 0$

    a. $(x-10)(x+2) = 0$

    $x = 10; \quad x = -2$

    b. $(x^2 - 8x) - 20 = 0$

    $(x^2 - 8x + 16) - 20 - 16 = 0$

    $(x-4)^2 - 36 = 0$

    $(x-4)^2 = 36$

    $x - 4 = \pm 6$

    $x = 4 \pm 6$

    $x = 10; \quad x = -2$

    c. $x = \dfrac{8 \pm \sqrt{(-8)^2 - 4(1)(-20)}}{2(1)}$

    $x = \dfrac{8 \pm \sqrt{64 + 80}}{2}$

    $x = \dfrac{8 \pm \sqrt{144}}{2}$

    $x = \dfrac{8 \pm 12}{2}$

    $x = 10; \quad x = -2$

85. $y = \dfrac{2}{3}x - 1$

2.6 Technology Highlight

Exercise 1: Shifted right 3 units; answers will vary.

2.6 Exercises

1. Stretch; compression

3. $(-5, -9)$; upward

5. Answers will vary.

7. $f(x) = x^2 + 4x$

    a. quadratic
    b. up/up, Vertex $(-2,-4)$,
       Axis of symmetry $x = -2$,
       x–intercepts $(-4, 0)$ and $(0,0)$,
       y–intercept $(0,0)$
    c. D: $x \in \mathbb{R}$, R: $y \in [-4, \infty)$

9. $p(x) = x^2 - 2x - 3$

    a. quadratic
    b. up/up, Vertex $(1,-4)$,
       Axis of symmetry $x = 1$,
       x–intercepts $(-1, 0)$ and $(3,0)$,
       y–intercept $(0,-3)$
    c. D: $x \in \mathbb{R}$, R: $y \in [-4, \infty)$

11. $f(x) = x^2 - 4x - 5$

    a. quadratic
    b. up/up, Vertex $(2,-9)$,
       Axis of symmetry $x = 2$,
       x–intercepts $(-1, 0)$ and $(5,0)$,
       y–intercept $(0,-5)$
    c. D: $x \in \mathbb{R}$, R: $y \in [-9, \infty)$

13. $p(x) = 2\sqrt{x+4} - 2$

    a. square root
    b. up to the right, Initial point $(-3,-4)$,
       x–intercept $(-3, 0)$,
       y–intercept $(0,2)$
    c. D: $x \in [-4, \infty)$, R: $y \in [-2, \infty)$

15. $r(x) = -3\sqrt{4-x} + 3$

    a. square root
    b. down to the left, Initial point $(4,3)$,
       x–intercept $(3, 0)$,
       y–intercept $(0,-3)$
    c. D: $x \in (-\infty, 4]$, R: $y \in (-\infty, 3]$

17. $g(x) = 2\sqrt{4-x}$

    a. square root
    b. up to the left, Initial point $(4,0)$,
       x–intercept $(4, 0)$,
       y–intercept $(0,4)$
    c. D: $x \in (-\infty, 4]$, R: $y \in [0, \infty)$

19. $p(x) = 2|x+1| - 4$

    a. absolute value
    b. up/up, Vertex $(-1,-4)$,
       Axis of symmetry $x = -1$,
       x–intercepts $(-3, 0)$ and $(1,0)$,
       y–intercept $(0,-2)$
    c. D: $x \in \mathbb{R}$, R: $y \in [-4, \infty)$

2.6 Exercises

21. $r(x) = -2|x+1| + 6$
    a. absolute value
    b. down/down, Vertex $(-1,6)$,
       Axis of symmetry $x = -1$,
       x–intercepts $(-4, 0)$ and $(2,0)$,
       y–intercept $(0,4)$
    c. D: $x \in \mathbf{i}$, R: $y \in (-\infty, 6]$

23. $g(x) = -3|x| + 6$
    a. absolute value
    b. down/down, Vertex $(0,6)$,
       Axis of symmetry $x = 0$,
       x–intercepts $(-2, 0)$ and $(2,0)$,
       y–intercept $(0,6)$
    c. D: $x \in \mathbf{i}$, R: $y \in (-\infty, 6]$

25. $f(x) = -(x-1)^3$
    a. cubic
    b. up/down, Inflection point $(1,0)$,
       x–intercept $(1, 0)$,
       y–intercept $(0,1)$
    c. D: $x \in \mathbf{i}$, R: $y \in \mathbf{i}$

27. $h(x) = x^3 + 1$
    a. cubic
    b. down/up, Inflection point $(0,1)$,
       x–intercept $(-1, 0)$,
       y–intercept $(0,1)$
    c. D: $x \in \mathbf{i}$, R: $y \in \mathbf{i}$

29. $q(x) = \sqrt[3]{x-1} - 1$
    a. cube root
    b. down/up, Inflection point $(1,-1)$,
       x–intercept $(2, 0)$,
       y–intercept $(0,-2)$
    c. D: $x \in \mathbf{i}$, R: $y \in \mathbf{i}$

31. Function family: Square root
    x–intercept: $(-3, 0)$
    y–intercept: $(0, 2)$
    Initial point: $(-4, -2)$
    End behavior: Up on right

33. Function family: Cubic
    x–intercept: $(-2, 0)$
    y–intercept: $(0, -2)$
    Inflection point: $(-1, -1)$
    End behavior: Up/down

35. $f(x) = \sqrt{x}$; $g(x) = \sqrt{x} + 2$; $h(x) = \sqrt{x} - 3$

x	f(x)	g(x)	h(x)
0	0	2	–3
4	2	4	–1
9	3	5	0
16	4	6	1
25	5	7	2

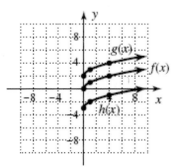

From the parent graph $f(x) = \sqrt{x}$, $g(x)$
   shifts
up 2 units and $h(x)$ shifts down 3 units.

37. $p(x) = |x|$; $q(x) = |x| - 5$; $r(x) = |x| + 2$

x	p(x)	q(x)	r(x)
–2	2	–3	4
–1	1	–4	3
0	0	–5	2
1	1	–4	3
2	2	–3	4

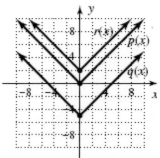

From the parent graph $p(x) = |x|$, $q(x)$ shifts
down 5 units and $r(x)$ shifts up 2 units.

39. $f(x) = x^3 - 2$

Shifts down 2 units.

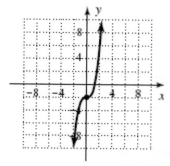

41. $h(x) = x^2 + 3$

Shifts up 3 units.

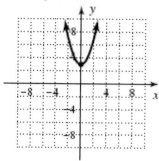

43. $p(x) = x^2$; $q(x) = (x+3)^2$

x	$p(x) = x^2$	$q(x) = (x+3)^2$
−5	25	4
−3	9	0
−1	1	4
1	1	16
3	9	36

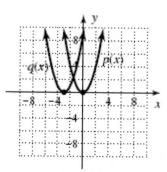

From the parent graph $p(x) = x^2$, $q(x)$ shifts left 3 units.

45. $Y_1 = |x|$; $Y_2 = |x - 1|$

x	$Y_1 = |x|$	$Y_2 = |x - 1|$
−2	2	3
−1	1	2
0	0	1
1	1	0
2	2	1

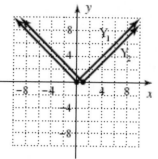

From the parent graph $Y_1 = |x|$, Y_2 shifts right 1 unit.

47. $p(x) = (x - 3)^2$

Shifts right 3 units.

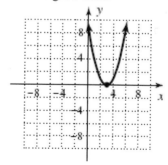

49. $h(x) = |x + 3|$

Shifts left 3 units.

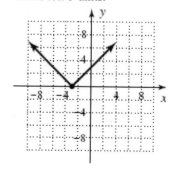

50. $f(x) = \sqrt[3]{x+2}$

Shifts left 2 units.

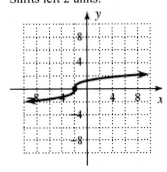

51. $g(x) = -|x|$

Reflects across the x–axis.

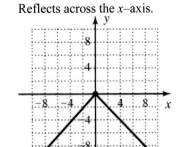

53. $f(x) = \sqrt[3]{-x}$

Reflects across the y–axis.

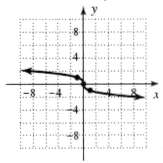

55. $p(x) = x^2$; $q(x) = 2x^2$; $r(x) = \dfrac{1}{2}x^2$

x	p(x)	q(x)	r(x)
−2	4	8	2
−1	1	2	½
0	0	0	0
1	1	2	½
2	4	8	2

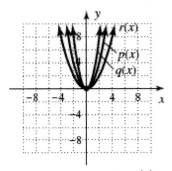

From the parent graph $p(x) = x^2$, $q(x)$ stretches upward and $r(x)$ compresses downward.

57. $Y_1 = |x|$; $Y_2 = 3|x|$; $Y_3 = \dfrac{1}{3}|x|$

x	Y_1	Y_2	Y_3
−2	2	6	2/3
−1	1	3	1/3
0	0	0	0
1	1	3	1/3
2	2	6	2/3

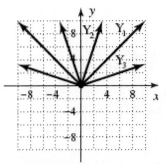

From the parent graph $Y_1 = |x|$, Y_2 stretches upward and Y_3 compresses downward.

59. $f(x) = 4\sqrt[3]{x}$

Stretches upward and downward.

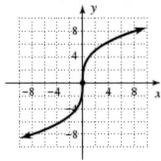

61. $p(x) = \frac{1}{3}x^3$

Compresses downward.

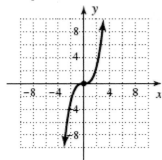

63. $f(x) = \frac{1}{2}x^3$; g

65. $f(x) = -(x-3)^2 + 2$; i

67. $f(x) = |x+4| + 1$; e

69. $f(x) = -\sqrt{x+6} - 1$; j

71. $f(x) = (x-4)^2 - 3$; l

73. $f(x) = \sqrt{x+3} - 1$; c

75. $f(x) = \sqrt{x+2} - 1$

Left 2, down 1
Initial point: $(-2, -1)$

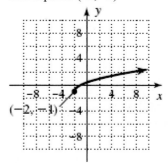

77. $h(x) = -(x+3)^2 - 2$

Left 3, reflected across x–axis, down 2
Vertex: $(-3, -2)$

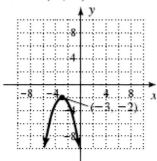

79. $p(x) = (x+3)^3 - 1$

Left 3, down 1
Inflection point: $(-3, -1)$

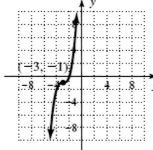

2.6 Exercises

81. $Y_1 = \sqrt[3]{x+1} - 2$

Left 1, down 2
Inflection point: $(-1, -2)$

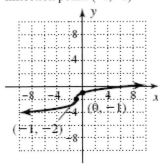

83. $f(x) = -|x+3| - 2$

Left 3, reflected across x–axis, down 2
Vertex: $(-3, -2)$

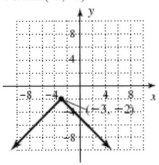

85. $h(x) = -2(x+1)^2 - 3$

Left 1, stretched vertically, reflected across
x–axis, down 3
Vertex: $(-1, -3)$

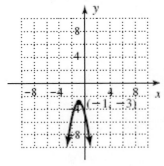

87. $p(x) = -\dfrac{1}{3}(x+2)^3 - 1$

Left 2, compressed vertically, reflected
across x–axis, down 1
Inflection point: $(-2, -1)$

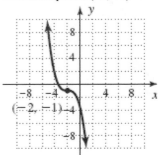

89. $Y_1 = -2\sqrt{-x-1} + 3$

Reflected across y–axis, left 1, reflected
across x–axis, stretched vertically, up 3
Initial point: $(-1, 3)$

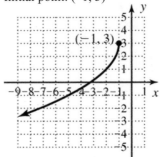

91. $h(x) = \dfrac{1}{5}(x-3)^2 + 1$

Right 3, compressed vertically, up 1
Vertex: $(3, 1)$

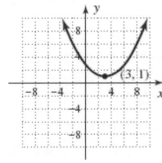

93. a. $f(x-2)$

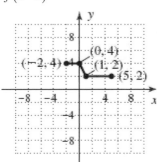

95. a. $h(x)+3$

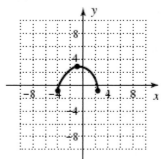

b. $-f(x)-3$

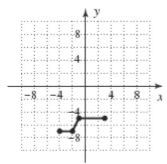

b. $-h(x-2)$

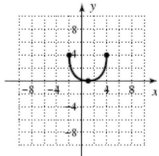

c. $\dfrac{1}{2}f(x+1)$

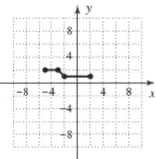

c. $h(x-2)-1$

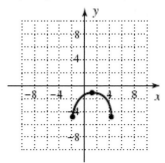

d. $f(-x)+1$

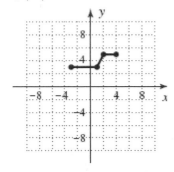

d. $\dfrac{1}{4}h(x)+5$

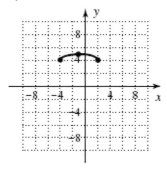

2.6 Exercises

97. Vertex: (2, 0)
Point: (0, –4)

$$y = a(x-h)^2 + k$$
$$-4 = a(0-2)^2 + 0$$
$$-4 = 4a$$
$$-1 = a;$$
$$y = -(x-2)^2$$

99. Node: (–3, 0)
Point: (6, 4.5)

$$y = a\sqrt{x-h} + k$$
$$4.5 = a\sqrt{6-(-3)} + 0$$
$$4.5 = 3a$$
$$1.5 = a;$$
$$y = 1.5\sqrt{x+3}$$

101. Vertex: (–4, 0)
Point: (1, 4)

$$y = a|x-h| + k$$
$$4 = a|1+4| + 0$$
$$4 = 5a$$
$$\frac{4}{5} = a;$$
$$y = \frac{4}{5}|x+4|$$

103. $V = \frac{4}{3}\pi r^3$

$$\frac{4}{3}\pi \approx 4.2$$

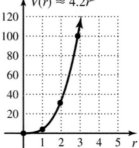

Volume estimate: 70 in^3

$$V = \frac{4}{3}\pi r^3$$
$$V = \frac{4}{3}\pi (2.5)^3$$
$$V = \frac{4}{3}\pi (15.625) \approx 65.4 \, \text{in}^3$$

Yes

105. $T(x) = \frac{1}{4}\sqrt{x}$

The graph can be obtained from $y = \sqrt{x}$ if it is compressed vertically.

$$T(81) = \frac{1}{4}\sqrt{81} = \frac{1}{4}(9) = 2.25 \, \text{sec}$$

This point is on the graph.

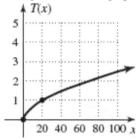

107. $P(v) = \dfrac{8}{125}v^3$

   a. The graph can be obtained from $y = v^3$ if it is compressed vertically.

   b.

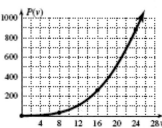

$$P(15) = \frac{8}{125}(15)^3 = 216 \text{ watts}$$

   c. About 15.6, 161.5, Power increases dramatically at higher windspeeds.

109. $d(t) = 2t^2$

   a. Vertical stretch by a factor of 2

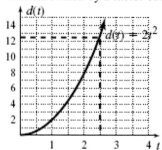

   b. $d(2.5) = 2(2.5)^2 = 2(6.25) = 12.5$ ft

   c. 5, 13, distance fallen per unit time increases very fast.

111. $f(x) = |x|$ and $g(x) = 2\sqrt{x}$

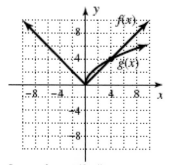

Interval: $x \in (0, 4)$

$x = 1$

$f(1) = |1| = 1$ and $g(1) = 2\sqrt{1} = 2$

$g(h) > f(h)$

Interval: $x \in (4, \infty)$

$x = 9$

$f(9) = |9| = 9$ and $g(1) = 2\sqrt{9} = 6$

$g(k) < f(k)$

113. $f(x) = x^2 - 4$

$F(x) = |x^2 - 4|$

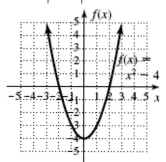

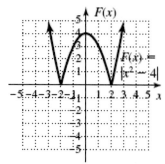

Any points in QIII and IV will reflected across the x–axis and thus move to QI and II.

115. $P = 32 + 32 + 38 + 24 + 6 + 8 = 140$ in.

$A = 32(32) + 24(6) = 1024 + 144 = 1168$ in^2

117. $f(x) = (x - 4)^2 + 3$

Quadratic, opens upward, Vertex (4,3)

$f(x) \downarrow : (-\infty, 4)$;

$f(x) \uparrow : (4, \infty)$

2.7 Technology Highlight

Exercise 1: They are approaching 4; not defined.

2.7 Exercises

1. Continuous

3. Smooth

5. Each piece must be continuous on the corresponding interval, and the function values at the endpoints of each interval must be equal. Answers will vary.

7. a. $f(x) = \begin{cases} x^2 - 6x + 10 & 0 \le x \le 5 \\ \dfrac{3}{2}x - \dfrac{5}{2} & 5 < x \le 9 \end{cases}$

    b. $y \in [1, 11]$

9. $h(x) = \begin{cases} -2 & x < -2 \\ |x| & -2 \le x < 3 \\ 5 & x \ge 3 \end{cases}$

    $h(-5) = -2$;

    $h(-2) = |-2| = 2$;

    $h\left(-\dfrac{1}{2}\right) = \left|-\dfrac{1}{2}\right| = \dfrac{1}{2}$;

    $h(0) = |0| = 0$;

    $h(2.999) = |2.999| = 2.999$;

    $h(3) = 5$

11. $p(x) = \begin{cases} 5 & x < -3 \\ x^2 - 4 & -3 \le x \le 3 \\ 2x + 1 & x > 3 \end{cases}$

    $p(-5) = 5$;

    $p(-3) = (-3)^2 - 4 = 9 - 4 = 5$;

    $p(-2) = (-2)^2 - 4 = 4 - 4 = 0$;

    $p(0) = (0)^2 - 4 = 0 - 4 = -4$;

    $p(3) = (3)^2 - 4 = 9 - 4 = 5$;

    $p(5) = 2(5) + 1 = 10 + 1 = 11$

13. $p(x) = \begin{cases} x + 2 & -6 \le x \le 2 \\ 2|x - 4| & x > 2 \end{cases}$

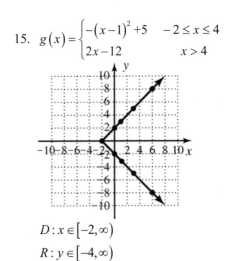

D: $x \in [-6, \infty)$

R: $y \in [-4, \infty)$

15. $g(x) = \begin{cases} -(x-1)^2 + 5 & -2 \le x \le 4 \\ 2x - 12 & x > 4 \end{cases}$

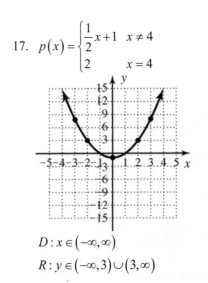

$D : x \in [-2, \infty)$

$R : y \in [-4, \infty)$

17. $p(x) = \begin{cases} \dfrac{1}{2}x + 1 & x \ne 4 \\ 2 & x = 4 \end{cases}$

$D : x \in (-\infty, \infty)$

$R : y \in (-\infty, 3) \cup (3, \infty)$

19. $H(x) = \begin{cases} -x+3 & x < 1 \\ -|x-5|+6 & 1 \le x < 9 \end{cases}$

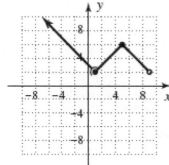

$D: x \in (-\infty, 9)$
$R: y \in [2, \infty)$

21. $f(x) = \begin{cases} -x-3 & x < -3 \\ 9-x^2 & -3 \le x < 2 \\ 4 & x \ge 2 \end{cases}$

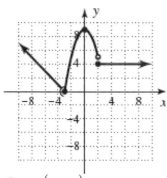

$D: x \in (-\infty, \infty)$
$R: y \in [0, \infty)$

23. $f(x) = \begin{cases} \dfrac{x^2-9}{x+3} & x \ne -3 \\ c & x = -3 \end{cases}$

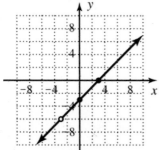

$D: x \in (-\infty, \infty)$
$R: y \in (-\infty, -6) \cup (-6, \infty)$
Discontinuity at $x = -3$
Redefine $f(x) = -6$ at $x = -3$; $c = -6$

25. $f(x) = \begin{cases} \dfrac{x^3-1}{x-1} & x \ne 1 \\ c & x = 1 \end{cases}$

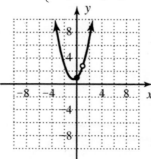

$D: x \in (-\infty, \infty)$
$R: y \in [0.75, \infty)$
Discontinuity at $x = 1$
Redefine $f(x) = 3$ at $x = 1$; $c = 3$

27. Left line contains the points $(-4, -3)$ and $(2, 0)$.
$$m = \frac{0-(-3)}{2-(-4)} = \frac{1}{2};$$
$$y - 0 = \frac{1}{2}(x-2)$$
$$y = \frac{1}{2}x - 1;$$
Right line contains the points $(2, 0)$ and $(3, 3)$.
$$m = \frac{3-0}{3-2} = 3;$$
$$y - 0 = 3(x-2)$$
$$y = 3x - 6;$$
$$f(x) = \begin{cases} \dfrac{1}{2}x-1 & -4 \le x < 2 \\ 3x-6 & x \ge 2 \end{cases}$$

29. The first equation is a quadratic with vertex $(-1, -4)$, opening up.
$$y = (x+1)^2 - 4$$
$$y = x^2 + 2x - 3;$$
The line is bounded by $(1, 2)$ and contains $(4, 5)$.
$$m = \frac{5-2}{4-1} = 1$$
$$y - 2 = 1(x-1)$$
$$y = x + 1;$$
$$p(x) = \begin{cases} x^2+2x-3 & x \le 1 \\ x+1 & x > 1 \end{cases}$$

2.7 Exercises

31. $|x| = \begin{cases} -x & x < 0 \\ x & x \geq 0 \end{cases}$

$f(x) = \dfrac{|x|}{x}$

Graph is discontinuous at $x = 0$.

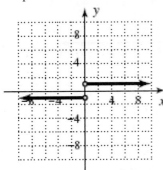

If $x < 0$, $f(x) = -1$.

If $x > 0$, $f(x) = 1$.

33. a. $S(t) = \begin{cases} -t^2 + 6t & 0 \leq t \leq 5 \\ 500 & t > 5 \end{cases}$

b. $S(t) \in [0, 9]$

35. $P(t) = \begin{cases} -0.03t^2 + 1.28t + 1.68 & 0 \leq t \leq 30 \\ 1.89t - 43.5 & t > 30 \end{cases}$

a. $P(5) = -0.03(5)^2 + 1.28(5) + 1.68 = 7.33$

$P(15) = -0.03(15)^2 + 1.28(15) + 1.68 = 14.13$

$P(25) = -0.03(25)^2 + 1.28(25) + 1.68 = 14.93$

$P(35) = 1.89(35) - 43.5 = 22.65$

$P(45) = 1.89(45) - 43.5 = 41.55$

$P(55) = 1.89(55) - 43.5 = 60.45$

b. Each piece gives a slightly different value due to rounding of coefficients in each model. At $t = 30$ we use the "first" piece: $P(30) = 13.08$.

37. $C(h) = \begin{cases} 0.09h & 0 \leq h \leq 1000 \\ 0.18h - 90 & h > 1000 \end{cases}$

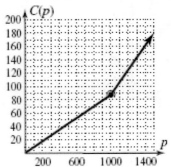

$C(1200) = 0.18(1200) - 90 = 216 - 90 = \$126$

39. $C(t) = \begin{cases} 0.75t & 0 \leq t \leq 25 \\ 1.5t - 18.75 & t > 25 \end{cases}$

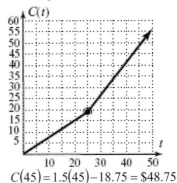

$C(45) = 1.5(45) - 18.75 = \$48.75$

41. $S(t) = \begin{cases} -1.35t^2 + 31.9t + 152 & 0 \le t \le 12 \\ 2.5t^2 - 80.6t + 950 & 12 < t \le 22 \end{cases}$

$S(25) = 2.5(25)^2 - 80.6(25) + 950$

$= 2.5(625) - 2015 + 950 = 497.5$

$\approx \$498$ billion;

$S(28) = 2.5(28)^2 - 80.6(28) + 950$

$= 2.5(784) - 2256.8 + 950 = 653.2$

$\approx \$653$ billion;

$S(30) = 2.5(30)^2 - 80.6(30) + 950$

$= 2.5(900) - 2418 + 950 = 782$

$\approx \$782$ billion

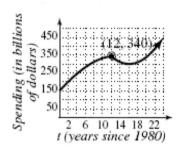

43. $C(m) = \begin{cases} 3.3m & 0 \le m \le 30 \\ 3.3(30) + 7(m - 30) & m > 30 \end{cases}$

$C(m) = \begin{cases} 3.3m & 0 \le m \le 30 \\ 7m - 111 & m > 30 \end{cases}$

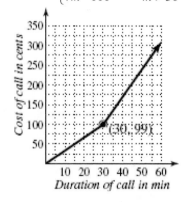

$C(46) = 7(46) - 111 = \$2.11$

45. $C(a) = \begin{cases} 0 & a < 2 \\ 2 & 2 \le a < 13 \\ 5 & 13 \le a < 20 \\ 7 & 20 \le a < 65 \\ 5 & a \ge 65 \end{cases}$

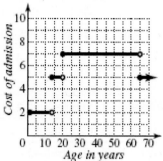

One grandparent:

$C(70) = 5$;

Two adults:

$C(44) = 7; C(45) = 7$;

Three teenagers:

$3 \cdot 5 = 15$;

Two children:

$2 \cdot 2 = 4$;

One infant: 0

Total Cost: $5 + 7 + 7 + 15 + 4 + 0 = \$38$

47. a. $C(w) = 17\lceil w - 1 \rceil + 80$

For an envelope weighing between 0 and 1 oz, the cost is $0.80. Each step interval increases by 0.17.

b. $0 < w \le 13$

c. 80 cents

d. 165 cents

e. 165 cents

f. 165 cents

g. 182 cents

49. $h(x) = |x-2| - |x+3|$

x	$h(x) = |x-2| - |x+3|$
-5	$h(-5) = |-5-2| - |-5+3| = 7 - 2 = 5$
-4	$h(-4) = |-4-2| - |-4+3| = 6 - 1 = 5$
-3	$h(-3) = |-3-2| - |-3+3| = 5 - 0 = 5$
-2	$h(-2) = |-2-2| - |-2+3| = 4 - 1 = 3$
-1	$h(-1) = |-1-2| - |-1+3| = 3 - 2 = 1$
0	$h(0) = |0-2| - |0+3| = 2 - 3 = -1$
1	$h(1) = |1-2| - |1+3| = 1 - 4 = -3$
2	$h(2) = |2-2| - |2+3| = 0 - 5 = -5$
3	$h(3) = |3-2| - |3+3| = 1 - 6 = -5$
4	$h(4) = |4-2| - |4+3| = 2 - 7 = -5$
5	$h(5) = |5-2| - |5+3| = 3 - 8 = -5$

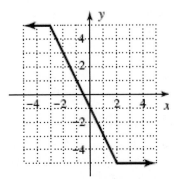

The function is continuous.

$$h(x) = \begin{cases} 5 & x \le -3 \\ -2x-1 & -3 < x < 2 \\ -5 & x \ge 2 \end{cases}$$

51. $Y_1 = \dfrac{x+2}{x+2}$, $Y_2 = \dfrac{|x+2|}{x+2}$

Y_1 has a removable discontinuity at $x = -2$.

Y_2 is discontinuous at $x = -2$.

53. $\dfrac{3}{x-2} + 1 = \dfrac{30}{x^2-4}$

$$\left(\dfrac{3}{x-2} + 1 = \dfrac{30}{(x-2)(x+2)} \right)(x-2)(x+2)$$

$$3(x+2) + 1(x-2)(x+2) = 30$$

$$3x + 6 + x^2 - 4 = 30$$

$$x^2 + 3x - 28 = 0$$

$$(x+7)(x-4) = 0$$

$$x = -7; \quad x = 4$$

55. a. $a^2 + b^2 = c^2$

$$8^2 + b^2 = 12^2$$

$$64 + b^2 = 144$$

$$b^2 = 80$$

$$b = 4\sqrt{5} \text{ cm}$$

b. $A = \dfrac{1}{2}bh$

$$A = \dfrac{1}{2}\left(4\sqrt{5}\right)(8)$$

$$A = 16\sqrt{5} \text{ cm}^2$$

c. $V = \left(\dfrac{bh}{2}\right)h$

$$V = \left(16\sqrt{5}\right)(20) = 320\sqrt{5} \text{ cm}^3$$

2.8 Technology Highlight

Exercise 1: $Y_1 = \sqrt{x}$ and $Y_2 = x + 7$
Yes, graph shifts 7 units to the left.

2.8 Exercises

1. $(f + g)(x);\ A \cap B$

3. Intersection; $g(x)$

5. Answers will vary.

7. a. Domain:
$f(x) = 2x^2 - x - 3; x \in \mathbb{R}$;
$g(x) = x^2 + 5x; x \in \mathbb{R}$;
$h(x) = f(x) - g(x); x \in \mathbb{R}$
$h(-2) = f(-2) - g(-2)$

   b. $= 2(-2)^2 - (-2) - 3 - \left((-2)^2 + 5(-2)\right)$
$= 7 - (-6) = 13$

9. $h(x) = f(x) - g(x)$
   a. $h(x) = 2x^2 - x - 3 - \left(x^2 + 5x\right)$
$= 2x^2 - x - 3 - x^2 - 5x$
$= x^2 - 6x - 3$
   b. $h(-2) = (-2)^2 - 6(-2) - 3 = 13$
   c. Same result

11. a. Domain of $f(x) = \sqrt{x-3}$
$x - 3 \geq 0$
$x \geq 3;\ [3, \infty)$
Domain of $g(x): x \in \mathbb{R}$;
Domain of $h(x): x \in [3, \infty)$
   b. $h(x) = (f + g)(x)$
$= f(x) + g(x)$
$= \sqrt{x-3} + 2x^3 - 54$
   c. $h(4) = \sqrt{4-3} + 2(4)^3 - 54 = 75$;
$h(2) = \sqrt{2-3} + 2(2)^3 - 54$
$= \sqrt{-1} + 16 - 54$
$\sqrt{-1}$ is not a real number;
2 is not in the domain of $h(x)$.

13. a. Domain of $p(x) = \sqrt{x+5}$
$x + 5 \geq 0$
$x \geq -5;\ x \in [-5, \infty)$
Domain of $q(x) = \sqrt{3-x}$
$3 - x \geq 0$
$-x \geq -3$
$x \leq 3;\ x \in (-\infty, 3]$
Domain of $r(x): x \in [-5, 3]$
   b. $r(x) = (p + q)(x)$
$= p(x) + q(x)$
$= \sqrt{x+5} + \sqrt{3-x}$
   c. $r(2) = \sqrt{2+5} + \sqrt{3-2} = \sqrt{7} + 1$
$r(4) = \sqrt{4+5} + \sqrt{3-4} = \sqrt{9} + \sqrt{-1}$
$\sqrt{-1}$ is not a real number;
4 is not in the domain of $r(x)$.

15. a. Domain of $f(x) = \sqrt{x+4}$
$x + 4 \geq 0$
$x \geq -4;\ x \in [-4, \infty)$
Domain of $g(x) = 2x + 3 : x \in \mathbb{R}$
Domain of $h(x): x \in [-4, \infty)$
   b. $h(x) = (f \cdot g)(x)$
$= f(x) \cdot g(x)$
$= \sqrt{x+4}\,(2x+3)$
   c. $h(-4) = \sqrt{-4+4}\,(2(-4)+3) = 0$;
$h(21) = \sqrt{21+4}\,(2(21)+3) = 225$

17. a. Domain of $p(x) = \sqrt{x+1}$
$x + 1 \geq 0$
$x \geq -1;\ x \in [-1, \infty)$
Domain of $q(x) = \sqrt{7-x}$
$7 - x \geq 0$
$-x \geq -7$
$x \leq 7;\ x \in (-\infty, 7]$
Domain of $r(x): x \in [-1, 7]$
   b. $r(x) = (p \cdot q)(x)$
$= p(x) \cdot q(x)$
$= \sqrt{x+1} \cdot \sqrt{7-x}$
$= \sqrt{-x^2 + 6x + 7}$

c. $r(15) = \sqrt{-(15)^2 + 6(15) + 7} = \sqrt{-128}$

$\sqrt{-128}$ is not a real number;

15 is not in the domain of $r(x)$.

$r(3) = \sqrt{-(3)^2 + 6(3) + 7} = \sqrt{16} = 4$

19. a. Domain of $f(x) = x^2 - 16 : x \in \square$

Domain of $g(x) = x + 4 : x \in \square$

Domain of $h(x) = \dfrac{x^2 - 16}{x + 4}, x \neq -4$

$x \in (-\infty, -4) \cup (-4, \infty)$

b. $h(x) = \dfrac{f}{g}(x) = \dfrac{x^2 - 16}{x + 4}$

$h(x) = \dfrac{(x+4)(x-4)}{x+4} = x - 4; \; x \neq -4$

21. a. Domain of

$f(x) = x^3 + 4x^2 - 2x - 8 : x \in \square$

Domain of $g(x) = x + 4, x \in \square$

Domain of

$h(x) = \dfrac{x^3 + 4x^2 - 2x - 8}{x + 4}, x \neq -4$

$x \in (-\infty, -4) \cup (-4, \infty)$

b. $h(x) = \dfrac{f}{g}(x) = \dfrac{x^3 + 4x^2 - 2x - 8}{x + 4}$

$h(x) = \dfrac{x^2(x+4) - 2(x+4)}{x+4}$

$= \dfrac{(x+4)(x^2 - 2)}{x+4} = x^2 - 2; \; x \neq -4$

23. a. Domain of $f(x) = x^3 - 7x^2 + 6x : x \in \square$

Domain of $g(x) = x - 1 : x \in \square$

Domain of

$h(x) = \dfrac{x^3 - 7x^2 + 6x}{x - 1}, x \neq 1$

$x \in (-\infty, 1) \cup (1, \infty)$

b. $h(x) = \dfrac{f}{g}(x) = \dfrac{x^3 - 7x^2 + 6x}{x - 1}$

$h(x) = \dfrac{x(x^2 - 7x + 6)}{x - 1}$

$= \dfrac{x(x-6)(x-1)}{x-1} = x(x-6)$

$= x^2 - 6x; \; x \neq 1$

25. a. Domain of $f(x) = x + 1 : x \in \square$

Domain of $g(x) = x - 5 : x \in \square$

Domain of

$h(x) = \dfrac{x+1}{x-5}, x \neq 5$

$x \in (-\infty, 5) \cup (5, \infty)$

b. $h(x) = \dfrac{f}{g}(x) = \dfrac{x+1}{x-5}; \; x \neq 1$

27. a. Domain of $p(x) = 2x - 3 : x \in \square$

Domain of $q(x) = \sqrt{-2 - x}$,

$-2 - x \geq 0$

$-x \geq 2$

$x \leq -2; \; x \in (-\infty, -2]$

Domain of $r(x) = \dfrac{2x - 3}{\sqrt{-2 - x}}$,

$-2 - x > 0$

$-x > 2$

$x < -2; \; x \in (-\infty, -2)$

b. $r(x) = \dfrac{p}{q}(x) = \dfrac{2x - 3}{\sqrt{-2 - x}}$

c. $r(6) = \dfrac{2(6) - 3}{\sqrt{-2 - 6}} = \dfrac{9}{\sqrt{-8}}$

$\sqrt{-8}$ is not a real number;

6 is not in the domain of $r(x)$.

$r(-6) = \dfrac{2(-6) - 3}{\sqrt{-2 + 6}} = \dfrac{-15}{\sqrt{4}} = -\dfrac{15}{2}$

29. a. Domain of $p(x) = x - 5 : x \in \mathbb{R}$

Domain of $q(x) = \sqrt{x-5}$,

$x - 5 \geq 0$

$x \geq 5; \; x \in [5, \infty)$

Domain of $r(x) = \dfrac{x-5}{\sqrt{x-5}}$,

$x - 5 > 0$

$x > 5; \; x \in (5, \infty)$

b. $r(x) = \dfrac{p}{q}(x) = \dfrac{x-5}{\sqrt{x-5}}$

c. $r(6) = \dfrac{6-5}{\sqrt{6-5}} = \dfrac{1}{\sqrt{1}} = 1$

$r(-6) = \dfrac{-6-5}{\sqrt{-6-5}} = \dfrac{-11}{\sqrt{-11}}$

$\sqrt{-11}$ is not a real number;

-6 is not in the domain of $r(x)$.

31. a. Domain of $p(x) = x^2 - 36 : x \in \mathbb{R}$

Domain of $q(x) = \sqrt{2x+13}$,

$2x + 13 \geq 0$

$2x \geq -13$

$x \geq -\dfrac{13}{2}; \; x \in \left[-\dfrac{13}{2}, \infty\right)$

Domain of $r(x) = \dfrac{x^2 - 36}{\sqrt{2x+13}}$,

$2x + 13 > 0$

$2x > -13$

$x > -\dfrac{13}{2}; \; x \in \left(-\dfrac{13}{2}, \infty\right)$

b. $r(x) = \dfrac{p}{q}(x) = \dfrac{x^2 - 36}{\sqrt{2x+13}}$

c. $r(6) = \dfrac{6^2 - 36}{\sqrt{2(6)+13}} = \dfrac{0}{\sqrt{25}} = 0$

$r(-6) = \dfrac{(-6)^2 - 36}{\sqrt{2(-6)+13}} = \dfrac{0}{\sqrt{1}} = 0$

33. a. $f(x) = \dfrac{6x}{x-3}, \; g(x) = \dfrac{3x}{x+2}$

$h(x) = \dfrac{f(x)}{g(x)} = \dfrac{\dfrac{6x}{x-3}}{\dfrac{3x}{x+2}}$

$= \dfrac{6x}{x-3} \div \dfrac{3x}{x+2} = \dfrac{6x}{x-3} \cdot \dfrac{x+2}{3x}$

$= \dfrac{2(x+2)}{x-3} = \dfrac{2x+4}{x-3}$

b. Domain of $h(x) = \dfrac{2x+4}{x-3}, x \neq 3$

$x \in (-\infty, 3) \cup (3, \infty)$

c. $x + 2 \neq 0$

$x \neq -2;$

$\dfrac{3x}{x+2} \neq 0$

$x \neq 0$

35. $f(x) = 2x + 3$ and $g(x) = x - 2$

Sum:

$f(x) + g(x) = 2x + 3 + x - 2 = 3x + 1$

Domain contains all values of x.

$D : x \in (-\infty, \infty)$

Difference:

$f(x) - g(x) = 2x + 3 - (x - 2)$

$= 2x + 3 - x + 2 = x + 5$

Domain contains all values of x.

$D : x \in (-\infty, \infty)$

Product:

$f(x) \cdot g(x) = (2x + 3)(x - 2)$

$= 2x^2 - 4x + 3x - 6$

$= 2x^2 - x - 6$

Domain contains all values of x.

$D : x \in (-\infty, \infty)$

Quotient:

$\dfrac{f(x)}{g(x)} = \dfrac{2x+3}{x-2}$

$x - 2 \neq 0$

$x \neq 2$

$D : x \in (-\infty, 2) \cup (2, \infty)$

2.8 Exercises

37. $f(x) = x^2 + 7$ and $g(x) = 3x - 2$

Sum:

$$f(x) + g(x) = x^2 + 7 + 3x - 2 = x^2 + 3x + 5$$

Domain contains all values of x.

$D : x \in (-\infty, \infty)$

Difference:

$$f(x) - g(x) = x^2 + 7 - (3x - 2)$$
$$= x^2 + 7 - 3x + 2$$
$$= x^2 - 3x + 9$$

Domain contains all values of x.

$D : x \in (-\infty, \infty)$

Product:

$$f(x) \cdot g(x) = (x^2 + 7)(3x - 2)$$
$$= 3x^3 - 2x^2 + 21x - 14$$

Domain contains all values of x.

$D : x \in (-\infty, \infty)$

Quotient:

$$\frac{f(x)}{g(x)} = \frac{x^2 + 7}{3x - 2}$$
$$3x - 2 \neq 0$$
$$3x \neq 2$$
$$x \neq \frac{2}{3}$$

$D : x \in \left(-\infty, \frac{2}{3}\right) \cup \left(\frac{2}{3}, \infty\right)$

39. $f(x) = x^2 + 2x - 3$ and $g(x) = x - 1$

Sum:

$$f(x) + g(x) = x^2 + 2x - 3 + x - 1$$
$$= x^2 + 3x - 4$$

Domain contains all values of x.

$D : x \in (-\infty, \infty)$

Difference:

$$f(x) - g(x) = x^2 + 2x - 3 - (x - 1)$$
$$= x^2 + 2x - 3 - x + 1$$
$$= x^2 + x - 2$$

Domain contains all values of x.

$D : x \in (-\infty, \infty)$

Product:

$$f(x) \cdot g(x) = (x^2 + 2x - 3)(x - 1)$$
$$= x^3 - x^2 + 2x^2 - 2x - 3x + 3$$
$$= x^3 + x^2 - 5x + 3$$

Domain contains all values of x.

$D : x \in (-\infty, \infty)$

Quotient:

$$\frac{f(x)}{g(x)} = \frac{x^2 + 2x - 3}{x - 1}$$
$$= \frac{(x + 3)(x - 1)}{x - 1} = x + 3$$
$$x - 1 \neq 0$$
$$x \neq 1$$

$D : x \in (-\infty, 1) \cup (1, \infty)$

41. $f(x) = 3x + 1$ and $g(x) = \sqrt{x - 3}$

Sum:

$$f(x) + g(x) = 3x + 1 + \sqrt{x - 3}$$
$$x - 3 \geq 0$$
$$x \geq 3$$

$D : x \in [3, \infty)$

Difference:

$$f(x) - g(x) = 3x + 1 - \sqrt{x - 3}$$
$$x - 3 \geq 0$$
$$x \geq 3$$

$D : x \in [3, \infty)$

Product:

$$f(x) \cdot g(x) = (3x + 1)\sqrt{x - 3}$$
$$x - 3 \geq 0$$
$$x \geq 3$$

$D : x \in [3, \infty)$

Quotient:

$$\frac{f(x)}{g(x)} = \frac{3x + 1}{\sqrt{x - 3}}$$
$$x - 3 > 0$$
$$x > 3$$

$D : x \in (3, \infty)$

43. $f(x) = 2x^2$ and $g(x) = \sqrt{x+1}$

Sum:

$f(x) + g(x) = 2x^2 + \sqrt{x+1}$

$x + 1 \geq 0$

$x \geq -1$

$D: x \in [-1, \infty)$

Difference:

$f(x) - g(x) = 2x^2 - \sqrt{x+1}$

$x + 1 \geq 0$

$x \geq -1$

$D: x \in [-1, \infty)$

Product:

$f(x) \cdot g(x) = 2x^2\sqrt{x+1}$

$x + 1 \geq 0$

$x \geq -1$

$D: x \in [-1, \infty)$

Quotient:

$\dfrac{f(x)}{g(x)} = \dfrac{2x^2}{\sqrt{x+1}}$

$x + 1 > 0$

$x > -1$

$D: x \in (-1, \infty)$

45. $f(x) = \dfrac{2}{x-3}$ and $g(x) = \dfrac{5}{x+2}$

Sum:

$f(x) + g(x) = \dfrac{2}{x-3} + \dfrac{5}{x+2}$

$= \dfrac{2(x+2) + 5(x-3)}{(x-3)(x+2)}$

$= \dfrac{2x+4+5x-15}{(x-3)(x+2)}$

$= \dfrac{7x-11}{(x-3)(x+2)}$

$x - 3 \neq 0 \quad x + 2 \neq 0$

$x \neq 3 \qquad x \neq -2$

$D: x \in (-\infty, -2) \cup (-2, 3) \cup (3, \infty)$

Difference:

$f(x) - g(x) = \dfrac{2}{x-3} - \dfrac{5}{x+2}$

$= \dfrac{2(x+2) - 5(x-3)}{(x-3)(x+2)}$

$= \dfrac{2x+4-5x+15}{(x-3)(x+2)}$

$= \dfrac{-3x+19}{(x-3)(x+2)}$

$x - 3 \neq 0 \quad x + 2 \neq 0$

$x \neq 3 \qquad x \neq -2$

$D: x \in (-\infty, -2) \cup (-2, 3) \cup (3, \infty)$

Product:

$f(x) \cdot g(x) = \left(\dfrac{2}{x-3}\right)\left(\dfrac{5}{x+2}\right)$

$= \dfrac{10}{(x-3)(x+2)}$

$= \dfrac{10}{x^2 - x - 6}$

$x - 3 \neq 0 \quad x + 2 \neq 0$

$x \neq 3 \qquad x \neq -2$

$D: x \in (-\infty, -2) \cup (-2, 3) \cup (3, \infty)$

Quotient:

$\dfrac{f(x)}{g(x)} = \dfrac{\dfrac{2}{x-3}}{\dfrac{5}{x+2}} = \left(\dfrac{2}{x-3}\right)\left(\dfrac{x+2}{5}\right)$

$= \dfrac{2(x+2)}{5(x-3)} = \dfrac{2x+4}{5x-15}$

$x - 3 \neq 0 \quad x + 2 \neq 0$

$x \neq 3 \qquad x \neq -2$

$D: x \in (-\infty, -2) \cup (-2, 3) \cup (3, \infty)$

47. $f(x) = x^2 - 5x - 14$

$f(-2) = (-2)^2 - 5(-2) - 14 = 4 + 10 - 14 = 0;$

$f(7) = (7)^2 - 5(7) - 14 = 49 - 35 - 14 = 0;$

$f(2) = (2)^2 - 5(2) - 14 = 4 - 10 - 14 = -20;$

$f(a-2) = (a-2)^2 - 5(a-2) - 14$

$= a^2 - 4a + 4 - 5a + 10 - 14$

$= a^2 - 9a$

49. $f(x) = \sqrt{x+3}$ and $g(x) = 2x - 5$

  (a) $h(x) = (f \circ g)(x) = f[g(x)]$

$$= \sqrt{g(x) + 3}$$
$$= \sqrt{(2x-5)+3}$$
$$= \sqrt{2x-2}$$

  (b) $H(x) = (g \circ f)(x) = g[f(x)]$

$$= 2(f(x)) - 5$$
$$= 2\sqrt{x+3} - 5$$

  (c) $2x - 2 \geq 0$

$$2x \geq 2$$
$$x \geq 1$$

Domain of $h : x \in [1, \infty)$

$$x + 3 \geq 0$$
$$x \geq -3$$

Domain of H: $x \in [-3, \infty)$

51. $f(x) = \sqrt{x-3}$ and $g(x) = 3x + 4$

  (a) $h(x) = (f \circ g)(x)$

$$h(x) = f[g(x)]$$
$$h(x) = \sqrt{g(x) - 3}$$
$$= \sqrt{3x + 4 - 3}$$
$$= \sqrt{3x + 1}$$

  (b) $H(x) = (g \circ f)(x)$

$$H(x) = g[f(x)]$$
$$H(x) = 3(f(x)) + 4$$
$$= 3\sqrt{x-3} + 4$$

  (c) $3x + 1 \geq 0$

$$3x \geq -1$$
$$x \geq -\frac{1}{3}$$

Domain of h: $\left\{ x \mid x \geq -\frac{1}{3} \right\}$

or $\left[-\frac{1}{3}, \infty \right)$;

$$x - 3 \geq 0$$
$$x \geq 3$$

Domain of H: $\{x \mid x \geq 3\}$

or $[3, \infty)$

53. $f(x) = x^2 - 3x$ and $g(x) = x + 2$

  (a) $h(x) = (f \circ g)(x)$

$$h(x) = f[g(x)]$$
$$h(x) = (g(x))^2 - 3(g(x))$$
$$= (x+2)^2 - 3(x+2)$$
$$= x^2 + 4x + 4 - 3x - 6$$
$$= x^2 + x - 2$$

  (b) $H(x) = (g \circ f)(x)$

$$H(x) = g[f(x)]$$
$$H(x) = (f(x)) + 2$$
$$= x^2 - 3x + 2$$

  (c) Domain of h: $(-\infty, \infty)$

Domain of H: $(-\infty, \infty)$

55. $f(x) = x^2 + x - 4$ and $g(x) = x + 3$

  (a) $h(x) = (f \circ g)(x)$

$$h(x) = f[g(x)]$$
$$h(x) = (g(x))^2 + g(x) - 4$$
$$= (x+3)^2 + x + 3 - 4$$
$$= x^2 + 6x + 9 + x - 1$$
$$= x^2 + 7x + 8$$

  (b) $H(x) = (g \circ f)(x)$

$$H(x) = g[f(x)]$$
$$H(x) = f(x) + 3$$
$$= x^2 + x - 4 + 3$$
$$= x^2 + x - 1$$

  (c) Domain of h: $(-\infty, \infty)$

Domain of H: $(-\infty, \infty)$

57. $f(x) = |x| - 5$ and $g(x) = -3x + 1$

  (a) $h(x) = (f \circ g)(x)$

    $h(x) = f[g(x)]$

    $h(x) = |g(x)| - 5$

    $= |-3x + 1| - 5$

  (b) $H(x) = (g \circ f)(x)$

    $H(x) = g[f(x)]$

    $H(x) = -3(f(x)) + 1$

    $= -3(|x| - 5) + 1$

    $= -3|x| + 15 + 1$

    $= -3|x| + 16$

  (c) Domain of h: $(-\infty, \infty)$

    Domain of H: $(-\infty, \infty)$

59. $f(x) = \dfrac{2x}{x+3}$ and $g(x) = \dfrac{5}{x}$

  (a)

    $(f \circ g)(x)$: For $g(x)$ to be defined, $x \neq 0$.

    For $f[g(x)] = \dfrac{2g(x)}{g(x)+3}$,

    $g(x) \neq -3$ so $x \neq -\dfrac{5}{3}$.

    Domain: $\left\{ x \mid x \neq 0, x \neq -\dfrac{5}{3} \right\}$

  (b)

    $(g \circ f)(x)$: For $f(x)$ to be defined, $x \neq -3$.

    For $g[f(x)] = \dfrac{5}{f(x)}$,

    $f(x) \neq 0$ so $x \neq 0$.

    Domain: $\{ x \mid x \neq 0, x \neq -3 \}$

  (c) $(f \circ g)(x) = f[g(x)]$

    $= \dfrac{2(g(x))}{g(x)+3} = \dfrac{2\left(\dfrac{5}{x}\right)}{\dfrac{5}{x}+3} = \dfrac{\dfrac{10}{x}}{\dfrac{5+3x}{x}}$

    $= \dfrac{10x}{x(5+3x)} = \dfrac{10}{5+3x}$

    $(g \circ f)(x) = g[f(x)]$

    $= \dfrac{5}{f(x)} = \dfrac{5}{\dfrac{2x}{x+3}} = \dfrac{5(x+3)}{2x} = \dfrac{5x+15}{2x}$

61. $f(x) = \dfrac{4}{x}$ and $g(x) = \dfrac{1}{x-5}$

  (a)

    $(f \circ g)(x)$: For $g(x)$ to be defined, $x \neq 5$.

    For $f[g(x)] = \dfrac{4}{g(x)}$,

    $g(x) \neq 0$ and $g(x)$ is never zero.

    Domain: $\{ x \mid x \neq 5 \}$

  (b)

    $(g \circ f)(x)$: For $f(x)$ to be defined, $x \neq 0$.

    For $g[f(x)] = \dfrac{1}{f(x)-5}$,

    $f(x) \neq 5$ so $x \neq \dfrac{4}{5}$.

    Domain: $\left\{ x \mid x \neq 0, x \neq \dfrac{4}{5} \right\}$

  (c) $h(x) = (f \circ g)(x)$

    $h(x) = f[g(x)]$

    $h(x) = \dfrac{4}{g(x)}$

    $= \dfrac{4}{\dfrac{1}{x-5}}$

    $= 4(x-5)$

    $= 4x - 20$

    $H(x) = (g \circ f)(x)$

    $H(x) = g[f(x)]$

    $H(x) = \dfrac{1}{f(x)-5}$

    $= \dfrac{1}{\dfrac{4}{x}-5}$

    $= \dfrac{1}{\dfrac{4-5x}{x}}$

    $= \dfrac{x}{4-5x}$

2.8 Exercises

63. $f(x) = x^2 - 8$ and $g(x) = x + 2$
    $h(x) = (f \circ g)(x)$

    a. $(f \circ g)(x) = f[g(x)]$
    $= (g(x)^2) - 8$
    $= (x+2)^2 - 8$
    $= x^2 + 4x + 4 - 8$
    $= x^2 + 4x - 4;$
    $h(x) = x^2 + 4x - 4$
    $h(5) = (5)^2 + 4(5) - 4 = 25 + 20 - 4 = 41$

    b. $g(5) = 5 + 2 = 7$
    $f[g(5)] = f(7)$
    $= (7)^2 - 8 = 49 - 8 = 41$

65. $h(x) = \left(\sqrt{x-2}+1\right)^3 - 5$

    Answers may vary.
    $g(x) = \sqrt{x-2}+1, \; f(x) = x^3 - 5$

67. $f(x) = 2x - 1, \; g(x) = x^2 - 1,$
    $h(x) = x + 4$

    a. $p(x) = f\left[g\left([h(x)]\right)\right]$
    $p(x) = f\left[(x+4)^2 - 1\right]$
    $= 2\left[(x+4)^2 - 1\right] - 1$
    $= 2(x+4)^2 - 2 - 1$
    $= 2(x+4)^2 - 3$

    b. $q(x) = g\left[f\left([h(x)]\right)\right]$
    $q(x) = g\left[2(x+4) - 1\right]$
    $= g[2x + 8 - 1] = g[2x + 7]$
    $= (2x+7)^2 - 1$

69. a. $C(5) = 6000$
    b. $T(8) = 3000$
    c. $C(9) + T(9) = 6000 + 2000 = 8000$
    d. $C(9) - T(9) = 6000 - 2000 = 4000$

71. a. $R(2) = \$1$ billion
    b. $C(8) = \$5$ billion
    c. $R(t) = C(t)$
    Broke even 2003, 2007, 2010
    d. $C(t) > R(t):$
    $t \in (2000, 2003) \cup (2007, 2010)$
    e. $R(t) > C(t), t \in (2003, 2007)$
    f. $R(5) - C(5) = 5 - 1 = \$4$ billion

73. a. $(f + g)(-4) = f(-4) + g(-4)$
    $= 5 + (-1) = 4$
    b. $(f \cdot g)(1) = f(1) \cdot g(1)$
    $= 0(3) = 0$
    c. $(f - g)(4) = f(4) - g(4)$
    $= 5 - 3 = 2$
    d. $(f + g)(0) = f(0) + g(0)$
    $= 1 + 2 = 3$
    e. $\left(\dfrac{f}{g}\right)(2) = \dfrac{f(2)}{g(2)} = \dfrac{-1}{3}$
    f. $(f \cdot g)(-2) = f(-2) \cdot g(-2)$
    $= 3(2) = 6$
    g. $(g \cdot f)(2) = g(2) \cdot f(2)$
    $= 3(-1) = -3$
    h. $(f - g)(-1) = f(-1) - g(-1)$
    $= 2 - 1 = 1$
    i. $(f + g)(8) = f(8) + g(8)$
    $= -1 + 2 = 1$
    j. $\left(\dfrac{f}{g}\right)(7) = \dfrac{f(7)}{g(7)} = \dfrac{1}{0}$ undefined
    k. $(g \circ f)(4) = g(f(4))$
    $= g(5) = \dfrac{1}{2} = 0.5$
    l. $(f \circ g)(4) = f(g(4))$
    $= f(3) = 2$

75. $h(x) = f(x) - g(x)$
    $= 5 - \left(\dfrac{2}{3}x + 1\right) = 5 - \dfrac{2}{3}x - 1$
    $= -\dfrac{2}{3}x + 4$

77. $h(x) = f(x) - g(x)$
    $= \left(5x - x^2\right) - x = 4x - x^2$

79. $A = 40\pi r + 2\pi r^2$

$A = 2\pi r(20 + r)$

$A(r) = (f \cdot g)(r)$

$f(r) = 2\pi r, g(r) = 20 + r$

$A(5) = 2\pi(5)(20 + 5) = 10\pi(25) = 250\pi$ units^2

81. Revenue: $R(x) = 40,000x$

Cost: $C(x) = 108,000 + 28,000x$

a. $P(x) = R(x) - C(x)$

$= 40,000x - 108,000 - 28,000x$

$= 12,000x - 108,000$

b. Break even when $P(x) = 0$

$12,000x - 108,000 = 0$

$12,000x = 108,000$

$x = 9$

9 boats must be sold to break even.

83. a. $P(n) = R(n) - C(n)$

$P(n) = 11.45n - 0.1n^2$

b. $P(12) = 11.45(12) - 0.1(12)^2$

$= 137.4 - 14.4 = \$123$

c. $P(60) = 11.45(60) - 0.1(60)^2$

$= 687 - 360 = \$327$

d. At $n = 115$, costs exceed revenue, $C(115) > R(115)$.

85. $f(x) = 0.5x - 14$; $g(x) = 2x + 23$

$h(x) = (f \circ g)(x) = f[g(x)]$

$h(x) = 0.5(g(x)) - 14$

$= 0.5(2x + 23) - 14$

$= x + 11.5 - 14$

$= x - 2.5$;

$h(13) = 13 - 2.5 = 10.5$

87. $T(x) = 41.6x$; $R(x) = 10.9x$

(a) $T(100) = 41.6(100) = 4160$ baht

(b) $R(4160) = 10.9(4160) = 45,344$

rRinggit

(c) $M(x) = (R \circ T)(x) = R[T(x)]$

$M(x) = 10.9(T(x))$

$= 10.9(41.6x)$

$= 453.44x$;

$M(100) = 453.44(100) = 45344$ ringgit

Parts B and C agree.

89. $r(t) = 3t$; $A = \pi r^2$

(a) $r(2) = 3(2) = 6$ ft

(b) $A(6) = \pi(6)^2 = 36\pi$ ft^2

(c) $A(t) = (A \circ r)(t) = A[r(t)]$

$A(t) = \pi(r(t))^2$

$= \pi(3t)^2$

$= 9\pi t^2$;

$A(2) = 9\pi(2)^2 = 36\pi$ ft^2

The answers do agree.

91. $C(x) = 0.0345x^4 - 0.8996x^3 + 7.5383x^2 - 21.7215x + 40$

$L(x) = -0.0345x^4 + 0.8996x^3 - 7.5383x^2 + 21.7215x + 10$

(a) Using the grapher, 1995 to 1996; 1999 to 2004

(b) Using the grapher, 30 seats; 1995

(c) Using the grapher, 20 seats; 1997

(d) Using the grapher, the total number in the senate (50); the number of additional seats held by the majority.

93. $f(x) = \sqrt{1-x}$ and $g(x) = \sqrt{x-2}$

Using the grapher,
$(f+g)(x)$ cannot be found because their
domains do not overlap.

95. $f(x) = (x-3)^2 + 2, g(x) = 4|x-3| - 5$

x	$f(x)$	$g(x)$	$(f\text{–}g)(x)$
−2	27	15	12
−1	18	11	7
0	11	7	4
1	6	3	3
2	3	−1	4
3	2	−5	7
4	3	−1	4
5	6	3	3
6	11	7	4
7	18	11	7
8	27	15	12

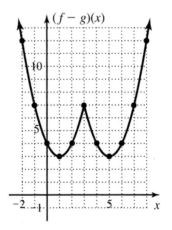

97. $f(x) = \sqrt{x}$; $g(x) = \sqrt[3]{x}$; $h(x) = |x|$

(a)

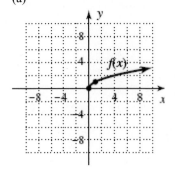

(b)

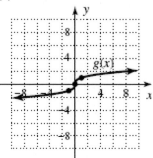

(c)

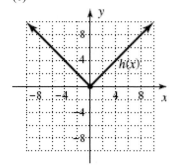

99. $-2x + 3y = 9$

$3y = 2x + 9$

$y = \dfrac{2}{3}x + 3$

$m = \dfrac{2}{3}$;

Slope of a line perpendicular is $-\dfrac{3}{2}$.

y– intercept (0,0);

Equation: $y = -\dfrac{3}{2}x$

Chapter 2 Summary and Review

1.

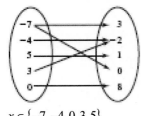

$$x \in \{-7, -4, 0, 3, 5\}$$
$$y \in \{-2, 0, 1, 3, 8\}$$

3. $(19, 25), (-14, -31)$
$$d = \sqrt{(-14-19)^2 + (-31-25)^2}$$
$$= \sqrt{1089 + 3136} = \sqrt{4225} = 65$$
65 miles

5. $x^2 + y^2 = 16$
Center (0,0), Radius 4

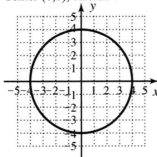

7. $(-3, 0)$ and $(0, 4)$
To find the center, find the midpoint.
$$\left(\frac{-3+0}{2}, \frac{0+4}{2} \right) = \left(-\frac{3}{2}, 2 \right)$$
To find the radius, find the distance between
$$\left(-\frac{3}{2}, 2 \right) \text{ and } (0, 4)$$
$$d = \sqrt{\left(0 - \left(-\frac{3}{2} \right) \right)^2 + (4-2)^2}$$
$$= \sqrt{\frac{9}{4} + 4} = \sqrt{\frac{25}{4}} = \frac{5}{2} = 2.5$$
Radius: 2.5
Equation: $\left(x + \frac{3}{2} \right)^2 + (y-2)^2 = 6.25$

9. a. L_1: $(-2, 0)$ and $(0, 6)$
$$m = \frac{6-0}{0-(-2)} = \frac{6}{2} = 3$$
L_2: $(1, 8)$ and $(0, 5)$
$$m = \frac{5-8}{0-1} = \frac{-3}{-1} = 3$$
Parallel

b. L_1: $(1, 10)$ and $(-1, 7)$
$$m = \frac{7-10}{-1-1} = \frac{-3}{-2} = \frac{3}{2}$$
L_2: $(-2, -1)$ and $(1, -3)$
$$m = \frac{-3-(-1)}{1-(-2)} = \frac{-2}{3}$$
Perpendicular

11. a. $2x + 3y = 6$
x–intercept: $(3, 0)$ y–intercept: $(0, 2)$
$$2x + 3(0) = 6 \qquad 2(0) + 3y = 6$$
$$2x = 6 \qquad\qquad 3y = 6$$
$$x = 3 \qquad\qquad\quad y = 2$$

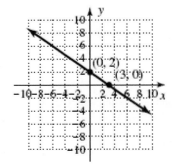

Summary and Review

b. $y = \dfrac{4}{3}x - 2$

x–intercept: $\left(\dfrac{3}{2}, 0\right)$ y–intercept: $(0, -2)$

$0 = \dfrac{4}{3}x - 2$ $y = \dfrac{4}{3}(0) - 2$

$2 = \dfrac{4}{3}x$ $y = -2$

$\dfrac{6}{4} = x$

$\dfrac{3}{2} = x$

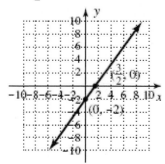

13. $(-5, -4)$ $(7, 2)$ $(0, 16)$

$m = \dfrac{16 - 2}{0 - 7} = \dfrac{14}{-7} = -2$;

$m = \dfrac{2 - (-4)}{7 - (-5)} = \dfrac{6}{12} = \dfrac{1}{2}$

Yes

15. a. $4x + 3y - 12 = 0$

$3y = -4x + 12$

$y = -\dfrac{4}{3}x + 4$

$m = -\dfrac{4}{3}$; y–intercept $(0, 4)$

b. $5x - 3y = 15$

$-3y = -5x + 15$

$y = \dfrac{5}{3}x - 5$

$m = \dfrac{5}{3}$; y–intercept $(0, -5)$

17. a. $m = \dfrac{2}{3}$; $(1, 4)$

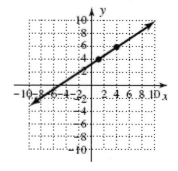

b. $m = -\dfrac{1}{2}$; $(-2, 3)$

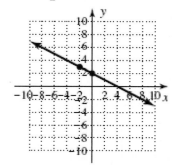

19. $(1, 2)$ and $(-3, 5)$

$m = \dfrac{5 - 2}{-3 - 1} = -\dfrac{3}{4}$

$y - 2 = -\dfrac{3}{4}(x - 1)$

$y - 2 = -\dfrac{3}{4}x + \dfrac{3}{4}$

$y = -\dfrac{3}{4}x + \dfrac{11}{4}$

Chapter 2: Relations, Functions and Graphs

21. $m = \dfrac{2}{5}$; y–intercept $(0, 2)$

    $y = \dfrac{2}{5}x + 2$

    When the rabbit population increases by 500, the wolf population increases by 200.

23. a. $f(x) = \sqrt{4x + 5}$

    $4x + 5 \geq 0$

    $4x \geq -5$

    $x \geq -\dfrac{5}{4}$

    $x \in \left[-\dfrac{5}{4}, \infty\right)$

    b. $g(x) = \dfrac{x - 4}{x^2 - x - 6}$

    $x^2 - x - 6 = 0$

    $(x - 3)(x + 2) = 0$

    $x - 3 = 0$ or $x + 2 = 0$

    $x = 3$ or $x = -2$

    These values must be excluded because they cause division by zero.

    $x \in (-\infty, -2) \cup (-2, 3) \cup (3, \infty)$

25. It is a function.

27. $D: x \in (-\infty, \infty)$

    $R: y \in [-5, \infty)$

    $f(x) \uparrow: x \in (2, \infty)$

    $f(x) \downarrow: x \in (-\infty, 2)$

    $f(x) > 0: x \in (-\infty, -1) \cup (5, \infty)$

    $f(x) < 0: x \in (-1, 5)$

29. $D: x \in (-\infty, \infty)$

    $R: y \in (-\infty, \infty)$

    $f(x) \uparrow: x \in (-\infty, -3) \cup (1, \infty)$

    $f(x) \downarrow: x \in (-3, 1)$

    $f(x) > 0: x \in (-5, -1) \cup (4, \infty)$

    $f(x) < 0: x \in (-\infty, -5) \cup (-1, 4)$

31. a. $\dfrac{f(x_2) - f(x_1)}{x_2 - x_1} = \dfrac{\sqrt{5 + 4} - \sqrt{-3 + 4}}{5 - (-3)}$

    $= \dfrac{3 - 1}{8} = \dfrac{1}{4}$

    Graph is rising to the right.

    b. $\dfrac{j(x + h) - j(x)}{h}$

    $= \dfrac{(x + h)^2 - (x + h) - (x^2 - x)}{h}$

    $= \dfrac{x^2 + 2xh + h^2 - x - h - x^2 + x}{h}$

    $= \dfrac{2xh + h^2 - h}{h} = 2x - 1 + h$

    $x = 2, h = 0.01$

    $2(2) - 1 + 0.01 = 3.01$

33. Squaring function
    a. up on left/up on the right
    b. x–intercepts: $(-4,0)$, $(0,0)$
       y–intercept: $(0,0)$
    c. Vertex: $(-2,-4)$
    d. $x \in (-\infty, \infty)$, $y \in [-4, \infty)$

35. Cubing function
    a. down on left/up on right
    b. x–intercepts: $(-2,0)$, $(-1,0)$, $(4,0)$
       y–intercept: $(0,2)$
    c. Inflection point: $(1,0)$
    d. $x \in (-\infty, \infty)$, $y \in (-\infty, \infty)$

Summary and Review

37. Cube root
    a. up on left/ down on right
    b. x–intercept: $(1,0)$
       y–intercept: $(0,1)$
    c. Inflection: $(1,0)$
    d. $x \in (-\infty, \infty), y \in (-\infty, \infty)$

39. $f(x) = 2|x+3|$; Absolute Value

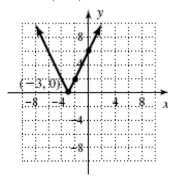

41. $f(x) = \sqrt{x-5} + 2$; Square Root

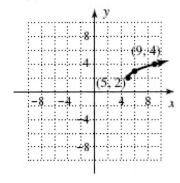

43. a. $f(x-2)$
       Right 2

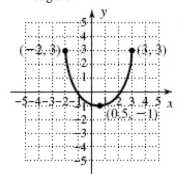

   b. $-f(x) + 4$
       Reflect, up 4

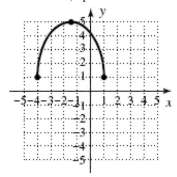

   c. $\dfrac{1}{2}f(x)$
       Compressed down

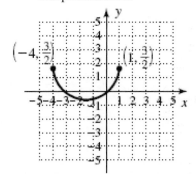

45. $h(x) = \begin{cases} \dfrac{x^2 - 2x - 15}{x + 3} & x \neq -3 \\ -6 & x = -3 \end{cases}$

$x \in (-\infty, \infty)$

$y \in (-\infty, -8) \cup (-8, \infty)$

Discontinuity at $x = -3$

Define $h(x) = -8$ at $x = -3$

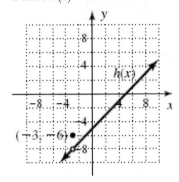

49. $f(x) = x^2 + 4x$ and $g(x) = 3x - 2$

$(f + g)(a) = f(a) + g(a)$

$= a^2 + 4a + 3a - 2$

$= a^2 + 7a - 2$

51. $f(x) = x^2 + 4x$ and $g(x) = 3x - 2$

$\left(\dfrac{f}{g}\right)(x) = \dfrac{x^2 + 4x}{3x - 2}$

$D : x \in \left(-\infty, \dfrac{2}{3}\right) \cup \left(\dfrac{2}{3}, \infty\right)$

53. $p(x) = 4x - 3$; $q(x) = x^2 + 2x$;

$(q \circ p)(3) = q[p(3)]$

$p(3) = 4(3) - 3 = 12 - 3 = 9$

$q(9) = (9)^2 + 2(9) = 81 + 18 = 99$

47. $q(x) = \begin{cases} 2\sqrt{-x-3} - 4 & x \leq -3 \\ -2|x| + 2 & -3 < x < 3 \\ 2\sqrt{x-3} - 4 & x \geq 3 \end{cases}$

55. $h(x) = \sqrt{3x - 2} + 1$;

$f(x) = \sqrt{x} + 1$;

$g(x) = 3x - 2$

57. $r(t) = 2t + 3$

$A(t) = \pi(2t + 3)^2$

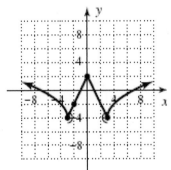

$D : x \in (-\infty, \infty)$

$R : y \in [-4, \infty)$

Chapter 2 Mixed Review

1. $4x + 3y = 12$

    $3y = -4x + 12$

    $y = -\dfrac{4}{3}x + 4$

3. a. $f(x) = \dfrac{x+1}{x^2 - 5x + 4}$

    $x^2 - 5x + 4 = 0$

    $(x-4)(x-1) = 0$

    $x - 4 = 0 \text{ or } x - 1 = 0$

    $x = 4 \quad \text{ or } x = 1$

    These values are restricted because they cause division by zero.

    Domain: $(-\infty, 1) \cup (1, 4) \cup (4, \infty)$

    b. $g(x) = \dfrac{1}{\sqrt{2x-3}}$

    Set the radicand greater than zero. (Zero must be excluded because the radical is in the denominator.

    $2x - 3 > 0$

    $2x > 3$

    $x > \dfrac{3}{2}$

    Domain: $\left(\dfrac{3}{2}, \infty \right)$

5. $m = -\dfrac{3}{2}$; y–intercept $(0, -2)$

    $y = -\dfrac{3}{2}x - 2$

7. $L_1 : (-3, 7), (2, 2)$

    Slope: $\dfrac{2-7}{2-(-3)} = -1$;

    $L_2 : (2, 2), (5, 5)$

    Slope: $\dfrac{5-2}{5-2} = 1$;

    Lines are perpendicular. Vertex of right angle is (2, 2).

    Radius: distance between (–3,7) and (2,2).

    $d = \sqrt{\left(2-(-3)\right)^2 + (2-7)^2} = \sqrt{50}$;

    Center (2,2), radius $\sqrt{50}$

    Equation: $(x-2)^2 + (y-2)^2 = 50$

9. $y = \dfrac{3}{5}x - 2$

    y–intercept $(0, -2)$

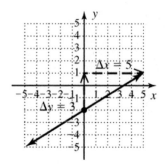

11. a. $p(x) = -2x^2 + 8x$

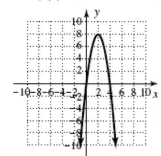

    Rate of change is positive in [–2, –1] since p is increasing in $(-\infty, 2)$.

    The rate of change in [1, 2] will be less than the rate of change in [–2, –1].

    $\dfrac{\Delta p}{\Delta x} = \dfrac{p(2) - p(1)}{2 - 1} = \dfrac{8 - 6}{1} = 2$;

    $\dfrac{\Delta p}{\Delta x} = \dfrac{p(-1) - p(-2)}{-1 - (-2)} = \dfrac{-10 - (-24)}{1} = 14$

    b. $A(t) = 1000e^{0.07t}$

    For [10, 10.01],

    $\dfrac{1000e^{0.07(10.01)} - 1000e^{0.07(10)}}{10.01 - 10} \approx 141.0$;

    For [15, 15.01],

    $\dfrac{1000e^{0.07(15.01)} - 1000e^{0.07(15)}}{15.01 - 15} \approx 200.1$;

    For [20, 20.01],

    $\dfrac{1000e^{0.07(20.01)} - 1000e^{0.07(20)}}{20.01 - 20} \approx 284.0$;

    In the interval: [15, 15.01]

13. $f(x) = \dfrac{3}{x^2 - 1}, g(x) = 3x - 2$

$(f \circ g)(x) = \dfrac{3}{(3x-2)^2 - 1}$

$= \dfrac{3}{9x^2 - 12x + 4 - 1} = \dfrac{3}{9x^2 - 12x + 3}$

$= \dfrac{3}{3(3x^2 - 4x + 1)} = \dfrac{1}{3x^2 - 4x + 1}$

To find domain, $3x^2 - 4x + 1 = 0$

$(3x - 1)(x - 1) = 0$

$x = \dfrac{1}{3}, x = 1$

Domain: $\left(-\infty, \dfrac{1}{3}\right) \cup \left(\dfrac{1}{3}, 1\right) \cup (1, \infty)$

15. $f(x) = x^2 + 1, g(x) = 3x - 2$

$\dfrac{f(x+h) - f(x)}{h}$

$\dfrac{\left[(x+h)^2 + 1\right] - \left[x^2 + 1\right]}{h}$

$= \dfrac{x^2 + 2xh + h^2 + 1 - x^2 - 1}{h}$

$= \dfrac{2xh + h^2}{h} = \dfrac{h(2x + h)}{h} = 2x + h \; ;$

$\dfrac{g(x+h) - g(x)}{h}$

$\dfrac{\left[3(x+h) - 2\right] - \left[3x - 2\right]}{h}$

$= \dfrac{3x + 3h - 2 - 3x + 2}{h} = \dfrac{3h}{h} = 3 \; ;$

For small h, $2x + h = 3$

when $x \approx \dfrac{3}{2}$

17. a. $D : x \in (-\infty, 6]$
    $R : y \in (-\infty, 3]$

b. Min: $(3, -3)$
    Max: $y = 3$ for $x \in (-6, -3); (6, 0)$

c. $g(x)\!\uparrow : x \in (-\infty, -6) \cup (3, 6)$
    $g(x)\!\downarrow : x \in (-3, 3)$
    $g(x)$ constant : $x \in (-6, -3)$

d. $g(x) > 0 : x \in (-7, -1)$
    $g(x) < 0 : x \in (-\infty, -7) \cup (-1, 6)$

19. x–intercepts: $(-1, 0)$, $(1.5, 0)$
    y–intercept: $(0, 3)$

$f(x) = a(x + 1)(x - 1.5)$

$f(x) = a\left(x^2 - \dfrac{1}{2}x - \dfrac{3}{2}\right)$

$3 = a\left(0^2 - \dfrac{1}{2}(0) - \dfrac{3}{2}\right)$

$3 = -\dfrac{3}{2}a$

$-2 = a;$

$f(x) = -2x^2 + x + 3$

Chapter 2 Practice Test

1. a. $x = y^2 + 2y$

    b. $y = \sqrt{5 - 2x}$

    c. $|y| + 1 = x$

    d. $y = x^2 + 2x$

    a and c are non–functions, do not pass the vertical line test.

3. $x + 4y = 8$

    $4y = -x + 8$

    $y = -\dfrac{1}{4}x + 2$

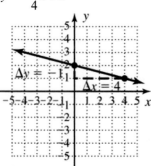

5. $6x + 5y = 3$

    $6x + 5y = 3$

    $5y = -6x + 3$

    $y = -\dfrac{6}{5}x + \dfrac{3}{5}$

    Slope: $-\dfrac{6}{5}$

    Point $(2, -2)$, slope $-\dfrac{6}{5}$

    $y - (-2) = -\dfrac{6}{5}(x - 2)$

    $y + 2 = -\dfrac{6}{5}x + \dfrac{12}{5}$

    $y = -\dfrac{6}{5}x + \dfrac{2}{5}$

7. $L_1: x = -3$

    $L_2: y = 4$

9. a. $W(24) = 300$

    b. $h = 30$ when $W(h) = 375$

    c. $(20, 250)$ and $(40, 500)$

    $$m = \frac{500 - 250}{40 - 20} = \frac{250}{20} = \frac{25}{2}$$

    $$W(h) = \frac{25}{2}h$$

    d. Wages are \$12.50 per hour.

    e. $h \in [0, 40]$

    $w \in [0, 500]$

11. $f(x) = \dfrac{2 - x^2}{x^2}$

    a. $f\left(\dfrac{2}{3}\right) = \dfrac{2 - \left(\dfrac{2}{3}\right)^2}{\left(\dfrac{2}{3}\right)^2} = \dfrac{2 - \left(\dfrac{4}{9}\right)}{\dfrac{4}{9}}$

    $$= \frac{\dfrac{14}{9}}{\dfrac{4}{9}} = \frac{14}{9} \div \frac{4}{9} = \frac{7}{2}$$

    b. $f(a + 3) = \dfrac{2 - (a + 3)^2}{(a + 3)^2}$

    $$= \frac{2 - (a^2 + 6a + 9)}{a^2 + 6a + 9} = \frac{2 - a^2 - 6a - 9}{a^2 + 6a + 9}$$

    $$= \frac{-a^2 - 6a - 7}{a^2 + 6a + 9}$$

    c. $f(1 + 2i) = \dfrac{2 - (1 + 2i)^2}{(1 + 2i)^2}$

    $$= \frac{2 - (1 + 4i + 4i^2)}{1 + 4i + 4i^2} = \frac{2 - 1 - 4i - 4i^2}{1 + 4i - 4}$$

    $$= \frac{1 - 4i + 4}{-3 + 4i} = \frac{5 - 4i}{-3 + 4i}$$

    $$= \frac{5 - 4i}{-3 + 4i} \cdot \frac{-3 - 4i}{-3 - 4i} = \frac{-15 - 20i + 12i + 16i^2}{9 - 16i^2}$$

    $$= \frac{-15 - 8i - 16}{9 + 16} = \frac{-31 - 8i}{25}$$

    $$= -\frac{31}{25} - \frac{8}{25}i$$

13. $S(t) = 2t^2 - 3t$

   a. No, new company and sales should be growing.

   b. For $[5,6], S(5) = 2(5)^2 - 3(5) = 35$

     $S(6) = 2(6)^2 - 3(6) = 54$

     Rate of Change: $\dfrac{54-35}{6-5} = 19$

     For $[6,7], S(7) = 2(7)^2 - 3(7) = 77$

     Rate of Change: $\dfrac{77-54}{7-6} = 23$

   c. $\dfrac{2(t+h)^2 - 3(t+h) - \left(2t^2 - 3t\right)}{h}$

     $= \dfrac{2\left(t^2 + 2th + h^2\right) - 3t - 3h - 2t^2 + 3t}{h}$

     $= \dfrac{2t^2 + 4th + 2h^2 - 3h - 2t^2}{h}$

     $= \dfrac{4th + 2h^2 - 3h}{h} = 4t - 3 + 2h$

     For small h:

     $4(10) - 3 = 37, 4(18) - 3 = 69,$

     $4(24) - 3 = 93$

     For small h, sales volume is approximately

     $\dfrac{37,000 \text{ units}}{1 \text{ mo}}$ in month 10,

     $\dfrac{69,000 \text{ units}}{1 \text{ mo}}$ in month 18,

     $\dfrac{93,000 \text{ units}}{1 \text{ mo}}$ in month 24

15. $g(x) = -(x+3)^2 - 2$

   Left 3, reflected across x–axis, down 2

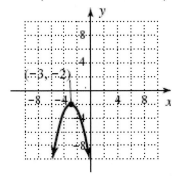

17. a. $D: x \in [-4, \infty)$

     $R: y \in [-3, \infty)$

   b. $f(-1) \approx 2.2$

   c. $f(x) < 0 : x \in (-4, -3)$

     $f(x) > 0 : x \in (-3, \infty)$

   d. $f(x) \uparrow : (-4, \infty)$

     $f(x) \downarrow$: none

   e. Parent graph: $y = \sqrt{x}$

     Graph shifts left 4, down 3

     $y = a\sqrt{x+4} - 3$

     $3 = a\sqrt{0+4} - 3$

     $3 = a\sqrt{4} - 3$

     $3 = 2a - 3$

     $6 = 2a$

     $3 = a$

     $y = 3\sqrt{x+4} - 3$

19.

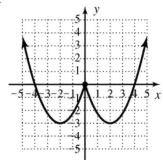

Ch 2 Calculator Exploration

Exercise 1: $y = -5(x+4)^2 + 6; (-4, 6)$

Ch. 2 Strengthening Core Skills

Exercise 1: $f(x) = x^2 - 8x - 12$

$$\frac{b}{2a} = \frac{-8}{2(1)} = -4$$

$$g(x) = x + 4$$

$$h(x) = f[g(x)] = f(x+4)$$

$$= (x+4)^2 - 8(x+4) - 12$$

$$= x^2 + 8x + 16 - 8x - 32 - 12$$

$$= x^2 - 28$$

$$x^2 - 28 = 0$$

$$x^2 = 28$$

$$x = \pm 2\sqrt{7} \;;$$

$$4 \pm 2\sqrt{7}$$

Exercise 3: $f(x) = 2x^2 - 10x + 11$

$$\frac{b}{2a} = \frac{-10}{2(2)} = -\frac{5}{2}$$

$$g(x) = x + \frac{5}{2}$$

$$h(x) = f[g(x)] = f\left(x + \frac{5}{2}\right)$$

$$= 2\left(x + \frac{5}{2}\right)^2 - 10\left(x + \frac{5}{2}\right) + 11$$

$$= 2\left(x^2 + 5x + \frac{25}{4}\right) - 10x - \frac{50}{2} + 11$$

$$= 2x^2 + 10x + \frac{50}{4} - 10x - \frac{50}{2} + 11$$

$$= 2x^2 - \frac{3}{2}$$

$$2x^2 - \frac{3}{2} = 0$$

$$2x^2 = \frac{3}{2}$$

$$x^2 = \frac{3}{4}$$

$$x = \pm\frac{\sqrt{3}}{2}$$

$$\frac{5}{2} \pm \frac{\sqrt{3}}{2}$$

Cumulative Review Chapters 1–2

1. $\left(x^3 - 5x^2 + 2x - 10\right) \div (x - 5)$

$$= \frac{x^2(x-5) + 2(x-5)}{x-5}$$

$$= \frac{(x-5)(x^2+2)}{x-5}$$

$$= x^2 + 2$$

3. $A = \pi r^2$

$$69 = \pi r^2$$

$$\frac{69}{\pi} = r^2$$

$$21.96 \approx r^2$$

$$4.686 \approx r;$$

$$C = 2\pi r$$

$$C = 2\pi(4.686)$$

$$C \approx 29.45 \text{ cm}$$

5. $-2(3-x) + 5x = 4(x+1) - 7$

$$-6 + 2x + 5x = 4x + 4 - 7$$

$$7x - 6 = 4x - 3$$

$$3x = 3$$

$$x = 1$$

7. a. $(-4, 7)$ and $(2, 5)$

$$m = \frac{7-5}{-4-2} = \frac{2}{-6} = -\frac{1}{3}$$

b. $3x - 5y = 20$

$$-5y = -3x + 20$$

$$y = \frac{3}{5}x - 4$$

$$m = \frac{3}{5}$$

9. $(-3, 2)$; $m = \dfrac{1}{2}$

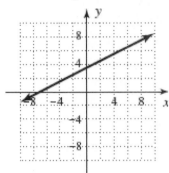

$$y - 2 = \frac{1}{2}(x + 3)$$

$$y - 2 = \frac{1}{2}x + \frac{3}{2}$$

$$y = \frac{1}{2}x + \frac{7}{2}$$

11. $f(x) = 3x^2 - 6x$ and $g(x) = x - 2$

$$(f \cdot g)(x) = f(x) \cdot g(x)$$

$$= (3x^2 - 6x)(x - 2)$$

$$= 3x^3 - 6x^2 - 6x^2 + 12x$$

$$= 3x^3 - 12x^2 + 12x$$

$$(f \div g)(x) = \frac{f(x)}{g(x)}$$

$$= \frac{3x^2 - 6x}{x - 2}$$

$$= \frac{3x(x - 2)}{x - 2}$$

$$= 3x; \quad x \neq 2;$$

$$(g \circ f)(-2) = g[f(-2)];$$

$$f(-2) = 3(-2)^2 - 6(-2) = 24;$$

$$g(24) = 24 - 2 = 22$$

13. $f(x) = \begin{cases} x^2 - 4 & x < 2 \\ x - 1 & 2 \leq x \leq 8 \end{cases}$

a. $D: x \in (-\infty, 8]$
   $R: y \in [-4, \infty)$

b. $f(-3) = (-3)^2 - 4 = 9 - 4 = 5$;
   $f(-1) = (-1)^2 - 4 = 1 - 4 = -3$;
   $f(1) = (1)^2 - 4 = 1 - 4 = -3$;
   $f(2) = 2 - 1 = 1$;
   $f(3) = 3 - 1 = 2$

c. $(-2, 0)$

d. $f(x) < 0: x \in (-2, 2)$
   $f(x) > 0: x \in (-\infty, -2) \cup [2, 8]$

e. Max: $(8, 7)$
   Min: $(0, -4)$

f. $f(x)\uparrow: x \in (0, 8)$
   $f(x)\downarrow: x \in (-\infty, 0)$

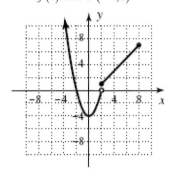

15. a. $\dfrac{-2}{x^2 - 3x - 10} + \dfrac{1}{x + 2}$

$$= \frac{-2}{(x - 5)(x + 2)} + \frac{1}{x + 2}$$

$$= \frac{-2}{(x - 5)(x + 2)} + \frac{1(x - 5)}{(x - 5)(x + 2)}$$

$$= \frac{-2 + x - 5}{(x - 5)(x + 2)}$$

$$= \frac{x - 7}{(x - 5)(x + 2)}$$

b. $\dfrac{b^2}{4a^2} - \dfrac{c}{a} = \dfrac{b^2}{4a^2} - \dfrac{4ac}{4a^2} = \dfrac{b^2 - 4ac}{4a^2}$

17. a. $N \subset Z \subset W \subset Q \subset R$
       False
    b. $W \subset N \subset Z \subset Q \subset R$
       False
    c. $N \subset W \subset Z \subset Q \subset R$
       True
    d. $N \subset R \subset Z \subset Q \subset W$
       False

19. $2x^2 + 49 = -20x$

$$2x^2 + 20x + 49 = 0$$

$$2x^2 + 20x = -49$$

$$x^2 + 10x = \frac{-49}{2}$$

$$x^2 + 10x + 25 = -\frac{49}{2} + 25$$

$$(x+5)^2 = \frac{1}{2}$$

$$x + 5 = \pm\sqrt{\frac{1}{2}}$$

$$x + 5 = \pm\frac{\sqrt{2}}{2}$$

$$x = -5 \pm \frac{\sqrt{2}}{2};$$

$$x \approx -4.293$$
$$x \approx -5.707$$

21. Let w represent the width.
    Let l represent the length.

$$A = lw$$

$$1457 = (w + 16)w$$

$$0 = w^2 + 16w - 1457$$

$$0 = (w - 31)(w + 47)$$

$$w = 31 \text{ cm}; \ l = 47 \text{ cm}$$

23. a. $6x^2 - 7x = 20$

$$6x^2 - 7x - 20 = 0$$

$$(3x + 4)(2x - 5) = 0$$

$$x = -\frac{4}{3}; \quad x = \frac{5}{2}$$

    b. $x^3 + 5x^2 - 15 = 3x$

$$x^3 + 5x^2 - 3x - 15 = 0$$

$$(x^3 + 5x^2) - (3x + 15) = 0$$

$$x^2(x + 5) - 3(x + 5) = 0$$

$$(x + 5)(x^2 - 3) = 0$$

$$x = -5; \quad x = \sqrt{3}; \quad x = -\sqrt{3}$$

25. $(-4, 5), (4, -1), (0, 8)$

$$d = \sqrt{(-4-4)^2 + (5+1)^2}$$

$$d = \sqrt{(-8)^2 + (6)^2}$$

$$d = \sqrt{100}$$

$$d = 10;$$

$$d = \sqrt{(4-0)^2 + (-1-8)^2}$$

$$d = \sqrt{(4)^2 + (-9)^2}$$

$$d = \sqrt{97}$$

$$d \approx 9.85;$$

$$d = \sqrt{(-4-0)^2 + (5-8)^2}$$

$$d = \sqrt{(-4)^2 + (-3)^2}$$

$$d = \sqrt{25}$$

$$d = 5;$$

$$P = 10 + \sqrt{97} + 5 = 15 + \sqrt{97}$$

$$\approx 15 + 9.85 \approx 24.85 \text{ units}$$

No it is not a right triangle.

$$5^2 + \left(\sqrt{97}\right)^2 \neq 10^2$$

3.1 Technology Highlight

Exercise 1: 1.35, 6.65

Exercise 3: $-2.87, 0.87$

3.1 Exercises

1. $\dfrac{25}{2}$

3. $0, f(x)$

5. Answers will vary.

7. $f(x) = x^2 + 4x - 5$

$f(x) = \left(x^2 + 4x + 4\right) - 5 - 4$

$f(x) = (x+2)^2 - 9;$

$x = \dfrac{-4 \pm \sqrt{(4)^2 - 4(1)(-5)}}{2(1)}$

$x = \dfrac{-4 \pm \sqrt{36}}{2}$

$x = \dfrac{-4 \pm 6}{2}$

$x = 1; \quad x = -5$

Left 2, down 9
x-intercepts: $(1, 0), (-5, 0)$
y-intercept: $(0, -5)$
Vertex: $(-2, -9)$

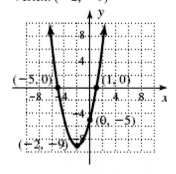

9. $h(x) = -x^2 + 2x + 3$

$h(x) = -\left(x^2 - 2x + 1\right) + 3 + 1$

$h(x) = -(x-1)^2 + 4;$

$x = \dfrac{-2 \pm \sqrt{(2)^2 - 4(-1)(3)}}{2(-1)}$

$x = \dfrac{-2 \pm \sqrt{16}}{-2}$

$x = \dfrac{-2 \pm 4}{-2}$

$x = -1; \quad x = 3$

Reflected in x-axis, right 1, up 4
x-intercepts: $(-1, 0), (3, 0)$
y-intercept: $(0, 3)$
Vertex: $(1, 4)$

3.1 Exercises

11. $Y_1 = 3x^2 + 6x - 5$

$Y_1 = 3(x^2 + 2x) - 5$

$Y_1 = 3(x^2 + 2x + 1) - 5 - 3$

$Y_1 = 3(x+1)^2 - 8;$

$x = \dfrac{-6 \pm \sqrt{(6)^2 - 4(3)(-5)}}{2(3)}$

$x = \dfrac{-6 \pm \sqrt{96}}{6}$

$x \approx 0.6; \quad x \approx -2.6$

Left 1, down 8, stretched vertically

x-intercepts: $(0.6, 0)$, $(-2.6, 0)$

y-intercept: $(0, -5)$

Vertex: $(-1, -8)$

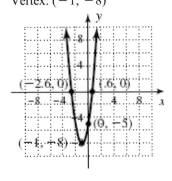

13. $f(x) = -2x^2 + 8x + 7$

$f(x) = -2(x^2 - 4x) + 7$

$f(x) = -2(x^2 - 4x + 4) + 7 + 8$

$f(x) = -2(x-2)^2 + 15;$

$x = \dfrac{-8 \pm \sqrt{(8)^2 - 4(-2)(7)}}{2(-2)}$

$x = \dfrac{-8 \pm \sqrt{120}}{-4}$

$x \approx -0.7 \quad x \approx 4.7$

Reflected in x-axis, right 2, up 15, stretched vertically

x-intercepts: $(-0.7, 0)$, $(4.7, 0)$

y-intercept: $(0, 7)$

Vertex: $(2, 15)$

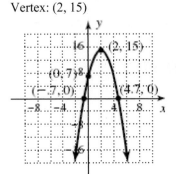

15. $p(x) = 2x^2 - 7x + 3$

$p(x) = 2\left(x^2 - \dfrac{7}{2}x\right) + 3$

$p(x) = 2\left(x^2 - \dfrac{7}{2}x + \dfrac{49}{16}\right) + 3 - \dfrac{49}{8}$

$p(x) = 2\left(x - \dfrac{7}{4}\right)^2 - \dfrac{25}{8};$

$x = \dfrac{7 \pm \sqrt{(-7)^2 - 4(2)(3)}}{2(2)}$

$x = \dfrac{7 \pm \sqrt{25}}{4}$

$x = \dfrac{7 \pm 5}{4}$

$x = 3; \quad x = \dfrac{1}{2}$

Right $\dfrac{7}{4}$, down $\dfrac{25}{8}$, stretched vertically

x-intercepts: $(3, 0)$, $\left(\dfrac{1}{2}, 0\right)$

y-intercept: $(0, 3)$

Vertex: $\left(\dfrac{7}{4}, -\dfrac{25}{8}\right)$

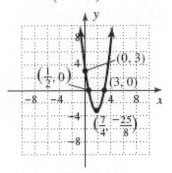

Chapter 3: Polynomial and Rational Functions

17. $f(x) = -3x^2 - 7x + 6$

$f(x) = -3\left(x^2 + \dfrac{7}{3}x\right) + 6$

$f(x) = -3\left(x^2 + \dfrac{7}{3}x + \dfrac{49}{36}\right) + 6 + \dfrac{49}{12}$

$f(x) = -3\left(x + \dfrac{7}{6}\right)^2 + \dfrac{121}{12};$

$x = \dfrac{7 \pm \sqrt{(-7)^2 - 4(-3)(6)}}{2(-3)}$

$x = \dfrac{7 \pm \sqrt{121}}{-6}$

$x = \dfrac{7 \pm 11}{-6}$

$x = -3; \quad x = \dfrac{2}{3}$

Reflected in x-axis, left $\dfrac{7}{6}$, up $\dfrac{121}{12}$, stretched vertically

x-intercepts: $(-3, 0)$, $\left(\dfrac{2}{3}, 0\right)$

y-intercept: $(0, 6)$

Vertex: $\left(-\dfrac{7}{6}, \dfrac{121}{12}\right)$

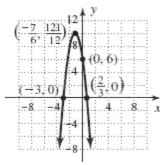

19. $p(x) = x^2 - 5x + 2$

$p(x) = \left(x^2 - 5x + \dfrac{25}{4}\right) + 2 - \dfrac{25}{4}$

$p(x) = \left(x - \dfrac{5}{2}\right)^2 - \dfrac{17}{4};$

$x = \dfrac{5 \pm \sqrt{(-5)^2 - 4(1)(2)}}{2(1)}$

$x = \dfrac{5 \pm \sqrt{17}}{2}$

$x \approx 4.56; \quad x \approx 0.44$

Right $\dfrac{5}{2}$, down $\dfrac{17}{4}$

x-intercepts: $(4.6, 0), (0.4, 0)$

y-intercept: $(0, 2)$

Vertex: $\left(\dfrac{5}{2}, -\dfrac{17}{4}\right)$

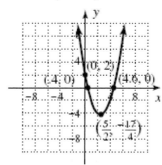

3.1 Exercises

21. $f(x) = x^2 + 2x - 6$

$f(x) = (x^2 + 2x) - 6$

$f(x) = (x^2 + 2x + 1) - 6 - 1$

$f(x) = (x + 1)^2 - 7;$

$(x + 1)^2 - 7 = 0$

$(x + 1)^2 = 7$

$x + 1 = \pm\sqrt{7}$

$x = -1 \pm \sqrt{7}$

$x \approx 1.6; \quad x \approx -3.6$

Left 1, down 7
x-intercepts: (1.6, 0), (-3.6, 0)
y-intercept: $(0, -6)$
Vertex: $(-1, -7)$

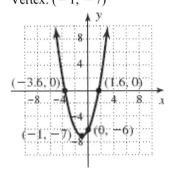

23. $h(x) = -x^2 + 4x + 2$

$h(x) = -(x^2 - 4x) + 2$

$h(x) = -(x^2 - 4x + 4) + 2 + 4$

$h(x) = -(x - 2)^2 + 6;$

$-(x - 2)^2 + 6 = 0$

$-(x - 2)^2 = -6$

$(x - 2)^2 = 6$

$x - 2 = \pm\sqrt{6}$

$x = 2 \pm \sqrt{6}$

$x \approx 4.4; \quad x \approx -0.4$

Reflected across x-axis, right 2, up 6,
x-intercepts: (4.4, 0), $(-0.4, 0)$
y-intercept: (0, 2)
Vertex: (2, 6)

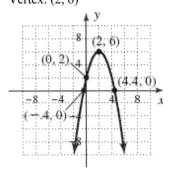

25. $Y_1 = 0.5x^2 + 3x + 7$

$Y_1 = 0.5(x^2 + 6x) + 7$

$Y_1 = 0.5(x^2 + 6x + 9) + 7 - 4.5$

$Y_1 = 0.5(x + 3)^2 + 2.5;$

$0.5(x + 3)^2 + 2.5 = 0$

$0.5(x + 3)^2 = -2.5$

$(x + 3)^2 = -5$

Left 3, up 2.5, compressed vertically
No x-intercepts
y-intercept: (0, 7)

Vertex: $\left(-3, \dfrac{5}{2}\right)$

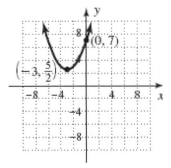

27. $Y_1 = -2x^2 + 10x - 7$

$Y_1 = -2(x^2 - 5x) - 7$

$Y_1 = -2\left(x^2 - 5x + \dfrac{25}{4}\right) - 7 + \dfrac{50}{4}$

$Y_1 = -2\left(x - \dfrac{5}{2}\right)^2 + \dfrac{11}{2}$;

$-2\left(x - \dfrac{5}{2}\right)^2 + \dfrac{11}{2} = 0$

$-2\left(x - \dfrac{5}{2}\right)^2 = -\dfrac{11}{2}$

$\left(x - \dfrac{5}{2}\right)^2 = \dfrac{11}{4}$

$x - \dfrac{5}{2} = \pm\dfrac{\sqrt{11}}{2}$

$x = \dfrac{5}{2} \pm \dfrac{\sqrt{11}}{2}$

$x \approx 4.2; \ x \approx 0.8$

Reflected across x-axis, right $\dfrac{5}{2}$, up $\dfrac{11}{2}$,

stretched vertically
x-intercepts: (4.2, 0), (0.8, 0)
y-intercept: (0, -7)

Vertex: $\left(\dfrac{5}{2}, \dfrac{11}{2}\right)$

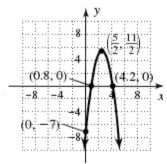

29. $f(x) = 4x^2 - 12x + 3$

$f(x) = 4(x^2 - 3x) + 3$

$f(x) = 4\left(x^2 - 3x + \dfrac{9}{4}\right) + 3 - 9$

$f(x) = 4\left(x - \dfrac{3}{2}\right)^2 - 6$;

$4\left(x - \dfrac{3}{2}\right)^2 - 6 = 0$

$4\left(x - \dfrac{3}{2}\right)^2 = 6$

$\left(x - \dfrac{3}{2}\right)^2 = \dfrac{3}{2}$

$x - \dfrac{3}{2} = \pm\dfrac{\sqrt{3}}{\sqrt{2}}$

$x = \dfrac{3}{2} \pm \dfrac{\sqrt{6}}{2}$

$x \approx 2.7; \ x \approx 0.3$

Right $\dfrac{3}{2}$, down 6, stretched vertically

x-intercepts: (2.7, 0), (0.3, 0)
y-intercept: (0, 3)

Vertex: $\left(\dfrac{3}{2}, -6\right)$

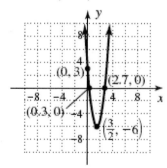

3.1 Exercises

31. $p(x) = \dfrac{1}{2}x^2 + 3x - 5$

$p(x) = \dfrac{1}{2}\left(x^2 + 6x\right) - 5$

$p(x) = \dfrac{1}{2}\left(x^2 + 6x + 9\right) - 5 - \dfrac{9}{2}$

$p(x) = \dfrac{1}{2}(x+3)^2 - \dfrac{19}{2};$

$\dfrac{1}{2}(x+3)^2 - \dfrac{19}{2} = 0$

$\dfrac{1}{2}(x+3)^2 = \dfrac{19}{2}$

$(x+3)^2 = 19$

$x+3 = \pm\sqrt{19}$

$x = -3 \pm \sqrt{19}$

$x \approx 1.4 \quad x \approx -7.4$

Left 3, down $\dfrac{19}{2}$, compressed vertically

x-intercepts: $(1.4, 0), (-7.4, 0)$
y-intercept: $(0, -5)$

Vertex: $\left(-3, \dfrac{-19}{2}\right)$

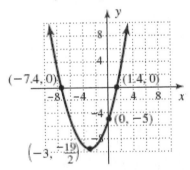

33. Compare to the graph of $y = x^2$, this graph is shifted right 2 and down 1: $y = a(x-2)^2 - 1$ where a is positive. Choose a point on the graph, using $(3, 0)$

$y = a(x-2)^2 - 1$
$0 = a(3-2)^2 - 1$
$0 = a - 1$
$1 = a$
Equation: $y = 1(x-2)^2 - 1$

35. Compare to the graph of $y = x^2$, this graph is reflected across the x-axis, shifted to the left 2 and up 4: $y = a(x+2)^2 - 4$.

Choose a point on the graph, using $(0, 0)$

$0 = a(0+2)^2 + 4$
$0 = a(4) + 4$
$-4 = 4a$
$a = -1$
Equation: $y = -1(x+2)^2 + 4$

37. Compare to the graph of $y = x^2$, this graph is reflected across the x-axis, shifted left 2 and up 3:
$y = a(x+2)^2 + 3$.

Choose a point on the graph, using $(0, -3)$

$-3 = a(0+2)^2 + 3$
$-3 = 4a + 3$
$-6 = 4a$
$\dfrac{-3}{2} = a$

Equation: $y = -\dfrac{3}{2}(x+2)^2 + 3$

39. i. a. $y = (x+3)^2 - 5$

$(x+3)^2 - 5 = 0$
$(x+3)^2 = 5$
$x+3 = \pm\sqrt{5}$
$x = -3 \pm \sqrt{5}$

b. $x = -3 \pm \sqrt{-\dfrac{-5}{1}}$

$x = -3 \pm \sqrt{5}$

ii. a. $y = -(x-4)^2 + 3$

$-(x-4)^2 + 3 = 0$
$-(x-4)^2 = -3$
$(x-4)^2 = 3$
$x-4 = \pm\sqrt{3}$
$x = 4 \pm \sqrt{3}$

b. $x = 4 \pm \sqrt{-\dfrac{3}{-1}}$

$x = 4 \pm \sqrt{3}$

iii. a. $y = 2(x+4)^2 - 7$

$2(x+4)^2 - 7 = 0$

$2(x+4)^2 = 7$

$(x+4)^2 = \dfrac{7}{2}$

$x + 4 = \pm\sqrt{\dfrac{7}{2}}$

$x = -4 \pm \dfrac{\sqrt{14}}{2}$

b. $x = -4 \pm \sqrt{-\dfrac{-7}{2}}$

$x = -4 \pm \sqrt{\dfrac{7}{2}}$

$x = -4 \pm \dfrac{\sqrt{14}}{2}$

iv. a. $y = -3(x-2)^2 + 6$

$-3(x-2)^2 + 6 = 0$

$-3(x-2)^2 = -6$

$(x-2)^2 = 2$

$x - 2 = \pm\sqrt{2}$

$x = 2 \pm \sqrt{2}$

b. $x = 2 \pm \sqrt{-\dfrac{6}{-3}}$

$x = 2 \pm \sqrt{2}$

v. a. $s(t) = 0.2(t+0.7)^2 - 0.8$

$0.2(t+0.7)^2 - 0.8 = 0$

$0.2(t+0.7)^2 = 0.8$

$(t+0.7)^2 = 4$

$t + 0.7 = \pm 2$

$t = -0.7 \pm 2$

$t = -2.7; \ t = 1.3$

b. $t = -0.7 \pm \sqrt{-\dfrac{-0.8}{0.2}}$

$t = -0.7 \pm \sqrt{4}$

$t = -0.7 \pm 2$

$t = -2.7; t = 1.3$

vi. a. $r(t) = -0.5(t-0.6)^2 + 2$

$-0.5(t-0.6)^2 + 2 = 0$

$-0.5(t-0.6)^2 = -2$

$(t-0.6)^2 = 4$

$t - 0.6 = \pm 2$

$t = 0.6 \pm 2$

$t = -1.4; \ t = 2.6$

b. $t = 0.6 \pm \sqrt{-\dfrac{2}{-0.5}}$

$t = 0.6 \pm \sqrt{4}$

$t = 0.6 \pm 2$

$t = -1.4; t = 2.6$

41. $P(x) = -10x^2 + 3500x - 66000$

a. (0, −66000) When no cars are produced, there is a profit loss of $66,000.

b. $P(x) = -10x^2 + 3500x - 66000$

$-10x^2 + 3500x - 66000 = 0$

$-10(x^2 - 350 + 6600) = 0$

$-10(x - 20)(x - 330) = 0$

$x = 20; \ x = 330$

(20, 0) and (330, 0)

No profit will be made if less than 20 cars or more than 330 cars are produced.

c. $x = \dfrac{-b}{2a} = \dfrac{-3500}{-20} = 175$ cars

d. $P(175) = -10(175)^2 + 3500(175) - 66000$

$P(175) = \$240{,}250$

43. $d(x) = x^2 - 12x$

a. $x = \dfrac{-b}{2a} = \dfrac{12}{2} = 6$ miles

b. $d(6) = (6)^2 - 12(6) = -36$
3600 feet

c. $d(4) = (4)^2 - 12(4) = -32$
3200 feet

d. $x^2 - 12x = 0$
$x(x - 12) = 0$

$x = 0; \quad x = 12$
12 miles

45. $P(x) = -0.5x^2 + 175x - 3300$

a. $(0, -3300)$ If no appliances are sold, the loss will be $3300.

b. $-0.5x^2 + 175x - 3300 = 0$

$-0.5\left(x^2 - 350x + 6600\right) = 0$

$-0.5(x - 20)(x - 330) = 0$

$x = 20; \quad x = 330$
$(20, 0)$ and $(330, 0)$
If less than 20 or more than 330 appliances are made and sold, there will be no profit.

c. $0 \le x \le 200$ Greatest number of appliances to be produced each day is 200.

d. $x = \dfrac{-b}{2a} = \dfrac{175}{1} = 175$;

$P(175) = -0.5(175)^2 + 175(175) - 3300$
$= \$12{,}012.50$

47. $h(t) = -16t^2 + 176t$

a.

$$h(2) = -16(2)^2 + 176(2) = 288 \text{feet}$$

b.

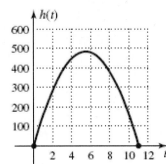

c. $x = \dfrac{-b}{2a} = \dfrac{-176}{-32} = 5.5 \sec$

$h(5.5) = -16(5.5)^2 + 176(5.5) = 484$ feet

d. $16t^2 - 176t = 0$
$16t(t - 11) = 0$

$t = 11$ sec

49. a. $h(t) = -16t^2 + 32t = 5$

b. $t = 0.5$
$h(0.5) = -16(0.5)^2 + 32(0.5) + 5$
$\qquad = 17$ ft
$t = 1.5$
$h(1.5) = -16(1.5)^2 + 32(1.5) + 5$
$\qquad = 17$ ft

c. The person is 17 ft height at 0.5 sec and is 17 feet height at 1.5 sec, so max height must occur between $t = 0.5$ and $t = 1.5$.

d. $h(t) = -16t^2 + 32t + 5$

$t = -\dfrac{32}{2(-16)}$

$= 1$ sec

e. $h(1) = -16(1)^2 + 32(1) + 5$
$= 21$ ft

f. $h(t) = -16t^2 + 32t + 5$

$$t = \frac{-32 \pm \sqrt{32^2 - 4(-16)(5)}}{2(-16)}$$

$$= \frac{-32 \pm \sqrt{1344}}{-32}$$

$$= \frac{-32 \pm (36.66)}{-32}$$

$$= \frac{-32 - (36.66)}{-32}$$

$$\approx 2.2 \text{ sec}$$

51. $C(x) = 16x - 63$

$R(x) = -x^2 + 326x - 7463$

$P(x) = R(x) - C(x)$

$P(x) = -x^2 + 326x - 7463 - 16x + 63$

$P(x) = -x^2 + 310x - 7400;$

$x = \dfrac{-b}{2a} = \dfrac{-310}{-2} = 155$

155,000 bottles;

$P(155) = -(155)^2 + 310(155) - 7400 = 16625$

$16,625

53. Let x represent the width.

Let $\dfrac{384 - 4x}{2}$ represent the length

$$x\left(\frac{384 - 4x}{2}\right) = 192x - 2x^2$$

a. Opens downward, x-coordinate of

vertex: $x = \dfrac{-192}{2(-2)} = 48$ ft;

$$\frac{384 - 4(48)}{2} = 96 \text{ ft}$$

b. $\dfrac{96}{3} = 32;$

32 ft x 48 ft

55. $x = 2 \pm 3i$

$f(x) = (x - (2 + 3i))(x - (2 - 3i))$

$f(x) = (x - 2 - 3i)(x - 2 + 3i)$

$f(x) = ((x-2)^2 - (3i)^2)$

$f(x) = x^2 - 4x + 4 - 9i^2$

$f(x) = x^2 - 4x + 4 + 9$

$f(x) = x^2 - 4x + 13$

57. a. radicand will be negative – two complex zeroes
    b. radicand will be positive – two real zeroes
    c. radicand is zero – one real zero
    d. two real, rational zeroes
    e. two real, irrational zeroes

59. $\dfrac{x^2 - 4x + 4}{x^2 + 3x - 10} \cdot \dfrac{x^2 - 25}{x^2 - 10x + 25}$

$x = \dfrac{(x-2)^2}{(x+5)(x-2)} \cdot \dfrac{(x-5)(x+5)}{(x-5)^2}$

$= \dfrac{x-2}{x-5}$

61. $f(x) = 3x^2 + 7x - 6; \quad f(x) \le 0$

$3x^2 + 7x - 6 = 0$

$(3x - 2)(x + 3) = 0$

$x = \dfrac{2}{3}; \quad x = -3$

Concave up

$x \in \left[-3, \dfrac{2}{3}\right]$

3.2 Exercises

1. synthetic; zero

3. $P(c)$, remainder

5. If polynomial P(x) is divided by the linear factor x-c, the remainder is identical to P(c).

7. $\dfrac{x^3 - 5x^2 - 4x + 23}{x - 2}$

$$
\begin{array}{r}
x^2 - 3x - 10 \\
x-2\overline{)x^3 - 5x^2 - 4x + 23} \\
\underline{x^3 - 2x^2} \\
-3x^2 - 4x \\
\underline{-3x^2 + 6x} \\
-10x + 23 \\
\underline{-10x + 20} \\
3
\end{array}
$$

$x^3 - 5x^2 - 4x + 23 = (x-2)(x^2 - 3x - 10) + 3$

9. $\left(2x^3 + 5x^2 + 4x + 17\right) \div \left(x + 3\right)$

$$
\begin{array}{r}
2x^2 - x + 7 \\
x+3\overline{)2x^3 + 5x^2 + 4x + 17} \\
\underline{2x^3 + 6x^2} \\
-1x^2 + 4x \\
\underline{-x^2 - 3x} \\
7x + 17 \\
\underline{7x + 21} \\
-4
\end{array}
$$

$2x^3 + 5x^2 + 4x + 17 = (x+3)(2x^2 - x + 7) - 4$

11. $(x^3 - 8x^2 + 11x + 20) \div (x - 5)$

$$
\begin{array}{r}
x^2 - 3x - 4 \\
x-5\overline{)x^3 - 8x^2 + 11x + 20} \\
\underline{x^3 - 5x^2} \\
-3x^2 + 11x \\
\underline{-3x^2 + 15x} \\
-4x + 20 \\
\underline{-4x + 20} \\
\end{array}
$$

$(x^3 - 8x^2 + 11x + 20) = (x-5)(x^2 - 3x - 4) + 0$

13. $\dfrac{2x^2 - 5x - 3}{x - 3}$

$$
\begin{array}{r}
3\underline{|2 \quad -5 \quad -3} \\
\quad 6 \quad 3 \\
\overline{2 \quad 1 \quad 0}
\end{array}
$$

a) $\dfrac{2x^2 - 5x - 3}{x - 3} = (2x + 1) + \dfrac{0}{x - 3}$

b) $2x^2 - 5x - 3 = (x-3)(2x+1) + 0$

15. $(x^3 - 7x^2 + 6x + 8) \div (x - 2)$

$$
\begin{array}{r}
2\underline{|1 \quad -7 \quad 6 \quad 8} \\
\quad 2 \quad -10 \quad -8 \\
\overline{1 \quad -5 \quad -4 \quad 0}
\end{array}
$$

a) $\dfrac{x^3 - 7x^2 + 16x + 8}{x - 2} = (x^2 - 5x - 4) + \dfrac{0}{x - 2}$

b) $x^3 - 7x^2 + 6x + 8 = (x-2)(x^2 - 5x - 4) + 0$

17. $\dfrac{x^3 - 5x^2 - 4x + 23}{x - 2}$

$$
\begin{array}{r}
2\underline{|1 \quad -5 \quad -4 \quad 23} \\
\quad 2 \quad -6 \quad -20 \\
\overline{1 \quad -3 \quad -10 \quad 3}
\end{array}
$$

a) $\dfrac{x^3 - 5x^2 - 4x + 23}{x - 2} = (x^2 - 3x - 10) + \dfrac{3}{x - 2}$

b) $x^3 - 5x^2 - 4x + 23 = (x-2)(x^2 - 3x - 10) + 3$

19. $(2x^3 - 5x^2 - 11x - 17) \div (x - 4)$

$$
\begin{array}{r}
4\underline{|2 \quad -5 \quad -11 \quad -17} \\
\quad 8 \quad 12 \quad 4 \\
\overline{2 \quad 3 \quad 1 \quad -13}
\end{array}
$$

a) $\dfrac{2x^3 - 5x^2 - 11x - 17}{x - 4} = (2x^2 + 3x + 1) - \dfrac{13}{x - 4}$

b) $2x^3 - 5x^2 - 11x - 17 = (x-4)(2x^2 + 3x + 1) - 13$

21. $(x^3 + 5x^2 + 7) \div (x + 1)$

$(x^3 + 5x^2 + 0x + 7) \div (x + 1)$

$$
\begin{array}{r}
-1\underline{|1 \quad 5 \quad 0 \quad 7} \\
\quad -1 \quad -4 \quad 4 \\
\overline{1 \quad 4 \quad -4 \quad 11}
\end{array}
$$

$x^3 + 5x^2 + 7 = (x+1)(x^2 + 4x - 4) + 11$

23. $(x^3 - 13x - 12) \div (x - 4)$

$(x^3 - 0x^2 - 13x - 12) \div (x - 4)$

$$\underline{4 |}\ 1 \quad 0 \quad -13 \quad -12$$
$$\phantom{\underline{4 |}\ 1}\quad 4 \quad 16 \quad 12$$
$$\overline{\phantom{\underline{4 |}}\ 1 \quad 4 \quad 3 \quad 0}$$

$x^3 - 13x - 12 = (x - 4)(x^2 + 4x + 3) + 0$

25. $\dfrac{3x^3 - 8x + 12}{x - 1}$

$(3x^3 + 0x^2 - 8x + 12) \div (x - 1)$

$$\underline{1 |}\ 3 \quad 0 \quad -8 \quad 12$$
$$\phantom{\underline{1 |}\ 3}\quad 3 \quad 3 \quad -5$$
$$\overline{\phantom{\underline{1 |}}\ 3 \quad 3 \quad -5 \quad 7}$$

$3x^3 - 8x + 12 = (x - 1)(3x^2 + 3x - 5) + 7$

27. $(n^3 + 27) \div (n + 3)$

$(n^3 + 0n^2 + 0n + 27) \div (n + 3)$

$$\underline{-3 |}\ 1 \quad 0 \quad 0 \quad 27$$
$$\phantom{\underline{-3 |}\ 1}\quad -3 \quad 9 \quad -27$$
$$\overline{\phantom{\underline{-3 |}}\ 1 \quad -3 \quad 9 \quad 0}$$

$n^3 + 27 = (n + 3)(n^2 - 3n + 9) + 0$

29. $(x^4 + 3x^3 - 16x - 8) \div (x - 2)$

$(x^4 + 3x^3 + 0x^2 - 16x - 8) \div (x - 2)$

$$\underline{2 |}\ 1 \quad 3 \quad 0 \quad -16 \quad -8$$
$$\phantom{\underline{2 |}\ 1}\quad 2 \quad 10 \quad 20 \quad 8$$
$$\overline{\phantom{\underline{2 |}}\ 1 \quad 5 \quad 10 \quad 4 \quad 0}$$

$x^4 + 3x^3 - 16x - 8$
$= (x - 2)(x^3 + 5x^2 + 10x + 4) + 0$

31. $\dfrac{2x^3 + 7x^2 - x + 26}{x^2 + 0x + 3}$

$$
\begin{array}{r}
2x + 7 \\
x^2 + 0x + 3 \overline{)2x^3 + 7x^2 - x + 26} \\
\underline{2x^3 + 0x^2 + 6x} \\
7x^2 - 7x + 26 \\
\underline{7x^2 + 0x + 21} \\
-7x + 5
\end{array}
$$

$\dfrac{2x^3 + 7x^2 - x + 26}{x^2 + 3} = (2x + 7) + \dfrac{-7x + 5}{x^2 + 3}$

33. $\dfrac{x^4 - 5x^2 - 4x + 7}{x^2 - 1}$

$$
\begin{array}{r}
x^2 - 4 \\
x^2 + 0x - 1 \overline{)x^4 + 0x^3 - 5x^2 - 4x + 7} \\
\underline{x^4 + 0x^3 - x^2} \\
-4x^2 - 4x + 7 \\
\underline{-4x^2 + 0x + 4} \\
-4x + 3
\end{array}
$$

$\dfrac{x^4 - 5x^2 - 4x + 7}{x^2 - 1} = (x^2 - 4) + \dfrac{-4x + 3}{x^2 - 1}$

35. $P(x) = x^3 - 6x^2 + 5x + 12$

    a. $P(-2) = -30$

$$\underline{-2 |}\ 1 \quad -6 \quad 5 \quad 12$$
$$\phantom{\underline{-2 |}\ 1}\quad -2 \quad 16 \quad -42$$
$$\overline{\phantom{\underline{-2 |}}\ 1 \quad -8 \quad 21 \quad \boxed{-30}}$$

    b. $P(5) = 12$

$$\underline{5 |}\ 1 \quad -6 \quad 5 \quad 12$$
$$\phantom{\underline{5 |}\ 1}\quad 5 \quad -5 \quad 0$$
$$\overline{\phantom{\underline{5 |}}\ 1 \quad -1 \quad 0 \quad \boxed{12}}$$

37. $P(x) = 2x^3 - x^2 - 19x + 4$

    a. $P(-3) = -2$

$$\underline{-3 |}\ 2 \quad -1 \quad -19 \quad 4$$
$$\phantom{\underline{-3 |}\ 2}\quad -6 \quad 21 \quad -6$$
$$\overline{\phantom{\underline{-3 |}}\ 2 \quad -7 \quad 2 \quad \boxed{-2}}$$

    b. $P(2) = -22$

$$\underline{2 |}\ 2 \quad -1 \quad -19 \quad 4$$
$$\phantom{\underline{2 |}\ 2}\quad 4 \quad 6 \quad -26$$
$$\overline{\phantom{\underline{2 |}}\ 2 \quad 3 \quad -13 \quad \boxed{-22}}$$

39. $P(x) = x^4 - 4x^2 + x + 1$

    a. $P(-2) = -1$

$$\underline{-2 |}\ 1 \quad 0 \quad -4 \quad 1 \quad 1$$
$$\phantom{\underline{-2 |}\ 1}\quad -2 \quad 4 \quad 0 \quad -2$$
$$\overline{\phantom{\underline{-2 |}}\ 1 \quad -2 \quad 0 \quad 1 \quad \boxed{-1}}$$

    b. $P(2) = 3$

$$\underline{2 |}\ 1 \quad 0 \quad -4 \quad 1 \quad 1$$
$$\phantom{\underline{2 |}\ 1}\quad 2 \quad 4 \quad 0 \quad 2$$
$$\overline{\phantom{\underline{2 |}}\ 1 \quad 2 \quad 0 \quad 1 \quad \boxed{3}}$$

41. $P(x) = 2x^3 - 7x + 33$

   a. $P(-2)$

$$\underline{-2|}\ \ 2\ \ \ 0\ \ -7\ \ \ 33$$
$$\underline{\ \ \ \ \ \ -4\ \ \ \ 8\ \ -2}$$
$$\ \ \ \ \ \ 2\ -4\ \ \ \ 1\ \ \ \ 31$$

   b. $P(-3) = 0$

$$\underline{-3|}\ \ 2\ \ \ 0\ \ -7\ \ \ 33$$
$$\underline{\ \ \ \ \ \ -6\ \ \ 18\ \ -33}$$
$$\ \ \ \ \ \ 2\ -6\ \ \ 11\ \ \ \ \ 0$$

43. $Px = 2x^3 + 3x^2 - 9x - 10$

   a. $P\left(\dfrac{3}{2}\right) = -10$

$$\underline{\tfrac{3}{2}|}\ \ 2\ \ \ 3\ \ -9\ \ -10$$
$$\underline{\ \ \ \ \ \ \ \ \ \ 3\ \ \ \ 9\ \ \ \ \ 0}$$
$$\ \ \ \ \ \ 2\ \ \ 6\ \ \ \ 0\ \ \underline{|-10}$$

   b. $P\left(-\dfrac{5}{2}\right) = 0$

$$\underline{-\tfrac{5}{2}|}\ \ 2\ \ \ \ 3\ \ -9\ \ -10$$
$$\underline{\ \ \ \ \ \ \ \ \ -5\ \ \ \ 5\ \ \ \ 10}$$
$$\ \ \ \ \ \ 2\ -2\ \ -4\ \ \ \underline{|\ 0}$$

45. $f(x) = x^3 - 3x^2 - 13x + 15$

$$\underline{-3|}\ \ 1\ \ -3\ \ -13\ \ \ \ 15$$
   a.
$$\underline{\ \ \ \ \ \ \ \ -3\ \ \ \ 18\ \ -15}$$
$$\ \ \ \ \ \ 1\ \ -6\ \ \ \ \ 5\ \ \ \ \ \ 0$$

yes, $(x + 3)$ is a factor since remainder is 0.

$$\underline{5|}\ \ 1\ \ -3\ \ -13\ \ \ \ 15$$
   b.
$$\underline{\ \ \ \ \ \ \ \ \ 5\ \ \ \ 10\ \ -15}$$
$$\ \ \ \ \ \ 1\ \ \ \ 2\ \ \ -3\ \ \ \ \ \ 0$$

yes, $(x - 5)$ is a factor since remainder is 0.

47. $f(x) = x^3 - 6x^2 + 3x + 10$

$$\underline{-2|}\ \ 1\ -6\ \ \ \ 3\ \ \ \ 10$$
   a.
$$\underline{\ \ \ \ \ \ \ -2\ \ \ 16\ \ -38}$$
$$\ \ \ \ \ \ 1\ -8\ \ \ 19\ \ -28$$

no, $(x + 2)$ is not a factor since remainder is not 0.

$$\underline{5|}\ \ 1\ -6\ \ \ \ 3\ \ \ \ 10$$
   b.
$$\underline{\ \ \ \ \ \ \ \ 5\ -5\ \ -10}$$
$$\ \ \ \ \ \ 1\ -1\ -2\ \ \ \ \ 0$$

yes, $(x - 5)$ is a factor since remainder is 0.

49. $f(x) = -x^3 + 7x - 6$

$$\underline{-3|}\ -1\ \ \ 0\ \ \ \ 7\ \ -6$$
   a.
$$\underline{\ \ \ \ \ \ \ \ \ \ 3\ -9\ \ \ \ 6}$$
$$\ \ \ \ \ -1\ \ \ 3\ \ -2\ \ \ \ 0$$

yes, $(x + 3)$ is a factor since remainder is 0.

$$\underline{2|}\ -1\ \ \ 0\ \ \ \ 7\ \ -6$$
   b.
$$\underline{\ \ \ \ \ \ \ \ -2\ -4\ \ \ \ 6}$$
$$\ \ \ \ \ -1\ -2\ \ \ \ 3\ \ \ \ 0$$

yes, $(x - 2)$ is a factor since remainder is 0.

51. $P(x) = x^3 + 2x^2 - 5x - 6;\ \ x = -3$

verified

$$\underline{-3|}\ \ 1\ \ \ \ 2\ \ -5\ \ -6$$
$$\underline{\ \ \ \ \ \ \ \ -3\ \ \ \ 3\ \ \ \ 6}$$
$$\ \ \ \ \ \ 1\ \ -1\ \ -2\ \ \underline{|\ 0}$$

$P(-3) = 0$

53. $P(x) = x^3 - 7x + 6;\ \ x = 2$

verified

$$\underline{2|}\ \ 1\ \ \ 0\ \ -7\ \ \ \ 6$$
$$\underline{\ \ \ \ \ \ \ \ \ 2\ \ \ \ 4\ \ -6}$$
$$\ \ \ \ \ \ 1\ \ \ 2\ \ -3\ \ \underline{|\ 0}$$

$P(2) = 0$

55. $P(x) = 9x^3 + 18x^2 - 4x - 8$

$$\underline{\tfrac{2}{3}|}\ 9\ \ \ 18\ \ -4\ \ -8$$
$$\underline{\ \ \ \ \ \ \ \ \ \ \ 6\ \ \ \ 16\ \ \ \ 8}$$
$$\ \ \ \ \ 9\ \ \ 24\ \ \ 12\ \ \ \ 0$$

$P\left(\dfrac{2}{3}\right) = 0$

57. $-2,\ 3,\ -5$; degree 3

$P(x) = (x+2)(x-3)(x+5);$

$P(x) = (x^2 - x - 6)(x+5)$

$P(x) = x^3 + 5x^2 - x^2 - 5x - 6x - 30$

$P(x) = x^3 + 4x^2 - 11x - 30$

59. $-2,\ \sqrt{3},\ -\sqrt{3}$; degree 3

$P(x) = (x+2)(x-\sqrt{3})(x+\sqrt{3});$

$P(x) = (x+2)(x^2 + \sqrt{3}x - \sqrt{3}x - 3)$

$P(x) = (x+2)(x^2 - 3)$

$P(x) = x^3 - 3x + 2x^2 - 6$

$P(x) = x^3 + 2x^2 - 3x - 6$

61. $-5,\ 2\sqrt{3},\ -2\sqrt{3}$; degree 3

$P(x) = (x+5)(x-2\sqrt{3})(x+2\sqrt{3});$

$P(x) = (x+5)(x^2 + 2\sqrt{3}x - 2\sqrt{3}x - 12)$

$P(x) = (x+5)(x^2 - 12)$

$P(x) = x^3 - 12x + 5x^2 - 60$

$P(x) = x^3 + 5x^2 - 12x - 60$

63. $1,\ -2,\ \sqrt{10},\ -\sqrt{10}$; degree 4

$P(x) = (x-1)(x+2)(x-\sqrt{10})(x+\sqrt{10});$

$P(x) = (x^2 + x - 2)(x^2 + \sqrt{10}x - \sqrt{10}x - 10)$

$P(x) = (x^2 + x - 2)(x^2 - 10)$

$P(x) = x^4 - 10x^2 + x^3 - 10x - 2x^2 + 20$

$P(x) = x^4 + x^3 - 12x^2 - 10x + 20$

65. $P(x) = x^3 - 5x^2 - 2x + 24$

$$\underline{-2|}\ \begin{array}{rrrr} 1 & -5 & -2 & 24 \\ & -2 & 14 & -24 \\ \hline 1 & -7 & 12 & 0 \end{array}$$

$P(x) = (x+2)(x^2 - 7x + 12)$

$P(x) = (x+2)(x-3)(x-4)$

67. $p(x) = x^4 + 2x^3 - 12x^2 - 18x + 27$

$$\underline{-3|}\ \begin{array}{rrrrr} 1 & 2 & -12 & -18 & 27 \\ & -3 & 3 & 27 & -27 \\ \hline 1 & -1 & -9 & 9 & 0 \end{array}$$

$p(x) = (x+3)(x^3 - x^2 - 9x + 9)$

$p(x) = (x+3)(x^2(x-1) - 9(x-1))$

$p(x) = (x+3)(x-1)(x^2 - 9)$

$p(x) = (x+3)(x-1)(x+3)(x-3)$

$p(x) = (x+3)^2(x-1)(x-3)$

69. $f(x) = 2x^3 + 11x^2 - x - 30$

$$\dfrac{3}{2}\bigg|\ \begin{array}{rrrr} 2 & 11 & -1 & -30 \\ & 3 & 21 & 30 \\ \hline 2 & 14 & 20 & 0 \end{array}$$

$f(x) = \left(x - \dfrac{3}{2}\right)(2x^2 + 14x + 20)$

$f(x) = \left(x - \dfrac{3}{2}\right)2(x^2 + 7x + 10)$

$f(x) = \left(x - \dfrac{3}{2}\right)2(x+2)(x+5)$

$f(x) = 2\left(x - \dfrac{3}{2}\right)(x+2)(x+5)$

71. $p(x) = x^3 - 3x^2 - 9x + 27$

$p(x) = x^2(x-3) - 9(x-3)$

$p(x) = (x-3)(x^2 - 9)$

$p(x) = (x-3)(x+3)(x-3)$

$p(x) = (x+3)(x-3)^2$

73. $p(x) = x^3 - 6x^2 + 12x - 8$

Possible Factors of 8, $\pm 1, \pm 2, \pm 4, \pm 8$

$p(x) = (x - 2)(x^2 - 4x + 4)$

$p(x) = (x - 2)(x - 2)(x - 2)$

$p(x) = (x - 2)^3$

75. $p(x) = (x^2 - 6x + 9)(x^2 - 9)$

$p(x) = (x - 3)(x - 3)(x + 3)(x - 3)$

$p(x) = (x + 3)(x - 3)^3$

77. $p(x) = (x^3 + 4x^2 - 9x - 36)(x^2 + x - 12)$

$p(x) = (x^2(x + 4) - 9(x + 4))(x + 4)(x - 3)$

$p(x) = (x + 4)(x^2 - 9)(x + 4)(x - 3)$

$p(x) = (x + 4)(x + 3)(x - 3)(x + 4)(x - 3)$

$p(x) = (x + 3)(x - 3)^2(x + 4)^2$

79. $640 = 4x^3 - 84x^2 + 432x$

$160 = x^3 - 21x^2 + 108x$

$0 = x^3 - 21x^2 + 108x - 160$

$$\begin{array}{r|rrrr} 4 & 1 & -21 & 108 & -160 \\ & & 4 & -68 & 160 \\ \hline & 1 & -17 & 40 & \boxed{0} \end{array}$$

$0 = (x - 4)(x^2 - 17x + 40)$

4 inch squares,

$24 - 8 = 16$ in.

$18 - 8 = 10$ in.

Dimensions of box: 16 in. x 10 in. x 4 in.

81. $P(w) = -0.1w^4 + 2w^3 - 14w^2 + 52w + 5$

a. week 10, 22.5 thousand

$P(5) = -0.1(5)^4 + 2(5)^3 - 14(5)^2 + 52(5) + 5 = 102.5$

$P(10) = -0.1(10)^4 + 2(10)^3 - 14(10)^2 + 52(10) + 5 = 125$

b. one week before closing, 36 thousand

$P(1) = -0.1(1)^4 + 2(1)^3 - 14(1)^2 + 52(1) + 5 = 44.9$

$P(11) = -0.1(11)^4 + 2(11)^3 - 14(11)^2 + 52(11) + 5 = 80.9$

c. week 9

$P(7) = -0.1(7)^4 + 2(7)^3 - 14(7)^2 + 52(7) + 5 = 128.9$

$P(8) = -0.1(8)^4 + 2(8)^3 - 14(8)^2 + 52(8) + 5 = 139.4$

$P(9) = -0.1(9)^4 + 2(9)^3 - 14(9)^2 + 52(9) + 5 = 140.9$

$P(10) = -0.1(10)^4 + 2(10)^3 - 14(10)^2 + 52(10) + 5 = 125$

83. $v(x) = x^3 + 11x^2 + 24x$

a. $v(3) = 198$

$$\begin{array}{r|rrrr} 3 & 1 & 11 & 24 & 0 \\ & & 3 & 42 & 198 \\ \hline & 1 & 14 & 66 & 198 \end{array}$$

Volume: 198 ft^3

b. $100 = x^3 + 11x^2 + 24x$

$0 = x^3 + 11x^2 + 24x - 100$

$$\begin{array}{r|rrrr} 2 & 1 & 11 & 24 & -100 \\ & & 2 & 26 & 100 \\ \hline & 1 & 13 & 50 & 0 \end{array}$$

Height: 2 ft

c. $y = x^3 + 11x^2 + 24x - 1000$

Use graphing calculator to find the zero: (6.8437621, 0) Depth: about 7 ft

85. $f(x) = x^3 - 3x^2 - 5x + k$

$$\begin{array}{r|rrrr} -2 & 1 & -3 & -5 & k \\ & & -2 & 10 & -10 \\ \hline & 1 & -5 & 5 & 0 \end{array}$$

$k - 10 = 0$

$k = 10$

87. $p(x) = x^3 - 3x^2 + k + 10$

$$\begin{array}{r|rrrr} 2 & 1 & -3 & k & 10 \\ & & 2 & -2 & 2k-4 \\ \hline & 1 & -1 & k-2 & 0 \end{array}$$

$2k - 4 = -10$

$2k = -6$

$k = -3$

89. $f(x) = (x - 2i)(x + 2i)(x - 3)$

$f(x) = (x^2 + 4)(x - 3)$

$f(x) = x^3 - 3x^2 + 4x - 12$

$$\underline{3}|\;1\quad -3\quad 4\quad -12$$
$$\qquad\quad 3\quad 0\quad 12$$
$$\overline{\qquad 1\quad 0\quad 4\quad\;\; 0}$$

$x^2 + 4 = 0$

$x^2 = -4$

$x = \pm 2i$

The theorems also apply to complex zeroes of polynomials.

91. (a) $1^3 + 2^3 + 3^3 = 1 + 8 + 27 = 36$

$S_3 = 36$

$$\underline{3}|\;1\quad 2\quad 1\quad 0\quad\;\; 0$$
$$\qquad\quad 3\quad 15\quad 48\quad 144$$
$$\overline{\qquad 1\quad 5\quad 16\quad 48\quad |\underline{144}}$$

$144 \div 4 = 36$

(b) $1^3 + 2^3 + 3^3 + 4^3 + 5^3$

$= 1 + 8 + 27 + 64 + 125 = 225$

$S_5 = 225$

$$\underline{5}|\;1\quad 2\quad 1\quad 0\quad\;\; 0$$
$$\qquad\quad 5\quad 35\quad 180\quad 900$$
$$\overline{\qquad 1\quad 7\quad 36\quad 180\quad |\underline{900}}$$

$900 \div 4 = 225$

93.

	D	r	t
John	1275	5	$\dfrac{1275}{5} = 255$
Rick	1025	4	$\dfrac{1025}{4} = 256.25$

John reaches the finish line in 25 secs while Rick reaches the finish line in 256.25 sec. Yes, John wins.

95. $(0, 5000), (5, 12000)$

$m = \dfrac{12000 - 5000}{5 - 0} = \dfrac{7000}{5} = 1400$

$G(t) = 1400t + 5000$ where t is the number of years since 2005.

3.3 Technology Highlight

1. They give an approximate location for each zero.

3. Outputs change sign at each zero.

3.3 Exercises

1. Coefficients

3. $a - bi$

5. b; 4 is not a factor of 6.

7. $P(x) = x^4 + 5x^2 - 36$

$P(x) = (x^2 - 4)(x^2 + 9)$

$P(x) = (x - 2)(x + 2)(x + 3i)(x - 3i)$

Zeroes: $x = 2, x = -2, x = 3i, x = -3i$

9. $Q(x) = x^4 - 16$

$Q(x) = (x^2 + 4)(x^2 - 4)$

$Q(x) = (x + 2i)(x - 2i)(x - 2)(x + 2)$

Zeroes: $x = -2, x = 2, x = 2i, x = -2i$

11. $P(x) = x^3 + x^2 - x - 1$

$P(x) = x^2(x + 1) - (x + 1)$

$P(x) = (x + 1)(x^2 - 1)$

$P(x) = (x + 1)(x + 1)(x - 1)$

Zeroes: $x = -1, x = -1, x = 1$

13. $Q(x) = x^3 - 5x^2 - 25x + 125$

$Q(x) = x^2(x - 5) - 25(x - 5)$

$Q(x) = (x - 5)(x^2 - 25)$

$Q(x) = (x - 5)(x + 5)(x - 5)$

Zeroes: $x = 5, x = -5, x = 5$

15. $p(x) = (x^2 - 10x + 25)(x^2 + 4x - 45)(x + 9)$

$p(x) = (x - 5)^2(x + 9)(x - 5)(x + 9)$

$p(x) = (x - 5)^3(x + 9)^2$

Zeroes: $x = 5$, multiplicity 3, $x = -9$, multiplicity 2

131

3.3 Exercises

17. $P(x) = (x^2 - 5x - 14)(x^2 - 49)(x + 2)$

$\quad P(x) = (x - 7)(x + 2)(x - 7)(x + 7)(x + 2)$

$\quad P(x) = (x - 7)^2 (x + 2)^2 (x + 7)$

$\quad$ Zeroes: $x = 7$, multiplicity 2,

$\quad x = -2$, multiplicity 2,

$\quad x = -7$, multiplicity 1

19. Degree 3, $x = 3, x = 2i$, $(x = -2i)$

$\quad P(x) = (x - 3)(x - 2i)(x + 2i)$

$\quad P(x) = (x - 3)(x^2 + 4)$

$\quad P(x) = x^3 - 3x^2 + 4x - 12$

21. Degree 4, $x = -1, x = 2, x = i$, $(x = -i)$

$\quad P(x) = (x + 1)(x - 2)(x - i)(x + i)$

$\quad P(x) = (x^2 - x - 2)(x^2 + 1)$

$\quad P(x) = x^4 - x^3 - 2x^2 + x^2 - x - 2$

$\quad P(x) = x^4 - x^3 - x^2 - x - 2$

23. Degree 4, $x = 3, x = 2i$, $(x = 3, x = -2i)$

$\quad P(x) = (x - 3)(x - 3)(x - 2i)(x + 2i)$

$\quad P(x) = (x^2 - 6x + 9)(x^2 + 4)$

$\quad P(x) = x^4 + 4x^2 - 6x^3 - 24x + 9x^2 + 36$

$\quad P(x) = x^4 - 6x^3 + 13x^2 - 24x + 36$

25. Degree 4, $x = -1, x = 1 + 2i$,

$\quad (x = -1, x = 1 - 2i)$

$\quad P(x) = (x + 1)(x + 1)\big(x - (1 + 2i)\big)\big(x - (1 - 2i)\big)$

$\quad P(x) = (x^2 + 2x + 1)\big((x - 1) - 2i\big)\big((x - 1) + 2i\big)$

$\quad P(x) = (x^2 + 2x + 1)\big((x - 1)^2 - 4i^2\big)$

$\quad P(x) = (x^2 + 2x + 1)\big((x - 1)^2 + 4\big)$

$\quad P(x) = (x^2 + 2x + 1)(x^2 - 2x + 1 + 4)$

$\quad P(x) = (x^2 + 2x + 1)(x^2 - 2x + 5)$

$\quad P(x) = x^4 - 2x^3 + 5x^2 + 2x^3 - 4x^2$

$\qquad\quad + 10x + x^2 - 2x + 5$

$\quad P(x) = x^4 + 2x^2 + 8x + 5$

27. Degree 4, $x = -3, x = 1 + i\sqrt{2}$,

$\quad (x = -3, x = 1 - i\sqrt{2})$

$\quad P(x) = (x + 3)(x + 3)\big(x - (1 + i\sqrt{2})\big)\big(x - (1 - i\sqrt{2})\big)$

$\quad P(x) = (x^2 + 6x + 9)\big((x - 1) - i\sqrt{2}\big)\big((x - 1) + i\sqrt{2}\big)$

$\quad P(x) = (x^2 + 6x + 9)\big((x - 1)^2 - i^2\sqrt{4}\big)$

$\quad P(x) = (x^2 + 6x + 9)\big((x - 1)^2 + 2\big)$

$\quad P(x) = (x^2 + 6x + 9)(x^2 - 2x + 1 + 2)$

$\quad P(x) = (x^2 + 6x + 9)(x^2 - 2x + 3)$

$\quad P(x) = x^4 - 2x^3 + 3x^2 + 6x^3 - 12x^2$

$\qquad\quad + 18x + 9x^2 - 18x + 27$

$\quad P(x) = x^4 + 4x^3 + 27$

29. $f(x) = x^3 + 2x^2 - 8x - 5$

$\quad$ a. $[-4, -3]$, yes

$\qquad f(-4)$

$\qquad = (-4)^3 + 2(-4)^2 - 8(-4) - 5 = -5$;

$\qquad f(-3)$

$\qquad = (-3)^3 + 2(-3)^2 - 8(-3) - 5 = 10$

$\quad$ b. $[2, 3]$, yes

$\qquad f(2) = (2)^3 + 2(2)^2 - 8(2) - 5 = -5$;

$\qquad f(3) = (3)^3 + 2(3)^2 - 8(3) - 5 = 16$

31. $h(x) = 2x^3 + 13x^2 + 3x - 36$

$\quad$ a. $[1, 2]$, yes

$\qquad h(1) = 2(1)^3 + 13(1)^2 + 3(1) - 36 = -18$

$\qquad h(2) = 2(2)^3 + 13(2)^2 + 3(2) - 36 = 38$

$\quad$ b. $[-3, -2]$, yes

$\qquad h(-3)$

$\qquad = 2(-3)^3 + 13(-3)^2 + 3(-3) - 36 = 18$

$\qquad h(-2)$

$\qquad = 2(-2)^3 + 13(-2)^2 + 3(-2) - 36 = -6$

33. $f(x) = 4x^3 - 19x - 15$

$\quad \dfrac{\{\pm 1, \pm 15, \pm 3, \pm 5\}}{\{\pm 1, \pm 4, \pm 2\}}$;

$\quad \left\{ \pm 1, \pm 15, \pm 3, \pm 5, \pm \dfrac{1}{4}, \pm \dfrac{15}{4}, \right.$

$\qquad \left. \pm \dfrac{3}{4}, \pm \dfrac{5}{4}, \pm \dfrac{1}{2}, \pm \dfrac{15}{2}, \pm \dfrac{3}{2}, \pm \dfrac{5}{2} \right\}$

35. $h(x) = 2x^3 - 5x^2 - 28x + 15$

$$\frac{\{\pm 1, \pm 15, \pm 3, \pm 5\}}{\{\pm 1, \pm 2\}};$$

$$\left\{\pm 1, \pm 15, \pm 3, \pm 5, \pm \frac{1}{2}, \pm \frac{15}{2}, \pm \frac{3}{2}, \pm \frac{5}{2}\right\}$$

37. $p(x) = 6x^4 - 2x^3 + 5x^2 - 28$

$$\frac{\{\pm 1, \pm 28, \pm 2, \pm 14, \pm 4, \pm 7\}}{\{\pm 1, \pm 6, \pm 2, \pm 3\}};$$

$$\left\{\pm 1, \pm 28, \pm 2, \pm 14, \pm 4, \pm 7, \pm \frac{1}{6}, \pm \frac{14}{3}, \pm \frac{1}{3}, \pm \frac{7}{3}, \pm \frac{2}{3}, \right.$$

$$\left. \pm \frac{7}{6}, \pm \frac{1}{2}, \pm \frac{7}{2}, \pm \frac{28}{3}, \pm \frac{4}{3} \right\}$$

39. $Y_1 = 32t^3 - 52t^2 + 17t + 3$

$$\frac{\{\pm 1, \pm 3\}}{\{\pm 1, \pm 32, \pm 2, \pm 16, \pm 4, \pm 8\}};$$

$$\left\{\pm 1, \pm \frac{1}{32}, \pm \frac{1}{2}, \pm \frac{1}{16}, \pm \frac{1}{4}, \pm \frac{1}{8}, \right.$$

$$\left. \pm 3, \pm \frac{3}{32}, \pm \frac{3}{2}, \pm \frac{3}{16}, \pm \frac{3}{4}, \pm \frac{3}{8} \right\}$$

41. $f(x) = x^3 - 13x + 12$

Possible rational zeroes:

$$\frac{\{\pm 1, \pm 12, \pm 2, \pm 6, \pm 3, \pm 4\}}{\{\pm 1\}};$$

$$\{\pm 1, \pm 12, \pm 2, \pm 6, \pm 3, \pm 4\}$$

$$\begin{array}{r|rrrr} -4 & 1 & 0 & -13 & 12 \\ & & -4 & 16 & -12 \\ \hline & 1 & -4 & 3 & \boxed{0} \end{array}$$

$f(x) = (x+4)(x^2 - 4x + 3)$
$f(x) = (x+4)(x-1)(x-3)$
$x = -4, 1, 3$

43. $h(x) = x^3 - 19x - 30$

Possible rational zeroes:

$$\frac{\{\pm 1, \pm 30, \pm 2, \pm 15, \pm 3, \pm 10, \pm 5, \pm 6\}}{\{\pm 1\}};$$

$$\{\pm 1, \pm 30, \pm 2, \pm 15, \pm 3, \pm 10, \pm 5, \pm 6\}$$

$$\begin{array}{r|rrrr} -3 & 1 & 0 & -19 & -30 \\ & & -3 & 9 & 30 \\ \hline & 1 & -3 & -10 & \boxed{0} \end{array}$$

$h(x) = (x+3)(x^2 - 3x - 10)$
$h(x) = (x+3)(x+2)(x-5)$
$x = -3, -2, 5$

45. $p(x) = x^3 - 2x^2 - 11x + 12$

Possible rational zeroes:

$$\frac{\{\pm 1, \pm 12, \pm 2, \pm 6, \pm 3, \pm 4\}}{\{\pm 1\}};$$

$$\{\pm 1, \pm 12, \pm 2, \pm 6, \pm 3, \pm 4\}$$

$$\begin{array}{r|rrrr} -3 & 1 & -2 & -11 & 12 \\ & & -3 & 15 & -12 \\ \hline & 1 & -5 & 4 & \boxed{0} \end{array}$$

$p(x) = (x+3)(x^2 - 5x + 4)$
$p(x) = (x+3)(x-1)(x-4)$
$x = -3, 1, 4$

47. $Y_1 = x^3 - 6x^2 - x + 30$

Possible rational zeroes:

$$\frac{\{\pm 1, \pm 30, \pm 2, \pm 15, \pm 3, \pm 10, \pm 5, \pm 6\}}{\{\pm 1\}};$$

$$\{\pm 1, \pm 30, \pm 2, \pm 15, \pm 3, \pm 10, \pm 5, \pm 6\}$$

$$\begin{array}{r|rrrr} -2 & 1 & -6 & -1 & 30 \\ & & -2 & 16 & -30 \\ \hline & 1 & -8 & 15 & \boxed{0} \end{array}$$

$Y_1 = (x+2)(x^2 - 8x + 15)$
$Y_1 = (x+2)(x-3)(x-5)$
$x = -2, 3, 5$

49. $Y_3 = x^4 - 15x^2 + 10x + 24$

Possible rational zeroes:

$$\frac{\{\pm 1, \pm 24, \pm 2, \pm 12, \pm 3, \pm 8, \pm 4, \pm 6\}}{\{\pm 1\}};$$

$$\{\pm 1, \pm 24, \pm 2, \pm 12, \pm 3, \pm 8, \pm 4, \pm 6\}$$

$$\begin{array}{r|rrrrr} -4 & 1 & 0 & -15 & 10 & 24 \\ & & -4 & 16 & -4 & -24 \\ \hline & 1 & -4 & 1 & 6 & \boxed{0} \end{array}$$

$$\begin{array}{r|rrrr} -1 & 1 & -4 & 1 & 6 \\ & & -1 & 5 & -6 \\ \hline & 1 & -5 & 6 & \boxed{0} \end{array}$$

$Y_3 = (x+4)(x+1)(x^2 - 5x + 6)$
$Y_3 = (x+4)(x+1)(x-2)(x-3)$
$x = -4, -1, 2, 3$

51. $f(x) = x^4 + 7x^3 - 7x^2 - 55x - 42$

Possible rational zeroes:

$$\frac{\{\pm 1, \pm 42, \pm 2, \pm 21, \pm 3, \pm 14, \pm 6, \pm 7\}}{\{\pm 1\}};$$

$$\{\pm 1, \pm 42, \pm 2, \pm 21, \pm 3, \pm 14, \pm 6, \pm 7\}$$

$$\begin{array}{r|rrrr} -7 & 1 & 7 & -7 & -55 & -42 \\ & & -7 & 0 & 49 & 42 \\ \hline & 1 & 0 & -7 & -6 & \underline{|0} \end{array}$$

$$\begin{array}{r|rrrr} -2 & 1 & 0 & -7 & -6 \\ & & -2 & 4 & 6 \\ \hline & 1 & -2 & -3 & \underline{|0} \end{array}$$

$f(x) = (x+7)(x+2)(x^2 - 2x - 3)$

$f(x) = (x+7)(x+2)(x+1)(x-3)$

$x = -7, -2, -1, 3$

53. $f(x) = 4x^3 - 7x + 3$

Possible rational zeroes: $\dfrac{\{\pm 1, \pm 3\}}{\{\pm 1, \pm 4, \pm 2\}}$

$$\begin{array}{r|rrrr} 1 & 4 & 0 & -7 & 3 \\ & & 4 & 4 & -3 \\ \hline & 4 & 4 & -3 & \underline{|0} \end{array}$$

$f(x) = (x-1)\left(4x^2 + 4x - 3\right)$

$f(x) = (x-1)(2x+3)(2x-1)$

$x = \dfrac{-3}{2}, \dfrac{1}{2}, 1$

55. $h(x) = 4x^3 + 8x^2 - 3x - 9$

Possible rational zeroes: $\dfrac{\{\pm 1, \pm 9, \pm 3\}}{\{\pm 1, \pm 4, \pm 2\}}$

$$\left\{\pm 1, \pm 9, \pm 3, \pm \frac{1}{4}, \pm \frac{9}{4}, \pm \frac{3}{4}, \pm \frac{1}{2}, \pm \frac{9}{2}, \pm \frac{3}{2}\right\}$$

$$\begin{array}{r|rrrr} 1 & 4 & 8 & -3 & -9 \\ & & 4 & 12 & 9 \\ \hline & 4 & 12 & 9 & \underline{|0} \end{array}$$

$h(x) = (x-1)\left(4x^2 + 12x + 9\right)$

$h(x) = (x-1)(2x+3)^2$

$x = \dfrac{-3}{2}, 1$

57. $Y_1 = 2x^3 - 3x^2 - 9x + 10$

Possible rational zeroes: $\dfrac{\{\pm 1, \pm 10, \pm 2, \pm 5\}}{\{\pm 1, \pm 2\}}$

$$\left\{\pm 1, \pm 10, \pm 2, \pm 5, \pm \frac{1}{2}, \pm \frac{5}{2}\right\}$$

$$\begin{array}{r|rrrr} 1 & 2 & -3 & -9 & 10 \\ & & 2 & -1 & -10 \\ \hline & 2 & -1 & -10 & \underline{|0} \end{array}$$

$Y_1 = (x-1)\left(2x^2 - x - 10\right)$

$Y_1 = (x-1)(x+2)(2x-5)$

$x = -2, 1, \dfrac{5}{2}$

59. $p(x) = 2x^4 + 3x^3 - 9x^2 - 15x - 5$

Possible rational zeroes: $\dfrac{\{\pm 1, \pm 5\}}{\{\pm 1, \pm 2\}}$

$$\left\{\pm 1, \pm 5, \pm \frac{1}{2}, \pm \frac{5}{2}\right\}$$

$$\begin{array}{r|rrrrr} -1 & 2 & 3 & -9 & -15 & -5 \\ & & -2 & -1 & 10 & 5 \\ \hline & 2 & 1 & -10 & -5 & \underline{|0} \end{array}$$

$p(x) = (x+1)\left(2x^3 + x^2 - 10x - 5\right)$

$p(x) = (x+1)\left(x^2(2x+1) - 5(2x+1)\right)$

$p(x) = (x+1)(2x+1)\left(x^2 - 5\right)$

$p(x) = (x+1)(2x+1)(x - \sqrt{5})(x + \sqrt{5})$

$x = -1, \dfrac{-1}{2}, \sqrt{5}, -\sqrt{5}$

61. $r(x) = 3x^4 - 5x^3 + 14x^2 - 20x + 8$

Possible rational zeroes: $\dfrac{\{\pm 1, \pm 8, \pm 2, \pm 4\}}{\{\pm 1, \pm 3\}}$

$\left\{ \pm 1, \pm 8, \pm 2, \pm 4, \pm\dfrac{1}{3}, \pm\dfrac{8}{3}, \pm\dfrac{2}{3}, \pm\dfrac{4}{3} \right\}$

$$\begin{array}{r|rrrrr} 1 & 3 & -5 & 14 & -20 & 8 \\ & & 3 & -2 & 12 & -8 \\ \hline & 3 & -2 & 12 & -8 & \underline{|0} \end{array}$$

$r(x) = (x-1)(3x^3 - 2x^2 + 12x - 8)$

$r(x) = (x-1)(x^2(3x-2) + 4(3x-2))$

$r(x) = (x-1)(3x-2)(x^2 + 4)$

$r(x) = (x-1)(3x-2)(x-2i)(x+2i)$

$x = 1, \dfrac{2}{3}, \pm 2i$

63. $f(x) = 2x^4 - 9x^3 + 4x^2 + 21x - 18$

Possible rational zeroes:

$\dfrac{\{\pm 1, \pm 18, \pm 2, \pm 9, \pm 3, \pm 6\}}{\{\pm 1, \pm 2\}}$

$\left\{ \pm 1, \pm 18, \pm 2, \pm 9, \pm 3, \pm 6, \pm\dfrac{1}{2}, \pm\dfrac{9}{2}, \pm\dfrac{3}{2} \right\}$

$$\begin{array}{r|rrrrr} 1 & 2 & -9 & 4 & 21 & -18 \\ & & 2 & -7 & -3 & 18 \\ \hline & 2 & -7 & -3 & 18 & \underline{|0} \end{array}$$

$$\begin{array}{r|rrrr} 2 & 2 & -7 & -3 & 18 \\ & & 4 & -6 & -18 \\ \hline & 2 & -3 & -9 & \underline{|0} \end{array}$$

$f(x) = (x-1)(x-2)(2x^2 - 3x - 9)$

$f(x) = (x-1)(x-2)(x-3)(2x+3)$

$x = 1, 2, 3, \dfrac{-3}{2}$

65. $h(x) = 3x^4 + 2x^3 - 9x^2 + 4$

Possible rational zeroes: $\dfrac{\{\pm 1, \pm 4, \pm 2\}}{\{\pm 1, \pm 3\}}$

$\left\{ \pm 1, \pm 4, \pm 2, \pm\dfrac{1}{3}, \pm\dfrac{4}{3}, \pm\dfrac{2}{3} \right\}$

$$\begin{array}{r|rrrrr} 1 & 3 & 2 & -9 & 0 & 4 \\ & & 3 & 5 & -4 & -4 \\ \hline & 3 & 5 & -4 & -4 & \underline{|0} \end{array}$$

$$\begin{array}{r|rrrr} -2 & 3 & 5 & -4 & -4 \\ & & -6 & 2 & 4 \\ \hline & 3 & -1 & -2 & \underline{|0} \end{array}$$

$h(x) = (x-1)(x+2)(3x^2 - x - 2)$

$h(x) = (x-1)(x+2)(3x+2)(x-1)$

$x = -2, 1, \dfrac{-2}{3}$

67. $P(x) = 2x^4 + 3x^3 - 24x^2 - 68x - 48$

Possible rational zeroes:

$\{\pm 1, \pm 48, \pm 2, \pm 24, \pm 3, \pm 16, \pm 4, \pm 12, \pm 6, \pm 8\}$ over $\{\pm 1, \pm 2\}$

$\left\{ \pm 1, \pm 48, \pm 2, \pm 24, \pm 3, \pm 16, \pm 4, \pm 12, \pm 6, \pm 8 \right.$

$\left. \pm\dfrac{1}{2}, \pm\dfrac{3}{2} \right\}$

$$\begin{array}{r|rrrrr} -2 & 2 & 3 & -24 & -68 & -48 \\ & & -4 & 12 & 44 & 48 \\ \hline & 2 & -1 & -22 & -24 & 0 \end{array}$$

$$\begin{array}{r|rrrr} 4 & 2 & -1 & -22 & -24 \\ & & 8 & 28 & 24 \\ \hline & 2 & 7 & 6 & 0 \end{array}$$

$2x^2 + 7x + 6 = (2x+3)(x+2)$

$P(x) = (x+2)^2 (x-4)(2x+3)$

Zeroes: $x = -2$, multiplicity 2

$x = 4$, multiplicity 1

$x = -\dfrac{3}{2}$, multiplicity 1

69. $r(x) = 3x^4 - 20x^3 + 34x^2 + 12x - 45$

Possible rational zeroes:

$$\frac{\{\pm 1, \pm 45, \pm 3, \pm 15, \pm 5, \pm 9\}}{\{\pm 1, \pm 3\}}$$

$$\left\{\pm 1, \pm 45, \pm 3, \pm 15, \pm 5, \pm 9\right.$$
$$\left.\pm \frac{1}{3}, \pm \frac{5}{3}\right\}$$

$$\begin{array}{r|rrrrr} -1 & 3 & -20 & 34 & 12 & -45 \\ & & -3 & 23 & -57 & 45 \\ \hline & 3 & -23 & 57 & -45 & 0 \end{array}$$

$$\begin{array}{r|rrrr} 3 & 3 & -23 & 57 & -45 \\ & & 9 & -42 & 45 \\ \hline & 3 & -14 & 15 & 0 \end{array}$$

$$3x^2 - 14x + 15 = (3x - 5)(x - 3)$$

$$r(x) = (x + 1)(x - 3)^2(3x - 5)$$

Zeroes: $x = 3$, multiplicity 2

$x = -1$, multiplicity 1

$x = \dfrac{5}{3}$, multiplicity 1

71. $Y_1 = x^5 + 6x^2 - 49x + 42$

Possible rational zeroes:

$$\frac{\{\pm 1, \pm 42, \pm 2, \pm 21, \pm 3, \pm 14, \pm 6, \pm 7\}}{\{\pm 1\}}$$

$$\{\pm 1, \pm 42, \pm 2, \pm 21, \pm 3, \pm 14, \pm 6, \pm 7\}$$

$$\begin{array}{r|rrrrrr} 1 & 1 & 0 & 0 & 6 & -49 & 42 \\ & & 1 & 1 & 1 & 7 & -42 \\ \hline & 1 & 1 & 1 & 7 & -42 & 0 \end{array}$$

$$\begin{array}{r|rrrrr} 2 & 1 & 1 & 1 & 7 & -42 \\ & & 2 & 6 & 14 & 42 \\ \hline & 1 & 3 & 7 & 21 & 0 \end{array}$$

$$Y_1 = (x - 1)(x - 2)(x^3 + 3x^2 + 7x + 21)$$
$$Y_1 = (x - 1)(x - 2)(x^2(x + 3) + 7(x + 3))$$
$$Y_1 = (x - 1)(x - 2)(x + 3)(x^2 + 7)$$
$$Y_1 = (x - 1)(x - 2)(x + 3)(x + \sqrt{7}\,i)(x - \sqrt{7}\,i)$$

$$x = 1, \ 2, \ -3, \ \pm\sqrt{7}\,i$$

73. $P(x) = 3x^5 + x^4 + x^3 + 7x^2 - 24x + 12$

Possible rational zeroes:

$$\frac{\{\pm 1, \pm 12, \pm 2, \pm 6, \pm 3, \pm 4\}}{\{\pm 1, \pm 3\}}$$

$$\left\{\pm 1, \pm 12, \pm 2, \pm 6, \pm 3, \pm 4, \pm \frac{1}{3}, \pm \frac{2}{3}, \pm \frac{4}{3}\right\}$$

$$\begin{array}{r|rrrrrr} 1 & 3 & 1 & 1 & 7 & -24 & 12 \\ & & 3 & 4 & 5 & 12 & -12 \\ \hline & 3 & 4 & 5 & 12 & -12 & 0 \end{array}$$

$$\begin{array}{r|rrrrr} -2 & 3 & 4 & 5 & 12 & -12 \\ & & -6 & 4 & -18 & 12 \\ \hline & 3 & -2 & 9 & -6 & 0 \end{array}$$

$$P(x) = (x - 1)(x + 2)(3x^3 - 2x^2 + 9x - 6)$$
$$P(x) = (x - 1)(x + 2)(x^2(3x - 2) + 3(3x - 2))$$
$$P(x) = (x - 1)(x + 2)(3x - 2)(x^2 + 3)$$
$$P(x) = (x - 1)(x + 2)(3x - 2)(x + \sqrt{3}\,i)(x - \sqrt{3}\,i)$$

$$x = -2, \frac{2}{3}, 1, \pm\sqrt{3}\,i$$

75. $Y_1 = x^4 - 5x^3 + 20x - 16$

Possible rational zeroes: $\dfrac{\{\pm 1, \pm 16, \pm 2, \pm 8, \pm 4\}}{\{\pm 1\}}$

$$\{\pm 1, \pm 16, \pm 2, \pm 8, \pm 4\}$$

$$\begin{array}{r|rrrrr} 1 & 1 & -5 & 0 & 20 & -16 \\ & & 1 & -4 & -4 & 16 \\ \hline & 1 & -4 & -4 & 16 & 0 \end{array}$$

$$Y_1 = (x - 1)(x^3 - 4x^2 - 4x + 16)$$
$$Y_1 = (x - 1)(x^2(x - 4) - 4(x - 4))$$
$$Y_1 = (x - 1)(x - 4)(x^2 - 4)$$
$$Y_1 = (x - 1)(x - 4)(x + 2)(x - 2)$$
$$x = 1, \ 2, \ 4, \ -2$$

77. $r(x) = x^4 + 2x^3 - 5x^2 - 4x + 6$

Possible rational zeroes: $\dfrac{\{\pm 1, \pm 6, \pm 2, \pm 3\}}{\{\pm 1\}}$

$\{\pm 1, \pm 6, \pm 2, \pm 3\}$

$$
\begin{array}{r|rrrrr}
1 & 1 & 2 & -5 & -4 & 6 \\
  &   & 1 & 3 & -2 & -6 \\
\hline
  & 1 & 3 & -2 & -6 & \underline{|\,0}
\end{array}
$$

$r(x) = (x-1)(x^3 + 3x^2 - 2x - 6)$

$r(x) = (x-1)(x^2(x+3) - 2(x+3))$

$r(x) = (x-1)(x+3)(x^2 - 2)$

$r(x) = (x+3)(x-1)(x+\sqrt{2})(x-\sqrt{2})$

$x = -3, 1, \pm\sqrt{2}$

79. $p(x) = 2x^4 - x^3 + 3x^2 - 3x - 9$

Possible rational zeroes: $\dfrac{\{\pm 1, \pm 9, \pm 3\}}{\{\pm 1, \pm 2\}}$

$\left\{\pm 1, \pm 9, \pm 3, \pm\dfrac{1}{2}, \pm\dfrac{9}{2}, \pm\dfrac{3}{2}\right\}$

$$
\begin{array}{r|rrrrr}
-1 & 2 & -1 & 3 & -3 & -9 \\
   &   & -2 & 3 & -6 & 9 \\
\hline
   & 2 & -3 & 6 & -9 & \underline{|\,0}
\end{array}
$$

$p(x) = (x+1)(2x^3 - 3x^2 + 6x - 9)$

$p(x) = (x+1)(x^2(2x-3) + 3(2x-3))$

$p(x) = (x+1)(2x-3)(x^2 + 3)$

$p(x) = (x+1)(2x-3)(x+\sqrt{3}\,i)(x-\sqrt{3}\,i)$

$x = -1, \dfrac{3}{2}, \pm\sqrt{3}\,i$

81. $f(x) = 2x^5 - 7x^4 + 13x^3 - 23x^2 + 21x - 6$

Possible rational zeroes: $\dfrac{\{\pm 1, \pm 6, \pm 2, \pm 3\}}{\{\pm 1, \pm 2\}}$

$\left\{\pm 1, \pm 6, \pm 2, \pm 3, \pm\dfrac{1}{2}, \pm\dfrac{3}{2}\right\}$

$$
\begin{array}{r|rrrrrr}
1 & 2 & -7 & 13 & -23 & 21 & -6 \\
  &   & 2 & -5 & 8 & -15 & 6 \\
\hline
  & 2 & -5 & 8 & -15 & 6 & \underline{|\,0}
\end{array}
$$

$$
\begin{array}{r|rrrrr}
2 & 2 & -5 & 8 & -15 & 6 \\
  &   & 4 & -2 & 12 & -6 \\
\hline
  & 2 & -1 & 6 & -3 & \underline{|\,0}
\end{array}
$$

$f(x) = (x-1)(x-2)(2x^3 - x^2 + 6x - 3)$

$f(x) = (x-1)(x-2)(x^2(2x-1) + 3(2x-1))$

$f(x) = (x-1)(x-2)(2x-1)(x^2 + 3)$

$f(x) = (x-1)(x-2)(2x-1)(x+\sqrt{3}\,i)(x-\sqrt{3}\,i)$

$x = \dfrac{1}{2}, 1, 2, \pm\sqrt{3}\,i$

83. $f(x) = x^4 - 2x^3 + 4x - 8$

   a. Possible rational zeroes:

   $\dfrac{\{\pm 1, \pm 8, \pm 2, \pm 4\}}{\{\pm 1\}}$

   $\{\pm 1, \pm 8, \pm 2, \pm 4\}$

   b. Zeroes of unity: none, neither 1 nor -1 is a zero

   $(1 - 2 + 4 - 8 \neq 0), (1 + 2 - 4 - 8 \neq 0)$

   c. # of positive zeroes: 3 or 1 zeroes
   # of negative zeroes: 1 root

   d. Bounds: zeroes must lie between -2 and 2.

   $$
   \begin{array}{r|rrrrr}
   -2 & 1 & -2 & 0 & 4 & -8 \\
      &   & -2 & 8 & -16 & 24 \\
   \hline
      & 1 & -4 & 8 & -12 & \underline{|\,16}
   \end{array}
   $$

   $$
   \begin{array}{r|rrrrr}
   2 & 1 & -2 & 0 & 4 & -8 \\
     &   & 2 & 0 & 0 & 8 \\
   \hline
     & 1 & 0 & 0 & 4 & \underline{|\,0}
   \end{array}
   $$

85. $h(x) = x^5 + x^4 - 3x^3 + 5x + 2$

   a. Possible rational zeroes: $\dfrac{\{\pm 1, \pm 2\}}{\{\pm 1\}}$

       $\{\pm 1, \pm 2\}$

   b. Zeroes of unity: -1 is a root
      $(1 + 1 - 3 + 5 + 2 \neq 0)$,
      $(-1 + 1 + 3 - 5 + 2 = 0)$

   c. # of positive zeroes: 2 or 0 zeroes
      # of negative zeroes: 3 or 1 zeroes

   d. Bounds: zeroes must lie between -3 and 2.

```
-3 | 1   1  -3   0   5    2
   |    -3   6  -9  27  -96
   ---------------------------
     1  -2   3  -9  32  |-94

 2 | 1   1  -3   0   5    2
   |     2   6   6  12   34
   ---------------------------
     1   3   3   6  17  | 36
```

87. $p(x) = x^5 - 3x^4 + 3x^3 - 9x^2 - 4x + 12$

   a. Possible rational zeroes:
      $\{\pm 1, \pm 12, \pm 2, \pm 6, \pm 3, \pm 4\}$

   b. Zeroes of unity: $x = 1$ and $x = -1$ are zeroes.
      $(1 - 3 + 3 - 9 - 4 + 12 = 0)$,
      $(-1 - 3 - 3 - 9 + 4 + 12 = 0)$

   c. # of positive zeroes: 4, 2, or 0 zeroes
      # of negative zeroes: 1 root

   d. Bounds: zeroes must lie between -1 and 4.

```
-1 | 1  -3   3  -9  -4   12
   |    -1   4  -7  16  -12
   ---------------------------
     1  -4   7 -16  12  | 0

 4 | 1  -3   3  -9  -4   12
   |     4   4  28  76  288
   ---------------------------
     1   1   7  19  72  |300
```

89. $r(x) = 2x^4 + 7x^2 + 11x - 20$

   a. Possible rational zeroes:

      $\dfrac{\{\pm 1, \pm 20, \pm 2, \pm 10, \pm 4, \pm 5\}}{\{\pm 1, \pm 2\}}$

      $\left\{ \pm 1, \pm 20, \pm 2, \pm 10, \pm 4, \pm 5, \pm \dfrac{1}{2}, \pm \dfrac{5}{2} \right\}$

   b. Zeroes of unity: $x = 1$ is a root.
      $(2 + 7 + 11 - 20 = 0)$,
      $(2 + 7 - 11 - 20 \neq 0)$

   c. # of positive zeroes: 1 root
      # of negative zeroes: 1 root

   d. Bounds: zeroes must lie between -2 and 1.

```
-2 | 2   0   7   11  -20
   |    -4   8  -30   38
   -----------------------
     2  -4  15  -19  |18

 1 | 2   0   7   11  -20
   |     2   2    9   20
   -----------------------
     2   2   9   20  | 0
```

91. $f(x) = 4x^3 - 16x^2 - 9x + 36$

Possible Positive zeroes	Possible Negative zeroes	Possible Complex zeroes	Total number of zeroes
2	1	0	3
0	1	2	3

Possible rational zeroes:

$\dfrac{\{\pm 1, \pm 36, \pm 2, \pm 18, \pm 3, \pm 12, \pm 4, \pm 9, \pm 6\}}{\{\pm 1, \pm 4, \pm 2\}}$;

$\left\{ \pm 1, \pm 36, \pm 2, \pm 18, \pm 3, \pm 12, \pm 4, \pm 9, \pm 6, \right.$
$\left. \pm \dfrac{1}{4}, \pm \dfrac{1}{2}, \pm \dfrac{9}{2}, \pm \dfrac{3}{4}, \pm \dfrac{9}{4}, \pm \dfrac{3}{2} \right\}$

```
4 | 4  -16  -9   36
  |     16   0  -36
  --------------------
    4    0  -9  | 0
```

$f(x) = (x - 4)(4x^2 - 9)$
$f(x) = (x - 4)(2x - 3)(2x + 3)$

$x = \dfrac{-3}{2}, \dfrac{3}{2}, 4$

93. $h(x) = 6x^3 - 73x^2 + 10x + 24$

Possible Positive zeroes	Possible Negative zeroes	Possible Complex zeroes	Total number of zeroes
2	1	0	3
0	1	2	3

Possible rational zeroes:

$$\frac{\{\pm 1, \pm 24, \pm 2, \pm 12, \pm 3, \pm 8, \pm 4, \pm 6\}}{\{\pm 1, \pm 6, \pm 2, \pm 3\}};$$

$$\left\{\pm 1, \pm 24, \pm 2, \pm 12, \pm 3, \pm 8, \pm 4, \pm 6, \pm \frac{1}{6}, \right.$$
$$\left. \pm \frac{1}{3}, \pm \frac{1}{2}, \pm \frac{4}{3}, \pm \frac{2}{3}, \pm \frac{3}{2}, \pm \frac{8}{3}\right\}$$

$$\begin{array}{r|rrrr} 12 & 6 & -73 & 10 & 24 \\ & & 72 & -12 & -24 \\ \hline & 6 & -1 & -2 & \underline{|0} \end{array}$$

$h(x) = (x-12)(6x^2 - x - 2)$

$h(x) = (x-12)(3x-2)(2x+1)$

$x = \dfrac{-1}{2}, \dfrac{2}{3}, 12$

95. $p(x) = 4x^4 + 40x^3 - 97x^2 - 10x + 24$

Possible Positive zeroes	Possible Negative zeroes	Possible Complex zeroes	Total number of zeroes
2	2	0	4
0	2	2	4
2	0	2	4
0	0	4	4

Possible rational zeroes:

$$\frac{\{\pm 1, \pm 24, \pm 2, \pm 12, \pm 3, \pm 8, \pm 4, \pm 6\}}{\{\pm 1, \pm 4, \pm 2\}}$$

$$\left\{\pm 1, \pm 24, \pm 2, \pm 12, \pm 3, \pm 8, \pm 4, \pm 6, \pm \frac{1}{4}, \pm \frac{1}{2}, \pm \frac{3}{4}, \pm \frac{3}{2}\right\}$$

$$\begin{array}{r|rrrrr} 2 & 4 & 40 & -97 & -10 & 24 \\ & & 8 & 96 & -2 & -24 \\ \hline & 4 & 48 & -1 & -12 & \underline{|0} \end{array}$$

$p(x) = (x-2)(4x^3 + 48x^2 - 1x - 12)$

$p(x) = (x-2)(4x^2(x+12) - 1(x+12))$

$p(x) = (x-2)(x+12)(4x^2 - 1)$

$p(x) = (x-2)(x+12)(2x-1)(2x+1)$

$x = -12, \dfrac{-1}{2}, \dfrac{1}{2}, 2$

97. $z = a + bi : |z| = \sqrt{a^2 + b^2}$

(a) $|3+4i| = \sqrt{(3)^2 + (4)^2} = 5$

(b) $|-5+12i| = \sqrt{(-5)^2 + (12)^2} = 13$

(c) $|1 + \sqrt{3}\, i| = \sqrt{(1)^2 + (\sqrt{3})^2} = 2$

99. $f(x) = 4x^3 - 12x^2 - 24x + 32$

Possible rational zeroes:

$$\frac{\{\pm 1, \pm 32, \pm 2, \pm 16, \pm 4, \pm 8\}}{\{\pm 1, \pm 4, \pm 2\}};$$

$$\left\{\pm 1, \pm 32, \pm 2, \pm 16, \pm 4, \pm 8, \pm \frac{1}{4}, \pm \frac{1}{2}\right\}$$

$$\begin{array}{r|rrrr} -2 & 4 & -12 & -24 & 32 \\ & & -8 & 40 & -32 \\ \hline & 4 & -20 & 16 & \underline{|0} \end{array}$$

$f(x) = (x+2)(4x^2 - 20x + 16)$

$f(x) = 4(x+2)(x^2 - 5x + 4)$

$f(x) = 4(x+2)(x-4)(x-1)$

$x = -2, 1, 4$

Yes; grapher shows maximum and minimum values occur at the zeroes of f.

101. $g(x) = 4x^3 - 18x^2 + 2x + 24$

Possible rational zeroes:

$$\frac{\{\pm 1, \pm 24, \pm 2, \pm 12, \pm 3, \pm 8, \pm 4, \pm 6\}}{\{\pm 1, \pm 4, \pm 2\}};$$

$$\left\{\pm 1, \pm 24, \pm 2, \pm 12, \pm 3, \pm 8, \pm 4, \pm 6, \pm \frac{1}{4}, \pm \frac{1}{2}, \pm \frac{3}{4}, \pm \frac{3}{2}\right\}$$

$$\begin{array}{r|rrrr} -1 & 4 & -18 & 2 & 24 \\ & & -4 & 22 & -24 \\ \hline & 4 & -22 & 24 & \underline{|0} \end{array}$$

$g(x) = (x+1)(4x^2 - 22x + 24)$

$g(x) = 2(x+1)(2x^2 - 11x + 12)$

$g(x) = 2(x+1)(2x-3)(x-4)$

$x = -1, \dfrac{3}{2}, 4$

Yes; grapher shows maximum and minimum values occur at the zeroes of g.

103. $v = x \cdot x \cdot (x-1) = x^3 - x^2$

   (a) $x^3 - x^2 = 48$

      $x^3 - x^2 - 48 = 0$

      Possible rational zeroes:

$$\frac{\{\pm 1, \pm 48, \pm 2, \pm 24, \pm 3, \pm 16, \pm 4, \pm 12, \pm 6, \pm 8\}}{\{\pm 1\}};$$

$$\{\pm 1, \pm 48, \pm 2, \pm 24, \pm 3, \pm 16, \pm 4, \pm 12, \pm 6, \pm 8\}$$

```
4|1  -1   0   -48
        4  12    48
    1   3  12     0
```

      $x = 4$

      4 cm $\times$ 4 cm $\times$ 4 cm

   (b) $x^3 - x^2 = 100$

      $x^3 - x^2 - 100 = 0$

      Possible rational zeroes:

      $\{\pm 1, \pm 100, \pm 2, \pm 50, \pm 4, \pm 25, \pm 5, \pm 20, \pm 10\}$

```
5 | 1  -1    0   -100
          5   20    100
      1   4   20    | 0
```

```
5|2  -4   0  -150
       10  30   150
    2   6  30    0
```

      $x = 5$

      5cm $\times$ 5 cm $\times$ 5 cm

105. V = LWH

    $2w(w)(w-2) = 150$

    $2w^3 - 4w^2 - 150 = 0$

    $2(w^3 - 2w^2 - 75) = 0$

    Possible rational zeroes:

    $\{\pm 1, \pm 75, \pm 3, \pm 25, \pm 5, \pm 15\}$

```
5 | 2  -4   0  -150
        10  30   150
     2   6  30   | 0
```

    $w = 5; \ 2w = 10; \ w - 2 = 3$

    length 10 in., width 5 in., height 3 in.

107. $f(x) = \frac{1}{4}x^4 - 6x^3 + 42x^2 - 72x - 64$

    $0 = \frac{1}{4}x^4 - 6x^3 + 42x^2 - 72x - 64$

    $0 = x^4 - 24x^3 + 168x^2 - 288x - 256$

    Using a grapher:

    $x = 4, 8$, between 12 and 13

    1994, 1998, 2002; 5 years

109. $f(x) = -0.4192x^4 + 18.9663x^3 - 319.9714x^2$
                   $+2384.2x - 6615.8$

    To solve $f(x) = 0$, find zeros using the graphing calculator.

    a. $x = 8.97$ m, $x = 11.29$ m,

        $x = 12.05$ m, $x = 12.94$ m

    b. Find maximum (9.70, 3.71)

        9.7 m will maximize the efficiency of the boat with rating 3.7.

111. $P(x) = 0.2x^3 - 0.24x^2 - 1.04x + 2.68$

    a.

```
-10|0.02  -0.24  -1.04    2.68
              0.2    4.4  -33.6
      0.02  -0.44   3.36  -30.92
```

    alternate signs, -10 is a lower bound

    b.

```
10|0.02  -0.24  -1.04    2.68
             0.2   -0.4  -14.4
     0.02  -0.04  -1.44  -11.72
```

    no

    c. about 14.88

Chapter 3: Polynomial and Rational Functions

113.A. a. $p(x) = x^2 + 25$

$p(x) = (x + 5i)(x - 5i)$

b. $q(x) = x^2 + 9$

$q(x) = (x + 3i)(x - 3i)$

c. $r(x) = x^2 + 7$

$r(x) = (x + i\sqrt{7})(x - i\sqrt{7})$

B. a. $x^2 - 7 = 0$

$(x + \sqrt{7})(x - \sqrt{7}) = 0$

$x + \sqrt{7} = 0$ or $x - \sqrt{7} = 0$

$x = -\sqrt{7}$ or $x = \sqrt{7}$

b. $x^2 - 12 = 0$

$(x + \sqrt{12})(x - \sqrt{12}) = 0$

$x + \sqrt{12} = 0$ or $x - \sqrt{12} = 0$

$x = -\sqrt{12}$ or $x = \sqrt{12}$

$x = -2\sqrt{3}$ or $x = 2\sqrt{3}$

c. $x^2 - 18 = 0$

$(x + \sqrt{18})(x - \sqrt{18}) = 0$

$x + \sqrt{18} = 0$ or $x - \sqrt{18} = 0$

$x = -\sqrt{18}$ or $x = \sqrt{18}$

$x = -3\sqrt{2}$ or $x = 3\sqrt{2}$

115.a. $C(z) = z^3 + (1 - 4i)z^2 + (-6 - 4i)z + 24i$

$z = 4i$;

$$\begin{array}{r|rrr} 4i & 1 & 1-4i & -6-4i & 24i \\ & & 4i & 4i & -24i \\ \hline & 1 & 1 & -6 & \boxed{0} \end{array}$$

$C(z) = (z - 4i)(z^2 + z - 6)$

$C(z) = (z - 4i)(z + 3)(z - 2)$

b. $C(x) = x^3 + (5 - 9i)x^2 + (4 - 45i)x - 36i$;

$z = 9i$;

$$\begin{array}{r|rrr} 9i & 1 & 5-9i & 4-45i & -36i \\ & & 9i & 45i & 36i \\ \hline & 1 & 5 & 4 & \boxed{0} \end{array}$$

$C(z) = (z - 9i)(z^2 + 5z + 4)$

$C(z) = (z - 9i)(z + 4)(z + 1)$

c. $C(z) = z^3 + (-2 - 3i)z^2 + (5 + 6i)z - 15i$;

$z = 3i$;

$$\begin{array}{r|rrr} 3i & 1 & -2-3i & 5+6i & -15i \\ & & 3i & -6i & 15i \\ \hline & 1 & -2 & 5 & \boxed{0} \end{array}$$

$C(z) = (z - 3i)(z^2 - 2z + 5)$;

$z = \dfrac{-(-2) \pm \sqrt{(-2)^2 - 4(1)(5)}}{2(1)}$

$= \dfrac{2 \pm \sqrt{-16}}{2} = 1 \pm 2i$;

$C(z) = (z - 3i)(z - 1 - 2i)(z - 1 + 2i)$

d. $C(z) = z^3 + (-4 - i)z^2 + (29 + 4i)z - 29i$

$z = i$;

$$\begin{array}{r|rrr} i & 1 & -4-i & 29+4i & -29i \\ & & i & -4i & 29i \\ \hline & 1 & -4 & 29 & \boxed{0} \end{array}$$

$C(z) = (z - i)(z^2 - 4z + 29)$;

$z = \dfrac{-(-4) \pm \sqrt{(-4)^2 - 4(1)(29)}}{2(1)}$

$= \dfrac{4 \pm \sqrt{-100}}{2} = 2 \pm 5i$;

$C(z) = (z - i)(z - 2 - 5i)(z - 2 + 5i)$

3.3 Exercises

e. $C(z) = z^3 + (-2-6i)z^2 + (4+12i)z - 24i$

$z = 6i$;

$6i$	1	$-2-6i$	$4+12i$	$-24i$
		$6i$	$-12i$	$24i$
	1	-2	4	$\boxed{0}$

$C(z) = (z-6i)(z^2 - 2z + 4)$;

$z = \dfrac{-(-2) \pm \sqrt{(-2)^2 - 4(1)(4)}}{2(1)}$

$= \dfrac{2 \pm \sqrt{-12}}{2} = 1 \pm \sqrt{3}\, i$;

$C(z) = (z-6i)(z-1-\sqrt{3}\, i)(z-1+\sqrt{3}\, i)$

f. $C(z) = z^3 + (-6+4i)z^2 + (11-24i)z + 44i$

$z = -4i$;

$-4i$	1	$-6+4i$	$11-24i$	$44i$
		$-4i$	$24i$	$-44i$
	1	-6	11	$\boxed{0}$

$C(z) = (z+4i)(z^2 - 6z + 11)$;

$z = \dfrac{-(-6) \pm \sqrt{(-6)^2 - 4(1)(11)}}{2(1)}$

$= \dfrac{6 \pm \sqrt{-8}}{2} = 3 \pm \sqrt{2}\, i$;

$C(z) = (z+4i)(z-3-\sqrt{2}\, i)(z-3+\sqrt{2}\, i)$

g. $C(z) = z^3 + (-2-i)z^2 + (5+4i)z + (-6+3i)$;

$z = 2-i$;

$2-i$	1	$-2-i$	$5+4i$	$-6+3i$
		$2-i$	$-2-4i$	$6-3i$
	1	$-2i$	3	$\boxed{0}$

$C(z) = (z-2+i)(z^2 - 2iz + 3)$;

$z = \dfrac{-(-2i) \pm \sqrt{(-2i)^2 - 4(1)(3)}}{2(1)}$

$= \dfrac{2i \pm \sqrt{-16}}{2} = i \pm 2i = 3i$ or $-i$;

$C(z) = (z-2+i)(z-3i)(z+i)$

h. $C(z) = z^3 - 2z^2 + (19+6i)z + (-20+30i)$;

$z = 2-3i$;

$2-3i$	1	-2	$19+6i$	$-20+30i$
		$2-3i$	$-9-6i$	$20-30i$
	1	$-3i$	10	$\boxed{0}$

$C(z) = (z-2+3i)(z^2 - 3iz + 10)$;

$z = \dfrac{-(-3i) \pm \sqrt{(-3i)^2 - 4(1)(10)}}{2(1)}$

$= \dfrac{3i \pm \sqrt{-49}}{2} = \dfrac{3i \pm 7i}{2} = 5i$ or $-2i$;

$C(z) = (z-2+3i)(z-5i)(z+2i)$

117. Let x represent the width.

Let $\dfrac{1200-4x}{2}$ represent the length

$x\left(\dfrac{1200-4x}{2}\right) = 600x - 2x^2$

a. Opens downward, x-coordinate of

vertex: $x = \dfrac{-600}{2(-2)} = 150$ ft;

$\dfrac{1200 - 4(150)}{2} = 300$ ft

b. $\dfrac{300}{3} = 100$;

$100(150) = 15{,}000$ ft^2

119. Node $(-4, -2)$

$r(x) = a\sqrt{x+4} - 2$

Passing through $(0, 2)$

$2 = a\sqrt{0+4} - 2$

$4 = 2a$

$2 = a$;

$r(x) = 2\sqrt{x+4} - 2$

3.4 Exercises

1. zero, m

3. Bounce, flatter

5. Answers will vary.

7. polynomial, degree 3

9. not a polynomial, sharp turns

11. polynomial, degree 2

13. up/down

15. down/down

17. down/up, $(0, -2)$

19. down/down, $(0, -6)$

21. up/down, $(0, 6)$

23. (a) even
    (b) -3, odd; -1, even; 3, odd
    (c) $f(x) = (x+3)(x+1)^2(x-3)$
    (d) D: $x \in \mathbb{R}$, R: $y \in [-9, \infty)$

25. (a) even
    (b) -3, odd; -1, odd; 2, odd; 4, odd
    (c) deg 4; $f(x) = -(x+3)(x+1)(x-2)(x-4)$
    (d) D: $x \in \mathbb{R}$, R: $y \in (-\infty, 25]$

27. (a) odd
    (b) -1, even; 3, odd
    (c) deg 3; $f(x) = -(x+1)^2(x-3)$
    (d) D: $x \in \mathbb{R}$, R: $y \in \mathbb{R}$

29. degree 6, up/up, $(0, -12)$

31. degree 5, up/down, $(0, -24)$

33. degree 6, up/up, $(0, -192)$

35. degree 5, up/down, $(0, 2)$

37. b

39. e

41. c

43. $f(x) = (x+3)(x+1)(x-2)$
    end behavior: down/up
    x-intercepts: $(-3,0), (-1,0),$ and $(2,0)$;
    crosses at all x-intercepts
    $f(0) = (0+3)(0+1)(0-2) = -6$
    y-intercept: $(0,-6)$

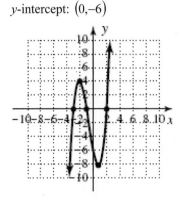

45. $p(x) = -(x+1)^2(x-3)$
    end behavior: up/down
    x-intercepts: $(-1,0)$ and $(3,0)$;
    crosses at $(3,0)$, bounces at $(-1,0)$
    $p(0) = -(0+1)^2(0-3) = 3$
    y-intercept: $(0,3)$

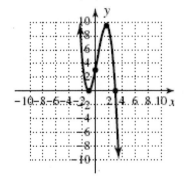

47. $Y_1 = (x+1)^2 (3x-2)(x+3)$
    end behavior: up/up
    x-intercepts: $(-1,0), \left(\dfrac{2}{3},0\right)$ and $(-3,0)$;

    crosses at $(-3,0)$ and $\left(\dfrac{2}{3},0\right)$,

    bounces at $(-1,0)$
    $Y_1 = (0+1)^2 (3(0)-2)(0+3) = -6$
    y-intercept: $(0,-6)$

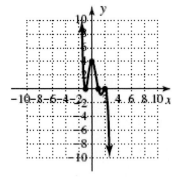

49. $r(x) = -(x+1)^2 (x-2)^2 (x-1)$
    end behavior: up/down
    x-intercepts: $(-1,0), (2,0)$ and $(1,0)$;
    crosses at $(1,0)$, bounces at $(-1,0)$ and $(2,0)$
    $r(0) = -(0+1)^2 (0-2)^2 (0-1) = 4$
    y-intercept: $(0,4)$

51. $f(x) = (2x+3)(x-1)^3$
    end behavior: up/up
    x-intercepts: $\left(-\dfrac{3}{2},0\right)$ and $(1,0)$;

    crosses at all x-intercepts
    $f(0) = (2(0)+3)(0-1)^3 = -3$
    y-intercept: $(0,-3)$

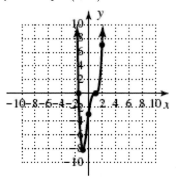

53. $h(x) = (x+1)^3 (x-3)(x-2)$
    end behavior: down/up
    x-intercepts: $(-1,0), (3,0)$ and $(2,0)$;
    crosses at all x-intercepts
    $h(0) = (0+1)^3 (0-3)(0-2) = 6$
    y-intercept: $(0,6)$

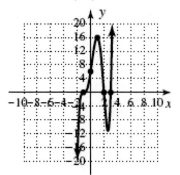

55. $Y_3 = (x+1)^3(x-1)^2(x-2)$

end behavior: up/up

x-intercepts: $(-1,0),(1,0)$ and $(2,0)$;

crosses at $(-1,0)$ and $(2,0)$, bounces at $(1,0)$;

$Y_3 = (0+1)^3(0-1)^2(0-2) = -2$

y-intercept: $(0,-2)$

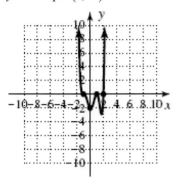

57. $y = x^3 + 3x^2 - 4$

end behavior: down/up

Possible rational roots: $\{\pm1,\pm4,\pm2\}$

$y = (x+2)^2(x-1)$

x-intercepts: $(-2,0)$ and $(1,0)$

crosses at $(1,0)$, bounces at $(-2,0)$;

$y = 0^3 + 3(0)^2 - 4 = -4$

y-intercept: $(0,-4)$

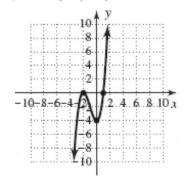

59. $f(x) = x^3 - 3x^2 - 6x + 8$

end behavior: down/up

Possible rational roots: $\{\pm1,\pm8,\pm2,\pm4\}$

$f(x) = (x+2)(x-1)(x-4)$

x-intercepts: $(-2,0),(1,0)$ and $(4,0)$

crosses at all x-intercepts;

$$f(0) = 0^3 - 3(0)^2 - 6(0) + 8 = 8$$

y-intercept: $(0,8)$

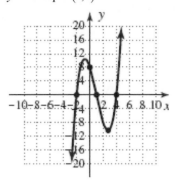

61. $h(x) = -x^3 - x^2 + 5x - 3$

end behavior: up/down

Possible rational roots: $\{\pm1,\pm3\}$

$h(x) = -1(x+3)(x-1)^2$

x-intercepts: $(-3,0)$ and $(1,0)$

crosses at $(-3,0)$, bounces at $(1,0)$;

$h(0) = -0^3 - (0)^2 + 5(0) - 3 = -3$

y-intercept: $(0,-3)$

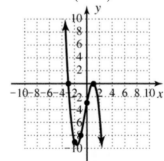

63. $p(x) = -x^4 + 10x^2 - 9$

    end behavior: down/down

    $p(x) = -1(x^2 - 9)(x^2 - 1)$

    $p(x) = -1(x+3)(x-3)(x+1)(x-1)$

    x-intercepts:

    $(-3,0), (3,0), (-1,0)$ and $(1,0)$

    crosses at all x-intercepts;

    $p(0) = -0^4 + 10(0)^2 - 9 = -9$

    y-intercept: $(0,-9)$

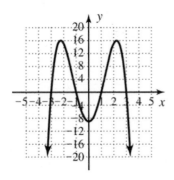

65. $r(x) = x^4 - 9x^2 - 4x + 12$

    end behavior: up/up
    Possible rational roots:
    $\{\pm1, \pm12, \pm2, \pm6, \pm3, \pm4\}$

    $r(x) = (x+2)^2(x-1)(x-3)$

    x-intercepts: $(-2,0), (1,0)$ and $(3,0)$

    crosses at $(1,0)$ and $(3,0)$, bounces at $(-2,0)$;

    $r(0) = 0^4 - 9(0)^2 - 4(0) + 12 = 12$

    y-intercept: $(0,12)$

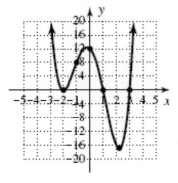

67. $Y_1 = x^4 - 6x^3 + 8x^2 + 6x - 9$

    end behavior: up/up
    Possible rational roots: $\{\pm1, \pm9, \pm3\}$

    $Y_1 = (x+1)(x-1)(x-3)^2$

    x-intercepts: $(-1,0), (1,0),$ and $(3,0)$

    crosses at $(-1,0)$ and $(1,0)$, bounces at $(3,0)$;

    $Y_1 = 0^4 - 6(0)^3 + 8(0)^2 + 6(0) - 9 = -9$

    y-intercept: $(0,-9)$

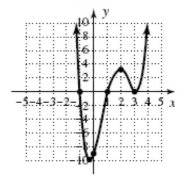

69. $Y_3 = 3x^4 + 2x^3 - 36x^2 + 24x + 32$

    end behavior: up/up
    Possible rational roots:
    $\{\pm1, \pm32, \pm2, \pm16, \pm4, \pm8\}$

    $Y_3 = (x+4)(3x+2)(x-2)^2$

    x-intercepts:

    $(-4,0), \left(-\dfrac{2}{3},0\right),$ and $(2,0)$

    crosses at $(-4,0)$ and $\left(-\dfrac{2}{3},0\right)$,

    bounces at $(2,0)$;

    $Y_3 = 3(0)^4 + 2(0)^3 - 36(0)^2 + 24(0) + 32 = 32$

    y-intercept: $(0,32)$

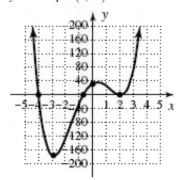

71. $F(x) = 2x^4 + 3x^3 - 9x^2$

$F(x) = x^2(2x^2 + 3x - 9)$

end behavior: up/up

Possible rational roots: $\dfrac{\{\pm 1, \pm 9, \pm 3\}}{\{\pm 1, \pm 2\}}$;

$\left\{ \pm 1, \pm 9, \pm 3, \pm \dfrac{1}{2}, \pm \dfrac{9}{2}, \pm \dfrac{3}{2} \right\}$

$F(x) = x^2(x+3)(2x-3)$

x-intercepts:

$(0,0), \ (-3,0), \ \text{and} \ \left(\dfrac{3}{2}, 0 \right)$

crosses at $(--3,0)$ and $\left(\dfrac{3}{2}, 0 \right)$ bounces at

$(0,0)$;

$F(0) = 2(0)^4 + 3(0)^3 - 9(0)^2 = 0$

y-intercept: $(0,0)$

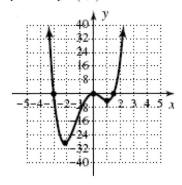

73. $f(x) = x^5 + 4x^4 - 16x^2 - 16x$

$f(x) = x(x^4 + 4x^3 - 16x - 16)$

end behavior: down/up

Possible rational roots: $\{\pm 1, \pm 16, \pm 2, \pm 8, \pm 4\}$

$f(x) = x(x+2)^3(x-2)$

x-intercepts: $(0,0), \ (-2,0), \ \text{and} \ (2,0)$

crosses at all x-intercepts;

$f(0) = (0)^5 + 4(0)^4 - 16(0)^2 - 16(0) = 0$

y-intercept: $(0,0)$

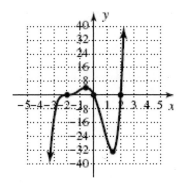

75. $h(x) = x^6 - 2x^5 - 4x^4 + 8x^3$

$h(x) = x^3(x^3 - 2x^2 - 4x + 8)$

$h(x) = x^3(x^2(x-2) - 4(x-2))$

$h(x) = x^3(x-2)(x^2 - 4)$

$h(x) = x^3(x-2)(x+2)(x-2)$

end behavior: up/up

x-intercepts: $(0,0), \ (-2,0) \ \text{and} \ (2,0)$

crosses at $(-2,0)$ and $(0,0)$, bounces at $(2,0)$;

$h(0) = (0)^6 - 2(0)^5 - 4(0)^4 + 8(0)^3 = 0$

y-intercept: $(0,0)$

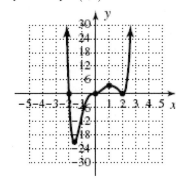

3.4 Exercises

77. $h(x) = x^5 + 4x^4 - 9x - 36$

$$h(x) = (x+4)(x-\sqrt{3})(x+\sqrt{3})(x^2+3)$$

$$h(x) = (x+4)(x-\sqrt{3})(x+\sqrt{3})(x-\sqrt{3}\,i)(x+\sqrt{3}\,i)$$

y-intercept: $(0, -36)$

79. $f(x) = 2x^5 + 5x^4 - 10x^3 - 25x^2 + 12x + 30$

$$f(x) = 2\left(x+\frac{5}{2}\right)(x-\sqrt{2})(x+\sqrt{2})(x-\sqrt{3})(x+\sqrt{3})$$

y-intercept: $(0, 30)$

81. $P(x) = a(x+4)(x-1)(x-3)$;

y-intercept; $(0,2)$

$$2 = a(0+4)(0-1)(0-3)$$

$$2 = 12a$$

$$\frac{1}{6} = a;$$

$$P(x) = \frac{1}{6}(x+4)(x-1)(x-3)$$

$$P(x) = \frac{1}{6}(x^3 - 13x + 12)$$

83. $P(x) = (x+3)(x+1)(x-2)(x-4)$

$$P(x) = (x^2 + 4x + 3)(x^2 - 6x + 8)$$

$$P(x) = x^4 - 6x^3 + 8x^2 + 4x^3 - 24x^2 + 32x$$

$$+ 3x^2 - 18x + 24$$

$$P(x) = x^4 - 2x^3 - 13x^2 + 14x + 24$$

85. $v(t) = -t^4 + 25t^3 - 192t^2 + 432t$

a. $v(2) = -(2)^4 + 25(2)^3 - 192(2)^2 + 432(2) = 280$

280 vehicles above average;

$v(6) = -(6)^4 + 25(6)^3 - 192(6)^2 + 432(6) = -216$

216 vehicles below average;

$v(11) = -(11)^4 + 25(11)^3 - 192(11)^2 + 432(11) = 154$

154 vehicles below average

b. $0 = -t^4 + 25t^3 - 192t^2 + 432t$

$$0 = -t(t^3 - 25t^2 + 192t - 432)$$

Possible rational zeroes:

$\{\pm 1, \pm 432, \pm 2, \pm 216, \pm 3, \pm 144, \pm 4, \pm 108, \pm 6, \pm 72,$

$\pm 8, \pm 54, \pm 9, \pm 48, \pm 12, \pm 36, \pm 16, \pm 27, \pm 18, \pm 24\}$

$$0 = t(t-4)(t-9)(t-12)$$

$$t = 0, t = 4, t = 9, t = 12$$

6 am, 10 am, 3 pm, 6 pm

c. $x \in [-2,13,1], y \in [-300,300,30]$

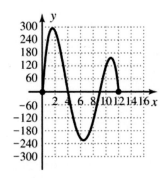

Chapter 3: Polynomial and Rational Functions

87. a. 3

b. $9 - 4 = 5$

c. $B(x) = a(x-4)(x-9)$;

y-intercept: $(1,6)$;

$6 = a(1-4)(1-9)$

$\dfrac{1}{4} = a$;

$B(x) = \dfrac{1}{4}x(x-4)(x-9)$

$B(8) = \dfrac{1}{4}(8)(8-4)(8-9) = -\$80{,}000$

89. a. $f(x) \to \infty, f(x) \to -\infty$

b. $g(x) \to \infty, g(x) \to -\infty$,

$x^4 \geq 0$ for all x .

91. $x^5 - x^4 - x^3 + x^2 - 2x + 3 = 0$

Possible rational roots: $\{\pm 1, \pm 3\}$

Testing these four roots by synthetic division shows there are no rational roots. Verified

93. $h(x) = (f \circ g)(x) = \left(\dfrac{1}{x}\right)^2 - 2\left(\dfrac{1}{x}\right)$

$= \dfrac{1}{x^2} - \dfrac{2}{x} = \dfrac{1-2x}{x^2}$;

$D : x \in \{x \mid x \neq 0\}$;

$H(x) = (g \circ f)(x) = \dfrac{1}{x^2 - 2x}$;

$D : x \in \{x \mid x \neq 0, x \neq 2\}$

95. a. $-(2x+5)-(6-x)+3 = x - 3(x+2)$

$-2x - 5 - 6 + x + 3 = x - 3x - 6$

$-x - 8 = -2x - 6$

$x = 2$

b. $\sqrt{x+1} + 3 = \sqrt{2x} + 2$

$\sqrt{x+1} = \sqrt{2x} - 1$

$\left(\sqrt{x+1}\right)^2 = \left(\sqrt{2x} - 1\right)^2$

$x + 1 = 2x - 2\sqrt{2x} + 1$

$-x = -2\sqrt{2x}$

$(-x)^2 = \left(-2\sqrt{2x}\right)^2$

$x^2 = 4(2x)$

$x^2 - 8x = 0$

$x(x-8) = 0$

$x = 0 \text{ or } x - 8 = 0$

$x = 8$

$x = 8 \ (x = 0 \text{ does not check})$

c. $\dfrac{2}{x-3} + 5 = \dfrac{21}{x^2 - 9} + 4$

$\dfrac{2}{x-3} + 5 = \dfrac{21}{(x+3)(x-3)} + 4$

$(x-3)(x+3)\left[\dfrac{2}{x-3} + 5 = \dfrac{21}{(x+3)(x-3)} + 4\right]$

$2(x+3) + 5(x^2 - 9) = 21 + 4(x^2 - 9)$

$2x + 6 + 5x^2 - 45 = 21 + 4x^2 - 36$

$2x + 5x^2 - 39 = 4x^2 - 15$

$x^2 + 2x - 24 = 0$

$(x+6)(x-4) = 0$

$x + 6 = 0 \text{ or } x - 4 = 0$

$x = -6 \qquad \text{or } x = 4$

Mid-Chapter Check

1. a. $x^3 + 8x^2 + 7x - 14 = \left(x^2 + 6x - 5\right)(x + 2) - 4$

$$
\begin{array}{r}
x^2 + 6x - 5 \\
x + 2 \overline{\smash{\big)}\ x^3 + 8x^2 + 7x - 14} \\
\underline{-\left(x^3 + 2x^2\right)} \\
6x^2 + 7x \\
\underline{-\left(6x^2 + 12x\right)} \\
-5x - 14 \\
\underline{-\left(-5x - 10\right)} \\
-4
\end{array}
$$

b. $\dfrac{x^3 + 8x^2 + 7x - 14}{x + 2} = x^2 + 6x - 5 - \dfrac{4}{x + 2}$;

3. $f(-2) = 7$

$$
\begin{array}{r|rrrrr}
-2 & -3 & 0 & 7 & -8 & 11 \\
   &   & 6 & -12 & 10 & -4 \\
\hline
   & -3 & 6 & -5 & 2 & \boxed{7}
\end{array}
$$

5. $g(2) = (2)^3 - 6(2) - 4 = -8$;

$g(3) = (3)^3 - 6(3) - 4 = 5$;

They have opposite signs.

7. $h(x) = x^4 + 3x^3 + 10x^2 + 6x - 20$

Possible Rational Roots:
$\{\pm 1, \pm 20, \pm 2, \pm 10, \pm 4, \pm 5\}$;

$h(x) = (x + 2)(x - 1)\left(x^2 + 2x + 10\right)$

$x = -2, x = 1, x = -1 \pm 3i$

9. $q(x) = x^3 + 5x^2 + 2x - 8$

end behavior: down/up

Possible rational roots: $\{\pm 1, \pm 8, \pm 2, \pm 4\}$

$q(x) = (x + 4)(x + 2)(x - 1)$

x-intercepts: $(-4, 0), (-2, 0), (1, 0)$

crosses at all x-intercepts;

$q(0) = 0^3 + 5(0)^2 + 2(0) - 8 = -8$

y-intercept: $(0, -8)$

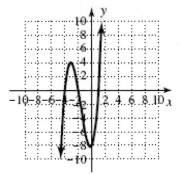

Reinforcing Basic Concepts

1. 1.532

3.5 Technology Highlight

1. $3,420,000; undefinded since cost becomes negative.

3. very closely

3.5 Exercises

1. as $x \rightarrow -\infty, y \rightarrow 2$

3. denominator, numerator

5. about $x = 98$

7. $V(x) = \dfrac{1}{(x-1)} + 2$

   a. as $x \rightarrow \infty, y \rightarrow 2$;
      as $x \rightarrow -\infty, y \rightarrow 2$;

   b. as $x \rightarrow -1^-, y \rightarrow -\infty$;
      as $x \rightarrow -1^+, y \rightarrow \infty$;

9. $Q(x) = \dfrac{1}{(x+2)^2} + 1$

   a. as $x \rightarrow \infty, y \rightarrow 1$;
      as $x \rightarrow -\infty, y \rightarrow 1$;

   b. as $x \rightarrow -2^-, y \rightarrow \infty$;
      as $x \rightarrow -2^+, y \rightarrow \infty$;

11. reciprocal quadratic,
    $S(x) = \dfrac{1}{(x+1)^2} - 2$

13. reciprocal function,
    $Q(x) = \dfrac{1}{(x+1)} - 2$

15. reciprocal quadratic,
    $f(x) = \dfrac{1}{(x+2)^2} - 5$

17. $y \rightarrow -2$

19. $y \rightarrow -\infty$

21. $x \rightarrow -1, \ y \rightarrow \pm\infty$

23. $x - 3 = 0$
    $x = 3$
    $D : x \in (-\infty, 3) \cup (3, \infty)$

25. $x^2 - 9 = 0$
    $x^2 = 9$
    $x = 3, x = -3$
    $D : x \in (-\infty, -3) \cup (-3, 3) \cup (3, \infty)$

27. $2x^2 + 3x - 5 = 0$
    $(2x + 5)(x - 1) = 0$
    $2x + 5 = 0$ or $x - 1 = 0$
    $2x = -5$ or $x = 1$
    $x = -\dfrac{5}{2}$ or $x = 1$
    $D : x \in \left(-\infty, -\dfrac{5}{2}\right) \cup \left(-\dfrac{5}{2}, 1\right) \cup (1, \infty)$

29. $x^2 + x + 1 = 0, b^2 - 4ac < 0$
    no vertical asymptotes
    $D : x \in (-\infty, \infty)$

31. $x^2 - x - 6 = 0$
    $(x - 3)(x + 2) = 0$
    $x - 3 = 0$ or $x + 2 = 0$
    $x = 3$ or $x = -2$
    yes   yes

33. $x^2 - 6x + 9 = 0$
    $(x - 3)(x - 3) = 0$
    $x - 3 = 0$
    $x = 3$
    no

35. $x^3 + 2x^2 - 4x - 8 = 0$
    $x^2(x + 2) - 4(x + 2) = 0$
    $(x + 2)(x^2 - 4) = 0$
    $(x + 2)(x + 2)(x - 2) = 0$
    $x + 2 = 0$ or $x - 2 = 0$
    $x = -2$ or $x = 2$
    no   yes

37. $Y_1 = \dfrac{2x-3}{x^2+1}$

    (a) $HA: y = 0$

    (b) $2x - 3 = 0$

    $x = \dfrac{3}{2}$

    crosses at $\left(\dfrac{3}{2}, 0\right)$

39. $r(x) = \dfrac{4x^2 - 9}{x^2 - 3x - 18}$

    (a) $HA: y = 4$

    (b) $4x^2 - 9 = 4\left(x^2 - 3x - 18\right)$

    $4x^2 - 9 = 4x^2 - 12x - 72$

    $12x = -63$

    $12x = -63$

    $x = -\dfrac{63}{12} = -\dfrac{21}{4}$

    crosses at $\left(-\dfrac{21}{4}, 4\right)$

41. $p(x) = \dfrac{3x^2 - 5}{x^2 - 1}$

    (a) $HA: y = 3$

    (b) $3x^2 - 5 = 3\left(x^2 - 1\right)$

    $3x^2 - 5 = 3x^2 - 3$

    does not cross

43. $f(x) = \dfrac{x^2 - 3x}{x^2 - 5}$

    $f(x) = \dfrac{x(x-3)}{x^2 - 5}$

    x-intercepts: (0, 0) cross, (3, 0) cross;
    y-intercept: (0, 0)

45. $g(x) = \dfrac{x^2 + 3x - 4}{x^2 - 1}$

    $g(x) = \dfrac{(x+4)(x-1)}{(x+1)(x-1)}$

    x-intercept: (−4, 0) cross;
    y-intercept: (0, 4)

47. $h(x) = \dfrac{x^3 - 6x^2 + 9x}{4 - x^2}$

    $h(x) = \dfrac{x\left(x^2 - 6x + 9\right)}{4 - x^2}$

    $h(x) = \dfrac{x(x-3)(x-3)}{(2+x)(2-x)}$

    x-intercepts: (0, 0) cross, (3, 0) bounce;
    y-intercept: (0, 0)

49. $f(x) = \dfrac{x+3}{x-1}$

    $f(0) = \dfrac{(0)+3}{(0)-1} = -3;$

    y-intercept: (0, −3);
    $x - 1 = 0$
    vertical asymptote: $x = 1$
    x-intercept: $(-3, 0)$
    horizontal asymptote: $y = 1$
    deg num = deg den

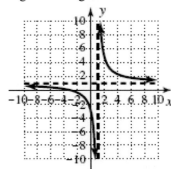

51. $F(x) = \dfrac{8x}{x^2 + 4}$

$F(0) = \dfrac{8(0)}{(0)^2 + 4} = 0;$

y-intercept: $(0,0)$;

$x^2 + 4 \neq 0$

vertical asymptote: none

x-intercept: $(0,0)$

horizontal asymptote: $y = 0$

deg num $<$ deg den

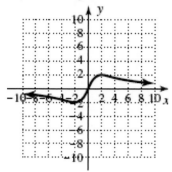

53. $p(x) = \dfrac{-2x^2}{x^2 - 4}$

$p(0) = \dfrac{-2(0)^2}{(0)^2 - 4} = 0;$

y-intercept: $(0,0)$;

$x^2 - 4 = 0$

$(x+2)(x-2) = 0$

vertical asymptote: $x = -2$ and $x = 2$

x-intercept: $(0,0)$

horizontal asymptote: $y = -2$

deg num $=$ deg den

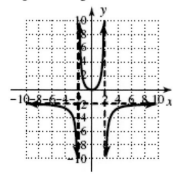

55. $q(x) = \dfrac{2x - x^2}{x^2 + 4x - 5}$

$q(0) = \dfrac{2(0) - (0)^2}{(0)^2 + 4(0) - 5} = 0;$

y-intercept: $(0,0)$;

$x^2 + 4x - 5 = 0$

$(x+5)(x-1) = 0$

vertical asymptotes: $x = -5$ and $x = 1$

$2x - x^2 = 0$

$x(2 - x) = 0$

$x = 0$ or $x = 2$

x-intercepts: $(0,0), (2,0)$

horizontal asymptote: $y = -1$

deg num $=$ deg den

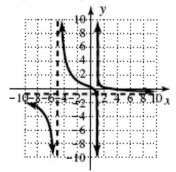

57. $h(x) = \dfrac{-3x}{x^2 - 6x + 9}$

$h(0) = \dfrac{-3(0)}{(0)^2 - 6(0) + 9} = 0;$

y-intercept: $(0,0)$;

$x^2 - 6x + 9 = 0$

$(x-3)(x-3) = 0$

vertical asymptote: $x = 3$

x-intercept: $(0,0)$

horizontal asymptote: $y = 0$

deg num $<$ deg den

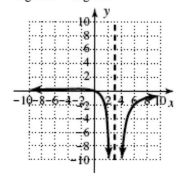

59. $Y_1 = \dfrac{x-1}{x^2 - 3x - 4}$

$Y_1 = \dfrac{(0)-1}{(0)^2 - 3(0) - 4} = \dfrac{1}{4}$;

y-intercept: $\left(0, \dfrac{1}{4}\right)$;

$x^2 - 3x - 4 = 0$
$(x-4)(x+1) = 0$

vertical asymptotes: $x = 4$ and $x = -1$
x-intercept: $(1,0)$
horizontal asymptote: $y = 0$
deg num $<$ deg den

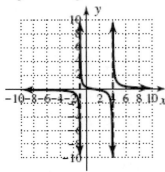

61. $s(x) = \dfrac{4x^2}{2x^2 + 4}$

$s(0) = \dfrac{4(0)^2}{2(0)^2 + 4} = 0$;

y-intercept: $(0,0)$;

$2x^2 + 4 \neq 0$
vertical asymptotes: none
x-intercept: $(0,0)$
horizontal asymptote: $y = 2$
deg num $=$ deg den

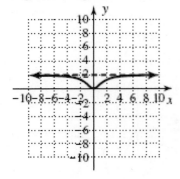

63. $Y_1 = \dfrac{x^2 - 4}{x^2 - 1}$

$Y_1 = \dfrac{(0)^2 - 4}{(0)^2 - 1} = 4$;

y-intercept: $(0,4)$;

$Y_1 = \dfrac{(x+2)(x-2)}{(x+1)(x-1)}$

vertical asymptotes: $x = -1$ and $x = 1$
x-intercepts: $(-2,0)$ and $(2,0)$
horizontal asymptote: $y = 1$
deg num $=$ deg den

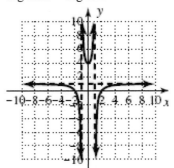

65. $v(t) = \dfrac{-2x}{x^3 + 2x^2 - 4x - 8}$

$v(0) = \dfrac{-2(0)}{(0)^3 + 2(0)^2 - 4(0) - 8} = 0$

y-intercept: $(0,0)$;

$v(t) = \dfrac{-2x}{x^2(x+2) - 4(x+2)}$

$v(t) = \dfrac{-2x}{(x+2)(x^2 - 4)} = \dfrac{-2x}{(x+2)^2(x-2)}$

vertical asymptotes: $x = -2$ and $x = 2$
x-intercepts: $(0,0)$
horizontal asymptote: $y = 0$
deg num $<$ deg den

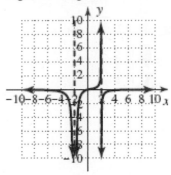

67. VA: $x = -2, x = 3$

HA: $y = 1$

$$f(x) = \frac{(x-4)(x+1)}{(x+2)(x-3)}$$

69. VA: $x = -3, x = 3$

HA: $y = -1$

$$f(x) = \frac{x^2 - 4}{9 - x^2}$$

71. $D(x) = \dfrac{63x}{x^2 + 20}$

a. Population density approaches zero far from town.

b. 10 miles, 20 miles

c. 4.5 miles, 704

73. $C(p) = \dfrac{80p}{100 - p}$

a. $C(20) = \dfrac{80(20)}{100 - 20} = 20$; $20,000

$C(50) = \dfrac{80(50)}{100 - 50} = 80$; $80,000

$C(80) = \dfrac{80(80)}{100 - 80} = 320$; $320,000

Cost increases dramatically

b.

c. As $p \to 100^-$, $C \to \infty$

75. $C(h) = \dfrac{2h^2 + h}{h^3 + 70}$

a. According to the graph, 5 hours; about 0.28

b. $\dfrac{\Delta C}{\Delta h} = \dfrac{C(10) - C(8)}{10 - 8} = \dfrac{0.196 - 0.234}{2}$

$\dfrac{\Delta C}{\Delta h} = -0.019$;

$\dfrac{\Delta C}{\Delta h} = \dfrac{C(22) - C(20)}{22 - 20} = \dfrac{0.0924 - 0.102}{2}$

$\dfrac{\Delta C}{\Delta h} = -0.005$

As number of hours increases, the rate of change decreases.

c. Horizontal asymptote:

As $h \to \infty$, $C \to 0^+$

77. $W(t) = \dfrac{6t + 40}{t}$

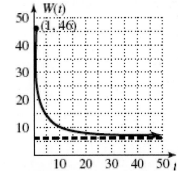

a. 2; 10

b. 10; 20

c. On the average, the number of words remembered for life is 6.

79. a. $C(x) = \dfrac{40 + 3x}{160 + 4x}$

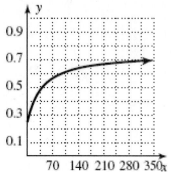

b. 35%, 62.5%, 160 gallons

c. 160 gallons; 200 gallons

d. 70%, 75%

81. $A(x) = \dfrac{125x + 50000}{x}$; [0,5000]

a. $C(500) = \$225$;

$C(1000) = \$175$

b. $150 = \dfrac{125x + 50000}{x}$

$150x = 125x + 50000$

$25x = 50000$

$x = 2000$ heaters

c. $137.50 = \dfrac{125x + 50000}{x}$

$137.50x = 125x + 50000$

$12.50x = 50000$

$x = 4000$ heaters

d. The horizontal asymptote at $y = 125$ means the average cost approaches $125 as monthly production gets very large. Due to the limitations on production (maximum of 5000 heaters) the average cost will never fall below A(5000)=135.

83. $G(n) = \dfrac{336 + n(95)}{4 + n}$

a. $90 = \dfrac{336 + n(95)}{4 + n}$

$90(4 + n) = 336 + n(95)$

$360 + 90n = 336 + 95n$

$24 = 5n$

$\dfrac{24}{5} = n$

5 tests

b. $93 = \dfrac{336 + n(95)}{4 + n}$

$93(4 + n) = 336 + n(95)$

$372 + 93n = 336 + 95n$

$36 = 2n$

$18 = n$

c. HA: $y = 95$

$95 = \dfrac{336 + n(95)}{4 + n}$

$95(4 + n) = 336 + n(95)$

$380 + 95n = 336 + 95n$

$380 \neq 336$

The horizontal asymptote at $y = 95$ means her average grade will approach 95 as the number of tests taken increases; no.

d. $93 = \dfrac{336 + n(100)}{4 + n}$

$93(4 + n) = 336 + n(100)$

$372 + 93n = 336 + 100n$

$36 = 7n$

$n \approx 6$

85. a.
$$\frac{\Delta C}{\Delta x} = \frac{\frac{250(61)}{100-61} - \frac{250(60)}{100-60}}{61-60} = 16.0;$$

$$\frac{\Delta C}{\Delta x} = \frac{\frac{250(71)}{100-71} - \frac{250(70)}{100-70}}{71-70} = 28.7;$$

$$\frac{\Delta C}{\Delta x} = \frac{\frac{250(81)}{100-81} - \frac{250(80)}{100-80}}{81-80} = 65.8;$$

$$\frac{\Delta C}{\Delta x} = \frac{\frac{250(91)}{100-91} - \frac{250(90)}{100-90}}{91-90} = 277.8;$$

b. 12.7, 37.1, 212.0

c.
$$\frac{\Delta C}{\Delta x} = \frac{\frac{350(61)}{100-61} - \frac{350(60)}{100-60}}{61-60} = 22.4;$$

$$\frac{\Delta C}{\Delta x} = \frac{\frac{350(71)}{100-71} - \frac{350(70)}{100-70}}{71-70} = 40.2;$$

$$\frac{\Delta C}{\Delta x} = \frac{\frac{350(81)}{100-81} - \frac{350(80)}{100-80}}{81-80} = 92.1;$$

$$\frac{\Delta C}{\Delta x} = \frac{\frac{350(91)}{100-91} - \frac{350(90)}{100-90}}{91-90} = 388.9;$$

17.8, 51.9, 296.8
Answers will vary.

87. a. $V(x) = \dfrac{3x^2 - 16x - 20}{x^2 - 3x - 10}$

$$\begin{array}{r} 3 \\ x^2 - 3x - 10 \overline{)\,3x^2 - 16x - 20} \\ \underline{3x^2 - 9x - 30} \\ -7x + 10 \end{array}$$

$q(x) = 3$, horizontal asymptote at $y = 3$.
$r(x) = -7x + 10$, graph crosses HA at $x = \dfrac{10}{7}$

b. $v(x) = \dfrac{-2x^2 + 4x + 13}{x^2 - 2x - 3}$

$$\begin{array}{r} -2 \\ x^2 - 2x - 3 \overline{)\,-2x^2 + 4x + 13} \\ \underline{-2x^2 + 4x + 6} \\ 7 \end{array}$$

$q(x) = -2$, horizontal asymptote at $y = -2$.
$r(x) = 7$, no zeroes-graph will not cross.

89. $3x - 4y = 12$
$-4y = -3x + 12$
$y = \dfrac{3}{4}x - 3$; slope is $\dfrac{3}{4}$;

Slope of perpendicular is $-\dfrac{4}{3}$;

$y - (-3) = -\dfrac{4}{3}(x - 2)$

$y + 3 = -\dfrac{4}{3}x + \dfrac{8}{3}$

$y = -\dfrac{4}{3}x - \dfrac{1}{3}$

91. $f(4) = 39$;

$$\begin{array}{r|rrrr} 4 & 2 & -7 & 5 & 3 \\ & & 8 & 4 & 36 \\ \hline & 2 & 1 & 9 & \underline{|39} \end{array}$$

$f\left(\dfrac{3}{2}\right) = \dfrac{3}{2}$;

$$\begin{array}{r|rrrr} \frac{3}{2} & 2 & -7 & 5 & 3 \\ & & 3 & -6 & -\frac{3}{2} \\ \hline & 2 & -4 & -1 & \underline{|\frac{3}{2}} \end{array}$$

$f(2) = 1$

$$\begin{array}{r|rrrr} 2 & 2 & -7 & 5 & 3 \\ & & 4 & -6 & -2 \\ \hline & 2 & -3 & -1 & \underline{|1} \end{array}$$

3.6 Technology Highlight

1. $(-2, -4)$

3. $(-1, 3)$

3.6 Exercises

1. Nonremovable

3. Two

5. Answers will vary.

7. $f(x) = \dfrac{(x+2)(x-2)}{x+2}$;

$x + 2 \neq 0$
$x \neq -2$;

$f(x) = \begin{cases} \dfrac{x^2 - 4}{x+2} ; & x \neq -2 \\ -4 & ; x = -2 \end{cases}$

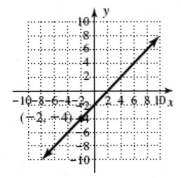

9. $g(x) = \dfrac{(x-3)(x+1)}{x+1}$;

$x + 1 \neq 0$
$x \neq -1$;

$g(x) = \begin{cases} \dfrac{x^2 - 2x - 3}{x+1} ; & x \neq -1 \\ -4 & ; x = -1 \end{cases}$

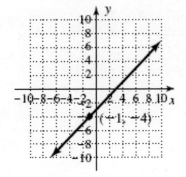

11. $h(x) = \dfrac{x(3 - 2x)}{2x - 3} = \dfrac{-x(2x - 3)}{2x - 3}$

$2x - 3 \neq 0$
$x \neq \dfrac{3}{2}$;

$h(x) = \begin{cases} \dfrac{3x - 2x^2}{2x - 3} ; & x \neq \dfrac{3}{2} \\ -\dfrac{3}{2} & ; x = \dfrac{3}{2} \end{cases}$

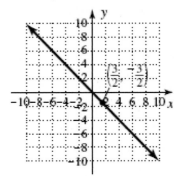

13. $p(x) = \dfrac{(x-2)(x^2 + 2x + 4)}{x - 2}$;

$x - 2 \neq 0$
$x \neq 2$;

$p(x) = \begin{cases} \dfrac{x^3 - 8}{x - 2} ; & x \neq 2 \\ 12 & ; x = 2 \end{cases}$

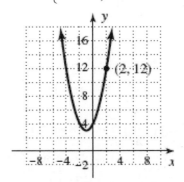

15. $q(x) = \dfrac{x^3 - 7x - 6}{x + 1}$;

$x + 1 \neq 0$

$x \neq -1$;

$$q(x) = \begin{cases} \dfrac{x^3 - 7x - 6}{x + 1} & x \neq -1 \\ \\ -4 & x = -1 \end{cases}$$

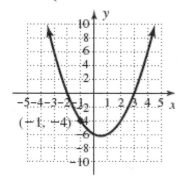

17. $r(x) = \dfrac{x^2(x+3) - (x+3)}{(x+3)(x-1)} = \dfrac{(x+3)(x^2-1)}{(x+3)(x-1)}$

$\qquad = \dfrac{(x+3)(x+1)(x-1)}{(x+3)(x-1)}$

$$r(x) = \begin{cases} \dfrac{x^3 + 3x^2 - x - 3}{x^2 + 2x - 3} & ; \quad x \neq -3, x \neq 1 \\ \\ -2 & ; \quad x = -3 \\ \\ 2 & ; \quad x = 1 \end{cases}$$

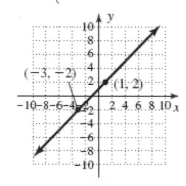

19. $Y_1 = \dfrac{x^2 - 4}{x}$

$x^2 - 4 = 0$

$x^2 = 4$

$x = \pm 2$

x-intercepts: $(-2, 0)$ and $(2, 0)$;

y-intercept: none;

$Y_1 = \dfrac{x^2}{x} - \dfrac{4}{x} = x - \dfrac{4}{x}$

$q(x) = x$

Oblique Asymptote: $y = x$

Vertical Asymptote: $x = 0$

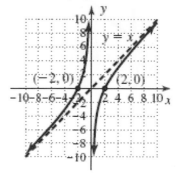

21. $v(x) = \dfrac{3 - x^2}{x}$

$3 - x^2 = 0$

$3 = x^2$

$\pm\sqrt{3} = x$

x-intercepts: $\left(-\sqrt{3}, 0\right)$ and $\left(\sqrt{3}, 0\right)$

y-intercept: none;

$v(x) = \dfrac{3}{x} - \dfrac{x^2}{x} = \dfrac{3}{x} - x$

$q(x) = -x$

Oblique Asymptote: $y = -x$

Vertical Asymptote: $x = 0$

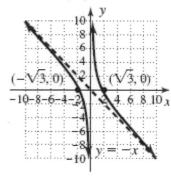

3.6 Exercises

23. $w(x) = \dfrac{x^2 + 1}{x}$

$x^2 + 1 \neq 0$ (complex solutions)

x-intercepts: none

y-intercept: none;

$w(x) = \dfrac{x^2}{x} + \dfrac{1}{x} = x + \dfrac{1}{x}$

$q(x) = x$

Oblique Asymptote: $y = x$

Vertical Asymptote: $x = 0$

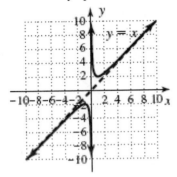

25. $h(x) = \dfrac{x^3 - 2x^2 + 3}{x^2}$

$x^3 - 2x^2 + 3 = 0$

Possible rational roots: $\dfrac{\{\pm 1, \pm 3\}}{\{\pm 1\}}$;

$\pm 1, \pm 3$

x-intercept: $(-1, 0)$

y-intercept: none;

$h(x) = \dfrac{x^3}{x^2} - \dfrac{2x^2}{x^2} + \dfrac{3}{x^2} = x - 2 + \dfrac{3}{x^2}$

$q(x) = x - 1$

Oblique Asymptote: $y = x - 2$

Vertical Asymptote: $x = 0$

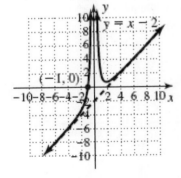

27. $Y_1 = \dfrac{x^3 + 3x^2 - 4}{x^2}$

$x^3 + 3x^2 - 4 = 0$

Possible rational roots: $\dfrac{\{\pm 1, \pm 4, \pm 2\}}{\{\pm 1\}}$;

$\pm 1, \pm 4, \pm 2$

x-intercepts: $(1, 0)$; $(-2, 0)$

y-intercept: none;

$Y_1 = \dfrac{x^3}{x^2} + \dfrac{3x^2}{x^2} - \dfrac{4}{x^2} = x + 3 - \dfrac{4}{x^2}$

$q(x) = x + 3$

Oblique Asymptote: $y = x + 3$

Vertical Asymptote: $x = 0$

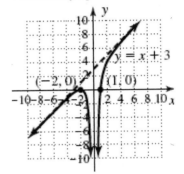

29. $f(x) = \dfrac{x^3 - 3x + 2}{x^2}$

$x^3 - 3x + 2 = 0$

Possible rational roots: $\dfrac{\{\pm 1, \pm 2\}}{\{\pm 1\}}$;

$\pm 1, \pm 2$

x-intercepts: $(-2, 0)$ and $(1, 0)$

y-intercept: none;

$f(x) = \dfrac{x^3}{x^2} - \dfrac{3x}{x^2} + \dfrac{2}{x^2} = x - \dfrac{3}{x} + \dfrac{2}{x^2}$

$q(x) = x$

Oblique Asymptote: $y = x$

Vertical Asymptote: $x = 0$

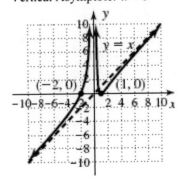

31. $Y_3 = \dfrac{x^3 - 5x^2 + 4}{x^2}$

$x^3 - 5x^2 + 4 = 0$

Possible rational roots: $\dfrac{\{\pm 1, \pm 4, \pm 2\}}{\{\pm 1\}}$;

$\pm 1, \pm 4, \pm 2$

$Y_3 = (x-1)(x^2 - 4x - 4)$;

$x = \dfrac{-(-4) \pm \sqrt{(-4)^2 - 4(1)(-4)}}{2(1)}$

$x = \dfrac{4 \pm \sqrt{32}}{2}$

$x = \dfrac{4 \pm 4\sqrt{2}}{2}$

$x = 2 \pm 2\sqrt{2}$

x-intercepts: $(1,0)$, $(2+\sqrt{2},0)$ and $(2-\sqrt{2},0)$

y-intercept: none;

$Y_3 = \dfrac{x^3}{x^2} - \dfrac{5x^2}{x^2} + \dfrac{4}{x^2} = x - 5 + \dfrac{4}{x^2}$

$q(x) = x - 5$

Oblique Asymptote: $y = x - 5$

Vertical Asymptote: $x = 0$

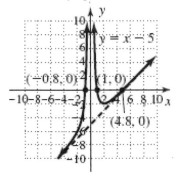

33. $r(x) = \dfrac{x^3 - x^2 - 4x + 4}{x^2}$

$x^3 - x^2 - 4x + 4 = 0$

$x^2(x-1) - 4(x-1) = 0$

$(x-1)(x^2 - 4) = 0$

$(x-1)(x+2)(x-2) = 0$

$x - 1 = 0$ or $x + 2 = 0$ or $x - 2 = 0$

$x = 1$ or $x = -2$ or $x = 2$

x-intercepts: $(-2,0)$ and $(1,0)$ and $(2,0)$

y-intercept: none;

$r(x) = \dfrac{x^3}{x^2} - \dfrac{x^2}{x^2} - \dfrac{4x}{x^2} + \dfrac{4}{x^2} = x - 1 - \dfrac{4}{x} + \dfrac{4}{x^2}$

$q(x) = x - 1$

Oblique Asymptote: $y = x - 1$

Vertical Asymptote: $x = 0$

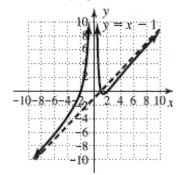

35. $g(x) = \dfrac{x^2 + 4x + 4}{x + 3}$

$x^2 + 4x + 4 = 0$

$(x+2)(x+2) = 0$

$x + 2 = 0$

$x = -2$

x-intercept: $(-2,0)$

y-intercept: none;

$$
\begin{array}{r}
x+1 \\
x+3\overline{\smash{\big)}\,x^2+4x+4} \\
\underline{-(x^2+3x)} \\
x+4 \\
\underline{-(x+3)} \\
1
\end{array}
$$

Oblique Asymptote: $y = x + 1$

$x + 3 = 0$

Vertical Asymptote: $x = -3$

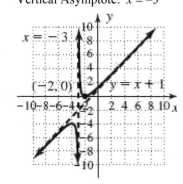

3.6 Exercises

37. $f(x) = \dfrac{x^2+1}{x+1}$

$x^2+1 \neq 0$ (complex solutions)

x-intercepts: none
y-intercept: $(0,1)$;

$$
\begin{array}{r}
x-1 \\
x+1{\overline{\smash{\big)}\,x^2+0x+1}} \\
-\underline{(x^2+x)} \\
-x+1 \\
-\underline{(-x-1)} \\
2
\end{array}
$$

Oblique Asymptote: $y = x-1$

$x+1 = 0$
Vertical Asymptote: $x = -1$

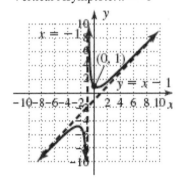

39. $Y_3 = \dfrac{x^2-4}{x+1}$

$x^2-4 = 0$

$x^2 = 4$

$x = \pm 2$

x-intercepts: $(-2,0)$ and $(2,0)$;
y-intercept: $(0,-4)$;

$$
\begin{array}{r}
x-1 \\
x+1{\overline{\smash{\big)}\,x^2+0x-4}} \\
-\underline{(x^2+x)} \\
-x-4 \\
-\underline{(-x-1)} \\
-3
\end{array}
$$

Oblique Asymptote: $y = x-1$

$x+1 = 0$
Vertical Asymptote: $x = -1$

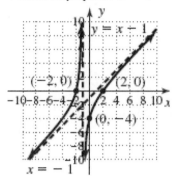

41. $v(x) = \dfrac{x^3-4x}{x^2-1}$

$x(x^2-4) = 0$

$x(x+2)(x-2) = 0$

$x = 0$ or $x+2 = 0$ or $x-2 = 0$

$x = 0$ or $x = -2$ or $x = 2$

x-intercepts: $(-2,0), (2,0)$ and $(0,0)$

y-intercept: $(0,0)$;

$$
\begin{array}{r}
x \\
x^2-1{\overline{\smash{\big)}\,x^3-4x}} \\
-\underline{(x^3-x)} \\
-3x
\end{array}
$$

Oblique Asymptote: $y = x$

$x^2-1 = 0$

$(x+1)(x-1) = 0$

$x+1 = 0$ or $x-1 = 0$

$x = -1$ or $x = 1$

Vertical Asymptote: $x = -1$ or $x = 1$

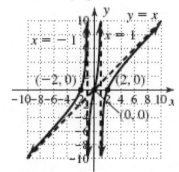

43. $w(x) = \dfrac{16x - x^3}{x^2 + 4}$

$x(16 - x^2) = 0$

$x(4 + x)(4 - x) = 0$

$x = 0$ or $4 + x = 0$ or $4 - x = 0$

$x = 0$ or $x = -4$ or $x = 4$

x-intercepts: $(-4, 0), (4, 0)$ and $(0, 0)$

y-intercept: $(0, 0)$;

$$
\begin{array}{r}
-x \\
x^2 + 4 \overline{) -x^3 + 16x} \\
-(-x^3 - 4x) \\
\hline
20x
\end{array}
$$

Oblique Asymptote: $y = -x$

$x^2 + 4 \neq 0$ (complex solutions)

Vertical Asymptote: none

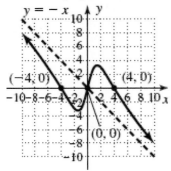

45. $W(x) = \dfrac{x^3 - 3x + 2}{x^2 - 9}$

$x^3 - 3x + 2 = 0$

Possible rational roots: $\dfrac{\pm 1, \pm 2}{\pm 1}$;

$\pm 1, \pm 2$

x-intercept: $(-2, 0)$ and $(1, 0)$

y-intercept: $\left(0, -\dfrac{2}{9}\right)$;

$$
\begin{array}{r}
x \\
x^2 - 9 \overline{) x^3 - 3x + 2} \\
-(x^3 - 9x) \\
\hline
6x + 2
\end{array}
$$

Oblique Asymptote: $y = x$

$x^2 - 9 = 0$

$(x + 3)(x - 3) = 0$

$x + 3 = 0$ or $x - 3 = 0$

$x = -3$ or $x = 3$

Vertical Asymptote: $x = -3$ or $x = 3$

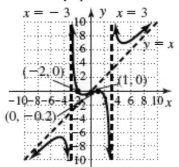

47. $p(x) = \dfrac{x^4 + 4}{x^2 + 1}$

$x^4 + 4 = 0$

Possible rational roots: $\dfrac{\pm 1, \pm 4, \pm 2}{\pm 1}$;

$\pm 1, \pm 4, \pm 2$

$x^4 + 4 \neq 0$ (complex solutions)

x-intercept: none

y-intercept: $(0, 4)$;

$$
\begin{array}{r}
x^2 - 1 \\
x^2 + 1 \overline{) x^4 + 0x^2 + 4} \\
-(x^4 + x^2) \\
\hline
-x^2 + 4 \\
-(-x^2 - 1) \\
\hline
5
\end{array}
$$

Oblique Asymptote: $y = x^2 - 1$

$x^2 + 1 \neq 0$ (complex solutions)

Vertical Asymptote: none

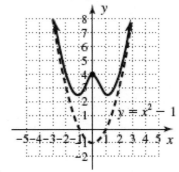

3.6 Exercises

49. $q(x) = \dfrac{10 + 9x^2 - x^4}{x^2 + 5}$

$10 + 9x^2 - x^4 = 0$

$\left(10 - x^2\right)\left(1 + x^2\right) = 0$

$10 - x^2 = 0$ or $1 + x^2 \neq 0$

$10 = x^2$

$\pm\sqrt{10} = x$

x-intercepts: $\left(-\sqrt{10}, 0\right)$ and $\left(\sqrt{10}, 0\right)$

y-intercept: $(0, 2)$;

$$\begin{array}{r} -x^2 + 14 \\ x^2 + 5 \overline{) -x^4 + 9x^2 + 10} \\ -\left(-x^4 - 5x^2\right) \\ \hline 14x^2 + 10 \\ -\left(14x^2 + 70\right) \\ \hline -60 \end{array}$$

Oblique Asymptote: $y = -x^2 + 14$

$x^2 + 5 \neq 0$ (complex solutions)

Vertical Asymptote: none

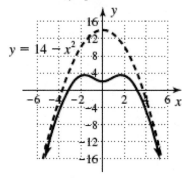

$y = 14 - x^2$

51. $f(x) = \dfrac{x^3}{x} + \dfrac{500}{x} = x^2 + \dfrac{500}{x}$

Oblique Asymptote: $y = x^2$

Minimum: 119.1

53. $A(a) = \dfrac{1}{2}\left(\dfrac{ka^2}{a - h}\right)$

$A(a) = \dfrac{1}{2}\left(\dfrac{6a^2}{a - 5}\right)$

$A(a) = \dfrac{3a^2}{a - 5}$

a. Oblique Asymptote: $y = 3a + 15$

$a - 5 = 0$

Vertical Asymptote: $a = 5$

b. $A(11) = \dfrac{3(11)^2}{11 - 5} = 60.5$

c. $(10, 0)$

55. a. $A(x) = \dfrac{4x^2 + 53x + 250}{x}$

Vertical Asymptote: $x = 0$

Oblique Asymptote: $q(x) = 4x + 53$

b. Cost: \$307, \$372, \$445

Avg Cost: \$307, \$186, \$148.33

c. 8, \$116.25

d. verified

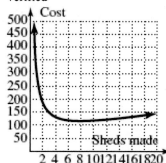

57. a. $S(x, y) = 2x^2 + 4xy$;

$V(x, y) = x^2 y$

b. $12 = x^2 y$

$\dfrac{12}{x^2} = y$;

$S(x) = 2x^2 + 4x\left(\dfrac{12}{x^2}\right)$

$= 2x^2 + \dfrac{48}{x} = \dfrac{2x^3 + 48}{x}$

c. $S(x)$ is asymptotic to $y = 2x^2$

d. $x = 2$ ft 3.5 in;

$y = 2$ ft 3.5 in

Chapter 3: Polynomial and Rational Functions

59. a. $A(x, y) = xy$;
    $R(x, y) = (x - 2.5)(y - 2)$
    b. $60 = (x - 2.5)(y - 2)$

    $\dfrac{60}{x - 2.5} = y - 2$

    $\dfrac{60}{x - 2.5} + 2 = y$

    $\dfrac{2x + 55}{x - 2.5} = y$;

    $A(x) = x\left(\dfrac{60}{x - 2.5} + 2\right)$

    $A(x) = \dfrac{60x}{x - 2.5} + 2x$

    $A(x) = \dfrac{60x}{x - 2.5} + 2x\left(\dfrac{x - 2.5}{x - 2.5}\right)$

    $A(x) = \dfrac{60x + 2x^2 - 5x}{x - 2.5} = \dfrac{2x^2 + 55x}{x - 2.5}$

    c. $A(x)$ is asymptotic to $y = 2x + 60$
    d. $x \approx 11.16$ in.;
       $y \approx 8.93$ in.

61. a. $V = \pi r^2 h$;

    $\dfrac{V}{\pi r^2} = h$

    b. $S = 2\pi r^2 + 2\pi r\left(\dfrac{V}{\pi r^2}\right) = 2\pi r^2 + \dfrac{2V}{r}$

    c. $S = 2\pi r^2 + \dfrac{2V}{r} = \dfrac{2\pi r^3 + 2V}{r}$

    d. $\dfrac{1200}{\pi r^2} = h$

    $r \approx 5.76$ cm, $h \approx 11.51$ cm;
    $S \approx 625.13\ \text{cm}^3$

63. Answers will vary.

65. $S = \dfrac{\pi r^3 + 2V}{r}$;

    $S = \dfrac{\pi r^3 + 180}{r}$;

    $90 = \dfrac{\pi r^3 + 180}{r}$

    $90r = \pi r^3 + 180$

    $0 = \pi r^3 - 90r + 180$

    Using grapher, $r \approx 3.1$ in., $h \approx 3.0$ in.

67. $-3x + 4y = -16$
    $4y = 3x - 16$

    $y = \dfrac{3}{4}x - 4$; $m = \dfrac{3}{4}$; $(0, -4)$

69. a. $\left(\overline{AB}\right)^2 = 12^2 + 5^2$

    $\overline{AB} = \sqrt{169} = 13$;
    Perimeter $= 12 + 5 + 13 = 30$ cm

    b. $\left(\overline{CB}\right)^2 = \overline{AB} \cdot \overline{DB}$

    $5^2 = 13 \cdot \overline{DB}$

    $\dfrac{25}{13} = \overline{DB}$;

    $\left(\overline{CD}\right)^2 + \left(\overline{DB}\right)^2 = 5^2$

    $\left(\overline{CD}\right)^2 + \left(\dfrac{25}{13}\right)^2 = 5^2$

    $\left(\overline{CD}\right)^2 = \dfrac{3600}{169}$

    $\overline{CD} = \dfrac{60}{13}$ cm

    c. $A = \dfrac{1}{2}(13)\left(\dfrac{60}{13}\right) = 30\ \text{cm}^2$

    d. $A_{BDC} = \dfrac{1}{2}\left(\dfrac{60}{13}\right)\left(\dfrac{25}{13}\right) = \dfrac{750}{169} \approx 4.4\ \text{cm}^2$;

    $A_{ADC} = 30 - 4.4 = \dfrac{4320}{169} \approx 25.6\ \text{cm}^2$

3.7 Technology Highlight

1. $P(x) < 0 : x \in (-3.1, -1.7) \cup (1.3, 2.4)$

3.7 Exercises

1. Vertical, multiplicity

3. Empty

5. Answers will vary.

7. $f(x) = -x^2 + 4x; \quad f(x) > 0$
   $-x^2 + 4x = 0$
   $-x(x-4) = 0$
   $x = 0; \quad x = 4$
   Concave down
   $x \in (0, 4)$

9. $h(x) = x^2 + 4x - 5; \quad h(x) \geq 0$
   $x^2 + 4x - 5 = 0$
   $(x+5)(x-1) = 0$
   $x = -5; \quad x = 1$
   Concave up
   $x \in (-\infty, -5] \cup [1, \infty)$

11. $q(x) = 2x^2 - 5x - 7; \quad q(x) < 0$
    $2x^2 - 5x - 7 = 0$
    $(2x-7)(x+1) = 0$
    $x = \dfrac{7}{2}; \quad x = -1$
    Concave up
    $x \in \left(-1, \dfrac{7}{2}\right)$

13. $7 \geq x^2$
    $7 = x^2$
    $\pm\sqrt{7} = x$
    $x \in \left[-\sqrt{7}, \sqrt{7}\right]$

15. $x^2 + 3x \leq 6$
    $x^2 + 3x - 6 = 0$
    $x = \dfrac{-3 \pm \sqrt{3^2 - 4(1)(-6)}}{2(1)}$
    $x = \dfrac{-3 \pm \sqrt{9 + 24}}{2}$
    $x = \dfrac{-3 \pm \sqrt{33}}{2}$
    Concave up
    $x \in \left(-\infty, \dfrac{-3-\sqrt{33}}{2}\right] \cup \left[\dfrac{-3+\sqrt{33}}{2}, \infty\right)$

17. $3x^2 \geq -2x + 5$
    $3x^2 + 2x - 5 = 0$
    $(3x+5)(x-1) = 0$
    $x = \dfrac{-5}{3}; \quad x = 1$
    Concave up
    $x \in \left(-\infty, \dfrac{-5}{3}\right] \cup [1, \infty)$

19. $s(x) = x^2 - 8x + 16; \quad s(x) \geq 0$
    $x^2 - 8x + 16 = 0$
    $(x-4)(x-4) = 0$
    $x = 4$
    Concave up
    $x \in (-\infty, \infty)$

21. $r(x) = 4x^2 + 12x + 9; \quad r(x) < 0$
    $4x^2 + 12x + 9 = 0$
    $(2x+3)(2x+3) = 0$
    $x = -\dfrac{3}{2}$
    Concave up
    No solution

23. $g(x) = -x^2 + 10x - 25; \quad g(x) < 0$

$$-x^2 + 10x - 25 = 0$$
$$-\left(x^2 - 10x + 25\right) = 0$$
$$-(x-5)(x-5) = 0$$
$$x = 5$$

Concave down
$$x \in (-\infty, 5) \cup (5, \infty)$$

25. $-x^2 > 2$

$$-x^2 = 2$$
$$x^2 = -2$$
$$x = \sqrt{-2}$$

No x-intercepts
Concave down
No solution

27. $x^2 - 2x > -5$

$$x^2 - 2x + 5 = 0$$
$$x = \frac{2 \pm \sqrt{(-2)^2 - 4(1)(5)}}{2(1)}$$
$$x = \frac{2 \pm \sqrt{4 - 20}}{2}$$
$$x = \frac{2 \pm \sqrt{-16}}{2}$$

No x-intercepts
Concave up
$$x \in (-\infty, \infty)$$

29. $p(x) = 2x^2 - 6x + 9; \quad p(x) \geq 0$

$$2x^2 - 6x + 9 = 0$$
$$x = \frac{6 \pm \sqrt{(-6)^2 - 4(2)(9)}}{2(2)}$$
$$x = \frac{6 \pm \sqrt{36 - 72}}{4}$$
$$x = \frac{6 \pm \sqrt{-36}}{4}$$

No x-intercepts
Concave up
$$x \in (-\infty, \infty)$$

31. $h(x) = \sqrt{x^2 - 25}$

$$x^2 - 25 \geq 0$$
To find zeroes, solve
$$x^2 = 25$$
$$x = \pm 5$$

Use a number line diagram, plot (-5, 0) and (5, 0). Sketch a parabola opening upward. The graph is above the x-axis when domain is $\quad x \in (-\infty, -5] \cup [5, \infty)$

33. $q(x) = \sqrt{x^2 - 5x}$

To find zeroes, solve
$$x^2 - 5x = 0$$
$$x(x - 5) = 0$$
$$x = 0 \text{ or } x = 5$$

Use a number line diagram, plot (0, 0) and (5, 0). Sketch a parabola opening upward. The graph is above the x-axis when domain is: $\quad x \in (-\infty, 0] \cup [5, \infty)$

35. $t(x) = \sqrt{-x^2 + 3x - 4}$

To find the zeroes, solve
$$-x^2 + 3x - 4 = 0$$
$$a = -1, b = 3, c = -4$$
$$x = \frac{-3 \pm \sqrt{(-3)^2 - 4(-1)(-4)}}{2(-1)}$$
$$x = \frac{-3 \pm \sqrt{-7}}{-1}$$

No solution

37. $(x + 3)(x - 5) < 0$

$$x \in (-3, 5)$$

39. $(x + 1)^2 (x - 4) \geq 0$

$$x \in [4, \infty) \cup \{-1\}$$

3.7 Exercises

41. $(x+2)^3(x-2)^2(x-4) \geq 0$

pos neg neg pos
-2   2   4

$x \in (-\infty, -2] \cup \{2\} \cup [4, \infty)$

42. $(x-1)^3(x+2)^2(x-3) \leq 0$

pos pos neg pos
-2   1   3

$x \in [1,3] \cup \{-2\}$

43. $x^2 + 4x + 1 < 0$;

$$x = \frac{-(4) \pm \sqrt{(4)^2 - 4(1)(1)}}{2(1)} = \frac{-4 \pm \sqrt{12}}{2}$$

$$= \frac{-4 \pm 2\sqrt{3}}{2} = -2 \pm \sqrt{3} ;$$

pos neg pos
$-2-\sqrt{3}$ $-2+\sqrt{3}$

$x \in \left(-2-\sqrt{3}, -2+\sqrt{3}\right)$

45. $x^3 + x^2 - 5x + 3 \leq 0$

Possible rational roots: $\dfrac{\{\pm 1, \pm 3\}}{\{\pm 1\}}$;

$\{\pm 1, \pm 3\}$

$$
\begin{array}{r|rrrr}
1 & 1 & 1 & -5 & 3 \\
  &   & 1 & 2 & -3 \\
\hline
  & 1 & 2 & -3 & \underline{|0}
\end{array}
$$

$(x-1)(x^2 + 2x - 3) \leq 0$

$(x-1)(x+3)(x-1) \leq 0$

$(x+3)(x-1)^2 \leq 0$

neg pos pos
-3   1

$x \in (-\infty, -3] \cup \{1\}$

47. $x^3 - 7x + 6 > 0$

Possible rational roots: $\dfrac{\{\pm 1, \pm 6, \pm 2, \pm 3\}}{\{\pm 1\}}$;

$\{\pm 1, \pm 6, \pm 2, \pm 3\}$

$$
\begin{array}{r|rrrr}
1 & 1 & 0 & -7 & 6 \\
  &   & 1 & 1 & -6 \\
\hline
  & 1 & 1 & -6 & \underline{|0}
\end{array}
$$

$(x-1)(x^2 + x - 6) > 0$

$(x-1)(x+3)(x-2) > 0$

neg pos neg pos
-3   1   2

$x \in (-3, 1) \cup (2, \infty)$

49. $x^4 - 10x^2 > -9$

$x^4 - 10x^2 + 9 > 0$

$(x^2 - 1)(x^2 - 9) > 0$

$(x+1)(x-1)(x+3)(x-3) > 0$

pos neg pos neg pos
-3   -1   1   3

$x \in (-\infty, -3) \cup (-1, 1) \cup (3, \infty)$

51. $x^4 - 9x^2 > 4x - 12$

$x^4 - 9x^2 - 4x + 12 > 0$

Possible rational roots:

$\dfrac{\{\pm 1, \pm 12, \pm 2, \pm 6, \pm 3, \pm 4\}}{\{\pm 1\}}$;

$\{\pm 1, \pm 12, \pm 2, \pm 6, \pm 3, \pm 4\}$

$$
\begin{array}{r|rrrrr}
1 & 1 & 0 & -9 & -4 & 12 \\
  &   & 1 & 1 & -8 & -12 \\
\hline
  & 1 & 1 & -8 & -12 & \underline{|0}
\end{array}
$$

$$
\begin{array}{r|rrrr}
3 & 1 & 1 & -8 & -12 \\
  &   & 3 & 12 & 12 \\
\hline
  & 1 & 4 & 4 & \underline{|0}
\end{array}
$$

$(x-1)(x-3)(x^2 + 4x + 4) > 0$

$(x-1)(x-3)(x+2)^2 > 0$

pos pos neg pos
-2   1   3

$x \in (-\infty, -2) \cup (-2, 1) \cup (3, \infty)$

53. $x^4 - 6x^3 \le -8x^2 - 6x + 9$

$x^4 - 6x^3 + 8x^2 + 6x - 9 \le 0$

Possible rational roots: $\dfrac{\{\pm 1, \pm 9, \pm 3\}}{\{\pm 1\}}$;

$\{\pm 1, \pm 9, \pm 3\}$

$$
\begin{array}{r|rrrrr}
-1 & 1 & -6 & 8 & 6 & -9 \\
   &   & -1 & 7 & -15 & 9 \\
\hline
   & 1 & -7 & 15 & -9 & \boxed{0}
\end{array}
$$

$$
\begin{array}{r|rrrr}
1 & 1 & -7 & 15 & -9 \\
  &   & 1 & -6 & 9 \\
\hline
  & 1 & -6 & 9 & \boxed{0}
\end{array}
$$

$(x+1)(x-1)(x^2 - 6x + 9) \le 0$

$(x+1)(x-1)(x-3)^2 \le 0$

$x \in [-1,1] \cup \{3\}$

55. $\dfrac{x+3}{x-2} \le 0$

$x \in [-3,2)$

57. $\dfrac{x+1}{x^2 + 4x + 4} < 0$

$\dfrac{x+1}{(x+2)^2} < 0$

$x \in (-\infty, -2) \cup (-2, -1)$

59. $\dfrac{2-x}{x^2 - x - 6} \ge 0$

$\dfrac{2-x}{(x-3)(x+2)} \ge 0$

$x \in (-\infty, -2) \cup [2,3)$

61. $\dfrac{2x - x^2}{x^2 + 4x - 5} < 0$

$\dfrac{x(2-x)}{(x+5)(x-1)} < 0$

$x \in (-\infty, -5) \cup (0,1) \cup (2, \infty)$

63. $\dfrac{x^2 - 4}{x^3 - 13x + 12} \ge 0$

Possible rational roots of denominator:

$\dfrac{\{\pm 1, \pm 12, \pm 2, \pm 6, \pm 3, \pm 4\}}{\{\pm 1\}}$;

$\{\pm 1, \pm 12, \pm 2, \pm 6, \pm 3, \pm 4\}$

$$
\begin{array}{r|rrrr}
1 & 1 & 0 & -13 & 12 \\
  &   & 1 & 1 & -12 \\
\hline
  & 1 & 1 & -12 & \boxed{0}
\end{array}
$$

$x^3 - 13x + 12 = (x-1)(x^2 + x - 12)$

$\qquad = (x-1)(x+4)(x-3)$;

$\dfrac{(x+2)(x-2)}{(x-1)(x+4)(x-3)} \ge 0$

$x \in (-4, -2] \cup (1,2] \cup (3, \infty)$

65. $\dfrac{x^2 + 5x - 14}{x^3 + x^2 - 5x + 3} > 0$

Possible rational roots of denominator:

$\dfrac{\{\pm 1, \pm 3\}}{\{\pm 1\}}$; $\{\pm 1, \pm 3\}$;

$$
\begin{array}{r|rrrr}
1 & 1 & 1 & -5 & 3 \\
  &   & 1 & 2 & -3 \\
\hline
  & 1 & 2 & -3 & \boxed{0}
\end{array}
$$

$x^3 + x^2 - 5x + 3 = (x-1)(x^2 + 2x - 3)$

$\qquad = (x-1)(x+3)(x-1)$;

$\dfrac{(x+7)(x-2)}{(x-1)^2(x+3)} > 0$

$x \in (-7, -3) \cup (2, \infty)$

3.7 Exercises

67. $\dfrac{2}{x-2} \le \dfrac{1}{x}$

$\dfrac{2}{x-2} - \dfrac{1}{x} \le 0$

$\dfrac{2x-x+2}{x(x-2)} \le 0$

$\dfrac{x+2}{x(x-2)} \le 0$

$$\begin{array}{ccccc} \text{neg} & \text{pos} & \text{neg} & \text{pos} \\ \bullet & \circ & \circ & \\ -2 & 0 & 2 & \end{array}$$

$x \in (-\infty, -2] \cup (0, 2)$

69. $\dfrac{x-3}{x+17} > \dfrac{1}{x-1}$

$\dfrac{x-3}{x+17} - \dfrac{1}{x-1} > 0$

$\dfrac{(x-3)(x-1) - 1(x+17)}{(x+17)(x-1)} > 0$

$\dfrac{x^2 - 4x + 3 - x - 17}{(x+17)(x-1)} > 0$

$\dfrac{x^2 - 5x - 14}{(x+17)(x-1)} > 0$

$\dfrac{(x-7)(x+2)}{(x+17)(x-1)} > 0$

$$\begin{array}{ccccccc} \text{pos} & \text{neg} & \text{pos} & \text{neg} & \text{pos} \\ \circ & \circ & \circ & \bullet & \\ -17 & -2 & 1 & 7 & \end{array}$$

$x \in (-\infty, -17) \cup (-2, 1) \cup (7, \infty)$

71. $\dfrac{x+1}{x-2} \ge \dfrac{x+2}{x+3}$

$\dfrac{x+1}{x-2} - \dfrac{x+2}{x+3} \ge 0$

$\dfrac{(x+1)(x+3) - (x+2)(x-2)}{(x-2)(x+3)} \ge 0$

$\dfrac{x^2 + 4x + 3 - x^2 + 4}{(x-2)(x+3)} \ge 0$

$\dfrac{4x+7}{(x-2)(x+3)} \ge 0$

$$\begin{array}{ccccc} \text{neg} & \text{pos} & \text{neg} & \text{pos} \\ \circ & \bullet & \circ & \\ -3 & -\frac{7}{4} & 2 & \end{array}$$

$x \in \left(-3, -\dfrac{7}{4}\right] \cup (2, \infty)$

72. $\dfrac{x-3}{x-6} \le \dfrac{x+1}{x+4}$

$\dfrac{x-3}{x-6} - \dfrac{x+1}{x+4} \le 0$

$\dfrac{(x-3)(x+4) - (x+1)(x-6)}{(x-6)(x+4)} \le 0$

$\dfrac{x^2 + x - 12 - x^2 + 5x + 6}{(x-6)(x+4)} \le 0$

$\dfrac{6x - 6}{(x-6)(x+4)} \le 0$

$\dfrac{6(x-1)}{(x-6)(x+4)} \le 0$

$$\begin{array}{ccccccc} \text{neg} & \text{pos} & \text{neg} & \text{pos} \\ \circ & \bullet & \circ & \\ -4 & 1 & 6 & \end{array}$$

$x \in (-\infty, -4) \cup [1, 6)$

73. $\dfrac{x+2}{x^2+9} > 0$

$x^2 + 9$ has no real roots

$$\begin{array}{cc} \text{neg} & \text{pos} \\ \circ & \\ -2 & \end{array}$$

$x \in (-2, \infty)$

75. $\dfrac{x^3+1}{x^2+1} > 0$

$\dfrac{(x+1)(x^2 - x + 1)}{x^2 + 1} > 0$

$x^2 - x + 1, \; x^2 + 1$ have no real roots

$$\begin{array}{cc} \text{neg} & \text{pos} \\ \circ & \\ -1 & \end{array}$$

$x \in (-1, \infty)$

77. $\dfrac{x^4 - 5x^2 - 36}{x^2 - 2x + 1} > 0$

$\dfrac{(x^2 - 9)(x^2 + 4)}{(x-1)^2} > 0$

$\dfrac{(x+3)(x-3)(x^2 + 4)}{(x-1)^2} > 0$

$x^2 + 4$ has no real roots

$$\begin{array}{ccccc} \text{pos} & \text{neg} & \text{neg} & \text{pos} \\ \circ & \circ & \circ & \\ -3 & 1 & 3 & \end{array}$$

$x \in (-\infty, -3) \cup (3, \infty)$

79. $x^2 - 2x \geq 15$

$x^2 - 2x - 15 \geq 0$

$(x-5)(x+3) \geq 0$

$x \in (-\infty, -3] \cup [5, \infty)$

81. $x^3 \geq 9x$

$x^3 - 9x \geq 0$

$x(x^2 - 9) \geq 0$

$x(x+3)(x-3) \geq 0$

$x \in [-3, 0] \cup [3, \infty)$

83. $-4x + 12 < -x^3 + 3x^2$

$x^3 - 3x^2 - 4x + 12 < 0$

$x^2(x-3) - 4(x-3) < 0$

$(x-3)(x^2-4) < 0$

$(x-3)(x+2)(x-2) < 0$

$x \in (-\infty, -2) \cup (2, 3)$

85. $\dfrac{x^2 - x - 6}{x^2 - 1} \geq 0$

$\dfrac{(x+2)(x-3)}{(x+1)(x-1)} \geq 0$

$x \in (-\infty, -2] \cup (-1, 1) \cup [3, \infty)$

87. b

89. b

91. a. $D = -(4p^3 + 27(p+1)^2)$

$D = -(4p^3 + 27(p^2 + 2p + 1))$

$D = -(4p^3 + 27p^2 + 54p + 27)$

verified

b. $-(4p^3 + 27p^2 + 54p + 27) = 0$

Possible rational roots: $\dfrac{\{\pm 1, \pm 3, \pm 9, \pm 27\}}{\{\pm 1, \pm 2, \pm 4\}}$

$D = -(p+3)^2\left(p + \dfrac{3}{4}\right)$

$p = -3, q = -3 + 1 = -2$

$p = -\dfrac{3}{4}, q = -\dfrac{3}{4} + 1 = \dfrac{1}{4}$

c. $-(p+3)^2\left(p + \dfrac{3}{4}\right) > 0$

$(-\infty, -3) \cup \left(-3, -\dfrac{3}{4}\right)$

d. Verified

93. $d(x) = k(x^3 - 192x + 1024)$

a. $\dfrac{k(x^3 - 3(8)^2 x + 2(8)^3)}{k} < 189$

$x^3 - 192x + 1024 < 189$

$x^3 - 192x + 835 < 0$

Possible rational roots:
$\pm 1, \pm 835, \pm 5, \pm 167$

$(x-5)(x^2 + 5x - 167) < 0$

$x \in (5, 8]$

b. $(4)^3 - 192(4) + 1024 = 320$ units

c. $\dfrac{k(x^3 - 3(8)^2 x + 2(8)^3)}{k} > 475$

$x^3 - 192x + 1024 > 475$

$x^3 - 192x + 549 > 0$

Possible rational roots:
$\dfrac{\{\pm 1, \pm 3, \pm 9, \pm 61, \pm 183, \pm 549\}}{\{\pm 1\}}$

$(x-3)(x^2 + 3x - 183) > 0$

$x \in [0, 3)$

d. $\dfrac{k(x^3 - 3(8)^2 x + 2(8)^3)}{k} \leq 648$

$x^3 - 192x + 1024 \leq 648$

$x^3 - 192x + 376 \leq 0$

Possible rational roots:
$\dfrac{\{\pm 1, \pm 2, \pm 4, \pm 8, \pm 47, \pm 94, \pm 188, \pm 376\}}{\{\pm 1\}}$

$(x-2)(x^2 + 2x - 188) \leq 0$

2 feet

95. a. $R = \dfrac{2D}{t_1 + t_2}$

$40 = \dfrac{2(80)}{t_1 + t_2}$

$1 = \dfrac{4}{t_1 + t_2}$

$1 = \dfrac{4}{\dfrac{80}{r_1} + \dfrac{80}{r_2}}$

$1 = \dfrac{4r_1 r_2}{80r_1 + 80r_2}$

$80r_1 + 80r_2 = 4r_1 r_2$

$20r_1 + 20r_2 = r_1 r_2$

$20r_2 - r_1 r_2 = -20r_1$

$r_2(20 - r_1) = -20r_1$

$r_2 = \dfrac{-20r_1}{20 - r_1}$

$r_2 = \dfrac{20r_1}{r_1 - 20}$

Verified

b. Horizontal: $r_2 = 20$, as r_1 increases, r_2 decreases to maintain $R = 40$.

Vertical: $r_1 = 20$, as r_1 decreases, r_2 increases to maintain $R = 40$.

c. $\dfrac{20r_1}{r_1 - 20} > r_1$

$\dfrac{20r_1}{r_1 - 20} - r_1 > 0$

$\dfrac{20r_1}{r_1 - 20} - \dfrac{r_1(r_1 - 20)}{r_1 - 20} > 0$

$\dfrac{20r_1 - r_1^2 + 20r_1}{r_1 - 20} > 0$

$\dfrac{40r_1 - r_1^2}{r_1 - 20} > 0$

$\dfrac{r_1(40 - r_1)}{r_1 - 20} > 0$

Critical points: 0, 20, 40

$r_1 \in (20, 40)$

97. $R(t) = 0.01t^2 + 0.1t + 30$

a. $0.01t^2 + 0.1t + 30 < 42$

$0.01t^2 + 0.1t - 12 < 0$

$t^2 + 10t - 1200 < 0$

$(t + 40)(t - 30) < 0$

$[0°, 30°)$

b. $R(t) = 0.01t^2 + 0.1t + 20$

$0.01t^2 + 0.1t + 30 > 36$

$0.01t^2 + 0.1t - 6 > 0$

$t^2 + 10t - 600 > 0$

$(t - 20)(t + 30) > 0$

$(20°, \infty)$

c. $0.01t^2 + 0.1t + 30 > 60$

$0.01t^2 + 0.1t - 30 > 0$

$t^2 + 10t - 3000 > 0$

$(t + 60)(t - 50) > 0$

$(50°, \infty)$

99. a. $\dfrac{2n^3 + 3n^2 + n}{6} \geq 30$

$2n^3 + 3n^2 + n \geq 180$

$2n^3 + 3n^2 + n - 180 \geq 0$

Possible rational roots:

$\{\pm 1, \pm 180, \pm 2, \pm 90, \pm 3, \pm 60, \pm 4, \pm 45, \pm 5, \pm 36,$

$\pm 6, \pm 30, \pm 9, \pm 20, \pm 10, \pm 18, \pm 12, \pm 15, \pm \dfrac{1}{2},$

$\pm \dfrac{3}{2}, \pm \dfrac{45}{2}, \pm \dfrac{5}{2}, \pm \dfrac{9}{2}, \pm \dfrac{15}{2}\}$

$(n - 4)(2n^2 + 11n + 45) \geq 0$

$n \geq 4$

b. $\dfrac{2n^3+3n^2+n}{6}\le 285$

$2n^3+3n^2+n\le 1710$

$2n^3+3n^2+n-1710\le 0$

Possible rational roots:

$\{\pm 1,\pm 1710,\pm 2,\pm 855,\pm 3,\pm 570,\pm 5,\pm 342,$

$\pm 6,\pm 285,\pm 9,\pm 190,\pm 10,\pm 171,\pm 15,$

$\pm 114,\pm 18,\pm 95,\pm 19,\pm 90,\pm 30,\pm 57,$

$\pm 38,\pm 45,\pm\dfrac{1}{2},\pm\dfrac{855}{2},\pm\dfrac{3}{2},\pm\dfrac{5}{2},$

$\pm\dfrac{285}{2},\pm\dfrac{9}{2},\pm\dfrac{171}{2},\pm\dfrac{15}{2},\pm\dfrac{95}{2},$

$\pm\dfrac{19}{2},\pm\dfrac{57}{2},\pm\dfrac{45}{2}\}$

$(n-9)(2n^2+21n+190)\le 0$

$n\le 9$

c. $\dfrac{2n^3+3n^2+n}{6}\le 999$

$2n^3+3n^2+n\le 5994$

$2n^3+3n^2+n-5994\le 0$

Possible rational roots:

$\{\pm 1,\pm 5994,\pm 2,\pm 2997,\pm 3,\pm 1998,\pm 6,\pm 999,\pm 9,$

$\pm 666,\pm 18,\pm 333,\pm 27,\pm 222,\pm 37,\pm 162,\pm 54,$

$\pm 111,\pm 74,\pm 81,\pm\dfrac{1}{2},\pm\dfrac{3}{2},\pm\dfrac{999}{2},\pm\dfrac{9}{2},\pm\dfrac{333}{2},$

$\pm\dfrac{27}{2},\pm\dfrac{37}{2},\pm\dfrac{111}{2},\pm\dfrac{81}{2};\}$

Not factorable

$\dfrac{2(13)^3+3(13)^2+(13)}{6}\le 999$

$819\le 999$

$n=13$

101. a. yes, $x^2\ge 0$

b. yes, $\dfrac{x^2}{x^2+1}\ge 0$

103. $x(x+2)(x-1)^2>0;\quad \dfrac{x(x+2)}{(x-1)^2}>0$

105. $R(x)=\dfrac{x^2-16x+28}{(x-8)^2}$

$R(x)=\dfrac{(x-14)(x-2)}{(x-8)^2}$

Solve: $\dfrac{(x-14)(x-2)}{(x-8)^2}<0$

$R(x)<0$ for $x\in(2,8)\cup(8,14)$

107. $f(x)=\dfrac{x^2+2x-8}{x+4}$

$f(x)=\dfrac{(x+4)(x-2)}{x+4}=x-2$

$f(x)=-6$ when $x=-4$

$F(x)=\begin{cases} f(x) & x\ne -4 \\ -6 & x=-4 \end{cases}$

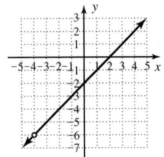

109. $3x+1<10$ and $x^2-3<1$

$3x<9$ and $x^2-4<0$

$x<3$ and $(x+2)(x-2)<0$

3.8 Exercises

1. Constant

3. $y = \dfrac{k}{x^2}$

5. Answers will vary.

7. $d = kr$

9. $F = ka$

11. $y = kx$

$0.6 = k(24)$

$0.025 = k$

$y = 0.025x$

x	$f(x) = 0.025x$
500	$f(500) = 0.025(500) = 12.5$
650	$16.25 = 0.025x$ $650 = x$
750	$f(750) = 0.025(750) = 18.75$

13. $w = kh$

$344.25 = k(37.5)$

$9.18 = k$;

$w = 9.18h$

$w = 9.18(35)$

$w = \$321.30$

k represents the hourly wage.

15. a. $s = kh$

$192 = k(47)$

$\dfrac{192}{47} = k$;

$s = \dfrac{192}{47}h$

b.

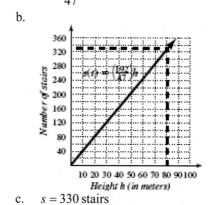

c. $s = 330$ stairs

d. $s = \dfrac{192}{47}(81) \approx 331$; Yes

17. $A = kS^2$

19. $P = kc^2$

21. $p = kq^2$

$280 = k(50)^2$

$\dfrac{280}{(50)^2} = k$

$0.112 = k$;

$p = 0.112q^2$

q	$p(q) = 0.112q^2$
45	$p(45) = 0.112(45)^2 = 226.8$
55	$338.8 = 0.112q^2$ $3025 = q^2$ $55 = q$
70	$p(70) = 0.112(70)^2 = 548.8$

23. $A = ks^2$

$3528 = k(14\sqrt{3})^2$

$3528 = 588k$

$\dfrac{3528}{588} = k$

$6 = k;$

$A = 6s^2;$

$A = 6(303,600)^2;$

$A = 553,037,760,000 \text{ cm}^2$

$A = 55,303,776 \text{ m}^2$

25. a. $d = kt^2$

$169 = k(3.25)^2$

$169 = 10.5625k$

$16 = k;$

$d = 16t^2$

b.

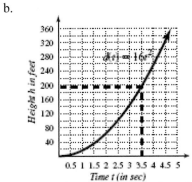

c. According to the graph, about 3.5 seconds

d. $196 = 16t^2$

$12.25 = t^2$

$3.5 \sec = t$

Yes, it was close.

e. $121 = 16t^2$

$7.5625 = t^2$

$2.75 = t$

2.75 seconds

27. $F = \dfrac{k}{d^2}$

29. $S = \dfrac{k}{L}$

31. $Y = \dfrac{k}{Z^2}$

$1369 = \dfrac{k}{3^2}$

$12321 = k;$

$Y = \dfrac{12321}{Z^2}$

Z	Y
37	$Y(37) = \dfrac{12321}{37^2} = 9$
74	$2.25 = \dfrac{12321}{Z^2}$ $2.25Z^2 = 12321$ $Z = 74$
111	$Y(111) = \dfrac{12321}{111^2} = 1$

33. $w = \dfrac{k}{r^2}$

$75 = \dfrac{k}{(6400)^2}$

$3072000000 = k;$

$w = \dfrac{3072000000}{r^2}$

$w = \dfrac{3072000000}{(8000)^2}$

$w = 48 \text{ kg}$

35. $I = krt$

37. $A = kh(B + b)$

39. $V = ktr^2$

41. $C = \dfrac{kR}{S^2}$

$21 = \dfrac{k(7)}{(1.5)^2}$

$47.25 = 7k$

$6.75 = k$;

$C = \dfrac{6.75R}{S^2}$

R	S	C
120	6	22.5
200	12.5	8.64
350	15	10.5

$22.5 = \dfrac{6.75(120)}{S^2}$

$22.5S^2 = 810$

$S = 6$;

$C = \dfrac{6.75(200)}{(12.5)^2} = \dfrac{1350}{156.25} = 8.64$;

$10.5 = \dfrac{6.75R}{(15)^2}$

$2362.5 = 6.75R$

$350 = R$

43. $E = kmv^2$

$200 = k(1)(20)^2$

$0.5 = k$;

$E = 0.5mv^2$;

$E = 0.5(1)(35)^2$

$E = 612.5$ joules

45. $R(A) = \sqrt[3]{A} - 1$

$f(x) = \sqrt[3]{x}$; Cube root family

Amount A	Rate r
1.0	0.0
1.05	0.016
1.10	0.032
1.15	0.048
1.20	0.063
1.25	0.077

$R(A) = \sqrt[3]{1.17} - 1 = 0.054 = 5.4\%$

Interest Rate: 5.4%

47. $T = \dfrac{k}{V}$

$4 = \dfrac{k}{12}$

$48 = k$;

$T = \dfrac{48}{V}$;

$T = \dfrac{48}{1.5}$

$T = 32$ volunteers

49. $M = kE$

$16 = k(96)$

$\dfrac{16}{96} = k$

$\dfrac{1}{6} = k$;

$M = \dfrac{1}{6}E$;

$M = \dfrac{1}{6}(250)$

$M \approx 41.7$ kg

51. $D = k\sqrt{S}$

$108 = k\sqrt{25}$

$21.6 = k$;

$D = 21.6\sqrt{S}$;

$D = 21.6\sqrt{45}$

$D \approx 144.9$ ft

53. $C = kLD$

$76.50 = k(36)\left(\dfrac{1}{4}\right)$

$76.50 = 9k$

$8.5 = k$;

$C = 8.5LD$;

$C = 8.5(24)\left(\dfrac{3}{8}\right)$

$C = \$76.50$

55. $C = \dfrac{kp_1p_2}{d^2}$

$300 = \dfrac{k(300000)(420000)}{430^2}$

$55470000 = 1.26\text{x}10^{11}k$

$4.4\text{x}10^{-4} = k$;

$C = \dfrac{\left(4.4\text{x}10^{-4}\right)p_1p_2}{d^2}$

$C = \dfrac{\left(4.4\text{x}10^{-4}\right)(170000)(550000)}{430^2} \approx 222.5$

about 223 calls

57. $V = k \cdot l \cdot w^2$

$12.27 = k \cdot (3.75) \cdot (2.50)^2$

$\dfrac{12.27}{(3.75) \cdot (2.50)^2} = k$;

a. $V = \dfrac{12.27}{3.75(2.50)^2} \cdot (4.65) \cdot (3.10)^2$

$V \approx 23.39$ cm^3

b. $\dfrac{23.39}{12.27} \approx 1.91$ or 191%

59. a. $M = k(w)h^2\left(\dfrac{1}{L}\right)$

b. $270 = k(18)(2)^2\left(\dfrac{1}{8}\right)$

$270 = 9k$

$30 = k$;

$M = 30(18)(2)^2\left(\dfrac{1}{12}\right) = 180$ lb

61. $f(x) = k\dfrac{1}{x}$

$\dfrac{\Delta y}{\Delta x} = \dfrac{f(0.6) - f(0.5)}{0.6 - 0.5} = -\dfrac{10}{3}$

$g(x) = k\dfrac{1}{x^2}$

$\dfrac{\Delta y}{\Delta x} = \dfrac{g(0.6) - g(0.5)}{0.6 - 0.5} = -\dfrac{110}{9}$

From $x = 0.7$ to $x = 0.8$, the rate of decrease will be less because as x approaches infinity, y approaches 0.

$x \to \infty, \quad y \to 0^+$

63. $I = \dfrac{k}{d^2}$

a. $I = \dfrac{k}{5^2}$

$25I = k$;

$2I = \dfrac{25I}{d^2}$

$d^2 = \dfrac{25I}{2I}$

$d^2 = \dfrac{25}{2}$

$d = \sqrt{\dfrac{25}{2}} \approx 3.5\ \text{ft}$

b. $I = \dfrac{k}{12^2}$

$144I = k$;

$3I = \dfrac{144I}{d^2}$

$d^2 = \dfrac{144I}{3I}$

$d^2 = \dfrac{144}{3}$

$d = \sqrt{\dfrac{144}{3}} \approx 6.9\ \text{ft}$

65. $x^3 + 4x^2 + 8x = 0$

$x(x^2 + 4x + 8) = 0$;

$a = 1, \quad b = 4, \quad c = 8$

$x = \dfrac{-4 \pm \sqrt{(4)^2 - 4(1)(8)}}{2(1)}$

$x = \dfrac{-4 \pm \sqrt{16 - 32}}{2}$

$x = \dfrac{-4 \pm \sqrt{-16}}{2}$

$x = \dfrac{-4 \pm 4i}{2}$;

$x = -2 \pm 2i; \quad x = 0$

67. $f(x) = -2|x - 3| + 5$

Right 3, up 5, reflected across the x-axis.

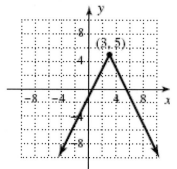

Chapter 3 Summary and Concept Review

1. $f(x) = x^2 + 8x + 15$

$0 = x^2 + 8x + 15$

$0 = \left(x^2 + 8x + 16\right) + 15 - 16$

$0 = (x+4)^2 - 1$

$0 = (x+4)^2 - 1$

$1 = (x+4)^2$

$\pm 1 = x + 4$

$x = -4 \pm 1$

x-intercepts: $(-5, 0)$ and $(-3, 0)$
Vertex: $(-4, -1)$

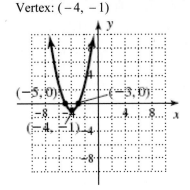

3. $f(x) = 4x^2 - 12x + 3$

$0 = 4x^2 - 12x + 3$

$0 = 4\left(x^2 - 3x\right) + 3$

$0 = 4\left(x^2 - 3x + \dfrac{9}{4}\right) + 3 - 9$

$0 = 4\left(x - \dfrac{3}{2}\right)^2 - 6$

$0 = 4\left(x - \dfrac{3}{2}\right)^2 - 6$

$6 = 4\left(x - \dfrac{3}{2}\right)^2$

$\dfrac{3}{2} = \left(x - \dfrac{3}{2}\right)^2$

$x - \dfrac{3}{2} = \sqrt{\dfrac{3}{2}}$

$x - \dfrac{3}{2} = \pm\dfrac{\sqrt{6}}{2}$

$x = \dfrac{3}{2} \pm \dfrac{\sqrt{6}}{2}$

x-intercepts: $(2.7, 0)$ and $(0.3, 0)$;
y-intercept: $(0, 3)$

$f(0) = 4(0)^2 - 12(0) + 3 = 3$

Vertex: $\left(\dfrac{3}{2}, -6\right)$

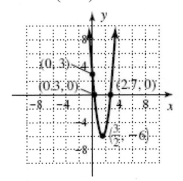

5. $\dfrac{x^3 + 4x^2 - 5x - 6}{x - 2}$

$$
\begin{array}{r}
x^2 + 6x + 7 \\
x-2 \overline{\smash{\big)}\, x^3 + 4x^2 - 5x - 6} \\
\underline{-\left(x^3 - 2x^2\right)} \\
6x^2 - 5x \\
\underline{-\left(6x^2 - 12x\right)} \\
7x - 6 \\
\underline{-\left(7x - 14\right)} \\
8
\end{array}
$$

$q(x) = x^2 + 6x + 7$

$R = 8$

7. Since $R = 0$, -7 is a root and $x + 7$ is a factor.

$$
\begin{array}{r|rrrrr}
-7 & 2 & 13 & -6 & 9 & 14 \\
& & -14 & 7 & -7 & -14 \\
\hline
& 2 & -1 & 1 & 2 & \underline{|\,0}
\end{array}
$$

9. $p(x) = x^3 + 2x^2 - 11x - 12$

Possible rational roots: $\pm 1, \pm 12, \pm 2, \pm 6, \pm 3, \pm 4$

$$
\begin{array}{r|rrrr}
-4 & 1 & 2 & -11 & -12 \\
& & -4 & 8 & 12 \\
\hline
& 1 & -2 & -3 & \underline{|\,0}
\end{array}
$$

$p(x) = (x+4)\left(x^2 - 2x - 3\right)$

$p(x) = (x+4)(x+1)(x-3)$

11. $P(x) = 4x^3 + 8x^2 - 3x - 1$

$$\frac{1}{2} \begin{array}{|rrrr} 4 & 8 & -3 & -1 \\ & 2 & 5 & 1 \\ \hline 4 & 10 & 2 & \boxed{0} \end{array}$$

Since $R = 0$, $\dfrac{1}{2}$ is a root and

$\left(x - \dfrac{1}{2}\right)$ is a factor.

13. $h(x) = x^3 + 9x^2 + 13x - 10$

$$-7 \begin{array}{|rrrr} 1 & 9 & 13 & -10 \\ & -7 & -14 & 7 \\ \hline 1 & 2 & -1 & \boxed{-3} \end{array}$$

$h(-7) = -3$

15. $C(x) = (x-1)^2(x+2i)(x-2i)$

$C(x) = (x^2 - 2x + 1)(x^2 - 4i^2)$

$C(x) = (x^2 - 2x + 1)(x^2 + 4)$

$C(x) = x^4 + 4x^2 - 2x^3 - 8x + x^2 + 4$

$C(x) = x^4 - 2x^3 + 5x^2 - 8x + 4$

17. $p(x) = 4x^3 - 16x^2 + 11x + 10$

Possible rational roots:

$\dfrac{\{\pm 1, \pm 10, \pm 2, \pm 5\}}{\{\pm 1, \pm 2, \pm 4\}}$;

$\left\{\pm 1, \pm 10, \pm 2, \pm 5, \pm \dfrac{1}{2}, \pm \dfrac{5}{2}, \pm \dfrac{1}{4}, \pm \dfrac{5}{4}\right\}$

19. $P(x) = 2x^3 - 3x^2 - 17x - 12$

Possible rational roots:

$\dfrac{\{\pm 1, \pm 12, \pm 2, \pm 6, \pm 3, \pm 4\}}{\{\pm 1\}}$

$$4 \begin{array}{|rrrr} 2 & -3 & -17 & -12 \\ & 8 & 20 & 12 \\ \hline 2 & 5 & 3 & 0 \end{array}$$

$P(x) = (x-4)(2x^2 + 5x + 3)$

$P(x) = (x-4)(x+1)(2x+3)$

21. $P(x) = x^4 - 3x^3 - 8x^2 + 12x + 6$

$[-2, -1]$

$P(-2) = (-2)^4 - 3(-2)^3 - 8(-2)^2 + 12(-2) + 6 = -10$;

$P(-1) = (-1)^4 - 3(-1)^3 - 8(-1)^2 + 12(-1) + 6 = -10$;

$[1, 2]$

$P(1) = (1)^4 - 3(1)^3 - 8(1)^2 + 12(1) + 6 = 8$;

$P(2) = (2)^4 - 3(2)^3 - 8(2)^2 + 12(2) + 6 = -10$;

$[2, 3]$

$P(3) = (3)^4 - 3(3)^3 - 8(3)^2 + 12(3) + 6 = -30$;

$[4, 5]$

$P(4) = (4)^4 - 3(4)^3 - 8(4)^2 + 12(4) + 6 = -10$;

$P(5) = (5)^4 - 3(5)^3 - 8(5)^2 + 12(5) + 6 = 116$

Sign changes in intervals: $[1, 2]$, $[4, 5]$, verified

23. $f(x) = -3x^5 + 2x^4 + 9x - 4$

$f(0) = -3(0)^5 + 2(0)^4 + 9(0) - 4 = -4$

degree 5; up/down; $(0, -4)$

25. $p(x) = (x+1)^3(x-2)^2$

end behavior: down/up

bounce at $(2, 0)$; cross at $(-1, 0)$

$p(0) = (0+1)^3(0-2)^2 = 4$

y-intercept: $(0, 4)$

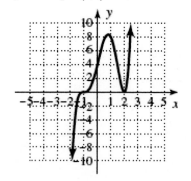

27. $h(x) = x^4 - 6x^3 + 8x^2 + 6x - 9$

end behavior: up/up

Possible rational roots: $\dfrac{\{\pm 1, \pm 9, \pm 3\}}{\{\pm 1\}}$

$h(x) = (x+1)(x-1)(x-3)^2$

bounce at $(3,0)$; cross at $(-1,0)$ and $(1,0)$

y-intercept: $(0,-9)$

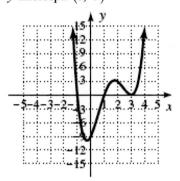

29. $V(x) = \dfrac{x^2 - 9}{x^2 - 3x - 4}$

$V(x) = \dfrac{(x+3)(x-3)}{(x-4)(x+1)}$

a. $\{x \mid x \in R, x \neq -1, 4\}$

b. HA: $y = 1$
(deg num = deg den)
VA: $x = -1, x = 4$

c. $V(0) = \dfrac{0^2 - 9}{0^2 - 3(0) - 4} = \dfrac{9}{4}$

y-intercept $\left(0, \dfrac{9}{4}\right)$;

x-intercepts : $(-3, 0)$ and $(3, 0)$

d. $V(1) = \dfrac{1^2 - 9}{1^2 - 3(1) - 4} = \dfrac{4}{3}$

31. $v(x) = \dfrac{x^2 - 4x}{x^2 - 4}$

$v(0) = \dfrac{(0)^2 - 4(0)}{(0)^2 - 4} = 0$;

y-intercept: $(0,0)$

$v(x) = \dfrac{x(x-4)}{(x+2)(x-2)}$;

vertical asymptotes: $x = -2$ and $x = 2$

x-intercepts: $(0,0)$ and $(4,0)$

horizontal asymptote: $y = 1$

(deg num = deg den)

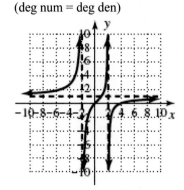

33. $V(x) = \dfrac{(x+3)(x-4)}{(x+2)(x-3)}$

$V(x) = \dfrac{x^2 - x - 12}{x^2 - x - 6}$;

$V(0) = \dfrac{(0)^2 - (0) - 12}{(0)^2 - (0) - 6} = 2$

35. $h(x) = \dfrac{x^3 - 2x^2 - 9x + 18}{x - 2}$

$h(x) = \dfrac{x^2(x-2) - 9(x-2)}{x - 2}$

$h(x) = \dfrac{(x-2)(x^2 - 9)}{x - 2}$

$h(x) = \dfrac{(x-2)(x+3)(x-3)}{x - 2}$

If $x = 2$, $x^2 - 9 = (2)^2 - 9 = -5$

Removable discontinuity at $(2, -5)$.

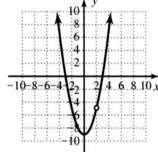

37. $h(x) = \dfrac{x^2 - 2x}{x - 3}$

$h(0) = \dfrac{(0)^2 - 2(0)}{(0) - 3} = 0;$

y-intercept: $(0,0)$

$h(x) = \dfrac{x(x-2)}{x-3};$

vertical asymptote: $x = 3$

x-intercepts: $(0,0)$ and $(2,0)$

horizontal asymptote: none
(deg num > deg den)

oblique asymptote: $y = x + 1$

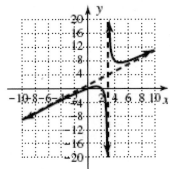

39. $A(x) = \dfrac{x^2 - 2x + 6}{x}$

a.

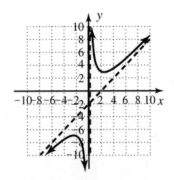

b. about 2450 favors

c. $A(2.45) \approx 2.90$
about \$2.90 each

41. $\dfrac{x^2 - 3x - 10}{x - 2} \geq 0$

$\dfrac{(x-5)(x+2)}{x-2} \geq 0$

neg   pos   neg   pos
-2   2   5

Outputs are positive for $x \in [-2, 2) \cup [5, \infty)$

43. $y = k\sqrt[3]{x};$

$52.5 = k\sqrt[3]{27}$

$\dfrac{52.5}{3} = k$

$17.5 = k;$

$y = 17.5\sqrt[3]{x};$

x	y
216	105
0.343	12.25
729	157.5

45. $t = \dfrac{kuv}{w};$

$30 = \dfrac{k(2)(3)}{5}$

$t = \dfrac{25uv}{w};$

$t = \dfrac{25(8)(12)}{15} = 160;$

Chapter 3: Polynomial and Rational Functions

Chapter 3 Mixed Review

1. Vertex $\left(\dfrac{1}{2},\dfrac{9}{2}\right)$

$y = a\left(x-\dfrac{1}{2}\right)^2+\dfrac{9}{2}$, $(2,0)$

$0 = a\left(2-\dfrac{1}{2}\right)^2+\dfrac{9}{2}$

$-\dfrac{9}{2}=a\left(\dfrac{3}{2}\right)^2$

$-2 = a$;

$y = -2\left(x-\dfrac{1}{2}\right)^2+\dfrac{9}{2}$

3. $C(s) = \dfrac{1}{180}s^2 - \dfrac{8}{9}s + \dfrac{680}{9}$

The s value of the vertex: $\dfrac{-\left(-\dfrac{8}{9}\right)}{2\left(\dfrac{1}{180}\right)} = 80$

$C(80) = \dfrac{1}{180}(80)^2 - \dfrac{8}{9}(80) + \dfrac{680}{9} = 40$

80 GB, $40.00

5. $\dfrac{x^4 - 3x^2 + 5x - 1}{x+2}$

$$\begin{array}{r|rrrrr} -2 & 1 & 0 & -3 & 5 & -1 \\ & & -2 & 4 & -2 & -6 \\ \hline & 1 & -2 & 1 & 3 & \underline{|-7} \end{array}$$

$q(x) = x^3 - 2x^2 + x + 3$

$R = -7$

7. $P(x) = 6x^3 - 23x^2 - 40x + 31$

(a) $P(-1) = 42$

$$\begin{array}{r|rrrr} -1 & 6 & -23 & -40 & 31 \\ & & -6 & 29 & 11 \\ \hline & 6 & -29 & -11 & \underline{|42} \end{array}$$

(b) $P(1) = -26$

$$\begin{array}{r|rrrr} 1 & 6 & -23 & -40 & 31 \\ & & 6 & -17 & -57 \\ \hline & 6 & -17 & -57 & \underline{|-26} \end{array}$$

(c) $P(5) = 6$

$$\begin{array}{r|rrrr} 5 & 6 & -23 & -40 & 31 \\ & & 30 & 35 & -25 \\ \hline & 6 & 7 & -5 & \underline{|\,6} \end{array}$$

9. a. $6x^3 + x^2 - 20x - 12 = 0$

Possible rational roots:

$\left\{\pm1,\pm12,\pm2,\pm6,\pm3,\pm4,\right.$
$\left.\pm\dfrac{1}{6},\pm\dfrac{1}{3},\pm\dfrac{1}{2},\pm\dfrac{2}{3},\pm\dfrac{3}{2},\pm\dfrac{4}{3}\right\}$

$x = 9$ and $x = \dfrac{8}{3}$ CANNOT be roots.

b. $P(x) = x^4 - x^3 + 7x^2 - 9x - 18$

Possible rational roots:
$\{\pm1,\pm18,\pm2,\pm9,\pm3,\pm6\}$

$$\begin{array}{r|rrrrr} 2 & 1 & -1 & 7 & -9 & -18 \\ & & 2 & 2 & 18 & 18 \\ \hline & 1 & 1 & 9 & 9 & \underline{|\,0} \end{array}$$

$$\begin{array}{r|rrrr} -1 & 1 & 1 & 9 & 9 \\ & & -1 & 0 & -9 \\ \hline & 1 & 0 & 9 & \underline{|\,0} \end{array}$$

$P(x) = (x-2)(x+1)(x^2+9)$
$P(x) = (x-2)(x+1)(x+3i)(x-3i)$
$x = 2, x = -1, x = -3i, x = 3i$

11. $p(x) = \dfrac{x^2 - 2x}{x^2 - 2x + 1}$

$p(0) = \dfrac{(0)^2 - 2(0)}{(0)^2 - 2(0) + 1} = 0$

y-intercept: $(0,0)$

$p(x) = \dfrac{x(x-2)}{(x-1)^2}$;

vertical asymptote: $x = 1$

x-intercepts: $(0,0)$ and $(2,0)$

horizontal asymptote: $y = 1$

13. $r(x) = \dfrac{x^3 - 13x + 12}{x^2}$

$r(0) = \dfrac{(0)^3 - 13(0) + 12}{(0)^2} =$ undefined

y-intercept: none

Possible rational roots:
$\{\pm 1, \pm 12, \pm 2, \pm 6, \pm 3, \pm 4\}$

$r(x) = \dfrac{(x+4)(x-1)(x-3)}{x^2}$;

vertical asymptote: $x = 0$

x-intercepts: $(-4,0), (1,0)$ and $(3,0)$

horizontal asymptote: none

$r(x) = x - \dfrac{13}{x} + \dfrac{12}{x^2}$

oblique asymptote: $y = x$

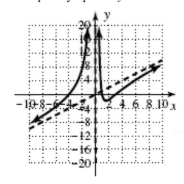

15. $x^3 - 4x < 12 - 3x^2$

$x^3 + 3x^2 - 4x - 12 < 0$

Possible rational roots:
$\{\pm 1, \pm 12, \pm 2, \pm 6, \pm 3, \pm 4\}$

$(x+2)(x-2)(x+3) < 0$

```
 neg  pos    neg    pos
 ──o────o──────────o────▶
   -3   -2          2
```

$x \in (-\infty, -3) \cup (-2, 2)$

17. a. $V(s) = (24 - 2x)(16 - 2x)(x)$

$= (384 - 80x + 4x^2)(x)$

$= 384x - 80x^2 + 4x^3$

$= 4x^3 - 80x^2 + 384x$

b. $512 = 4x^3 - 80x^2 + 384x$

$0 = 4x^3 - 80x^2 + 384x - 512$

$0 = x^3 - 20x^2 + 96x - 128$

c. For $0 < x < 8$, possible rational zeroes
are: 1, 2, and 4

d. $x = 4$ inches

$$
\begin{array}{r|rrr}
4 & 1 & -20 & 96 & -128 \\
  &   & 4 & -64 & 128 \\
\hline
  & 1 & -16 & 32 & 0
\end{array}
$$

e. $(x - 4)(x^2 - 16x + 32)$

$x = \dfrac{-(-16) \pm \sqrt{(-16)^2 - 4(1)(32)}}{2(1)}$

$x = \dfrac{16 \pm \sqrt{128}}{2} = 8 - 4\sqrt{2} \approx 2.34$ inches

19. $R = kL\left(\dfrac{1}{A}\right)$

Chapter 3 Practice Test

1. a. $f(x) = -x^2 + 10x - 16$

$f(x) = -\left(x^2 - 10x\right) - 16$

$f(x) = -\left(x^2 - 10x + 25\right) - 16 + 25$

$f(x) = -(x - 5)^2 + 9$;

Vertex: (5, 9), opens downward.

$f(0) = -0^2 + 10(0) - 16 = -16$

y-intercept $(0, \,^-16)$

$0 = -x^2 + 10x - 16$

$x^2 - 10x + 16 = 0$

$(x - 8)(x - 2) = 0$

$x = 8$ or $x = 2$

x-intercepts $(8,0)$ and $(2,0)$

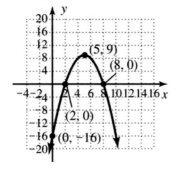

b. $g(x) = \dfrac{1}{2}x^2 + 4x + 16$

$g(x) = \dfrac{1}{2}\left(x^2 + 8x\right) + 16$

$g(x) = \dfrac{1}{2}\left(x^2 + 8x + 16\right) + 16 - 8$

$g(x) = \dfrac{1}{2}(x + 4)^2 + 8$;

Vertex: $(-4, 8)$, opens upward.

$g(0) = \dfrac{1}{2}(0)^2 + 4(0) + 16 = 16$

y-intercept $(0, 16)$

$0 = \dfrac{1}{2}x^2 + 4x + 16$

$0 = x^2 + 8x + 32$

$b^2 - 4ac = 64 - 4(1)(32) < 0$

No x-intercepts

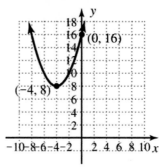

3. $d(t) = t^2 - 14t$

a. $d(4) = (4)^2 - 14(4) = -40$

40 ft

$d(6) = (6)^2 - 14(6) = -48$

48 ft

b. $d(t) = t^2 - 14t + 49 - 49$

$d(t) = (t - 7)^2 - 49$

49 ft

c. $2(7) = 14$ seconds

5. $\dfrac{x^3 + 4x^2 - 5x - 20}{x+2} = x^2 + 2x - 9 + \dfrac{-2}{x+2}$

$$
\begin{array}{r|rrrr}
-2 & 1 & 4 & -5 & -20 \\
   &   & -2 & -4 & 18 \\
\hline
   & 1 & 2 & -9 & \underline{|-2}
\end{array}
$$

7. $f(x) = 2x^3 + 4x^2 - 5x + 2$

$f(-3) = -1$

$$
\begin{array}{r|rrrr}
-3 & 2 & 4 & -5 & 2 \\
   &   & -6 & 6 & -3 \\
\hline
   & 2 & -2 & 1 & \underline{|-1}
\end{array}
$$

9. $Q(x) = \left(x^2 - 3x + 2\right)\left(x^3 - 2x^2 - x + 2\right)$

$Q(x) = (x-2)(x-1)\left(x^2(x-2) - (x-2)\right)$

$Q(x) = (x-2)(x-1)(x-2)\left(x^2 - 1\right)$

$Q(x) = (x-2)(x-1)(x-2)(x+1)(x-1)$

$Q(x) = (x-2)^2 (x-1)^2 (x+1)$

2 multiplicity 2

1 multiplicity 2, -1 multiplicity 1

11. $f(x) = \dfrac{1}{2}x^3 - 7x^2 + 28x - 32$

(a) $0 = \dfrac{1}{2}x^3 - 7x^2 + 28x - 32$

$0 = x^3 - 14x^2 + 56x - 64$

Possible rational roots:

$\left\{ \pm 1, \pm 64, \pm 2, \pm 32, \pm 4, \pm 16, \pm 8 \right\}$

$$
\begin{array}{r|rrrr}
2 & 1 & -14 & 56 & -64 \\
  &   & 2 & -24 & 64 \\
\hline
  & 1 & -12 & 32 & 0
\end{array}
$$

$0 = (x-2)\left(x^2 - 12x + 32\right)$

$0 = (x-2)(x-4)(x-8)$

$x = 2,\ x = 4,\ x = 8$

1992, 1994, 1998

(b) 4 years (1992-1994, 1998-2000)

(c) surplus of $2.5 million

13. $g(x) = x^4 - 9x^2 - 4x + 12$

end behavior: up/up

Possible rational roots:

$\dfrac{\left\{ \pm 1, \pm 12, \pm 2, \pm 6, \pm 3, \pm 4 \right\}}{\left\{ \pm 1 \right\}}$

$$
\begin{array}{r|rrrrr}
-2 & 1 & 0 & -9 & -4 & 12 \\
   &   & -2 & 4 & 10 & -12 \\
\hline
   & 1 & -2 & -5 & 6 & \underline{|0}
\end{array}
$$

$$
\begin{array}{r|rrrr}
-2 & 1 & -2 & -5 & 6 \\
   &   & -2 & 8 & -6 \\
\hline
   & 1 & -4 & 3 & \underline{|0}
\end{array}
$$

$g(x) = (x+2)^2 \left(x^2 - 4x + 3\right)$

$g(x) = (x+2)^2 (x-1)(x-3)$

-2 multiplicity 2, 1 multiplicity 1, 3 multiplicity 1

bounce at $(-2, 0)$; cross at $(1, 0)$ and $(3, 0)$

$g(0) = 0^4 - 9(0)^2 - 4(0) + 12 = 12$

y-intercept: $(0, 12)$

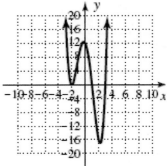

15. $C(x) = \dfrac{300x}{100 - x}$

   a. VA: $x = 100$; removal of 100% of the contaminants

   b. From 80% to 85%:

$$C(85) = \frac{300(85)}{100 - (85)} = 1700 ;$$

$1,700,000;

$$C(80) = \frac{300(80)}{100 - (80)} = 1200 ;$$

$1700 - 1200 = 500$, \$500,000;

From 90% to 95 %:

$$C(95) = \frac{300(95)}{100 - (95)} = 5700 ;$$

$5,700,000

$$C(90) = \frac{300(90)}{100 - (90)} = 2700 ;$$

$5700 - 2700 = 3000$; \$3,000,000;

It becomes cost prohibitive to remove all the contaminants.

   c. $2200 = \dfrac{300x}{100 - x}$

$$2200(100 - x) = 300x$$
$$220000 - 2200x = 300x$$
$$220000 = 2500x$$
$$88 = x$$
$$x = 88\%$$

17. $\overline{C(x)} = \dfrac{2x^2 + 25x + 128}{x}$

Using grapher: $x = 8$; 800 items
Minimizes costs

19. $C(h) = \dfrac{2h^2 + 5h}{h^3 + 55}$

   a.

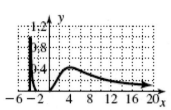

   b. $h^3 + 55 = 0$

$$h^3 = -55$$
$$h = -\sqrt[3]{55} , \text{ no}$$

   c. $C(2) = \dfrac{2(2)^2 + 5(2)}{(2)^3 + 55} \approx 0.286 = 28.6\% ;$

$$C(8) = \frac{2(8)^2 + 5(8)}{(8)^3 + 55} \approx 0.296 = 29.6\%$$

   d. $\dfrac{2h^2 + 5h}{h^3 + 55} < 0.2$

Using grapher: ≈ 11.7 hours

   e. Using grapher: 4 hours, 43.7%

   f. A trace amount of the chemical will remain in the bloodstream.

Chapter 3 Calculator Exploration

1. $Y_1 = \left(x^3 - 6x^2 + 32\right)\left(x^2 + 1\right)$
$$= (x - 4)^2 (x + 2)\left(x^2 + 1\right)$$

$Y_2 = x^3 - 6x^2 + 32 = (x - 4)^2 (x + 2)$

$Y_3 = x + 2$

3. They do not affect the solution.

Chapter 3 Strengthening Core Skills

1. $x^3 - 3x - 18 \le 0$

$$\frac{\{\pm 1, \pm 18, \pm 2, \pm 9, \pm 3, \pm 6\}}{\{\pm 1\}}$$

$$\begin{array}{r|rrr}
3 & 1 & 0 & -3 & -18 \\
& & 3 & 9 & 18 \\
\hline
& 1 & 3 & 6 & \boxed{0}
\end{array}$$

$(x-3)(x^2 + 3x + 6) \le 0$

$x \in (-\infty, 3]$

3. $x^3 - 13x + 12 < 0$

$$\frac{\{\pm 1, \pm 12, \pm 2, \pm 6, \pm 3, \pm 4\}}{\{\pm 1\}}$$

$$\begin{array}{r|rrrr}
-4 & 1 & 0 & -13 & 12 \\
& & -4 & 16 & -12 \\
\hline
& 1 & -4 & 3 & \boxed{0}
\end{array}$$

$(x+4)(x^2 - 4x + 3) < 0$

$(x-3)(x-1)(x+4) < 0$

$x \in (-\infty, -4) \cup (1, 3)$

5. $x^4 - x^2 - 12 > 0$

$(x^2 - 4)(x^2 + 3) > 0$

$(x-2)(x+2)(x^2 + 3) > 0$

$(x^2 + 3)$ does not affect the solution set.

$x \in (-\infty, -2) \cup (2, \infty)$

Cumulative Review Chapters R-3

1. $\dfrac{1}{R} = \dfrac{1}{R_1} + \dfrac{1}{R_2}$

$RR_1 R_2 \left[\dfrac{1}{R} \right] = \left[\dfrac{1}{R_1} + \dfrac{1}{R_2} \right] RR_1 R_2$

$R_1 R_2 = RR_2 + RR_1$

$R_1 R_2 = R(R_2 + R_1)$

$\dfrac{R_1 R_2}{R_1 + R_2} = R$

3. a. $x^3 - 1$

$= (x-1)(x^2 + x + 1)$

b. $x^3 - 3x^2 - 4x + 12$

$= x^2(x-3) - 4(x-3)$

$= (x-3)(x^2 - 4)$

$= (x-3)(x+2)(x-2)$

5. $x + 3 < 5$ or $5 - x < 4$

$x < 2$ or $-x < -1$

$x < 2$ or $x > 1$

$x \in (-\infty, \infty)$

7. $(2-3i)^2 - 4(2-3i) + 13 = 0$

$4 - 12i + 9i^2 - 8 + 12i + 13 = 0$

$4 - 12i - 9 - 8 + 12i + 13 = 0$

$0 = 0$

Verified

9. $(1, 17), (61, 28)$

$m = \dfrac{28 - 17}{61 - 1} = \dfrac{11}{60};$

$y - 17 = \dfrac{11}{60}(x - 1)$

$y - 17 = \dfrac{11}{60}x - \dfrac{11}{60}$

$y = \dfrac{11}{60}x + \dfrac{1009}{60};$

$y = \dfrac{11}{60}(121) + \dfrac{1009}{60} = 39$ minutes;

Driving time increases 11 minutes every 60 days.

11. $y = 1.18x^2 - 10.99x + 4.6$;

Using grapher, the profit is first earned in the 9th month.

13. $f(x) = \sqrt[3]{2x-3}$;

$x = \sqrt[3]{2y-3}$

$x^3 = 2y-3$

$x^3 + 3 = 2y$

$\dfrac{x^3+3}{2} = y$

$f^{-1}(x) = \dfrac{x^3+3}{2}$

$(f \circ f^{-1})(x) = \sqrt[3]{2\left(\dfrac{x^3+3}{2}\right)-3}$

$= \sqrt[3]{x^3+3-3} = \sqrt[3]{x^3} = x$;

$(f^{-1} \circ f)(x) = \dfrac{\left(\sqrt[3]{2x-3}\right)^3 + 3}{2}$

$= \dfrac{2x-3+3}{2} = \dfrac{2x}{2} = x$;

Verified

15. $F(x) = -f(x+1)+2$

Reflected in x-axis, left 1, up 2

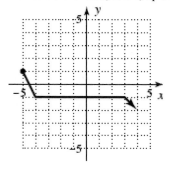

17. $Y = \dfrac{kX}{Z^2}$;

$10 = \dfrac{k(32)}{(4)^2}$

$5 = k$;

$Y = \dfrac{5X}{Z^2}$;

$1.4 = \dfrac{5x}{(15)^2}$

$x = 63$

19. $f(x) = x^3 - 3x^2 - 6x + 8$

Possible rational roots: $\{\pm 1, \pm 8, \pm 2, \pm 4\}$

$$
\begin{array}{r|rrrr}
1 & 1 & -3 & -6 & 8 \\
  &   & 1 & -2 & -8 \\
\hline
  & 1 & -2 & -8 & 0
\end{array}
$$

$f(x) = (x-1)(x^2 - 2x - 8)$

$f(x) = (x-1)(x+2)(x-4)$

$f(x) = (x-1)(x+2)(x-4)$

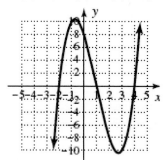

4.1 Technology Highlight

1. a. $f(x) = 2x + 1$;

    $x = 2y + 1$

    $x - 1 = 2y$

    $\dfrac{x-1}{2} = y$

    $f^{-1}(x) = \dfrac{x-1}{2}$

    b, c verified using a graphing calculator

    d.

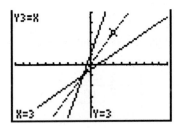

3. a. $h(x) = \dfrac{x}{x+1}$;

    $x = \dfrac{y}{y+1}$

    $x(y+1) = y$

    $xy + x = y$

    $xy - y = -x$

    $y(x-1) = -x$

    $y = \dfrac{-x}{x-1}$

    $f^{-1}(x) = \dfrac{-x}{x-1} = \dfrac{x}{1-x}$

    b, c verified using a graphing calculator

    d.

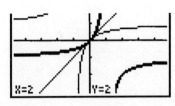

4.1 Exercises

1. Second; one

3. $(-11, -2), (-5, 0), (1, 2), (19, 4)$

5. False, answers will vary.

7. One-to-one

9. One-to-one

11. Not a function

13. One-to-one

15. Not one-to-one

17. Not one-to-one; $y = 7$ is paired with $x = -2$ and $x = 2$.

19. One-to-one

21. One-to-one

23. Not one-to-one; for $p(t) > 5$, one y corresponds to two x-values.

25. One-to-one

27. One-to-one

29. $f(x) = \{(-2, 1), (-1, 4), (0, 5), (2, 9), (5, 15)\}$
    $f^{-1}(x) = \{(1, -2), (4, -1), (5, 0), (9, 2), (15, 5\}$

31. $v(x) = \{(-4, 3), (-3, 2), (0, 1), (5, 0), (12, -1), (21, -2), (32, -3)\}$
    $v^{-1}(x) = \{(3, -4), (2, -3), (1, 0), (0, 5), (-1, 12), (-2, 21), (-3, 32)\}$

33. $f(x) = x + 5$
    $y = x + 5$;
    $x = y + 5$
    $x - 5 = y$
    $f^{-1}(x) = x - 5$

35. $p(x) = -\dfrac{4}{5}x$

$y = -\dfrac{4}{5}x;$

$x = -\dfrac{4}{5}y$

$-\dfrac{5}{4}x = y$

$p^{-1}(x) = -\dfrac{5}{4}x$

37. $f(x) = 4x + 3$

Multiply by 4, add 3
Inverse:
Subtract 3, divide by 4

$f^{-1}(x) = \dfrac{x-3}{4}$

39. $Y_1 = \sqrt[3]{x-4}$

Subtract 4, take cube root
Inverse:
Cube x, add 4
$Y_1^{-1} = x^3 + 4$

41. $f(x) = \sqrt[3]{x-2}$

a. $f(10) = \sqrt[3]{10-2} = \sqrt[3]{8} = 2$;

$f(-6) = \sqrt[3]{-6-2} = \sqrt[3]{-8} = -2$;

$f(1) = \sqrt[3]{1-2} = \sqrt[3]{-1} = -1$;

$(-6,-2),(1,-1),(10,2)$

b. $f(x) = \sqrt[3]{x-2}$;

Interchange x and y to find the inverse.

$x = \sqrt[3]{y-2}$

$x^3 = y - 2$

$x^3 + 2 = y$

$f^{-1}(x) = x^3 + 2$

c. $f^{-1}(-2) = (-2)^3 + 2 = -6$;

$f^{-1}(-1) = (-1)^3 + 2 = 1$;

$f^{-1}(2) = (2)^3 + 2 = 10$

43. $f(x) = x^3 + 1$

a. $f(0) = (0)^3 + 1 = 1$;

$f(1) = (1)^3 + 1 = 2$;

$f(-1) = (-1)^3 + 1 = 0$;

$(0,1),(1,2),(-1,0)$

b. $f(x) = x^3 + 1$;
Interchange x and y to find the inverse.

$x = y^3 + 1$

$x - 1 = y^3$

$\sqrt[3]{x-1} = y$

$f^{-1}(x) = \sqrt[3]{x-1}$

c. $f^{-1}(1) = \sqrt[3]{1-1} = 0$;

$f^{-1}(2) = \sqrt[3]{2-1} = 1$;

$f^{-1}(0) = \sqrt[3]{0-1} = -1$

45. $f(x) = \dfrac{8}{x+2}$

a. $f(0) = \dfrac{8}{0+2} = 4$;

$f(2) = \dfrac{8}{2+2} = 2$;

$f(6) = \dfrac{8}{6+2} = 1$;

$(0,4),(2,2),(6,1)$

b. $f(x) = \dfrac{8}{x+2}$;

Interchange x and y to find the inverse.

$x = \dfrac{8}{y+2}$

$x(y+2) = 8$

$xy + 2x = 8$

$xy = 8 - 2x$

$y = \dfrac{8}{x} - 2$

$f^{-1}(x) = \dfrac{8}{x} - 2$

4.1 Exercises

c. $f^{-1}(4) = \dfrac{8}{4} - 2 = 0$;

$f^{-1}(2) = \dfrac{8}{2} - 2 = 2$;

$f^{-1}(1) = \dfrac{8}{1} - 2 = 6$

47. $f(x) = \dfrac{x}{x+1}$

a. $f(0) = \dfrac{0}{0+1} = 0$;

$f(1) = \dfrac{1}{1+1} = \dfrac{1}{2}$;

$f(-2) = \dfrac{-2}{-2+1} = 2$;

$(0,0), \left(1, \dfrac{1}{2}\right), (-2,2)$

b. $f(x) = \dfrac{x}{x+1}$;

Interchange x and y to find the inverse.

$x = \dfrac{y}{y+1}$

$x(y+1) = y$

$xy + x = y$

$xy - y = -x$

$y(x-1) = -x$

$y = \dfrac{-x}{x-1}$

$y = \dfrac{x}{1-x}$

$f^{-1}(x) = \dfrac{x}{1-x}$

c. $f^{-1}(0) = \dfrac{0}{1-0} = 0$;

$f^{-1}\left(\dfrac{1}{2}\right) = \dfrac{\dfrac{1}{2}}{1 - \dfrac{1}{2}} = 1$;

$f^{-1}(2) = \dfrac{2}{1-2} = -2$

49. $f(x) = (x+5)^2$

a. Parabola with vertex (-5,0)
Restricting domain to $x \geq -5$ leaves
right branch of $f(x) = (x+5)^2$ with
range $y \geq 0$

b. For $x \geq -5$

$f(x) = (x+5)^2$

$y = (x+5)^2$

Interchange x and y to find the inverse.

$x = (y+5)^2$

$\pm\sqrt{x} = \sqrt{(y+5)^2}$

$\pm\sqrt{x} = y+5$

use $\sqrt{x}$ since $x \geq -5$

$\sqrt{x} - 5 = y$

$f^{-1}(x) = \sqrt{x} - 5$,

Domain $x \in [0,\infty)$, Range $y \in [-5,\infty)$

51. $v(x) = \dfrac{8}{(x-3)^2}$

a. Restricting domain to $x > 3$, range
$y > 0$

b. $v(x) = \dfrac{8}{(x-3)^2}$

Interchange x and y to find the inverse.

$x = \dfrac{8}{(y-3)^2}$

$x(y-3)^2 = 8$

$(y-3)^2 = \dfrac{8}{x}$

$y - 3 = \pm\sqrt{\dfrac{8}{x}}$

use $+\sqrt{\dfrac{8}{x}}$ since $x > 3$

$y = 3 + \sqrt{\dfrac{8}{x}}$

$v^{-1}(x) = 3 + \sqrt{\dfrac{8}{x}}$,

Domain $x \in (0,\infty)$, Range $y \in (3,\infty)$

53. $p(x) = (x+4)^2 - 2$

    a. Restricting domain to $x \geq -4$, range
        $y \geq -2$

    b. $p(x) = (x+4)^2 - 2$

        Interchange x and y to find the inverse.

        $x = (y+4)^2 - 2$

        $x + 2 = (y+4)^2$

        $\pm\sqrt{x+2} = y+4$

        use $\sqrt{x+2}$ since $x \geq -4$

        $\sqrt{x+2} - 4 = y$

        $p^{-1}(x) = \sqrt{x+2} - 4$

        Domain $x \in [-2, \infty)$,, Range $y \in [-4, \infty)$

55. $f(x) = -2x + 5$; $g(x) = \dfrac{x-5}{-2}$

$(f \circ g)(x) = f[g(x)]$

$= -2(g(x)) + 5$

$= -2\left(\dfrac{x-5}{-2}\right) + 5 = x - 5 + 5 = x;$

$(g \circ f)(x) = g[f(x)]$

$= \dfrac{f(x) - 5}{-2}$

$= \dfrac{-2x + 5 - 5}{-2}$

$= \dfrac{-2x}{-2}$

$= x$

57. $f(x) = \sqrt[3]{x+5}$; $g(x) = x^3 - 5$

$(f \circ g)(x) = f[g(x)]$

$= \sqrt[3]{g(x) + 5}$

$= \sqrt[3]{x^3 - 5 + 5}$

$= \sqrt[3]{x^3}$

$= x;$

$(g \circ f)(x) = g[f(x)]$

$= (f(x))^3 - 5$

$= \left(\sqrt[3]{x+5}\right)^3 - 5$

$= x + 5 - 5$

$= x$

59. $f(x) = \dfrac{2}{3}x - 6$; $g(x) = \dfrac{3}{2}x + 9$

$(f \circ g)(x) = f[g(x)]$

$= \dfrac{2}{3}g(x) - 6$

$= \dfrac{2}{3}\left(\dfrac{3}{2}x + 9\right) - 6$

$= x + 6 - 6$

$= x;$

$(g \circ f)(x) = g[f(x)]$

$= \dfrac{3}{2}f(x) + 9$

$= \dfrac{3}{2}\left(\dfrac{2}{3}x - 6\right) + 9$

$= x - 9 + 9$

$= x$

61. $f(x) = x^2 - 3$; $x \geq 0$; $g(x) = \sqrt{x+3}$

$(f \circ g)(x) = f[g(x)]$

$= (g(x))^2 - 3$

$= \left(\sqrt{x+3}\right)^2 - 3$

$= x + 3 - 3$

$= x;$

$(g \circ f)(x) = g[f(x)]$

$= \sqrt{f(x) + 3}$

$= \sqrt{x^2 - 3 + 3}$

$= \sqrt{x^2}$

$= x$

4.1 Exercises

63. $f(x) = 3x - 5$

$y = 3x - 5$

$x = 3y - 5$

$x + 5 = 3y$

$\dfrac{x+5}{3} = y$

$f^{-1}(x) = \dfrac{x+5}{3}$

$\left(f \circ f^{-1}\right)(x) = f\left[f^{-1}(x)\right]$

$= 3\left(f^{-1}(x)\right) - 5$

$= 3\left(\dfrac{x+5}{3}\right) - 5$

$= x + 5 - 5$

$= x;$

$\left(f^{-1} \circ f\right)(x) = f^{-1}[f(x)]$

$= \dfrac{f(x)+5}{3}$

$= \dfrac{3x-5+5}{3}$

$= \dfrac{3x}{3}$

$= x$

65. $f(x) = \dfrac{x-5}{2}$

$y = \dfrac{x-5}{2}$

$x = \dfrac{y-5}{2}$

$2x = y - 5$

$2x + 5 = y$

$f^{-1}(x) = 2x + 5$

$\left(f \circ f^{-1}\right)(x) = f\left[f^{-1}(x)\right]$

$= \dfrac{f^{-1}(x)-5}{2}$

$= \dfrac{2x+5-5}{2}$

$= \dfrac{2x}{2}$

$= x;$

$\left(f^{-1} \circ f\right)(x) = f^{-1}[f(x)]$

$= 2(f(x)) + 5$

$= 2\left(\dfrac{x-5}{2}\right) + 5$

$= x - 5 + 5$

$= x$

67. $f(x) = \dfrac{1}{2}x - 3$

$y = \dfrac{1}{2}x - 3$

$x = \dfrac{1}{2}y - 3$

$x + 3 = \dfrac{1}{2}y$

$2x + 6 = y$

$f^{-1}(x) = 2x + 6$

$\left(f \circ f^{-1}\right)(x) = f\left[f^{-1}(x)\right]$

$= \dfrac{1}{2}\left(f^{-1}(x)\right) - 3$

$= \dfrac{1}{2}(2x+6) - 3$

$= x + 3 - 3$

$= x;$

$\left(f^{-1} \circ f\right)(x) = f^{-1}[f(x)]$

$= 2(f(x)) + 6$

$= 2\left(\dfrac{1}{2}x - 3\right) + 6$

$= x - 6 + 6$

$= x$

69. $f(x) = x^3 + 3$

$\quad y = x^3 + 3$

$\quad x = y^3 + 3$

$\quad x - 3 = y^3$

$\quad \sqrt[3]{x-3} = y$

$\quad f^{-1}(x) = \sqrt[3]{x-3}$

$\quad \left(f \circ f^{-1}\right)(x) = f\left[f^{-1}(x)\right]$

$\quad = \left(f^{-1}(x)\right)^3 + 3$

$\quad = \left(\sqrt[3]{x-3}\right)^3 + 3$

$\quad = x - 3 + 3$

$\quad = x;$

$\quad \left(f^{-1} \circ f\right)(x) = f^{-1}\left[f(x)\right]$

$\quad = \sqrt[3]{f(x) - 3}$

$\quad = \sqrt[3]{x^3 + 3 - 3}$

$\quad = \sqrt[3]{x^3}$

$\quad = x$

71. $f(x) = \sqrt[3]{2x+1}$

$\quad y = \sqrt[3]{2x+1}$

$\quad x = \sqrt[3]{2y+1}$

$\quad x^3 = 2y + 1$

$\quad x^3 - 1 = 2y$

$\quad \dfrac{x^3 - 1}{2} = y$

$\quad f^{-1}(x) = \dfrac{x^3 - 1}{2}$

$\quad \left(f \circ f^{-1}\right)(x) = f\left[f^{-1}(x)\right]$

$\quad = \sqrt[3]{2\left(f^{-1}(x)\right) + 1}$

$\quad = \sqrt[3]{2\left(\dfrac{x^3 - 1}{2}\right) + 1}$

$\quad = \sqrt[3]{x^3 - 1 + 1}$

$\quad = \sqrt[3]{x^3}$

$\quad = x;$

$\left(f^{-1} \circ f\right)(x) = f^{-1}\left[f(x)\right]$

$\quad = \dfrac{(f(x))^3 - 1}{2}$

$\quad = \dfrac{\left(\sqrt[3]{2x+1}\right)^3 - 1}{2}$

$\quad = \dfrac{2x + 1 - 1}{2}$

$\quad = \dfrac{2x}{2}$

$\quad = x$

73. $f(x) = \dfrac{(x-1)^3}{8}$

$\quad y = \dfrac{(x-1)^3}{8}$

$\quad x = \dfrac{(y-1)^3}{8}$

$\quad 8x = (y-1)^3$

$\quad \sqrt[3]{8x} = y - 1$

$\quad 2\sqrt[3]{x} + 1 = y$

$\quad f^{-1}(x) = 2\sqrt[3]{x} + 1$

$\quad \left(f \circ f^{-1}\right)(x) = f\left[f^{-1}(x)\right]$

$\quad = \dfrac{\left(f^{-1}(x) - 1\right)^3}{8}$

$\quad = \dfrac{\left(2\sqrt[3]{x} + 1 - 1\right)^3}{8}$

$\quad = \dfrac{\left(2\sqrt[3]{x}\right)^3}{8}$

$\quad = \dfrac{8x}{8}$

$\quad = x;$

$\quad \left(f^{-1} \circ f\right)(x) = f^{-1}\left[f(x)\right]$

$\quad = 2\sqrt[3]{f(x)} + 1$

$\quad = 2\sqrt[3]{\dfrac{(x-1)^3}{8}} + 1$

$\quad = 2\left(\dfrac{x-1}{2}\right) + 1$

$\quad = x - 1 + 1$

$\quad = x$

4.1 Exercises

75. $f(x) = \sqrt{3x+2}$,

$x \in \left[-\dfrac{2}{3}, \infty\right), y \in [0, \infty)$

$y = \sqrt{3x+2}$

$x = \sqrt{3y+2}$

$x^2 = 3y+2$

$x^2 - 2 = 3y$

$\dfrac{x^2-2}{3} = y$

$f^{-1}(x) = \dfrac{x^2-2}{3}; x \geq 0; \ y \in \left[-\dfrac{2}{3}, \infty\right)$

$\left(f \circ f^{-1}\right)(x) = f\left(f^{-1}(x)\right)$

$= \sqrt{3\left(f^{-1}(x)\right)+2}$

$= \sqrt{3\left(\dfrac{x^2-2}{3}\right)+2}$

$= \sqrt{x^2 - 2 + 2}$

$= \sqrt{x^2}$

$= |x|$

$= x; \ \text{ since } x \geq 0$

$\left(f^{-1} \circ f\right)(x) = f^{-1}[f(x)]$

$= \dfrac{(f(x))^2 - 2}{3}$

$= \dfrac{\left(\sqrt{3x+2}\right)^2 - 2}{3}$

$= \dfrac{3x + 2 - 2}{3}$

$= \dfrac{3x}{3}$

$= x$

77. $p(x) = 2\sqrt{x-3}$

$x \in [3, \infty), y \in [0, \infty)$

$y = 2\sqrt{x-3}$

$x = 2\sqrt{y-3}$

$\dfrac{x}{2} = \sqrt{y-3}$

$\dfrac{x^2}{4} = y - 3$

$\dfrac{x^2}{4} + 3 = y$

$p^{-1}(x) = \dfrac{x^2}{4} + 3; x \geq 0; \ y \in [3, \infty)$

$\left(p \circ p^{-1}\right)(x) = p\left(p^{-1}(x)\right)$

$= 2\sqrt{p^{-1}(x) - 3}$

$= 2\sqrt{\dfrac{x^2}{4} + 3 - 3}$

$= 2\sqrt{\dfrac{x^2}{4}}$

$= 2\left(\dfrac{x}{2}\right)$

$= x;$

$\left(p^{-1} \circ p\right)(x) = p^{-1}[p(x)]$

$= \dfrac{(p(x))^2}{4} + 3$

$= \dfrac{\left(2\sqrt{x-3}\right)^2}{4} + 3$

$= \dfrac{4(x-3)}{4} + 3$

$= x - 3 + 3$

$= x$

79. $v(x) = x^2 + 3; x \geq 0, y \in [3, \infty)$

$y = x^2 + 3$

$x = y^2 + 3$

$x - 3 = y^2$

$\sqrt{x-3} = y$

$v^{-1}(x) = \sqrt{x-3}$; $x \geq 3, y \in [0, \infty)$

$\left(v \circ v^{-1}\right)(x) = v\left[v^{-1}(x)\right]$

$= \left(v^{-1}(x)\right)^2 + 3$

$= \left(\sqrt{x-3}\right)^2 + 3$

$= x - 3 + 3$

$= x;$

$\left(v^{-1} \circ v\right)(x) = v^{-1}[v(x)]$

$= \sqrt{v(x) - 3}$

$= \sqrt{x^2 + 3 - 3}$

$= \sqrt{x^2}$

$= |x|$

$= x;$ since $x \geq 0$

81. $f(x) = 4x + 1$; $f^{-1}(x) = \dfrac{x-1}{4}$

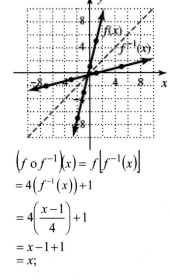

$\left(f \circ f^{-1}\right)(x) = f\left[f^{-1}(x)\right]$

$= 4\left(f^{-1}(x)\right) + 1$

$= 4\left(\dfrac{x-1}{4}\right) + 1$

$= x - 1 + 1$

$= x;$

$\left(f^{-1} \circ f\right)(x) = f^{-1}[f(x)]$

$= \dfrac{f(x) - 1}{4}$

$= \dfrac{4x + 1 - 1}{4}$

$= \dfrac{4x}{4}$

$= x$

83. $f(x) = \sqrt[3]{x+2}$; $f^{-1}(x) = x^3 - 2$

$\left(f \circ f^{-1}\right)(x) = f\left[f^{-1}(x)\right]$

$= \sqrt[3]{f^{-1}(x) + 2}$

$= \sqrt[3]{x^3 - 2 + 2}$

$= \sqrt[3]{x^3}$

$= x;$

$\left(f^{-1} \circ f\right)(x) = f^{-1}[f(x)]$

$= (f(x))^3 - 2$

$= \left(\sqrt[3]{x+2}\right)^3 - 2$

$= x + 2 - 2$

$= x$

4.1 Exercises

85. $f(x) = 0.2x + 1$; $f^{-1}(x) = 5x - 5$

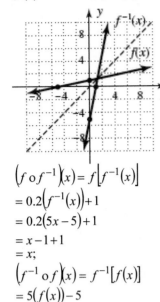

$$\left(f \circ f^{-1}\right)(x) = f\left[f^{-1}(x)\right]$$
$$= 0.2\left(f^{-1}(x)\right) + 1$$
$$= 0.2(5x - 5) + 1$$
$$= x - 1 + 1$$
$$= x;$$
$$\left(f^{-1} \circ f\right)(x) = f^{-1}\left[f(x)\right]$$
$$= 5(f(x)) - 5$$
$$= 5(0.2x + 1) - 5$$
$$= x + 5 - 5$$
$$= x$$

87. $f(x) = (x + 2)^2; x \geq -2$; $f^{-1}(x) = \sqrt{x} - 2$

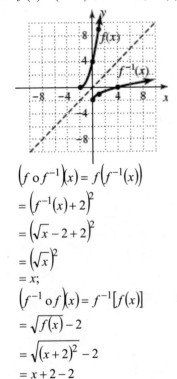

$$\left(f \circ f^{-1}\right)(x) = f\left(f^{-1}(x)\right)$$
$$= \left(f^{-1}(x) + 2\right)^2$$
$$= \left(\sqrt{x} - 2 + 2\right)^2$$
$$= \left(\sqrt{x}\right)^2$$
$$= x;$$
$$\left(f^{-1} \circ f\right)(x) = f^{-1}\left[f(x)\right]$$
$$= \sqrt{f(x)} - 2$$
$$= \sqrt{(x + 2)^2} - 2$$
$$= x + 2 - 2$$
$$= x$$

89. $f(x)$ $f^{-1}(x)$

$D: x \in [0, \infty)$ $D: x \in [-2, \infty)$

$R: y \in [-2, \infty)$ $R: y \in [0, \infty)$

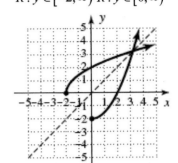

91. $f(x)$ $f^{-1}(x)$

$D: x \in (0, \infty)$ $D: x \in (-\infty, \infty)$

$R: y \in (-\infty, \infty)$ $R: y \in (0, \infty)$

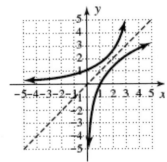

93. $f(x)$ $f^{-1}(x)$

$D: x \in (-\infty, 4]$ $D: x \in (-\infty, 4]$

$R: y \in (-\infty, 4]$ $R: y \in (-\infty, 4]$

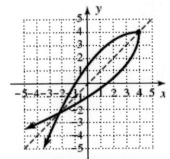

95. $f(x) = \frac{1}{2}x - 8.5$

   a. $f(80) = \frac{1}{2}(80) - 8.5 = 31.5$ cm

   b. $y = \frac{1}{2}x - 8.5$

   $x = \frac{1}{2}y - 8.5$

   $x + 8.5 = \frac{1}{2}y$

   $2x + 17 = y$

   $f^{-1}(x) = 2x + 17$

   $f^{-1}(31.5) = 2(31.5) + 17 = 80$

   It gives the distance of the projector from screen.

97. $f(x) = -\frac{7}{2}x + 59$

   a. $f(35) = -\frac{7}{2}(35) + 59 = -63.5°F$

   b. $y = -\frac{7}{2}x + 59$

   $x = -\frac{7}{2}y + 59$

   $x - 59 = -\frac{7}{2}y$

   $-\frac{2}{7}(x - 59) = y$

   $f^{-1}(x) = -\frac{2}{7}(x - 59)$

   Independent: temperature
   Dependent: altitude

   c. $f^{-1}(-18) = -\frac{2}{7}(-18 - 59)$

   $= -\frac{2}{7}(-77) = 22$

   The approximate altitude is 22000 feet.

99. $f(x) = 16x^2; x \geq 0$

   a. $f(3) = 16(3)^2 = 16(9) = 144$ ft

   b. $y = 16x^2$

   $x = 16y^2$

   $\frac{x}{16} = y^2$

   $\frac{\sqrt{x}}{4} = y$

   $f^{-1}(x) = \frac{\sqrt{x}}{4};$

   Independent: distance fallen
   Dependent: time fallen

   c. $f^{-1}(784) = \frac{\sqrt{784}}{4} = \frac{28}{4} = 7$ sec

101. $f(x) = \frac{1}{3}\pi x^3$

   a. $f(30) = \frac{1}{3}\pi(30)^3 = 9000\pi \approx 28260$ ft³

   b. $y = \frac{1}{3}\pi x^3$

   $x = \frac{1}{3}\pi y^3$

   $3x = \pi y^3$

   $\frac{3x}{\pi} = y^3$

   $\sqrt[3]{\frac{3x}{\pi}} = y$

   $f^{-1}(x) = \sqrt[3]{\frac{3x}{\pi}};$

   Independent: volume
   Dependent: height

   c. $f^{-1}(763.02) = \sqrt[3]{\frac{3(763.02)}{\pi}} = 9$ ft

4.1 Exercises

103. $f(x) = \{(x,y) \mid y = 3x - 6\}$

   a. $f(2) = 3(2) - 6 = 0, \ (2,0);$

       $f(0) = 3(0) - 6 = -6, \ (0,-6);$

       $f(1) = 3(1) - 6 = -3, \ (1,-3);$

       $f(3) = 3(3) - 6 = 3, \ (3,3);$

       $f(-1) = 3(-1) - 6 = -9, \ (-1,-9)$

   b. $(0,2),(-6,0),(-3,1),(3,3),(-9,-1)$

$$f^{-1}(x) = \left\{(x,y) \mid y = \frac{x}{3} + 2\right\}$$

$$f^{-1}(0) = \frac{0}{3} + 2 = 2, \ (0,2);$$

$$f^{-1}(-6) = \frac{-6}{3} + 2 = 0, \ (-6,0);$$

$$f^{-1}(-3) = \frac{-3}{3} + 2 = 1, \ (-3,1);$$

$$f^{-1}(3) = \frac{3}{3} + 2 = 3, \ (3,3);$$

$$f^{-1}(-9) = \frac{-9}{3} + 2 = -1, \ (-9,-1)$$

105. $f(x) = \dfrac{2}{3}\left(x - \dfrac{1}{2}\right)^5 + \dfrac{4}{5}$

$$y = \frac{2}{3}\left(x - \frac{1}{2}\right)^5 + \frac{4}{5}$$

$$x = \frac{2}{3}\left(y - \frac{1}{2}\right)^5 + \frac{4}{5}$$

$$x - \frac{4}{5} = \frac{2}{3}\left(y - \frac{1}{2}\right)^5$$

$$\frac{3}{2}\left(x - \frac{4}{5}\right) = \left(y - \frac{1}{2}\right)^5$$

$$\sqrt[5]{\frac{3}{2}\left(x - \frac{4}{5}\right)} = y - \frac{1}{2}$$

$$\sqrt[5]{\frac{3}{2}\left(x - \frac{4}{5}\right)} + \frac{1}{2} = y$$

   d. $f^{-1}(x) = \sqrt[5]{\dfrac{3}{2}\left(x - \dfrac{4}{5}\right)} + \dfrac{1}{2}$

107. $f(x) = x^2 - x - 2; \ f(x) \le 0$

$$x^2 - x - 2 = 0$$
$$(x-2)(x+1) = 0$$

$$x = 2; \ x = -1$$

Concave up

$$x \in [-1, 2]$$

109. a. Perimeter of a rectangle:
      $P = 2l + 2w$

   b. Area of a circle:
      $A = \pi r^2$

   c. Volume of a cylinder:
      $V = \pi r^2 h$

   d. Volume of a cone:
      $V = \dfrac{1}{3}\pi r^2 h$

   e. Circumference of a circle:
      $C = 2\pi r$

   f. Area of a triangle:
      $A = \dfrac{1}{2}bh$

   g. Area of a trapezoid:
      $A = \dfrac{1}{2}(b_1 + b_2)h$

   h. Volume of a sphere:
      $V = \dfrac{4}{3}\pi r^3$

   i. Pythagorean Theorem:
      $a^2 + b^2 = c^2$

4.2 Technology Highlight

1. $3^x = 22$; $x \approx 2.8$

3. $e^{x-1} = 9$; $x \approx 3.2$

4.2 Exercises

1. b^x, b, b, x

3. $a, 1$

5. False; for $|b| < 1$ and $x_2 > x_1, b^{x_2} < b^{x_1}$
    so the function is decreasing.

7. $P(t) = 2500 \cdot 4^t$;
    $P(2) = 2500 \cdot 4^2 = 40000$;
    $P\left(\dfrac{1}{2}\right) = 2500 \cdot 4^{\frac{1}{2}} = 5000$;
    $P\left(\dfrac{3}{2}\right) = 2500 \cdot 4^{\frac{3}{2}} = 20000$;
    $P\left(\sqrt{3}\right) = 2500 \cdot 4^{\sqrt{3}} \approx 27589.162$

9. $f(x) = 0.5 \cdot 10^x$;
    $f(3) = 0.5 \cdot 10^3 = 500$;
    $f\left(\dfrac{1}{2}\right) = 0.5 \cdot 10^{\frac{1}{2}} \approx 1.581$;
    $f\left(\dfrac{2}{3}\right) = 0.5 \cdot 10^{\frac{2}{3}} \approx 2.321$;
    $f\left(\sqrt{7}\right) = 0.5 \cdot 10^{\sqrt{7}} \approx 221.168$

11. $V(n) = 10,000\left(\dfrac{2}{3}\right)^n$;

    $V(0) = 10,000\left(\dfrac{2}{3}\right)^0 = 10000$;

    $V(4) = 10,000\left(\dfrac{2}{3}\right)^4 \approx 1975.309$;

    $V(4.7) = 10,000\left(\dfrac{2}{3}\right)^{4.7} \approx 1487.206$;

    $V(5) = 10,000\left(\dfrac{2}{3}\right)^5 \approx 1316.872$

13. $y = 3^x$
    y-intercept: $(0, 1)$

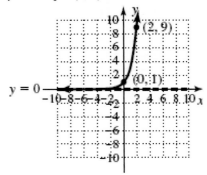

increasing

15. $y = \left(\dfrac{1}{3}\right)^x$
    y-intercept: $(0, 1)$

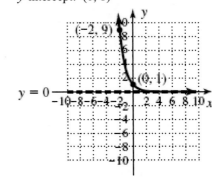

decreasing

17. $y = 3^x + 2$

    up 2

    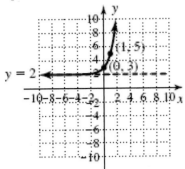

19. $y = 3^{x+3}$

    left 3

    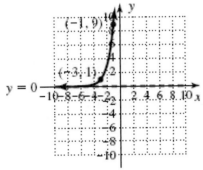

21. $y = 2^{-x}$

    reflected in the y-axis

    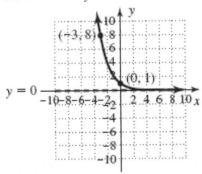

23. $y = 2^{-x} + 3$

    reflected in the y-axis, up 3

    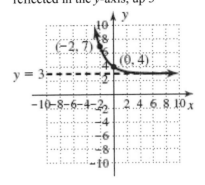

25. $y = 2^{x+1} - 3$

    left 1, down 3

    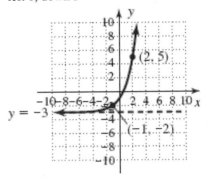

27. $y = \left(\dfrac{1}{3}\right)^x + 1$

    up 1

    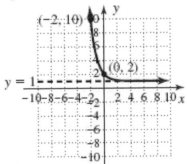

29. $y = \left(\dfrac{1}{3}\right)^{x-2}$

   right 2

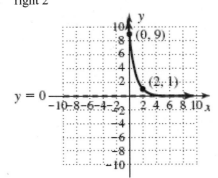

31. $y = \left(\dfrac{1}{3}\right)^{x} - 2$

   down 2

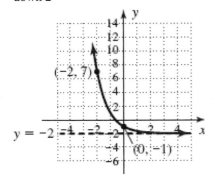

33. e; $y = 5^{-x}$

35. a; $y = 3^{-x+1}$

37. b; $y = 2^{x+1} - 2$

39. $e^{1} \approx 2.718282$

41. $e^{2} \approx 7.389056$

43. $e^{1.5} \approx 4.481689$

45. $e^{\sqrt{2}} \approx 4.113250$

47. $f(x) = e^{x+3} - 2$

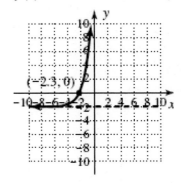

49. $r(t) = -e^{t} + 2$

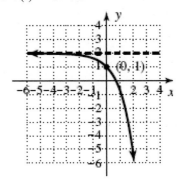

51. $p(x) = e^{-x+2} - 1$

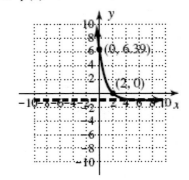

4.2 Exercises

53. $10^x = 1000$
$10^x = 10^3$
$x = 3$

55. $25^x = 125$
$5^{2x} = 5^3$
$2x = 3$
$x = \dfrac{3}{2}$

57. $8^{x+2} = 32$
$2^{3(x+2)} = 2^5$
$3x + 6 = 5$
$3x = -1$
$x = -\dfrac{1}{3}$

59. $32^x = 16^{x+1}$
$2^{5x} = 2^{4(x+1)}$
$5x = 4x + 4$
$x = 4$

61. $\left(\dfrac{1}{5}\right)^x = 125$
$\left(\dfrac{1}{5}\right)^x = \left(\dfrac{1}{5}\right)^{-3}$
$x = -3$

63. $\left(\dfrac{1}{3}\right)^{2x} = 9^{x-6}$
$\left(\dfrac{1}{3}\right)^{2x} = \left(\dfrac{1}{3}\right)^{-2(x-6)}$
$2x = -2x + 12$
$4x = 12$
$x = 3$

65. $\left(\dfrac{1}{9}\right)^{x-5} = 3^{3x}$
$\left(\dfrac{1}{3}\right)^{2(x-5)} = \left(\dfrac{1}{3}\right)^{-1(3x)}$
$2x - 10 = -3x$
$5x = 10$
$x = 2$

67. $25^{3x} = 125^{x-2}$
$5^{6x} = 5^{3(x-2)}$
$6x = 3x - 6$
$3x = -6$
$x = -2$

69. $\dfrac{e^4}{e^{2-x}} = e^3 e^1$
$e^{4-(2-x)} = e^4$
$4 - (2 - x) = 4$
$4 - 2 + x = 4$
$2 + x = 4$
$x = 2$

71. $\left(e^{2x-4}\right)^3 = \dfrac{e^{x+5}}{e^2}$
$e^{6x-12} = e^{x+5-2}$
$6x - 12 = x + 3$
$5x - 12 = 3$
$5x = 15$
$x = 3$

73. $P(t) = 1000 \cdot 3^t$

   (a) 12 hr $= \dfrac{1}{2}$ day

   $P\left(\dfrac{1}{2}\right) = 1000 \cdot 3^{\frac{1}{2}} \approx 1732$;

   $P(1) = 1000 \cdot 3^1 = 3000$;

   $P\left(\dfrac{3}{2}\right) = 1000 \cdot 3^{\frac{3}{2}} \approx 5196$;

   $P(2) = 1000 \cdot 3^2 = 9000$

   (b) yes

   (c) as $t \to \infty, P \to \infty$

   (d)

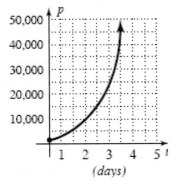

   (days)

75. $T(x) = T_R + (T_0 - T_R)e^{kx}$

   $T_R = 73°, T_0 = -10°, k \approx -0.031$

   $T(x) = 73 + (-10 - 73)e^{-0.031x}$

   $35 = 73 + (-10 - 73)e^{-0.031x}$

   Using calculator and table,

   $t \approx 25$ min , $25 - 15 = 10$ minutes after guests arrive

77. $V(t) = V_0 \cdot \left(\dfrac{4}{5}\right)^t$

   (a) $V(1) = 125000 \cdot \left(\dfrac{4}{5}\right)^1 = \$100{,}000$

   (b) $64000 = 125000 \cdot \left(\dfrac{4}{5}\right)^t$

   $\dfrac{64}{125} = \left(\dfrac{4}{5}\right)^t$

   $\left(\dfrac{4}{5}\right)^3 = \left(\dfrac{4}{5}\right)^t$

   $t = 3$ yr

79. $V(t) = V_0 \cdot \left(\dfrac{5}{6}\right)^t$

   (a) $V(5) = 216000 \cdot \left(\dfrac{5}{6}\right)^5 \approx \$86{,}806$

   (b) $125000 = 216000 \cdot \left(\dfrac{5}{6}\right)^t$

   $\dfrac{125}{216} = \left(\dfrac{5}{6}\right)^t$

   $\left(\dfrac{5}{6}\right)^3 = \left(\dfrac{5}{6}\right)^t$

   $t = 3$ yr

81. $R(t) = R_0 \cdot 2^t$

   (a) $R(4) = 2.5 \cdot 2^4 = \$40$ million

   (b) $320 = 2.5 \cdot 2^t$

   $128 = 2^t$

   $2^7 = 2^t$

   $t = 7$ yr

83. $T(x) = 0.85^x$

   $T(7) = 0.85^7 = 0.32058 \approx 32\%$,

   transparent

85. $T(11) = 0.85^{11} = 0.167734 \approx 17\%$,

   transparent

4.2 Exercises

87. $P(t) = P_0(1.05)^t$

$P(10) = 20000(1.05)^{10} \approx \$32,578$

89. $Q(t) = Q_0\left(\dfrac{1}{2}\right)^{\frac{t}{h}}$

(a) $Q(24) = 64\left(\dfrac{1}{2}\right)^{\frac{24}{8}} = 8 \text{ grams}$

(b) $1 = 64\left(\dfrac{1}{2}\right)^{\frac{t}{8}}$

$\dfrac{1}{64} = \left(\dfrac{1}{2}\right)^{\frac{t}{8}}$

$\left(\dfrac{1}{2}\right)^6 = \left(\dfrac{1}{2}\right)^{\frac{t}{8}}$

$6 = \dfrac{t}{8}$

$t = 48 \text{ minutes}$

91. $f(20) = \left(\dfrac{1}{2}\right)^{20} = 9.5 \times 10^{-7}$;

Answers will vary.

93. $5^{3x} = 27$, find 5^{2x}

$\left(5^x\right)^3 = 3^3$

$5^x = 3^1$

$\left(5^x\right)^2 = 3^2$

$5^{2x} = 9$

95. $\left(\dfrac{1}{2}\right)^{x+1} = \dfrac{1}{3}$, find $\left(\dfrac{1}{2}\right)^{-x}$

$\dfrac{1}{2}\left(\dfrac{1}{2}\right)^x = \dfrac{1}{3}$

$\left(\dfrac{1}{2}\right)^x = \dfrac{2}{3}$

$\left(\left(\dfrac{1}{2}\right)^x\right)^{-1} = \left(\dfrac{2}{3}\right)^{-1}$

$\left(\dfrac{1}{2}\right)^{-x} = \dfrac{3}{2}$

97. $\left(1 + \dfrac{1}{x}\right)^x$

a. $\dfrac{f(x+0.01) - f(x)}{x+0.01 - x}$

$= \dfrac{\left(1 + \dfrac{1}{x+0.01}\right)^{x+0.01} - \left(1 + \dfrac{1}{x}\right)^x}{0.01}$

TABLE:
At $x = 1$, 0.3842;
At $x = 4$, 0.0564;
At $x = 10$, 0.0114
At $x = 20$, 0.0031
The rate of growth seems to be approaching 0.

b. Using TABLE At $e = 2.718281828...$
$x = 16,608$

c. yes, the secant lines are becoming virtually horizontal $(y = e)$

99. $D : x \in [-2, \infty), R : y \in [-1, \infty)$

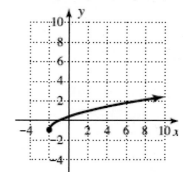

101.a. Volume of a sphere: $\dfrac{4}{3}\pi r^3$

b. Area of a triangle: $\dfrac{1}{2}bh$

c. Volume of a rectangular prism: lwh

d. Pythagorean theorem: $a^2 + b^2 = c^2$

Chapter 4: Exponential and Logarithmic Functions

4.3 Exercises

1. $\log_b x$, b, b, greater

3. $(1,0)$; 0

5. 5; answers will vary

7. $3 = \log_2 8$
   $2^3 = 8$

9. $-1 = \log_7 \dfrac{1}{7}$
   $7^{-1} = \dfrac{1}{7}$

11. $0 = \log_9 1$
    $9^0 = 1$

13. $\dfrac{1}{3} = \log_8 2$
    $8^{\frac{1}{3}} = 2$

15. $1 = \log_2 2$
    $2^1 = 2$

17. $\log_7 49 = 2$
    $7^2 = 49$

19. $\log_{10} 100 = 2$
    $10^2 = 100$

21. $\log_e(54.598) \approx 4$
    $e^4 \approx 54.598$

23. $4^3 = 64$
    $\log_4 64 = 3$

25. $3^{-2} = \dfrac{1}{9}$
    $\log_3\left(\dfrac{1}{9}\right) = -2$

27. $e^0 = 1$
    $0 = \log_e 1$

29. $\left(\dfrac{1}{3}\right)^{-3} = 27$
    $\log_{\frac{1}{3}} 27 = -3$

31. $10^3 = 1000$
    $\log 1000 = 3$

33. $10^{-2} = \dfrac{1}{100}$
    $\log \dfrac{1}{100} = -2$

35. $4^{\frac{3}{2}} = 8$
    $\log_4 8 = \dfrac{3}{2}$

37. $4^{\frac{-3}{2}} = \dfrac{1}{8}$
    $\log_4 \dfrac{1}{8} = \dfrac{-3}{2}$

39. $\log_4 4$
    $= 1$

41. $\log_{11} 121 = x$
    $11^x = 121$
    $11^x = 11^2$
    $x = 2$

43. $\log_e e = \log_e e^1 = 1$

45. $\log_4 2 = x$
    $4^x = 2$
    $2^{2x} = 2^1$
    $2x = 1$
    $x = \dfrac{1}{2}$

47. $\log_7 \dfrac{1}{49} = x$
    $7^x = \dfrac{1}{49}$
    $7^x = 7^{-2}$
    $x = -2$

49. $\log_e \dfrac{1}{e^2} = \log_e e^{-2} = -2$

51. $\log 50 = 1.6990$

53. $\ln 1.6 = 0.4700$

55. $\ln 225 = 5.4161$

57. $\log \sqrt{37} = 0.7841$

59. $f(x) = \log_2 x + 3$
    Shift up 3

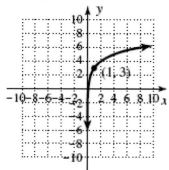

61. $h(x) = \log_2 (x-2) + 3$
    Shift right 2, up 3

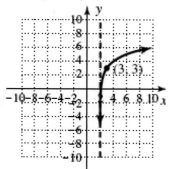

63. $q(x) = \ln(x+1)$
    Shift left 1

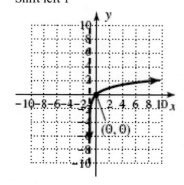

65. $Y_1 = -\ln(X+1)$
    Reflected across x–axis, shift left 1

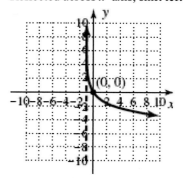

67. $y = \log_b (x+2)$, II

69. $y = 1 - \log_b x$, VI

71. $y = \log_b x + 2$, V

73. $y = \log_6 \left(\dfrac{x+1}{x-3} \right)$

    $\dfrac{x+1}{x-3} > 0, x \neq 3$

    critical values: -1 and 3

    pos　neg　pos

    -1　　3

    $x \in (-\infty, -1) \cup (3, \infty)$

75. $y = \log_5 \sqrt{2x-3}$
    $2x - 3 > 0$
    $2x > 3$
    $x > \dfrac{3}{2}$
    $x \in \left(\dfrac{3}{2}, \infty \right)$

77. $y = \log\left(9 - x^2\right)$
    $9 - x^2 > 0$;
    $(3+x)(3-x) > 0$
    critical values: 3 and -3

    neg　pos　neg

    -3　　3

    $x \in (-3, 3)$

79. $f(x) = -\log_{10} x$

$x = 7.94 \times 10^{-5}$

$f(7.94 \times 10^{-5}) = -\log_{10}(7.94 \times 10^{-5})$

pH ≈ 4.1; acid

81. $M(I) = \log\left(\dfrac{I}{I_0}\right)$

a. $I = 50,000 I_0$

$M(50000 I_0) = \log\left(\dfrac{50000 I_0}{I_0}\right)$

$M(50000 I_0) = \log(50000) \approx 4.7$

b. $I = 75,000 I_0$

$M(75000 I_0) = \log\left(\dfrac{75000 I_0}{I_0}\right)$

$M(75000 I_0) = \log(75000) \approx 4.9$

83. $M(I) = \log\left(\dfrac{I}{I_0}\right)$

1989: $6.2 = \log\left(\dfrac{I}{I_0}\right)$

$10^{6.2} = \dfrac{I}{I_0}$

$I = 10^{6.2} I_0$;

2006: $6.7 = \log\left(\dfrac{I}{I_0}\right)$

$10^{6.7} = \dfrac{I}{I_0}$

$I = 10^{6.7} I_0$;

Comparing the 1989 earthquakes to the 2006 earthquakes: $\dfrac{10^{6.7} I_0}{10^{6.2} I_0} \approx 3.2$ times

85. $M(I) = 6 - 2.5 \cdot \log\left(\dfrac{I}{I_0}\right)$

a. $I = 27 \cdot I_0$

$M(27 I_0) = 6 - 2.5 \cdot \log\left(\dfrac{27 I_0}{I_0}\right)$

≈ 2.4

b. $I = 85 \cdot I_0$

$M(85 I_0) = 6 - 2.5 \cdot \log\left(\dfrac{85 I_0}{I_0}\right)$

≈ 1.2

87. $D(I) = 10 \cdot \log\left(\dfrac{I}{I_0}\right)$

a. $I = 10^{-14}$

$D(10^{-14}) = 10 \cdot \log\left(\dfrac{10^{-14}}{10^{-16}}\right) = 20$ dB

b. $I = 10^{-4}$

$D(10^{-4}) = 10 \cdot \log\left(\dfrac{10^{-4}}{10^{-16}}\right) = 120$ dB

89. $D(I) = 10 \cdot \log\left(\dfrac{I}{I_0}\right)$;

Aircompressor: $D(I) = 110$

$110 = 10 \cdot \log\left(\dfrac{I}{I_0}\right)$

$\dfrac{110}{10} = \log\left(\dfrac{I}{I_0}\right)$

$11 = \log\left(\dfrac{I}{I_0}\right)$

$10^{11} = \dfrac{I}{I_0}$

$I = 10^{11} I_0$;

Hair Dryer: $D(I) = 75$

$75 = 10 \cdot \log\left(\dfrac{I}{I_0}\right)$

$\dfrac{75}{10} = \log\left(\dfrac{I}{I_0}\right)$

$7.5 = \log\left(\dfrac{I}{I_0}\right)$

$10^{7.5} = \dfrac{I}{I_0}$

$I = 10^{7.5} I_0$;

Comparison: $\dfrac{10^{11} I_0}{10^{7.5} I_0} \approx 3162$ times

4.3 Exercises

91. $H = (30T + 8000) \cdot \ln\left(\dfrac{P_0}{P}\right)$

 $T = -10, P = 34, P_0 = 76$

 $H = (30(-10) + 8000) \cdot \ln\left(\dfrac{76}{34}\right)$

 $H \approx 6194$ meters

93. $H = (30T + 8000) \cdot \ln\left(\dfrac{P_0}{P}\right)$

 a. $T = 8, P = 39.3, P_0 = 76$

 $H = (30(8) + 8000) \cdot \ln\left(\dfrac{76}{39.3}\right)$

 $H \approx 5,434$ meters

 b. $T = 12, P = 47.1, P_0 = 76$

 $H = (30(12) + 8000) \cdot \ln\left(\dfrac{76}{47.1}\right)$

 $H \approx 4,000$ meters

95. $N(A) = 1500 + 315 \cdot \ln(A)$

 (a) $N(10) = 1500 + 315 \cdot \ln(10)$
 $= 2225$ items

 (b) $N(50) = 1500 + 315 \cdot \ln(50)$
 $= 2732$ items

 (c) $\approx \$117,000$

 (d) $\dfrac{VN}{VA} = \dfrac{8}{1}$

 $\dfrac{N(39.4) - N(39.3)}{39.4 - 39.3} = \dfrac{8}{1}$

97. $C(x) = 42 \ln x - 270$

 (a) $C(2500) = 42 \ln 2500 - 270$
 ≈ 58.6 cfm

 (b) $40 = 42 \ln x - 270$
 $310 = 42 \ln x$
 $\dfrac{310}{42} = \ln x$
 $e^{\frac{310}{42}} = x$
 $x \approx 1,605$ ft^2

99. $P(x) = 95 - 14 \cdot \log_2(x)$

 a. 1 day
 $P(1) = 95 - 14 \cdot \log_2(1) = 95\%$

 b. 4 days
 $P(4) = 95 - 14 \cdot \log_2(4) = 67\%$

 c. 16 days
 $P(16) = 95 - 14 \cdot \log_2(16) = 39\%$

101. $f(x) = -\log_{10} x$

 $x = 5.1 \times 10^{-5}$
 $f(5.1 \times 10^{-5}) = -\log_{10}(5.1 \times 10^{-5})$
 pH ≈ 4.3 ; acid

103. a. Threshold of audibility
 0 dB

 b. Lawn Mower
 90 dB

 c. Whisper
 15 dB

 d. Loud rock concert
 120 dB

 e. Lively party
 100 dB

 f. Jet engine
 140 dB

 Many sources give the threshold of pain as 120dB; answers will vary.

Chapter 4: Exponential and Logarithmic Functions

105.a. $\log_{64}\dfrac{1}{16}=x$

Convert to exponential form:

$64^x=\dfrac{1}{16}$

$4^{3x}=4^{-2}$

$3x=-2$

$x=\dfrac{-2}{3}$

b. $\log_{\frac{4}{9}}\left(\dfrac{27}{8}\right)=x$

Convert to exponential form:

$\left(\dfrac{4}{9}\right)^x=\dfrac{27}{8}$

$\left(\dfrac{2}{3}\right)^{2x}=\left(\dfrac{2}{3}\right)^{-3}$

$2x=-3$

$x=\dfrac{-3}{2}$

c. $\log_{0.25}32=x$

$(0.25)^x=32$

$\left(\dfrac{1}{4}\right)^x=2^5$

$\left(4^{-1}\right)^x=2^5$

$\left(2^{-2}\right)^x=2^5$

$2^{-2x}=2^5$

$-2x=5$

$x=\dfrac{-5}{2}$

107. $g(x)=\sqrt[3]{x+2}-1$

D: $x\in R$

R: $y\in R$

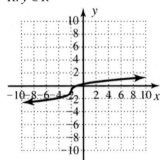

108.a. x^3-8

$=(x-2)\left(x^2+2x+4\right)$

b. a^2-49

$=(a+7)(a-7)$

c. $n^2-10n+25$

$=(n-5)(n-5)$

$=(n-5)^2$

d. $2b^2-7b+6$

$=(2b-3)(b-2)$

109. $x\in(-\infty,-5)$;

$f(x)=(x+5)(x-4)^2$

$f(x)=(x+5)\left(x^2-8x+16\right)$

$f(x)=x^3-8x^2+16x+5x^2-40x+80$

$f(x)=x^3-3x^2-24x+80$

Chapter 4 Mid-Chapter Check

1. a. $27^{\frac{2}{3}} = 9$

   $\frac{2}{3} = \log_{27} 9$

   b. $81^{\frac{5}{4}} = 243$

   $\frac{5}{4} = \log_{81} 243$

3. a. $4^{2x} = 32^{x-1}$

   $\left(2^2\right)^{2x} = \left(2^5\right)^{x-1}$

   $2^{4x} = 2^{5x-5}$

   $4x = 5x - 5$

   $x = 5$

   b. $\left(\frac{1}{3}\right)^{4b} = 9^{2b-5}$

   $\left(3^{-1}\right)^{4b} = \left(3^2\right)^{2b-5}$

   $3^{-4b} = 3^{4b-10}$

   $-4b = 4b - 10$

   $-8b = -10$

   $b = \frac{5}{4}$

5. $V(t) = V_0\left(\frac{9}{8}\right)^t$

   a. $V(3) = 50,000\left(\frac{9}{8}\right)^3 = \$71,191.41$

   b. 6 yr

7. $f(x) = \sqrt{x-3} + 1$

   $D: x \in [3, \infty); R: y \in [1, \infty)$;

   $f(x) = \sqrt{x-3} + 1$

   Interchange x and y.

   $x = \sqrt{y-3} + 1$

   $x - 1 = \sqrt{y-3}$

   $(x-1)^2 = y - 3$

   $(x-1)^2 + 3 = y$

   $f^{-1}(x) = (x-1)^2 + 3$

   $D: x \in [1, \infty); R: y \in [3, \infty)$

9. (a) $\frac{2}{3} = \log_{27} 9$

   $27^{\frac{2}{3}} = 9$

   $\left(3^3\right)^{\frac{2}{3}} = 9$

   $3^2 = 9$, verified

   (b) $1.4 \approx \ln 4.0552$

   $e^{1.4} \approx 4.0552$,

   verified on calculator

Reinforcing Basic Concepts

1. $14 - 11.8 = 2.2; \ 10^{2.2} \approx 158$

   About 158 times

3. $7.5 - 3.4 = 4.1; \ 10^{4.1} \approx 12,589$

   About 12,589 times

5. $9.1 - 4.5 = 4.6; \ 10^{4.6} \approx 39,811$

   About 39,811 times

4.4 Exercises

1. e

3. Extraneous

5. $\ln(4x+3)+\ln 2 = 3.2$
   $\ln 2(4x+3) = 3.2$
   $\ln(8x+6) = 3.2$
   $e^{3.2} = 8x+6$
   $e^{3.2}-6 = 8x$
   $\dfrac{e^{3.2}-6}{8} = x$
   $x = 2.316566275$

7. $\ln x = 3.4$
   $e^{\ln x} = e^{3.4}$
   $x = e^{3.4}$
   $x \approx 29.964$

9. $\log x = \dfrac{1}{4}$
   $10^{\log x} = 10^{\frac{1}{4}}$
   $x = 10^{\frac{1}{4}}$
   $x \approx 1.778$

11. $e^x = 9.025$
    $\ln e^x = \ln 9.025$
    $x = \ln 9.025$
    $x \approx 2.200$

13. $10^x = 18.197$
    $\log 10^x = \log 18.197$
    $x = \log 18.197$
    $x \approx 1.260$

15. $4e^{x-2}+5 = 70$
    $4e^{x-2} = 65$
    $e^{x-2} = 16.25$
    $\ln e^{x-2} = \ln 16.25$
    $x-2 = \ln 16.25$
    $x = 2+\ln 16.25$
    $x \approx 4.7881$

17. $10^{x+5}-228 = -150$
    $10^{x+5} = 78$
    $\log 10^{x+5} = \log 78$
    $x+5 = \log 78$
    $x = -5+\log 78$
    $x \approx -3.1079$

19. $-150 = 290.8 - 190e^{-0.75x}$
    $-440.8 = -190e^{-0.75x}$
    $\dfrac{-440.8}{-190} = e^{-0.75x}$
    $\dfrac{58}{25} = e^{-0.75x}$
    $\ln\left(\dfrac{58}{25}\right) = \ln e^{-0.75x}$
    $\ln\left(\dfrac{58}{25}\right) = -0.75x$
    $\dfrac{\ln\left(\dfrac{58}{25}\right)}{-0.75} = x$
    $x \approx -1.1221$

21. $3\ln(x+4)-5 = 3$
    $3\ln(x+4) = 8$
    $\ln(x+4) = \dfrac{8}{3}$
    $x+4 = e^{\frac{8}{3}}$
    $x = e^{\frac{8}{3}}-4$
    $x \approx 10.3919$

23. $-1.5 = 2\log(5-x)-4$
    $2.5 = 2\log(5-x)$
    $1.25 = \log(5-x)$
    $10^{1.25} = 10^{\log(5-x)}$
    $10^{1.25} = 5-x$
    $x = 5-10^{1.25}$
    $x \approx -12.7828$

4.4 Exercises

25. $\dfrac{1}{2}\ln(2x+5)+3=3.2$

$\dfrac{1}{2}\ln(2x+5)=0.2$

$\ln(2x+5)=0.4$

$e^{\ln(2x+5)}=e^{0.4}$

$2x+5=e^{0.4}$

$2x=e^{0.4}-5$

$x=\dfrac{e^{0.4}-5}{2}$

$x\approx-1.7541$

27. $\ln(2x)+\ln(x-7)$

$=\ln(2x(x-7))$

$=\ln(2x^2-14x)$

29. $\log(x+1)+\log(x-1)$

$=\log((x+1)(x-1))$

$=\log(x^2-1)$

31. $\log_3 28-\log_3 7$

$=\log_3\left(\dfrac{28}{7}\right)$

$=\log_3(4)$

33. $\log x-\log(x+1)$

$=\log\left(\dfrac{x}{x+1}\right)$

35. $\ln(x-5)-\ln x$

$=\ln\left(\dfrac{x-5}{x}\right)$

37. $\ln(x^2-4)-\ln(x+2)$

$=\ln\left(\dfrac{x^2-4}{x+2}\right)$

$=\ln\left(\dfrac{(x+2)(x-2)}{x+2}\right)$

$=\ln(x-2)$

39. $\log_2 7+\log_2 6$

$=\log_2(7\cdot6)$

$=\log_2 42$

41. $\log_5(x^2-2x)+\log_5 x^{-1}$

$=\log_5\left(x^{-1}(x^2-2x)\right)$

$=\log_5(x-2)$

43. $\log 8^{x+2}=(x+2)\log 8$

45. $\ln 5^{2x-1}=(2x-1)\ln 5$

47. $\log\sqrt{22}=\log 22^{\frac{1}{2}}=\dfrac{1}{2}\log 22$

49. $\log_5 81=\log_5 3^4=4\log_5 3$

51. $\log(a^3 b)=\log a^3+\log b=3\log a+\log b$

53. $\ln\left(x\sqrt[4]{y}\right)=\ln x+\ln y^{\frac{1}{4}}$

$=\ln x+\dfrac{1}{4}\ln y$

55. $\ln\left(\dfrac{x^2}{y}\right)=\ln x^2-\ln y$

$=2\ln x-\ln y$

57. $\log\left(\sqrt{\dfrac{x-2}{x}}\right)=\log\left(\dfrac{x-2}{x}\right)^{\frac{1}{2}}$

$=\dfrac{1}{2}\log\left(\dfrac{x-2}{x}\right)$

$=\dfrac{1}{2}\left[\log(x-2)-\log x\right]$

59. $\ln\left(\dfrac{7x\sqrt{3-4x}}{2(x-1)^3}\right)$

$= \ln\left(7x\sqrt{3-4x}\right) - \ln\left(2(x-1)^3\right)$

$= \ln 7x + \ln\sqrt{3-4x} - \left[\ln 2 + \ln(x-1)^3\right]$

$= \ln 7x + \ln(3-4x)^{\frac{1}{2}} - \ln 2 - \ln(x-1)^3$

$= \ln 7 + \ln x + \dfrac{1}{2}\ln(3-4x) - \ln 2 - 3\ln(x-1)$

61. $\log_7 60 = \dfrac{\ln 60}{\ln 7} = 2.104076884$

63. $\log_5 152 = \dfrac{\ln 152}{\ln 5} \approx 3.121512475$

65 $\log_3 1.73205 = \dfrac{\log 1.73205}{\log 3}$

≈ 0.499999576

67. $\log_{0.5} 0.125 = \dfrac{\log 0.125}{\log 0.5} = 3$

69. $f(x) = \log_3 x = \dfrac{\log x}{\log 3}$;

$f(5) = \dfrac{\log 5}{\log 3} \approx 1.4650$;

$f(15) = \dfrac{\log 15}{\log 3} \approx 2.4650$;

$f(45) = \dfrac{\log 45}{\log 3} \approx 3.4650$;

Outputs increase by 1; $f\left(3^3 \cdot 5\right) \approx 4.465$

71. $h(x) = \log_9 x = \dfrac{\log x}{\log 9}$;

$h(2) = \dfrac{\log 2}{\log 9} \approx 0.3155$;

$h(4) = \dfrac{\log 4}{\log 9} \approx 0.6309$;

$h(8) = \dfrac{\log 8}{\log 9} \approx 0.9464$;

Outputs are multiples of 0.3155;
$h\left(2^4\right) = 4(0.3155) \approx 1.2619$

73. $\log 4 + \log(x-7) = 2$

$\log 4(x-7) = 2$

$4x - 28 = 10^2$

$4x = 128$

$x = 32$;

Check:

$\log 4 + \log(32-7) = 2$

$\log 4 + \log 25 = 2$

$\log 100 = 2$

$\log 10^2 = 2$

$2 = 2$

75. $\log(2x-5) - \log 78 = -1$

$\log\left(\dfrac{2x-5}{78}\right) = -1$

$\dfrac{2x-5}{78} = 10^{-1}$

$\dfrac{2x-5}{78} = \dfrac{1}{10}$

$2x - 5 = \dfrac{78}{10}$

$2x = 12.8$

$x = 6.4$;

Check:

$\log(2(6.4)-5) - \log 78 = -1$

$\log(7.8) - \log 78 = -1$

$\log\left(\dfrac{7.8}{78}\right) = -1$

$\log(0.1) = -1$

$\log 10^{-1} = -1$

$-1 = -1$

77. $\log(x-15)-2=-\log x$

$\log(x-15)+\log x=2$

$\log x(x-15)=2$

$x(x-15)=10^2$

$x^2-15x=100$

$x^2-15x-100=0$

$(x-20)(x+5)=0$

$x-20=0$ or $x+5=0$

$x=20$ or $x=-5$;

Check $x=20$:

$\log(20-15)-2=-\log 20$

$\log(5)-2=-\log 20$

$-1.3010=-1.3010$;

Check $x=-5$:

$\log(-5-15)-2=-\log(-5)$

-5 is not in the domain;

$x=20, x=-5$ is extraneous

79. $\log(2x+1)=1-\log x$

$\log(2x+1)+\log x=1$

$\log x(2x+1)=1$

$x(2x+1)=10^1$

$2x^2+x=10$

$2x^2+x-10=0$

$(2x+5)(x-2)=0$

$2x+5=0$ or $x-2=0$

$x=\dfrac{-5}{2}$ or $x=2$;

Check $x=-\dfrac{5}{2}$:

$\log\left(2\left(-\dfrac{5}{2}\right)+1\right)=1-\log\left(-\dfrac{5}{2}\right)$

$\log(-4)=1-\log\left(-\dfrac{5}{2}\right)$

$-\dfrac{5}{2}$ is not in the domain;

Check $x=2$:

$\log(2(2)+1)=1-\log(2)$

$\log(5)=1-\log(2)$

$0.69897=0.69897$;

$x=2, x=-\dfrac{5}{2}$ is extraneous

81. $\log(5x+2)=\log 2$

$5x+2=2$

$5x=0$

$x=0$

83. $\log_4(x+2)-\log_4 3=\log_4(x-1)$

$\log_4\left(\dfrac{x+2}{3}\right)=\log_4(x-1)$

$\dfrac{x+2}{3}=x-1$

$x+2=3x-3$

$-2x=-5$

$x=\dfrac{5}{2}$

85. $\ln(8x-4)=\ln 2+\ln x$

$\ln(8x-4)=\ln(2x)$

$8x-4=2x$

$6x=4$

$x=\dfrac{2}{3}$

87. $\log(2x-1)+\log 5=1$

Write in exponential form:

$\log(10x-5)=1$

$10^1=10x-5$

$15=10x$

$x=\dfrac{3}{2}$

89. $\log_2 9+\log_2(x+3)=3$

Write in exponential form:

$\log_2(9x+27)=3$

$2^3=9x+27$

$8=9x+27$

$-19=9x$

$x=\dfrac{-19}{9}$

91. $\ln(x+7)+\ln 9=2$

Write in exponential form:

$\ln(9x+63)=2$

$e^2=9x+63$

$e^2-63=9x$

$x=\dfrac{e^2-63}{9}$

93. $\log(x+8) + \log x = \log(x+18)$

Write in exponential form:

$\log(x^2 + 8x) = \log(x + 18)$

$x^2 + 8x = x + 18$

$x^2 + 7x - 18 = 0$

$(x + 9)(x - 2) = 0$

$x + 9 = 0 \ \text{ or } \ x - 2 = 0$

$x = -9 \quad \text{ or } \ x = 2$

$x = 2, \ -9$ is extraneous

95. $\ln(2x + 1) = 3 + \ln 6$

$e^{3+\ln 6} = 2x + 1$

$e^{3+\ln 6} - 1 = 2x$

$\dfrac{e^{3+\ln 6} - 1}{2} = x$

$\dfrac{e^3 e^{\ln 6} - 1}{2} = x$

$\dfrac{e^3(6) - 1}{2} = x$

$\dfrac{6e^3 - 1}{2} = x$

$x = 3e^3 - \dfrac{1}{2}$

$x \approx 59.75661077$

97. $\log(-x - 1) = \log(5x) - \log x$

$\log(-x - 1) = \log 5$

$-x - 1 = 5$

$-x = 6$

$x = -6$

$x = --6$ is extraneous, No Solution

99. $\ln(2t + 7) = \ln(3) - \ln(t + 1)$

$\ln(2t + 7) = \ln\left(\dfrac{3}{t+1}\right)$

$2t + 7 = \dfrac{3}{t+1}$

$(t + 1)(2t + 7) = \left(\dfrac{3}{t+1}\right)(t+1)$

$2t^2 + 9t + 7 = 3$

$2t^2 + 9t + 4 = 0$

$(2t + 1)(t + 4) = 0$

$2t + 1 = 0 \ \text{ or } \ t + 4 = 0$

$2t = -1 \quad \text{ or } \ t = -4$

$t = -\dfrac{1}{2}, \ -4$ is extraneous

101. $\log(x - 1) - \log x = \log(x - 3)$

$\log \dfrac{x-1}{x} = \log(x - 3)$

$\dfrac{x-1}{x} = x - 3$

$x - 1 = x(x - 3)$

$x - 1 = x^2 - 3x$

$0 = x^2 - 4x + 1$

$a = 1, b = -4, c = 1$

$x = \dfrac{-(-4) \pm \sqrt{(-4)^2 - 4(1)(1)}}{2(1)}$

$x = \dfrac{4 \pm \sqrt{12}}{2}$

$x = \dfrac{4 \pm 2\sqrt{3}}{2}$

$x = 2 \pm \sqrt{3}$

$x = 2 + \sqrt{3}, x = 2 - \sqrt{3}$ is extraneous

103. $7^{x+2} = 231$

$\ln 7^{x+2} = \ln 231$

$(x + 2)\ln 7 = \ln 231$

$x + 2 = \dfrac{\ln 231}{\ln 7}$

$x = \dfrac{\ln 231}{\ln 7} - 2$

$x \approx 0.7968$

105. $5^{3x-2} = 128,965$

$\ln 5^{3x-2} = \ln 128965$

$(3x - 2)\ln 5 = \ln 128965$

$3x - 2 = \dfrac{\ln 128695}{\ln 5}$

$3x = \dfrac{\ln 128695}{\ln 5} + 2$

$x = \dfrac{\ln 128965}{3\ln 5} + \dfrac{2}{3}$

$x \approx 3.1038$

107. $2^{x+1} = 3^x$

$\ln 2^{x+1} = \ln 3^x$

$(x+1)\ln 2 = x\ln 3$

$x\ln 2 + \ln 2 = x\ln 3$

$x\ln 2 - x\ln 3 = -\ln 2$

$x(\ln 2 - \ln 3) = -\ln 2$

$x = \dfrac{-\ln 2}{\ln 2 - \ln 3}$

$x = \dfrac{\ln 2}{\ln 3 - \ln 2}$

$x \approx 1.7095$

109. $5^{2x+1} = 9^{x+1}$

$\ln 5^{2x+1} = \ln 9^{x+1}$

$(2x+1)\ln 5 = (x+1)\ln 9$

$2x\ln 5 + \ln 5 = x\ln 9 + \ln 9$

$2x\ln 5 - x\ln 9 = \ln 9 - \ln 5$

$x(2\ln 5 - \ln 9) = \ln 9 - \ln 5$

$x = \dfrac{\ln 9 - \ln 5}{2\ln 5 - \ln 9}$

$x \approx 0.5753$

111. $\dfrac{250}{1+4e^{-0.06x}} = 200$

$250 = 200\left(1+4e^{-0.06x}\right)$

$\dfrac{250}{200} = 1+4e^{-0.06x}$

$\dfrac{1}{4} = 4e^{-0.06x}$

$\dfrac{1}{16} = e^{-0.06x}$

$\ln\dfrac{1}{16} = \ln e^{-0.06x}$

$\ln\dfrac{1}{16} = -0.06x$

$\dfrac{\ln\dfrac{1}{16}}{-0.06} = x$

$x \approx 46.2$

113. $P = \dfrac{C}{1+ae^{-kt}}$

$P\left(1+ae^{-kt}\right) = C$

$1+ae^{-kt} = \dfrac{C}{P}$

$ae^{-kt} = \dfrac{C}{P} - 1$

$e^{-kt} = \dfrac{\dfrac{C}{P} - 1}{a}$

$\ln e^{-kt} = \ln\left(\dfrac{\dfrac{C}{P} - 1}{a}\right)$

$-kt = \ln\left(\dfrac{\dfrac{C}{P} - 1}{a}\right)$

$t = \dfrac{\ln\left(\dfrac{\dfrac{C}{P} - 1}{a}\right)}{-k}$;

$C = 450,\ a = 8,\ P = 400,\ k = 0.075$

$t = \dfrac{\ln\left(\dfrac{\dfrac{450}{400} - 1}{8}\right)}{-0.075} \approx 55.45$

115. $P(t) = \dfrac{750}{1 + 24e^{-0.075t}}$

a. $P(0) = \dfrac{750}{1 + 24e^{-0.075(0)}}$

$P(0) = 30$ fish

b. $300 = \dfrac{750}{1 + 24e^{-0.075t}}$

$300\left(1 + 24e^{-0.075t}\right) = 750$

$1 + 24e^{-0.075t} = \dfrac{750}{300}$

$24e^{-0.075t} = \dfrac{3}{2}$

$e^{-0.075t} = \dfrac{1}{16}$

$\ln e^{-0.075t} = \ln \dfrac{1}{16}$

$-0.075t = \ln \dfrac{1}{16}$

$t = \dfrac{\ln \dfrac{1}{16}}{-0.075}$

$t \approx 37$ months

117. $H = (30T + 8{,}000)\ln\left(\dfrac{P_0}{P}\right),\ P_0 = 76$

a. $H = 18250, T = -75$

$18{,}250 = (30(-75) + 8{,}000)\ln\left(\dfrac{76}{P}\right)$

$18{,}250 = 5{,}750\ln\left(\dfrac{76}{P}\right)$

$\dfrac{18{,}250}{5{,}750} = \ln\left(\dfrac{76}{P}\right)$

$e^{\frac{73}{23}} = \dfrac{76}{P}$

$Pe^{\frac{73}{23}} = 76$

$P = \dfrac{76}{e^{\frac{73}{23}}}$

$P \approx 3.2$ cmHg

119. $T = T_R + (T_0 - T_R)e^{-kh}$

$32 = -20 + (75 - (-20))e^{-0.012h}$

$52 = 95e^{-0.012h}$

$\dfrac{52}{95} = e^{-0.012h}$

$\ln \dfrac{52}{95} = \ln e^{-0.012h}$

$\ln \dfrac{52}{95} = -0.012h$

$\dfrac{\ln \dfrac{52}{95}}{-0.012} = h$

$h \approx 50.2$ min

121. $T = k \ln \dfrac{V_n}{V_f}$

$3 = 5 \ln \dfrac{28500}{V_f}$

$\dfrac{3}{5} = \ln \dfrac{28500}{V_f}$

$e^{\frac{3}{5}} = \dfrac{28500}{V_f}$

$V_f e^{\frac{3}{5}} = 285000$

$V_f = \dfrac{28500}{e^{\frac{3}{5}}}$

$V_f = \$15{,}641$

123. $T(p) = \dfrac{-\ln p}{k}$

a. $k = 0.072$

$T(0.65) = \dfrac{-\ln 0.65}{0.072} \approx 5.98$

About 6 hours

b. $24 = \dfrac{-\ln P}{0.072}$

$24(0.072) = -\ln P$

$1.728 = -\ln P$

$-1.728 = \ln P$

$e^{-1.728} = P$

$P \approx 0.1776$ or 18.0%

4.4 Exercises

125. $V_s = V_e \ln\left(\dfrac{M_s}{M_s - M_f}\right)$

$6 = 8\ln\left(\dfrac{100}{100 - M_f}\right)$

$\dfrac{6}{8} = \ln\left(\dfrac{100}{100 - M_f}\right)$

$\dfrac{3}{4} = \ln\left(\dfrac{100}{100 - M_f}\right)$

$e^{\frac{3}{4}} = \dfrac{100}{100 - M_f}$

$e^{\frac{3}{4}}\left(100 - M_f\right) = 100$

$100e^{\frac{3}{4}} - M_f e^{\frac{3}{4}} = 100$

$100e^{\frac{3}{4}} = 100 + M_f e^{\frac{3}{4}}$

$100e^{\frac{3}{4}} - 100 = M_f e^{\frac{3}{4}}$

$\dfrac{100e^{\frac{3}{4}} - 100}{e^{\frac{3}{4}}} = M_f$

$M_f = 52.76$ tons

127. $P(t) = 5.9 + 12.6\ln t$

a. $P(5) = 5.9 + 12.6\ln 5 = 26$ planes

b. $34 = 5.9 + 12.6\ln t$

$28.1 = 12.6\ln t$

$\dfrac{28.1}{12.6} = \ln t$

$e^{\frac{28.1}{12.6}} = t$

$t \approx 9$ days

129. Answers will vary.

131. a. d
  b. e
  c. b
  d. f
  e. a
  f. c

133. $3e^{2x} - 4e^x - 7 = -3$

Let $u = e^x$

$3u^2 - 4u - 4 = 0$

$(3u + 2)(u - 2) = 0$

$3u + 2 = 0$ or $u - 2 = 0$

$3u = -2$ or $u = 2$

$u = -\dfrac{2}{3}$ or $u = 2$

$e^x = -\dfrac{2}{3}$ or $e^x = 2$

$\ln e^x = \ln\left(-\dfrac{2}{3}\right)$ or $\ln e^x = \ln 2$

$x \neq \ln\left(-\dfrac{2}{3}\right)$ or $x = \ln 2$

$x \approx 0.69314718$

135. a. $f(x) = 3^{x-2}$; $g(x) = \log_3 x + 2$

$(f \circ g)(x) = 3^{(\log_3 x + 2) - 2} = 3^{\log_3 x} = x$;

$(g \circ f)(x) = \log_3\left(3^{x-2}\right) + 2 =$

$(x - 2)\log_3 3 + 2 = x - 2 + 2 = x$

b. $f(x) = e^{x-1}$; $g(x) = \ln x + 1$

$(f \circ g)(x) = e^{(\ln x + 1) - 1} = e^{\ln x} = x$;

$(g \circ f)(x) = \ln\left(e^{x-1}\right) + 1 =$

$(x - 1)\ln e + 1 = x - 1 + 1 = x$

137.a. $y = e^{x \ln 2} = e^{\ln 2^x} = 2^x$;

$y = 2^x$

$\ln y = x \ln 2$

$e^{\ln y} = e^{x \ln 2}$

$y = e^{x \ln 2}$

b. $y = b^x$

$\ln y = \ln b^x$

$\ln y = x \ln b$

$e^{\ln y} = e^{x \ln b}$

$y = e^{xr}$ for $r = \ln b$

139. Answers will vary.

141. b

143. $r(x) = \dfrac{x^2 - 4}{x - 1}$

$r(x) = \dfrac{(x + 2)(x - 2)}{x - 1}$

VA: $x = 1$

HA: none (deg num > deg den)

$$\begin{array}{r} x + 1 \\ x - 1 \overline{) x^2 + 0x - 4} \\ -\underline{(x^2 - x)} \\ x - 4 \\ -\underline{(x - 1)} \\ -3 \end{array}$$

Oblique Asymptote: $y = x + 1$

x-intercepts: $(-2, 0)$ and $(2, 0)$;

$r(0) = \dfrac{0^2 - 4}{0 - 1} = 4$

y-intercept: $(0, 4)$

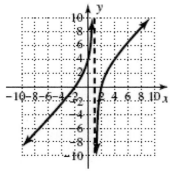

4.5 Technology Highlight

1. $A = P\left(1 + \dfrac{r}{n}\right)^{nt}$

Doubling time, find x when $y = 2000$

$Y_1 = 1000\left(1 + \dfrac{0.08}{4}\right)^{4x} \approx 8.75$ yr;

$Y_1 = 1000\left(1 + \dfrac{0.08}{12}\right)^{12x} \approx 8.69$ yr;

$Y_1 = 1000\left(1 + \dfrac{0.08}{365}\right)^{365x} \approx 8.665$ yr;

$Y_1 = 1000\left(1 + \dfrac{0.08}{365 \cdot 24}\right)^{365 \cdot 24x} \approx 8.664$ yr

8.75 yr compounded quarterly; 8.69 yr compounded monthly; 8.665 yr compounded daily; 8.664 yr compounded hourly

3. No. Examples will vary.

4.5 Exercises

1. Compound

3. $Q_0 e^{-rt}$

5. Answers will vary.

7. $I = prt$;

   9 months $= \dfrac{3}{4}$ year;

   $229.50 = p(0.0625)(0.75)$

   $\dfrac{229.50}{(0.0625)(0.75)} = p$

   $\$4896 = p$

9. $I = prt$

   $297.50 - 260 = 260r\left(\dfrac{3}{52}\right)$

   $37.5 = 15r$

   $\dfrac{37.5}{15} = r$

   $2.50 = r$

   $r = 250\%$

11. $A = p(1+rt)$

    $2500 = p\left(1 + 0.0625\left(\dfrac{31}{12}\right)\right)$

    $2500 = p\left(\dfrac{223}{192}\right)$

    $p \approx \$2152.47$

13. $A = p(1+rt)$

    $149925 = 120000(1+0.0475t)$

    $\dfrac{1999}{1600} = 1 + 0.0475t$

    $\dfrac{399}{1600} = 0.0475t$

    $5.25 \text{ years} = t$

15. $I = prt$

    $40 = 200r\left(\dfrac{13}{52}\right)$

    $40 = 50r$

    $0.80 = r$

    $r = 80\%$

17. $A = p(1+r)^t$

    $48428 = 38000(1+0.0625)^t$

    $\dfrac{12107}{9500} = (1+0.0625)^t$

    $\ln\dfrac{12107}{9500} = \ln(1+0.0625)^t$

    $\ln\dfrac{12107}{9500} = t\ln(1+0.0625)$

    $\dfrac{\ln\dfrac{12107}{9500}}{\ln(1+0.0625)} = t$

    $t \approx 4 \text{ years}$

19. $A = p(1+r)^t$

    $4575 = 1525(1+0.071)^t$

    $3 = (1+0.071)^t$

    $\ln 3 = \ln(1+0.071)^t$

    $\ln 3 = t\ln(1+0.071)$

    $\dfrac{\ln 3}{\ln(1+0.071)} = t$

    $t \approx 16 \text{ years}$

21. $P = \dfrac{A}{(1+r)^t}$

    $P = \dfrac{10000}{(1+0.0575)^5}$

    $P \approx \$7561.33$

23. $A = p\left(1+\dfrac{r}{n}\right)^{nt}$

    $129500 = 90000\left(1+\dfrac{0.07125}{52}\right)^{52t}$

    $\dfrac{259}{180} = \left(1+\dfrac{0.07125}{52}\right)^{52t}$

    $\ln\left(\dfrac{259}{180}\right) = \ln(1.001370192)^{52t}$

    $\ln\left(\dfrac{259}{180}\right) = 52t\ln(1.001370192)$

    $\dfrac{\ln\left(\dfrac{259}{180}\right)}{52\ln(1.001370192)} = t$

    $t \approx 5 \text{ years}$

25. $A = p\left(1 + \dfrac{r}{n}\right)^{nt}$

$10000 = 5000\left(1 + \dfrac{0.0925}{365}\right)^{365t}$

$2 = (1.000253425)^{365t}$

$\ln 2 = \ln(1.000253425)^{365t}$

$\ln 2 = 365t \ln(1.000253425)$

$\dfrac{\ln 2}{365 \ln(1.000253425)} = t$

$t \approx 7.5$ years

27. $A = p\left(1 + \dfrac{r}{n}\right)^{nt}$

$A = 10\left(1 + \dfrac{0.10}{10}\right)^{10(10)} \approx \$27.04$, No

29. $A = p\left(1 + \dfrac{r}{n}\right)^{nt}$

(a) $A = 175000\left(1 + \dfrac{0.0875}{2}\right)^{2(4)}$

$\approx \$246496.05$, No

(b) $r \approx 9.12\%$

31. $A = pe^{rt}$

$2500 = 1750e^{0.045t}$

$\dfrac{10}{7} = e^{0.045t}$

$\ln\left(\dfrac{10}{7}\right) = \ln e^{0.045t}$

$\ln\left(\dfrac{10}{7}\right) = 0.045t \ln e$

$\ln\left(\dfrac{10}{7}\right) = 0.045t$

$\dfrac{\ln\left(\dfrac{10}{7}\right)}{0.045} = t$

$t \approx 7.9$ years

33. $A = pe^{rt}$

$10000 = 5000e^{0.0925t}$

$2 = e^{0.0925t}$

$\ln 2 = \ln e^{0.0925t}$

$\ln 2 = 0.0925t \ln e$

$\dfrac{\ln 2}{0.0925} = t$

$t \approx 7.5$ years

35. $A = pe^{rt}$

(a) $A = 12500e^{0.086(5)} = 19215.72$ euros, No

(b) $20000 = 12500e^{r(5)}$

$\dfrac{8}{5} = e^{5r}$

$\ln\left(\dfrac{8}{5}\right) = \ln e^{5r}$

$\ln\left(\dfrac{8}{5}\right) = 5r \ln e$

$\dfrac{\ln\left(\dfrac{8}{5}\right)}{5} = r$

$r \approx 9.4\%$

37. $A = pe^{rt}$

(a) $A = 12000e^{0.055(7)} \approx 17635.37$ euros, No

(b) $20000 = Pe^{0.055(7)}$

$\dfrac{20000}{e^{0.055(7)}} = P$

$P \approx 13,609$ euros

39. $T = \dfrac{1}{r} \cdot \ln\left(\dfrac{A}{P}\right)$

$8 = \dfrac{1}{0.05} \cdot \ln\left(\dfrac{A}{200000}\right)$

$0.4 = \ln\left(\dfrac{A}{200000}\right)$

By definition, $\ln x = y$ iff $e^y = x$

$e^{0.4} = \dfrac{A}{200000}$

$200000e^{0.4} = A$

No, \$298364.94;

$8 = \dfrac{1}{0.05} \cdot \ln\left(\dfrac{350000}{P}\right)$

$0.4 = \ln\left(\dfrac{350000}{P}\right)$

By definition, $\ln x = y$ iff $e^y = x$

$e^{0.4} = \dfrac{350000}{P}$

$\dfrac{350000}{e^{0.4}} = P$

$P = \$234{,}612.01$

41. $A = \dfrac{p\left[(1+R)^{nt} - 1\right]}{R}$

$10000 = \dfrac{90\left[\left(1 + \dfrac{0.0775}{12}\right)^{12t} - 1\right]}{\dfrac{0.0775}{12}}$

$\dfrac{775}{12} = 90\left[\left(1 + \dfrac{0.0775}{12}\right)^{12t} - 1\right]$

$\dfrac{155}{216} = \left(1 + \dfrac{0.0775}{12}\right)^{12t} - 1$

$\dfrac{371}{216} = \left(1 + \dfrac{0.0775}{12}\right)^{12t}$

$\ln\left(\dfrac{371}{216}\right) = \ln\left(1 + \dfrac{0.0775}{12}\right)^{12t}$

$\ln\left(\dfrac{371}{216}\right) = 12t \ln\left(1 + \dfrac{0.0775}{12}\right)$

$\dfrac{\ln\left(\dfrac{371}{216}\right)}{12 \ln\left(1 + \dfrac{0.0775}{12}\right)} = t$

≈ 7 years

43. $A = \dfrac{p\left[(1+R)^{nt} - 1\right]}{R}$

$30000 = \dfrac{50\left[\left(1 + \dfrac{0.062}{12}\right)^{12t} - 1\right]}{\dfrac{0.062}{12}}$

$155 = 50\left[\left(1 + \dfrac{0.062}{12}\right)^{12t} - 1\right]$

$3.1 = \left(1 + \dfrac{0.062}{12}\right)^{12t} - 1$

$4.1 = \left(1 + \dfrac{0.062}{12}\right)^{12t}$

$\ln 4.1 = \ln\left(1 + \dfrac{0.062}{12}\right)^{12t}$

$\ln(4.1) = 12t \ln\left(1 + \dfrac{0.062}{12}\right)$

$\dfrac{\ln(4.1)}{12 \ln\left(1 + \dfrac{0.062}{12}\right)} = t$

≈ 23 years

45. $A = \dfrac{p\left[(1+R)^{nt} - 1\right]}{R}$

(a) $A = \dfrac{250\left[\left(1+\dfrac{0.085}{12}\right)^{12(5)} - 1\right]}{\dfrac{0.085}{12}}$

$\approx \$18610.61$, No

(b) $22500 = \dfrac{p\left[\left(1+\dfrac{0.085}{12}\right)^{12(5)} - 1\right]}{\dfrac{0.085}{12}}$

$159.375 = p\left[\left(1+\dfrac{0.085}{12}\right)^{12(5)} - 1\right]$

$\dfrac{159.375}{\left[\left(1+\dfrac{0.085}{12}\right)^{12(5)} - 1\right]} = p$

$p \approx \$302.25$

47. $A = p + prt$

a. $A - p = prt$

$\dfrac{A-p}{pr} = t$

b. $A = p(1+rt)$

$\dfrac{A}{1+rt} = p$

49. $A = P\left(1+\dfrac{r}{n}\right)^{nt}$

a. $\dfrac{A}{P} = \left(1+\dfrac{r}{n}\right)^{nt}$

$\sqrt[nt]{\dfrac{A}{P}} = 1 + \dfrac{r}{n}$

$\sqrt[nt]{\dfrac{A}{P}} - 1 = \dfrac{r}{n}$

$n\left(\sqrt[nt]{\dfrac{A}{P}} - 1\right) = r$

b. $\ln\left(\dfrac{A}{P}\right) = \ln\left(1+\dfrac{r}{n}\right)^{nt}$

$\ln\left(\dfrac{A}{P}\right) = nt\ln\left(1+\dfrac{r}{n}\right)$

$\dfrac{\ln\left(\dfrac{A}{P}\right)}{n\ln\left(1+\dfrac{r}{n}\right)} = t$

51. $Q(t) = Q_0 e^{rt}$

a. $\dfrac{Q(t)}{e^{rt}} = Q_0$

b. $\dfrac{Q(t)}{Q_0} = e^{rt}$

$\ln\left(\dfrac{Q(t)}{Q_0}\right) = \ln e^{rt}$

$\ln\left(\dfrac{Q(t)}{Q_0}\right) = rt\ln e$

$\dfrac{\ln\left(\dfrac{Q(t)}{Q_0}\right)}{r} = t$

53. $P = \dfrac{AR}{1 - (1+R)^{-nt}}$

$P = \dfrac{125000\left(\dfrac{0.055}{12}\right)}{1 - \left(1+\left(\dfrac{0.055}{12}\right)\right)^{-12(30)}}$

$P \approx \$709.74$

55. $Q(t) = Q_0 e^{rt}$

    (a) $2000 = 1000 e^{r(12)}$

    $2 = e^{12r}$

    $\ln 2 = \ln e^{12r}$

    $\ln 2 = 12r \ln e$

    $\dfrac{\ln 2}{12} = r$

    $r \approx 5.78\%$

    (b) $200000 = 1000 e^{(0.0578)t}$

    $200 = e^{(0.0578)t}$

    $\ln 200 = \ln e^{(0.0578)t}$

    $\ln 200 = 0.0578t \ln e$

    $\dfrac{\ln 200}{0.0578} = t$

    $t \approx 91.67$ hours

57. $r = \dfrac{\ln 2}{t}$

    $r = \dfrac{\ln 2}{8}$

    $r \approx 0.087$ or $r \approx 8.7\%$;

    $Q(t) = Q_0 e^{-rt}$

    $0.5 = Q_0 e^{-0.087(3)}$

    $\dfrac{0.5}{e^{-0.087(3)}} = Q_0$

    $Q_0 \approx 0.65$ grams

59. $r = \dfrac{\ln 2}{t}$

    $r = \dfrac{\ln 2}{432}$;

    $Q(t) = Q_0 e^{-rt}$

    $2.7 = 10 e^{-\frac{\ln 2}{432}t}$

    $0.27 = e^{-\frac{\ln 2}{432}t}$

    $\ln 0.27 = \ln e^{-\frac{\ln 2}{432}t}$

    $\ln 0.27 = -\dfrac{\ln 2}{432}t \ln e$

    $\dfrac{\ln 0.27}{-\dfrac{\ln 2}{432}} = t$

    ≈ 816 years

61. $T = -8267 \cdot \ln p$

    $17255 = -8267 \cdot \ln p$

    $\dfrac{17255}{-8267} = \ln p$

    $e^{-\frac{17255}{8267}} = e^{\ln p}$

    $e^{-\frac{17255}{8267}} = p$

    $p \approx 0.124$

    About 12.4 %

63. $A = pe^{rt}$

    $A = 10000 e^{0.062(120)} = \$17,027,502.21$

    Answers will vary.

65. $A = p\left(1 + \dfrac{r}{n}\right)^{nt}$

    $25000 = 6000\left(1 + \dfrac{r}{365}\right)^{365(18)}$

    $\dfrac{25}{6} = \left(1 + \dfrac{r}{365}\right)^{6570}$

    $\sqrt[6570]{\dfrac{25}{6}} = 1 + \dfrac{r}{365}$

    $\sqrt[6570]{\dfrac{25}{6}} - 1 = \dfrac{r}{365}$

    $365\left[\sqrt[6570]{\dfrac{25}{6}} - 1\right] = r$

    $r \approx 7.93\%$

67. $2000^2 + 1580^2 = x^2$

    $x \approx 2548.8$ meters

69. $P(x) = (x-3)(x+1)(x-(1+2i))(x-(1-2i))$

    $P(x) = \left(x^2 - 2x - 3\right)\left((x-1)^2 - 4i^2\right)$

    $P(x) = \left(x^2 - 2x - 3\right)\left(x^2 - 2x + 1 + 4\right)$

    $P(x) = \left(x^2 - 2x - 3\right)\left(x^2 - 2x + 5\right)$

    $P(x) = x^4 - 2x^3 + 5x^2 - 2x^3 + 4x^2$

    $\qquad -10x - 3x^2 + 6x - 15$

    $P(x) = x^4 - 4x^3 + 6x^2 - 4x - 15$

Chapter 4 Summary and Concept Review

1. $h(x) = -|x-2| + 3$; No

3. $s(x) = \sqrt{x-1} + 5$; Yes

5. $f(x) = x^2 - 2, \quad x \geq 0$

$$y = x^2 - 2$$
$$x = y^2 - 2$$
$$x + 2 = y^2$$
$$\sqrt{x+2} = y$$
$$f^{-1}(x) = \sqrt{x+2}$$
$$\left(f \circ f^{-1}\right)(x) = f\left[f^{-1}(x)\right]$$
$$= \left(f^{-1}(x)\right)^2 - 2$$
$$= \left(\sqrt{x+2}\right)^2 - 2$$
$$= x + 2 - 2$$
$$= x;$$
$$\left(f^{-1} \circ f\right)(x) = f^{-1}[f(x)]$$
$$= \sqrt{f(x) + 2}$$
$$= \sqrt{x^2 - 2 + 2}$$
$$= \sqrt{x^2}$$
$$= x$$

7. $f(x)$:
$$\begin{cases} D : x \in [-4, \infty) \\ R : y \in [0, \infty) \end{cases}$$
$f^{-1}(x)$:
$$\begin{cases} D : x \in [0, \infty) \\ R : y \in [-4, \infty) \end{cases}$$

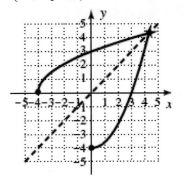

9. $f(x)$:
$$\begin{cases} D : x \in (-\infty, \infty) \\ R : y \in (0, \infty) \end{cases}$$
$f^{-1}(x)$:
$$\begin{cases} D : x \in (0, \infty) \\ R : y \in (-\infty, \infty) \end{cases}$$

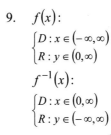

11. $y = 2^x + 3$

Asymptote: $y = 3$

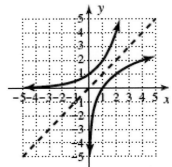

13. $y = -e^{x+1} - 2$

Left 1, reflected across the x-axis, down 2,

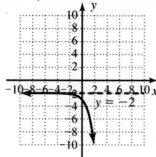

15. $4^x = \dfrac{1}{16}$

$4^x = 4^{-2}$

$x = -2$

17. $20000 = 142000 \cdot (0.85)^t$

$\dfrac{10}{71} = 0.85^t$

$\ln\left(\dfrac{10}{71}\right) = \ln 0.85^t$

$\ln\left(\dfrac{10}{71}\right) = t \ln 0.85$

$\dfrac{\ln\left(\dfrac{10}{71}\right)}{\ln 0.85} = t$

About 12.1 years

19. $\log_5 \dfrac{1}{125} = -3$

$5^{-3} = \dfrac{1}{125}$

21. $5^2 = 25$

$\log_5 25 = 2$

23. $3^4 = 81$

$\log_3 81 = 4$

25. $\ln \dfrac{1}{e} = x$

$e^x = \dfrac{1}{e}$

$e^x = e^{-1}$

$x = -1$

27. $f(x) = \log_2 x$

Asymptote: $x = 0$

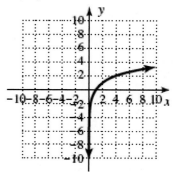

29. $f(x) = 2 + \ln(x-1)$

Asymptote: $x = 1$

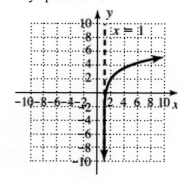

31. $g(x) = \log \sqrt{2x+3}$

$2x + 3 > 0$

$2x > -3$

$x > -\dfrac{3}{2}$

Domain: $x \in (-\dfrac{3}{2}, \infty)$

33. a. $\ln x = 32$

$e^{32} = x$

b. $\log x = 2.38$

$10^{2.38} = x$

c. $e^x = 9.8$

$\ln e^x = \ln 9.8$

$x = \ln 9.8$

d. $10^x = \sqrt{7}$

$\log 10^x = \log \sqrt{7}$

$x = \log \sqrt{7}$

35. a. $\ln 7 + \ln 6$

$\ln 42$

b. $\log_9 2 + \log_9 15$

$\log_9 30$

c. $\ln(x+3) - \ln(x-1)$

$\ln\left(\dfrac{x+3}{x-1}\right)$

d. $\log x + \log(x+1)$

$\log(x^2 + x)$

37. a. $\ln\left(x\sqrt[4]{y}\right)$

$= \ln x + \ln y^{\frac{1}{4}}$

$= \ln x + \dfrac{1}{4}\ln y$

b. $\ln\left(\sqrt[3]{pq}\right)$

$= \ln p^{\frac{1}{3}} + \ln q$

$= \dfrac{1}{3}\ln p + \ln q$

c. $\log\left(\dfrac{\sqrt[3]{x^5 y^4}}{\sqrt{x^5 y^3}}\right)$

$= \log\left(\sqrt[3]{x^5 y^4}\right) - \log\sqrt{x^5 y^3}$

$= \log x^{\frac{5}{3}} y^{\frac{4}{3}} - \log x^{\frac{5}{2}} y^{\frac{3}{2}}$

$= \log x^{\frac{5}{3}} + \log y^{\frac{4}{3}} - \log x^{\frac{5}{2}} - \log y^{\frac{3}{2}}$

$= \dfrac{5}{3}\log x + \dfrac{4}{3}\log y - \dfrac{5}{2}\log x - \dfrac{3}{2}\log y$

d. $\log\left(\dfrac{4\sqrt[3]{p^5 q^4}}{\sqrt{p^3 q^2}}\right)$

$= \log 4\sqrt[3]{p^5 q^4} - \log\sqrt{p^3 q^2}$

$= \log 4 p^{\frac{5}{3}} q^{\frac{4}{3}} - \log p^{\frac{3}{2}} q$

$= \log 4 + \log p^{\frac{5}{3}} + \log q^{\frac{4}{3}} - \left(\log p^{\frac{3}{2}} + \log q\right)$

$= \log 4 + \dfrac{5}{3}\log p + \dfrac{4}{3}\log q - \dfrac{3}{2}\log p - \log q$

39. $2^x = 7$

$\ln 2^x = \ln 7$

$x \ln 2 = \ln 7$

$x = \dfrac{\ln 7}{\ln 2}$

41. $e^{x-2} = 3^x$

$\ln e^{x-2} = \ln 3^x$

$x - 2 = x \ln 3$

$x - x \ln 3 = 2$

$x(1 - \ln 3) = 2$

$x = \dfrac{2}{1 - \ln 3}$

43. $\log x + \log(x - 3) = 1$

$\log x(x - 3) = 1$

$10^1 = x(x - 3)$

$0 = x^2 - 3x - 10$

$0 = (x - 5)(x + 2)$

$x = 5 \text{ or } x = -2$

$5,\ -2$ is extraneous

45. $R(h) = \dfrac{\ln(2)}{h}$

   a. $R(3.9) = \dfrac{\ln(2)}{3.9}$

        $\approx 17.77\%$

   b. $0.0289 = \dfrac{\ln(2)}{h}$

       $0.0289h = \ln 2$

       $h = \dfrac{\ln 2}{0.0289}$

       About 23.98 days

47. $I = \mathrm{P}rt$

$27.75 = 600r\left(\dfrac{3}{12}\right)$

$4\left(\dfrac{27.75}{600}\right) = r$

$r = 0.185$

18.5%

49. $A = \dfrac{p\left[(1 + R)^{nt} - 1\right]}{R}$

  (a) $A = \dfrac{260\left[\left(1 + \dfrac{0.075}{12}\right)^{12(4)} - 1\right]}{\dfrac{0.075}{12}}$

      $A \approx \$14501.72$,

      No

  (b) $15000 = \dfrac{p\left[\left(1 + \dfrac{0.075}{12}\right)^{12(4)} - 1\right]}{\dfrac{0.075}{12}}$

     $93.75 = p\left[\left(1 + \dfrac{0.075}{12}\right)^{12(4)} - 1\right]$

     $\dfrac{93.75}{\left[\left(1 + \dfrac{0.075}{12}\right)^{12(4)} - 1\right]} = p$

     $p \approx \$268.93$

Chapter 4 Mixed Review

1. a. $\log_2 30$

   $\dfrac{\log 30}{\log 2} \approx 4.9069$

   b. $\log_{0.25} 8$

   $\dfrac{\log 8}{\log 0.25} = -1.5$

   c. $\log_8 2$

   $\dfrac{\log 2}{\log 8} = \dfrac{1}{3}$

3. a. $\log_{10} 20^2$

   $= 2\log_{10} 20$

   b. $\log 10^{0.05x}$

   $= 0.05x \log 10$

   $= 0.05x$

   c. $\ln 2^{x-3}$

   $= (x-3)\ln 2$

5. $y = 5 \cdot 2^{-x}$

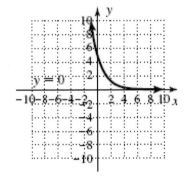

7. $y = \log_2(-x) - 4$

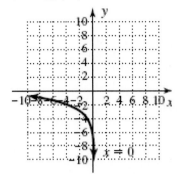

9. a. $\log_5 625 = 4$

   $5^4 = 625$

   b. $\ln 0.15x = 0.45$

   $e^{0.45} = 0.15x$

   c. $\log(0.1 \times 10^8) = 7$

   $10^7 = 0.1 \times 10^8$

11. $g(x) = \sqrt{x-1} + 2$

   a. $D: x \in [1, \infty), R: y \in [2, \infty)$

   b. $g(x) = \sqrt{x-1} + 2$

   Interchange x and y.

   $x = \sqrt{y-1} + 2$

   $x - 2 = \sqrt{y-1}$

   $(x-2)^2 = y - 1$

   $(x-2)^2 + 1 = y$

   $g^{-1}(x) = (x-2)^2 + 1$

   $D: x \in [2, \infty), R: y \in [1, \infty)$

   c. Answers will vary.

13. $10^{x-4} = 200$

$\log 10^{x-4} = \log 200$

$(x-4)\log 10 = \log 2 \cdot 10^2$

$x - 4 = \log 2 + \log 10^2$

$x - 4 = \log 2 + 2\log 10$

$x - 4 = \log 2 + 2$

$x = 6 + \log 2$

15. $\log_2(2x-5) + \log_2(x-2) = 4$

$\log_2(2x-5)(x-2) = 4$

$2^4 = (2x-5)(x-2)$

$16 = 2x^2 - 9x + 10$

$0 = 2x^2 - 9x - 6$

$a = 2, b = -9, c = -6$

$x = \dfrac{-(-9) \pm \sqrt{(-9)^2 - 4(2)(-6)}}{2(2)}$

$x = \dfrac{9 \pm \sqrt{129}}{4}$

$x = \dfrac{9 + \sqrt{129}}{4}$; $x = \dfrac{9 - \sqrt{129}}{4}$ is extraneous

17. $6.5 = \log\left(\dfrac{I}{2\times 10^{11}}\right)$

$10^{6.5} = \dfrac{I}{2\times 10^{11}}$

$10^{6.5}\left(2\times 10^{11}\right) = I$

$2 \cdot 10^{0.5} \cdot 10^{17} = I$

$2 \cdot \sqrt{10} \cdot 10^{17} = I$

$I \approx 6.3 \times 10^{17}$

19. $r(n) = 2(0.8)^n$

$r(6) = 2(0.8)^6 = 0.524$

0.52 m;

n	$r(n) = 2(0.8^n)$
1	1.6 m
2	1.28 m
3	1.02 m
4	0.82 m
5	0.66 m
6	0.52 m

Chapter 4 Practice Test

1. $\log_3 81 = 4$
    $3^4 = 81$

3. $\log_b \left(\dfrac{\sqrt{x^5} \, y^3}{z} \right)$

    $= \log_b \sqrt{x^5} \, y^3 - \log_b z$

    $= \log_b x^{\frac{5}{2}} y^3 - \log_b z$

    $= \log_b x^{\frac{5}{2}} + \log_b y^3 - \log_b z$

    $= \dfrac{5}{2} \log_b x + 3 \log_b y - \log_b z$

5. $5^{x-7} = 125$
    $5^{x-7} = 5^3$
    $x - 7 = 3$
    $x = 10$

7. $\log_a 45$
    $= \log_a \left(3^2 \cdot 5 \right)$
    $= \log_a 3^2 + \log_a 5$
    $= 2 \log_a 3 + \log_a 5$
    $= 2(0.48) + 1.72$
    $= 2.68$

9. $g(x) = -2^{x-1} + 3$
    HA: $y = 3$

    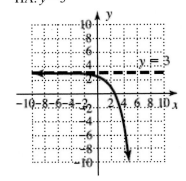

11. a. $\log_3 100$

    $= \dfrac{\log 100}{\log 3}$

    $= \dfrac{\log 10^2}{\log 3}$

    $= \dfrac{2 \log 10}{\log 3}$

    $= \dfrac{2}{\log 3}$

    ≈ 4.19

    b. $\log_6 0.235$

    $= \dfrac{\log 0.235}{\log 6}$

    ≈ -0.81

13. $3^{x-1} = 89$
    $\ln 3^{x-1} = \ln 89$
    $(x-1)\ln 3 = \ln 89$

    $x - 1 = \dfrac{\ln 89}{\ln 3}$

    $x = 1 + \dfrac{\ln 89}{\ln 3}$

15. $3000 = 8000(0.82)^t$

    $\dfrac{3}{8} = (0.82)^t$

    $\ln \left(\dfrac{3}{8} \right) = \ln (0.82)^t$

    $\ln \left(\dfrac{3}{8} \right) = t \ln (0.82)$

    $\dfrac{\ln \left(\dfrac{3}{8} \right)}{\ln 0.82} = t$

    $t \approx 5 \text{ years}$

17. $Q(t) = -2600 + 1900 \ln t$

$3000 = -2600 + 1900 \ln t$

$5600 = 1900 \ln t$

$\dfrac{56}{19} = \ln t$

$e^{\frac{56}{19}} = e^{\ln t}$

$e^{\frac{56}{19}} = t$

$t \approx 19.1$ months

19. $A = \dfrac{p\left[(1+R)^{nt} - 1\right]}{R}$

(a) $A = \dfrac{50\left[\left(1 + \dfrac{0.0825}{12}\right)^{12(5)} - 1\right]}{\dfrac{0.0825}{12}}$

$A \approx \$3697.88$

No

(b) $4000 = \dfrac{p\left[\left(1 + \dfrac{0.0825}{12}\right)^{12(5)} - 1\right]}{\dfrac{0.0825}{12}}$

$27.5 = p\left[\left(1 + \dfrac{0.0825}{12}\right)^{60} - 1\right]$

$\dfrac{27.5}{\left[\left(1 + \dfrac{0.0825}{12}\right)^{60} - 1\right]} = p$

$p \approx \$54.09$

Chapter 4: Calculator Exploration and Discovery

1. $a = 25$, $b = 0.5$, $c = 2500$

3. b

5. $b = 0.6$

7. Verified

Strengthening Core Skills

1. Answers will vary.

3. Answers will vary.

Cumulative Review Chapters 1 to 4

1. $x^2 - 4x + 53 = 0$

$a = 1, b = -4, c = 53$

$x = \dfrac{-(-4) \pm \sqrt{(-4)^2 - 4(1)(53)}}{2(1)}$

$x = \dfrac{4 \pm \sqrt{-196}}{2}$

$x = \dfrac{4 \pm 14i}{2}$

$x = 2 \pm 7i$

3. $(4 + 5i)^2 - 8(4 + 5i) + 41 = 0$

$-9 + 40i - 32 - 40i + 41 = 0$

$0 = 0$

5. $f(x) = x^3 - 2$, $g(x) = \sqrt[3]{x + 2}$;

$f(g(x)) = \left(\sqrt[3]{x + 2}\right)^3 - 2 = x + 2 + x = x$;

$g(f(x)) = \sqrt[3]{x^3 - 2 + 2} = \sqrt[3]{x^3} = x$

Since $(f \circ g)(x) = (g \circ f)(x)$, they are inverse functions.

7. 1991 → year 1

(a) $(1,3100), (9,6740)$

$$m = \frac{6740 - 3100}{9 - 1} = 455$$

$$y - 3100 = 455(x - 1)$$

$$y - 3100 = 455x - 455$$

$$y = 455x + 2645$$

$$T(t) = 455t + 2645$$

(b) $\dfrac{\Delta T}{\Delta t} = \dfrac{455}{1}$, triple births increase by 455 each year.

(c) In 1996, $T(6) = 455(6) + 2645 = 5375$

sets of triplets

In 2007,

$t = 17, T(17) = 455(17) + 2645 = 10{,}380$

sets of triplets

9.
$$h(x) = \begin{cases} -4 & -10 \le x < -2 \\ -x^2 & -2 \le x < 3 \\ 3x - 18 & x \ge 3 \end{cases}$$

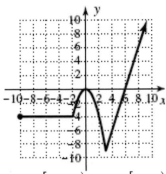

$D: x \in [-10, \infty), R: y \in [-9, \infty)$;

$h(x) \uparrow: (-2, 0) \cup (3, \infty)$

$h(x) \downarrow: (0, 3)$

11. $f(x) = x^4 - 3x^3 - 12x^2 + 52x - 48$

Possible rational roots:

$$\frac{\{\pm 1, \pm 48, \pm 2, \pm 24, \pm 3, \pm 16, \pm 4, \pm 12, \pm 6, \pm 8\}}{\{\pm 1\}} \; ;$$

$$\{\pm 1, \pm 48, \pm 2, \pm 24, \pm 3, \pm 16, \pm 4, \pm 12, \pm 6, \pm 8\}$$

$$\begin{array}{r|rrrr} 3 & 1 & -3 & -12 & 52 & -48 \\ & & 3 & 0 & -36 & 48 \\ \hline & 1 & 0 & -12 & 16 & \underline{|\,0} \end{array}$$

$$\begin{array}{r|rrrr} 2 & 1 & 0 & -12 & 16 \\ & & 2 & 4 & -16 \\ \hline & 1 & 2 & -8 & \underline{|\,0} \end{array}$$

$$f(x) = (x - 3)(x - 2)(x^2 + 2x - 8)$$

$$f(x) = (x - 3)(x - 2)(x - 2)(x + 4)$$

$x = 3, x = 2$ (multiplicity 2), $x = -4$

13. $V = \dfrac{1}{2}\pi b^2 a$

$$\frac{2V}{\pi a} = b^2$$

$$\sqrt{\frac{2V}{\pi a}} = b$$

15. a) $f(x) = \dfrac{2x + 3}{5}$

$$y = \frac{2x + 3}{5}$$

$$x = \frac{2y + 3}{5}$$

$$5x = 2y + 3$$

$$5x - 3 = 2y$$

$$\frac{5x - 3}{2} = y$$

$$f^{-1}(x) = \frac{5x - 3}{2}$$

b) graph

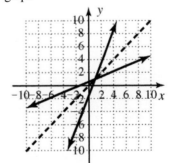

c) $f\left(f^{-1}(x)\right) = f\left(\dfrac{5x-3}{2}\right)$

$= \dfrac{2\left(\dfrac{5x-3}{2}\right)+3}{5}$

$= \dfrac{5x-3+3}{5}$

$= \dfrac{5x}{5}$

$f\left(f^{-1}(x)\right) = x$

$f^{-1}\left(f(x)\right) = f^{-1}\left(\dfrac{2x+3}{5}\right)$

$= \dfrac{5\left(\dfrac{2x+3}{5}\right)-3}{2}$

$= \dfrac{2x+3-3}{2}$

$= \dfrac{2x}{2}$

$f^{-1}\left(f(x)\right) = x$

17. $\ln(x+3) + \ln(x-2) = \ln 24$

$\ln(x+3)(x-2) = \ln 24$

$(x+3)(x-2) = 24$

$x^2 + x - 6 = 24$

$x^2 + x - 30 = 0$

$(x+6)(x-5) = 0$

$x+6 = 0 \ \text{ or } \ x-5 = 0$

$x = -6 \ \ \text{ or } \ x = 5$

$x = 5, x = -6$ is an extraneous root

19. a) Sportwagon:
$H(3000) = 123\ln(3000) - 897$
$\approx 88 \text{ hp}$

Minivan:
$H(3000) = 193\ln(3000) - 1464$
$\approx 81 \text{ hp}$

b) $123\ln r - 897 = 193\ln r - 1464$
$123\ln r + 567 = 193\ln r$
$567 = 193\ln r - 123\ln r$
$567 = 70\ln r$
$\dfrac{567}{70} = \ln r$
$e^{\frac{567}{70}} = r$
$r \approx 3294 \text{ rpm}$

c) Sportwagon:
$H(5600) = 123\ln(5600) - 897$
$\approx 164.6 \text{ hp}$

Minivan:
$H(5800) = 193\ln(5800) - 1464$
$\approx 208.46 \text{ hp}$

Minivan, 208 hp @ 5800 rpm

Chapter 5: Introduction to Trigonometric Functions

5.1 Exercises

1. Complementary; 180; less; greater

3. $r\theta$; $\frac{1}{2}r^2\theta$; radians

5. Answers will vary.

7. a. Complement $= 90° - 12.5° = 77.5°$
   b. Supplement $= 180° - 149.2° = 30.8°$

9. $\alpha = 90° - 37° = 53°$

11. $42°30' = 42° + \left(\frac{30}{60}\right)^{\circ} = 42.5°$

13. $67°33'19'' = 67° + \left(\frac{33}{60}\right)^{\circ} + \left(\frac{19}{3,600}\right)^{\circ}$
    $= 67.555°$

15. $285°00'09'' = 285° + \left(\frac{09}{3,600}\right)^{\circ} = 285.0025°$

17. $45°45'45'' = 45° + \left(\frac{45}{60}\right)^{\circ} + \left(\frac{45}{3,600}\right)^{\circ}$
    $= 45.7625°$

19. $20.25° = 20° + 0.25(60)' = 20°15'00''$

21. $67.307° = 67° + 0.307(60)' = 67°18.42'$
    $= 67°18' + 0.42(60) = 67°18'25.2''$

23. $275.33° = 275° + 0.33(60)' = 275°19.8'$
    $= 275°19' + 0.8(60)'' = 275°19'48''$

25. $5.4525° = 5° + 0.4525(60)' = 5° + 27.15'$
    $= 5° + 27' + 0.15(60)'' = 5°27'9''$

27. No; $19 + 16 < 40$.

29. $\alpha = 180° - (53° + 58°) = 69°$

31. $\angle A = 180° - (90° + 65°) = 25°$

33. Let x be the height of the helicopter.

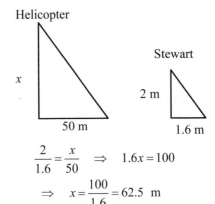

$\frac{2}{1.6} = \frac{x}{50}$ $\Rightarrow$ $1.6x = 100$

$\Rightarrow$ $x = \frac{100}{1.6} = 62.5$ m

35. $82 = \sqrt{2} \cdot a$; $a = \frac{82}{\sqrt{2}} = 41\sqrt{2} \approx 58$ ft

    Height of the firetruck: 10 ft
    Total height: $58 + 10 = 68$ ft

37. $\theta = 75°$; $\theta + 360k$
    $k = -2; 75 + 360(-2) = -645°$;
    $k = -1; 75 + 360(-1) = -285°$;
    $k = 1; 75 + 360(1) = 435°$;
    $k = 2; 75 + 360(2) = 795°$;
    $-645°, -285°, 435°, 795°$

39. $\theta = -45°$; $\theta + 360k$
    $k = -2; -45 + 360(-2) = -765°$;
    $k = -1; -45 + 360(-1) = -405°$;
    $k = 1; -45 + 360(1) = 315°$;
    $k = 2; -45 + 360(2) = 675°$;
    $-765°, -405°, 315°, 675°$

41. $s = r\theta = 280(3.5) = 980$ m

43. $s = r\theta$; $2,007 = 2,676 \cdot \theta$
    $\theta = \frac{2,007}{2,676} = 0.75$ rad

45. $s = r\theta$; $4,146.9 = r \cdot \frac{3\pi}{4}$
    $r = \frac{4,146.9}{\frac{3\pi}{4}} = 4,146.9 \cdot \frac{4}{3\pi} \approx 1,760$ yd

5.1 Exercises

47. $s = r\theta = 2 \cdot \dfrac{4\pi}{3} = \dfrac{8\pi}{3}$ mi

49. $s = r\theta$; $252.35 = 980 \cdot \theta$

   $\theta = \dfrac{252.35}{980} = 0.2575$ rad

51. Convert 320° to radians first:

   $320° \cdot \dfrac{\pi \text{ rad}}{180°} = \dfrac{16\pi}{9}$ rad

   $s = r\theta$; $52.5 = r \cdot \dfrac{16\pi}{9}$

   $r = \dfrac{52.5}{\dfrac{16\pi}{9}} \approx 9.4$ km

53. $A = \dfrac{1}{2}r^2\theta = \dfrac{1}{2}(6.8)^2(5) = 115.6$ km^2

55. $A = \dfrac{1}{2}r^2\theta$; $1080 = \dfrac{1}{2}(60)^2 \cdot \theta = 1800 \cdot \theta$

   $\theta = \dfrac{1080}{1800} = 0.6$ rad

57. $A = \dfrac{1}{2}r^2\theta$; $16.5 = \dfrac{1}{2}r^2 \cdot \dfrac{7\pi}{6} = \dfrac{7\pi}{12}r^2$

   $r^2 = \dfrac{16.5}{\dfrac{7\pi}{12}} \approx 9.004$; $r \approx 3$ m

   (We discard the negative answer since r is a distance.)

59. $r = 5$ cm; $\theta = 1.5$ rad
   $s = r\theta$; $s = 5(1.5) = 7.5$ cm
   $A = \dfrac{1}{2}r^2\theta$; $A = \dfrac{1}{2}(5)^2(1.5) = 18.75$ cm^2

61. $r = 10$ m; $s = 43$ m
   $s = r\theta$; $43 = 10 \cdot \theta$; $\theta = \dfrac{43}{10} = 4.3$ rad
   $A = \dfrac{1}{2}r^2\theta$; $A = \dfrac{1}{2}(10)^2(4.3) = 215$ m^2

63. $A = 864$ mm^2; $\theta = 3$ rad
   $A = \dfrac{1}{2}r^2\theta$; $864 = \dfrac{1}{2}r^2 \cdot 3 = \dfrac{3}{2}r^2$
   $r^2 = 864 \cdot \dfrac{2}{3} = 576$; $r = 24$ mm;
   $s = r\theta = (24)(3) = 72$ mm

65. $360° \cdot \dfrac{\pi \text{ rad}}{180°} = 2\pi$ rad

67. $45° \cdot \dfrac{\pi \text{ rad}}{180°} = \dfrac{\pi}{4}$ rad

69. $210° \cdot \dfrac{\pi \text{ rad}}{180°} = \dfrac{7\pi}{6}$ rad

71. $-120° \cdot \dfrac{\pi \text{ rad}}{180°} = -\dfrac{2\pi}{3}$ rad

73. $27° \cdot \dfrac{\pi \text{ rad}}{180°} \approx 0.4712$ rad

75. $227.9° \cdot \dfrac{\pi \text{ rad}}{180°} \approx 3.9776$ rad

77. $\dfrac{\pi}{3}$ rad $\cdot \dfrac{180°}{\pi \text{ rad}} = 60°$

79. $\dfrac{\pi}{6}$ rad $\cdot \dfrac{180°}{\pi \text{ rad}} = 30°$

81. $\dfrac{2\pi}{3}$ rad $\cdot \dfrac{180°}{\pi \text{ rad}} = 120°$

83. 4π rad $\cdot \dfrac{180°}{\pi \text{ rad}} = 720°$

85. $\dfrac{11\pi}{12}$ rad $\cdot \dfrac{180°}{\pi \text{ rad}} = 165°$

87. 3.2541 rad $\cdot \dfrac{180°}{\pi \text{ rad}} \approx 186.4°$

89. 3 rad $\cdot \dfrac{180°}{\pi \text{ rad}} \approx 171.9°$

91. -2.5 rad $\cdot \dfrac{180°}{\pi \text{ rad}} \approx -143.2°$

93. $a = 15, b = 8, c = 17$

   $h = \dfrac{8(15)}{17} \approx 7.06$ cm; $m = \dfrac{8^2}{17} \approx 3.76$ cm

   $n = \dfrac{15^2}{17} \approx 13.24$ cm

95. $40.3° - 26.4° = 13.9°$, so $13.9°$ separates the cities. Convert to radians:

$13.9° \cdot \dfrac{\pi \text{ rad}}{180°} = 0.2426 \text{ rad}$

Find arc length if $r = 3{,}960$, $\theta = 0.2426$:

$s = r\theta$; $s = (3{,}960)(0.2426) = 960.7$

miles apart

97. a. $r = 12 \text{ m}$; $\theta = 40° \cdot \dfrac{\pi \text{ rad}}{180°} \approx 0.698 \text{ rad}$

$A = \dfrac{1}{2}r^2\theta$; $A = \dfrac{1}{2}(12)^2(0.698) \approx 50.3 \text{ m}^2$

b. For $A = 100.6 \text{ m}^2$ and $r = 12 \text{ m}$, find θ.

$100.6 = \dfrac{1}{2}(12)^2 \cdot \theta = 72\theta$

$\theta = \dfrac{100.6}{72} \approx 1.4 \text{ rad} \cdot \dfrac{180°}{\pi \text{ rad}} \approx 80°$

c. For $A = 100.6 \text{ m}^2$ and $\theta = 0.698 \text{ rad}$, find r.

$100.6 = \dfrac{1}{2}r^2(0.698)$; $r^2 = \dfrac{100.6}{0.349} = 288.3$

$r \approx 17 \text{ m}$ (We discard the negative answer since r is a distance.)

99. a. $\omega = \dfrac{\frac{3}{4}\text{ rev}}{\text{sec}} \cdot \dfrac{2\pi \text{ rad}}{1 \text{ rev}} = 1.5\pi \dfrac{\text{rad}}{\text{sec}}$

b. $V = r\omega$;

$V = (56 \text{ } in.)\left(1.5\pi \dfrac{\text{rad}}{\text{sec}}\right) \approx 263.9 \dfrac{\text{in.}}{\text{sec}}$

Convert to mi./hr.

$\left(\dfrac{263.9 \text{ in.}}{\text{sec}}\right)\left(\dfrac{1 \text{ mi}}{5{,}280(12)\text{in}}\right) \cdot \left(\dfrac{3{,}600 \text{ sec}}{1 \text{ hr}}\right)$

$\approx 15 \dfrac{\text{mi}}{\text{hr}}$

101.a. $\omega = \dfrac{20 \text{ rev}}{\text{min}} \cdot \dfrac{2\pi \text{ rad}}{\text{rev}} = 40\pi \dfrac{\text{rad}}{\text{min}}$

b. $V = r\omega$

$V = 3 \text{ in.} \cdot \dfrac{40\pi \text{ rad}}{\text{min}} = 120\pi \dfrac{\text{in.}}{\text{min}}$

Convert to ft/sec

$\left(\dfrac{120\pi \text{ in.}}{\text{sec}}\right)\left(\dfrac{1 \text{ min}}{60 \text{ sec}}\right)\left(\dfrac{1 \text{ ft}}{12 \text{ in.}}\right)$

$\approx 0.52 \dfrac{\text{ft}}{\text{sec}}$

c. $\text{dist.} = \text{speed} \cdot \text{time}$

$6 \text{ ft} = 0.52 \dfrac{\text{ft}}{\text{sec}} \cdot t$

$t = \dfrac{6}{0.52} \approx 11.5 \text{ sec}$

103.a. Each concentric line represents 250 m in elevation, and 4 lines separate A and B, so the change in elevation is $4(250) = 1{,}000$ m.

b. $\dfrac{1 \text{ cm}}{625 \text{ m}} = \dfrac{1.6 \text{ cm}}{x \text{ m}}$

$x = 1.6(625) = 1{,}000 \text{ m}$

c.

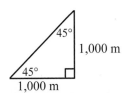

Trail length $= 1{,}000\sqrt{2} \approx 1{,}414.2$ m

105. In the next 0.5 hour, each plane will go 50 miles. The angle between their paths is $90°$.

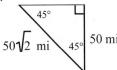

This is a 45-45-90 triangle, so the distance between them is $50\sqrt{2}$ mi or about 70.7 mi apart.

107.a. $\left(\dfrac{1\ \text{rev}}{7.15\ \text{days}}\right)\left(\dfrac{2\pi\ \text{rad}}{1\ \text{rev}}\right)\approx 0.8788\dfrac{\text{rad}}{\text{day}}$;

$\left(\dfrac{0.8788\ \text{rad}}{\text{day}}\right)\left(\dfrac{180°}{\pi\ \text{rad}}\right)\approx 50.3°/\text{day}$

b. $\left(\dfrac{0.8788\ \text{rad}}{\text{day}}\right)\left(\dfrac{1\ \text{day}}{24\ \text{hr}}\right)\approx 0.0366\dfrac{\text{rad}}{\text{hr}}$

c. $V=r\omega=\left(656,000\ \text{mi}\right)\left(0.0366\dfrac{\text{rad}}{\text{hr}}\right)$

$\cdot\left(\dfrac{1\ \text{hr}}{3,600\ \text{sec}}\right)\approx 6.67\dfrac{\text{mi}}{\text{sec}}$

109. Answers will vary.

111.a. <u>Adult bike:</u>
Linear velocity of pedal sprocket:

$\left(\dfrac{50\ \text{rev}}{\text{min}}\right)\left(\dfrac{2\pi\ \text{rad}}{\text{rev}}\right)=100\pi\dfrac{\text{rad}}{\text{min}}$

$V_p=(4\ \text{in.})\left(\dfrac{100\pi\ \text{rad}}{\text{min}}\right)=400\pi\dfrac{\text{in.}}{\text{min}}$

This will be the same as the linear velocity of the wheel sprocket, so we can use it to find the angular velocity:

$400\pi\dfrac{\text{in.}}{\text{min}}=(2\ \text{in.})\omega_w$

$\omega_w=200\pi\dfrac{\text{rad}}{\text{min}}$

This is the same as the angular velocity of the tire, so we can use it to find the tire's linear velocity, which is the speed of the bike.

$V_t=(13\ \text{in.})\left(\dfrac{200\pi\ \text{rad}}{\text{min}}\right)=2,600\pi\dfrac{\text{in.}}{\text{min}}$

<u>Kid's bike:</u> (Same calculations)

$V_b=(2.5\ \text{in})\left(\dfrac{100\pi\ \text{rad}}{\text{min}}\right)=250\pi\dfrac{\text{in}}{\text{min}}$

$250\pi\dfrac{\text{in.}}{\text{min}}=(1.5\ \text{in.})\omega_w$

$\omega_w=\dfrac{250\pi}{1.5}=\dfrac{500\pi\ \text{rad}}{3\ \text{min}}$

$V_t=(9\ \text{in.})\left(\dfrac{500\pi\ \text{rad}}{3\ \text{min}}\right)=1,500\pi\dfrac{\text{in.}}{\text{min}}$

The difference in speeds between the bikes is $1,100\pi$ in/min, so in 2 minutes, the adult bike will go $2,200\pi$ inches further.

$\left(2,200\pi\ \text{in.}\right)\left(\dfrac{1\ \text{yd}}{36\ \text{in.}}\right)\approx 192\text{yd}$

b. We need to basically do the kid's bike calculations from part a backwards, starting with a linear velocity of $2,600\pi$ in./min.

$2,600\pi\dfrac{\text{in.}}{\text{min}}=(9\ \text{in.})\omega_w$

$\omega_w=\dfrac{2,600\pi}{9}=907.5\dfrac{\text{rad}}{\text{min}}$

Linear velocity of wheel sprocket:

$V_w=(1.5\ \text{in.})\left(907.5\dfrac{\text{rad}}{\text{min}}\right)=1,361.25\dfrac{\text{in.}}{\text{min}}$

This equals the linear velocity of the pedal sprocket so we can use it to find the angular velocity:

$1,361.25\dfrac{\text{in.}}{\text{min}}=(2.5\ \text{in.})\omega_p$

$\omega_p=\dfrac{1,361.25}{2.5}=544.5\dfrac{\text{rad}}{\text{min}}$

Convert to revolutions per min:

$\left(\dfrac{544.5\ \text{rad}}{\text{min}}\right)\left(\dfrac{1\ \text{rev}}{2\pi\ \text{rad}}\right)\approx 86.7\ \text{rpm}$

113. Use $A=P\left(1+\dfrac{r}{n}\right)^{nt}$ with $A=1,500$,
$P=1,000$, $n=12$, and $t=5$. Solve for r.

$1,500=1,000\left(1+\dfrac{r}{12}\right)^{60}$

$1.5=\left(1+\dfrac{r}{12}\right)^{60}$

$1.5^{1/60}=1+\dfrac{r}{12}$

$1.5^{1/60}-1=\dfrac{r}{12}$

$r=12\left(1.5^{1/60}-1\right)\approx 0.0814$

Interest rate is 8.14%.

115. The vertex of the parabola is $(2,-4)$, so the equation is $f(x)=a(x-2)^2-4$.
$(0,-3)$ is on the graph, so $f(0)=-3$:

$f(0)=a(0-2)^2-4=-3$

$4a-4=-3;\quad 4a=1;\quad a=\dfrac{1}{4}$

$f(x)=\dfrac{1}{4}(x-2)^2-4$

5.2 Exercises

1. x; y; origin

3. x; y; $\dfrac{y}{x}$; $\sec t$; $\csc t$; $\cot t$

5. Answers will vary.

7. $x^2 + (-0.8)^2 = 1$; $x^2 + 0.64 = 1$
   $x^2 = 1 - 0.64 = 0.36$
   $x = \pm 0.6$; QIII, so choose $x = -0.6$
   $(-0.6, -0.8)$

9. $\left(\dfrac{5}{13}\right)^2 + y^2 = 1$; $\dfrac{25}{169} + y^2 = 1$

   $y^2 = 1 - \dfrac{25}{169} = \dfrac{144}{169}$

   $y = \pm\dfrac{12}{13}$; QIV, so choose $y = -\dfrac{12}{13}$

   $\left(\dfrac{5}{13}, -\dfrac{12}{13}\right)$

11. $\left(\dfrac{\sqrt{11}}{6}\right)^2 + y^2 = 1$; $\dfrac{11}{36} + y^2 = 1$

    $y^2 = 1 - \dfrac{11}{36} = \dfrac{25}{36}$

    $y = \pm\dfrac{5}{6}$; QI, so choose $y = \dfrac{5}{6}$

    $\left(\dfrac{\sqrt{11}}{6}, \dfrac{5}{6}\right)$

13. $\left(-\dfrac{\sqrt{11}}{4}\right)^2 + y^2 = 1$; $\dfrac{11}{16} + y^2 = 1$

    $y^2 = 1 - \dfrac{11}{16} = \dfrac{5}{16}$

    $y = \pm\dfrac{\sqrt{5}}{4}$; QII, so choose $y = \dfrac{\sqrt{5}}{4}$

    $\left(-\dfrac{\sqrt{11}}{4}, \dfrac{\sqrt{5}}{4}\right)$

15. $x^2 + (-0.2137)^2 = 1$; $x^2 + 0.0457 = 1$
    $x^2 = 1 - 0.0457$; $x = \pm\sqrt{1 - 0.0457}$
    $x = \pm 0.9769$; QIII, so choose $x = -0.9769$
    $(-0.9769, -0.2137)$

17. $x^2 + (0.1198)^2 = 1$; $x^2 + 0.0144 = 1$
    $x^2 = 1 - 0.0144$; $x = \pm\sqrt{1 - 0.0144}$
    $x = \pm 0.9928$; QII, so choose $x = -0.9928$
    $(-0.9928, 0.1198)$

19. $\left(-\dfrac{\sqrt{3}}{2}\right)^2 + \left(\dfrac{1}{2}\right)^2 = \dfrac{3}{4} + \dfrac{1}{4} = \dfrac{4}{4} = 1$

    Other points: $\left(\dfrac{\sqrt{3}}{2}, \dfrac{1}{2}\right)$ (QI),

    $\left(-\dfrac{\sqrt{3}}{2}, -\dfrac{1}{2}\right)$ (QIII), $\left(\dfrac{\sqrt{3}}{2}, -\dfrac{1}{2}\right)$ (QIV)

21. $\left(\dfrac{\sqrt{11}}{6}\right)^2 + \left(-\dfrac{5}{6}\right)^2 = \dfrac{11}{36} + \dfrac{25}{36} = \dfrac{36}{36} = 1$

    Other points: $\left(\dfrac{\sqrt{11}}{6}, \dfrac{5}{6}\right)$ (QI)

    $\left(-\dfrac{\sqrt{11}}{6}, \dfrac{5}{6}\right)$ (QII), $\left(-\dfrac{\sqrt{11}}{6}, -\dfrac{5}{6}\right)$ (QIII)

23. $(0.3325)^2 + (0.9431)^2 = 0.1106 + 0.8894 = 1$
    Other points: $(-0.3325, 0.9431)$ (QII)
       $(-0.3325, -0.9431)$ (QIII),
       $(0.3325, -0.9431)$ (QIV)

25. $(0.9937)^2 + (-0.1121)^2 = 0.9874 + 0.0126 = 1$
    Other points: $(0.9937, 0.1121)$ (QI),
    $(-0.9937, 0.1121)$ (QII)
    $(-0.9937, -0.1121)$ (QIII)

27.

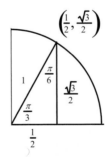

29. $\dfrac{5\pi}{4} - \pi = \dfrac{5\pi}{4} - \dfrac{4\pi}{4} = \dfrac{\pi}{4}$; $\dfrac{5\pi}{4}$ is in QIII, so

the point is $\left(-\dfrac{\sqrt{2}}{2}, -\dfrac{\sqrt{2}}{2}\right)$.

31. $-\dfrac{5\pi}{6} - (-\pi) = -\dfrac{5\pi}{6} + \dfrac{6\pi}{6} = \dfrac{\pi}{6}$; $-\dfrac{5\pi}{6}$ is in

QIII, so the point is $\left(-\dfrac{\sqrt{3}}{2}, -\dfrac{1}{2}\right)$

33. $3\pi - \dfrac{11\pi}{4} = \dfrac{12\pi}{4} - \dfrac{11\pi}{4} = \dfrac{\pi}{4}$; $\dfrac{11\pi}{4}$ is in

QII, so the point is $\left(-\dfrac{\sqrt{2}}{2}, \dfrac{\sqrt{2}}{2}\right)$

35. $\dfrac{25\pi}{6} - 4\pi = \dfrac{25\pi}{6} - \dfrac{24\pi}{6} = \dfrac{\pi}{6}$; $\dfrac{25\pi}{6}$ is in

Q1, so the point is $\left(\dfrac{\sqrt{3}}{2}, \dfrac{1}{2}\right)$.

37. The reference angle for each angle is $\dfrac{\pi}{4}$.

a. $\sin\left(\dfrac{\pi}{4}\right) = \dfrac{\sqrt{2}}{2}$ (QI)

b. $\sin\left(\dfrac{3\pi}{4}\right) = \dfrac{\sqrt{2}}{2}$ (QII)

c. $\sin\left(\dfrac{5\pi}{4}\right) = -\dfrac{\sqrt{2}}{2}$ (QIII)

d. $\sin\left(\dfrac{7\pi}{4}\right) = -\dfrac{\sqrt{2}}{2}$ (QIV)

e. $\sin\left(\dfrac{9\pi}{4}\right) = \dfrac{\sqrt{2}}{2}$ (QI)

f. $\sin\left(-\dfrac{\pi}{4}\right) = -\dfrac{\sqrt{2}}{2}$ (QIV)

g. $\sin\left(-\dfrac{5\pi}{4}\right) = \dfrac{\sqrt{2}}{2}$ (QII)

h. $\sin\left(-\dfrac{11\pi}{4}\right) = -\dfrac{\sqrt{2}}{2}$ (QIII)

39. Note that these are all quadrantal angles.

a. $\cos \pi = -1$

b. $\cos 0 = 1$

c. $\cos\left(\dfrac{\pi}{2}\right) = 0$

d. $\cos\left(\dfrac{3\pi}{2}\right) = 0$

41. The reference arc for each number is $\dfrac{\pi}{6}$.

a. $\cos\left(\dfrac{\pi}{6}\right) = \dfrac{\sqrt{3}}{2}$ (QI)

b. $\cos\left(\dfrac{5\pi}{6}\right) = -\dfrac{\sqrt{3}}{2}$ (QII)

c. $\cos\left(\dfrac{7\pi}{6}\right) = -\dfrac{\sqrt{3}}{2}$ (QIII)

d. $\cos\left(\dfrac{11\pi}{6}\right) = \dfrac{\sqrt{3}}{2}$ (QIV)

e. $\cos\left(\dfrac{13\pi}{6}\right) = \dfrac{\sqrt{3}}{2}$ (QI)

f. $\cos\left(-\dfrac{\pi}{6}\right) = \dfrac{\sqrt{3}}{2}$ (QIV)

g. $\cos\left(-\dfrac{5\pi}{6}\right) = -\dfrac{\sqrt{3}}{2}$ (QIII)

h. $\cos\left(-\dfrac{23\pi}{6}\right) = \dfrac{\sqrt{3}}{2}$ (QI)

43. Note that these are all quadrantal angles.

a. $\tan \pi = \dfrac{0}{-1} = 0$

b. $\tan 0 = \dfrac{0}{1} = 0$

c. $\tan\left(\dfrac{\pi}{2}\right) = \dfrac{1}{0}$ This is undefined.

d. $\tan\left(\dfrac{3\pi}{2}\right) = \dfrac{-1}{0}$ This is undefined.

45. $\sin t = 0.6$; $\cos t = -0.8$;

$\tan t = \dfrac{0.6}{-0.8} = -0.75$; $\cot t = \dfrac{-0.8}{0.6} = -1.\overline{3}$

$\sec t = \dfrac{1}{-0.8} = -1.25$; $\csc t = \dfrac{1}{0.6} = 1.\overline{6}$

47. $\sin t = -\dfrac{12}{13}$; $\cos t = \dfrac{-5}{13}$;

$\tan t = \dfrac{-\frac{12}{13}}{\frac{-5}{13}} = \dfrac{12}{5}$; $\cot t = \dfrac{1}{\frac{12}{5}} = \dfrac{5}{12}$;

$\sec t = \dfrac{1}{\frac{-5}{13}} = -\dfrac{13}{5}$; $\csc t = \dfrac{1}{\frac{-12}{13}} = -\dfrac{13}{12}$

49. $\sin t = \dfrac{\sqrt{11}}{6}$; $\cos t = \dfrac{5}{6}$;

$\tan t = \dfrac{\frac{\sqrt{11}}{6}}{\frac{5}{6}} = \dfrac{\sqrt{11}}{5}$;

$\cot t = \dfrac{1}{\frac{\sqrt{11}}{5}} = \dfrac{5}{\sqrt{11}} = \dfrac{5\sqrt{11}}{11}$;

$\sec t = \dfrac{1}{\frac{5}{6}} = \dfrac{6}{5}$;

$\csc t = \dfrac{1}{\frac{\sqrt{11}}{6}} = \dfrac{6}{\sqrt{11}} = \dfrac{6\sqrt{11}}{11}$

51. $\sin t = \dfrac{\sqrt{21}}{5}$; $\cos t = -\dfrac{2}{5}$;

$\tan t = \dfrac{\frac{\sqrt{21}}{5}}{\frac{-2}{5}} = -\dfrac{\sqrt{21}}{2}$;

$\cot t = \dfrac{1}{\frac{-\sqrt{21}}{2}} = -\dfrac{2}{\sqrt{21}} = \dfrac{-2\sqrt{21}}{21}$;

$\sec t = \dfrac{1}{\frac{-2}{5}} = -\dfrac{5}{2}$;

$\csc t = \dfrac{1}{\frac{\sqrt{21}}{5}} = \dfrac{5}{\sqrt{21}} = \dfrac{5\sqrt{21}}{21}$

53. $\sin t = -\dfrac{2\sqrt{2}}{3}$; $\cos t = -\dfrac{1}{3}$;

$\tan t = \dfrac{\frac{-2\sqrt{2}}{3}}{\frac{-1}{3}} = 2\sqrt{2}$; $\cot t = \dfrac{1}{2\sqrt{2}} = \dfrac{\sqrt{2}}{4}$;

$\sec t = \dfrac{1}{\frac{-1}{3}} = -3$;

$\csc t = \dfrac{1}{\frac{-2\sqrt{2}}{3}} = -\dfrac{3}{2\sqrt{2}} = -\dfrac{3\sqrt{2}}{4}$

55. $\sin t = \dfrac{\sqrt{3}}{2}$; $\cos t = \dfrac{1}{2}$;

$\tan t = \dfrac{\frac{\sqrt{3}}{2}}{\frac{1}{2}} = \sqrt{3}$; $\cot t = \dfrac{1}{\sqrt{3}} = \dfrac{\sqrt{3}}{3}$;

$\sec t = \dfrac{1}{\frac{1}{2}} = 2$; $\csc t = \dfrac{1}{\frac{\sqrt{3}}{2}} = \dfrac{2}{\sqrt{3}} = \dfrac{2\sqrt{3}}{3}$

57. $\sin t = \dfrac{\sqrt{2}}{2}$; $\cos t = -\dfrac{\sqrt{2}}{2}$;

$\tan t = \dfrac{\frac{\sqrt{2}}{2}}{\frac{-\sqrt{2}}{2}} = -1$; $\cot t = \dfrac{1}{-1} = -1$;

$\sec t = \dfrac{1}{\frac{-\sqrt{2}}{2}} = -\dfrac{2}{\sqrt{2}} = -\dfrac{2\sqrt{2}}{2} = -\sqrt{2}$;

$\csc t = \dfrac{1}{\frac{\sqrt{2}}{2}} = \dfrac{2}{\sqrt{2}} = \dfrac{2\sqrt{2}}{2} = \sqrt{2}$

59. QI; $\sin 0.75 \approx 0.7$

61. QIV; $\cos 0.75 \approx 0.7$

63. QI; $\tan 0.8 = \dfrac{\sin 0.8}{\cos 0.8} \approx 1$

65. QII; $\csc 2.0 = \dfrac{1}{\sin 2.0} \approx \dfrac{1}{0.9} \approx 1.1$

67. $\cos\left(\dfrac{5\pi}{8}\right) \approx -0.4$; QII

69. $\tan\left(\dfrac{8\pi}{5}\right) \approx -3.1$; QIV

71. $\sin\left(\dfrac{2\pi}{3}\right) = \dfrac{\sqrt{3}}{2}$

73. $\cos\left(\dfrac{7\pi}{6}\right) = -\dfrac{\sqrt{3}}{2}$

75. $\tan\left(\dfrac{2\pi}{3}\right) = \dfrac{\sin\left(\dfrac{2\pi}{3}\right)}{\cos\left(\dfrac{2\pi}{3}\right)} = \dfrac{\dfrac{\sqrt{3}}{2}}{\dfrac{-1}{2}} = -\sqrt{3}$

77. $\sin\left(\dfrac{\pi}{2}\right) = 1$

79. $\sec t = -\sqrt{2} \;\Rightarrow\; \cos t = -\dfrac{1}{\sqrt{2}} = -\dfrac{\sqrt{2}}{2}$

This occurs at $t = \dfrac{3\pi}{4}$ and $\dfrac{5\pi}{4}$.

81. $\tan t$ undefined $\;\Rightarrow\; \cos t = 0$

This occurs at $t = \dfrac{\pi}{2}$ and $\dfrac{3\pi}{2}$.

83. $\cos t = -\dfrac{\sqrt{2}}{2}$ occurs at $t = \dfrac{3\pi}{4}$ and $\dfrac{5\pi}{4}$

85. $\sin t = 0$ occurs at $t = 0,\; \pi$

87. a. Since $\cos t > 0$ and $\sin t < 0$, t is in QIV;

$-t$ is in QI and has coordinates $\left(\dfrac{3}{4}, \dfrac{4}{5}\right)$.

b. $t + \pi$ is in QII and has coordinates $\left(-\dfrac{3}{4}, \dfrac{4}{5}\right)$.

89. 0.8 is in QI. The additional value is in QII, and is $\pi - 0.8 = 2.3416$.

91. 4.5 is in QIII. The additional value is in QII, and is $2\pi - 4.5 = 1.7832$.

93. 0.4 is in QI. The additional value is in QIII, and is $\pi + 0.4 = 3.5416$.

95. $(x, y, r) = \left(\dfrac{x}{r}, \dfrac{y}{r}, 1\right)$

a. $\left(\dfrac{x}{r}, \dfrac{y}{r}, 1\right) = \left(\dfrac{5}{13}, \dfrac{12}{13}, 1\right)$

$\left(\dfrac{5}{13}\right)^2 + \left(\dfrac{12}{13}\right)^2 = \dfrac{25}{169} + \dfrac{144}{169} = 1$

$\sin t = \dfrac{12}{13}$; $\cos t = \dfrac{5}{13}$;

$\tan t = \dfrac{12}{5}$; $\csc t = \dfrac{13}{12}$;

$\sec t = \dfrac{13}{5}$; $\cot t = \dfrac{5}{12}$

b. $\left(\dfrac{x}{r}, \dfrac{y}{r}, 1\right) = \left(\dfrac{7}{25}, \dfrac{24}{25}, 1\right)$

$\left(\dfrac{7}{25}\right)^2 + \left(\dfrac{24}{25}\right)^2 = \dfrac{49}{625} + \dfrac{576}{625} = 1$

$\sin t = \dfrac{24}{25}$; $\cos t = \dfrac{7}{25}$;

$\tan t = \dfrac{24}{7}$; $\csc t = \dfrac{25}{24}$;

$\sec t = \dfrac{25}{7}$; $\cot t = \dfrac{7}{24}$

c. $\left(\dfrac{x}{r}, \dfrac{y}{r}, 1\right) = \left(\dfrac{12}{37}, \dfrac{35}{37}, 1\right)$

$\left(\dfrac{12}{37}\right)^2 + \left(\dfrac{35}{37}\right)^2 = \dfrac{144}{1369} + \dfrac{1225}{1369} = 1$

$\sin t = \dfrac{35}{37}$; $\cos t = \dfrac{12}{37}$;

$\tan t = \dfrac{35}{12}$; $\csc t = \dfrac{37}{35}$;

$\sec t = \dfrac{37}{12}$; $\cot t = \dfrac{12}{35}$

d. $\left(\dfrac{x}{r}, \dfrac{y}{r}, 1\right) = \left(\dfrac{9}{41}, \dfrac{40}{41}, 1\right)$

$\left(\dfrac{9}{41}\right)^2 + \left(\dfrac{40}{41}\right)^2 = \dfrac{81}{1681} + \dfrac{1600}{1681} = 1$

$\sin t = \dfrac{40}{41}$; $\cos t = \dfrac{9}{41}$;

$\tan t = \dfrac{40}{9}$; $\csc t = \dfrac{41}{40}$;

$\sec t = \dfrac{41}{9}$; $\cot t = \dfrac{9}{40}$

97. a. The circumference of the roller is $2\pi r =$ 2π ft, so 2π ft corresponds to 1 revolution, or 2π radians.

$$5\,\text{ft} \cdot \frac{2\pi\,\text{rad}}{2\pi\,\text{ft}} = 5\,\text{rad}$$

b. 5 rad corresponds to 5 ft, so 30 rad corresponds to 30 ft.

99. a. The circumference of the spool is $2\pi r =$ 2π dm, so 2π dm corresponds to 1 revolution, or 2π radians.

$$5\,\text{rad} \cdot \frac{2\pi\,\text{dm}}{2\pi\,\text{rad}} = 5\,\text{dm}$$

b. If 5 rad corresponds to 5 dm, then 2π rad corresponds to 2π (≈ 6.28) dm.

101. a. Use $s = r\theta$ with $r = 1$ AU, $\theta = 2.5$ rad.
$s = (1\ \text{AU})(2.5\ \text{rad}) = 2.5$ AU.

b. If 2.5 AU corresponds to 2.5 rad, then 1 revolution (2π rad) corresponds to 2π, or about 6.28 AU.

103. Yes, the distance equals the circumference.

105. Since sine and cosine are determined by coordinates of points on a circle of radius 1, they have to be between −1 and 1, so the range for both is [−1, 1].

107. a. $2t \approx 2.2$
b. $t \approx 1.1$, which is in QI.
c. $\cos(1.1) \approx 0.5$
d. $\cos(2t) = -0.6$; $2\cos t = 2(0.5)$; No

109. $(-3, -4), (5, 2)$

a. $d = \sqrt{(-3-5)^2 + (-4-2)^2}$
$d = \sqrt{(-8)^2 + (-6)^2} = \sqrt{64 + 36} = \sqrt{100} = 10$

b. $\left(\dfrac{-3+5}{2}, \dfrac{-4+2}{2}\right) = (1, -1)$

c. $m = \dfrac{-4-2}{-3-5} = \dfrac{-6}{-8} = \dfrac{3}{4}$

111. a. $2|x+1| - 3 = 7$
$2|x+1| = 10$
$|x+1| = 5$
$x+1 = -5$ or $x+1 = 5$
$x = -6$ or $x = 4$

b. $2\sqrt{x+1} - 3 = 7$
$2\sqrt{x+1} = 10$
$\sqrt{x+1} = 5$
$\left(\sqrt{x+1}\right)^2 = 5^2$
$x+1 = 25$
$x = 24$

Technology Highlights

1. $Y_2(0.00025) \approx 0.8639$
Xmin = 0
Xmax = 1/1336
Xscl = 1/113,360
Ymin = −1
Ymax = 1
Yscl = 0.25

2. $Y_1(550) \approx 362.58$
Xmin = 0
Xmax = 1256.6
Xscl = 120
Ymin = −1000
Ymax = 1000
Yscl = 200

5.3 Exercises

1. increasing

3. $(-\infty,\infty)$; $[-1, 1]$

5. Answers will vary.

7.

t	$\cos t$
π	-1
$\dfrac{7\pi}{6}$	$\dfrac{-\sqrt{3}}{2}$
$\dfrac{5\pi}{4}$	$\dfrac{-\sqrt{2}}{2}$
$\dfrac{4\pi}{3}$	$\dfrac{-1}{2}$
$\dfrac{3\pi}{2}$	0
$\dfrac{5\pi}{3}$	$\dfrac{1}{2}$
$\dfrac{7\pi}{4}$	$\dfrac{\sqrt{2}}{2}$
$\dfrac{11\pi}{6}$	$\dfrac{\sqrt{3}}{2}$
2π	1

9. a. $t = \left(\dfrac{\pi}{6}+10\pi\right)$, II

   b. $t = -\dfrac{\pi}{4}$, V

   c. $t = -\dfrac{15\pi}{4}$; IV

   d. $t = 13\pi$; I

   e. $t = \dfrac{21\pi}{2}$; III

11. $y = \sin t$ for $t \in \left[-\dfrac{3\pi}{2}, \dfrac{\pi}{2}\right]$

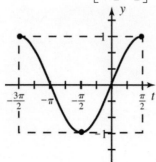

13. $y = \cos t$ for $t \in \left[-\dfrac{\pi}{2}, 2\pi\right]$

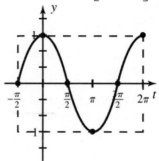

15. $y = 3\sin t$
    Amplitude 3, Period 2π

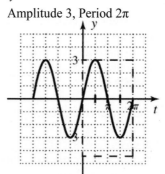

17. $y = -2\cos t$
    Amplitude 2, Period 2π

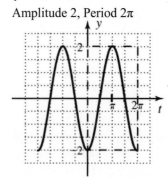

19. $y = \dfrac{1}{2}\sin t$

Amplitude $\dfrac{1}{2}$, Period 2π

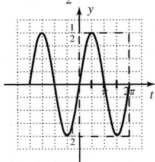

21. $y = -\sin(2t)$

Amplitude 1, Period $\dfrac{2\pi}{2} = \pi$

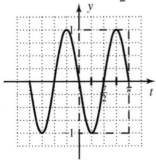

23. $y = 0.8\cos(2t)$

Amplitude 0.8, Period $\dfrac{2\pi}{2} = \pi$

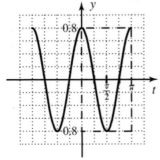

25. $f(t) = 4\cos\left(\dfrac{1}{2}t\right)$

Amplitude 4, Period $\dfrac{2\pi}{\dfrac{1}{2}} = 4\pi$

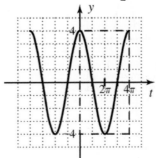

27. $f(t) = 3\sin(4\pi t)$

Amplitude 3, Period $\dfrac{2\pi}{4\pi} = \dfrac{1}{2}$

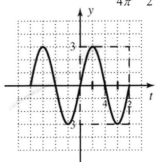

29. $y = 4\sin\left(\dfrac{5\pi}{3}t\right)$

Amplitude 4, Period $\dfrac{2\pi}{\dfrac{5\pi}{3}} = 2\pi \cdot \dfrac{3}{5\pi} = \dfrac{6}{5}$

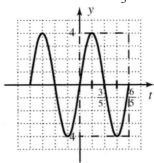

5.3 Exercises

31. $f(t) = 2\sin(256\pi t)$

Amplitude 2, Period $\dfrac{2\pi}{256\pi} = \dfrac{1}{128}$

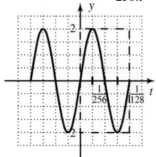

33. $y = 3\csc t$

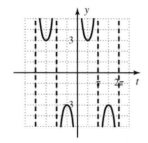

35. $y = 2\sec t$

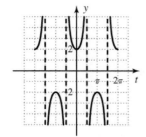

37. $y = -2\cos(4t)$

Amplitude 2, Period $\dfrac{2\pi}{4} = \dfrac{\pi}{2}$. Graph goes through $(0, -2)$ and matches graph k.

39. $y = 3\sin(2t)$

Amplitude 3, Period $\dfrac{2\pi}{2} = \pi$. Graph goes through $(0, 0)$ and matches graph f.

41. $y = 2\csc\left(\dfrac{1}{2}t\right)$

No amplitude, Period $\dfrac{2\pi}{\dfrac{1}{2}} = 4\pi$. Graph has max's and mins at height 2 and matches graph h.

43. $f(t) = \dfrac{3}{4}\cos(0.4t)$

Amplitude $\dfrac{3}{4}$, Period $\dfrac{2\pi}{0.4} = 5\pi$. Graph goes through $\left(0, \dfrac{3}{4}\right)$ and matches graph b.

45. $y = \sec(8\pi t)$

No amplitude, Period $\dfrac{2\pi}{8\pi} = \dfrac{1}{4}$. Graph has max's and mins at height 1 and matches graph j.

47. $y = 4\sin(144\pi t)$

Amplitude 4, Period $\dfrac{2\pi}{144\pi} = \dfrac{1}{72}$. Graph goes through $(0, 0)$ and matches graph d.

49. The amplitude is $\frac{3}{4}$, and the period is $\frac{\pi}{4}$.

    $\frac{\pi}{4} = \frac{2\pi}{B} \;\Rightarrow\; \pi B = 8\pi \;\Rightarrow\; B = 8$. This is a

    cosine graph that goes through $\left(0, -\frac{3}{4}\right)$, so

    $A = -\frac{3}{4}$. $\; y = -\frac{3}{4}\cos(8t)$

51. The max's and mins are at height 0.2, and the first portion to the right of the y-axis is below the x-axis, so $A = -0.2$. The period is

    4π, so $\frac{2\pi}{B} = 4\pi \;\Rightarrow\; 4\pi B = 2\pi$

    $\Rightarrow\; B = \frac{2\pi}{4\pi} = \frac{1}{2}$. $\; y = -0.2\csc\left(\frac{1}{2}t\right)$

53. The amplitude is 6 and the graph goes through $(0, 6)$, so $A = 6$. The period is 3, so

    $\frac{2\pi}{B} = 3 \;\Rightarrow\; 3B = 2\pi \;\Rightarrow\; B = \frac{2\pi}{3}$.

    $y = 6\cos\left(\frac{2\pi}{3}t\right)$

55. The red graph is $y = -\cos x$, and the blue is

    $y = \sin x$. The graphs cross at $x = \frac{3\pi}{4}$ and

    $\frac{7\pi}{4}$.

57. The red graph is $y = -2\cos x$, and the blue is
    $y = 2\sin(3x)$. The graphs cross at

    $x = \frac{3\pi}{8}, \frac{3\pi}{4}, \frac{7\pi}{8}, \frac{11\pi}{8}, \frac{7\pi}{4}$ and $\frac{15\pi}{8}$.

59. $\sin^2\theta + \cos^2\theta = 1$; $\left(\frac{15}{113}\right)^2 + \cos^2\theta = 1$

    $\frac{225}{12{,}769} + \cos^2\theta = 1$

    $\cos^2\theta = 1 - \frac{225}{12{,}769} = \frac{12{,}544}{12{,}769}$

    $\cos\theta = \pm\sqrt{\frac{12{,}544}{12{,}769}} = \pm\frac{112}{113}$

    QI, so choose $\cos t = \frac{112}{113}$. The Pythagorean triple is (15, 112, 113).

61. a. The height from crest to trough is 3 ft.
    b. The first cycle ends at $x = 80$, so the wavelength is 80 mi.
    c. The amplitude is 1.5, the period is 80,

       so $A = 1.5$, $\frac{2\pi}{B} = 80 \;\Rightarrow\; 80B = 2\pi$

       $\Rightarrow\; B = \frac{\pi}{40}$. The equation is

       $h = 1.5\cos\left(\frac{\pi}{40}x\right)$

63. a. The amplitude is 4 and the graph goes through $(0, -4)$, so $A = -4$. The period is 24, so

       $\frac{2\pi}{B} = 24 \;\Rightarrow\; 24B = 2\pi \;\Rightarrow\; B = \frac{\pi}{12}$.

       The equation is $D = -4\cos\left(\frac{\pi}{12}t\right)$.

    b. $D(11) = -4\cos\left(\frac{11\pi}{12}\right) \approx 3.86$

    c. Midnight corresponds to $t = 18$. At $t = 18$, the deviation is 0, so the temperature is 72°.

65. a. The amplitude is 15, so $A = 15$. The period is 2, so $\frac{2\pi}{B} = 2 \Rightarrow B = \pi$. The equation is $D = 15\cos(\pi t)$.

   b. The period is 2, so the height at $t = 6.5$ is the same as at $t = 4.5$, which is zero. The tail is at center.

   c. Only one cycle is completed every two seconds, so the shark is probably swimming at a leisurely pace.

67. a. Graph a: the energy is highest when closest to the Sun, and this is at $t = 0$ for graph a.

   b. This graph is at height 62.5 at about $t = 76$ days.

   c. The period is 96 days.

69. a. Wavelength:

   $\dfrac{2\pi}{\dfrac{\pi}{240}} = 2\pi \cdot \dfrac{240}{\pi} = 480\,\text{nm}$. This is in the blue range.

   b. $\dfrac{2\pi}{\dfrac{\pi}{310}} = 2\pi \cdot \dfrac{310}{\pi} = 620\,\text{nm}$

   This is in the orange range.

71. $A = 30$, period = 1/25, so $\dfrac{2\pi}{\omega} = \dfrac{1}{25}$
   $\Rightarrow \omega = 50\pi$. The equation is
   $I = 30\sin(50\pi t)$.
   $I(0.045) = 30\sin(50\pi(0.045)) \approx 21.2$ amps

73. All the functions graphed in this section have average value zero.

t	0	$\frac{\pi}{2}$	π	$\frac{3\pi}{2}$	2π
y	3	5	3	1	3

The average value is $\dfrac{(1+5)}{2} = 3$. The graph is shifted up by 3 because of the +3 on the end of the function. The average value of $y = -2\cos t + 1$ is 1. The amplitude is "centered" on the average value.

75. The function with the largest coefficient of t has the shortest period: $g(t)$.

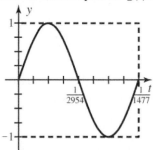

77. This is a 30-60-90 triangle, so the longer leg is $\sqrt{3}$ times the shorter. The shorter leg is $\dfrac{100}{\sqrt{3}}$ yd. The hypotenuse is twice as long, or $\dfrac{200}{\sqrt{3}} \approx 115.5$ yd.

79. a. $z_1 + z_2 = (1+i) + (2-5i) = 3 - 4i$

   b. $z_1 - z_2 = (1+i) - (2-5i) = -1 + 6i$

   c. $z_1 z_2 = (1+i)(2-5i) = 2 - 3i - 5i^2$
   $= 2 - 3i - 5(-1) = 2 - 3i + 5 = 7 - 3i$

   d. $\dfrac{z_2}{z_1} = \dfrac{2-5i}{1+i} \cdot \dfrac{1-i}{1-i} = \dfrac{2-7i+5i^2}{1-i^2}$
   $= \dfrac{2-7i+5(-1)}{1-(-1)} = \dfrac{-3-7i}{2}$

Chapter 5: Introduction to Trigonometric Functions

Technology Highlights

1. There is asymptotic behavior at these zeroes.

5.4 Exercises

1. π; $P = \dfrac{\pi}{B}$

3. odd; $-f(t)$; -0.268

5. a. Use the formula $\cot t = \dfrac{\cos t}{\sin t}$
   b. They are the reciprocals of $\tan t$ values.

7.

t	π	$\dfrac{7\pi}{6}$	$\dfrac{5\pi}{4}$	$\dfrac{4\pi}{3}$	$\dfrac{3\pi}{2}$
$\tan t$	0	$\dfrac{1}{\sqrt{3}}$	1	$\sqrt{3}$	Undefined

9. $\dfrac{\pi}{2} \approx 1.6$; $\dfrac{\pi}{4} \approx 0.8$; $\dfrac{\pi}{6} \approx 0.5$; $\sqrt{2} \approx 1.4$

   $\dfrac{\sqrt{2}}{2} \approx 0.7$; $\dfrac{2}{\sqrt{3}} \approx 1.2$

11. a. $\tan\left(-\dfrac{\pi}{4}\right) = \dfrac{\dfrac{\sqrt{2}}{2}}{\dfrac{-\sqrt{2}}{2}} = -1$ (QIV)

    b. $\cot\left(\dfrac{\pi}{6}\right) = \dfrac{\dfrac{\sqrt{3}}{2}}{\dfrac{1}{2}} = \sqrt{3}$ (Q1)

    c. $\cot\left(\dfrac{3\pi}{4}\right) = \dfrac{\dfrac{-\sqrt{2}}{2}}{\dfrac{\sqrt{2}}{2}} = -1$ (QII)

    d. $\tan\left(\dfrac{\pi}{3}\right) = \dfrac{\dfrac{\sqrt{3}}{2}}{\dfrac{1}{2}} = \sqrt{3}$ (Q1)

13. a. $\tan\left(\dfrac{7\pi}{4}\right) = -1$, so $\tan^{-1}(-1) = \dfrac{7\pi}{4}$

    b. $\cot\left(\dfrac{7\pi}{6}\right) = \sqrt{3}$, so $\cot^{-1}\sqrt{3} = \dfrac{7\pi}{6}$

    c. $\cot\left(\dfrac{5\pi}{3}\right) = -\dfrac{1}{\sqrt{3}}$, so $\cot^{-1}\left(-\dfrac{1}{\sqrt{3}}\right) = \dfrac{5\pi}{3}$

    d. $\tan\left(\dfrac{3\pi}{4}\right) = -1$, so $\tan^{-1}(-1) = \dfrac{3\pi}{4}$

15.

t	π	$\dfrac{7\pi}{6}$	$\dfrac{5\pi}{4}$	$\dfrac{4\pi}{3}$	$\dfrac{3\pi}{2}$
$\cot t$	Undefined	$\sqrt{3}$	1	$\dfrac{1}{\sqrt{3}}$	0

17. There are many different choices: three nearby are $\dfrac{11\pi}{24} - \pi = -\dfrac{13\pi}{24}$, $\dfrac{11\pi}{24} + \pi = \dfrac{35\pi}{24}$, and $\dfrac{11\pi}{24} + 2\pi = \dfrac{59\pi}{24}$.

19. There are many different choices: three nearby are $1.5 - \pi \approx -1.6$, $1.5 + \pi \approx 4.6$, and $1.5 + 2\pi \approx 7.8$.

21. $t = \left(\dfrac{\pi}{10}\right)$; $\tan t \approx 0.3249$. The period is π, so all solutions are given by $\dfrac{\pi}{10} + \pi k, k \in Z$.

23. $t = \left(\dfrac{\pi}{12}\right)$; $\cot t \approx 3.732$; $2 + \sqrt{3} \approx 3.732$

    The period is π, so all solutions are given by $\dfrac{\pi}{12} + \pi k, k \in Z$.

25. $f(t) = 2 \tan t; [-2\pi, 2\pi]$

Period $= \pi$. Asymptotes at $x = -\dfrac{3\pi}{2}, -\dfrac{\pi}{2}, \dfrac{\pi}{2}$

and $\dfrac{3\pi}{2}$. Zeros at $x = -2\pi, -\pi, 0, \pi, -2\pi$.

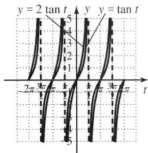

27. $h(t) = 3 \cot t; [-2\pi, 2\pi]$

Period $= \pi$. Asymptotes at $x = -2\pi, -\pi, 0,$

$\pi, -2\pi$. Zeros at $x = -\dfrac{3\pi}{2}, -\dfrac{\pi}{2}, \dfrac{\pi}{2}, \dfrac{3\pi}{2}$.

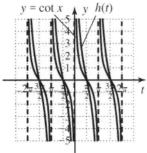

29. $y = \tan(2t); \left[-\dfrac{\pi}{2}, \dfrac{\pi}{2}\right]$

Period $= \dfrac{\pi}{2}$. Asymptotes at $x = -\dfrac{\pi}{4}, \dfrac{\pi}{4}$.

Zeros at $x = -\dfrac{\pi}{2}, 0, \dfrac{\pi}{2}$.

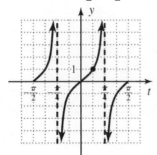

31. $y = \cot(4t); \left[-\dfrac{\pi}{4}, \dfrac{\pi}{4}\right]$

Period $= \dfrac{\pi}{4}$. Asymptotes at $x = -\dfrac{\pi}{4}, 0, \dfrac{\pi}{4}$.

Zeros at $x = -\dfrac{\pi}{8}, \dfrac{\pi}{8}$.

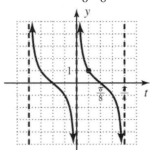

33. $y = 2 \tan(4t); \left[-\dfrac{\pi}{4}, \dfrac{\pi}{4}\right]$

Period $= \dfrac{\pi}{4}$. Asymptotes at $x = -\dfrac{\pi}{8}, \dfrac{\pi}{8}$.

Zeros at $x = -\dfrac{\pi}{4}, 0, \dfrac{\pi}{4}$.

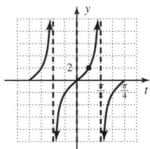

35. $y = 5 \cot\left(\dfrac{1}{3}t\right); [-3\pi, 3\pi]$

Period $= \dfrac{\pi}{\frac{1}{3}} = 3\pi$. Asymptotes at

$x = -3\pi, 0, 3\pi$. Zeros at $x = -\dfrac{3\pi}{2}, \dfrac{3\pi}{2}$

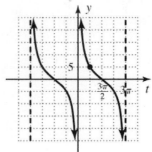

37. $y = 3\tan(2\pi t); \left[-\dfrac{1}{2}, \dfrac{1}{2}\right]$

Period $= \dfrac{\pi}{2\pi} = \dfrac{1}{2}$. Asymptotes at

$x = -\dfrac{1}{4}, \dfrac{1}{4}.$

Zeros at $x = -\dfrac{1}{2}, 0, \dfrac{1}{2}.$

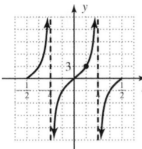

39. $f(t) = 2\cot(\pi t); [-1, 1]$

Period $= \dfrac{\pi}{\pi} = 1$. Asymptotes at $x = -1, 0, 1$.

Zeros at $x = -\dfrac{1}{2}, \dfrac{1}{2},$

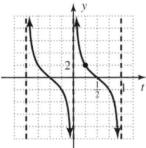

41. The period is 2π, so $\dfrac{\pi}{B} = 2\pi \Rightarrow B = \dfrac{1}{2}$.

So the equation is $y = A\tan\left(\dfrac{1}{2}t\right)$, and

$\left(\dfrac{\pi}{2}, 3\right)$ is on the graph, so $3 = A\tan\left(\dfrac{1}{2}\cdot\dfrac{\pi}{2}\right)$

$\Rightarrow 3 = A\tan\left(\dfrac{\pi}{4}\right) = A$. The equation is

$y = 3\tan\left(\dfrac{1}{2}t\right)$.

43. The period is $\dfrac{3}{2}$, so $\dfrac{\pi}{B} = \dfrac{3}{2} \Rightarrow B = \dfrac{2\pi}{3}$.

The equation is $y = A\cot\left(\dfrac{2\pi}{3}t\right)$, and

$\left(\dfrac{1}{4}, 2\sqrt{3}\right)$ is on the graph, so

$2\sqrt{3} = A\cot\left(\dfrac{2\pi}{3}\cdot\dfrac{1}{4}\right) = A\cot\left(\dfrac{\pi}{6}\right) = \sqrt{3}A$ and

$A = 2$. $y = 2\cot\left(\dfrac{2\pi}{3}t\right)$ is the equation.

45. Graphing $Y_1 = \cos(3t)$ and $Y_2 = \tan t$, we see

the graphs intersect at $t = \dfrac{\pi}{8}$ (≈ 0.3926) and

$\dfrac{3\pi}{8}$ (≈ 1.1781).

47. Plug in $u = 40°$, $v = 65°$, $d = 100$.

$h = \dfrac{100}{\cot 40° - \cot 65°} \approx 137.8\,\text{ft}$

49. The asymptotes occur where the output is infinite, so the two asymptotes provided by the table are at –6 and 6. This means $P = 12$,

and $\dfrac{\pi}{B} = 12 \Rightarrow B = \dfrac{\pi}{12}$. The asymptotes

will occur at $6 + 12k$, $k \in Z$. The equation

will look like $y = A\tan\left(\dfrac{\pi}{12}x\right)$. If we use the

point (3, 5.2) to find A:

$5.2 = A\tan\left(\dfrac{\pi}{12}\cdot 3\right) \Rightarrow A = 5.2$. (Note that if

you choose a different point, the result will differ somewhat.) So the equation is

$y = 5.2\tan\left(\dfrac{\pi}{12}x\right)$. Using this model,

$y(2) = 5.2\tan\left(\dfrac{\pi}{12}\cdot 2\right) \approx 3.002$ and

$y(-2) = 5.2\tan\left(\dfrac{\pi}{12}\cdot -2\right) \approx -3.002$. These

results agree well with results from the table.

51. The asymptote will be at 90° and the zero is at $\theta = 0$, so the period is 180°, and
$$\frac{180°}{B} = 180° \Rightarrow B = 1. \quad y = A\tan\theta. \text{ Using}$$
the point (30°, 6.9), we get
$$6.9 = A\tan 30° = A \cdot \frac{1}{\sqrt{3}} \Rightarrow A = 11.95.$$
The equation is $y = 11.95\tan\theta$. The asymptotes are at $90° + 180°k$, $k \in Z$. Note that $y(45°) = 11.95\tan 45° = 11.95$. When the pen is at that angle, it forms a 45° -45° -90° triangle with the other leg equal to the length of the pen, so the pen is about 12 cm in length.

53. a. Perimeter (circumference) of a circle is
$$P = 2\pi r = 2\pi(10) = 20\pi \text{ cm} \approx 62.8 \text{ cm}.$$

   b. When $n = 4$ it's a square; the radius of the circle is half the length of a side, so the sides have length 20 cm and the perimeter is 4(20) = 80 cm.

   c. Plug $r = 10$ and each value of n into the given formula.

n	10	20	30	100
P	64.984	63.354	63.063	62.853

   As n gets large, the perimeter approaches 20π, about 62.8 cm.

55. a.

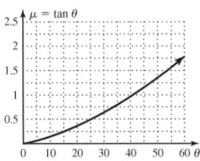

   The block will not slide at 30°. The smallest angle is about 35°.

   b. $\mu = \tan 46.5° = 1.05$

   c. This would require an angle steeper than 68.2°, which is quite steep. Something like soft rubber on sandstone might have a coefficient that high.

57. a. $\tan(80°) = 5.67$ units

   b. $\tan^{-1} 16.35 = 86.5°$

   c. Yes, it can be any length. The range of tangent is $(-\infty, \infty)$.

   d. As θ gets close to 90°, tangent of the angle increases without bound, as does the length of the line segment.

59. Make a table of values for the function.
$$D = 5\tan\left(\frac{\pi}{8}t\right).$$

t	2	3	3.5	3.8
D	5	12.1	25.1	63.5

t	3.9	3.99
D	127.3	1,273.23

[2, 3]: $\dfrac{D(3) - D(2)}{3 - 2} = 12.1 - 5 \approx 7.1 \dfrac{\text{m}}{\text{sec}}$;

[3, 3.5]:
$$\frac{D(3.5) - D(3)}{3.5 - 3} = \frac{25.1 - 12.1}{.5} \approx 26 \frac{\text{m}}{\text{sec}} ;$$

[3.5, 3.8]:
$$\frac{D(3.8) - D(3.5)}{3.8 - 3.5} = \frac{63.5 - 25.1}{0.3} \approx 128 \frac{\text{m}}{\text{sec}}$$

The velocity of the beam is increasing dramatically.

[3.9, 3.99]:
$$\frac{D(3.99) - D(3.9)}{3.99 - 3.9} = \frac{1,273.23 - 127.3}{0.09}$$
$$\approx 12,733 \frac{\text{m}}{\text{sec}}$$

61. a. $h(0) = 0$; y-intercept: (0,0)
   $3x^2 - 9x = 0$; $3x(x - 3) = 0$; $x = 0, 3$
   x-intercepts: (0,0), (3,0)
   $2x^2 - 8 = 0$; $2x^2 = 8$; $x^2 = 4$; $x = \pm 2$
   Vertical asymptotes: $x = 2, -2$
   Horizontal asymptote: $y = 3/2$

   b. $t(0)$ is undefined: no y-intercept
   $x + 1 = 0$; $x = -1$; x-intercept: (-1,0)
   $x^2 - 4x = 0$; $x(x - 4) = 0$; $x = 0, 4$
   Vertical asymptotes: $x = 0, 4$
   Horizontal asymptote: $y = 0$

   c. $p(0) = -\dfrac{1}{2}$; y-intercept $\left(0, -\dfrac{1}{2}\right)$.

   $x^2 - 1 = 0$; $x = -1, 1$;
   x-intercepts: (-1,0), (1,0)
   $x + 2 = 0$; $x = -2$;
   Vertical asymptote: $x = -2$
   Horizontal asymptote: none
   $$\frac{x^2 - 1}{x + 2} = x - 2 + \frac{3}{x + 2}$$
   Slant asymptote: $y = x - 2$

Chapter 5: Introduction to Trigonometric Functions

Chapter 5 Mid-Chapter Check

1. a. $36°06'36'' = \left(36 + \dfrac{6}{60} + \dfrac{36}{3,600}\right)^{\circ}$

    $= 36.11°\text{N}$

    $115°04'48'' = \left(115 + \dfrac{4}{60} + \dfrac{48}{3,600}\right)^{\circ}$

    $= 115.08° \text{ W}$

    b. $s = r\theta;\ s = 3,960(36.11°)\left(\dfrac{\pi \text{ rad}}{180°}\right)$

    $\approx 2,495.7 \text{ mi}$

3. a. $\cot 60° = \dfrac{\cos 60°}{\sin 60°} = \dfrac{1/2}{\sqrt{3}/2} = \dfrac{1}{\sqrt{3}}$

    b. $\sin\left(\dfrac{7\pi}{4}\right) = -\dfrac{\sqrt{2}}{2}$ (QIV)

5. $\left(-\dfrac{\sqrt{5}}{3}\right)^2 + y^2 = 1;\ y^2 = 1 - \dfrac{5}{9} = \dfrac{4}{9}$

    $y = \pm\sqrt{\dfrac{4}{9}} = \pm\dfrac{2}{3}$. QIII, so choose $-\dfrac{2}{3}$.

    $\sin\theta = -\dfrac{2}{3};\quad \cos\theta = -\dfrac{\sqrt{5}}{3}$

    $\tan\theta = \dfrac{-2/3}{-\sqrt{5}/3} = \dfrac{2}{\sqrt{5}};\quad \cot\theta = \dfrac{\sqrt{5}}{2}$

    $\sec\theta = \dfrac{1}{-\sqrt{5}/3} = -\dfrac{3}{\sqrt{5}};\quad \csc\theta = -\dfrac{3}{2}$

7. $y = 3\tan\left(\dfrac{\pi}{2}t\right)$

    $P = \dfrac{\pi}{\dfrac{\pi}{2}} = 2$. One zero is $t = 0$, which is

    halfway between asymptotes, so $t = -1$ and 1 are asymptotes. The others are $t = -3$, -5, 3, and 5.

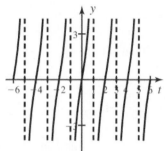

9. a. QIV

    b. $2\pi - 5.94 \approx 0.343$

    c. $\sin t$ and $\tan t$ are negative in QIV.

Reinforcing Basic Concepts

1. $\left(-\dfrac{1}{2}, \dfrac{\sqrt{3}}{2}\right), \cos t = -\dfrac{1}{2}, \sin t = \dfrac{\sqrt{3}}{2}$

3. QIV, negative since $y < 0$

Technology Highlights

1. $y = -2\cos(\pi t) + 1$

    $t = \dfrac{1}{3},\ \dfrac{5}{3},\ \dfrac{7}{3}$

3. $y = \dfrac{3}{2}\tan(2x) - 1$

    $t \approx 0.2940,\ 1.8648$

5.5 Exercises

1. $y = A\sin(Bt + C) + D$;
   $y = A\cos(Bt + C) + D$

3. $0 \leq Bt + C < 2\pi$

5. Answers will vary.

7. a. $A = 50,\ P = 24$
   b. $f(14) \approx -25$
   c. $f(x) \geq 20$ on $[1.6, 10.4]$.

9. a. $A = 200, P = 3$
   b. $f(2) \approx -175$
   c. $f(x) \leq -100$ on $[1.75, 2.75]$.

11. $A = \dfrac{100 - 20}{2} = 40$; $D = \dfrac{100 + 20}{2} = 60$

   $B = \dfrac{2\pi}{30} = \dfrac{\pi}{15}$; $y = 40\sin\left(\dfrac{\pi}{15}t\right) + 60$

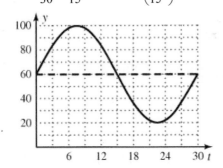

13. $A = \dfrac{20 - 4}{2} = 8$; $D = \dfrac{20 + 4}{2} = 12$

   $B = \dfrac{2\pi}{360} = \dfrac{\pi}{180}$; $y = 8\sin\left(\dfrac{\pi}{180}t\right) + 12$

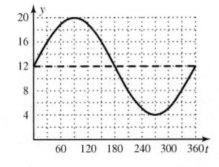

15. a. $A = \dfrac{39 - 29}{2} = 5$; $D = \dfrac{39 + 29}{2} = 34$

   $B = \dfrac{2\pi}{24} = \dfrac{\pi}{12}$; $y = 5\sin\left(\dfrac{\pi}{12}t\right) + 34$

   b.

   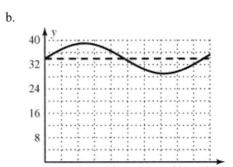

   c. At about $t = 13.5$ (1:30 A.M.) and $t = 22.5$ (10:30 A.M.)

17. a. $A = \dfrac{18.8 - 6}{2} = 6.4$; $D = \dfrac{18.8 + 6}{2} = 12.4$

   $B = \dfrac{2\pi}{12} = \dfrac{\pi}{6}$ ($P = 12$ since 12 months

   in a year.) $y = -6.4\cos\left(\dfrac{\pi}{6}t\right) + 12.4$

   Note that we want a cosine function since the low occurs at $t = 0$.

   b.

   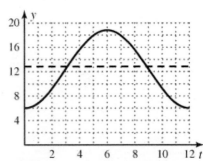

   c. The graph is above height 15 from about $t = 3.8$ to $t = 8.2$, which is a span of 4.4 months, or about 134 days.

19. a. $P = \dfrac{2\pi}{\dfrac{2\pi}{11}} = 11$ years

    b.

    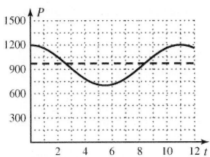

    c. The maximum is 1,200 and the minimum is 700.
    d. The height of the graph is below 740 from about $t = 4.5$ to 6.5, a span of 2 years.

21. The period is 11 years, and the average value occurs $\dfrac{1}{4}$ way through a period. This is at $t = 2.75$, so we need to shift 2.75 units to the right.

    $$P(t) = 250\cos\left[\dfrac{2\pi}{11}(t - 2.75)\right] + 950$$

    or $P(t) = 250\sin\left(\dfrac{2\pi}{11}t\right) + 950$

23. $A = 120$; avg. value $= 0$; $P = \dfrac{2\pi}{\dfrac{\pi}{12}} = 24$; no

    vertical shift; horizontal shift is 6 units right;

    PI: $\quad 0 \le \dfrac{\pi}{12}(t - 6) < 2\pi$
    $\quad\quad 0 \le t - 6 < 24$
    $\quad\quad 6 \le t < 30$

25. $A = 1$; avg. value $= 0$; $P = \dfrac{2\pi}{\dfrac{\pi}{6}} = 12$; no

    vertical shift; horizontal shift is $\dfrac{\dfrac{\pi}{3}}{\dfrac{\pi}{6}} = 2$

    units right;

    PI: $\quad 0 \le \dfrac{\pi}{6}t - \dfrac{\pi}{3} < 2\pi$
    $\quad\quad \dfrac{\pi}{3} \le \dfrac{\pi}{6}t < \dfrac{7\pi}{3}$
    $\quad\quad 2 \le t < 14$

27. $A = 1$; avg. value $= 0$; $P = \dfrac{2\pi}{\dfrac{\pi}{4}} = 8$; no

    vertical shift; horizontal shift is $\dfrac{\dfrac{\pi}{6}}{\dfrac{\pi}{4}} = \dfrac{2}{3}$

    unit right;

    PI: $\quad 0 \le \dfrac{\pi}{4}t - \dfrac{\pi}{6} < 2\pi$
    $\quad\quad \dfrac{\pi}{6} \le \dfrac{\pi}{4}t < \dfrac{13\pi}{6}$
    $\quad\quad \dfrac{2}{3} \le t < \dfrac{26}{3}$

29. $A = 24.5$; avg. value $= 15.5$; $P = \dfrac{2\pi}{\dfrac{\pi}{10}} = 20$;

    vertical shift is 15.5 units up; horizontal shift is 2.5 units right;

    PI: $\quad 0 \le \dfrac{\pi}{10}(t - 2.5) < 2\pi$
    $\quad\quad 0 \le t - 2.5 < 20$
    $\quad\quad 2.5 \le t < 22.5$

31. $A = 28$; avg. value = 92; $P = \dfrac{2\pi}{\dfrac{\pi}{6}} = 12$;

vertical shift is 92 units up; horizontal shift

is $\dfrac{\dfrac{5\pi}{12}}{\dfrac{\pi}{6}} = \dfrac{5}{2}$ units right;

PI: $\quad 0 \le \dfrac{\pi}{6}t - \dfrac{5\pi}{12} < 2\pi$

$\dfrac{5\pi}{12} \le \dfrac{\pi}{6}t < \dfrac{29\pi}{12}$

$\dfrac{5}{2} \le t < \dfrac{29}{2}$

33. $A = 2{,}500$; avg. value = 3,150; $P = \dfrac{2\pi}{\dfrac{\pi}{4}} = 8$;

vertical shift is 3,150 units up; horizontal

shift is $\dfrac{\dfrac{\pi}{12}}{\dfrac{\pi}{4}} = \dfrac{1}{3}$ unit left;

PI: $\quad 0 \le \dfrac{\pi}{4}t + \dfrac{\pi}{12} < 2\pi$

$-\dfrac{\pi}{12} \le \dfrac{\pi}{4}t < \dfrac{23\pi}{12}$

$-\dfrac{1}{3} \le t < \dfrac{23}{3}$

35. Max and min are 600 and 100, so

$A = \dfrac{600 - 100}{2} = 250$. Avg. value is 350.

$P = 24$, so $B = \dfrac{2\pi}{24} = \dfrac{\pi}{12}$.

$y = 250\sin\left(\dfrac{\pi}{12}t\right) + 350$

37. Max and min are 18 and 8, so

$A = \dfrac{18 - 8}{2} = 5$. The primary cycle would

start at -25, so the horizontal shift is 25 units left. The avg. value is 13. $P = 100$, so

$B = \dfrac{2\pi}{100} = \dfrac{\pi}{50}$.

$y = 5\sin\left(\dfrac{\pi}{50}(t + 25)\right) + 13$

$= 5\sin\left(\dfrac{\pi}{50}t + \dfrac{\pi}{2}\right) + 13$

39. Max and min are 3 and 11, so

$A = \dfrac{11 - 3}{2} = 4$. The primary cycle begins at -45, so the horizontal shift is 45 units left. The avg. value is 7. $P = 360$, so

$B = \dfrac{2\pi}{360} = \dfrac{\pi}{180}$.

$y = 4\sin\left(\dfrac{\pi}{180}(t + 45)\right) + 7$

$= 4\sin\left(\dfrac{\pi}{180}t + \dfrac{\pi}{4}\right) + 7$

41. $f(t) = 25\sin\left[\dfrac{\pi}{4}(t - 2)\right] + 55$

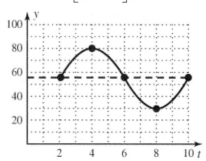

43. $h(t) = 3\sin(4t - \pi)$

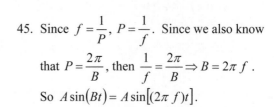

45. Since $f = \dfrac{1}{P}$, $P = \dfrac{1}{f}$. Since we also know

that $P = \dfrac{2\pi}{B}$, then $\dfrac{1}{f} = \dfrac{2\pi}{B} \Rightarrow B = 2\pi f$.

So $A\sin(Bt) = A\sin\left[(2\pi f)t\right]$.

Chapter 5: Introduction to Trigonometric Functions

47. a. $P = \dfrac{2\pi}{\dfrac{\pi}{2}} = 4$ sec; $f = \dfrac{1}{P} = \dfrac{1}{4}$ cycle/sec

    b. $d(2.5) = 6\sin\left(\dfrac{\pi}{2}(2.5)\right) = -4.24$ cm. 2.5

    is just past halfway through the period, so it's moving away from the equilibrium point.

    c. $d(3.5) = 6\sin\left(\dfrac{\pi}{2}(3.5)\right) = -4.24$ cm. 3.5

    is just short of the end of the period, so it's moving toward the equilibrium point.

    d. $\left|d(1.5) - d(1)\right| = \left|4.24 - 6\right| = 1.76$ cm .
    The average velocity is
    $\dfrac{1.76\,\text{cm}}{0.5\,\text{sec}} = 3.52\dfrac{\text{cm}}{\text{sec}}$. The weight
    reaches the equilibrium point at $t = 2$, so it's still speeding up between 1.5 and 2.

49. The amplitude is 15 and the period is 1.6, so
    $B = \dfrac{2\pi}{1.6} = 1.25\pi = \dfrac{5\pi}{4}$. $d(t) = 15\cos\left(\dfrac{5\pi}{4}t\right)$
    Note that we use cosine because the pendulum is at max distance at time 0.

51. Red: The period appears to be about 0.0068, so $f = \dfrac{1}{0.0068} \approx 147$. According to the chart, this looks like D_3.

    Blue: $P \approx 0.0088$, so $f = \dfrac{1}{0.0088} \approx 114$.
    According to the chart, this looks like $A\#_3$.

53. D_3: $f = 146.84$ and $P = \dfrac{1}{146.84} \approx 0.0068$
    sec. $y = \sin\left[146.84(2\pi t)\right]$

    G_4: $f = 392$ and $P = \dfrac{1}{392} \approx 0.00255$ sec.
    $y = \sin\left[392(2\pi t)\right]$

55. a. Caracas:
    $$D(15) = \dfrac{1.3}{2}\sin\left(\dfrac{2\pi}{365}(15 - 79)\right) + 12$$
    ≈ 11.4 hr
    Tokyo:
    $$D(15) = \dfrac{4.8}{2}\sin\left(\dfrac{2\pi}{365}(15 - 79)\right) + 12$$
    ≈ 9.9 hr

    b.

    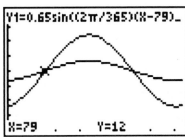

    i) The graphs intersect at $t = 79$ and 261.5, so they have the same number of hours on the 79th and 261st days.

    ii) The graph for Caracas is below 11.5 for $t < 28$ and $t > 312$, so there are 28 + (365–312) = 81 days with less than 11.5 hours. For Tokyo, it's $t < 67$ and $t > 274$, so there are 67 + (365–274) = 158 days.

57. a. Adds 12 hours. The sinusoidal behavior is actually on hours more/less than average of 12 hours of light.

    b. Means 12 hours of light and dark on March 20th. (Solstice)

    c. How many extra hours deviation from average. In the north, the planet is tilted closer toward the sun or farther from the Sun, depending on date. Variations will be greater.

59. 3.7 is in QIII; $3.7 - \pi \approx 0.5584$

61. $-1 + i\sqrt{5} + \left(-1 - i\sqrt{5}\right) = -2$

    $-1 + i\sqrt{5} - \left(-1 - i\sqrt{5}\right) = 2i\sqrt{5}$

    $\left(-1 + i\sqrt{5}\right)\left(-1 - i\sqrt{5}\right) = 1 - 5i^2 = 6$

    $\dfrac{-1 + i\sqrt{5}}{-1 - i\sqrt{5}} \cdot \dfrac{-1 + i\sqrt{5}}{-1 + i\sqrt{5}} = \dfrac{1 - 2i\sqrt{5} + 5i^2}{6}$

    $= \dfrac{-4 - 2i\sqrt{5}}{6} = -\dfrac{2}{3} - \dfrac{i\sqrt{5}}{3}$

5.6 Exercises

1. $\theta = \tan^{-1} x$

3. opposite; hypotenuse

5. To find the measure of all three angles and all three sides

Note that the diagrams in 7 – 12 are not drawn to scale.

7.

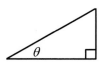

$$x^2 + 5^2 = 13^2; \quad x = \sqrt{169 - 25} = 12$$

$$\sec\theta = \frac{13}{5}; \quad \sin\theta = \frac{12}{13}; \quad \csc\theta = \frac{13}{12};$$

$$\tan\theta = \frac{12}{5}; \quad \cot\theta = \frac{5}{12}$$

9.

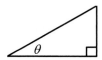

$$x^2 = 13^2 + 84^2 = 7{,}225; \quad x = 85$$

$$\cot\theta = \frac{13}{84}; \quad \sin\theta = \frac{84}{85}; \quad \csc\theta = \frac{85}{84}$$

$$\cos\theta = \frac{13}{85}; \quad \sec\theta = \frac{85}{13}$$

11.

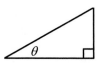

$$x^2 = 11^2 + 2^2; \quad x = \sqrt{121 + 4} = \sqrt{125} = 5\sqrt{5}$$

$$\tan\theta = \frac{11}{2}; \quad \sin\theta = \frac{11}{5\sqrt{5}}; \quad \csc\theta = \frac{5\sqrt{5}}{11}$$

$$\cos\theta = \frac{2}{5\sqrt{5}}; \quad \sec\theta = \frac{5\sqrt{5}}{2}$$

13. $B = 90° - 30° = 60°$

$$\sin 30° = \frac{a}{196}; \quad a = 196\sin 30° = 98 \text{ cm}$$

$$\cos 30° = \frac{b}{196}; \quad b = 196\cos 30° = 98\sqrt{3}$$

Angle	Side
$A = 30°$	$a = 98$ cm
$B = 60°$	$b = 98\sqrt{3}$ cm
$C = 90°$	$c = 196$ cm

15. $B = 90° - 45° = 45°$

$$\sin 45° = \frac{9.9}{c}; \quad c\sin 45° = 9.9; \quad c = \frac{9.9}{\sin 45°}$$

$$c = \frac{9.9}{\sqrt{2}/2} = \frac{19.8}{\sqrt{2}} = \frac{19.8\sqrt{2}}{2} = 9.9\sqrt{2} \text{ mm}$$

$$\cos 45° = \frac{a}{9.9\sqrt{2}}; \quad a = 9.9\sqrt{2}\cos 45°$$

$$a = 9.9\sqrt{2}\,\frac{\sqrt{2}}{2} = 9.9$$

Angle	Side
$A = 45°$	$a = 9.9$ mm
$B = 45°$	$b = 9.9$ mm
$C = 90°$	$c = 9.9\sqrt{2}$ mm

17. $B = 90° - 22° = 68°$

$$\sin 22° = \frac{14}{c}; \quad c\sin 22° = 14; \quad c = \frac{14}{\sin 22°}$$

$$c \approx 37.37 \text{ m}$$

$$\tan 22° = \frac{14}{b}; \quad b\tan 22° = 14; \quad b = \frac{14}{\tan 22°}$$

$$b \approx 34.65 \text{ m}$$

Angle	Side
$A = 22°$	$a = 14$ m
$B = 68°$	$b \approx 34.65$ m
$C = 90°$	$c \approx 37.37$ m

19. $A = 90° - 58° = 32°$

$$\cos 58° = \frac{5.6}{c}; \quad c\cos 58° = 5.6$$

$$c = \frac{5.6}{\cos 58°}; \quad c \approx 10.57 \text{ mi}$$

$$\tan 58° = \frac{b}{5.6}; \quad b = 5.6\tan 58° \approx 8.96 \text{ mi}$$

Angle	Side
$A = 32°$	$a = 5.6$ mi
$B = 58°$	$b \approx 8.96$ mi
$C = 90°$	$c \approx 10.57$ mi

21. $B = 90° - 65° = 25°$

    $\sin 65° = \dfrac{625}{c}$; $c \sin 65° = 625$; $c = \dfrac{625}{\sin 65°}$

    $c \approx 689.61$ mm

    $\tan 65° = \dfrac{625}{b}$; $b \tan 65° = 625$; $b = \dfrac{625}{\tan 65°}$

    $b \approx 291.44$ mm

Angle	Side
$A = 65°$	$a = 625$ mm
$B = 25°$	$b \approx 291.44$ mm
$C = 90°$	$c \approx 689.61$ mm

23. $\sin 27° = 0.4540$

25. $\tan 40° = 0.8391$

27. $\sec 40.9° = 1.3230$

29. $\sin 65° = 0.9063$

31. $A = \sin^{-1}(0.4540) \approx 27°$

33. $\theta = \tan^{-1}(0.8390) \approx 40°$

35. $B = \cos^{-1}\left(\dfrac{1}{1.3230}\right) \approx 40.9°$

37. $A = \sin^{-1}(0.9063) \approx 65°$

39. $\alpha = \tan^{-1}(0.9896) \approx 44.7°$

41. $\alpha = \sin^{-1}(0.3453) \approx 20.2°$

43. $\tan \theta = \dfrac{6}{18}$; $\theta = \tan^{-1}\left(\dfrac{1}{3}\right) \approx 18.4°$

45. $\tan \gamma = \dfrac{19.5}{18.7}$; $\gamma = \tan^{-1}\left(\dfrac{19.5}{18.7}\right) \approx 46.2°$

47. $\cos B = \dfrac{20}{42}$; $B = \cos^{-1}\left(\dfrac{20}{42}\right) \approx 61.6°$

49.

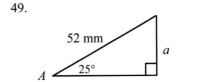

$\sin 25° = \dfrac{a}{52}$; $a = 52 \sin 25° \approx 21.98$ mm

51.

$\tan 32° = \dfrac{1.9}{b}$; $b \tan 32° = 1.9$; $b = \dfrac{1.9}{\tan 32°}$

$b \approx 3.04$ mi

53. $\cos 62.3° = \dfrac{82.5}{c}$; $c \cos 62.3° = 82.5$

    $c = \dfrac{82.5}{\cos 62.3°} \approx 177.48$ furlongs

55. $\sin 25° \approx 0.4266$; $\cos 65° \approx 0.4266$
    They have like values.

57. $\tan 5° \approx 0.0875$; $\cot 85° \approx 0.0875$
    They have like values.

59. $\sin 47° = \cos 43°$

61. $\cot 69° = \tan 21°$

63.

θ	30°		θ	30°
$\sin \theta$	$\frac{1}{2}$		$\tan(90 - \theta)$	$\sqrt{3}$
$\cos \theta$	$\frac{\sqrt{3}}{2}$		$\csc \theta$	2
$\tan \theta$	$\frac{\sqrt{3}}{3}$		$\sec \theta$	$\frac{2\sqrt{3}}{3}$
$\sin(90 - \theta)$	$\frac{\sqrt{3}}{2}$		$\cot \theta$	$\sqrt{3}$
$\cos(90 - \theta)$	$\frac{1}{2}$			

5.6 Exercises

65. $\sqrt{6}\csc 15° = \sqrt{6}\sec 75° = \sqrt{6}\left(\sqrt{6}+\sqrt{2}\right)$

$= 6+\sqrt{12} = 6+2\sqrt{3}$

67. $\cot^2 15° = \tan^2 75° = \left(2+\sqrt{3}\right)^2$

$= 4+4\sqrt{3}+3 = 7+4\sqrt{3}$

69. $\sin\theta = \dfrac{2A}{ab}$

$\sin\theta = \dfrac{2(38.9)}{(17)(24)}$; $\theta = \sin^{-1}\left(\dfrac{2(38.9)}{(17)(24)}\right)$

$\theta \approx 11.0°$
Repeat for β using 24 and 8 for a and b.

$\sin\beta = \dfrac{2(38.9)}{8(24)}$; $\beta = \sin^{-1}\left(\dfrac{2(38.9)}{8(24)}\right) \approx 23.9°$

$\gamma = 180° - (11.0° + 23.9°) = 145.1°$

71.

71.6°   100 m h

$\tan 71.6° = \dfrac{h}{100}$; $h = 100\tan 71.6°$

$h \approx 300.6\text{ m}$

73.

89°   25.9 ft h

$\tan 89° = \dfrac{h}{25.9}$; $h = 25.9\tan 89°$
$h \approx 1,483.8\text{ ft}$

75. $\tan 83° = \dfrac{d}{50}$; $d = 50\tan 83° \approx 407.22\text{ ft}$

$\dfrac{407.22\text{ ft}}{2.35\text{ sec}} \cdot \dfrac{1\text{ mi}}{5,280\text{ ft}} \cdot \dfrac{3,600\text{ sec}}{1\text{ hr}} = 118.1\text{ mph}$

77. Let's first find the distances $h_s, h_1,$ and h_2 in the diagram below, then answer the questions.

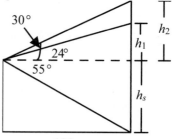

$\tan 55° = \dfrac{h_s}{175}$; $h_s = 175\tan 55° \approx 250\text{ yd}$

$\tan 24° = \dfrac{h_1}{175}$; $h_1 = 175\tan 24° \approx 77.9\text{ yd}$

$\tan 30° = \dfrac{h_2}{175}$; $h_2 = 175\tan 30° \approx 101\text{ yd}$

a. $h_s \approx 250\text{ yd}$ is the height of the south rim.

b. $h_s + h_2 \approx 351\text{ yd}$ is the height of the north rim.

c. $h_2 - h_1 \approx 23.1\text{ yd}$ is how far the climbers have to go to the top.

79. Let h_t be the height of the tower and h_r be the height of the restaurant.

$\tan 74.6° = \dfrac{h_t}{500}$; $h_t = 500\tan 74.6°$
$h_t \approx 1,815.2\text{ ft}$;

$\tan 66.5° = \dfrac{h_r}{500}$; $h_r = 500\tan 66.5°$
$\approx 1,149.9\text{ ft}$
Difference: 665.3 ft.

81. $\cos 34° = \dfrac{320}{Z}$; $Z\cos 34° = 320$

$Z = \dfrac{320}{\cos 34°} \approx 386.0\,\Omega$

83. a. Five contour lines, so change in elevation is $5(175) = 875$ m.

   b. $\dfrac{2.4 \text{ cm}}{x \text{ m}} = \dfrac{1 \text{ cm}}{500 \text{ m}}$; $\quad x = 2.4(500)$
   $= 1200$ m

   c.

   $\tan\theta = \dfrac{875}{1200}$; $\quad \theta = \tan^{-1}\left(\dfrac{875}{1200}\right) \approx 36.1°$

   $d^2 = 1200^2 + 875^2$; $\quad d = \sqrt{2,205,625}$

   $d \approx 1485$ m

85. $\tan 42° = \dfrac{h}{500}$; $\quad h = 500\tan 42° \approx 450$ ft

87. a. The triangle at the base of the box is isosceles with sides x, so the dotted diagonal on the bottom has length $x\sqrt{2}$ (45-45-90 triangle). The triangle formed by that diagonal, the diagonal across the box, and one edge of the box is right, and we can apply the Pythagorean Theorem:

   $\left(x\sqrt{2}\right)^2 + x^2 = 35^2$; $\quad 2x^2 + x^2 = 1225$;

   $3x^2 = 1225$; $\quad x^2 \approx 408.3$; $\quad x \approx 20.2$ cm

   b. Now we can find the angle using cosine:

   $\cos\theta = \dfrac{20.2\sqrt{2}}{35}$;

   $\theta = \cos^{-1}\left(\dfrac{20.2\sqrt{2}}{35}\right) \approx 35.3°$

89. $\cot u = \dfrac{x}{h}$; $\quad x = h\cot u$

   $\cot v = \dfrac{x-d}{h}$; $\quad \cot v = \dfrac{h\cot u - d}{h}$

   $h\cot v = h\cot u - d$
   $d = h\cot u - h\cot v = h(\cot u - \cot v)$

   $h = \dfrac{d}{\cot u - \cot v}$

91. a. If $\theta = 39.5°$, the complementary angle at the bottom of the right triangle is $50.5°$.

   $\sin 50.5° = \dfrac{r}{3960}$; $\quad r = 3960\sin 50.5°$

   $r \approx 3055.6$ mi

   b. The closest measure between the longitudes is $169°$: $(180 - 116) + (180 - 75)$.

   $s = 3,055.6(169°)\left(\dfrac{\pi \text{ rad}}{180°}\right) \approx 9012.8$ mi

   c. $9012.8 \text{ mi} \cdot \dfrac{1 \text{ hr}}{1250 \text{ mi}} = 7.21$ hr, or about

   7 hr, 13 min

93. a. Local maximums: $(-5, 2), (2, 3)$
   Local minimums: $(-7, -2), (-2, -1),$
   $(6, -3)$

   b. Zeros: $x = -6, -3, -1, 4$

   c. $T(x)\downarrow$ on $(-5, -2)$ and $(2, 6)$
   $T(x)\uparrow$ on $(-7, -5)$ and $(-2, 2)$

   d. $T(x) > 0$ on $(-6, -3)$ and $(-1, 4)$
   $T(x) < 0$ on $[-7, -6), (-3, -1)$ and $(4, 6]$

95. The diagonal of one side forms a 45-45-90 triangle with legs 38 in, so the hypotenuse (diagonal) is $38\sqrt{2} \approx 53.74$ in.
   The diagonal through the center of the box forms a right triangle with legs 38 in and $38\sqrt{2}$ in, so we can use the Pythagorean Theorem to find D:

   $D^2 = 38^2 + \left(38\sqrt{2}\right)^2 = 4,332$

   $D = \sqrt{4,332} \approx 65.82$ in.

5.7 Exercises

1. origin; *x*-axis

3. positive; clockwise

5. Answers will vary.

7.

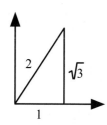

The lengths were obtained using the special triangle relationships for 30-60-90. The slope of the line is $\sqrt{3}$ since the rise is $\sqrt{3}$ and the run is 1. The equation is $y = \sqrt{3}x$. We'll choose the point $\left(3, 3\sqrt{3}\right)$. Then

$$r = \sqrt{3^2 + (3\sqrt{3})^2} = \sqrt{36} = 6.$$

$$\sin 60° = \frac{3\sqrt{3}}{6} = \frac{\sqrt{3}}{2}; \quad \cos 60° = \frac{3}{6} = \frac{1}{2}$$

$$\tan 60° = \frac{3\sqrt{3}}{3} = \sqrt{3}. \text{ These match our}$$
known values.

9.

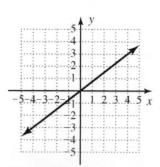

QI and QIII. For QI we chose (4, 3). Then
$$r = \sqrt{3^2 + 4^2} = \sqrt{25} = 5.$$
$$\sin\theta = \frac{3}{5}; \quad \cos\theta = \frac{4}{5}; \quad \tan\theta = \frac{3}{4}$$
For QIII, we chose (−4, −3); *r* is still 5.
$$\sin\theta = -\frac{3}{5}; \quad \cos\theta = -\frac{4}{5}; \quad \tan\theta = \frac{-3}{-4} = \frac{3}{4}$$

11.

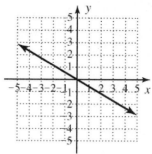

QII and QIV. For QII, we chose $\left(-3, \sqrt{3}\right)$.

$$r = \sqrt{(-3)^2 + (\sqrt{3})^2} = \sqrt{12} = 2\sqrt{3}$$

$$\sin\theta = \frac{\sqrt{3}}{2\sqrt{3}} = \frac{1}{2}; \quad \cos\theta = \frac{-3}{2\sqrt{3}} = -\frac{\sqrt{3}}{2}$$

$$\tan\theta = \frac{\sqrt{3}}{-3} = -\frac{1}{\sqrt{3}}$$

For QIV, we chose $\left(3, -\sqrt{3}\right)$; *r* is $2\sqrt{3}$.

$$\sin\theta = \frac{-\sqrt{3}}{2\sqrt{3}} = -\frac{1}{2}; \quad \cos\theta = \frac{3}{2\sqrt{3}} = \frac{\sqrt{3}}{2}$$

$$\tan\theta = \frac{-\sqrt{3}}{3} = -\frac{1}{\sqrt{3}}$$

13. $r = \sqrt{8^2 + 15^2} = \sqrt{289} = 17$
$$\sin\theta = \frac{15}{17}; \quad \csc\theta = \frac{17}{15}; \quad \cos\theta = \frac{8}{17}$$
$$\sec\theta = \frac{17}{8}; \quad \tan\theta = \frac{15}{8}; \quad \cot\theta = \frac{8}{15}$$

15. $r = \sqrt{(-20)^2 + 21^2} = \sqrt{841} = 29$
$$\sin\theta = \frac{21}{29}; \quad \csc\theta = \frac{29}{21}; \quad \cos\theta = -\frac{20}{29}$$
$$\sec\theta = -\frac{29}{20}; \quad \tan\theta = -\frac{21}{20}; \quad \cot\theta = -\frac{20}{21}$$

17. $r = \sqrt{(7.5)^2 + (-7.5)^2} = \sqrt{2(7.5)^2} = 7.5\sqrt{2}$
$$\sin\theta = -\frac{7.5}{7.5\sqrt{2}} = -\frac{\sqrt{2}}{2}; \quad \csc\theta = -\frac{2}{\sqrt{2}}$$
$$\cos\theta = \frac{7.5}{7.5\sqrt{2}} = \frac{\sqrt{2}}{2}; \quad \sec\theta = \frac{2}{\sqrt{2}}$$
$$\tan\theta = -1; \quad \cot\theta = -1$$

Chapter 5: Introduction to Trigonometric Functions

19. $r = \sqrt{4^2 + \left(\dfrac{4\sqrt{3}}{3}\right)^2} = \sqrt{16 + \dfrac{48}{9}} = \sqrt{\dfrac{64}{3}} = \dfrac{8}{\sqrt{3}}$

$\sin\theta = \dfrac{4\sqrt{3}}{8/\sqrt{3}} = \dfrac{1}{2}; \ \csc\theta = 2$

$\cos\theta = \dfrac{4}{8/\sqrt{3}} = \dfrac{\sqrt{3}}{2}; \ \sec\theta = \dfrac{2}{\sqrt{3}}$

$\tan\theta = \dfrac{1}{\sqrt{3}}; \ \cot\theta = \sqrt{3}$

21. $r = \sqrt{2^2 + 8^2} = \sqrt{68} = 2\sqrt{17}$

$\sin\theta = \dfrac{8}{2\sqrt{17}} = \dfrac{4}{\sqrt{17}}; \ \csc\theta = \dfrac{\sqrt{17}}{4}$

$\cos\theta = \dfrac{2}{2\sqrt{17}} = \dfrac{1}{\sqrt{17}}; \ \sec\theta = \sqrt{17}$

$\tan\theta = \dfrac{8}{2} = 4; \ \cot\theta = \dfrac{2}{8} = \dfrac{1}{4}$

23. Based on similar triangles, we can multiply both coordinates by 4 to clear decimals:
$(-15, -10)$

$r = \sqrt{(-15)^2 + (-10)^2} = \sqrt{325} = 5\sqrt{13}$

$\sin\theta = -\dfrac{10}{5\sqrt{13}} = -\dfrac{2}{\sqrt{13}}; \ \csc\theta = -\dfrac{\sqrt{13}}{2}$

$\cos\theta = -\dfrac{15}{5\sqrt{13}} = -\dfrac{3}{\sqrt{13}}; \ \sec\theta = -\dfrac{\sqrt{13}}{3}$

$\tan\theta = \dfrac{2}{3}; \ \cot\theta = \dfrac{3}{2}$

25. Based on similar triangles, we can multiply both coordinates by 9 to clear fractions:
$(-5, 6)$. $r = \sqrt{(-5)^2 + 6^2} = \sqrt{61}$

$\sin\theta = \dfrac{6}{\sqrt{61}}; \ \csc\theta = \dfrac{\sqrt{61}}{6}$

$\cos\theta = -\dfrac{5}{\sqrt{61}}; \ \sec\theta = -\dfrac{\sqrt{61}}{5}$

$\tan\theta = -\dfrac{6}{5}; \ \cot\theta = -\dfrac{5}{6}$

27. Based on similar triangles, we can multiply both coordinates by 4 to clear fractions:
$(1, -2\sqrt{5})$. $r = \sqrt{1^2 + (-2\sqrt{5})^2} = \sqrt{21}$

$\sin\theta = -\dfrac{2\sqrt{5}}{\sqrt{21}}; \ \csc\theta = -\dfrac{\sqrt{21}}{2\sqrt{5}}$

$\cos\theta = \dfrac{1}{\sqrt{21}}; \ \sec\theta = \sqrt{21}$

$\tan\theta = -2\sqrt{5}; \ \cot\theta = -\dfrac{1}{2\sqrt{5}}$

29. Every point on the terminal side looks like $(0, k)$, and $r = \sqrt{0^2 + k^2} = |k| = k$, $k > 0$
since $\theta = 90°$

$\sin 90° = \dfrac{k}{k} = 1; \ \cos 90° = \dfrac{0}{k} = 0;$

$\tan 90° = \dfrac{k}{0}$ Undefined

$\csc 90° = 1; \sec 90° = \dfrac{k}{0}$ Undefined;

$\cot 90° = \dfrac{0}{k} = 0$

31. QII: $\theta_r = 180° - 120° = 60°$

33. QII: $\theta_r = 180° - 135° = 45°$

35. QIV: $\theta_r = 0 - (-45°) = 45°$

37. QII: $\theta_r = 180° - 112° = 68°$

39. QII: Coterminal with 140°.
$\theta_r = 180° - 140° = 40°$

41. $-168.4° + 360° = 191.6°$ QIII
$\theta_r = 191.6° - 180° = 11.6°$

43. QII

45. QII

47. QIV: $\theta_r = 360° - 330° = 30°$

$\sin\theta = -\dfrac{1}{2}; \ \cos\theta = \dfrac{\sqrt{3}}{2}; \ \tan\theta = -\dfrac{1}{\sqrt{3}}$

49. QIV: $\theta_r = 0 - (-45°) = 45°$

$\sin\theta = -\dfrac{\sqrt{2}}{2}; \ \cos\theta = \dfrac{\sqrt{2}}{2}; \ \tan\theta = -1$

5.7 Exercises

51. QIII: $\theta_r = 240° - 180° = 60°$

$$\sin\theta = -\frac{\sqrt{3}}{2}; \quad \cos\theta = -\frac{1}{2}; \quad \tan\theta = \sqrt{3}$$

53. $-150° + 360° = 210°$. QIII: $210° - 180° = 30°$.

$$\sin\theta = -\frac{1}{2}; \quad \cos\theta = -\frac{\sqrt{3}}{2}; \quad \tan\theta = \frac{1}{\sqrt{3}}$$

55. $x = 4, r = 5$; $4^2 + y^2 = 5^2$; $y = \pm\sqrt{25-16}$ $y = \pm 3$. Since $\sin\theta < 0$, $y = -3$. QIV

$$\sin\theta = -\frac{3}{5}; \quad \csc\theta = -\frac{5}{3}; \quad \sec\theta = \frac{5}{4};$$

$$\tan\theta = -\frac{3}{4}; \quad \cot\theta = -\frac{4}{3}$$

57. $r = 37, y = -35$. Since $\csc\theta < 0$, we know $\sin\theta < 0$, and $\tan\theta > 0$ tells us we're in QIII.

$$x^2 + (-35)^2 = 37^2; \quad x = \pm\sqrt{37^2 - (-35)^2}$$
$$= \pm\sqrt{144} = \pm 12 \quad \text{Choose } x = -12$$
$$\sin\theta = -\frac{35}{37}; \quad \cos\theta = -\frac{12}{37}; \quad \sec\theta = -\frac{37}{12}$$
$$\tan\theta = \frac{35}{12}; \quad \cot\theta = \frac{12}{35}$$

59. $\csc\theta > 0$, so $\sin\theta > 0$, and $\cos\theta > 0$ as well, so QI. $y = 1, r = 3$.

$$x^2 + 1^2 = 3^2; \quad x = \sqrt{9-1} = 2\sqrt{2}$$
$$\sin\theta = \frac{1}{3}; \quad \cos\theta = \frac{2\sqrt{2}}{3}; \quad \sec\theta = \frac{3}{2\sqrt{2}}$$
$$\tan\theta = \frac{1}{2\sqrt{2}}; \quad \cot\theta = 2\sqrt{2}$$

61. $\sin\theta < 0$ and $\sec\theta < 0$, and $\cos\theta < 0$ so we're in QIII. $y = -7, r = 8$.

$$x^2 + (-7)^2 = 8^2; \quad x = -\sqrt{64-49} = -\sqrt{15}$$
$$\csc\theta = -\frac{8}{7}; \quad \cos\theta = -\frac{\sqrt{15}}{8}; \quad \sec\theta = -\frac{8}{\sqrt{15}}$$
$$\tan\theta = \frac{7}{\sqrt{15}}; \quad \cot\theta = \frac{\sqrt{15}}{7}$$

63. $52° + 360°k$
$52° + 360° = 412°$; $52° + 720° = 772°$
$52° - 360° = -308°$; $52° - 720° = -668°$

65. $87.5° + 360°k$
$87.5° + 360° = 447.5°$; $87.5° + 720°$
$= 807.5°$; $87.5° - 360° = -272.5°$;
$87.5° - 720° = -632.5°$

67. $225° + 360°k$
$225° + 360° = 585°$; $225° + 720° = 945°$;
$225° - 360° = -135°$; $225° - 720° = -495°$

69. $-107° + 360°k$
$-107° + 360° = 253°$; $-107° + 720° = 613°$;
$-107° - 360° = -467°$; $-107° - 720° = -827°$

71. $\sin 120° = \frac{\sqrt{3}}{2}$ (QII, Ref. angle = 60°);

$-240° + 360° = 120°$; $\cos(-240°) = -\frac{1}{2}$;

$480° - 360° = 120°$; $\tan 480° = -\sqrt{3}$

73. $\sin -30° = -\frac{1}{2}$; (QIV, Ref. angle = 30°);

$-390° + 360° = -30°$; $\cos -390° = \frac{\sqrt{3}}{2}$;

$690° - 720° = -30°$; $\tan 690° = -\frac{1}{\sqrt{3}}$

75. $600° - 360° = 240°$; QIII, Ref. angle = 60°;

$$\sin\theta = -\frac{\sqrt{3}}{2}; \quad \cos\theta = -\frac{1}{2}; \quad \tan\theta = \sqrt{3}$$

77. $-840° + 3(360)° = 240°$; QIII, Ref. angle = 60°;

$$\sin\theta = -\frac{\sqrt{3}}{2}; \quad \cos\theta = -\frac{1}{2}; \quad \tan\theta = \sqrt{3}$$

79. $570° - 360° = 210°$; QIII, ref. angle = 30°

$$\sin\theta = -\frac{1}{2}; \quad \cos\theta = -\frac{\sqrt{3}}{2}; \quad \tan\theta = \frac{1}{\sqrt{3}}$$

81. $-1230° + 4(360°) = 210°$; QIII, ref. angle = 30°

$$\sin\theta = -\frac{1}{2}; \quad \cos\theta = -\frac{\sqrt{3}}{2}; \quad \tan\theta = \frac{1}{\sqrt{3}}$$

83. $719° - 360° = 359°$; QIV, negative
$\sin 719° = -0.0175$

85. $-419° + 720° = 301°$; QIV, negative
$\tan(-419°) = -1.6643$

86. $-621° + 720° = 99°$; QII, negative

$\sec(-621°) = -6.3925$

87. $681° - 360° = 321°$; QIV, negative

$\csc 681° = -1.5890$

89. $805° - 720° = 85°$; QI, positive

$\cos 805° = 0.0872$

91. a. $A = ab \sin \theta = (9)(21) \sin 50°$

≈ 144.78 units ^{2}

b. Enter the function $Y_1 = 9 * 21 \sin x$ and set up a table with TblStart = 50 and ΔTbl = 1: The first value over 150 is at 53°.

c. For 90°, the parallelogram is a rectangle with area $A = ab \sin 90° = ab$.

d. Divide the parallelogram in half with a diagonal to get a triangle. The area given two sides and the angle between them is then $A = \frac{1}{2} ab \sin \theta = \frac{ab}{2} \sin \theta$.

93. $\theta = 60° + 360°k$; $\theta = 300° + 360°k$

95. $\theta = 240° + 360°k$; $\theta = 300° + 360°k$

97. $\sin^{-1} 0.8754 = 61.1°$; $180° - 61.1° = 118.8°$

$\theta = 61.1° + 360°k$; $\theta = 118.9° + 360°k$

99. $\tan^{-1}(-2.3512) = -67.0°$

$-67.0° + 180° = 113°$; $113° + 180° = 293°$

$\theta = 113.0° + 360°k$; $\theta = 293.0° + 360°k$

101. Five complete turns: 5(360°) = 1800°

Back to 3 o'clock: 90°. Total: 1890°

All coterminal angles: $90° + 360°k$

103. Assuming he was on his feet at the start, after 2.5 revolutions, he'd go in the water head first. $2(360°) + 180° = 900°$.

105. The angle between the terminal side and the x-axis can be found using a right triangle with opposite 2 and adjacent 6:

$\tan \theta = \frac{2}{6}$; $\theta = \tan^{-1}\left(\frac{1}{3}\right) \approx 18.4°$

The total revolution is this much shy of two full turns: $720° - 18.4° = 701.6°$

107. Area of triangle:

$A_t = \frac{1}{2} ab \sin \theta = \frac{1}{2}(18)(18) \sin 150° = 81$

Area of sector: (shaded plus triangle)

$A_s = \frac{1}{2} r^2 \theta = \frac{1}{2}(18)^2 \left(150° \cdot \frac{\pi \text{ rad}}{180°}\right) \approx 424.12$

Shaded area is difference between the two:

$A = 424.12 - 81 = 343.12$ in^2

109. Answers will vary.

111. a. 3 sec = 36 revolutions = 36(360°) = 12,960°.

b. $C = 2\pi r = 2\pi(20) = 40\pi \approx 125.66$ in.

c. 10 sec = 120 revolutions

$120 \text{ rev} \cdot \frac{125.66 \text{ in.}}{\text{rev}} \approx 15,080$ in.

d. $\frac{15,080 \text{ in.}}{10 \text{ sec}} \cdot \frac{3600 \text{ sec}}{1 \text{ hr}} \cdot \frac{1 \text{ mi}}{12(5280) \text{ in.}}$

≈ 85.68 mph

113. $\tan 78° = \frac{x}{117}$

$117 \tan 78° = x$

$550.4 \approx x$;

Approximate height of monument:

$550.4 + 5 = 555.4$ feet

115. $4x - 5y = 15$

$y = \frac{4}{5}x - 3$

Slope $= -\frac{5}{4}$,

$y - (-3) = -\frac{5}{4}(x - 4)$

$y + 3 = -\frac{5}{4}x + 5$

$y = -\frac{5}{4}x + 2$

Summary and Concept Review

1. $147 + \left(\dfrac{36}{60}\right) + \dfrac{48}{3600} = 147.61\overline{3}°$

3. Let x = the length of the bottom, y = the length of the other leg. Note that the hypotenuse of the smallest triangle is 5. (Pythagorean triple).

$\dfrac{5}{3} = \dfrac{16.875}{y}$; $5y = 50.625$; $y = 10.125$.

$16.875^2 = 10.125^2 + x^2$; $x = \sqrt{182.25} = 13.5$

10.125 by 13.5 by 16.875

5. $\dfrac{2\pi}{3} \cdot \dfrac{180°}{\pi \text{ rad}} = 120°$

7. $s = r\theta$; $s = 5(57°)\left(\dfrac{\pi \text{ rad}}{180°}\right) \approx 4.97$ units

9. $s = r\theta = 15(1.7) = 25.5$ cm;

$A = \dfrac{1}{2}r^2\theta = \dfrac{1}{2}(15)^2(1.7) = 191.25$ cm^2

11. $A = \dfrac{1}{2}r^2\theta$; $152 = \dfrac{1}{2}(8)^2\theta = 32\theta$;

$\theta = \dfrac{152}{32} = 4.75$ rad;

$s = r\theta = 8(4.75) = 38$ m

13. $y^2 + \left(\dfrac{\sqrt{13}}{7}\right)^2 = 1$; $y^2 = 1 - \dfrac{13}{49} = \dfrac{36}{49}$

$y = -\dfrac{6}{7}$; $\left(-\dfrac{\sqrt{13}}{7}, -\dfrac{6}{7}\right)$,

$\left(-\dfrac{\sqrt{13}}{7}, \dfrac{6}{7}\right)$, $\left(\dfrac{\sqrt{13}}{7}, \dfrac{6}{7}\right)$

15. $\csc t = \dfrac{2}{\sqrt{3}}$ $\Rightarrow$ $\sin t = \dfrac{\sqrt{3}}{2}$; $t = \dfrac{\pi}{3}, \dfrac{2\pi}{3}$

17. a. The circumference of the drum is $2\pi r = 2\pi$ yd, so 6π ft corresponds to 1 revolution or 2π radians.

$59\text{ft} \cdot \dfrac{2\pi \text{ rad}}{6\pi \text{ ft}} = \dfrac{59}{3} \approx 19.6667$ rad

   b. Each radian corresponds to 3 feet, so 25 rad.

19. $y = 3\sec t$; No amplitude, $P = 2\pi$

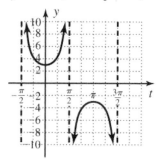

21. $y = 1.7\sin(4t)$; $A = 1.7$, $P = \dfrac{2\pi}{4} = \dfrac{\pi}{2}$

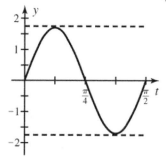

23. $g(t) = 3\sin(398\pi t)$; $A = 3$, $P = \dfrac{2\pi}{398\pi} = \dfrac{1}{199}$

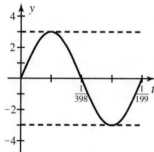

Chapter 5: Introduction to Trigonometric Functions

25. The max's and mins are at height 4 and –4, so $A = 4$. $P = \dfrac{2}{3}$, so $B = \dfrac{2\pi}{\frac{2}{3}} = 3\pi$.

$$y = 4\csc(3\pi\, t)$$

27. $\tan\left(\dfrac{7\pi}{4}\right) = \dfrac{\frac{\sqrt{2}}{2}}{\frac{-\sqrt{2}}{2}} = -1$;

$\cot\left(\dfrac{\pi}{3}\right) = \dfrac{\frac{1}{2}}{\frac{\sqrt{3}}{2}} = \dfrac{1}{\sqrt{3}}$

29. $y = 6\tan\left(\dfrac{1}{2}t\right)$

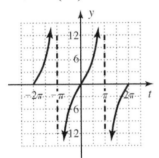

31. The period is π, so three additional solutions are $1.55 + \pi \approx 4.69$, $1.55 + 2\pi \approx 7.83$ and $1.55 + 3\pi \approx 10.97$. There are many others. $1.55 + k\pi$ radians, $k \in Z$

33. $h = \dfrac{144}{\cot 25° - \cot 40°} \approx 151.14\ \text{m}$

35. a. $A = 240$; vert. shift = 520 units up

$P = \dfrac{2\pi}{\frac{\pi}{6}} = 12$; horiz. shift = 3 units right

b.

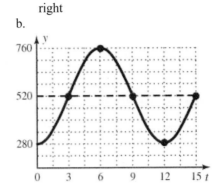

37. $A = 125$; vertical shift = 175 units up $P = 24$ (half cycle from 9 to 21), so

$B = \dfrac{2\pi}{24} = \dfrac{\pi}{12}$. Horiz. shift = 3 units right

for cosine. $y = 125\cos\left[\dfrac{\pi}{12}(t-3)\right] + 175$

39. a. $A = \dfrac{2.26 - 0.44}{2} = 0.91$;

$D = \dfrac{2.26 + 0.4}{2} = 1.35$

$B = \dfrac{2\pi}{12} = \dfrac{\pi}{6}$ ($P = 12$ since 12 months per year.)

$P(t) = 0.91\sin\left(\dfrac{\pi}{6}t\right) + 1.35$

b. $P(5) = 0.91\sin\left(\dfrac{5\pi}{6}\right) + 1.35 \approx 1.81$ in

(August)

$P(9) = 0.91\sin\left(\dfrac{9\pi}{6}\right) + 1.35 \approx 0.44$ in

(December)

41. a. $\tan 57.4° = \cot(90° - 57.4°) = \cot 32.6°$

  b. $\sin(19°30'15'') =$
  $\cos(90° - 19°30'15'') = \cos(70°29'45'')$

43. $\tan A = \dfrac{20}{21}$; $A = \tan^{-1}\left(\dfrac{20}{21}\right) \approx 43.6°$

  $B \approx 90° - 43.6° = 46.4°$
  $c^2 = 20^2 + 21^2$; $c = \sqrt{400 + 441} = 29$

Angles	Sides
$A \approx 43.6°$	$a = 20$ m
$B \approx 46.4°$	$b = 21$ m
$C = 90°$	$c = 29$ m

45. a.

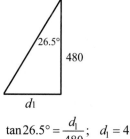

  $\tan 26.5° = \dfrac{d_1}{480}$; $d_1 = 480 \tan 26.5°$

  $d_1 \approx 239.32$ m

  b.

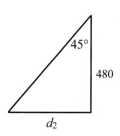

  $\tan 45° = \dfrac{d_2}{480}$; $d_2 = 480 \tan 45°$

  $d_2 = 480$ m
  The boats are about $480 - 239.32 =$
  240.68 m apart.

47. $207° + 360°k$; answers will vary
  $207° + 360° = 567°$; $207° + 720 = 927°$;
  $207° - 360° = -153°$; $207° - 720° = -513°$

49. a. $r = \sqrt{(-12)^2 + 35^2} = \sqrt{1369} = 37$

  $\sin \theta = \dfrac{35}{37}$; $\csc \theta = \dfrac{37}{35}$; $\cos \theta = -\dfrac{12}{37}$

  $\sec \theta = -\dfrac{37}{12}$; $\tan \theta = -\dfrac{35}{12}$; $\cot \theta = -\dfrac{12}{35}$

  b. $r = \sqrt{12^2 + (-18)^2} = \sqrt{468} = 6\sqrt{13}$

  $\sin \theta = -\dfrac{18}{6\sqrt{13}} = -\dfrac{3}{\sqrt{13}}$; $\csc \theta = -\dfrac{\sqrt{13}}{3}$

  $\cos \theta = \dfrac{12}{6\sqrt{13}} = \dfrac{2}{\sqrt{13}}$; $\sec \theta = \dfrac{\sqrt{13}}{2}$

  $\tan \theta = -\dfrac{18}{12} = -\dfrac{3}{2}$; $\cot \theta = -\dfrac{2}{3}$

51. a. $\theta = 135° + 180°k$

  b. $\theta = 30° + 360°k$ or $330° + 360°k$

  c. $\theta = \tan^{-1} 4.0108 \approx 76° + 180°k$

  d. $\theta = \sin^{-1}(-0.4540) \approx -27° + 360°k$
  or $\theta = 207° + 360°k$

Mixed Review

1. a. $A = 10$
  b. avg. value $= 15$
  c. $P = 6$
  d. $f(4) = 20$

3. $t = \dfrac{2\pi}{3}, \dfrac{4\pi}{3}$

5. $220° + 0.813\overline{8}(60)' = 220° + 48.830\overline{3}'$
  $= 220° + 48' + 0.830\overline{3}(60)'' \approx 220°48'50''$

7. 45-45-90 triangle: cuts are $12\sqrt{2} \approx 16.97''$
  Wall is 5 of these lengths across, so
  $60\sqrt{2}'' \approx 84.9''$

9. $r = \sqrt{\left(-4\sqrt{3}\right)^2 + (-4)^2} = \sqrt{64} = 8$

Let α = the angle between the terminal side and the negative x-axis

$\tan \alpha = \dfrac{4}{4\sqrt{3}} = \dfrac{1}{\sqrt{3}} \Rightarrow \alpha = \dfrac{\pi}{6}$

Then $\theta = \pi + \dfrac{\pi}{6} = \dfrac{7\pi}{6}$

$s = r\theta = 8\left(\dfrac{7\pi}{6}\right) = \dfrac{56\pi}{6} = \dfrac{28\pi}{3} \approx 29.3$ units

$A = \dfrac{1}{2}r^2\theta = \dfrac{1}{2}(8)^2\left(\dfrac{7\pi}{6}\right) = \dfrac{224\pi}{6}$

$= \dfrac{112\pi}{3} \approx 117.3$ units^2

11. $86 + \dfrac{54}{60} + \dfrac{54}{3600} = 86.915°$

13. $r^2 = 15^2 + (-8)^2 = 289; \quad r = 17$

$\sin \theta = -\dfrac{8}{17}; \quad \csc \theta = -\dfrac{17}{8}; \quad \cos \theta = \dfrac{15}{17}$

$\sec \theta = \dfrac{17}{15}; \quad \tan \theta = -\dfrac{8}{15}; \quad \cot \theta = -\dfrac{15}{8}$

15. $\sin \theta = \dfrac{100}{115.47}; \quad \theta = \sin^{-1}\left(\dfrac{100}{115.47}\right) \approx 60°$

17. a. $\omega = \dfrac{3 \text{ rev}}{\min} \cdot \dfrac{2\pi \text{ rad}}{\text{rev}} = 6\pi \, \tfrac{\text{rad}}{\sec}$

b. $V = r\omega = 20(6\pi) = 120\pi \approx 377 \, \tfrac{\text{cm}}{\sec}$

19. a. $A = 5, P = \dfrac{2\pi}{2} = \pi$, vertical shift 8 units down, no horizontal shift, primary interval $= [0, \pi)$

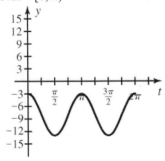

b. $A = \dfrac{7}{2}, P = \dfrac{2\pi}{\pi/2} = 4$, no vertical shift, horizontal shift = 1 unit right, primary interval $= [1, 5)$

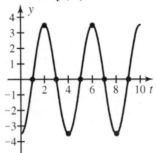

c. $P = \dfrac{\pi}{1/4} = 4\pi$, no horizontal shift, no amplitude, no vertical shift, $(-2\pi, 2\pi)$

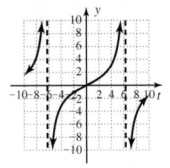

d. $P = 2\pi$, horizontal shift $= \dfrac{\pi}{2}$ right, no vertical shift, no amplitude, PI: $(-\pi, \pi)$

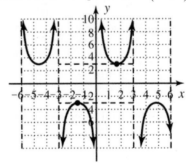

Practice Test

1. Complement: $90° - 35° = 55°$
   Supplement: $180° - 35° = 145°$

3. $30° + 360°k, k \in Z;$
   $30° + 360° = 390°;$ $30° + 720° = 750°$
   $30° - 360° = -330°;$ $30° - 720° = -690°$

5. Let d_1 and d_2 be the distance from Four
   Corners to point P and the length of
   Colorado's southern border respectively.

   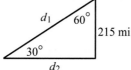

   a. $\sin 30° = \dfrac{215}{d_1};$ $d_1 = \dfrac{215}{\sin 30°} \approx 430$ mi

   b. $\tan 60° = \dfrac{d_2}{215};$ $d_2 = 215\tan 60° \approx 372$ mi

7. $x = 2, r = 5;$ $y^2 + 2^2 = 5^2;$ $y = -\sqrt{21}$ (QIV)
   $\sec\theta = \dfrac{5}{2};$ $\sin\theta = -\dfrac{\sqrt{21}}{5};$ $\csc\theta = -\dfrac{5}{\sqrt{21}}$

   $\tan\theta = -\dfrac{\sqrt{21}}{2};$ $\cot\theta = -\dfrac{2}{\sqrt{21}}$

9. a. $s = r\theta = 75(172.5°)\left(\dfrac{\pi\ \text{rad}}{180°}\right) \approx 225.8'$
      $225.8' = 225' + 0.8(12)'' = 225'\,9.6''$

   b. $\omega = \dfrac{172.5°}{20\ \text{sec}} \cdot \dfrac{\pi\ \text{rad}}{180°} = \dfrac{23\pi}{480} \approx 0.1505\tfrac{\text{rad}}{\text{sec}}$

   c. $V = r\omega = 75(0.1505) \approx 11.29\tfrac{\text{ft}}{\text{sec}} \approx 7.7\tfrac{\text{mi}}{\text{hr}}$

11. $d^2 + 57^2 = 88^2;$ $d = \sqrt{88^2 - 57^2} \approx 67$ cm
    $\cos\theta = \dfrac{57}{88};$ $\theta = \cos^{-1}\left(\dfrac{57}{88}\right) \approx 49.6°$

13. a. $t = \dfrac{7\pi}{6}$

    b. $\sec t = \dfrac{2\sqrt{3}}{3} \Rightarrow \cos t = \dfrac{\sqrt{3}}{2};$ $t = \dfrac{11\pi}{6}$

    c. $t = \dfrac{3\pi}{4}$

15. a. Domain: all real numbers
       Range: $y \in [-2, 2]$
       $P = \dfrac{2\pi}{\pi/5} = 10;$ $A = 2$

       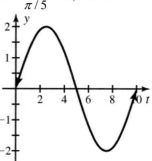

    b. Domain: $t : t \ne \dfrac{\pi}{2}(2k + 1)$ where k is
       any integer.
       Range: $(-\infty, -1] \cup [1, \infty)$
       $P = 2\pi$, no amplitude

       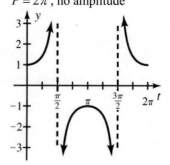

c. Domain: $3t \neq \dfrac{\pi}{2}(2k+1)$

so $t \neq \dfrac{\pi}{6}(2k+1)$ for any integer k

Range: all real numbers

$P = \dfrac{\pi}{3}$, no amplitude

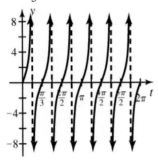

17. $3.5(360°) = 1260°$

19. Max and min are 20 and 5, so

$A = \dfrac{20-5}{2} = 7.5$, $D = \dfrac{20+5}{2} = 12.5$

$P = 12$, so $B = \dfrac{2\pi}{12} = \dfrac{\pi}{6}$. Max would ordinarily occur ¼ way through a period, which is 3, and the max is at 6, so the horiz. shift is 3 units right.

$y = 7.5 \sin\left[\dfrac{\pi}{6}(t-3)\right] + 12.5$

or $y = 7.5 \sin\left(\dfrac{\pi}{6}t - \dfrac{\pi}{2}\right) + 12.5$

a. $\sin^{-1}(-0.7568) \approx -0.86$. In QIII, $\pi + 0.86 \approx 4$

b. $\sec t = -1.5 \Rightarrow \cos t = -\dfrac{1}{1.5} = -\dfrac{2}{3}$

$\cos^{-1}\left(-\dfrac{2}{3}\right) \approx 2.3$

20. a. $\sin^{-1}(-0.7568) \approx -0.86$. In QIII, $\pi + 0.86 \approx 4$

b. $\sec t = -1.5 \Rightarrow \cos t = -\dfrac{1}{1.5} = -\dfrac{2}{3}$

$\cos^{-1}\left(-\dfrac{2}{3}\right) \approx 2.3$

Calculator Exploration & Discovery

1. Answers will vary.

Strengthening Core Skills

1. These values can all be found in Chapter 6.

t	0	$\dfrac{\pi}{6}$	$\dfrac{\pi}{4}$	$\dfrac{\pi}{3}$	$\dfrac{\pi}{2}$
$\sin t = y$	0	$\dfrac{1}{2}$	$\dfrac{\sqrt{2}}{2}$	$\dfrac{\sqrt{3}}{2}$	1
$\cos t = x$	1	$\dfrac{\sqrt{3}}{2}$	$\dfrac{\sqrt{2}}{2}$	$\dfrac{1}{2}$	0
$\tan t = \dfrac{y}{x}$	0	$\dfrac{\sqrt{3}}{3}$	1	$\sqrt{3}$	—
$\dfrac{2\pi}{3}$	$\dfrac{3\pi}{4}$	$\dfrac{5\pi}{6}$	π	$\dfrac{7\pi}{6}$	$\dfrac{5\pi}{4}$
$\dfrac{\sqrt{3}}{2}$	$\dfrac{\sqrt{2}}{2}$	$\dfrac{1}{2}$	0	$\dfrac{-1}{2}$	$\dfrac{-\sqrt{2}}{2}$
$\dfrac{-1}{2}$	$\dfrac{-\sqrt{2}}{2}$	$\dfrac{-\sqrt{3}}{2}$	-1	$\dfrac{-\sqrt{3}}{2}$	$\dfrac{-\sqrt{2}}{2}$
$-\sqrt{3}$	-1	$\dfrac{-\sqrt{3}}{3}$	0	$\dfrac{\sqrt{3}}{3}$	1

3. a. $\sqrt{6}\sin t - 2 = 1$

$\sqrt{6}\sin t = 3$

$\sin t = \dfrac{3}{\sqrt{6}} > 1$, no solution

b. $-3\sqrt{2}\cos t + \sqrt{2} = 0$

$-3\sqrt{2}\cos t = -\sqrt{2}$

$\cos t = \dfrac{1}{3}, t \approx 1.2310, t \approx 5.0522$

c. $3\tan t + \dfrac{1}{2} = -\dfrac{1}{4}$

$3\tan t = -\dfrac{3}{4}$

$\tan t = -\dfrac{1}{4}, t \approx 6.0382, t \approx 2.8966$

d. $2\sec t = -5; \ \sec t = -\dfrac{5}{2}; \ \cos t = -\dfrac{2}{5}$

$t = \cos^{-1}\left(-\dfrac{2}{5}\right) \approx 1.9823$ or

$t = 2\pi - 1.9823 \approx 4.3009$

Cumulative Review: Chap 1–5

1. $2|x+1|-3<5$; $2|x+1|<8$; $|x+1|<4$
   $-4<x+1<4$; $-5<x<3$

3.

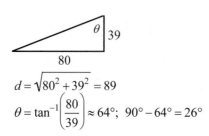

$d=\sqrt{80^2+39^2}=89$

$\theta=\tan^{-1}\left(\dfrac{80}{39}\right)\approx 64°$; $90°-64°=26°$

5. $\sin t=-\dfrac{\sqrt{7}}{4}$; $\csc t=-\dfrac{4}{\sqrt{7}}=-\dfrac{4\sqrt{7}}{7}$;

   $\cos t=\dfrac{3}{4}$; $\sec t=\dfrac{4}{3}$;

   $\tan t=-\dfrac{\sqrt{7}}{3}$; $\cot t=-\dfrac{3}{\sqrt{7}}=-\dfrac{3\sqrt{7}}{7}$

7. a. $2x-3\geq 0$; $2x\geq 3$; $x\geq\dfrac{3}{2}$

   $D:x\in\left[\dfrac{3}{2},\infty\right)$; $R:y\in[0,\infty)$

   b. $x^2-49=0$; $x^2=49$; $x=\pm 7$
   $D:\{x\,|\,x\in\mathbb{R},x\neq\pm 7\}$ or
   $(-\infty,-7)\cup(-7,7)\cup(7,\infty)$
   $R:y\in\mathbb{R}$

9. a. Local max: (–2, 4); Endpoint max:
   (4, 0); Local min: (2, –4)
   Endpoint min: (–4, 0)

   b. $f\geq 0$ for $x\in[-4,0]\cup\{4\}$
   $f<0$ for $x\in(0,4)$

   c. $f\uparrow$ for $x\in(-4,-2)\cup(2,4)$
   $f\downarrow$ for $x\in(-2,2)$

   d. f is odd (symmetric about origin)

11. $\tan 60°=\dfrac{h}{66}$; $h=66\tan 60°\approx 114.3$ ft

13. Shift the graph of $y=\dfrac{1}{x}$ one unit left and
    two units down.

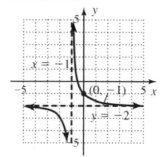

15. $r=\sqrt{(-9)^2+40^2}=41$

    $\sin\theta=\dfrac{40}{41}$; $\csc\theta=\dfrac{41}{40}$; $\cos\theta=-\dfrac{9}{41}$

    $\sec\theta=-\dfrac{41}{9}$; $\tan\theta=-\dfrac{40}{9}$; $\cot\theta=-\dfrac{9}{40}$

    $\theta=\cos^{-1}\left(-\dfrac{9}{41}\right)\approx 102.7°$

17. a. $s=r\theta=15(1.2)=18$

    b. $A=\dfrac{1}{2}(15)^2(1.2)=135$ m^2

19. Max and min are 2 and –1, so

$$A = \frac{2-(-1)}{2} = \frac{3}{2}, \quad D = \frac{2+(-1)}{2} = \frac{1}{2}$$

$P = \frac{\pi}{2}$, so $B = \frac{2\pi}{\pi/2} = 4$. Max would ordinarily occur ¼ way through a period, or at $\frac{\pi}{8}$. It actually occurs at $\frac{\pi}{4}$, so the horiz. shift is $\frac{\pi}{8}$ to the right.

$$y = \frac{3}{2}\sin\left[4\left(t - \frac{\pi}{8}\right)\right] + \frac{1}{2} \text{ or}$$

$$y = \frac{3}{2}\sin\left(4t - \frac{\pi}{2}\right) + \frac{1}{2}$$

21.

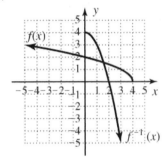

23. $3x - 4y = 8$; $-4y = 8 - 3x$; $y = \frac{8}{-4} - \frac{3}{-4}x$

$y = -2 + \frac{3}{4}x = \frac{3}{4}x - 2$; $m = \frac{3}{4}$, y-intercept: (0, –2)

25. $2275 = 1000e^{r(12)}$; $e^{12r} = \frac{2275}{1000} = 2.275$

$12r = \ln 2.275$; $r = \frac{\ln 2.275}{12} \approx 0.0685$

About 6.85%

Connections to Calculus - Chapter 5

1. $\left(\sqrt{x^2 + 8x}\right)^2 + 4^2 = (\text{hyp})^2$

$x^2 + 8x + 16 = (\text{hyp})^2$

$(x + 4)^2 = (\text{hyp})^2$

$\text{hyp} = x + 4$

$\sin\theta = \frac{4}{x+4}$; $\cos\theta = \frac{\sqrt{x^2+8x}}{x+4}$

$\csc\theta = \frac{x+4}{4}$; $\sec\theta = \frac{x+4}{\sqrt{x^2+8x}}$

$\tan\theta = \frac{4}{\sqrt{x^2+8x}}$; $\cot\theta = \frac{\sqrt{x^2+8x}}{4}$

3. $x = 4\tan\theta$

$\tan\theta = \frac{x}{4}$

$x^2 + 4^2 = (\text{hyp})^2$

$x^2 + 16 = (\text{hyp})^2$

$\sqrt{x^2 + 16} = \text{hyp}$

$\sin\theta = \frac{x}{\sqrt{x^2+16}}$; $\cos\theta = \frac{4}{\sqrt{x^2+16}}$

$\csc\theta = \frac{\sqrt{x^2+16}}{x}$; $\sec\theta = \frac{\sqrt{x^2+16}}{4}$

$\cot\theta = \frac{4}{x}$

Connections to Calculus

5. $\csc\theta = \dfrac{\sqrt{u^2+169}}{u}$

$u^2+(\text{adj})^2 = \left(\sqrt{u^2+169}\right)^2$

$u^2+(\text{adj})^2 = u^2+169$

$(\text{adj})^2 = 169$

$\text{adj} = \sqrt{169} = 13$

$\cot\theta = \dfrac{13}{u}$; $\sec\theta = \dfrac{\sqrt{u^2+169}}{13}$

7. $y=2$

$x = r\cos\theta$; $y = r\sin\theta$;

$x^2+y^2 = r^2$

$y=2$

$r\sin\theta = 2$

$r = \dfrac{2}{\sin\theta}$

9. $y=-2x+3$

$x = r\cos\theta$; $y = r\sin\theta$;

$x^2+y^2 = r^2$

$y=-2x+3$

$r\sin\theta = -2(r\cos\theta)+3$

$r\sin\theta + 2r\cos\theta = 3$

$r(\sin\theta + 2\cos\theta) = 3$

$r = \dfrac{3}{\sin\theta + 2\cos\theta}$

11. $x = r\cos\theta$; $y = r\sin\theta$; $x^2+y^2 = r^2$

$\dfrac{x}{r} = \cos\theta$; $\dfrac{y}{r} = \sin\theta$

$r = 5\left(\dfrac{y}{r}\right)$

$r^2 = 5y$

$x^2+y^2 = 5y$

$x^2+y^2-5y = 0$

circle

13. $x = r\cos\theta$; $y = r\sin\theta$; $x^2+y^2 = r^2$

$r(3\cos\theta - 2\sin\theta) = 6$

$3(r\cos\theta) - 2(r\sin\theta) = 6$

$3x-2y = 6$; line

6.1 Exercises

1. $\sin\theta,\ \sec\theta,\ \cos\theta$

3. One, false

5. $\dfrac{\cos x}{\sin x} - \dfrac{\sin x}{\sec x}$

$= \dfrac{\cos x\sec x - \sin^2 x}{\sin x\sec x} = \dfrac{1-\sin^2 x}{\sin x\sec x}$;

Answers will vary.

7. Answers may vary.

$\tan x = \dfrac{\sin x}{\cos x}$;

$\tan x = \dfrac{\sec x}{\csc x}$;

$\dfrac{\sin x}{\cos x} = \dfrac{\sec x}{\csc x}$;

$\dfrac{1}{\cot x} = \dfrac{\sec x}{\csc x}$;

$\dfrac{1}{\cot x} = \dfrac{\sin x}{\cos x}$

9. $1 + \tan^2 x = \sec^2 x$;

$1 = \sec^2 x - \tan^2 x$;

$\tan^2 x = \sec^2 x - 1$;

$1 = (\sec x + \tan x)(\sec x - \tan x)$

$\tan x = \pm\sqrt{\sec^2 x - 1}$

11. $\sin x\cot x = \cos x$

$\sin x\left(\dfrac{\cos x}{\sin x}\right) = \cos x$

$\cos x = \cos x$

13. $\sec^2 x\cot^2 x = \csc^2 x$

$\dfrac{1}{\cos^2 x}\left(\dfrac{\cos^2 x}{\sin^2 x}\right) = \csc^2 x$

$\dfrac{1}{\sin^2 x} = \csc^2 x$

$\csc^2 x = \csc^2 x$

15. $\cos x(\sec x - \cos x) = \sin^2 x$

$\cos x\sec x - \cos^2 x = \sin^2 x$

$\cos x \cdot \dfrac{1}{\cos x} - \cos^2 x = \sin^2 x$

$1 - \cos^2 x = \sin^2 x$

$\sin^2 x = \sin^2 x$

17. $\sin x(\csc x - \sin x) = \cos^2 x$

$\sin x\csc x - \sin^2 x = \cos^2 x$

$\sin x \cdot \dfrac{1}{\sin x} - \sin^2 x = \cos^2 x$

$1 - \sin^2 x = \cos^2 x$

$\cos^2 x = \cos^2 x$

19. $\tan x(\csc x + \cot x) = \sec x + 1$

$\tan x\csc x + \tan x\cot x = \sec x + 1$

$\dfrac{\sin x}{\cos x}\cdot\dfrac{1}{\sin x} + \dfrac{\sin x}{\cos x}\cdot\dfrac{\cos x}{\sin x} = \sec x + 1$

$\dfrac{1}{\cos x} + 1 = \sec x + 1$

$\sec x + 1 = \sec x + 1$

21. $\tan^2 x\csc^2 x - \tan^2 x = 1$

$\tan^2 x(\csc^2 x - 1) = 1$

$\tan^2 x(\cot^2 x) = 1$

$\dfrac{\sin^2 x}{\cos^2 x}\left(\dfrac{\cos^2 x}{\sin^2 x}\right) = 1$

$1 = 1$

23. $\dfrac{\sin x\cos x + \sin x}{\cos x + \cos^2 x} = \tan x$

$\dfrac{\sin x(\cos x + 1)}{\cos x(1 + \cos x)} = \tan x$

$\dfrac{\sin x}{\cos x} = \tan x$

$\tan x = \tan x$

25. $\dfrac{1 + \sin x}{\cos x + \cos x\sin x} = \sec x$

$\dfrac{1(1 + \sin x)}{\cos x(1 + \sin x)} = \sec x$

$\dfrac{1}{\cos x} = \sec x$

$\sec x = \sec x$

27. $\dfrac{\sin x \tan x + \sin x}{\tan x + \tan^2 x} = \cos x$

$\dfrac{\sin x(\tan x + 1)}{\tan x(1 + \tan x)} = \cos x$

$\dfrac{\sin x}{\tan x} = \cos x$

$\dfrac{\sin x}{\dfrac{\sin x}{\cos x}} = \cos x$

$\dfrac{\sin x \cos x}{\sin x} = \cos x$

$\cos x = \cos x$

29. $\dfrac{(\sin x + \cos x)^2}{\cos x} = \sec x + 2 \sin x$

$\dfrac{\sin^2 x + 2 \sin x \cos x + \cos^2 x}{\cos x} =$

$\dfrac{\sin^2 x + \cos^2 x + 2 \sin x \cos x}{\cos x} =$

$\dfrac{1 + 2 \sin x \cos x}{\cos x} =$

$\dfrac{1}{\cos x} + \dfrac{2 \sin x \cos x}{\cos x} =$

$\dfrac{1}{\cos x} + 2 \sin x =$

$\sec x + 2 \sin x = \sec x + 2 \sin x$

31. $(1 + \sin x)[1 + \sin(-x)] = \cos^2 x$

$(1 + \sin x)(1 - \sin x) = \cos^2 x$

$1 - \sin^2 x = \cos^2 x$

$\cos^2 x = \cos^2 x$

33. $\dfrac{(\csc x - \cot x)(\csc x + \cot x)}{\tan x} = \cot x$

$\dfrac{\csc^2 x - \cot^2 x}{\tan x} = \cot x$

$\dfrac{1}{\tan x} = \cot x$

$\cot x = \cot x$

35. $\dfrac{\cos^2 x}{\sin x} + \dfrac{\sin x}{1} = \csc x$

$\dfrac{\cos^2 x}{\sin x} + \dfrac{\sin^2 x}{\sin x} = \csc x$

$\dfrac{\cos^2 x + \sin^2 x}{\sin x} = \csc x$

$\dfrac{1}{\sin x} = \csc x$

$\csc x = \csc x$

37. $\dfrac{\tan x}{\csc x} - \dfrac{\sin x}{\cos x} = \dfrac{\sin x - 1}{\cot x}$

$\dfrac{\tan x \cos x}{\csc x \cos x} - \dfrac{\sin x \csc x}{\csc x \cos x} = \dfrac{\sin x - 1}{\cot x}$

$\dfrac{\tan x \cos x - \sin x \csc x}{\csc x \cos x} = \dfrac{\sin x - 1}{\cot x}$

$\dfrac{\dfrac{\sin x}{\cos x} \cdot \cos x - \sin x \cdot \dfrac{1}{\sin x}}{\csc x \cos x} = \dfrac{\sin x - 1}{\cot x}$

$\dfrac{\sin x - 1}{\csc x \cos x} = \dfrac{\sin x - 1}{\cot x}$

$\dfrac{\sin x - 1}{\dfrac{1}{\sin x} \cdot \cos x} = \dfrac{\sin x - 1}{\cot x}$

$\dfrac{\sin x - 1}{\dfrac{\cos x}{\sin x}} = \dfrac{\sin x - 1}{\cot x}$

$\dfrac{\sin x - 1}{\cot x} = \dfrac{\sin x - 1}{\cot x}$

39. $\dfrac{\sec x}{\sin x} - \dfrac{\csc x}{\sec x} = \tan x$

$\dfrac{\sec^2 x}{\sin x \sec x} - \dfrac{\csc x \sin x}{\sin x \sec x} = \tan x$

$\dfrac{\sec^2 x - \csc x \sin x}{\sin x \sec x} = \tan x$

$\dfrac{\sec^2 x - \dfrac{1}{\sin x} \cdot \sin x}{\sin x \sec x} = \tan x$

$\dfrac{\sec^2 x - 1}{\sin x \sec x} = \tan x$

$\dfrac{\tan^2 x}{\sin x \sec x} = \tan x$

$\dfrac{\tan^2 x}{(\sin x) \cdot \dfrac{1}{\cos x}} = \tan x$

$\dfrac{\tan^2 x}{\tan x} = \tan x$

$\tan x = \tan x$

41. $\tan x$ in terms of $\sin x$

$\tan x = \dfrac{\sin x}{\cos x};$

Since $\cos^2 x = 1 - \sin^2 x$

$\cos x = \pm\sqrt{1 - \sin^2 x}$

$\tan x = \dfrac{\sin x}{\pm\sqrt{1 - \sin^2 x}}$

43. $\sec x$ in terms of $\cot x$

Since $\sec^2 x = \tan^2 x + 1$

$\sec^2 x = \dfrac{1}{\cot^2 x} + 1$

$\sec x = \pm\sqrt{\dfrac{1}{\cot^2 x} + 1}$

45. $\cot x$ in terms of $\sin x$

Since $\cot^2 x = \csc^2 x - 1$

$\cot x = \pm\sqrt{\csc^2 x - 1}$

$\cot x = \pm\sqrt{\dfrac{1}{\sin^2 x} - 1}$

$\cot x = \pm\sqrt{\dfrac{1 - \sin^2 x}{\sin^2 x}}$

$\cot x = \dfrac{\pm\sqrt{1 - \sin^2 x}}{\sin x}$

47. $\cos\theta = -\dfrac{20}{29}$ with θ in QII

$29^2 = 20^2 + y^2$

$21 = y;$

$\sec\theta = -\dfrac{29}{20}$, since cosine and secant are reciprocals.

$\tan^2\theta = \sec^2\theta - 1$

$\tan^2\theta = \left(\dfrac{-29}{20}\right)^2 - 1$

$\tan^2\theta = \dfrac{841}{400} - \dfrac{400}{400}$

$\tan^2\theta = \dfrac{441}{400}$

$\tan\theta = \pm\dfrac{21}{20},$

since $\tan\theta$ is negative in QII we choose

$\tan\theta = \dfrac{-21}{20}$ so $\cot\theta = \dfrac{-20}{21};$

$\tan\theta = \dfrac{\sin\theta}{\cos\theta}$

$\dfrac{-21}{20} = \dfrac{\sin\theta}{\dfrac{-20}{29}}$

$\dfrac{-21}{20} \cdot \dfrac{-20}{29} = \sin\theta$

$\sin\theta = \dfrac{21}{29};$

$\csc\theta = \dfrac{29}{21}$

6.1 Exercises

49. $\tan \theta = \dfrac{15}{8}$ with θ in QIII

$h^2 = 15^2 + 8^2$
$h = 17;$

$\cot \theta = \dfrac{8}{15};$

$\sec^2 \theta = 1 + \tan^2 \theta$

$\sec^2 \theta = 1 + \left(\dfrac{15}{8}\right)^2$

$\sec^2 \theta = 1 + \dfrac{225}{64}$

$\sec^2 \theta = \dfrac{289}{64}$

$\sec \theta = \pm \dfrac{17}{8},$

Since $\sec \theta$ is negative in QIII $\sec \theta = \dfrac{-17}{8}$

So $\cos \theta = \dfrac{-8}{17};$

$\sin^2 \theta = 1 - \cos^2 \theta$

$\sin^2 \theta = 1 - \left(\dfrac{-8}{17}\right)^2$

$\sin^2 \theta = 1 - \dfrac{64}{289}$

$\sin^2 \theta = \dfrac{225}{289}$

$\sin \theta = \pm \dfrac{15}{17},$ $\sin \theta$ is negative in QIII

$\sin \theta = \dfrac{-15}{17}$ so $\csc \theta = \dfrac{-17}{15}$

51. $\cot \theta = \dfrac{x}{5}$ with θ in QI

$x^2 + 5^2 = h^2$
$\sqrt{x^2 + 25} = h;$

$\tan \theta = \dfrac{5}{x};$

$\sec^2 \theta = \tan^2 \theta + 1$

$\sec^2 \theta = \left(\dfrac{5}{x}\right)^2 + 1$

$\sec^2 \theta = \dfrac{25}{x^2} + 1$

$\sec^2 \theta = \dfrac{25 + x^2}{x^2}$

$\sec \theta = \pm \dfrac{\sqrt{25 + x^2}}{x},$

since $\sec \theta$ is positive in QI

$\sec \theta = \dfrac{\sqrt{25 + x^2}}{x},$ $\cos \theta = \dfrac{x}{\sqrt{25 + x^2}};$

$\csc \theta = 1 + \cot^2 \theta$

$\csc \theta = 1 + \left(\dfrac{x}{5}\right)^2$

$\csc \theta = 1 + \dfrac{x^2}{25}$

$\csc^2 \theta = \dfrac{25 + x^2}{25}$

$\csc \theta = \pm \dfrac{\sqrt{25 + x^2}}{5}$

Since $\csc \theta$ is positive in QI

$\csc \theta = \dfrac{\sqrt{25 + x^2}}{5};$

$\sin \theta = \dfrac{5}{\sqrt{25 + x^2}}$

53. $\sin\theta = -\dfrac{7}{13}$ with θ in QIII

$\sqrt{13^2 - 7^2} = x$

$x = 2\sqrt{30}$;

$\csc\theta = -\dfrac{13}{7}$;

$\cos^2\theta = 1 - \sin^2\theta$

$\cos^2\theta = 1 - \left(\dfrac{-7}{13}\right)^2$

$\cos^2\theta = \dfrac{120}{169}$

$\cos\theta = \pm\dfrac{2\sqrt{30}}{13}$,

since $\cos\theta$ is negative in QIII

$\cos\theta = -\dfrac{2\sqrt{30}}{13}$,

$\sec\theta = \dfrac{-13}{2\sqrt{30}}$;

$\tan\theta = \dfrac{\sin\theta}{\cos\theta}$

$\tan\theta = \dfrac{\dfrac{-7}{13}}{\dfrac{-2\sqrt{30}}{13}}$

$\tan\theta = \dfrac{-7}{13}\cdot\dfrac{13}{-2\sqrt{30}} = \dfrac{7}{2\sqrt{30}}$

$\cot\theta = \dfrac{2\sqrt{30}}{7}$

55. $\sec\theta = -\dfrac{9}{7}$ with θ in QII

$y = \sqrt{9^2 - 7^2}$

$y = 4\sqrt{2}$;

$\cos\theta = -\dfrac{7}{9}$;

$\sin^2\theta = 1 - \cos^2\theta$

$\sin^2\theta = 1 - \left(\dfrac{-7}{9}\right)^2$

$\sin^2\theta = \dfrac{32}{81}$

$\sin\theta = \pm\dfrac{4\sqrt{2}}{9}$, $\sin\theta$ is positive in QII

$\sin\theta = \dfrac{4\sqrt{2}}{9}$, $\csc\theta = \dfrac{9}{4\sqrt{2}}$;

$\tan\theta = \dfrac{\sin\theta}{\cos\theta}$

$\tan\theta = \dfrac{\dfrac{4\sqrt{2}}{9}}{\dfrac{-7}{9}} = \dfrac{4\sqrt{2}}{9}\cdot\dfrac{-9}{7} = = \dfrac{-4\sqrt{2}}{7}$;

$\cot\theta = \dfrac{-7}{4\sqrt{2}}$

57. $\cos\left(\dfrac{\pi}{4}\right) + \cos\theta \neq \cos\left(\dfrac{\pi}{4} + \theta\right)$

Answers will vary.
We will substitute a convenient value to prove the equation is false, namely $\theta = \dfrac{\pi}{4}$.

$\cos\left(\dfrac{\pi}{4}\right) + \cos\left(\dfrac{\pi}{4}\right) \neq \cos\left(\dfrac{\pi}{4} + \dfrac{\pi}{4}\right)$

$\dfrac{\sqrt{2}}{2} + \dfrac{\sqrt{2}}{2} \neq \cos\left(\dfrac{\pi}{2}\right)$

$\sqrt{2} \neq 0$

59. $\tan(2\theta) \neq 2\tan\theta$
Answers will vary.
We will substitute a convenient value to prove the equation is false, namely $\theta = \dfrac{\pi}{4}$.

$\tan\left(2\cdot\dfrac{\pi}{4}\right) \neq 2\tan\left(\dfrac{\pi}{4}\right)$

undefined $\neq 2$

61. $\cos^2\theta - \sin^2\theta \neq -1$
Answers will vary.
We will substitute a convenient value to prove the equation is false, namely $\theta = 0$.
$\cos^2 0 - \sin^2 0 \neq -1$
$1 - 0 \neq -1$
$1 \neq -1$

63. $E = \dfrac{I\cos\theta}{r^2}$; $90° - 40° = 50°$

$E = \dfrac{800\cos 50°}{(2)^2}$

$E = \dfrac{800\cos 50°}{4}$

$E \approx 128.6$ lumens/m^2

65. $\cos^3 x = (\cos x)(\cos^2 x)$
    $\quad = \cos x(1 - \sin^2 x)$

67. $\tan x + \tan^3 x = \tan x(1 + \tan^2 x)$
    $\quad = \tan x(\sec^2 x)$

69. $\tan^2 x \sec x - 4\tan^2 x$
    $\quad = \tan^2 x(\sec x - 4)$
    $\quad = (\sec x - 4)(\tan^2 x)$
    $\quad = (\sec x - 4)(\sec^2 x - 1)$
    $\quad = (\sec x - 4)(\sec x - 1)(\sec x + 1)$

71. $\cos^2 x \sin x - \cos^2 x$
    $\quad = \cos^2 x(\sin x - 1)$
    $\quad = (1 - \sin^2 x)(\sin x - 1)$
    $\quad = (1 + \sin x)(1 - \sin x)(\sin x - 1)$
    $\quad = (1 + \sin x)(1 - \sin x)(-1)(1 - \sin x)$
    $\quad = -1(1 + \sin x)(1 - \sin x)^2$

73. $A = nr^2 \, \dfrac{\sin\left(\dfrac{\pi}{n}\right)}{\cos\left(\dfrac{\pi}{n}\right)}$

    a. $A = nr^2 \tan\left(\dfrac{\pi}{n}\right)$

    b. $A = 4(4)^2 \tan\left(\dfrac{\pi}{4}\right)$
       $A = 4(16)(1)$
       $A = 64 \, \text{m}^2$

    c. $A = 12(4)^2 \tan\left(\dfrac{\pi}{12}\right)$
       $A = 12(16)\tan\left(\dfrac{\pi}{12}\right)$
       $A \approx 51.45 \, \text{m}^2$

75. $(m_2 - m_1)\cos\theta = \sin\theta + m_1 m_2 \sin\theta$
    $(m_2 - m_1)\cos\theta = \sin\theta(1 + m_1 m_2)$
    $\dfrac{m_2 - m_1}{1 + m_1 m_2} = \dfrac{\sin\theta}{\cos\theta}$
    $\tan\theta = \dfrac{m_2 - m_1}{1 + m_1 m_2}$

77. $\tan\theta = \dfrac{m_2 - m_1}{1 + m_1 m_2}$
    $\tan\theta = \dfrac{-2 - 3}{1 + 3(-2)}$
    $\tan\theta = 1$
    $\theta = 45°$

79. $f(\theta) = -2\sin^4\theta + \sqrt{3}\sin^3\theta + 2\sin^2\theta - \sqrt{3}\sin\theta$
    $\quad = \sin\theta\left(-2\sin^3\theta + \sqrt{3}\sin^2\theta + 2\sin\theta - \sqrt{3}\right)$
    $\quad = \sin\theta\left(-2\sin^3\theta + 2\sin\theta + \sqrt{3}\sin^2\theta - \sqrt{3}\right)$
    $\quad = \sin\theta\left[(-2\sin\theta)(\sin^2\theta - 1) + \sqrt{3}(\sin^2\theta - 1)\right]$
    $\quad = \sin\theta\left[(\sin^2\theta - 1)(-2\sin\theta) + \sqrt{3})\right]$
    $\quad = \sin\theta(\sin\theta + 1)(\sin\theta - 1)(-2\sin\theta + \sqrt{3})$

    $\sin\theta = 0, \sin\theta = -1, \sin\theta = 1, \sin\theta = \dfrac{\sqrt{3}}{2}$

    x-intercepts in $[0, 2\pi)$ at :

    $0, \dfrac{\pi}{3}, \dfrac{\pi}{2}, \dfrac{2\pi}{3}, \pi, \dfrac{3\pi}{2}$

81. $\tan 77° = \dfrac{h}{265}$
    $265\tan 77° = h$
    About $1148 \, \text{ft}$

83. $y = 2\sin(2t)$ for $t \in [0, 2\pi)$
    Amplitude: 2
    Period: $\dfrac{2\pi}{2} = \pi$
    Ref Rect: $2A = 4$ by $P = \pi$ units
    Since $P_0 = \pi$, $t = 0, \dfrac{\pi}{4}, \dfrac{\pi}{2}, \dfrac{3\pi}{4}$, and π .

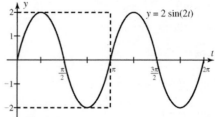

6.2 Exercises

Chapter 6: Trigonometric Identities, Inverses and Equations

1. Substituted

3. Complicated, simplify, build

5. Because we do not know if the equation is true.

7. $\sec x + \tan x$

$$= \frac{1}{\cos x} + \frac{\sin x}{\cos x}$$

$$= \frac{1 + \sin x}{\cos x}$$

9. $\left(1 - \sin^2 x\right)\sec x$

$$= \cos^2 x\left(\frac{1}{\cos x}\right)$$

$$= \cos x$$

11. $\dfrac{\sin x - \sin x \cos x}{\sin^2 x}$

$$= \frac{\sin x(1 - \cos x)}{\sin^2 x}$$

$$= \frac{1 - \cos x}{\sin x}$$

13. $\cos^2 x \tan^2 x = 1 - \cos^2 x$

$$\cos^2 x\left(\frac{\sin^2 x}{\cos^2 x}\right) =$$

$$\sin^2 x =$$

$$1 - \cos^2 x = 1 - \cos^2 x$$

15. $\tan x + \cot x = \sec x \csc x$

$$\frac{\sin x}{\cos x} + \frac{\cos x}{\sin x} =$$

$$\frac{\sin^2 x}{\cos x \sin x} + \frac{\cos^2 x}{\cos x \sin x} =$$

$$\frac{\sin^2 x + \cos^2 x}{\cos x \sin x} =$$

$$\frac{1}{\cos x \sin x} =$$

$$\frac{1}{\cos x} \cdot \frac{1}{\sin x} =$$

$$\sec x \csc x = \sec x \csc x$$

17. $\dfrac{\cos x}{\tan x} = \csc x - \sin x$

$$= \frac{1}{\sin x} - \sin x$$

$$= \frac{1 - \sin^2 x}{\sin x}$$

$$= \frac{\cos^2 x}{\sin x}$$

$$= \frac{\cos x \cos x}{\sin x}$$

$$= \frac{\cos x}{\dfrac{\sin x}{\cos x}}$$

$$= \frac{\cos x}{\tan x}$$

19. $\dfrac{\cos \theta}{1 - \sin \theta} = \sec \theta + \tan \theta$

$$= \frac{1}{\cos \theta} + \frac{\sin \theta}{\cos \theta}$$

$$= \frac{1 + \sin \theta}{\cos \theta}$$

$$= \frac{1 + \sin \theta}{\cos \theta} \cdot \frac{1 - \sin \theta}{1 - \sin \theta}$$

$$= \frac{1 - \sin^2 \theta}{\cos \theta(1 - \sin \theta)}$$

$$= \frac{\cos^2 \theta}{\cos \theta(1 - \sin \theta)}$$

$$= \frac{\cos \theta}{1 - \sin \theta}$$

21. $\dfrac{1 - \sin x}{\cos x} = \dfrac{\cos x}{1 + \sin x}$

$$\frac{1 - \sin x}{\cos x} \cdot \frac{(1 + \sin x)}{(1 + \sin x)} =$$

$$\frac{1 - \sin^2 x}{\cos x(1 + \sin x)} =$$

$$\frac{\cos^2 x}{\cos x(1 + \sin x)} =$$

$$\frac{\cos x}{1 + \sin x} = \frac{\cos x}{1 + \sin x}$$

23. $\dfrac{\csc x}{\cos x} - \dfrac{\cos x}{\csc x} = \dfrac{\cot^2 x + \sin^2 x}{\cot x}$

$\dfrac{\csc^2 x - \cos^2 x}{\cos x \csc x} =$

$\dfrac{\csc^2 x - \left(1 - \sin^2 x\right)}{\cos x \left(\dfrac{1}{\sin x}\right)} =$

$\dfrac{\csc^2 x - 1 + \sin^2 x}{\cot x} =$

$\dfrac{\left(\csc^2 x - 1\right) + \sin^2 x}{\cot x} =$

$\dfrac{\cot^2 x + \sin^2 x}{\cot x} = \dfrac{\cot^2 x + \sin^2 x}{\cot x}$

25. $\dfrac{\sin x}{1 + \sin x} - \dfrac{\sin x}{1 - \sin x} = -2 \tan^2 x$

$\dfrac{\sin x(1 - \sin x) - \sin x(1 + \sin x)}{(1 + \sin x)(1 - \sin x)} =$

$\dfrac{\sin x - \sin^2 x - \sin x - \sin^2 x}{1 - \sin^2 x} =$

$\dfrac{-2 \sin^2 x}{\cos^2 x} =$

$-2 \tan^2 x = -2 \tan^2 x$

27. $\dfrac{\cot x}{1 + \csc x} - \dfrac{\cot x}{1 - \csc x} = 2 \sec x$

$\dfrac{\cot x(1 - \csc x) - \cot x(1 + \csc x)}{(1 + \csc x)(1 - \csc x)} =$

$\dfrac{\cot x - \cot x \csc x - \cot x - \cot x \csc x}{1 - \csc^2 x} =$

$\dfrac{-2 \cot x \csc x}{-\cot^2 x} =$

$\dfrac{2 \cot x \csc x}{\cot^2 x} =$

$\dfrac{2 \csc x}{\cot x} =$

$\dfrac{2}{\sin x} \div \dfrac{\cos x}{\sin x} =$

$\dfrac{2}{\sin x} \cdot \dfrac{\sin x}{\cos x} =$

$\dfrac{2}{\cos x} =$

$2 \sec x = 2 \sec x$

29. $\dfrac{\sec^2 x}{1 + \cot^2 x} = \tan^2 x$

$\dfrac{\sec^2 x}{\csc^2 x} =$

$\dfrac{\dfrac{1}{\cos^2 x}}{\dfrac{1}{\sin^2 x}} =$

$\dfrac{1}{\cos^2 x} \div \dfrac{1}{\sin^2 x} =$

$\dfrac{1}{\cos^2 x} \cdot \dfrac{\sin^2 x}{1} =$

$\dfrac{\sin^2 x}{\cos^2 x} =$

$\tan^2 x = \tan^2 x$

31. $\sin^2 x\left(\cot^2 x - \csc^2 x\right) = -\sin^2 x$

$\sin^2 x \cot^2 x - \sin^2 x \csc^2 x =$

$\sin^2 x \cdot \dfrac{\cos^2 x}{\sin^2 x} - \sin^2 x \cdot \dfrac{1}{\sin^2 x} =$

$\cos^2 x - 1 =$

$-1\left(1 - \cos^2 x\right) =$

$-\sin^2 x = -\sin^2 x$

33. $\cos x \cot x + \sin x = \csc x$

$\cos x \cdot \dfrac{\cos x}{\sin x} + \sin x =$

$\dfrac{\cos^2 x}{\sin x} + \sin x =$

$\dfrac{\cos^2 x + \sin^2 x}{\sin x} =$

$\dfrac{1}{\sin x} =$

$\csc x = \csc x$

35. $\dfrac{\sec x}{\cot x + \tan x} = \sin x$

$$\dfrac{\dfrac{1}{\cos x}}{\dfrac{\cos x}{\sin x} + \dfrac{\sin x}{\cos x}} =$$

$$\dfrac{\dfrac{1}{\cos x}(\sin x)(\cos x)}{\left(\dfrac{\cos x}{\sin x} + \dfrac{\sin x}{\cos x}\right)(\sin x)(\cos x)} =$$

$$\dfrac{\sin x}{\cos^2 x + \sin^2 x} =$$

$$\dfrac{\sin x}{1} =$$

$$\sin x = \sin x$$

37. $\dfrac{\sin x - \csc x}{\csc x} = -\cos^2 x$

$$\dfrac{\sin x}{\csc x} - \dfrac{\csc x}{\csc x} =$$

$$\dfrac{\sin x}{\dfrac{1}{\sin x}} - 1 =$$

$$\sin^2 x - 1 =$$

$$-\cos^2 x = -\cos^2 x$$

39. $\dfrac{1}{\csc x - \sin x} = \tan x \sec x$

$$\dfrac{1}{\dfrac{1}{\sin x} - \sin x} =$$

$$\dfrac{1}{\dfrac{1}{\sin x} - \sin x} \cdot \dfrac{(\sin x)}{(\sin x)} =$$

$$\dfrac{\sin x}{1 - \sin^2 x} =$$

$$\dfrac{\sin x}{\cos^2 x} =$$

$$\dfrac{\sin x}{\cos x} \cdot \dfrac{1}{\cos x} =$$

$$\tan x \sec x = \tan x \sec x$$

41. $\dfrac{1+\sin x}{1-\sin x} = (\tan x + \sec x)^2$

$$\dfrac{1+\sin x}{1-\sin x} \cdot \dfrac{1+\sin x}{1+\sin x} =$$

$$\dfrac{1 + 2\sin x + \sin^2 x}{1 - \sin^2 x} =$$

$$\dfrac{1 + 2\sin x + \sin^2 x}{\cos^2 x} =$$

$$\dfrac{1}{\cos^2 x} + 2\dfrac{\sin x}{\cos x} \cdot \dfrac{1}{\cos x} + \dfrac{\sin^2 x}{\cos^2 x} =$$

$$\sec^2 x + 2\tan x \sec x + \tan^2 x =$$

$$(\sec x + \tan x)(\sec x + \tan x) =$$

$$(\sec x + \tan x)^2 =$$

$$(\tan x + \sec x)^2 = (\tan x + \sec x)^2$$

43. $\dfrac{\cos x - \sin x}{1 - \tan x} = \dfrac{\cos x + \sin x}{1 + \tan x}$

$$\dfrac{\cos x - \sin x}{1 - \tan x} \cdot \dfrac{\cos x + \sin x}{\cos x + \sin x} =$$

$$\dfrac{(\cos x - \sin x)(\cos x + \sin x)}{\cos x + \sin x - \sin x - \dfrac{\sin^2 x}{\cos x}} =$$

$$\dfrac{(\cos x - \sin x)(\cos x + \sin x)}{\cos x\left(1 - \dfrac{\sin^2 x}{\cos^2 x}\right)} =$$

$$\dfrac{(\cos x - \sin x)(\cos x + \sin x)}{\cos x(1 - \tan^2 x)} =$$

$$\dfrac{(\cos x - \sin x)(\cos x + \sin x)}{\cos x(1 - \tan x)(1 + \tan x)} =$$

$$\dfrac{(\cos x - \sin x)(\cos x + \sin x)}{(\cos x - \sin x)(1 + \tan x)} =$$

$$\dfrac{\cos x + \sin x}{1 + \tan x} = \dfrac{\cos x + \sin x}{1 + \tan x}$$

45. $\dfrac{\tan^2 x - \cot^2 x}{\tan x - \cot x} = \csc x \sec x$

$\dfrac{(\tan x + \cot x)(\tan x - \cot x)}{(\tan x - \cot x)} =$

$\tan x + \cot x =$

$\dfrac{\sin x}{\cos x} + \dfrac{\cos x}{\sin x} =$

$\dfrac{\sin^2 x + \cos^2 x}{\cos x \sin x} =$

$\dfrac{1}{\cos x \sin x} =$

$\dfrac{1}{\cos x} \cdot \dfrac{1}{\sin x} =$

$\sec x \csc x =$

$\csc x \sec x = \csc x \sec x$

47. $\dfrac{\cot x}{\cot x + \tan x} = 1 - \sin^2 x$

$\dfrac{\dfrac{\cos x}{\sin x}}{\dfrac{\cos x}{\sin x} + \dfrac{\sin x}{\cos x}} =$

$\dfrac{\dfrac{\cos x}{\sin x}}{\dfrac{\cos x}{\sin x} + \dfrac{\sin x}{\cos x}} \cdot \dfrac{(\cos x)(\sin x)}{(\cos x)(\sin x)} =$

$\dfrac{\cos^2 x}{\cos^2 x + \sin^2 x} =$

$\dfrac{\cos^2 x}{1} =$

$1 - \sin^2 x = 1 - \sin^2 x$

49. $\dfrac{\sec^4 x - \tan^4 x}{\sec^2 x + \tan^2 x} = 1$

$\dfrac{(\sec^2 x + \tan^2 x)(\sec^2 x - \tan^2 x)}{(\sec^2 x + \tan^2 x)} =$

$\sec^2 x - \tan^2 x =$

$1 = 1$

51. $\dfrac{\cos^4 x - \sin^4 x}{\cos^2 x} = 2 - \sec^2 x$

$\dfrac{(\cos^2 x - \sin^2 x)(\cos^2 x + \sin^2 x)}{\cos^2 x} =$

$\dfrac{(\cos^2 x - \sin^2 x)(1)}{\cos^2 x} =$

$\dfrac{\cos^2 x}{\cos^2 x} - \dfrac{\sin^2 x}{\cos^2 x} =$

$1 - \tan^2 x =$

$1 - (\sec^2 x - 1) =$

$1 - \sec^2 x + 1 =$

$2 - \sec^2 x = 2 - \sec^2 x$

53. $(\sec x + \tan x)^2 = \dfrac{(\sin x + 1)^2}{\cos^2 x}$

$\sec^2 x + 2 \sec x \tan x + \tan^2 x =$

$\dfrac{1}{\cos^2 x} + 2\left(\dfrac{1}{\cos x}\right)\left(\dfrac{\sin x}{\cos x}\right) + \dfrac{\sin^2 x}{\cos^2 x} =$

$\dfrac{1}{\cos^2 x} + \dfrac{2 \sin x}{\cos^2 x} + \dfrac{\sin^2 x}{\cos^2 x} =$

$\dfrac{1 + 2 \sin x + \sin^2 x}{\cos^2 x} =$

$\dfrac{(1 + \sin x)^2}{\cos^2 x} =$

$\dfrac{(\sin x + 1)^2}{\cos^2 x} = \dfrac{(\sin x + 1)^2}{\cos^2 x}$

55. $\dfrac{\cos x}{\sin x} + \dfrac{\sin x}{\cos x} + \dfrac{\csc x}{\sec x} = \dfrac{\sec x + \cos x}{\sin x}$

$\dfrac{\cos^2 x \sec x + \sin^2 x \sec x + \csc x \sin x \cos x}{\sin x \cos x \sec x} =$

$\dfrac{\sec x(\cos^2 x + \sin^2 x) + (1)\cos x}{\sin x \cos x \sec x} =$

$\dfrac{\sec x + \cos x}{\sin x \cos x \sec x} =$

$\dfrac{\sec x + \cos x}{\sin x} = \dfrac{\sec x + \cos x}{\sin x}$

57. $\dfrac{\sin^4 x - \cos^4 x}{\sin^3 x + \cos^3 x} = \dfrac{\sin x - \cos x}{1 - \sin x \cos x}$

Factor numerator as difference of two squares, denominator as sum of two cubes

$\dfrac{\left(\sin^2 x + \cos^2 x\right)\left(\sin^2 x - \cos^2 x\right)}{\left(\sin x + \cos x\right)\left(\sin^2 x - \sin x \cos x + \cos^2 x\right)} =$

$\dfrac{\left(1\right)\left(\sin x + \cos x\right)\left(\sin x - \cos x\right)}{\left(\sin x + \cos x\right)\left(\sin^2 x + \cos^2 x - \sin x \cos x\right)} =$

$\dfrac{\sin x - \cos x}{1 - \sin x \cos x} = \dfrac{\sin x - \cos x}{1 - \sin x \cos x}$

59. a. $d^2 = \left(20 + x\cos\theta\right)^2 + \left(20 - x\sin\theta\right)^2$

$= 400 + 40x\cos\theta + x^2 \cos^2\theta$
$\qquad + 400 - 40x\sin\theta + x^2 \sin^2\theta$
$= 800 + 40x\left(\cos\theta - \sin\theta\right)$
$\qquad + x^2\left(\cos^2\theta + \sin^2\theta\right)$
$= 800 + 40x\left(\cos\theta - \sin\theta\right) + x^2$

b. $d^2 = \left(20 + x\cos\theta\right)^2 + \left(20 - x\sin\theta\right)^2$

The distance between the first row and the 8th row is $3 \cdot 7 = 21$ feet.

$d^2 = \left(20 + 21\cos 18°\right)^2$
$\qquad + \left(20 - 21\sin 18°\right)^2$
$d^2 \approx 1{,}780.313$
$d \approx 42.2$ ft

61. a. $h^2 = \left(\sqrt{\cot x}\right)^2 + \left(\sqrt{\tan x}\right)^2$

$h^2 = \cot x + \tan x$
$h = \sqrt{\cot x + \tan x}$
$h = \sqrt{\cot 1.5 + \tan 1.5}$
$h \approx 3.76$ units

b. $h^2 = \cot x + \tan x$

$h^2 = \dfrac{\cos x}{\sin x} + \dfrac{\sin x}{\cos x}$

$h^2 = \dfrac{\cos^2 x + \sin^2 x}{\sin x \cos x}$

$h^2 = \dfrac{1}{\sin x \cos x}$

$h^2 = \dfrac{1}{\sin x} \cdot \dfrac{1}{\cos x}$

$h^2 = \csc x \sec x$

$h = \sqrt{\csc x \sec x}$;

$h = \sqrt{\csc(1.5)\sec(1.5)}$

$h \approx 3.76$ units ; yes

63. Using Pythagorean Theorem:

$D^2 = \left(20 + x\cos\theta\right)^2 + \left(x\sin\theta\right)^2$
$D^2 = 400 + 40x\cos\theta + x^2 \cos^2\theta + x^2 \sin^2\theta$
$D^2 = 400 + 40x\cos\theta + x^2\left(\cos^2\theta + \sin^2\theta\right)$
$D^2 = 400 + 40x\cos\theta + x^2$

The opposite side of θ can be represented by $x\sin\theta$, which is equivalent to the base of the triangle that contains side D.

$D^2 = 400 + 40\left(21\right)\cos\left(18°\right) + 21^2$
$D^2 = 1639.89$
$D = 40.5$ ft

65. $\sin\alpha = \dfrac{I_1 \cos\theta}{\sqrt{\left(I_1 \cos\theta\right)^2 + \left(I_2 \sin\theta\right)^2}}$

$\sin\alpha = \dfrac{I_1 \cos\theta}{\sqrt{\left(I_1 \cos\theta\right)^2 + \left(I_1 \sin\theta\right)^2}}$

$\sin\alpha = \dfrac{I_1 \cos\theta}{\sqrt{I_1^2 \cos^2\theta + I_1^2 \sin^2\theta}}$

$= \dfrac{I_1 \cos\theta}{\sqrt{I_1^2\left(\cos^2\theta + \sin^2\theta\right)}}$

$= \dfrac{I_1 \cos\theta}{\sqrt{I_1^2\left(1\right)}}$

$= \dfrac{I_1 \cos\theta}{I_1}$

$= \cos\theta$

67. Answers will vary.

69. $\sin^4 x + 2\sin^2 x \cos^2 x + \cos^4 x = 1$

$\left(\sin^2 x + \cos^2 x\right)^2 = 1$

$\left(1\right)^2 = 1$

$1 = 1$

71. $\left(\dfrac{\sqrt{7}}{4}\right)^2 + \left(\dfrac{3}{4}\right)^2$

$= \dfrac{7}{16} + \dfrac{9}{16}$

$= \dfrac{16}{16}$

$= 1;$

$\sin t = \dfrac{3}{4}; \cos t = \dfrac{\sqrt{7}}{4}; \tan t = \dfrac{3}{\sqrt{7}}$

73. $f(x) = -2|x-3| + 6$

Right 3, Reflected in x axis, stretched by a factor of 2, Up 6

Vertex: $(3,6)$

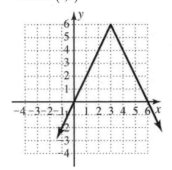

6.3 Exercises

1. False; QII

3. Repeat, opposite

5. Answers will vary.

7. $\cos 105° = \cos(45° + 60°)$

$= \cos 45° \cos 60° - \sin 45° \sin 60°$

$= \left(\dfrac{\sqrt{2}}{2}\right)\left(\dfrac{1}{2}\right) - \left(\dfrac{\sqrt{2}}{2}\right)\left(\dfrac{\sqrt{3}}{2}\right) = \dfrac{\sqrt{2}-\sqrt{6}}{4}$

9. $\cos\left(\dfrac{7\pi}{12}\right) = \cos\left(\dfrac{\pi}{3} + \dfrac{\pi}{4}\right)$

$= \cos\left(\dfrac{\pi}{3}\right)\cos\left(\dfrac{\pi}{4}\right) - \sin\left(\dfrac{\pi}{3}\right)\sin\left(\dfrac{\pi}{4}\right)$

$= \dfrac{1}{2}\left(\dfrac{\sqrt{2}}{2}\right) - \left(\dfrac{\sqrt{3}}{2}\right)\left(\dfrac{\sqrt{2}}{2}\right) = \dfrac{\sqrt{2}-\sqrt{6}}{4}$

11. a. $\cos(45° + 30°)$

$= \cos 45° \cos 30° - \sin 45° \sin 30°$

$= \dfrac{\sqrt{2}}{2} \cdot \dfrac{\sqrt{3}}{2} - \dfrac{\sqrt{2}}{2} \cdot \dfrac{1}{2}$

$= \dfrac{\sqrt{6} - \sqrt{2}}{4}$

b. $\cos(120° - 45°)$

$= \cos 120° \cos 45° + \sin 120° \sin 45°$

$= \dfrac{-1}{2} \cdot \dfrac{\sqrt{2}}{2} + \dfrac{\sqrt{3}}{2} \cdot \dfrac{\sqrt{2}}{2}$

$= \dfrac{-\sqrt{2} + \sqrt{6}}{4}$

$= \dfrac{\sqrt{6} - \sqrt{2}}{4}$

13. $\cos(7\theta)\cos(2\theta) + \sin(7\theta)\sin(2\theta)$

$= \cos(7\theta - 2\theta)$

$= \cos(5\theta)$

15. $\cos 183° \cos 153° + \sin 183° \sin 153°$

$= \cos(183° - 153°)$

$= \cos(30°)$

$= \dfrac{\sqrt{3}}{2}$

17. $\sin \alpha = \dfrac{-4}{5}, \tan \beta = \dfrac{-5}{12}$

$\sqrt{5^2 - 4^2} = 3; \cos \alpha = \dfrac{3}{5};$

$\sqrt{5^2 + 12^2} = 13; \sin \beta = \dfrac{5}{13}, \cos \beta = \dfrac{-12}{13};$

$\cos(\alpha + \beta) = \cos \alpha \cos \beta - \sin \alpha \sin \beta$

$= \dfrac{3}{5} \cdot \left(\dfrac{-12}{13}\right) - \left(\dfrac{-4}{5}\right)\left(\dfrac{5}{13}\right)$

$= \dfrac{-36}{65} + \dfrac{20}{65} = \dfrac{-16}{65}$

19. $\cos 57° = \sin(90° - 57°) = \sin 33°$

Recall: $\left(\sin\left(\dfrac{\pi}{2}\right) - t\right) = \cos t$

21. $\tan\left(\dfrac{5\pi}{12}\right) = \cot\left(\dfrac{\pi}{2} - \dfrac{5\pi}{12}\right) = \cot\left(\dfrac{\pi}{12}\right)$

    Recall: $\left(\cot\left(\dfrac{\pi}{2}\right) - t\right) = \tan t$

23. $\sin\left(\dfrac{\pi}{6} - \theta\right) = \cos\left(\dfrac{\pi}{2} - \left(\dfrac{\pi}{6} - \theta\right)\right)$

    $= \cos\left(\dfrac{\pi}{2} - \dfrac{\pi}{6} + \theta\right)$

    $= \cos\left(\dfrac{\pi}{3} + \theta\right)$

25. $\sin(3x)\cos(5x) + \cos(3x)\sin(5x)$

    $= \sin(3x + 5x)$

    $= \sin(8x)$

27. $\dfrac{\tan(5\theta) - \tan(2\theta)}{1 + \tan(5\theta)\tan(2\theta)}$

    $= \tan(5\theta - 2\theta)$

    $= \tan(3\theta)$

29. $\sin 137° \cos 47° - \cos 137° \sin 47°$

    $= \sin(137° - 47°)$

    $= \sin 90°$

    $= 1$

31. $\dfrac{\tan\left(\dfrac{11\pi}{21}\right) - \tan\left(\dfrac{4\pi}{21}\right)}{1 + \tan\left(\dfrac{11\pi}{21}\right)\tan\left(\dfrac{4\pi}{21}\right)}$

    $= \tan\left(\dfrac{11\pi}{21} - \dfrac{4\pi}{21}\right) = \tan\left(\dfrac{\pi}{3}\right) = \sqrt{3}$

33. $\cos\alpha = \dfrac{-7}{25}$, $\cot\beta = \dfrac{15}{8}$;

    $y = \sqrt{25^2 - 7^2} = 24$; $h = \sqrt{15^2 + 8^2} = 17$;

    $\sin\alpha = \dfrac{24}{25}$; $\sin\beta = \dfrac{-8}{17}$; $\cos\beta = \dfrac{-15}{17}$;

    $\tan\alpha = \dfrac{-24}{7}$; $\tan\beta = \dfrac{8}{15}$

    a. $\sin(\alpha + \beta) = \sin\alpha\cos\beta + \cos\alpha\sin\beta$

    $= \left(\dfrac{24}{25}\right)\left(\dfrac{-15}{17}\right) + \left(\dfrac{-7}{25}\right)\left(\dfrac{-8}{17}\right) = \dfrac{-304}{425}$

    b. $\tan(\alpha + \beta) = \dfrac{\tan\alpha + \tan\beta}{1 - \tan\alpha\tan\beta}$

    $= \dfrac{\dfrac{-24}{7} + \dfrac{8}{15}}{1 - \left(\dfrac{-24}{7}\right)\left(\dfrac{8}{15}\right)} = \dfrac{-304}{297}$

35. $\sin 105° = \sin(45° + 60°)$

    $= \sin 45° \cos 60° + \cos 45° \sin 60°$

    $= \left(\dfrac{\sqrt{2}}{2}\right)\left(\dfrac{1}{2}\right) + \left(\dfrac{\sqrt{2}}{2}\right)\left(\dfrac{\sqrt{3}}{2}\right)$

    $= \dfrac{\sqrt{2} + \sqrt{6}}{4} = \dfrac{\sqrt{6} + \sqrt{2}}{4}$

37. $\sin\left(\dfrac{5\pi}{12}\right) = \sin\left(\dfrac{\pi}{6} + \dfrac{\pi}{4}\right)$

    $= \sin\left(\dfrac{\pi}{6}\right)\cos\left(\dfrac{\pi}{4}\right) + \cos\left(\dfrac{\pi}{6}\right)\sin\left(\dfrac{\pi}{4}\right)$

    $= \left(\dfrac{1}{2}\right)\left(\dfrac{\sqrt{2}}{2}\right) + \dfrac{\sqrt{3}}{2}\left(\dfrac{\sqrt{2}}{2}\right)$

    $= \dfrac{\sqrt{2} + \sqrt{6}}{4} = \dfrac{\sqrt{6} + \sqrt{2}}{4}$

6.3 Exercises

39. $\tan 150° = \tan(180° - 30°)$

$$= \frac{\tan 180° - \tan 30°}{1 + \tan 180° \tan 30°}$$

$$= \frac{0 - \dfrac{\sqrt{3}}{3}}{1 + 0\left(\dfrac{\sqrt{3}}{3}\right)} = \frac{-\sqrt{3}}{3}$$

41. $\tan\left(\dfrac{2\pi}{3}\right) = \tan\left(\dfrac{\pi}{3} + \dfrac{\pi}{3}\right)$

$$= \frac{\tan\left(\dfrac{\pi}{3}\right) + \tan\left(\dfrac{\pi}{3}\right)}{1 - \tan\left(\dfrac{\pi}{3}\right)\tan\left(\dfrac{\pi}{3}\right)}$$

$$= \frac{\sqrt{3} + \sqrt{3}}{1 - \left(\sqrt{3}\right)\left(\sqrt{3}\right)} = \frac{2\sqrt{3}}{-2} = -\sqrt{3}$$

43. a. $\sin(45° - 30°)$

$$= \sin 45° \cos 30° - \cos 45° \sin 30°$$

$$= \left(\frac{\sqrt{2}}{2}\right)\left(\frac{\sqrt{3}}{2}\right) - \left(\frac{\sqrt{2}}{2}\right)\left(\frac{1}{2}\right)$$

$$= \frac{\sqrt{6} - \sqrt{2}}{4}$$

b. $\sin(135° - 120°)$

$$= \sin 135° \cos 120° - \cos 135° \sin 120°$$

$$= \left(\frac{\sqrt{2}}{2}\right)\left(\frac{-1}{2}\right) - \left(\frac{-\sqrt{2}}{2}\right)\left(\frac{\sqrt{3}}{2}\right)$$

$$= \frac{-\sqrt{2}}{4} + \frac{\sqrt{6}}{4} = \frac{\sqrt{6} - \sqrt{2}}{4}$$

45. Recall from # 35,

$$\cos 105° = \frac{\sqrt{2} - \sqrt{6}}{4}$$

and, $\sin 105° = \dfrac{\sqrt{6} + \sqrt{2}}{4}$;

$\sin 255° = \sin(150° + 105°)$

$= \sin 150° \cos 105° + \cos 150° \sin 105°$

$$= \left(\frac{1}{2}\right)\left(\frac{\sqrt{2} - \sqrt{6}}{4}\right) + \left(\frac{-\sqrt{3}}{2}\right)\left(\frac{\sqrt{6} + \sqrt{2}}{4}\right)$$

$$= \frac{\sqrt{2}}{8} - \frac{\sqrt{6}}{8} - \frac{\sqrt{18}}{8} - \frac{\sqrt{6}}{8}$$

$$= \frac{\sqrt{2}}{8} - \frac{\sqrt{6}}{8} - \frac{3\sqrt{2}}{8} - \frac{\sqrt{6}}{8}$$

$$= \frac{-2\sqrt{2}}{8} - \frac{-2\sqrt{6}}{8} = \frac{-\sqrt{2} - \sqrt{6}}{4}$$

47. $\sin \alpha = \dfrac{12}{13}$, $\tan \beta = \dfrac{35}{12}$;

$x = \sqrt{13^2 - 12^2} = 5$; $h = \sqrt{35^2 + 12^2} = 37$;

$\cos \alpha = \dfrac{5}{13}$; $\tan \alpha = \dfrac{12}{5}$; $\cos \beta = \dfrac{12}{37}$;

$\sin \beta = \dfrac{35}{37}$;

a. $\sin(\alpha + \beta) = \sin \alpha \cos \beta + \cos \alpha \sin \beta$

$$= \frac{12}{13}\left(\frac{12}{37}\right) + \frac{5}{13}\left(\frac{35}{37}\right) = \frac{319}{481}$$

b. $\cos(\alpha - \beta) = \cos \alpha \cos \beta + \sin \alpha \sin \beta$

$$= \frac{5}{13}\left(\frac{12}{37}\right) + \frac{12}{13}\left(\frac{35}{37}\right) = \frac{480}{481}$$

c. $\tan(\alpha + \beta) = \dfrac{\tan \alpha + \tan \beta}{1 - \tan \alpha \tan \beta}$

$$= \frac{\dfrac{12}{5} + \dfrac{35}{12}}{1 - \dfrac{12}{5}\left(\dfrac{35}{12}\right)} = \frac{-319}{360}$$

49. $\sin \alpha = \dfrac{28}{53}, \cos \beta = \dfrac{-13}{85}$;

$x = \sqrt{53^2 - 28^2} = 45; \, y = \sqrt{85^2 - 13^2} = 84$

$\cos \alpha = \dfrac{-45}{53}$; $\tan \alpha = \dfrac{-28}{45}$; $\sin \beta = \dfrac{84}{85}$;

$\tan \beta = \dfrac{-84}{13}$;

a. $\sin(\alpha - \beta) = \sin \alpha \cos \beta - \cos \alpha \sin \beta$

$= \dfrac{28}{53}\left(\dfrac{-13}{85}\right) - \left(\dfrac{-45}{53}\right)\left(\dfrac{84}{85}\right) = \dfrac{3416}{4505}$

b. $\cos(\alpha + \beta) = \cos \alpha \cos \beta - \sin \alpha \sin \beta$

$= \dfrac{-45}{53}\left(\dfrac{-13}{85}\right) - \left(\dfrac{28}{53}\right)\left(\dfrac{84}{85}\right) = \dfrac{-1767}{4505}$

c. $\tan(\alpha - \beta) = \dfrac{\tan \alpha - \tan \beta}{1 + \tan \alpha \tan \beta}$

$= \dfrac{\dfrac{-28}{45} - \dfrac{-84}{13}}{1 + \left(\dfrac{-28}{45}\right)\left(\dfrac{-84}{13}\right)} = \dfrac{3416}{2937}$

51. $h = \sqrt{12^2 + 5^2} = 13$

a. $\sin A = \sin(30° + \theta)$

$= \sin 30° \cos \theta + \cos 30° \sin \theta$

$= \dfrac{1}{2}\left(\dfrac{12}{13}\right) + \left(\dfrac{\sqrt{3}}{2}\right)\left(\dfrac{5}{13}\right) = \dfrac{12 + 5\sqrt{3}}{26}$

b. $\cos A = \cos(30° + \theta)$

$= \cos 30° \cos \theta - \sin 30° \sin \theta$

$= \dfrac{\sqrt{3}}{2}\left(\dfrac{12}{13}\right) - \dfrac{1}{2}\left(\dfrac{5}{13}\right) = \dfrac{12\sqrt{3} - 5}{26}$

c. $\tan A = \dfrac{\sin A}{\cos A}$

$= \dfrac{\dfrac{12 + 5\sqrt{3}}{26}}{\dfrac{12\sqrt{3} - 5}{26}} = \dfrac{12 + 5\sqrt{3}}{12\sqrt{3} - 5}$

53. Show $\theta = \alpha + \beta$

Third angle of 1st triangle: $90 - \alpha$

Third angle of the 3rd triangle: $90 - \beta$

Supplementary angles:

$90 - \alpha + \theta + 90 - \beta = 180$

$\theta = \alpha + \beta$;

$h_1 = \sqrt{32^2 + 24^2} = 40$

$h_2 = \sqrt{45^2 + 28^2} = 53$

$\sin \alpha = \dfrac{24}{40}$; $\cos \alpha = \dfrac{32}{40}$; $\sin \beta = \dfrac{28}{53}$;

$\cos \beta = \dfrac{45}{53}$;

a. $\sin \theta = \sin(\alpha + \beta)$

$= \sin \alpha \cos \beta + \cos \alpha \sin \beta$

$= \dfrac{24}{40}\left(\dfrac{45}{53}\right) + \left(\dfrac{32}{40}\right)\left(\dfrac{28}{53}\right) = \dfrac{247}{265}$

b. $\cos \theta = \cos(\alpha + \beta)$

$= \cos \alpha \cos \beta - \sin \alpha \sin \beta$

$= \left(\dfrac{32}{40}\right)\left(\dfrac{45}{53}\right) - \left(\dfrac{24}{40}\right)\left(\dfrac{28}{53}\right) = \dfrac{96}{265}$

c. $\tan \theta = \dfrac{\sin \theta}{\cos \theta} = \dfrac{\dfrac{247}{265}}{\dfrac{96}{265}} = \dfrac{247}{96}$

55. $\sin(\pi - \alpha) = \sin \alpha$
$\sin \pi \cos \alpha - \cos \pi \sin \alpha =$
$0 \cos \alpha - (-1)\sin \alpha =$
$\sin \alpha = \sin \alpha$

57. $\cos\left(x + \dfrac{\pi}{4}\right) = \dfrac{\sqrt{2}}{2}(\cos x - \sin x)$

$\cos x \cos\left(\dfrac{\pi}{4}\right) - \sin x \sin\left(\dfrac{\pi}{4}\right) =$

$(\cos x)\left(\dfrac{\sqrt{2}}{2}\right) - (\sin x)\left(\dfrac{\sqrt{2}}{2}\right) =$

$\dfrac{\sqrt{2}}{2}(\cos x - \sin x) = \dfrac{\sqrt{2}}{2}(\cos x - \sin x)$

59. $\tan\left(x + \dfrac{\pi}{4}\right) = \dfrac{1 + \tan x}{1 - \tan x}$

$\dfrac{\tan x + \tan\left(\dfrac{\pi}{4}\right)}{1 - \tan x \tan\left(\dfrac{\pi}{4}\right)} =$

$\dfrac{\tan x + 1}{1 - \tan x} =$

$\dfrac{1 + \tan x}{1 - \tan x} = \dfrac{1 + \tan x}{1 - \tan x}$

61. $\cos(\alpha + \beta) + \cos(\alpha - \beta) = 2\cos \alpha \cos \beta$
$\cos \alpha \cos \beta - \sin \alpha \sin \beta$
$\quad + \cos \alpha \cos \beta + \sin \alpha \sin \beta =$
$2\cos \alpha \cos \beta = 2\cos \alpha \cos \beta$

63. $\cos(2t) = \cos^2 t - \sin^2 t$
$\cos(t + t) =$
$\cos t \cos t - \sin t \sin t =$
$\cos^2 t - \sin^2 t = \cos^2 t - \sin^2 t$

65. $\sin(3t) = -4\sin^3 t + 3\sin t$
$\sin(2t + t) =$
$\sin(2t)\cos t + \cos(2t)\sin t =$
$(2\sin t \cos t)\cos t + (\cos^2 t - \sin^2 t)\sin t =$
$2\sin t \cos^2 t + \sin t \cos^2 t - \sin^3 t =$
$3\sin t \cos^2 t - \sin^3 t =$
$3\sin t(1 - \sin^2 t) - \sin^3 t =$
$3\sin t - 3\sin^3 t - \sin^3 t =$
$-4\sin^3 t + 3\sin t = -4\sin^3 t + 3\sin t$

67. $\cos\left(x - \dfrac{\pi}{4}\right) = \dfrac{\sqrt{2}}{2}(\cos x + \sin x)$

$\cos x \cos\left(\dfrac{\pi}{4}\right) + \sin x \sin\left(\dfrac{\pi}{4}\right) =$

$\cos x\left(\dfrac{\sqrt{2}}{2}\right) + \sin x\left(\dfrac{\sqrt{2}}{2}\right) =$

$\dfrac{\sqrt{2}}{2}(\cos x + \sin x) = \dfrac{\sqrt{2}}{2}(\cos x + \sin x)$

69. $F = \dfrac{Wk}{c}\tan(p-\theta)$

$F = \dfrac{Wk}{c}\tan\left(\dfrac{\pi}{6}-\dfrac{\pi}{4}\right)$

$F = \dfrac{Wk}{c}\cdot\dfrac{\tan\left(\dfrac{\pi}{6}\right)-\tan\left(\dfrac{\pi}{4}\right)}{1+\tan\left(\dfrac{\pi}{6}\right)\tan\left(\dfrac{\pi}{4}\right)}$

$F = \dfrac{Wk}{c}\cdot\dfrac{\dfrac{1}{\sqrt{3}}-1}{1+\left(\dfrac{1}{\sqrt{3}}\right)(1)}$

$F = \dfrac{Wk}{c}\cdot\dfrac{1-\sqrt{3}}{\sqrt{3}+1}$

$F = \dfrac{Wk}{c}\cdot\dfrac{1-\sqrt{3}}{1+\sqrt{3}}$

71. $R = \dfrac{\cos s\cos t}{\omega C\sin(s+t)}$

$= \dfrac{\cos s\cos t}{\omega C(\sin s\cos t+\cos s\sin t)}$

$= \dfrac{\cos s\cos t\cdot\dfrac{1}{\cos s\cos t}}{\omega C(\sin s\cos t+\cos s\sin t)\cdot\dfrac{1}{\cos s\cos t}}$

$= \dfrac{1}{\omega C\left(\dfrac{\sin s\cos t}{\cos s\cos t}+\dfrac{\cos s\sin t}{\cos s\cos t}\right)}$

$= \dfrac{1}{\omega C\left(\dfrac{\sin s}{\cos s}+\dfrac{\sin t}{\cos t}\right)}$

$= \dfrac{1}{\omega C(\tan s+\tan t)}$

73. $\dfrac{A}{B} = \dfrac{\tan\theta}{\tan(90°-\theta)}$

$\dfrac{A}{B} = \tan\theta\cdot\dfrac{1}{\tan(90°-\theta)}$

$\dfrac{A}{B} = \dfrac{\sin\theta}{\cos\theta}\cdot\dfrac{1}{\dfrac{\sin(90°-\theta)}{\cos(90°-\theta)}}$

$\dfrac{A}{B} = \dfrac{\sin\theta}{\cos\theta}\cdot\dfrac{\cos(90°-\theta)}{\sin(90°-\theta)}$

$\dfrac{A}{B} = \dfrac{\sin\theta(\cos90°\cos\theta+\sin90°\sin\theta)}{\cos\theta(\sin90°\cos\theta-\cos90°\sin\theta)}$

$\dfrac{A}{B} = \dfrac{\sin\theta(0\cos\theta+1\sin\theta)}{\cos\theta(1\cos\theta-0\sin\theta)}$

$\dfrac{A}{B} = \dfrac{\sin\theta(\sin\theta)}{\cos\theta(\cos\theta)}$

$\dfrac{A}{B} = \dfrac{\sin^2\theta}{\cos^2\theta}$

$\dfrac{A}{B} = \tan^2\theta$

75. $P(t) = A\sin(2\pi\ ft)+A\sin\left(2\pi\ ft+\dfrac{\pi}{2}\right)$

$= A\left[\sin(2\pi\ ft)+\sin\left(2\pi\ ft+\dfrac{\pi}{2}\right)\right]$

$= A\left[\sin(2\pi\ ft)+\cos(2\pi\ ft)\right]$

Verified using sum identity for sine

77. $f(x) = \sin x$

$$\frac{f(x+h) - f(x)}{h} = \frac{\sin(x+h) - \sin x}{h}$$

$$= \frac{\sin x \cos h + \cos x \sin h - \sin x}{h}$$

$$= \frac{\sin x \cos h - \sin x + \cos x \sin h}{h}$$

$$= \frac{\sin x(\cos h - 1) + \cos x \sin h}{h}$$

$$= \sin x \left(\frac{\cos h - 1}{h}\right) + \cos x \frac{(\sin h)}{h}$$

79. $\cos 1665°; \dfrac{1665}{360} = 4\dfrac{5}{8}$

4 multiples of $360°$, with

$\dfrac{5}{8}(360°) = 225°$ remaining

$\cos 1665° = \cos(225° + 360°(4))$

$= \cos(225°) = \dfrac{-\sqrt{2}}{2}$

81. $\sin\left(\dfrac{41\pi}{6}\right); \dfrac{41\pi}{6} \div 2\pi = \dfrac{41}{12} = 3\dfrac{5}{12}$

3 multiples of 2π, with

$\dfrac{5}{12}(2\pi) = \dfrac{5\pi}{6}$ remaining

$\sin\left(\dfrac{41\pi}{6}\right) = \sin\left(\dfrac{5\pi}{6} + 2\pi(3)\right)$

$= \sin\left(\dfrac{5\pi}{6}\right) = \dfrac{1}{2}$

83. $D = d$, so $D^2 = d^2$

Using the distance formula between the points $(\cos\alpha, \sin\alpha)$ and $(\cos\beta, \sin\beta)$:

$D^2 = (\cos\alpha - \cos\beta)^2 + (\sin\alpha - \sin\beta)^2$

$\begin{aligned} D^2 &= \cos^2\alpha - 2\cos\alpha\cos\beta + \cos^2\beta \\ &\quad + \sin^2\alpha - 2\sin\alpha\sin\beta + \sin^2\beta \end{aligned}$

$\begin{aligned} D^2 &= \cos^2\alpha + \sin^2\alpha + \cos^2\beta + \sin^2\beta \\ &\quad -2\cos\alpha\cos\beta - 2\sin\alpha\sin\beta \end{aligned}$

$D^2 = 2 - 2\cos\alpha\cos\beta - 2\sin\alpha\sin\beta$;

Using the distance formula between the points $(\cos(\alpha - \beta), \sin(\alpha - \beta))$ and $(1,0)$:

$d^2 = \sin^2(\alpha - \beta) + [\cos(\alpha - \beta) - 1]^2$

$\begin{aligned} d^2 &= \sin^2(\alpha - \beta) + \cos^2(\alpha - \beta) \\ &\quad - 2\cos(\alpha - \beta) + 1 \end{aligned}$

$d^2 = 1 - 2\cos(\alpha - \beta) + 1$

$d^2 = 2 - 2\cos(\alpha - \beta);$

$D^2 = d^2$ so

$2 - 2\cos\alpha\cos\beta - 2\sin\alpha\sin\beta = 2 - 2\cos(\alpha - \beta)$

$-2\cos\alpha\cos\beta - 2\sin\alpha\sin\beta = -2\cos(\alpha - \beta)$

$\dfrac{-2\cos\alpha\cos\beta - 2\sin\alpha\sin\beta}{-2} = \dfrac{-2\cos(\alpha - \beta)}{-2}$

$\cos\alpha\cos\beta + \sin\alpha\sin\beta = \cos(\alpha - \beta)$

85. a. $y = 3\sin\left(\dfrac{\pi}{8}x - \dfrac{\pi}{3}\right)$

Period $= \dfrac{2\pi}{\dfrac{\pi}{8}} = 2\pi \cdot \dfrac{8}{\pi} = 16$

The graph will go through $\dfrac{\pi}{8}$ of a complete cycle every 2π units.

b. $y = 4\tan\left(2x + \dfrac{\pi}{4}\right)$

Period $= \dfrac{\pi}{2}$

The graph will go through 2 times of a complete cycle every π units.

87. $\sin 40° = \dfrac{h}{30}$

$30\sin 40° = h$

$19.3\,\text{ft} \approx h$

Chapter 6: Trigonometric Identities, Inverses and Equations

6.4 Exercises

1. Sum, $\alpha = \beta$

3. $2x, x$

5. Answers will vary.

7. $\sin \theta = \dfrac{5}{13}$; θ in QII

$\cos(2\theta) = 1 - 2\sin^2\theta$

$= 1 - 2\left(\dfrac{5}{13}\right)^2 = 1 - \dfrac{50}{169}$

$\cos(2\theta) = \dfrac{119}{169}$;

$\dfrac{119}{169} = 2\cos^2\theta - 1$

$\dfrac{288}{169} = 2\cos^2\theta$

$\dfrac{144}{169} = \cos^2\theta$

$\pm \dfrac{12}{13} = \cos\theta$;

Since QII, $\cos\theta = \dfrac{-12}{13}$;

$\sin(2\theta) = 2\sin\theta\cos\theta$

$= 2\left(\dfrac{5}{13}\right)\left(\dfrac{-12}{13}\right)$

$\sin(2\theta) = \dfrac{-120}{169}$;

$\tan(2\theta) = \dfrac{2\tan\theta}{1 - \tan^2\theta}$

$= \dfrac{2\left(\dfrac{-5}{12}\right)}{1 - \left(\dfrac{-5}{12}\right)^2} = \dfrac{\dfrac{-10}{12}}{1 - \dfrac{25}{144}} = \dfrac{\dfrac{-10}{12}}{\dfrac{119}{144}} = \dfrac{-120}{119}$

9. $\cos\theta = \dfrac{-9}{41}$; θ in QII

$\cos(2\theta) = 2\cos^2\theta - 1$

$= 2\left(\dfrac{-9}{41}\right)^2 - 1 = 2\left(\dfrac{81}{1681}\right) - 1 = \dfrac{162}{1681} - 1$

$\cos(2\theta) = \dfrac{-1519}{1681}$;

$\cos(2\theta) = 1 - 2\sin^2\theta$

$\dfrac{-1519}{1681} = 1 - 2\sin^2\theta$

$\dfrac{-3200}{1681} = -2\sin^2\theta$

$\dfrac{1600}{1681} = \sin^2\theta$

$\pm \dfrac{40}{41} = \sin\theta$, QII $\to + \dfrac{40}{41}$;

$\sin(2\theta) = 2\sin\theta\cos\theta$

$= 2\left(\dfrac{40}{41}\right)\left(\dfrac{-9}{41}\right)$

$\sin(2\theta) = \dfrac{-720}{1681}$;

$\tan(2\theta) = \dfrac{2\tan\theta}{1 - \tan^2\theta}$

$= \dfrac{2\left(\dfrac{40}{-9}\right)}{1 - \left(\dfrac{40}{-9}\right)^2} = \dfrac{-\dfrac{80}{9}}{1 - \dfrac{1600}{81}} = \dfrac{-\dfrac{80}{9}}{\dfrac{-1519}{81}} = \dfrac{720}{1519}$

295

11. $\tan \theta = \dfrac{13}{84}$, θ in QIII

$\tan(2\theta) = \dfrac{2\tan\theta}{1-\tan^2\theta}$

$= \dfrac{2\left(\dfrac{13}{84}\right)}{1-\left(\dfrac{13}{84}\right)^2} = \dfrac{\dfrac{13}{42}}{\dfrac{6887}{7056}} = \dfrac{2184}{6887}$;

Using the identity

$1+\tan^2\theta = \sec^2\theta$

$1+\left(\dfrac{13}{84}\right)^2 = \sec^2\theta$

$\dfrac{7225}{7056} = \sec^2\theta$

$\pm\dfrac{85}{84} = \sec\theta$, QIII $\rightarrow \dfrac{-85}{84}$

And $\cos\theta = \dfrac{1}{\sec\theta}$

$= \dfrac{1}{\dfrac{-85}{84}} = \dfrac{-84}{85}$;

$\cos(2\theta) = 2\cos^2\theta - 1$

$= 2\left(\dfrac{-84}{85}\right)^2 - 1 = \dfrac{6887}{7225}$;

$\cos(2\theta) = 1 - 2\sin^2\theta$

$\dfrac{6887}{7225} = 1 - 2\sin^2\theta$

$-\dfrac{338}{7225} = -2\sin^2\theta$

$\dfrac{169}{7225} = \sin^2\theta$

$\pm\dfrac{13}{85} = \sin\theta$; QIII $\rightarrow -\dfrac{13}{85}$;

$\sin(2\theta) = 2\sin\theta\cos\theta$

$= 2\left(\dfrac{-13}{85}\right)\left(\dfrac{-84}{85}\right) = \dfrac{2184}{7225}$

13. $\sin \theta = \dfrac{48}{73}$; $\cos\theta < 0$, so θ in QII

$\cos(2\theta) = 1 - 2\sin^2\theta = 1 - 2\left(\dfrac{48}{73}\right)^2 = \dfrac{721}{5329}$

$\cos(2\theta) = 2\cos^2\theta - 1$

$\dfrac{721}{5329} = 2\cos^2\theta - 1$

$\dfrac{6050}{5329} = 2\cos^2\theta$

$\dfrac{3025}{5329} = \cos^2\theta$

$\pm\dfrac{55}{73} = \cos\theta$, QIII $\rightarrow \dfrac{-55}{73}$;

$\sin(2\theta) = 2\sin\theta\cos\theta$

$= 2\left(\dfrac{48}{73}\right)\left(\dfrac{-55}{73}\right) = \dfrac{-5280}{5329}$

$\tan(2\theta) = \dfrac{2\tan\theta}{1-\tan^2\theta}$

$= \dfrac{2\left(\dfrac{48}{-55}\right)}{1-\left(\dfrac{48}{-55}\right)^2} = \dfrac{\dfrac{96}{-55}}{\dfrac{721}{3025}} = -\dfrac{5280}{721}$

15. $\csc\theta = \dfrac{5}{3}$; $\sec\theta < 0$, so θ in QII

$\sin(\theta) = \dfrac{1}{\csc\theta} = \dfrac{1}{\frac{5}{3}}$; $\sin\theta = \dfrac{3}{5}$;

$\cos(2\theta) = 1 - 2\sin^2\theta = 1 - 2\left(\dfrac{3}{5}\right)^2 = \dfrac{7}{25}$;

$\cos(2\theta) = 2\cos^2\theta - 1$

$\dfrac{7}{25} = 2\cos^2\theta - 1$

$\dfrac{32}{25} = 2\cos^2\theta$

$\dfrac{16}{25} = \cos^2\theta$

$\pm\dfrac{4}{5} = \cos\theta$, QII $\rightarrow \dfrac{-4}{5}$;

$\sin(2\theta) = 2\sin\theta\cos\theta$

$\quad = 2\left(\dfrac{3}{5}\right)\left(\dfrac{-4}{5}\right) = \dfrac{-24}{25}$;

$\tan(2\theta) = \dfrac{2\tan\theta}{1 - \tan^2\theta}$

$\quad = \dfrac{2\left(\dfrac{3}{-4}\right)}{1 - \left(\dfrac{3}{-4}\right)^2} = \dfrac{\dfrac{-3}{2}}{\dfrac{7}{16}} = -\dfrac{24}{7}$

17. $\sin(2\theta) = \dfrac{24}{25}$; 2θ in QII

$\sqrt{25^2 - 24^2} = 7$; $\cos(2\theta) = \dfrac{-7}{25}$, (QII)

For 2θ in QII we have

$\dfrac{\pi}{2} < 2\theta < \pi$

$\dfrac{\pi}{4} < \theta < \dfrac{\pi}{2}$, θ in QI

$\sin\left(\dfrac{2\theta}{2}\right) = +\sqrt{\dfrac{1 - \cos 2\theta}{2}}$

$\sin\theta = \sqrt{\dfrac{1 - \left(\dfrac{-7}{25}\right)}{2}} = \sqrt{\dfrac{\frac{32}{25}}{2}} = \dfrac{4}{5}$;

$\cos\left(\dfrac{2\theta}{2}\right) = +\sqrt{\dfrac{1 + \cos 2\theta}{2}}$

$\cos\theta = \sqrt{\dfrac{1 + \left(\dfrac{-7}{25}\right)}{2}} = \sqrt{\dfrac{\frac{18}{25}}{2}} = \dfrac{3}{5}$;

$\tan\theta = \dfrac{\frac{4}{5}}{\frac{3}{5}} = \dfrac{4}{3}$

19. $\cos(2\theta) = -\dfrac{41}{841}$; 2θ in QII

For 2θ in QII, we have

$\dfrac{\pi}{2} < 2\theta < \pi$

$\dfrac{\pi}{4} < \theta < \dfrac{\pi}{2}$, θ in QI

$\cos(2\theta) = 2\cos^2\theta - 1$

$\dfrac{-41}{841} = 2\cos^2\theta - 1$

$\dfrac{800}{841} = 2\cos^2\theta$

$\dfrac{400}{841} = \cos^2\theta$

$\pm\dfrac{20}{29} = \cos\theta$, QI $\rightarrow \dfrac{20}{29}$;

$\cos(2\theta) = 1 - 2\sin^2\theta$

$\dfrac{-41}{841} = 1 - 2\sin^2\theta$

$\dfrac{-882}{841} = -2\sin^2\theta$

$\dfrac{441}{841} = \sin^2\theta$

$\pm\dfrac{21}{29} = \sin\theta$, QI $\rightarrow \dfrac{21}{29}$;

$\tan\theta = \dfrac{\frac{21}{29}}{\frac{20}{29}} = \dfrac{21}{20}$

21. $\sin(3\theta) = 3\sin\theta - 4\sin^3\theta$

$\sin(2\theta + \theta) =$

$\sin(2\theta)\cos\theta + \cos(2\theta)\sin\theta =$

$(2\sin\theta\cos\theta)\cos\theta +$

$\quad (\cos^2\theta - \sin^2\theta)\sin\theta =$

$2\sin\theta\cos^2\theta + \cos^2\theta\sin\theta - \sin^3\theta =$

$2\sin\theta(1 - \sin^2\theta)$

$\quad + (1 - \sin^2\theta)\sin\theta - \sin^3\theta =$

$2\sin\theta - 2\sin^3\theta + \sin\theta - \sin^3\theta - \sin^3\theta =$

$3\sin\theta - 4\sin^3\theta = 3\sin\theta - 4\sin^3\theta$

Verified.

23. $\cos 75°\sin 75°$

$\cos\alpha\sin\beta$

$= \frac{1}{2}[\sin(\alpha + \beta) - \sin(\alpha - \beta)]$

$\cos 75°\sin 75°$

$= \frac{1}{2}[\sin(75° + 75°) - \sin(75° - 75°)]$

$= \frac{1}{2}[\sin(150°) - \sin(0°)] = \frac{1}{2}\left[\frac{1}{2} - 0\right] = \frac{1}{4}$

25. $1 - 2\sin^2\left(\frac{\pi}{8}\right)$

$\cos(2\theta) = 1 - 2\sin^2\theta$

$\cos\left(\frac{\pi}{4}\right) = 1 - 2\sin^2\left(\frac{\pi}{8}\right) = \frac{\sqrt{2}}{2}$

27. $\dfrac{2\tan 22.5°}{1 - \tan^2 22.5°}$

$\tan(2\theta) = \dfrac{2\tan\theta}{1 - \tan^2\theta}$

$\tan 45° = \dfrac{2\tan 22.5°}{1 - \tan^2 22.5°} = 1$

29. $9\sin(3x)\cos(3x)$

$= \frac{9}{2}[2\sin(3x)\cos(3x)]$

$= \frac{9}{2}\sin[2(3x)]$

$= \frac{9}{2}\sin(6x)$

$= 4.5\sin(6x)$

31. $\sin^2 x\cos^2 x$

$\dfrac{1 - \cos(2x)}{2}\cdot\dfrac{1 + \cos(2x)}{2} = \dfrac{1 - \cos^2(2x)}{4}$

$= \frac{1}{4}(1 - \cos^2(2x))$

$= \frac{1}{4}\left(1 - \dfrac{1 + \cos(4x)}{2}\right)$

$= \frac{1}{4} - \dfrac{1 + \cos(4x)}{8}$

$= \frac{1}{4} - \frac{1}{8} - \dfrac{\cos(4x)}{8}$

$= \frac{1}{8} - \dfrac{\cos(4x)}{8}$

$= \frac{1}{8} - \frac{1}{8}\cos(4x)$

33. $3\cos^4 x = 3\left[\dfrac{1 + \cos(2x)}{2}\right]^2$

$= \frac{3}{4}[1 + 2\cos(2x) + \cos^2(2x)]$

$= \frac{3}{4}\left[1 + 2\cos(2x) + \dfrac{1 + \cos(4x)}{2}\right]$

$= \frac{3}{4} + \frac{3}{2}\cos(2x) + \frac{3}{8} + \dfrac{3\cos(4x)}{8}$

$= \frac{9}{8} + \frac{3}{2}\cos(2x) + \frac{3}{8}\cos(4x)$

35. $2\sin^6 x = 2\left[\dfrac{1-\cos(2x)}{2}\right]^3$

$= \dfrac{1}{4}[1-\cos(2x)][1-\cos(2x)][1-\cos(2x)]$

$= \dfrac{1}{4}[1-2\cos(2x)+\cos^2(2x)][1-\cos(2x)]$

$= \dfrac{1}{4}\left[1-2\cos(2x)+\dfrac{1+\cos(4x)}{2}\right][1-\cos(2x)]$

$= \dfrac{1}{4}\left[\dfrac{3}{2}-2\cos(2x)+\dfrac{1}{2}\cos(4x)\right][1-\cos(2x)]$

$= \dfrac{1}{4}\Big[\dfrac{3}{2}-2\cos(2x)+\dfrac{1}{2}\cos(4x)-\dfrac{3}{2}\cos(2x)$

$\qquad +2\cos^2(2x)-\dfrac{1}{2}\cos(2x)\cos(4x)\Big]$

$= \dfrac{1}{4}\Big[\dfrac{3}{2}-\dfrac{7}{2}\cos(2x)+\dfrac{1}{2}\cos(4x)+\dfrac{2(1+\cos(4x))}{2}$

$\qquad -\dfrac{1}{2}\cos(2x)\cos(4x)\Big]$

$= \dfrac{1}{4}\Big[\dfrac{5}{2}-\dfrac{7}{2}\cos(2x)+\dfrac{3}{2}\cos(4x)$

$\qquad -\dfrac{1}{2}\cos(2x)\cos(4x)\Big]$

$= \dfrac{5}{8}-\dfrac{7}{8}\cos(2x)+\dfrac{3}{8}\cos(4x)$

$\qquad -\dfrac{1}{8}\cos(2x)\cos(4x)$

37. $\theta = 22.5°$

$\sin 22.5° = \sin\left(\dfrac{45°}{2}\right) = \sqrt{\dfrac{1-\cos 45°}{2}}$

$= \sqrt{\dfrac{1-\dfrac{\sqrt{2}}{2}}{2}} = \dfrac{\sqrt{2-\sqrt{2}}}{2}$;

$\cos 22.5° = \cos\left(\dfrac{45°}{2}\right) = \sqrt{\dfrac{1+\cos 45°}{2}}$

$= \sqrt{\dfrac{1+\dfrac{\sqrt{2}}{2}}{2}} = \dfrac{\sqrt{2+\sqrt{2}}}{2}$;

$\tan 22.5° = \tan\left(\dfrac{45°}{2}\right) = \dfrac{1-\cos 45°}{\sin 45°}$

$= \dfrac{1-\dfrac{\sqrt{2}}{2}}{\dfrac{\sqrt{2}}{2}} = \dfrac{2-\sqrt{2}}{\sqrt{2}} = \dfrac{2\sqrt{2}-2}{2} = \sqrt{2}-1$

39. $\theta = \dfrac{\pi}{12}$

$\sin\left(\dfrac{\pi}{12}\right) = \sin\dfrac{\dfrac{\pi}{6}}{2} = \sqrt{\dfrac{1-\cos\left(\dfrac{\pi}{6}\right)}{2}}$

$= \sqrt{\dfrac{1-\dfrac{\sqrt{3}}{2}}{2}} = \dfrac{\sqrt{2-\sqrt{3}}}{2}$;

$\cos\left(\dfrac{\pi}{12}\right) = \cos\dfrac{\dfrac{\pi}{6}}{2} = \sqrt{\dfrac{1+\cos\left(\dfrac{\pi}{6}\right)}{2}}$

$= \sqrt{\dfrac{1+\dfrac{\sqrt{3}}{2}}{2}} = \dfrac{\sqrt{2+\sqrt{3}}}{2}$;

$\tan\left(\dfrac{\pi}{12}\right) = \tan\dfrac{\dfrac{\pi}{6}}{2} = \dfrac{1-\cos\left(\dfrac{\pi}{6}\right)}{\sin\left(\dfrac{\pi}{6}\right)}$

$= \dfrac{1-\dfrac{\sqrt{3}}{2}}{\dfrac{1}{2}} = \dfrac{2-\sqrt{3}}{1} = 2-\sqrt{3}$

41. $\theta = 67.5°$

$\sin 67.5° = \sin\left(\dfrac{135°}{2}\right) = \sqrt{\dfrac{1-\cos 135°}{2}}$

$= \sqrt{\dfrac{1-\dfrac{(-\sqrt{2})}{2}}{2}} = \sqrt{\dfrac{1+\dfrac{\sqrt{2}}{2}}{2}} = \dfrac{\sqrt{2+\sqrt{2}}}{2}$;

$\cos 67.5° = \cos\left(\dfrac{135°}{2}\right) = \sqrt{\dfrac{1+\cos 135°}{2}}$

$= \sqrt{\dfrac{1+\left(\dfrac{-\sqrt{2}}{2}\right)}{2}} = \dfrac{\sqrt{2-\sqrt{2}}}{2}$;

$\tan 67.5° = \tan\left(\dfrac{135°}{2}\right) = \dfrac{1-\cos 135°}{\sin 135°}$

$= \dfrac{1-\left(\dfrac{-\sqrt{2}}{2}\right)}{\dfrac{\sqrt{2}}{2}} = \dfrac{2+\sqrt{2}}{\sqrt{2}} = \dfrac{2\sqrt{2}+2}{2}$

$= \sqrt{2}+1$

6.4 Exercises

43. $\theta = \dfrac{3\pi}{8}$

$$\sin\left(\frac{3\pi}{8}\right) = \sin\left(\frac{\frac{3\pi}{4}}{2}\right) = \sqrt{\frac{1-\left(\frac{-\sqrt{2}}{2}\right)}{2}}$$

$$= \sqrt{\frac{1+\frac{\sqrt{2}}{2}}{2}} = \frac{\sqrt{2+\sqrt{2}}}{2};$$

$$\cos\left(\frac{3\pi}{8}\right) = \cos\left(\frac{\frac{3\pi}{4}}{2}\right) = \sqrt{\frac{1+\left(\frac{-\sqrt{2}}{2}\right)}{2}}$$

$$= \sqrt{\frac{1-\frac{\sqrt{2}}{2}}{2}} = \frac{\sqrt{2-\sqrt{2}}}{2};$$

$$\tan\left(\frac{3\pi}{8}\right) = \tan\left(\frac{\frac{3\pi}{4}}{2}\right) = \frac{1-\left(\frac{-\sqrt{2}}{2}\right)}{\frac{\sqrt{2}}{2}}$$

$$= \frac{1+\frac{\sqrt{2}}{2}}{\frac{\sqrt{2}}{2}} = \frac{2+\sqrt{2}}{\sqrt{2}} = \sqrt{2}+1$$

45. $\sin 11.25° = \sin\left(\dfrac{22.5°}{2}\right)$

$$= \sqrt{\frac{1-\cos 22.5°}{2}} = \frac{\sqrt{2-\sqrt{2+\sqrt{2}}}}{2}$$

Recall from problem #37:

$$\cos 22.5° = \cos\left(\frac{45°}{2}\right) = \sqrt{\frac{1+\cos 45°}{2}}$$

$$= \sqrt{\frac{1+\frac{\sqrt{2}}{2}}{2}} = \frac{\sqrt{2+\sqrt{2}}}{2}$$

47. $\sin\left(\dfrac{\pi}{24}\right) = \sin\left(\dfrac{\frac{\pi}{12}}{2}\right) = \sqrt{\dfrac{1-\frac{\sqrt{2+\sqrt{3}}}{2}}{2}}$

$$= \sqrt{\frac{2-\sqrt{2+\sqrt{3}}}{4}} = \frac{\sqrt{2-\sqrt{2+\sqrt{3}}}}{2}$$

Recall from problem #39:

$$\cos\left(\frac{\pi}{12}\right) = \cos\left(\frac{\frac{\pi}{6}}{2}\right) = \sqrt{\frac{1+\cos\left(\frac{\pi}{6}\right)}{2}}$$

$$= \sqrt{\frac{1+\frac{\sqrt{3}}{2}}{2}} = \frac{\sqrt{2+\sqrt{3}}}{2}$$

49. $\sqrt{\dfrac{1+\cos 30°}{2}} = \cos\left(\dfrac{30°}{2}\right) = \cos 15°$

51. $\sqrt{\dfrac{1-\cos(4\theta)}{1+\cos(4\theta)}} = \tan\left(\dfrac{4\theta}{2}\right) = \tan(2\theta)$

53. $\dfrac{\sin(2x)}{1+\cos(2x)} = \tan\left(\dfrac{2x}{2}\right) = \tan x$

55. $\sin\theta = \dfrac{12}{13}, \theta$ is obtuse

$$\cos^2\theta = 1 - \sin^2\theta = 1 - \left(\dfrac{12}{13}\right)^2 = \dfrac{25}{169}$$

$$\cos\theta = -\dfrac{5}{13} \text{ since QII;}$$

$$\sin\left(\dfrac{\theta}{2}\right) = \sqrt{\dfrac{1 - \left(\dfrac{-5}{13}\right)}{2}}$$

$$= \sqrt{\dfrac{\dfrac{18}{13}}{2}} = \sqrt{\dfrac{9}{13}} = \dfrac{3}{\sqrt{13}};$$

$$\cos\left(\dfrac{\theta}{2}\right) = \sqrt{\dfrac{1 + \left(\dfrac{-5}{13}\right)}{2}}$$

$$= \sqrt{\dfrac{\dfrac{8}{13}}{2}} = \sqrt{\dfrac{4}{13}} = \dfrac{2}{\sqrt{13}};$$

$$\tan\left(\dfrac{\theta}{2}\right) = \dfrac{\sin\left(\dfrac{\theta}{2}\right)}{\cos\left(\dfrac{\theta}{2}\right)} = \dfrac{\dfrac{3}{\sqrt{13}}}{\dfrac{2}{\sqrt{13}}} = \dfrac{3}{2}$$

57. $\cos\theta = -\dfrac{4}{5}, \theta$ in QII

$$\dfrac{\pi}{2} < \theta < \pi$$

$$\dfrac{\pi}{4} < \dfrac{\theta}{2} < \dfrac{\pi}{2}; \text{ QI}$$

$$\cos\left(\dfrac{\theta}{2}\right) = \sqrt{\dfrac{1 + \left(\dfrac{-4}{5}\right)}{2}}$$

$$= \sqrt{\dfrac{\dfrac{1}{5}}{2}} = \sqrt{\dfrac{1}{10}} = \dfrac{1}{\sqrt{10}};$$

$$\sin\left(\dfrac{\theta}{2}\right) = \sqrt{\dfrac{1 - \left(\dfrac{-4}{5}\right)}{2}}$$

$$= \sqrt{\dfrac{\dfrac{9}{5}}{2}} = \dfrac{3}{\sqrt{10}};$$

$$\tan\left(\dfrac{\theta}{2}\right) = \dfrac{\sin\left(\dfrac{\theta}{2}\right)}{\cos\left(\dfrac{\theta}{2}\right)} = \dfrac{\dfrac{3}{\sqrt{10}}}{\dfrac{1}{\sqrt{10}}} = 3$$

59. $\tan\theta = \dfrac{-35}{12}, \theta$ in QII

$$\sqrt{35^2 + 12^2} = 37;$$

$$\dfrac{\pi}{2} < \theta < \pi$$

$$\dfrac{\pi}{4} < \dfrac{\theta}{2} < \dfrac{\pi}{2}; \text{ QI}$$

$$\sin\left(\dfrac{\theta}{2}\right) = \sqrt{\dfrac{1 - \left(\dfrac{-12}{37}\right)}{2}} = \sqrt{\dfrac{49}{74}} = \dfrac{7}{\sqrt{74}};$$

$$\cos\left(\dfrac{\theta}{2}\right) = -\sqrt{\dfrac{1 + \left(\dfrac{-12}{37}\right)}{2}} = -\sqrt{\dfrac{25}{37}{2}} = \dfrac{5}{\sqrt{74}};$$

$$\tan\left(\dfrac{\theta}{2}\right) = \dfrac{\sin\left(\dfrac{\theta}{2}\right)}{\cos\left(\dfrac{\theta}{2}\right)} = \dfrac{\dfrac{7}{\sqrt{74}}}{\dfrac{5}{\sqrt{74}}} = \dfrac{7}{5}$$

61. $\sin\theta = \dfrac{15}{113}; \theta$ is acute

$$\cos^2\theta = 1 - \sin^2\theta$$

$$\cos^2\theta = 1 - \left(\dfrac{15}{113}\right)^2 = \dfrac{12544}{12769}$$

$$\cos\theta = \dfrac{112}{113};$$

$$\cos\left(\dfrac{\theta}{2}\right) = \sqrt{\dfrac{1 + \dfrac{112}{113}}{2}} = \sqrt{\dfrac{\dfrac{225}{113}}{2}} = \dfrac{15}{\sqrt{226}};$$

$$\sin\left(\dfrac{\theta}{2}\right) = \sqrt{\dfrac{1 - \dfrac{112}{113}}{2}} = \sqrt{\dfrac{\dfrac{1}{113}}{2}} = \dfrac{1}{\sqrt{226}};$$

$$\tan\left(\dfrac{\theta}{2}\right) = \dfrac{\sin\left(\dfrac{\theta}{2}\right)}{\cos\left(\dfrac{\theta}{2}\right)} = \dfrac{\dfrac{1}{\sqrt{226}}}{\dfrac{15}{\sqrt{226}}} = \dfrac{1}{15}$$

6.4 Exercises

63. $\cot\theta = \dfrac{21}{20}; \pi < \theta < \dfrac{3\pi}{2}$

$\dfrac{\pi}{2} < \dfrac{\theta}{2} < \dfrac{3\pi}{4}$

$\sqrt{21^2 + 20^2} = 29$;

$\sin\left(\dfrac{\theta}{2}\right) = \sqrt{\dfrac{1 - \left(\dfrac{-21}{29}\right)}{2}}$

$= \sqrt{\dfrac{\dfrac{50}{29}}{2}} = \sqrt{\dfrac{25}{29}} = \dfrac{5}{\sqrt{29}}$;

$\cos\left(\dfrac{\theta}{2}\right) = -\sqrt{\dfrac{1 + \left(\dfrac{-21}{29}\right)}{2}} = -\sqrt{\dfrac{\dfrac{8}{29}}{2}} = \dfrac{-2}{\sqrt{29}}$;

$\tan\left(\dfrac{\theta}{2}\right) = \dfrac{\sin\left(\dfrac{\theta}{2}\right)}{\cos\left(\dfrac{\theta}{2}\right)} = \dfrac{\dfrac{5}{\sqrt{29}}}{\dfrac{-2}{\sqrt{29}}} = \dfrac{-5}{2}$

65.

$\sin(-4\theta)\sin(8\theta) = \dfrac{1}{2}\left[\cos(-4\theta - 8\theta) - \cos(-4\theta + 8\theta)\right]$

$= \dfrac{1}{2}\left[\cos(-12\theta) - \cos(4\theta)\right]$

Recall $\cos(-x) = \cos x$

$= \dfrac{1}{2}\left[\cos(12\theta) - \cos(4\theta)\right]$

67. $2\cos\left(\dfrac{7t}{2}\right)\cos\left(\dfrac{3t}{2}\right)$

$= 2\left(\dfrac{1}{2}\right)\left[\cos\left(\dfrac{7t}{2} - \dfrac{3t}{2}\right) + \cos\left(\dfrac{7t}{2} + \dfrac{3t}{2}\right)\right]$

$= \cos(2t) + \cos(5t)$

69. $2\cos(1979\pi t)\cos(439\pi t)$

$= 2\left(\dfrac{1}{2}\right)\left[\cos(1979\pi t - 439\pi t) + \cos(1979\pi t + 439\pi t)\right]$

$= \cos(1540\pi t) + \cos(2418\pi t)$

71. $2\cos 15° \sin 135°$

$= 2\left(\dfrac{1}{2}\right)\left[\sin(15° + 135°) - \sin(15° - 135°)\right]$

$= \sin(150°) - \sin(-120°)$

$= \sin(150°) + \sin(120°)$

$= \dfrac{1}{2} + \dfrac{\sqrt{3}}{2}$

$= \dfrac{1 + \sqrt{3}}{2}$

73. $\sin\left(\dfrac{7\pi}{12}\right)\sin\left(-\dfrac{\pi}{12}\right)$

$= \dfrac{1}{2}\left[\cos\left(\dfrac{7\pi}{12} - \left(\dfrac{-\pi}{12}\right)\right) - \cos\left(\dfrac{7\pi}{12} + \left(\dfrac{-\pi}{12}\right)\right)\right]$

$= \dfrac{1}{2}\left[\cos\left(\dfrac{2\pi}{3}\right) - \cos\left(\dfrac{\pi}{2}\right)\right]$

$= \dfrac{1}{2}\left[\dfrac{-1}{2} - 0\right]$

$= \dfrac{-1}{4}$

75. $\sin(14k) + \sin(41k)$

$= 2\sin\left(\dfrac{14k + 41k}{2}\right)\cos\left(\dfrac{14k - 41k}{2}\right)$

$= 2\sin\left(\dfrac{55k}{2}\right)\cos\left(\dfrac{-27k}{2}\right)$

$= 2\sin\left(\dfrac{55k}{2}\right)\cos\left(\dfrac{27k}{2}\right)$

77. $\cos\left(\dfrac{7x}{6}\right) - \cos\left(\dfrac{5x}{6}\right)$

$= -2\sin\left(\dfrac{\dfrac{7x}{6} + \dfrac{5x}{6}}{2}\right)\sin\left(\dfrac{\dfrac{7x}{6} - \dfrac{5x}{6}}{2}\right)$

$= -2\sin\left(\dfrac{6x}{6}\right)\sin\left(\dfrac{x}{6}\right)$

$= -2\sin x \sin\left(\dfrac{x}{6}\right)$

79. $\cos(852\pi t) + \cos(1209\pi t)$

$= 2\cos\left(\dfrac{852\pi t + 1209\pi t}{2}\right)\cos\left(\dfrac{852\pi t - 1209\pi t}{2}\right)$

$= 2\cos\left(\dfrac{2061\pi t}{2}\right)\cos\left(\dfrac{-357\pi t}{2}\right)$

$= 2\cos\left(\dfrac{2061\pi t}{2}\right)\cos\left(\dfrac{357\pi t}{2}\right)$

81. $\sin\left(\dfrac{17}{12}\pi\right) - \sin\left(\dfrac{13\pi}{12}\right)$

$= 2\cos\left(\dfrac{\dfrac{17}{12}\pi + \dfrac{13}{12}\pi}{2}\right)\sin\left(\dfrac{\dfrac{17}{12}\pi - \dfrac{13}{12}\pi}{2}\right)$

$= 2\cos\left(\dfrac{5}{4}\pi\right)\sin\left(\dfrac{1}{6}\pi\right)$

$= 2\left(\dfrac{-\sqrt{2}}{2}\right)\left(\dfrac{1}{2}\right) = \dfrac{-\sqrt{2}}{2}$

83. $\dfrac{2\sin x \cos x}{\cos^2 x - \sin^2 x} = \tan(2x)$

$\dfrac{\sin 2x}{\cos 2x} =$

$\tan(2x) = \tan(2x)$

Verified.

85. $(\sin x + \cos x)^2 = 1 + \sin(2x)$

$\sin^2 x + 2\sin x \cos x + \cos^2 x =$

$\sin^2 x + \cos^2 x + 2\sin x \cos x =$

$1 + 2\sin x \cos x =$

$1 + \sin(2x) = 1 + \sin(2x)$

Verified.

87. $\cos(8\theta) = \cos^2(4\theta) - \sin^2(4\theta)$

$\cos(2 \cdot 4\theta) =$

$\cos^2(4\theta) - \sin^2(4\theta) = \cos^2(4\theta) - \sin^2(4\theta)$

Verified.

89. $\dfrac{\cos(2\theta)}{\sin^2\theta} = \cot^2\theta - 1$

$\dfrac{\cos^2\theta - \sin^2\theta}{\sin^2\theta} =$

$\dfrac{\cos^2\theta}{\sin^2\theta} - 1 =$

$\cot^2\theta - 1 = \cot^2\theta - 1$

Verified.

91. $\tan(2\theta) = \dfrac{2}{\cot\theta - \tan\theta}$

$\dfrac{(2\tan\theta)\left(\dfrac{1}{\tan\theta}\right)}{(1 - \tan^2\theta)\dfrac{1}{\tan\theta}} =$

$\dfrac{2}{\dfrac{1}{\tan\theta} - \tan\theta} =$

$\dfrac{2}{\cot\theta - \tan\theta} =$

Verified.

93. $\tan x + \cot x = 2\csc(2x)$

$= \dfrac{2}{\sin(2x)}$

$= \dfrac{2}{2\sin x \cos x}$

$= \dfrac{1}{\sin x \cos x}$

$= \dfrac{\sin^2 x + \cos^2 x}{\sin x \cos x}$

$= \dfrac{\sin^2 x}{\sin x \cos x} + \dfrac{\cos^2 x}{\sin x \cos x}$

$= \dfrac{\sin x}{\cos x} + \dfrac{\cos x}{\sin x}$

$= \tan x + \cot x$

Verified.

6.4 Exercises

95. $\cos^2\left(\dfrac{x}{2}\right) - \sin^2\left(\dfrac{x}{2}\right) = \cos x$

$\cos\left(2 \cdot \left(\dfrac{x}{2}\right)\right) =$

$\cos x = \cos x$

Verified.

97. $1 - \sin^2(2\theta) = 1 - 4\sin^2\theta + 4\sin^4\theta$

$= \left(1 - 2\sin^2\theta\right)^2$

$= \left(\cos(2\theta)\right)^2$

$= \cos^2(2\theta)$

$= 1 - \sin^2(2\theta)$

Verified.

99. $\dfrac{\sin(120\pi\,t) + \sin(80\pi\,t)}{\cos(120\pi\,t) - \cos(80\pi\,t)} = -\cot(20\pi\,t)$

$\dfrac{2\sin(100\pi\,t)\cos(20\pi\,t)}{-2\sin(100\pi\,t)\sin(20\pi\,t)} =$

$-\cot(20\pi\,t) = -\cot(20\pi\,t)$

Verified.

101. $\sin^2\alpha + (1 - \cos\alpha)^2 = \left[2\sin\left(\dfrac{\alpha}{2}\right)\right]^2$

$\sin^2\alpha + 1 - 2\cos\alpha + \cos^2\alpha =$

$\sin^2\alpha + \cos^2\alpha + 1 - 2\cos\alpha =$

$1 + 1 - 2\cos\alpha =$

$2 - 2\cos\alpha =$

$2(1 - \cos\alpha) =$

$4\left(\dfrac{1 - \cos\alpha}{2}\right) =$

$4\sin^2\left(\dfrac{\alpha}{2}\right) =$

$\left[2\sin\left(\dfrac{\alpha}{2}\right)\right]^2 = \left[2\sin\left(\dfrac{\alpha}{2}\right)\right]^2$

103. $\alpha = \beta$

$\sin(2\alpha) = \sin(\alpha + \alpha)$

$= \sin\alpha\cos\alpha + \cos\alpha\sin\alpha$

$= \sin\alpha\cos\alpha + \sin\alpha\cos\alpha$

$= 2\sin\alpha\cos\alpha\ ;$

$\tan(2\alpha) = \tan(\alpha + \alpha)$

$= \dfrac{\tan\alpha + \tan\alpha}{1 - \tan\alpha\tan\alpha}$

$= \dfrac{2\tan\alpha}{1 - \tan^2\alpha}$

105. Subtract the identities:

$\begin{cases} \cos\alpha\cos\beta + \sin\alpha\sin\beta = \cos(\alpha - \beta) \\ \cos\alpha\cos\beta - \sin\alpha\sin\beta = \cos(\alpha + \beta) \end{cases}$

$2\sin\alpha\sin\beta = \cos(\alpha - \beta) - \cos(\alpha + \beta)$

Divide by 2:

$\sin\alpha\sin\beta = \dfrac{1}{2}\left[\cos(\alpha - \beta) - \cos(\alpha + \beta)\right]$

107. $M = \csc\left(\dfrac{\theta}{2}\right)$

a. $\csc\left(\dfrac{30°}{2}\right) = \dfrac{1}{\sin\left(\dfrac{30°}{2}\right)}$

$= \dfrac{1}{\sqrt{1 - \dfrac{\sqrt{3}}{2}}} = \dfrac{1}{\dfrac{\sqrt{2 - \sqrt{3}}}{2}}$

$M = \dfrac{2}{\sqrt{2 - \sqrt{3}}}\ ;\ M \approx 3.9$

b. $\csc\left(\dfrac{45°}{2}\right) = \dfrac{1}{\sin\left(\dfrac{45°}{2}\right)}$

$= \dfrac{1}{\sqrt{1 - \dfrac{\sqrt{2}}{2}}} = \dfrac{1}{\dfrac{\sqrt{2 - \sqrt{2}}}{2}}$

$M = \dfrac{2}{\sqrt{2 - \sqrt{2}}}\ ;\ M \approx 2.6$

Chapter 6: Trigonometric Identities, Inverses and Equations

c. $2 = \csc\left(\dfrac{\theta}{2}\right)$

$2 = \dfrac{1}{\sin\left(\dfrac{\theta}{2}\right)}$

$2\sin\left(\dfrac{\theta}{2}\right) = 1$

$\sin\left(\dfrac{\theta}{2}\right) = \dfrac{1}{2}$

$\sqrt{\dfrac{1-\cos\theta}{2}} = \dfrac{1}{2}$

$\dfrac{1-\cos\theta}{2} = \dfrac{1}{4}$

$1-\cos\theta = \dfrac{1}{2}$

$-\cos\theta = -\dfrac{1}{2}$

$\cos\theta = \dfrac{1}{2}$

$\theta = 60°$

109. $r(\theta) = \dfrac{1}{32} v^2 \sin(2\theta)$

a. $r(22.5°) = \dfrac{1}{32}(96)^2 \sin 45°$

$= \dfrac{1}{32}(9216)\sin 45°$

$= 288 \sin 45°$

$= 288\left(\dfrac{1}{\sqrt{2}}\right) = \dfrac{288\sqrt{2}}{2} = 144\sqrt{2}$;

$r(45°) = \dfrac{1}{32}(96)^2 \sin 90°$

$= 288$

Number of ft short of maximum:

$288 - 144\sqrt{2}$ ft ≈ 84.3 ft

b. $r(67.5°) = \dfrac{1}{32}(96)^2 \sin 135°$

$= 288 \sin 135°$

$= 288\left(\dfrac{\sqrt{2}}{2}\right) = 144\sqrt{2}$

Number of ft short of maximum:

$288 - 144\sqrt{2}$ ft ≈ 84.3 ft

111. $y(t) = 2\cos(2150\pi t)\cos(268\pi t)$

$\cos\alpha \cos\beta = \dfrac{1}{2}\left[\cos(\alpha-\beta)+\cos(\alpha+\beta)\right];$

$= 2 \cdot \dfrac{1}{2}\left[\cos(2150\pi t - 268\pi t) + \cos(2150\pi t + 268\pi t)\right]$

$= \left[\cos(1882\pi t) + \cos(2418\pi t)\right]$

$= \cos(2418\pi t) + \cos(1882\pi t)$

$= \cos\left[2\pi(1209)t\right] + \cos\left[2\pi(941)t\right]$

The * key

113. $d(t) = \left|6\sin\left(\dfrac{\pi t}{60}\right)\right|$

$= \left|6\sin\left(\dfrac{1}{2}\dfrac{\pi t}{30}\right)\right| = \left|(6)\left(\pm\sqrt{\dfrac{1-\cos\left(\dfrac{\pi t}{30}\right)}{2}}\right)\right|$

$= 6\left(\sqrt{\dfrac{1-\cos\left(\dfrac{\pi t}{30}\right)}{2}}\right) = \sqrt{36 \cdot \dfrac{1-\cos\left(\dfrac{\pi t}{30}\right)}{2}}$

$= \sqrt{18\left[1-\cos\left(\dfrac{\pi t}{30}\right)\right]}$

$d(t) = \sqrt{18\left[1-\cos\left(\dfrac{\pi t}{30}\right)\right]}$

115.a. $f(\theta) = \sin(2\theta - 90°) + 1$

$= \sin(2\theta)\cos 90° - \cos(2\theta)\sin(90°) + 1$

$= 0 - \cos(2\theta) + 1$

$= 1 - \cos(2\theta)$

b. $g(\theta) = 2\sin^2\theta = 1 - \cos(2\theta)$

$= 1 - \left(\cos^2\theta - \sin^2\theta\right)$

$= 1 - \cos^2\theta + \sin^2\theta$

$= \sin^2\theta + \sin^2\theta$

$= 2\sin^2\theta$

$= 1 - \cos(2\theta)$

c. $k(\theta) = 1 + \sin^2\theta - \cos^2\theta$

$= 1 - \cos^2\theta + \sin^2\theta$

$= \sin^2\theta + \sin^2\theta$

$= 2\sin^2\theta$

$= 1 - \cos(2\theta)$

d. $h(\theta) = 1 - \cos(2\theta)$

6.4 Exercises

117. $\cos 15°$

Half-angle identity:

$$\cos\left(\frac{30°}{2}\right) = \sqrt{\frac{1+\cos 30°}{2}}$$

$$= \sqrt{\frac{1+\dfrac{\sqrt{3}}{2}}{2}} = \frac{\sqrt{2+\sqrt{3}}}{2} \; ;$$

Difference identity:

$$\cos 15° = \cos(45° - 30°)$$

$$= \cos 45° \cos 30° + \sin 45° \sin 30°$$

$$= \frac{\sqrt{2}}{2} \cdot \frac{\sqrt{3}}{2} + \frac{\sqrt{2}}{2} \cdot \frac{1}{2}$$

$$= \frac{\sqrt{6}}{4} - \frac{\sqrt{2}}{4} = \frac{\sqrt{6}+\sqrt{2}}{4}$$

a. $\dfrac{\sqrt{2+\sqrt{3}}}{2} \approx 0.9659$

$\dfrac{\sqrt{6}+\sqrt{2}}{4} \approx 0.9659$

b. $\dfrac{\sqrt{2+\sqrt{3}}}{2} = \dfrac{\sqrt{6}+\sqrt{2}}{4}$

$$\left(\frac{\sqrt{2+\sqrt{3}}}{2}\right)^2 = \left(\frac{\sqrt{6}+\sqrt{2}}{4}\right)^2$$

$$\frac{2+\sqrt{3}}{4} = \frac{6+2\sqrt{12}+2}{16}$$

$$\frac{2+\sqrt{3}}{4} = \frac{8+4\sqrt{3}}{16}$$

$$\frac{2+\sqrt{3}}{4} = \frac{2+\sqrt{3}}{4}$$

119. Must be a unit circle with θ in radians. Must use a right triangle definition of tangent:

$$\tan\left(\frac{\theta}{2}\right) = \frac{\text{opposite side}}{\text{adjacent side}} = \frac{\sin\theta}{1+\cos\theta}$$

121. $x^4 + x^3 - 8x^2 - 6x + 12 = 0$

Possible rational roots:

$$\frac{\{\pm 1, \pm 12, \pm 2, \pm 6, \pm 3, \pm 4, \}}{\{\pm 1\}}$$

$$\begin{array}{r|rrrr}
-2 & 1 & 1 & -8 & -6 & 12 \\
 & & -2 & 2 & 12 & -12 \\
\hline
 & 1 & -1 & -6 & 6 & \underline{|0} \\
\end{array}$$

$$\begin{array}{r|rrrr}
1 & 1 & -1 & -6 & 6 \\
 & & 1 & 0 & -6 \\
\hline
 & 1 & 0 & -6 & \underline{|0} \\
\end{array}$$

$$(x+2)(x-1)(x^2-6) = 0$$

$$x^2 - 6 = 0$$

$$x^2 = 6$$

$$x = \pm\sqrt{6}, x = 1, x = -2$$

123. $\left(\dfrac{16}{65}\right)^2 + \left(\dfrac{63}{65}\right)^2 = \dfrac{256}{4225} + \dfrac{3969}{4225} = \dfrac{4225}{4225} = 1$

$$\tan\theta = \frac{63}{16}; \sec\theta = \frac{65}{16}$$

$$1 + \tan^2\theta = \sec^2\theta$$

$$1 + \left(\frac{63}{16}\right)^2 = \left(\frac{65}{16}\right)^2$$

$$1 + \frac{3969}{256} = \frac{4225}{256}$$

$$\frac{256}{256} + \frac{3969}{256} = \frac{4225}{256}$$

Chapter 6 Mid-Chapter Check

1. $\sin x(\csc x - \sin x) = \cos^2 x$

$\sin x \csc x - \sin^2 x =$

$\dfrac{\sin x}{\sin x} - (1 - \cos^2 x) =$

$1 - 1 + \cos^2 x =$

$\cos^2 x = \cos^2 x$

3. $\dfrac{2 \sin x}{\sec x} - \dfrac{\cos x}{\csc x} = \cos x \sin x$

$\dfrac{2 \sin x}{\dfrac{1}{\cos x}} - \dfrac{\cos x}{\dfrac{1}{\sin x}} =$

$2 \sin x \cos x - \sin x \cos x =$

$\sin x \cos x =$

$\cos x \sin x = \cos x \sin x$

5. a. $\dfrac{\sin^3 x + \cos^3 x}{\sin x + \cos x} = 1 - \sin x \cos x$

$\dfrac{(\sin x + \cos x)(\sin^2 x - \sin x \cos x + \cos^2 x)}{(\sin x + \cos x)} =$

$1 - \sin x \cos x = 1 - \sin x \cos x$

b. $\dfrac{1 + \sec s}{\csc x} - \dfrac{1 + \cos x}{\cot x} = 0$

$\dfrac{1 + \dfrac{1}{\cos x}}{\dfrac{1}{\sin x}} - \dfrac{1 + \cos x}{\dfrac{\cos x}{\sin x}} = 0$

$\dfrac{\sin x \cos x + \sin x}{\cos x} - \dfrac{\sin x + \sin x \cos x}{\cos x} =$

$\dfrac{\sin x \cos x + \sin x - \sin x - \sin x \cos x}{\cos x} =$

$0 = 0$

7. $\sin \alpha = \dfrac{56}{65}$, $\tan \beta = \dfrac{-80}{39}$

$\sin^2 \alpha = 1 - \cos^2 \alpha$

$\left(\dfrac{56}{65}\right)^2 - 1 = -\cos^2 \alpha$

$\dfrac{-1089}{4225} = -\cos^2 \alpha$

$\cos^2 \alpha = \dfrac{1089}{4225}$

$\cos \alpha = \pm \dfrac{33}{65}$

$\cos \alpha = \dfrac{-33}{65}$;

$\tan \alpha = \dfrac{\sin \alpha}{\cos \alpha}$

$\tan \alpha = \dfrac{\dfrac{56}{65}}{\dfrac{-33}{65}} = \dfrac{-56}{33}$;

$\tan^2 \beta + 1 = \sec^2 \beta$

$\left(\dfrac{-80}{39}\right)^2 + 1 = \dfrac{1}{\cos^2 \beta}$

$\dfrac{7921}{1521} = \dfrac{1}{\cos^2 \beta}$

$7921 \cos 2\beta = 1521$

$\cos^2 \beta = \dfrac{1521}{7921}$

$\cos \beta = \pm \dfrac{39}{89}$

$\cos \beta = \dfrac{-39}{89}$;

$\sin^2 \beta = 1 - \left(\dfrac{-39}{89}\right)^2$

$\sin^2 \beta = \dfrac{6400}{7921}$

$\sin \beta = \pm \dfrac{80}{89}$

$\sin \beta = \dfrac{80}{89}$

a. $\sin(\alpha - \beta) = \sin \alpha \cos \beta - \cos \alpha \sin \beta$

$= \left(\dfrac{56}{65}\right)\left(\dfrac{-39}{89}\right) - \left(\dfrac{-33}{65}\right)\left(\dfrac{80}{89}\right) = \dfrac{456}{5785}$

b. $\cos(\alpha + \beta) = \cos \alpha \cos \beta - \sin \alpha \sin \beta$

$= \left(\dfrac{-33}{65}\right)\left(\dfrac{-39}{89}\right) - \left(\dfrac{56}{65}\right)\left(\dfrac{80}{89}\right) = -\dfrac{3193}{5785}$

c. $\tan(\alpha - \beta) = \dfrac{\left(\dfrac{-56}{33}\right) - \left(\dfrac{-80}{39}\right)}{1 + \left(\dfrac{-56}{33}\right)\left(\dfrac{-80}{39}\right)} = \dfrac{456}{5767}$

9. $\cos\theta = \dfrac{-15}{17}$, θ in QII

$$\frac{\pi}{2} < \theta < \pi$$

$$\frac{\pi}{4} < \frac{\theta}{2} < \frac{\pi}{2} \text{ in QI}$$

$$\sin\left(\frac{\theta}{2}\right) = \pm\sqrt{\frac{1 - \cos\theta}{2}}$$

$$= \sqrt{\frac{1 - \left(\frac{-15}{17}\right)}{2}} = \sqrt{\frac{\frac{32}{17}}{2}} = \sqrt{\frac{16}{17}} = \frac{4}{\sqrt{17}};$$

In QI

$$\cos\left(\frac{\theta}{2}\right) = +\sqrt{\frac{1 + \cos\theta}{2}}$$

$$= \sqrt{\frac{1 + \frac{-15}{17}}{2}} = \sqrt{\frac{\frac{2}{17}}{2}} = \frac{1}{\sqrt{17}}$$

Reinforcing Basic Concepts

1. $\sin^2\alpha + \cos^2\alpha = 1$

$$\frac{\sin^2\alpha + \cos^2\alpha}{\sin^2\alpha} = \frac{1}{\sin^2\alpha}$$

$$\frac{\sin^2\alpha}{\sin^2\alpha} + \frac{\cos^2\alpha}{\sin^2\alpha} = \frac{1}{\sin^2\alpha}$$

$$1 + \cot^2\alpha = \csc^2\alpha;$$

$$\sin^2\alpha + \cos^2\alpha = 1$$

$$\frac{\sin^2\alpha + \cos^2\alpha}{\cos^2\alpha} = \frac{1}{\cos^2\alpha}$$

$$\frac{\sin^2\alpha}{\cos^2\alpha} + \frac{\cos^2\alpha}{\cos^2\alpha} = \frac{1}{\cos^2\alpha}$$

$$\tan^2\alpha + 1 = \sec^2\alpha$$

6.5 Technology Highlight

Exercise 1: Using grapher:
$$Y1 = \cos x;$$
$$Y2 = \cos^{-1} x$$

6.5 Exercises

1. Horizontal line, one, one

3. $[-1,1]$, $\left[-\dfrac{\pi}{2}, \dfrac{\pi}{2}\right]$

5. $\cos^{-1}\left(\dfrac{1}{5}\right)$

7. $\sin^{-1}(0) = 0$; $\sin\left(\dfrac{\pi}{6}\right) = \sin 30° = \dfrac{1}{2}$;

$$\sin^{-1}\left(-\frac{1}{2}\right) = -\frac{\pi}{6};$$

$$\sin^{-1}(-1) = -\frac{\pi}{2}$$

9. $\sin^{-1}\left(\dfrac{\sqrt{2}}{2}\right)$:

y is the number or angle whose sine is $\dfrac{\sqrt{2}}{2}$

$$\sin^{-1}\left(\frac{\sqrt{2}}{2}\right) = y \text{ where } -\frac{\pi}{2} \le y \le \frac{\pi}{2}$$

$$\sin y = \frac{\sqrt{2}}{2}$$

$$y = \frac{\pi}{4}$$

11. $\sin^{-1} 1$:
y is the number or angle whose sine is 1
$\sin y = 1$

$$\sin^{-1} 1 = \frac{\pi}{2}$$

13. $\arcsin 0.8892 = 1.0956$

$$1.0956\left(\frac{180}{\pi}\right) \approx 62.8°$$

15. $\sin^{-1}\left(\dfrac{1}{\sqrt{7}}\right) = 0.3876$

$$0.3876\left(\frac{180}{\pi}\right) = 22.2°$$

17. $\sin\left[\sin^{-1}\left(\dfrac{\sqrt{2}}{2}\right)\right] = \dfrac{\sqrt{2}}{2}$

    since $\dfrac{\sqrt{2}}{2} \in [-1,1]$

19. $\arcsin\left[\sin\left(\dfrac{\pi}{3}\right)\right] = \dfrac{\pi}{3}$

    since $\dfrac{\pi}{3} \in \left[\dfrac{-\pi}{2}, \dfrac{\pi}{2}\right]$

21. $\sin^{-1}(\sin 135°) = 45°$

    since $45° \in [-90°, 90°]$ and
    $\sin 135° = \sin 45°$

23. $\sin\left(\sin^{-1} 68205\right) = 0.8205$
    since $0.8205 \in [-1,1]$

25. $\cos^{-1} 1 = 0$; $\cos\left(\dfrac{\pi}{6}\right) = \dfrac{\sqrt{3}}{2}$;

    $\arccos\left(-\dfrac{1}{2}\right) = 120°$;

    $\cos^{-1}(-1) = \pi$

27. $\cos^{-1}\left(\dfrac{1}{2}\right)$:

    y is the number or angle whose cosine is $\dfrac{1}{2}$

    $\cos^{-1} y = \dfrac{1}{2}$

    $\cos^{-1}\left(\dfrac{1}{2}\right) = \dfrac{\pi}{3}$

29. $\cos^{-1}(-1)$:

    y is the # or angle whose cosine is -1
    $\cos y = -1$
    $\cos^{-1}(-1) = \pi$

31. $\arccos 0.1352 \approx 82.2°$

    $82.2\left(\dfrac{\pi}{180}\right) = 1.4352$

33. $\cos^{-1}\left(\dfrac{\sqrt{5}}{3}\right) = 41.8°$

    $41.8\left(\dfrac{\pi}{180}\right) = 0.7297$

35. $\arccos\left[\cos\left(\dfrac{\pi}{4}\right)\right] = \dfrac{\pi}{4}$

    since $\dfrac{\pi}{4} \in [0, \pi]$

37. $\cos\left(\cos^{-1} 0.5560\right) = 0.5560$
    since $0.5560 \in [-1,1]$

39. $\cos\left[\cos^{-1}\left(\dfrac{-\sqrt{2}}{2}\right)\right] = \dfrac{-\sqrt{2}}{2}$

    since $\dfrac{-\sqrt{2}}{2} \in [-1,1]$

41. $\cos^{-1}\left[\cos\left(\dfrac{5\pi}{4}\right)\right] = \dfrac{3\pi}{4}$

    since $\dfrac{3\pi}{4} \in [0, \pi]$

    $\cos\dfrac{5\pi}{4} = -\cos\dfrac{\pi}{4} = \cos\dfrac{3\pi}{4}$

43.

$\tan 0 = 0$	$\tan^{-1} 0 \quad \underline{\quad 0 \quad}$
$\tan\left(-\dfrac{\pi}{3}\right) = \underline{\dfrac{-\sqrt{3}}{\quad}}$	$\arctan(-\sqrt{3}) = -\dfrac{\pi}{3}$
$\tan 30° = \dfrac{\sqrt{3}}{3}$	$\arctan\left(\dfrac{\sqrt{3}}{3}\right) = \underline{\quad 30° \quad}$
$\tan\left(\dfrac{\pi}{3}\right) = \underline{\quad \sqrt{3} \quad}$	$\tan^{-1}(\sqrt{3}) = \underline{\quad \dfrac{\pi}{3} \quad}$

6.5 Exercises

45. $\tan^{-1}\left(\dfrac{-\sqrt{3}}{3}\right) = -30° = \dfrac{-\pi}{6}$

47. $\arctan\left(\sqrt{3}\right) = 60° = \dfrac{\pi}{3}$

49. $\tan^{-1}(-2.05) \approx -64.0$

   $-64.0\left(\dfrac{\pi}{180}\right) = -1.1170$

51. $\arctan\left(\dfrac{29}{24}\right) = 54.1°$

   $54.1\left(\dfrac{\pi}{180}\right) = 0.9441$

53. $\sin^{-1}\left[\cos\left(\dfrac{2\pi}{3}\right)\right] = \dfrac{-\pi}{6}$

   $\cos\left(\dfrac{2\pi}{3}\right) = \dfrac{-1}{2}$

   $\sin^{-1}\left(\dfrac{-1}{2}\right) = \dfrac{-\pi}{6}$

55. $\tan\left[\arccos\left(\dfrac{\sqrt{3}}{2}\right)\right]$

   $\arccos\left(\dfrac{\sqrt{3}}{2}\right) = \dfrac{\pi}{6}$

   $\tan\left(\dfrac{\pi}{6}\right) = \dfrac{1}{\sqrt{3}} = \dfrac{\sqrt{3}}{3}$

57. $\csc\left[\sin^{-1}\left(\dfrac{\sqrt{2}}{2}\right)\right]$

   $\sin^{-1}\left(\dfrac{\sqrt{2}}{2}\right) = 45°$ or $\dfrac{\pi}{4}$

   $\csc\left(\dfrac{\pi}{4}\right) = \dfrac{1}{\sin\left(\dfrac{\pi}{4}\right)} = \sqrt{2}$

59. $\arccos\left[\sin(-30°)\right]$

   $\sin(-30°) = \dfrac{-1}{2}$

   $\arccos\left(\dfrac{-1}{2}\right) = 120°$

61. $\tan\left(\sin^{-1}1\right)$

   $\sin^{-1}1 = \dfrac{\pi}{2}$

   $\tan\left(\dfrac{\pi}{2}\right)$ cannot be evaluated because

   $x = \dfrac{\pi}{2}$ is a vertical asymptote for $\tan x$. $\dfrac{\pi}{2}$
   is not in the domain of $\tan x$.

63. $\sin^{-1}\left(\csc\dfrac{\pi}{4}\right)$

   $\csc\dfrac{\pi}{4} = \sqrt{2} > 1$;

   Not in domain of $\sin^{-1}x$

65. a) $\sin\theta = \dfrac{0.3}{0.5} = \dfrac{3}{5}$

   b) $\cos\theta = \dfrac{0.4}{0.5} = \dfrac{4}{5}$

   c) $\tan\theta = \dfrac{0.3}{0.4} = \dfrac{3}{4}$

67. a) $\sin\theta = \dfrac{\sqrt{x^2-36}}{x}$

   b) $\cos\theta = \dfrac{6}{x}$

   c) $\tan\theta = \dfrac{\sqrt{x^2-36}}{6}$

69. $\sin\left[\cos^{-1}\left(\dfrac{-7}{25}\right)\right] = \dfrac{24}{25}$

$x^2 + 7^2 = 25^2$

$x^2 + 49 = 625$

$x^2 = 576$

$x = 24$

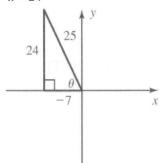

71. $\sin\left[\tan^{-1}\left(\dfrac{\sqrt{5}}{2}\right)\right] = \dfrac{\sqrt{5}}{3}$

$\sqrt{5}^2 + 2^2 = x^2$

$5 + 4 = x^2$

$9 = x^2$

$3 = x$

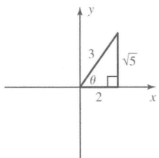

73. $\cot\left[\arcsin\left(\dfrac{3x}{5}\right)\right] = \dfrac{\sqrt{25 - 9x^2}}{3x}$

$y^2 + (3x)^2 = 5^2$

$y^2 + 9x^2 = 25$

$y^2 = 25 - 9x^2$

$y = \sqrt{25 - 9x^2}$

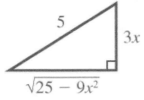

75. $\cos\left[\sin^{-1}\left(\dfrac{x}{\sqrt{12 + x^2}}\right)\right]$

$= \dfrac{\sqrt{12}}{\sqrt{25 + x^2}} = \sqrt{\dfrac{12}{12 + x^2}}$;

$x^2 + y^2 = \sqrt{12 + x^2}^2$

$y^2 = 12 + x^2 - x^2$

$y^2 = 12$

$y = \sqrt{12}$

77. $\sec^{-1} 1 = 0$;

$\sec\left(\dfrac{\pi}{3}\right) = 2$;

$\operatorname{arc\,sec}\left(\dfrac{2}{\sqrt{3}}\right) = 30°$;

$\sec(\pi) = -1$

79. $\operatorname{arc\,csc} x = \arcsin\left(\dfrac{1}{x}\right)$

$\operatorname{arc\,csc} 2 = \sin^{-1}\left(\dfrac{1}{2}\right) = 30° = \dfrac{\pi}{6}$

81. $\cot^{-1} \sqrt{3} = \tan^{-1}\left(\dfrac{1}{\sqrt{3}}\right) = 30° = \dfrac{\pi}{6}$

83. $\operatorname{arc\,sec} 5.789 = \cos^{-1}\left(\dfrac{1}{5.789}\right) = 80.1°$

85. $\sec^{-1} \sqrt{7} = \cos^{-1} \dfrac{1}{\sqrt{7}} = 67.8°$

87. $F_N = m g \cos \theta$

$m = 225 \text{ g} = 0.225 \text{ kg}$

a) Find F_N for $\theta = 15°$ and $\theta = 45°$

$F_N = 0.225(9.8)\cos 15°$

$= 2.13 \text{ N};$

$F_N = 0.225(9.8)\cos 45°$

1.56 N

b) Find θ for F_N for 1 N and $F_N = 2$ N

$1 = 0.225(9.8)\cos\theta \quad 2 = 0.225(9.8)\cos\theta$

$1 = 2.205 \cos \theta \quad\quad 2 = 2.205\cos\theta$

$\dfrac{1}{2.205} = \cos \theta \quad\quad \dfrac{2}{2.205} = \cos \theta$

$\cos^{-1}\left(\dfrac{1}{2.205}\right) = \theta \quad \cos^{-1}\left(\dfrac{2}{2.205}\right) = \theta$

$63° = \theta \quad\quad\quad 24.9° = \theta$

89. $\tan \theta = \dfrac{5.35}{20}$

$\theta = \tan^{-1}\left(\dfrac{5.35}{20}\right)$

$\theta \approx 14.98$

$2\theta \approx 30°$

91. $\tan \theta = \dfrac{150}{48}$

$\theta = \tan^{-1}\left(\dfrac{150}{48}\right)$

$\theta \approx 72.3°$

Straight line distance:

$48^2 + 150^2 = c^2$

$2304 + 22500 = c^2$

$24804 = c^2$

$157.5 \text{ yds} \approx c$

93. a) $\tan \beta = \dfrac{1.5}{x}$

$\beta = \tan^{-1}\left(\dfrac{1.5}{x}\right);$

$\tan \alpha = \dfrac{4}{x}$

$\alpha = \tan^{-1}\left(\dfrac{4}{x}\right)$

b) $\theta = \alpha - \beta = \tan^{-1}\left(\dfrac{4}{x}\right) - \tan^{-1}\left(\dfrac{1.5}{x}\right)$

c) $\theta \approx 27.0°$ at $x \approx 2.5$ ft

95. a) $\theta = \alpha - \beta$

$\tan \alpha = \dfrac{94}{d}$

$\alpha = \tan^{-1}\left(\dfrac{94}{d}\right);$

$\tan \beta = \dfrac{70}{d}$

$\beta = \tan^{-1}\left(\dfrac{70}{d}\right);$

$\theta = \tan^{-1}\left(\dfrac{94}{d}\right) - \tan^{-1}\left(\dfrac{70}{d}\right)$

b) $\theta \approx 8.4°$ at $d \approx 81.1$ ft

97. $\cos \theta = \dfrac{3960}{150 + 3960}$

a) $\theta = \cos^{-1}\left(\dfrac{3960}{150 + 3960}\right) = 15.5°$ or

$\approx 0.2705 \text{ rad}$

b) $d^2 = (x + r)^2 - r^2$

$= x^2 + 2xr + r^2 - r^2 = x^2 + 2xr$

$d \approx 1100 \text{ miles}$

$s = 3960 \cdot 0.2705 = 1071.2$

$d - s \approx 29 \text{ miles}$

99. a) $x = v_0 \cos \theta \, t$

$x = 70 \cos 10(6)$

$x \approx 413.6 \text{ ft away}$

b) $y = 0 + 70 \sin(10)(6) - 16(6)^2$

$= 420 \sin 10 - 576$

$\approx -503 \text{ ft}$

c) $503^2 + 413.6^2 = c^2$

$\sqrt{253009 + 171064.96} = c$

$c \approx 651.2 \text{ ft}$

101. $\sin 2\theta = 2 \sin \theta \cos \theta$

$= 2\left(\dfrac{6}{\sqrt{85}}\right)\left(\dfrac{7}{\sqrt{85}}\right)$

$= \dfrac{84}{85}$

103. $f(x) \le 0$

$f(x) = x^3 - 9x$

$= x(x^2 - 9)$

$= x(x - 3)(x + 3)$

$x = 0; x = 3; x = -3$

$(-\infty, -3] \cup [0, 3]$

6.6 Technology Highlight

Exercise 1:

$Y_1 = (1 + \sin x)^2 + \cos(2x);$

$Y_2 = 4\cos x(1 + \sin x);$

$x \approx -5.1126 + 2\pi k;$

$x \approx -2.1545 + 2\pi k$

6.6 Exercises

1. Principal, $[0, 2\pi)$, real

3. $\dfrac{\pi}{4}; \dfrac{\pi}{4}; \dfrac{3\pi}{4}; \dfrac{\pi}{4} + 2\pi k;, \dfrac{3\pi}{4} + 2\pi k$

5. Answers will vary.

7. $\sin x = \dfrac{-3}{4}$

    $y = \sin x$

    $y = \dfrac{-3}{4}$

    a) Principal root: QIV
    b) 2 roots

9. $y = \tan x$

    $y = -1.5$

    a) Principal root: QIV
    b) 2 roots

11.

θ	$\sin \theta$	$\cos \theta$	$\tan \theta$
0	0	1	0
$\dfrac{\pi}{6}$	$\dfrac{1}{2}$	$\dfrac{\sqrt{3}}{2}$	$\dfrac{\sqrt{3}}{3}$
$\dfrac{\pi}{3}$	$\dfrac{\sqrt{3}}{2}$	$\dfrac{1}{2}$	$\sqrt{3}$
$\dfrac{\pi}{2}$	1	0	Und
$\dfrac{2\pi}{3}$	$\dfrac{\sqrt{3}}{2}$	$-\dfrac{1}{2}$	$-\sqrt{3}$
$\dfrac{5\pi}{6}$	$\dfrac{1}{2}$	$-\dfrac{\sqrt{3}}{2}$	$-\dfrac{\sqrt{3}}{3}$
π	0	-1	0
$\dfrac{7\pi}{6}$	$-\dfrac{1}{2}$	$-\dfrac{\sqrt{3}}{2}$	$\dfrac{\sqrt{3}}{3}$
$\dfrac{4\pi}{3}$	$-\dfrac{\sqrt{3}}{2}$	$-\dfrac{1}{2}$	$\sqrt{3}$

13. $2\cos x = \sqrt{2}$

    $\cos x = \dfrac{\sqrt{2}}{2}$

    $x = \cos^{-1} \dfrac{\sqrt{2}}{2}$

    $x = \dfrac{\pi}{4}$

15. $-4\sin x = 2\sqrt{2}$

    $\sin x = \dfrac{-\sqrt{2}}{2}$

    $x = \sin^{-1}\left(\dfrac{-\sqrt{2}}{2}\right)$

    $x = \dfrac{-\pi}{4}$

17. $\sqrt{3}\tan x = 1$

    $\tan x = \dfrac{1}{\sqrt{3}}$

    $x = \tan^{-1} \dfrac{1}{\sqrt{3}}$

    $x = \dfrac{\pi}{6}$

6.6 Exercises

19. $2\sqrt{3}\sin x = -3$

$\sin x = \dfrac{-3}{2\sqrt{3}}$

$\sin x = \dfrac{-3\sqrt{3}}{6}$

$\sin x = \dfrac{-\sqrt{3}}{2}$

$x = \sin^{-1}\left(\dfrac{-\sqrt{3}}{2}\right)$

$x = \dfrac{-\pi}{3}$

21. $-6\cos x = 6$

$\cos x = -1$

$x = \cos^{-1}(-1)$

$x = \pi$

23. $\dfrac{7}{8}\cos x = \dfrac{7}{16}$

$\cos x = \dfrac{\frac{7}{16}}{\frac{7}{8}}$

$x = \cos^{-1}\left(\dfrac{1}{2}\right)$

$x = \dfrac{\pi}{3}$

25. $2 = 4\sin\theta$

$\dfrac{1}{2} = \sin\theta$

$\sin^{-1}\left(\dfrac{1}{2}\right) = \theta$

$\dfrac{\pi}{6} = \theta$

27. $-5\sqrt{3} = 10\cos\theta$

$\dfrac{-\sqrt{3}}{2} = \cos\theta$

$\cos^{-1}\left(\dfrac{-\sqrt{3}}{2}\right) = \theta$

$\dfrac{5\pi}{6} = \theta$

29. $9\sin x - 3.5 = 1$

$9\sin x = 4.5$

$\sin x = \dfrac{1}{2}$, QI or QII

$x = \sin^{-1}\left(\dfrac{1}{2}\right)$

$x = \dfrac{\pi}{6}$ or $\pi - \dfrac{\pi}{6} = \dfrac{5\pi}{6}$

$x = \dfrac{\pi}{6}, \dfrac{5\pi}{6}$

31. $8\tan x + 7\sqrt{3} = -\sqrt{3}$

$8\tan x = -8\sqrt{3}$

$\tan x = -\sqrt{3}$, QII or QIV

$x = \tan^{-1}\left(-\sqrt{3}\right)$

$x = \pi - \dfrac{\pi}{3} = \dfrac{2\pi}{3}$ or $2\pi - \dfrac{\pi}{3} = \dfrac{5\pi}{3}$

$x = \dfrac{2\pi}{3}, \dfrac{5\pi}{3}$

33. $\dfrac{2}{3}\cot x - \dfrac{5}{6} = \dfrac{-3}{2}$

$\dfrac{2}{3}\cot x = \dfrac{-2}{3}$

$\cot x = -1$, QII or QIV

$x = \cot^{-1}(-1)$

$x = \pi - \dfrac{\pi}{4} = \dfrac{3\pi}{4}$, $x = 2\pi - \dfrac{\pi}{4} = \dfrac{7\pi}{4}$

$x = \dfrac{3\pi}{4}, \dfrac{7\pi}{4}$

35. $4\cos^2 x = 3$

$\cos^2 x = \dfrac{3}{4}$

$\cos x = \dfrac{\pm\sqrt{3}}{2}$

$x = \cos^{-1}\left(\pm\dfrac{\sqrt{3}}{2}\right)$

$x = \dfrac{\pi}{6}, \dfrac{5\pi}{6}, \dfrac{7\pi}{6}, \dfrac{11\pi}{6}$

37. $-7\tan^2 x = -21$

$\tan^2 x = 3$

$\tan x = \pm\sqrt{3}$

$x = \tan^{-1}\left(\pm\sqrt{3}\right)$

$x = \dfrac{\pi}{3}, \dfrac{2\pi}{3}, \dfrac{4\pi}{3}, \dfrac{5\pi}{3}$

39. $-4\csc^2 x = -8$

$\csc^2 x = 2$

$\csc x = \pm\sqrt{2}$

$x = \csc^{-1}\left(\pm\sqrt{2}\right)$

$x = \dfrac{\pi}{4}, \dfrac{3\pi}{4}, \dfrac{5\pi}{4}, \dfrac{7\pi}{4}$

41. $4\sqrt{2}\sin^2 x = 4\sqrt{2}$

$\sin^2 x = 1$

$\sin x = \pm 1$

$x = \sin^{-1}(1)$ or $x = \sin^{-1}(-1)$

$x = \dfrac{\pi}{2}, \dfrac{3\pi}{2}$

43. $3\cos^2\theta + 14\cos\theta - 5 = 0$

Let $u = \cos\theta, u^2 = \cos^2 0$

$3u^2 + 14u - 5 = 0$

$(3u - 1)(u + 5) = 0$

$u = \dfrac{1}{3}$ or $u = -5$

$\cos\theta = \dfrac{1}{3}$ $\cos\theta = -5$

Extraneous because
$-1 \le \cos\theta \le 1$

$\theta = \cos^{-1}\left(\dfrac{1}{3}\right)$

$\theta = 1.2310 + 2\pi k$ or
$\theta = 5.0522 + 2\pi k$

45. $2\cos x \sin x - \cos x = 0$

$\cos x(2\sin x - 1) = 0$

$\cos x = 0$ or $2\sin x - 1 = 0$

$x = \dfrac{\pi}{2}$ or $x = \dfrac{3\pi}{2}$ $2\sin x = 1$

$x = \dfrac{\pi}{2} + \pi k$ $\sin x = \dfrac{1}{2}$

$x = \dfrac{\pi}{6}$ or $x = \dfrac{5\pi}{6}$

$x = \dfrac{\pi}{6} + 2\pi k$

or $x = \dfrac{5\pi}{6} + 2\pi k$

47. $\sec^2 x - 6\sec x = 16$

$\sec^2 x - 6\sec x - 16 = 0$

$(\sec x + 2)(\sec x - 8) = 0$

$\sec x + 2 = 0$ or $\sec x - 8 = 0$

$\sec^2 x = -2$ $\sec x = 8$

$x = \sec^{-1}(-2)$ $x = \sec^{-1}(8)$

$x = \cos^{-1}\left(\dfrac{-1}{2}\right)$ $x = \cos^{-1}\left(\dfrac{1}{8}\right)$

$x = \dfrac{2\pi}{3} + 2\pi k$ or $x = 1.4455 + 2\pi k$

$x = \dfrac{4\pi}{3} + 2\pi k$ $x = 4.8377 + 2\pi k$

49. $4\sin^2 x - 1 = 0$

$(2\sin x - 1)(2\sin x + 1) = 0$

$2\sin x = 1$ or $2\sin x + 1 = 0$

$\sin x = \dfrac{1}{2}$ $2\sin x = -1$

$x = \sin^{-1}\left(\dfrac{1}{2}\right)$ $\sin x = \dfrac{-1}{2}$

        $x = \sin^{-1}\left(\dfrac{-1}{2}\right)$

$x = \dfrac{\pi}{6} + 2\pi k$ $x = \dfrac{11\pi}{6} + 2\pi k$;

or $x = \dfrac{5\pi}{6} + 2\pi k$ $x = \dfrac{7\pi}{6} + 2\pi k$;

$x = \dfrac{\pi}{6} + \pi k$; $x = \dfrac{5\pi}{6} + \pi k$

6.6 Exercises

51. $-2\sin x = \sqrt{2}$

$$\sin x = \frac{-\sqrt{2}}{2}$$

$$\sin^{-1}(\sin x) = \sin^{-1}\left(\frac{-\sqrt{2}}{2}\right)$$

$$x = \frac{5\pi}{4} + 2\pi k \quad \text{or} \quad x = \frac{7\pi}{4} + 2\pi k$$

53. $-4\cos x = 2\sqrt{2}$

$$\cos x = \frac{-\sqrt{2}}{2}$$

$$\cos^{-1}(\cos x) = \cos^{-1}\left(\frac{-\sqrt{2}}{2}\right)$$

$$x = \cos^{-1}\left(\frac{-\sqrt{2}}{2}\right)$$

$$x = \frac{3\pi}{4} + 2\pi k \text{, QII}$$

$$x = \frac{5\pi}{4} + 2\pi k \text{, QIII}$$

55. $\sqrt{3}\tan x = -\sqrt{3}$

$\tan x = -1$

$\tan^{-1}(\tan x) = \tan^{-1}(-1)$

$$x = \frac{3\pi}{4} + \pi k \text{, QII or QIV}$$

57. $6\cos(2x) = -3$

$$\cos(2x) = \frac{-1}{2}$$

$$\cos^{-1}(\cos 2x) = \cos^{-1}\left(\frac{-1}{2}\right)$$

$$2x = \frac{2\pi}{3} + 2\pi k \text{, QII}$$

$$x = \frac{\pi}{3} + \pi k \text{ ;}$$

$$2x = \frac{4\pi}{3} + 2\pi k \text{, QIII}$$

$$x = \frac{2\pi}{3} + \pi k$$

59. $\sqrt{3}\tan 2x = -\sqrt{3}$

$\tan 2x = -1$

$\tan^{-1}(\tan 2x) = \tan^{-1}(-1)$

$$2x = \frac{3\pi}{4} + \pi k \text{, QII or QIV}$$

$$x = \frac{3\pi}{8} + \frac{\pi}{2} k$$

61. $-2\sqrt{3}\cos\left(\frac{1}{3}x\right) = 2\sqrt{3}$

$$\cos\left(\frac{1}{3}x\right) = -1$$

$$\cos^{-1}\left[\cos\left(\frac{1}{3}x\right)\right] = \cos^{-1}(-1)$$

$$\frac{1}{3}x = \pi + 2\pi k$$

$$x = 3\pi + 6\pi k$$

63. $\sqrt{2}\cos x \sin(2x) - 3\cos x = 0$

$$\cos x\left(\sqrt{2}\sin(2x) - 3\right) = 0$$

$$\cos x = 0 \qquad \text{or} \qquad \sqrt{2}\sin(2x) = 3$$

$$\sin(2x) = \frac{3}{\sqrt{2}} > 1$$

Extraneous

$$x = \frac{\pi}{2} + 2\pi k$$

$$\text{or} \quad \frac{3\pi}{2} + 2\pi k$$

Can be combined

$$x = \frac{\pi}{2} + \pi k, k \in Z$$

65. $\cos(3x)\csc(2x) - 2\cos(3x) = 0$

$\cos(3x)(\csc(2x) - 2) = 0$

$\cos(3x) = 0$

$3x = \dfrac{\pi}{2} + 2\pi k$ or $\quad 3x = \dfrac{3\pi}{2}x + 2\pi k$

$x = \dfrac{\pi}{6} + \dfrac{2}{3}\pi k \qquad x = \dfrac{\pi}{2} + \dfrac{2}{3}\pi k$;

Can be combined

$x = \dfrac{\pi}{6} + \dfrac{\pi}{3}k$;

OR

$\csc(2x) - 2 = 0$

$\csc(2x) = 2$

$\dfrac{1}{\sin(2x)} = 2$

$\sin(2x) = \dfrac{1}{2}$

$2x = \dfrac{\pi}{6} + 2\pi k$ or $\quad 2x = \dfrac{5\pi}{6} + 2\pi k$

$x = \dfrac{\pi}{12} + \pi k \qquad x = \dfrac{5\pi}{12} + \pi k$

67. $3\cos x = 1$

$\cos x = \dfrac{1}{3}$

$\cos^{-1}(\cos x) = \cos^{-1}\left(\dfrac{1}{3}\right)$

   a. $\quad x \approx 1.2310$

   b. $\quad 1.2310 + 2\pi k$, QI
   $2\pi - 1.2310 + 2\pi k = 5.0522 + 2\pi k$, QIV

69. $\sqrt{2}\sec x + 3 = 7$

$\sec x = \dfrac{4}{\sqrt{2}}$

$\cos x = \dfrac{\sqrt{2}}{4}$

$\cos^{-1}(\cos x) = \cos^{-1}\left(\dfrac{\sqrt{2}}{4}\right)$

   a. $\quad x \approx 1.2094$

   b. $\quad 1.2094 + 2\pi k$, QI
   $2\pi - 1.2094 = 5.0738$;
   $5.0738 + 2\pi k$, QIV

71. $\dfrac{1}{2}\sin(2\theta) = \dfrac{1}{3}$

$\sin(2\theta) = \dfrac{2}{3}$

$\sin^{-1}(\sin(2\theta)) = \sin^{-1}\left(\dfrac{2}{3}\right)$

   a. $\quad 2\theta \approx 0.7297$
   $\theta \approx 0.3649$

   b. $\quad 2\theta = 0.7297 + 2\pi k$
   $\theta = 0.3649 + \pi k$, QI;
   $\pi - 0.7297 = 2.4119$
   $2\theta = 2.4119 + 2\pi k$
   $\theta = 1.2059 + \pi k$

73. $-5\cos(2\theta) - 1 = 0$

$\cos(2\theta) = \dfrac{-1}{5}$

$\cos^{-1}(\cos(2\theta)) = \cos^{-1}\left(\dfrac{-1}{5}\right)$

   a. $\quad 2\theta \approx 1.7722$
   $\theta \approx 0.8861$

   b. $\quad 2\theta = 1.7722 + 2\pi k$
   $\theta = 0.8861 + \pi k$, QII;
   $\pi + \cos^{-1}(0.2) = 4.5110$, QIII;
   $2\theta = 4.5110 + 2\pi k$
   $\theta = 2.2555 + \pi k$

75. $\cos^2 x - \sin^2 x = \dfrac{1}{2}$

$\cos(2x) = \dfrac{1}{2}$

$2x = \cos^{-1}\left(\dfrac{1}{2}\right)$

$2x = \dfrac{\pi}{3} + 2\pi k$

$x = \dfrac{\pi}{6} + \pi k$, QI and QIII;

$2\pi - \dfrac{\pi}{3} = \dfrac{5\pi}{3}$;

$2x = \dfrac{5\pi}{3} + 2\pi k$

$x = \dfrac{5\pi}{6} + \pi k$, QII and QIV

77. $2\cos\left(\dfrac{1}{2}x\right)\cos x - 2\sin\left(\dfrac{1}{2}x\right)\sin x = 1$

$2\left[\cos\left(\dfrac{1}{2}x\right)\cos x - \sin\left(\dfrac{1}{2}x\right)\sin x\right] = 1$

$\cos\left(\dfrac{1}{2}x + x\right) = \dfrac{1}{2}$

$\cos^{-1}\left[\cos\left(\dfrac{3}{2}x\right)\right] = \cos^{-1}\left(\dfrac{1}{2}\right)$

$\dfrac{3}{2}x = \dfrac{\pi}{3}$;

$\dfrac{3}{2}x = \dfrac{\pi}{3} + 2\pi k$

$x = \dfrac{2\pi}{9} + \dfrac{4\pi}{3}k$, QI

$\dfrac{3}{2}x = \dfrac{5\pi}{3} + 2\pi k$

$x = \dfrac{10\pi}{9} + \dfrac{4\pi}{3}k$, QIV

79. $(\cos\theta + \sin\theta)^2 = 1$

$\cos^2\theta + 2\sin\theta\cos\theta + \sin^2\theta = 1$

$1 + 2\sin\theta\cos\theta = 1$

$2\sin\theta\cos\theta = 0$

$\sin(2\theta) = 0$

$\sin^{-1}[\sin(2\theta)] = \sin^{-1}(0)$

$2\theta = 0 + 2\pi k$ or $\quad 2\theta = \pi + 2\pi k$

$\theta = \pi k \qquad\qquad \theta = \dfrac{\pi}{2} + \pi k$

These can be combined as $\theta = \dfrac{\pi}{2}k$

81. $\cos(2\theta) + 2\sin^2\theta - 3\sin\theta = 0$

$1 - 2\sin^2\theta + 2\sin^2\theta - 3\sin\theta = 0$

$-3\sin\theta = -1$

$\sin\theta = \dfrac{1}{3}$

$\sin^{-1}(\sin\theta) = \sin^{-1}\left(\dfrac{1}{3}\right)$

$\theta \approx 0.3398 + 2\pi k$, QI;

$\pi - 0.3398 = 2.8018$

$\theta \approx 2.8018 + 2\pi k$, QII

83. $5\cos x - x = 3$

$Y_1 = 5\cos x - x$

$Y_2 = 3$

$x \approx 0.7290$

85. $\cos^2(2x) + x = 3$

$Y_1 = \cos^2(2x) + x$

$Y_2 = 3$

$x \approx 2.6649$

87. $x^2 + \sin(2x) = 1$

$Y_1 = x^2 + \sin(2x)$

$Y_2 = 1$

$x \approx 0.4566$

89. $R = \dfrac{5}{49}v^2\sin(2\theta)$

$16 = \dfrac{5}{49}(15)^2\sin(2\theta)$

$\dfrac{784}{1125} = \sin(2\theta)$

$\sin^{-1}\left(\dfrac{784}{1125}\right) = \sin^{-1}(\sin(2\theta))$

$44.2° \approx 2\theta$;

$2\theta = 44.2$

$\theta = 22.1°$, QI;

$180° - 44.2° = 135.8°$

$2\theta = 135.8°$

$\theta = 67.9°$, QII

91. $A(\theta) = 9.8\sin\theta$

$0 = 9.8\sin\theta$

$0 = \sin\theta$

$0° = \theta$

The ramp is horizontal.

93. $A(\theta) = 9.8\sin\theta$

$5 = 9.8\sin\theta$

$\dfrac{25}{49} = \sin\theta$

$\sin^{-1}\left(\dfrac{25}{49}\right) = \sin^{-1}(\sin\theta)$

$30.7° \approx \theta$;

$4.5 = 9.8\sin\theta$

$\dfrac{45}{98} = \sin\theta$

$27.3° \approx \theta$

Smaller

95. $\sin \alpha = k \sin \beta$

$\alpha = 90° - 55° = 35°$;

$\sin 35° = 1.33 \sin \beta$

$0.4313 \approx \sin \beta$

$\sin^{-1}(0.4313) \approx \sin^{-1}(\sin \beta)$

$25.5° \approx \beta$

97. $\sin \alpha = k \sin \beta$

$\alpha = (90° - 40°)$

$\sin(50°) = k \sin 34.3°$

$1.36 \approx k$;

$\sin \alpha = 1.36 \sin 15°$

$\sin \alpha \approx 0.3520$

$\sin^{-1}(\sin \alpha) \approx \sin^{-1}(0.3520)$

$\alpha = 20.6°$

99. $y = 5 \sin\left(\dfrac{1}{2} x\right) + 7$

   a. Since distance $x = 0$,
      y-int: 7 inches

   b. $9.5 = 5 \sin\left(\dfrac{1}{2} x\right) + 7$

   $2.5 = 5 \sin\left(\dfrac{1}{2} x\right)$

   $\dfrac{1}{2} = \sin\left(\dfrac{1}{2} x\right)$

   $\sin^{-1}\left(\dfrac{1}{2}\right) = \sin^{-1}\left[\sin\left(\dfrac{1}{2} x\right)\right]$

   QI

   $\dfrac{\pi}{6} = \dfrac{1}{2} x$

   $\dfrac{\pi}{3} = x$

   $x \approx 1.05$ in.

   QII

   $\dfrac{1}{2} x = \dfrac{5\pi}{6}$

   $x = \dfrac{5\pi}{3}$

   $x \approx 5.24$ in.

101. $A = \dfrac{1}{2} r^2 (\theta - \sin \theta)$

   $12 = \dfrac{1}{2} (10)^2 (\theta - \sin \theta)$

   $Y_1 = 12$

   $Y_2 = \dfrac{1}{2} (10)^2 (\theta - \sin \theta)$

   $\theta \approx 1.1547$

103. $5 \cos x - x = -x$

   $Y_1 = 5 \cos x - x$

   $Y_2 = -x$,

   $5 \cos x - x = -x$

   $5 \cos x = 0$

   $\cos x = 0$

   $x = \dfrac{\pi}{2} + \pi k$

   Intersection method; zero method.
   Explanations will vary.

105. $f(2 + i) = (2 + i)^2 - 4(2 + i) + 5$

   $= 4 + 4i + i^2 - 8 - 4i + 5$

   $= 4 + 4i - 1 - 8 - 4i + 5$

   $= 0$

107.a. $\tan\left[\sin^{-1}\left(-\dfrac{1}{2}\right)\right] = \tan(-30°) = -\dfrac{1}{\sqrt{3}}$

   b. $\sin\left[\tan^{-1}(-1)\right] = \dfrac{-\sqrt{2}}{2}$

   $\sin\left(-\dfrac{\pi}{4}\right) = -\dfrac{\sqrt{2}}{2}$

6.7 Exercises

1. $\sin^2 x + \cos^2 x = 1$, $1 + \tan^2 x = \sec^2 x$,
    $1 + \cot^2 x = \csc^2 x$

3. Factor, grouping

5. Answers will vary.

7. $\sin x + \cos x = \dfrac{\sqrt{6}}{2}$

    $(\sin x + \cos x)^2 = \left(\dfrac{\sqrt{6}}{2}\right)^2$

    $\sin^2 x + 2\sin x \cos x + \cos^2 x = \dfrac{3}{2}$

    $\sin^2 x + \cos^2 x + 2\sin x \cos x = \dfrac{3}{2}$

    $1 + 2\sin x \cos x = \dfrac{3}{2}$

    $2\sin x \cos x = \dfrac{1}{2}$

    $\sin 2x = \dfrac{1}{2}$

    Quadrant I:

    $2x = \dfrac{\pi}{6} + 2\pi k$

    $x = \dfrac{\pi}{12} + \pi k$

    $k = 0, x = \dfrac{\pi}{12}; k = 1, x = \dfrac{13\pi}{12}$

    Quadrant II:

    $2x = \dfrac{5\pi}{6} + 2\pi k$

    $x = \dfrac{5\pi}{12} + \pi k$

    $k = 0, x = \dfrac{5\pi}{12}; k = 1, x = \dfrac{17\pi}{12};$

    $x = \dfrac{\pi}{12}, x = \dfrac{5\pi}{12}$

    $\left(x = \dfrac{13\pi}{12}, x = \dfrac{17\pi}{12} \text{ are extraneous}\right)$

9. $\tan x - \sec x = -1$
    $\tan x = \sec x - 1$
    $\tan^2 x = (\sec x - 1)^2$
    $\tan^2 x = \sec^2 x - 2\sec x + 1$
    $\tan^2 x = 1 + \tan^2 x - 2\sec x + 1$
    $-2 = -2\sec x$
    $\sec x = 1$
    $\cos x = 1$
    $x = 0$

11. $\cos x + \sin x = \dfrac{4}{3}$

    $(\cos x + \sin x)^2 = \left(\dfrac{4}{3}\right)^2$

    $\cos^2 x + 2\cos x \sin x + \sin^2 x = \dfrac{16}{9}$

    $1 + 2\cos x \sin x = \dfrac{16}{9}$

    $2\cos x \sin x = \dfrac{7}{9}$

    $\sin 2x = \dfrac{7}{9}$

    $2x = \sin^{-1}\left(\dfrac{7}{9}\right)$

    $2x = 0.8911, \text{QI} \quad$ or $\quad 2x = \pi - 0.89112, \text{QII}$
    $x = 0.4456 \qquad\qquad\qquad 2x = 2.2505$
    $\qquad\qquad\qquad\qquad\qquad x = 1.1252$

Chapter 6: Trigonometric Identities, Inverses and Equations

13. $\cot x \csc x - 2\cot x - \csc x + 2 = 0$

$\left(\cot x \csc x - 2\cot x\right) - \left(\csc x - 2\right) = 0$

$\cot x\left(\csc x - 2\right) - \left(\csc x - 2\right) = 0$

$\left(\csc x - 2\right)\left(\cot x - 1\right) = 0$

$\csc x = 2 \qquad \text{or} \qquad \cot x = 1$

QI: $x = \dfrac{\pi}{6}$ \qquad\qquad QI: $x = \dfrac{\pi}{4}$

QII: $x = \pi - \dfrac{\pi}{6} = \dfrac{5\pi}{6}$; QIII $x = \pi + \dfrac{\pi}{4} = \dfrac{5\pi}{4}$

$\dfrac{\pi}{4}, \dfrac{5\pi}{4}, \dfrac{\pi}{6}, \dfrac{5\pi}{6}$

15. $3\tan^2 x \cos x - 3\cos x + 2 = 2\tan^2 x$

$3\tan^2 x \cos x - 3\cos x + 2 - 2\tan^2 x = 0$

$3\cos x\left(\tan^2 x - 1\right) + 2\left(1 - \tan^2 x\right) = 0$

$3\cos x\left(\tan^2 x - 1\right) - 2\left(\tan^2 x - 1\right) = 0$

$\left(\tan^2 x - 1\right)\left(3\cos x - 2\right) = 0$

First factor:

$\tan^2 x = 1$

$\tan x = \pm 1$

QI: $x = \dfrac{\pi}{4}$

QII: $x = \dfrac{3\pi}{4}$

QIII: $x = \dfrac{3\pi}{2} - \dfrac{\pi}{4} = \dfrac{5\pi}{4}$

QIV: $x = 2\pi - \dfrac{\pi}{4} = \dfrac{7\pi}{4}$

Second factor:

$3\cos x = 2$

$\cos x = \dfrac{2}{3} > 0$ in QI and QIV

QI: $x = \cos^{-1}\left(\dfrac{2}{3}\right) = 0.8411$

QIV: $x = 2\pi - 0.8411 = 5.4421$

$\dfrac{\pi}{4}, \dfrac{3\pi}{4}, \dfrac{5\pi}{4}, \dfrac{7\pi}{4}, 0.8411, 5.4421$

17. $\dfrac{1 + \cot^2 x}{\cot^2 x} = 2$

$\dfrac{\csc^2 x}{\cot^2 x} = 2$

$\dfrac{\csc x}{\cot x} = \pm\sqrt{2}$

$\dfrac{\dfrac{1}{\sin x}}{\dfrac{\cos x}{\sin x}} = \pm\sqrt{2}$

$\dfrac{1}{\cos x} = \pm\sqrt{2}$

$\cos x = \pm\dfrac{\sqrt{2}}{2}$, QI, QII, QIII, QIV

$x = \cos^{-1}\left(\pm\dfrac{\sqrt{2}}{2}\right)$

$x = \dfrac{\pi}{4}, \dfrac{5\pi}{4}$; Positive in QI, QIV

$x = \dfrac{3\pi}{4}, \dfrac{7\pi}{4}$; Negative in QII, QIII

$\dfrac{\pi}{4}, \dfrac{3\pi}{4}, \dfrac{5\pi}{4}, \dfrac{7\pi}{4}$

19. $3\cos(2x) + 7\sin x - 5 = 0$

$3\left(1 - 2\sin^2 x\right) + 7\sin x - 5 = 0$

$3 - 6\sin^2 x + 7\sin x - 5 = 0$

$-6\sin^2 x + 7\sin x - 2 = 0$

$6\sin^2 x - 7\sin x + 2 = 0$

$\left(2\sin x - 1\right)\left(3\sin x - 2\right) = 0$

First factor:

$2\sin x - 1 = 0$

$2\sin x = 1$

$\sin x = \dfrac{1}{2} > 0$ in QI and QII

$x = \dfrac{\pi}{6}, \dfrac{5\pi}{6}$;

Second factor:

$3\sin x - 2 = 0$

$\sin x = \dfrac{2}{3} > 0$ in QI and QII

QI: $x = 0.7297$

QII: $x = \pi - 0.7297 = 2.4103$

$\dfrac{\pi}{6}, \dfrac{5\pi}{6}, 0.7297, 2.4119$

21. $2\sin^2\left(\dfrac{x}{2}\right) - 3\cos\left(\dfrac{x}{2}\right) = 0$

$2\left(1 - \cos^2\left(\dfrac{x}{2}\right)\right) - 3\cos\left(\dfrac{x}{2}\right) = 0$

$2 - 2\cos^2\left(\dfrac{x}{2}\right) - 3\cos\left(\dfrac{x}{2}\right) = 0$

$2\cos^2\left(\dfrac{x}{2}\right) + 3\cos\left(\dfrac{x}{2}\right) - 2 = 0$

$\left(2\cos\left(\dfrac{x}{2}\right) - 1\right)\left(\cos\left(\dfrac{x}{2}\right) + 2\right) = 0$

First factor:

$2\cos\left(\dfrac{x}{2}\right) = 1$

$\cos\left(\dfrac{x}{2}\right) = \dfrac{1}{2}$

$\cos\left(\dfrac{x}{2}\right) > 0$ in QI and QIV

$\dfrac{x}{2} = \dfrac{\pi}{3}$

QI: $x = \dfrac{2\pi}{3}$

QIV: $\dfrac{x}{2} = 2\pi - \dfrac{\pi}{3} = \dfrac{5\pi}{3}$

$x = \dfrac{10\pi}{3}$ Not in interval.

Second factor:

$\cos\left(\dfrac{x}{2}\right) = -2$

Undefined

$\left\{\dfrac{2\pi}{3}\right\}$

23. $\cos(3x) + \cos(5x)\cos(2x)$
$\qquad + \sin(5x)\sin(2x) - 1 = 0$

$\cos(3x) + \cos(5x - 2x) - 1 = 0$

$\cos(3x) + \cos(3x) - 1 = 0$

$2\cos(3x) = 1$

$\cos(3x) = \dfrac{1}{2} > 0$ in QI and QIV

QI: $3x = \dfrac{\pi}{3}$

$x = \dfrac{\pi}{9}$

QIV: $3x = \dfrac{5\pi}{3}$

$x = \dfrac{5\pi}{9}$

$\dfrac{\pi}{9} + \dfrac{2\pi}{3}k, \dfrac{5\pi}{9} + \dfrac{2\pi}{3}k; k = 0, 1, 2$

25. $\sec^4 x - 2\sec^2 x \tan^2 x + \tan^4 x = \tan^2 x$

$\left(\sec^2 x - \tan^2 x\right)^2 = \tan^2 x$

$1^2 = \tan^2 x$

$\tan x = \pm 1$

$x = \tan^{-1}(\pm 1)$

$x = \dfrac{\pi}{4}, \dfrac{3\pi}{4}, \dfrac{5\pi}{4}, \dfrac{7\pi}{4}$; QI, QII, QIII, QIV

27. $250\sin\left(\dfrac{\pi}{6}x+\dfrac{\pi}{3}\right)-125=0$

$250\sin\left(\dfrac{\pi}{6}x+\dfrac{\pi}{3}\right)=125$

$\sin\left(\dfrac{\pi}{6}x+\dfrac{\pi}{3}\right)=\dfrac{1}{2}$

Let $u=\left(\dfrac{\pi}{6}x+\dfrac{\pi}{3}\right)$

Period: $\dfrac{2\pi}{\dfrac{\pi}{6}}=12\rightarrow[0,12)$;

$\sin u=\dfrac{1}{2}>0$ in QI and QII

$u=\sin^{-1}\left(\dfrac{1}{2}\right)$

$u=\dfrac{\pi}{6},\dfrac{5\pi}{6}$;

QI: $\dfrac{\pi}{6}x+\dfrac{\pi}{3}=\dfrac{\pi}{6}+2\pi k$

$\dfrac{\pi}{6}x=\dfrac{-\pi}{6}+2\pi k$

$x=-1+12k$

$k=0,x=-1$

$k=1,x=11$;

$\dfrac{\pi}{6}x+\dfrac{\pi}{3}=\dfrac{5\pi}{6}+2\pi k$, QII

$\dfrac{\pi}{6}x=\dfrac{\pi}{2}+2\pi k$

$x=3+12k$

If $k=0,x=3$;

Period: $\dfrac{2\pi}{\dfrac{\pi}{6}}=12\rightarrow[0,12)$

$x=3,x=11$

29. $1235\cos\left(\dfrac{\pi}{12}x-\dfrac{\pi}{4}\right)+772=1750$

$1235\cos\left(\dfrac{\pi}{12}x-\dfrac{\pi}{4}\right)=970$

$\cos\left(\dfrac{\pi}{12}x-\dfrac{\pi}{4}\right)=0.7919$

Let $u=\left(\dfrac{\pi}{12}x-\dfrac{\pi}{4}\right)$

$\cos u=0.7919$

$\cos^{-1}(\cos u)=\cos^{-1}(0.7919)$

$u=\cos^{-1}(0.7919)$

$u\approx0.6569$;

$\dfrac{\pi}{12}x-\dfrac{\pi}{4}\approx0.6569$ QI

$\dfrac{\pi}{12}x\approx1.4423$

$x\approx5.5091$;

QIV

$\dfrac{\pi}{12}x-\dfrac{\pi}{4}\approx-\cos^{-1}(0.7919)$

$x\approx0.4909$

Period: $\dfrac{2\pi}{\dfrac{\pi}{12}}=24\rightarrow[0,24)$;

$x\approx0.4909$, $x\approx5.5091$

31. $\cos x - \sin x = \dfrac{\sqrt{2}}{2}$

$\cos^2 x - 2\sin x \cos x + \sin^2 x = \dfrac{1}{2}$

$1 - 2\sin x \cos x = \dfrac{1}{2}$

$-2\sin x \cos x = \dfrac{-1}{2}$

$2\sin x \cos x = \dfrac{1}{2}$

$\sin 2x = \dfrac{1}{2} > 0$; QI and QII

$2x = \sin^{-1}\left(\dfrac{1}{2}\right)$

$2x = \dfrac{\pi}{6} + 2\pi k$ or $2x = \dfrac{5\pi}{6} + 2\pi k$

$x = \dfrac{\pi}{12} + \pi k$ $x = \dfrac{5\pi}{12} + \pi k$

If $k = 0, x = \dfrac{\pi}{12}$ If $k = 0, x = \dfrac{5\pi}{12}$

If $k = 1, x = \dfrac{13\pi}{12}$ If $k = 0, x = \dfrac{17\pi}{12}$

Extraneous: $\dfrac{13\pi}{12}, \dfrac{5\pi}{12}$

$\dfrac{\pi}{12}, \dfrac{17\pi}{12}$

33. $\dfrac{1 - \cos^2 x}{\tan^2 x} = \dfrac{\sqrt{3}}{2}$

$\dfrac{\sin^2 x}{\tan^2 x} = \dfrac{\sqrt{3}}{2}$

$\dfrac{\sin^2 x}{\dfrac{\sin^2 x}{\cos^2 x}} = \dfrac{\sqrt{3}}{2}$

$\cos^2 x = \dfrac{\sqrt{3}}{2}$

$\cos x = \pm\sqrt{\dfrac{\sqrt{3}}{2}}$; QI, QII, QIII, QIV

$x = \cos^{-1}\left(\pm\sqrt{\dfrac{\sqrt{3}}{2}}\right)$

QI: $x = 0.3747$
QII: $x = 2.7669$
QIII: $\pi + 0.3747 = 3.5163$
QIV: $2\pi - 0.3747 = 5.9085$

35. $\csc x + \cot x = 1$

$\csc x = 1 - \cot x$

$(\csc x)^2 = (1 - \cot x)^2$

$\csc^2 x = 1 - 2\cot x + \cot^2 x$

$1 + \cot^2 x = 1 - 2\cot x + \cot^2 x$

$0 = -2\cot x$

$2\cot x = 0$

$\cot x = 0$

$\dfrac{\pi}{2}; \left(\dfrac{3\pi}{2} \text{ is extraneous}\right)$

37. $\sec x \cos\left(\dfrac{\pi}{2} - x\right) = -1$

$\sec x \sin x = -1$

$\dfrac{1}{\cos x}\sin x = -1$

$\tan x = -1$ in QII and QIV

QII: $\pi - \dfrac{\pi}{4} = \dfrac{3\pi}{4}$

QIV: $2\pi - \dfrac{\pi}{4} = \dfrac{7\pi}{4}$

39. $\sec^2 x \tan\left(\dfrac{\pi}{2} - x\right) = 4$

$\dfrac{1}{\cos^2 x} \cdot \cot x = 4$

$\dfrac{1}{\cos^2 x} \cdot \dfrac{\cos x}{\sin x} = 4$

$\dfrac{1}{\cos x \sin x} = 4$

$\cos x \sin x = \dfrac{1}{4}$

$2\cos x \sin x = \dfrac{1}{2}$

$\sin 2x = \dfrac{1}{2} > 0$ in QI and QII

QI: $2x = \dfrac{\pi}{6} + 2\pi k$; $x = \dfrac{\pi}{12} + \pi k$

QII: $2x = \dfrac{5\pi}{6} + 2\pi k$; $x = \dfrac{5\pi}{12} + \pi k$

$x = \dfrac{\pi}{12} + \pi = \dfrac{13\pi}{12}$

$x = \dfrac{5\pi}{12} + \pi = \dfrac{17\pi}{12}$;

$k = 0, x = \dfrac{\pi}{12}, x = \dfrac{5\pi}{12}$

$k = 1, x = \dfrac{13\pi}{12}, x = \dfrac{17\pi}{12}$

$\dfrac{\pi}{12}, \dfrac{5\pi}{12}, \dfrac{13\pi}{12}, \dfrac{17\pi}{12}$

41. $y = \dfrac{D - x\cos\theta}{\sin\theta}$

I. $L_1 : y = -x + 5$

$L_2 : y = x$

a. Point of intersection: $\left(\dfrac{5}{2}, \dfrac{5}{2}\right)$

b. $D = \sqrt{a^2 + b^2}$

$D = \sqrt{2.5^2 + 2.5^2}$

$D = \sqrt{6.25 + 6.25}$

$D = \sqrt{12.5}$;

$\theta = \tan^{-1}\left(\dfrac{b}{a}\right)$

$\theta = \tan^{-1}\left(\dfrac{2.5}{2.5}\right)$

$\theta = \tan^{-1}(1)$

$\theta = \dfrac{\pi}{4}$;

$y = \dfrac{\sqrt{12.5} - x\cos\left(\dfrac{\pi}{4}\right)}{\sin\left(\dfrac{\pi}{4}\right)}$

c. Verified.

II. $L_1 : y = -\dfrac{1}{2}x + 5$

$L_2 : y = 2x$

a. Point of Intersection: $(2, 4)$

b. $D = \sqrt{2^2 + 4^2}$

$D = \sqrt{4 + 16}$

$D = \sqrt{20}$

$D = 2\sqrt{5}$;

$\theta = \tan^{-1}\left(\dfrac{4}{2}\right)$

$\theta = \tan^{-1}(2)$

$\theta \approx 1.1071$

$y = \dfrac{2\sqrt{5} - x\cos 1.1071}{\sin 1.1071}$

c. Verified.

III. $L_1 : y = \dfrac{-\sqrt{3}}{3}x + \dfrac{4\sqrt{3}}{3}$

$L_2 : y = \sqrt{3}x$

a. Point of intersection: $\left(1, \sqrt{3}\right)$

b. $D = \sqrt{(1)^2 + \left(\sqrt{3}\right)^2}$

$D = \sqrt{1 + 3}$

$D = \sqrt{4}$

$D = 2$;

$\theta = \tan^{-1}\left(\dfrac{\sqrt{3}}{1}\right)$

$\theta = \dfrac{\pi}{3}$;

$y = \dfrac{2 - x\cos\left(\dfrac{\pi}{3}\right)}{\sin\left(\dfrac{\pi}{3}\right)}$

c. Verified.

43. $v = \pi r^2 h \sin \theta$

$\theta = \dfrac{\pi}{2}$

$90° - \alpha = \theta$

$90 - 5° = \theta$

$r = 10, h = 25,$ angle of deflection: $x = 5°$

a. $V = \pi r^2 h \sin \theta$

$V = \pi (10)^2 (25) \sin\left(\dfrac{\pi}{2}\right)$

$V = 2500\pi \sin\left(\dfrac{\pi}{2}\right)$

$V = 2500\pi \ \text{ft}^3$ or $V \approx 7853.9816 \ \text{ft}^3$

b. $V = \pi (10)^2 (25) \sin(85)$

$V = 2500\pi \sin(85)$

$V \approx 7824.09 \ \text{ft}^3$

c. 98% of $7853.98 = 7696.90$

$7696.90 = 2500\pi \sin(\theta)$

$0.98 = \sin(\theta)$

$\sin^{-1}(0.98) = \theta$

$78.5° = \theta$

45. $D(t) = 36 \sin\left(\dfrac{\pi}{4} t - \dfrac{9}{4}\right) + 44$

a. Mid Sept. corresponds to $t = 4.5$

$D(4.5) = 36 \sin\left(\dfrac{\pi}{4}(4.5) - \dfrac{9}{4}\right) + 44$

$D(4.5) \approx 78.53 \ \text{m}^3/\text{sec}$

b. Using a grapher: August to November
Algebraically:

$50 = 36 \sin\left(\dfrac{\pi}{6} t - \dfrac{9}{4}\right) + 44$

$6 = 36 \sin\left(\dfrac{\pi}{4} t - \dfrac{9}{4}\right)$

$\dfrac{1}{6} = \sin\left(\dfrac{\pi}{4} t - \dfrac{9}{4}\right)$; Period = 12

Let $u = \dfrac{\pi}{4} t - \dfrac{9}{4}$

$\dfrac{1}{6} = \sin u$

$\sin u > 0$ in QI and QII

$\sin^{-1}\left(\dfrac{1}{6}\right) = \sin^{-1}(\sin u)$

$u = 0.1674$ or $u = \pi - 0.1674$

$u = 2.9741$

$\dfrac{\pi}{4} t - \dfrac{9}{4} = 0.1674$ $\qquad$ $\dfrac{\pi}{4} t - \dfrac{9}{4} = 2.9741$

$t = 3.0779$ (Aug) $\qquad$ $t = 6.6515$ (Nov)

For June through February, the discharge rate is over $50 \ \text{m}^3/\text{sec}$, in Aug., Sept., Oct., and Nov.

47. $S(x) = 1600 \cos\left(\dfrac{\pi}{6} x - \dfrac{\pi}{12}\right) + 5100$

a. $S(7) = 1600 \cos\left(\dfrac{\pi}{6}(7) - \dfrac{\pi}{12}\right) + 5100$

$S(7) \approx \$3554.52$

b. Using a grapher: May, June, July and August
Algebraically:

$4000 = 1600 \cos\left(\dfrac{\pi}{6} x - \dfrac{\pi}{12}\right) + 5100$

$-1100 = 1600 \cos\left(\dfrac{\pi}{6} x - \dfrac{\pi}{12}\right)$

$\dfrac{-11}{16} = \cos\left(\dfrac{\pi}{6} x - \dfrac{\pi}{12}\right)$

Let $u = \dfrac{\pi}{6} x - \dfrac{\pi}{12}$

$\cos u < 0$ in QII and QIII

$\dfrac{\pi}{6} x - \dfrac{\pi}{12} = 2.3288$

$x \approx 4.9478$

$x = 5, x = 6,7,8$

Or

$\dfrac{\pi}{6} x - \dfrac{\pi}{12} = \pi + 0.8128$

$x \approx 8.0522$

May, June, July, August

Chapter 6: Trigonometric Identities, Inverses and Equations

49. $T(x) = 9\cos\left(\dfrac{\pi}{6}x\right) + 15$

    a. Mid March corresponds to $x = 3.5$

    $T(3.5) = 9\cos\left(\dfrac{\pi}{6}(3.5)\right) + 15$

    $T(3.5) \approx 12.67$ in.

    b. Using a grapher,

    $10.5 = 9\cos\left(\dfrac{\pi}{6}x\right) + 15$

    $-4.5 = 9\cos\left(\dfrac{\pi}{6}x\right)$

    $\dfrac{-1}{2} = \cos\left(\dfrac{\pi}{6}x\right)$

    Let $u = \dfrac{\pi}{6}x$

    $\cos u < 0$ in QII and QIII

    $u = \dfrac{2\pi}{3}$ or $u = \dfrac{4\pi}{3}$

    $\dfrac{\pi}{6}x = \dfrac{2\pi}{3}$ $\qquad$ $\dfrac{\pi}{6}x = \dfrac{4\pi}{3}$

    $x = 4$ $\qquad\qquad$ $x = 8$

    $x = 4, x = 5, x = 6, x = 7, x = 8$

    The average thickness is at most 10.5 inches in April, May, June, July and August.

51. $G(x) = 21\cos\left(\dfrac{2\pi}{365}x + \dfrac{\pi}{2}\right) + 29$

    a. March 21 corresponds to $x = 80$

    $G(80) = 21\cos\left[\dfrac{2\pi}{365}(80) + \dfrac{\pi}{2}\right] + 29$

    $G(80) \approx 8.39$ gallons

    b. Using a grapher,

    $40 = 21\cos\left(\dfrac{2\pi}{365}x + \dfrac{\pi}{2}\right) + 29$

    $\dfrac{11}{21} = \cos\left(\dfrac{2\pi}{365}x + \dfrac{\pi}{2}\right)$

    Let $u = \dfrac{2\pi}{365}x + \dfrac{\pi}{2}$

    $\cos u > 0$ in QI and QIV

$u = 1.0195$

$\dfrac{2\pi}{365}x + \dfrac{\pi}{2} = 1.0195$

$x = -32.0257 + 365k$

At $k = 1$, x = 332.9743

Or

$u = 2\pi - 1.0195 = 5.2637$

$\dfrac{2\pi}{365}x + \dfrac{\pi}{2} = 5.2637$

$x = 214.5257$

$x \approx 214$ to day 333

53. $B(x) = 58\cos\left(\dfrac{\pi}{6}x + \pi\right) + 126$

    a. $B(0) = 58\cos\left[\dfrac{\pi}{6}(0) + \pi\right] + 126$

    $B(x) = 68$ bpm

    b. $B(5) = 58\cos\left[\dfrac{\pi}{6}(5) + \pi\right] + 126$

    $B(5) \approx 176.2$ bpm

    c. Using a grapher,

    $Y_1 = 58\cos\left(\dfrac{\pi}{6}x + \pi\right) + 126$

    $Y_2 = 170$

    From about 4.6 min to 7.4 min.

55. Answers will vary.

    (a) For example: $y = 19\cos\left(\pi - \dfrac{\pi}{6}x\right) + 53$

    (b) For example: $y = -21\sin\left(\dfrac{2\pi}{365}x\right) + 29$

327

57. Option I:
h = height (side opposite θ)
x = length and width (square)
Surface Area:
$2x^2 + 4xh = 1288$
$x^2 + 2xh = 644$;
Edges:
$4h + 4x + 4x = 176$
$4h + 8x = 176$
$h + 2x = 44$
$h = 44 - 2x$;
$x^2 + 2x(44 - 2x) = 644$
$x^2 + 88x - 4x^2 = 644$
$-3x^2 + 88x = 644$
$0 = 3x^2 - 88x + 644$
$0 = (3x - 46)(x - 14)$
$x = \dfrac{46}{3}$ or
$x = 14$; $h = 44 - 2(14) = 16$
14 cm x 14 cm x 16 cm
Option II:
Choosing base to be on the "side", length of base diagonal:
Base diagonal:
$\sqrt{14^2 + 16^2} = \sqrt{452} = 2\sqrt{113}$
$\cos\theta = \dfrac{2\sqrt{113}}{18\sqrt{2}}$
$\theta \approx 33.4°$
(a) Length of base diagonal:
$\sqrt{14^2 + 14^2} = 14\sqrt{2}$;
$\sqrt{\left(14\sqrt{2}\right)^2 + 16^2} = d$
$\sqrt{648} = d$
$18\sqrt{2} = d$
$25.5\,\text{cm} \approx d$
(b) $\cos\theta = \dfrac{14\sqrt{2}}{18\sqrt{2}}$
$\cos\theta = \dfrac{7}{9}$
$\theta \approx 38.9°$

59. $f(x) = x^4 - 3x^3 + 4x$
End behavior: up/up
$f(x) = x\left(x^3 - 3x^2 + 4\right)$
Possible rational roots:
$\dfrac{\{\pm 1, \pm 4, \pm 2\}}{\{\pm 1\}}$

$$\begin{array}{r|rrrr} -1 & 1 & -3 & 0 & 4 \\ & & -1 & 4 & -4 \\ \hline & 1 & -4 & 4 & \boxed{0} \end{array}$$

$f(x) = x(x + 1)\left(x^2 - 4x + 4\right)$
$f(x) = x(x + 1)(x - 2)^2$
Bounce at (2,0) (multiplicity 2),
cut at (0,0) and (−1,0)

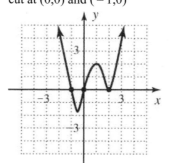

61. Let β represent the angle formed from the base of the tower to the top of the tower.
$\tan\beta = \dfrac{1450}{1000}$
$\tan^{-1}(\tan\beta) = \tan^{-1}\left(\dfrac{1450}{1000}\right)$
$\beta = 55.41°$;
Let α represent the angle formed from the base of the tower to the top of the antenna.
$\tan\alpha = \dfrac{1730}{1000}$
$\tan^{-1}(\tan\alpha) = \tan^{-1}\left(\dfrac{1730}{1000}\right)$
$\alpha = 59.97°$;
$\theta = \alpha - \beta = 59.97 - 55.41 = 4.56°$

Chapter 6: Trigonometric Identities, Inverses and Equations

<u>Summary and Concept Review</u>

1. $\sin x(\csc x - \sin x) = \cos^2 x$

$\sin x \csc x - \sin^2 x =$

$\sin x\left(\dfrac{1}{\sin x}\right) - \sin^2 x =$

$1 - \sin^2 x =$

$\cos^2 x = \cos^2 x$

3. $\dfrac{(\sec x - \tan x)(\sec x + \tan x)}{\csc x} = \sin x$

$\dfrac{\sec^2 x + \sec x \tan x - \sec x \tan x - \tan^2 x}{\csc x} =$

$\dfrac{\sec^2 x - \tan^2 x}{\csc x} =$

$\dfrac{1 + \tan^2 x - \tan^2 x}{\csc x} =$

$\dfrac{1}{\csc x} =$

$\sin x = \sin x$

5. $\cos\theta = \dfrac{-12}{37}; \ \theta$ in QIII

$\sqrt{37^2 - (-12)^2} = 35$

$\sin\theta = -\dfrac{35}{37} \ (\theta$ in QIII$);$

$\tan\theta = \dfrac{-35}{-12} = \dfrac{35}{12};$

$\cot\theta = \dfrac{12}{35};$

$\sec\theta = \dfrac{1}{\cos\theta} = \dfrac{1}{\dfrac{-12}{37}} = \dfrac{-37}{12};$

$\csc\theta = \dfrac{1}{\sin\theta} = \dfrac{1}{\dfrac{-35}{37}} = \dfrac{-37}{35}$

7. $\csc x + \cot x$

$= \dfrac{1}{\sin x} + \dfrac{\cos x}{\sin x}$

$= \dfrac{1}{\sin x} + \dfrac{\cos x}{\sin x}$

$= \dfrac{1 + \cos x}{\sin x}$

Reverse steps. Answers will vary.

9. $\dfrac{\csc^2 x\left(1 - \cos^2 x\right)}{\tan^2 x} = \cot^2 x$

$\dfrac{\csc^2 x\left(\sin^2 x\right)}{\tan^2 x} =$

$\dfrac{1}{\tan^2 x} =$

$\cot^2 x = \cot^2 x$

11. $\dfrac{\sin^4 x - \cos^4 x}{\sin x \cos x} = \tan x - \cot x$

$\dfrac{\left(\sin^2 x - \cos^2 x\right)\left(\sin^2 x + \cos^2 x\right)}{\sin x \cos x} =$

$\dfrac{\left(\sin^2 x - \cos^2 x\right)(1)}{\sin x \cos x} =$

$\dfrac{\sin^2 x}{\sin x \cos x} - \dfrac{\cos^2 x}{\sin x \cos x} =$

$\dfrac{\sin x}{\cos x} - \dfrac{\cos x}{\sin x} =$

$\tan x - \cot x = \tan x - \cot x$

13. a. $\cos 75° = \cos(30° + 45°)$

$= \cos 30° \cos 45° - \sin 30° \sin 45°$

$= \dfrac{\sqrt{3}}{2} \cdot \dfrac{\sqrt{2}}{2} - \dfrac{1}{2} \cdot \dfrac{\sqrt{2}}{2}$

$= \dfrac{\sqrt{6}}{4} - \dfrac{\sqrt{2}}{4} = \dfrac{\sqrt{6} - \sqrt{2}}{4}$

b. $\tan\left(\dfrac{\pi}{12}\right) = \tan\left(\dfrac{\pi}{3} - \dfrac{\pi}{4}\right)$

$= \dfrac{\tan\left(\dfrac{\pi}{3}\right) - \tan\left(\dfrac{\pi}{4}\right)}{1 + \tan\left(\dfrac{\pi}{3}\right)\tan\left(\dfrac{\pi}{4}\right)}$

$= \dfrac{\sqrt{3} - 1}{1 + \left(\sqrt{3}\right)(1)} = \dfrac{\sqrt{3} - 1}{\sqrt{3} + 1} \cdot \dfrac{\sqrt{3} - 1}{\sqrt{3} - 1}$

$= \dfrac{3 - \sqrt{3} - \sqrt{3} + 1}{3 - \sqrt{3} + \sqrt{3} - 1} = \dfrac{4 - 2\sqrt{3}}{2} = 2 - \sqrt{3}$

Summary and Concept Review

15. a. $\cos 109° \cos 71° - \sin 109° \sin 71°$
$= \cos(109° + 71°)$
$\cos 180° = -1$

 b. $\sin 139° \cos 19° - \cos 139° \sin 19°$
$= \sin(139° - 19°)$
$= \sin 120°$
$= \dfrac{\sqrt{3}}{2}$

17. a. $\cos 1170°$
$1170° - 3(360°) = 1170° - 1080° = 90°$
$\cos 1170° = \cos(1080° + 90°)$
$= \cos(1080°)\cos(90°) - \sin 1080° \sin 90°$
$= \cos 90° = 0$

 b. $\sin\left(\dfrac{57\pi}{4}\right)$
$\dfrac{57\pi}{4} = \dfrac{56\pi}{4} + \dfrac{\pi}{4} = 7(2\pi) + \dfrac{\pi}{4}$
$\sin\left(14\pi + \dfrac{\pi}{4}\right) =$
$= \sin(14\pi)\cos\left(\dfrac{\pi}{4}\right) + \cos(14\pi)\sin\left(\dfrac{\pi}{4}\right)$
$= \sin\dfrac{\pi}{4} = \dfrac{\sqrt{2}}{2}$

19. $\tan 15° = \tan(45° - 30°)$
$= \dfrac{\tan 45° - \tan 30°}{1 + \tan 45° \tan 30°}$
$= \dfrac{1 - \dfrac{1}{\sqrt{3}}}{1 + 1\left(\dfrac{1}{\sqrt{3}}\right)} = \dfrac{\dfrac{\sqrt{3}-1}{\sqrt{3}}}{\dfrac{\sqrt{3}+1}{\sqrt{3}}}$
$= \dfrac{\sqrt{3}-1}{\sqrt{3}} \cdot \dfrac{\sqrt{3}}{\sqrt{3}+1} = \dfrac{\sqrt{3}-1}{\sqrt{3}+1}$;
$\tan 15° = \tan(135° - 120°)$
$= \dfrac{\tan 135° - \tan 120°}{1 + \tan 135° \tan 120°}$
$= \dfrac{-1+\sqrt{3}}{1+(-1)(-\sqrt{3})} = \dfrac{-1+\sqrt{3}}{1+\sqrt{3}} = \dfrac{\sqrt{3}-1}{\sqrt{3}+1}$

Both expressions yield the same results.

21. a. $\cos\theta = \dfrac{13}{85}; \theta$ in QIV
$13^2 + y^2 = 85^2$
$y^2 = 85^2 - 13^2$
$y^2 = 7056$
$y = 84$;
$\sin(2\theta) = 2\sin\theta\cos\theta$
$= 2\left(\dfrac{-84}{85}\right)\left(\dfrac{13}{85}\right) = \dfrac{-2184}{7225}$;
$\cos(2\theta) = 2\cos^2\theta - 1$
$= 2\left(\dfrac{13}{85}\right)^2 - 1 = \dfrac{-6887}{7225}$;
$\tan(2\theta) = \dfrac{\sin 2\theta}{\cos 2\theta}$
$= \dfrac{\dfrac{-2184}{7225}}{\dfrac{-6887}{7225}} = \dfrac{2184}{6887}$

 b. $\csc\theta = \dfrac{-29}{20}; \theta$ in QIII
$x^2 + (-20)^2 = 29^2$
$x^2 + 400 = 841$
$x^2 = 441$
$x = -21$;
$\sin(2\theta) = 2\sin\theta\cos\theta$
$= 2\left(\dfrac{-20}{29}\right)\left(\dfrac{-21}{29}\right) = \dfrac{840}{841}$;
$\cos(2\theta) = \cos^2\theta - \sin^2\theta$
$= \left(\dfrac{-21}{29}\right)^2 - \left(\dfrac{-20}{29}\right)^2$
$= \dfrac{441}{841} - \dfrac{400}{841} = \dfrac{41}{841}$;
$\tan(2\theta) = \dfrac{2\tan\theta}{1 - \tan^2\theta}$
$= \dfrac{2\left(\dfrac{-20}{-21}\right)}{1 - \left(\dfrac{-20}{-21}\right)^2} = \dfrac{\dfrac{40}{21}}{1 - \dfrac{400}{441}}$
$= \dfrac{\left(\dfrac{40}{21}\right)}{\dfrac{441-400}{441}} = \dfrac{40}{21} \cdot \dfrac{21^2}{41}$
$= \dfrac{40 \cdot 21}{41} = \dfrac{840}{41}$

23. a. $\cos^2 22.5^2 - \sin^2 22.5°$
    $= \cos[2(22.5°)]$

    $= \cos 45° = \dfrac{\sqrt{2}}{2}$

   b. $1 - 2\sin^2\left(\dfrac{\pi}{12}\right) = \cos\left(2 \cdot \dfrac{\pi}{12}\right)$

    $= \cos\left(\dfrac{\pi}{6}\right) = \dfrac{\sqrt{3}}{2}$

25. a. $\cos\theta = \dfrac{24}{25}; 0° < \theta < 360°$, θ in QIV

    $270° < \theta < 360°$

    $\dfrac{270°}{2} < \dfrac{\theta}{2} < \dfrac{360°}{2}$

    $135° < \dfrac{\theta}{2} < 180°$

    $\cos\left(\dfrac{\theta}{2}\right) = -\sqrt{\dfrac{1+\cos\theta}{2}}$

    $= -\sqrt{\dfrac{1+\dfrac{24}{25}}{2}} = -\sqrt{\dfrac{49}{25} \cdot \dfrac{1}{2}}$

    $= -\sqrt{\dfrac{49}{50}} = \dfrac{-7}{5\sqrt{2}}, \left(\dfrac{\theta}{2} \text{ in QII}\right);$

    $\sin\left(\dfrac{\theta}{2}\right) = \sqrt{\dfrac{1-\cos\theta}{2}}$

    $= \sqrt{\dfrac{1-\dfrac{24}{25}}{2}} = \sqrt{\dfrac{1}{25} \cdot \dfrac{1}{2}}$

    $= \sqrt{\dfrac{1}{50}} = \dfrac{1}{5\sqrt{2}}, \dfrac{\theta}{2} \text{ in QII}$

   b. $\csc\theta = \dfrac{-65}{33}$; θ in QIV

    $-90° < \theta < 0°$

    $-\dfrac{90°}{2} < \dfrac{\theta}{2} < \dfrac{0°}{2}$

    $-45° < \dfrac{\theta}{2} < 0°$;

    $\sin\theta = -\dfrac{33}{65}$;

    $\sin^2\theta + \cos^2\theta = 1$

    $\left(\dfrac{-33}{65}\right)^2 + \cos^2\theta = 1$

$\cos^2\theta = 1 - \left(\dfrac{-33}{65}\right)^2 = \dfrac{3136}{65^2}$

$\cos\theta = \dfrac{56}{65}$ (in QII);

$\sin\left(\dfrac{\theta}{2}\right) = -\sqrt{\dfrac{1-\cos\theta}{2}}$

$= \sqrt{\dfrac{1-\dfrac{56}{65}}{2}} = -\sqrt{\dfrac{9}{65} \cdot \dfrac{1}{2}}$

$= -\sqrt{\dfrac{9}{130}} = \dfrac{-3}{\sqrt{130}}, \left(\dfrac{\theta}{2} \text{ in QIV}\right);$

$\cos\left(\dfrac{\theta}{2}\right) = \sqrt{\dfrac{1+\cos\theta}{2}} = \sqrt{\dfrac{1+\dfrac{56}{65}}{2}}$

$= \sqrt{\dfrac{121}{65} \cdot \dfrac{1}{2}} = \sqrt{\dfrac{121}{130}} = \dfrac{11}{\sqrt{130}}$

$\left(\dfrac{\theta}{2} \text{ in QIV}\right)$

27. $\cos(3x) + \cos x = 0$

$2\left[\cos\left(\dfrac{3x+x}{2}\right)\cos\left(\dfrac{3x-x}{2}\right)\right] = 0$

$2\cos(2x)\cos x = 0$

$2\cos 2x = 0 \qquad \text{or} \qquad \cos x = 0$

$\cos(2x) = 0 \qquad\qquad x = \dfrac{\pi}{2} + \pi k; k \in Z$

$2x = \dfrac{\pi}{2} + \pi k; k \in Z$

$x = \dfrac{\pi}{4} + \dfrac{\pi}{2}k; k \in Z$

29. $y = \sin^{-1}\left(\dfrac{\sqrt{2}}{2}\right)$

    $y = \dfrac{\pi}{4}$ or $45°$

31. $y = \arccos\left(-\dfrac{\sqrt{3}}{2}\right)$

    $y = \dfrac{5\pi}{6}$ or $150°$

33. $y = \sin^{-1}(0.8892)$
    $y = 1.0956$ or $62.8°$

35. $\sin\left[\sin^{-1}\left(\dfrac{1}{2}\right)\right] = \dfrac{1}{2}$

37. $\cos\left[\cos^{-1}(2)\right] =$ undefined

39. $\arccos[\cos(-60°)]$
    $= \arccos\left(\dfrac{1}{2}\right) = \dfrac{\pi}{3}$ or $60°$

41. $\sin\left[\cos^{-1}\left(\dfrac{12}{37}\right)\right] = \dfrac{35}{37}$
    $12^2 + y^2 = 37^2$
    $y = \sqrt{37^2 - 12^2} = 35$

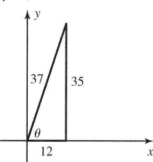

43. $\cot\left[\sin^{-1}\left(\dfrac{x}{\sqrt{81+x^2}}\right)\right] = \dfrac{9}{x}$
    $x^2 + a^2 = \left(\sqrt{81+x^2}\right)^2$
    $a^2 = 81 + x^2 - x^2$
    $a^2 = 81$
    $a = 9$

45. $7\sqrt{3}\sec\theta = x$
    $\sec\theta = \dfrac{x}{7\sqrt{3}}$
    $\theta = \sec^{-1}\left(\dfrac{x}{7\sqrt{3}}\right)$

47. $2\sin x = \sqrt{2}$
    $\sin x = \dfrac{\sqrt{2}}{2}$

    a) Principal root: $\dfrac{\pi}{4}$
       $\sin^{-1}(\sin x) = \sin^{-1}\left(\dfrac{\sqrt{2}}{2}\right)$
       $x = \dfrac{\pi}{4}$

    b) $[0, 2\pi); \left\{\dfrac{\pi}{4}, \dfrac{3\pi}{4}\right\}$
       Because $\dfrac{\sqrt{2}}{2}$ is positive, angles are in QI & QII.

    c) All real roots: $x = \dfrac{\pi}{4} + 2\pi k, k \in Z$ or
       $x = \dfrac{3\pi}{4} + 2\pi k, k \in Z$

49. $8\tan x + 7\sqrt{3} = -\sqrt{3}$
    $8\tan x = -8\sqrt{3}$
    $\tan x = -\sqrt{3}$
    $\tan^{-1}(\tan x) = \tan^{-1}\left(-\sqrt{3}\right)$
    $x = \dfrac{2\pi}{3}$

    a) Principal root: $\dfrac{-\pi}{3}$
       $\dfrac{2\pi}{3} - \pi = \dfrac{-\pi}{3}$

    b) $[0, 2\pi); \left\{\dfrac{2\pi}{3}, \dfrac{5\pi}{3}\right\}$
       Because -3 is negative, angles are in QII & QIV.
       $2\pi - \dfrac{\pi}{3} = \dfrac{6\pi}{3} - \dfrac{\pi}{3} = \dfrac{5\pi}{3}$.

    c) All real roots: $\dfrac{2\pi}{3} + k\pi, k \in Z$

Chapter 6: Trigonometric Identities, Inverses and Equations

51. $\dfrac{2}{5}\sin(2\theta) = \dfrac{1}{4}$

$\sin(2\theta) = \dfrac{5}{8}$

$\sin^{-1}(2\theta) = \sin^{-1}\left(\dfrac{5}{8}\right)$

Because $\dfrac{5}{8}$ is positive, angles are in QI &

QII.

$2\theta = 0.6751$

$\theta = 3.376$;

$2\theta = \pi - 0.6751$

$\theta = \dfrac{\pi - 0.6751}{2}$

$\theta = 1.2332$

a) Principal root: 0.3376

b) $\{0.3376, 1.2332, 3.4792, 4.3748\}$

   $\pi + 0.3376 = 3.4792$;

   $\pi + 1.2332 = 4.3748$

c) All real roots:

   $0.3376 + \pi k$ or $1.2332 + \pi k; k \in Z$

53. $A = \dfrac{1}{2}r^2\left(\theta - \sin\theta\right)$

$12 = \dfrac{1}{2}(10)^2\left(\theta - \sin\theta\right)$

$\dfrac{6}{25} = \theta - \sin\theta$

By grapher, $\theta = 1.1547$

55. $3\cos(2x) + 7\sin x - 5 = 0$

$3\left(1 - 2\sin^2 x\right) + 7\sin x - 5 = 0$

$3 - 6\sin^2 x + 7\sin x - 5 = 0$

$-6\sin^2 x + 7\sin x - 2 = 0$

$6\sin^2 x - 7\sin x + 2 = 0$

$\left(3\sin x - 2\right)\left(2\sin x - 1\right) = 0$

$3\sin x = 2$ or $2\sin x = 1$

$\sin x = \dfrac{2}{3}$ $\qquad$ $\sin x = \dfrac{1}{2}$

$\sin^{-1}(\sin x) = \sin^{-1}\left(\dfrac{2}{3}\right)$

$x = 0.7297$;

OR

sin x is positive, angle is in QI and QII.

$\sin x = \dfrac{1}{2}$

$\sin^{-1}(\sin x) = \sin^{-1}\left(\dfrac{1}{2}\right)$

$x = \dfrac{\pi}{6}$

$\pi - \dfrac{\pi}{6} = \dfrac{5\pi}{6}$;

$[0, 2\pi)$; $\pi - 0.7297 = 2.4119$;

$\left\{0.7297, 2.4119, \dfrac{\pi}{6}, \dfrac{5\pi}{6}\right\}$

57. $\csc x + \cot x = 1$

$\left(\csc x\right)^2 = \left(1 - \cot x\right)^2$

$\csc^2 x = 1 - 2\cot x + \cot^2 x$

$1 + \cot^2 x = 1 - 2\cot x + \cot^2 x$

$0 = -2\cot x$

$\cot x = 0$

$x = \dfrac{\pi}{2}$

59. $80\cos\left(\dfrac{\pi}{3}x + \dfrac{\pi}{4}\right) - 40\sqrt{2} = 0$

Period: $\dfrac{2\pi}{\dfrac{\pi}{3}} = 6 \rightarrow [0,6)$

$\cos\left(\dfrac{\pi}{3}x + \dfrac{\pi}{4}\right) = \dfrac{40\sqrt{2}}{80}$

QI or QIV $\left(2\pi - \dfrac{\pi}{4} = \dfrac{7\pi}{4}\right)$

$\cos^{-1}\left[\cos\left(\dfrac{\pi}{3}x + \dfrac{\pi}{4}\right)\right] = \cos^{-1}\left(\dfrac{\sqrt{2}}{2}\right)$

$\dfrac{\pi}{3}x + \dfrac{\pi}{4} = \dfrac{\pi}{4}$ $\quad$ or $\quad$ $\dfrac{\pi}{3}x + \dfrac{\pi}{4} = \dfrac{7\pi}{4}$

$\dfrac{\pi}{3}x = 0$ $\qquad\qquad$ $\dfrac{\pi}{3}x = \dfrac{3\pi}{2}$

$x = 0$ $\qquad\qquad\qquad$ $x = \dfrac{9}{2}$

$\left\{0, \dfrac{9}{2}\right\}$

Chapter 6 Mixed Review

1. $\csc\theta = \dfrac{\sqrt{117}}{6}$; θ in QII

$$\sqrt{\left(\sqrt{117}\right)^2 - 6^2} = 9$$

$$\sin\theta = \dfrac{6}{\sqrt{117}}, \sec\theta = \dfrac{-\sqrt{117}}{9},$$

$$\tan\theta = \dfrac{-6}{9} = \dfrac{-2}{3}, \cos\theta = \dfrac{-9}{\sqrt{117}},$$

$$\csc\theta = \dfrac{\sqrt{117}}{6}, \cot\theta = \dfrac{-9}{6} = \dfrac{-3}{2}$$

3. $\tan 255° = \tan(225° + 30°)$

$$= \dfrac{\tan 225° + \tan 30°}{1 - \tan 225° \tan 30°}$$

$$= \dfrac{1 + \dfrac{1}{\sqrt{3}}}{1 - (1)\left(\dfrac{1}{\sqrt{3}}\right)} = \dfrac{\sqrt{3}+1}{\sqrt{3}-1} \cdot \dfrac{\sqrt{3}+1}{\sqrt{3}+1}$$

$$= \dfrac{3 + 2\sqrt{3} + 1}{3 - 1} = \dfrac{4 + 2\sqrt{3}}{2} = 2 + \sqrt{3}$$

5. $\tan\left[\text{arc csc}\left(\dfrac{10}{x}\right)\right]$

$$\tan\theta = \dfrac{x}{\sqrt{100 - x^2}}$$

7. $-100\sin\left(\dfrac{\pi}{4}x - \dfrac{\pi}{6}\right) + 80 = 100$

$$-100\sin\left(\dfrac{\pi}{4}x - \dfrac{\pi}{6}\right) = 20; [0, 8)$$

$$\sin\left(\dfrac{\pi}{4}x - \dfrac{\pi}{6}\right) = \dfrac{-1}{5}$$

$$\sin^{-1}\left(\sin\left(\dfrac{\pi}{4}x - \dfrac{\pi}{6}\right)\right) = \sin^{-1}\left(\dfrac{-1}{5}\right)$$

$$= \dfrac{\pi}{4}x - \dfrac{\pi}{6} = -0.2014$$

$$\dfrac{\pi}{4}x = 0.3222$$

$$x \approx 0.4103$$

QIII or QIV

Or

QIII: $\pi + 0.2014 = 3.3429$

$$\dfrac{\pi}{4}x - \dfrac{\pi}{6} = 3.3429$$

$$\dfrac{\pi}{4}x = 3.8665$$

$$x \approx 4.9230$$

9. a) $R = \dfrac{1}{16}v^2 \sin\theta \cos\theta$

$$R = \dfrac{2}{2} \cdot \dfrac{1}{16}v^2 \sin\theta \cos\theta$$

$$R = 2 \cdot \dfrac{1}{32}v^2 \sin\theta \cos\theta$$

$$R = \dfrac{1}{32}v^2 \sin(2\theta)$$

b) $\sin(2\theta) = \sin\left[2\left(90° - \theta\right)\right]$

$$\sin(2\theta) = \sin\left(180° - 2\theta\right)$$

$$= \sin\left(180°\right)\cos\left(2\theta\right) - \cos\left(180°\right)\sin\left(2\theta\right)$$

$$= 0 - (-1)\sin\left(2\theta\right)$$

$$= \sin\left(2\theta\right)$$

11. $2\cos^2\left(\dfrac{\pi}{12}\right) - 1$

$$= \cos\left(2\left(\dfrac{\pi}{12}\right)\right) = \cos\left(\dfrac{\pi}{6}\right) = \dfrac{\sqrt{3}}{2}$$

13. $\dfrac{(\cos t + \sin t)^2}{\tan t} = \cot t + 2\cos^2 t$

$\dfrac{\cos^2 t + 2\cos t \sin t + \sin^2 t}{\tan t} =$

$\dfrac{1 + 2\cos t \sin t}{\tan t} =$

$\dfrac{1}{\tan t} + \dfrac{2\cos t \sin t}{\tan t} =$

$\cot t + \dfrac{2\cos t \sin t}{\dfrac{\sin t}{\cos t}} =$

$\cot t + 2\cos^2 t = \cot t + 2\cos^2 t$

15. $y = \operatorname{arc\,sec}\left(-\sqrt{2}\right) = \arccos\left(\dfrac{-1}{\sqrt{2}}\right)$

$135°$ or $\dfrac{3\pi}{4}$

17. $y = \arctan \sqrt{3}$

$60°$ or $\dfrac{\pi}{3}$

19. $\dfrac{x}{10} = \tan\theta$

$\tan^{-1}\left(\dfrac{x}{10}\right) = \tan^{-1}\left(\tan\theta\right)$

$\theta = \tan^{-1}\left(\dfrac{x}{10}\right)$

21. a) $D(t) = \left|8\sin\left(\dfrac{\pi t}{12}\right)\right| + 2$

Using a grapher,
6ft: 2 am, 2 pm and 10 am, 10 pm
10ft : 6 am, 6 pm

b) $D(4) = \left|8\sin\left(\dfrac{4\pi}{12}\right)\right| + 2$

≈ 8.9 ft

23. $\sin 172.5° - \sin 52.5°$

$\begin{cases} A + B = 172.5 \\ A - B = 52.5 \end{cases}$

$\overline{2A = 225}$

$A = \dfrac{225°}{2}$; $B = 172.5 - \dfrac{225}{2} = 60°$

a) $\sin 172.5° - \sin 52.5°$

$= \left[\sin\left(\dfrac{225}{2} + 60\right) - \sin\left(\dfrac{225}{2} - 60\right)\right]$

$= 2 \cdot \dfrac{1}{2}\left[\sin\left(\dfrac{225}{2} + 60\right) - \sin\left(\dfrac{225}{2} - 60\right)\right]$

$= 2\cos\left(\dfrac{225°}{2}\right)\sin 60°$

$= 2\left(-\sqrt{\dfrac{1 + \cos 225}{2}}\right)\left(\dfrac{\sqrt{3}}{2}\right)$

$= 2\left(-\sqrt{\dfrac{1 - \dfrac{\sqrt{2}}{2}}{2}}\right)\left(\dfrac{\sqrt{3}}{2}\right)$

$= \dfrac{-\sqrt{3}\sqrt{2 - \sqrt{2}}}{2}$

b) $\cos 172.5° + \cos 52.5°$

From part a, $A = \dfrac{225°}{2}, B = 60°$;

$2\cos\left(\dfrac{225°}{2}\right)\cos 60°$

$= 2\left(-\sqrt{\dfrac{1 - \dfrac{\sqrt{2}}{2}}{2}}\right)\left(\dfrac{1}{2}\right) = -\sqrt{\dfrac{2 - \sqrt{2}}{4}}$

$= \dfrac{-\sqrt{2 - \sqrt{2}}}{2}$

25. a) $\sin\left(\dfrac{13\pi}{24}\right)\cos\left(\dfrac{7\pi}{24}\right)$

$=\dfrac{1}{2}\left[\sin\left(\dfrac{13\pi}{24}+\dfrac{7\pi}{24}\right)+\sin\left(\dfrac{13\pi}{24}-\dfrac{7\pi}{24}\right)\right]$

$=\dfrac{1}{2}\left[\sin\left(\dfrac{5\pi}{6}\right)+\sin\left(\dfrac{\pi}{4}\right)\right]$

$=\dfrac{1}{2}\left[\dfrac{1}{2}+\dfrac{\sqrt{2}}{2}\right]=\dfrac{1+\sqrt{2}}{4}$

b) $\sin\left(\dfrac{13\pi}{24}\right)\sin\left(\dfrac{7\pi}{24}\right)$

$=\dfrac{1}{2}\left[\cos\left(\dfrac{13\pi}{24}-\dfrac{7\pi}{24}\right)-\cos\left(\dfrac{13\pi}{24}+\dfrac{7\pi}{24}\right)\right]$

$=\dfrac{1}{2}\left[\cos\left(\dfrac{\pi}{4}\right)-\cos\left(\dfrac{5\pi}{6}\right)\right]$

$=\dfrac{1}{2}\left[\dfrac{\sqrt{2}}{2}-\left(\dfrac{-\sqrt{3}}{2}\right)\right]$

$=\dfrac{\sqrt{2}}{4}+\dfrac{\sqrt{3}}{4}=\dfrac{\sqrt{2}+\sqrt{3}}{4}$

Chapter 6 Practice Test

1. $\dfrac{(\csc x-\cot x)(\csc x+\cot x)}{\sec x}=\cos x$

$\dfrac{\csc^2 x+\csc x\cot x-\csc x\cot x-\cot^2 x}{\sec x}=$

$\dfrac{\csc^2 x-\cot^2 x}{\sec x}=$

$\dfrac{(1+\cot^2 x)-\cot^2 x}{\sec x}=$

$\dfrac{1}{\sec x}=$

$\cos x=\cos x$

3. $\cos\theta=\dfrac{48}{73},\theta$ in QIV

$\sqrt{73^2-48^2}=55$;

$\sin\theta=\dfrac{-55}{73},\sec\theta=\dfrac{73}{48},\cot\theta=\dfrac{-48}{55}$

$\tan\theta=\dfrac{-55}{48},\csc\theta=\dfrac{-73}{55}$

5. $\cos 81°\cos 36°+\sin 81°\sin 36°$
$=\cos(81°-36°)$
$=\cos(45°)$
$=\dfrac{\sqrt{2}}{2}$

7. $\sin\left(x+\dfrac{\pi}{4}\right)-\sin\left(x-\dfrac{\pi}{4}\right)=\sqrt{2}\cos x$

$\sin x\cos\left(\dfrac{\pi}{4}\right)+\cos x\sin\left(\dfrac{\pi}{4}\right)$

$-\left[\sin x\cos\left(\dfrac{\pi}{4}\right)-\cos x\sin\left(\dfrac{\pi}{4}\right)\right]=$

$\sin x\cos\left(\dfrac{\pi}{4}\right)+\cos x\sin\left(\dfrac{\pi}{4}\right)$

$-\sin x\cos\left(\dfrac{\pi}{4}\right)+\cos x\sin\left(\dfrac{\pi}{4}\right)=$

$2\cos x\sin\left(\dfrac{\pi}{4}\right)=$

$2\cos x\left(\dfrac{\sqrt{2}}{2}\right)=$

$\sqrt{2}\cos x=\sqrt{2}\cos x$

9. $2\cos^2 75°-1$

$=\cos 2(75°)=\cos 150°=\dfrac{-\sqrt{3}}{2}$

11. $A=\dfrac{1}{2}bc\sin\alpha$

$A=\dfrac{1}{2}(8)(10)\sin 22.5°$

$A=40\sin\left(\dfrac{45°}{2}\right)$

$A=40\left(\sqrt{\dfrac{1-\cos 45°}{2}}\right)$

$A=40\sqrt{\dfrac{1-\dfrac{\sqrt{2}}{2}}{2}}$

$A=40\sqrt{\dfrac{2-\sqrt{2}}{4}}$

$A=40\cdot\dfrac{\sqrt{2-\sqrt{2}}}{2}$

$A=20\sqrt{2-\sqrt{2}}$

13. a) $y = \tan^{-1}\left(\dfrac{1}{\sqrt{3}}\right)$

   $y = 30°$

   b) $y = \sin\left[\sin^{-1}\left(\dfrac{1}{2}\right)\right]$

   $y = \dfrac{1}{2}$

   c) $y = \arccos(\cos 30°)$

   $y = 30°$

15. $\cos\left[\tan^{-1}\left(\dfrac{56}{33}\right)\right]$

   $\sqrt{56^2 + 33^3} = 65$

   $\cos\theta = \dfrac{33}{65}$

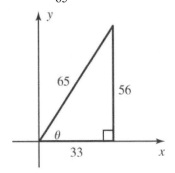

17. I. $8\cos x = -4\sqrt{2}$

   $\cos x = \dfrac{-\sqrt{2}}{2}$ QII or QIII

   a. $\cos^{-1}\left(\dfrac{-\sqrt{2}}{2}\right) = \dfrac{3\pi}{4}$

   b. $\left\{\dfrac{3\pi}{4}, \dfrac{5\pi}{4}\right\}$

   c. $\left\{\dfrac{3\pi}{4} + 2k\pi\right\} \cup \left\{\dfrac{5\pi}{4} + 2k\pi\right\}$

II. $\sqrt{3}\sec x + 2 = 4$

   $\sec x = \dfrac{2}{\sqrt{3}}$

   $\dfrac{1}{\cos x} = \dfrac{2}{\sqrt{3}}$

   $\cos x = \dfrac{\sqrt{3}}{2}$ (QI or QIV)

   a. $\cos^{-1}\left(\dfrac{\sqrt{3}}{2}\right) = \dfrac{\pi}{6}$

   b. $\left\{\dfrac{\pi}{6}, \dfrac{11\pi}{6}\right\}$

   c. $\left\{\dfrac{\pi}{6} + 2k\pi\right\} \cup \left\{\dfrac{11\pi}{6} + 2k\pi\right\}$

19. a. $Y_1 = 3\cos(2x - 1)$

   $Y_2 = \sin x, x \in [-\pi, \pi]$

   $\{-1.6875, -0.3413, 1.1321, 2.8967\}$

   b. $Y_1 = 2\sqrt{x} - 1$

   $Y_2 = 3\cos^2 x$

   $x \in [0, 2\pi]$

   $\{0.9671, 2.6110, 3.4538\}$

21. $3\sin(2x) + \cos x = 0$

   $3(2\sin x\cos x) + \cos x = 0$

   $\cos x[6\sin x + 1] = 0$

   $\cos x = 0 \quad$ or $\qquad 6\sin x + 1 = 0$

   $x = \cos^{-1}(0) \qquad\qquad 6\sin x = -1$

   $x = \dfrac{\pi}{2}$ or $\dfrac{3\pi}{2} \qquad \sin x = -\dfrac{1}{6}$

   $\qquad\qquad\qquad\qquad x = -0.1674$

   Principal root: 0.1674

   QIII or QIV

   $\pi + 0.1674 = 3.3090$

   $2\pi - 0.1674 = 6.1158$

   $\left\{\dfrac{\pi}{2}, 3.3090, \dfrac{3\pi}{2}, 6.1158\right\}$

23. $R(x) = 7.5 \cos\left(\dfrac{\pi}{6}x + \dfrac{4\pi}{3}\right) + 12.5$

   a. $R(9) = 7.5 \cos\left(\dfrac{\pi}{6} \cdot 9 + \dfrac{4\pi}{3}\right)$

       $+ 12.5 = 6$ or \$6,000

   b. Using grapher, January $x = 1$ through July $(x = 7)$

25. $2\cos(1979\pi t)\cos(439\pi t)$

    $= 2 \cdot \dfrac{1}{2}[\cos(1979\pi t + 439\pi t)$

    $+ \cos(1979\pi t - 439\pi t)]$

    $= \cos(2418\pi t) + \cos(1540\pi t)$

Calculator Exploration & Discovery

Exercise 1:

$Y_1 = \cos(14t); Y_2 = \cos(8t)$

$\cos(14t) + \cos(8t)$

$= 2\cos\left(\dfrac{14t + 8t}{2}\right)\cos\left(\dfrac{14t - 8t}{2}\right)$

   a. $Y_R = 2\cos(11t)\cos(3t)$

   b. $14 - 8 = 6$ beats

   c. $\dfrac{14 - 8}{2} = 3$, use $Y_2 = \pm 2\cos(3x)$

Exercise 3:

$Y_1 = \cos(14t); Y_2 = \cos(6t)$

$\cos(14t) + \cos(6t)$

$= 2\cos\left(\dfrac{14t + 6t}{2}\right)\cos\left(\dfrac{14t - 6t}{2}\right)$

   a. $Y_R = 2\cos(10t)\cos(4t)$

   b. $14 - 6 = 8$ beats

   c. $\dfrac{14 - 6}{2} = 4$, use $Y_2 = \pm 2\cos(4x)$

Strengthening Core Skills

Exercise 1:

$f(x) = 3\sin x + 2; f(x) > 3.7, x \in [0, 2\pi]$

Sine wave, amplitude 3, shifted 2 units up, sketch $y = 3.7$, solutions to $f(x)$ occur in QI and QIII, with solutions to $f(x) > 3.7$ between these solutions. Substitute 3.7 for $f(x)$ and isolating the sine function.

$3.7 = 3\sin x + 2$

$1.7 = 3\sin x$

$0.5667 \approx \sin x$

QI, $x = \sin^{-1}(0.5667) \approx 0.6025$

QIII, $x = \left[\pi - \sin^{-1}(0.5667)\right]$

$x = 2.5391$

$x \in (0.6025, 2.5391)$

Exercise 3

$h(x) = 125\sin\left(\dfrac{\pi}{6}x - \dfrac{\pi}{2}\right) + 175;$

$h(x) \le 150, x \in [0, 12]$

Sine wave, period $\dfrac{2\pi}{\left(\dfrac{\pi}{6}\right)} = 12$ days,

amplitude 125, shifted $\dfrac{-C}{B} = -\dfrac{-\dfrac{\pi}{2}}{\dfrac{\pi}{6}} = 3$

units to the right and 175 units up Sketch $y = 150$, solutions to $h(x) \le 150$ occur in QI and QIII, with solutions. outside this interval. Substitute 150 for $h(x)$ and isolate the sine function.

$150 = 125\sin\left(\dfrac{\pi}{6}x - \dfrac{\pi}{2}\right) + 175$

$-0.2 = \sin\left(\dfrac{\pi}{6}x - \dfrac{\pi}{2}\right)$

QI, $x = \left(\sin^{-1}(-0.2) + \dfrac{\pi}{2}\right)\left(\dfrac{6}{\pi}\right) = 2.6154$

QIII

$x = \left(\pi - \sin^{-1}(-0.2) + \dfrac{\pi}{2}\right)\left(\dfrac{6}{\pi}\right) = 9.3846$

$x \in [0, 2.6154] \cup [9.3847, 12]$

Chapter 6: Trigonometric Identities, Inverses and Equations

Cumulative Review: Chapters 1-6

1. $P(-13, 84)$

   $\sqrt{84^2 + 13^2} = 85$;

   $\sin\theta = \dfrac{84}{85}$, $\csc\theta = \dfrac{85}{84}$, $\cos\theta = \dfrac{-13}{85}$,

   $\sec\theta = \dfrac{-85}{13}$, $\tan\theta = \dfrac{-84}{13}$, $\cot\theta = \dfrac{-13}{84}$

3. $g(x) = x^2 - 4x + 1$

   $g(2 + \sqrt{3}) = (2 + \sqrt{3})^2 - 4(2 + \sqrt{3}) + 1$

   $= 4 + 4\sqrt{3} + 3 - 8 - 4\sqrt{3} + 1$

   $= 0$

   Verified

5. $\tan(36°56') = \dfrac{x}{26400}$

   $26400\tan(36°56') = x$

   $19846 \approx x$;

   $20,320 - 19846 = 474$

   About 474 ft

7. $h(x) = \dfrac{x-1}{x^2-4}$

   $h(x) = \dfrac{x-1}{(x+2)(x-2)}$

   V.A.: $x = -2, x = 2$

   H.A.: $y = 0$

   (deg num < deg den)

   $h(0) = \dfrac{0-1}{0^2-4} = \dfrac{1}{4}$

   y-intercept: $\left(0, \dfrac{1}{4}\right)$

   x-intercept: $(1, 0)$

9. $r = 45$ cm, $\dfrac{5\,\text{rev}}{\text{sec}}$ (1 rev = 2π radians)

   $\omega = 5(2\pi) = 10\pi$

   $v = r \cdot \omega$

   $= 45(10\pi) = 450\pi$ cm per sec

   $= 450(\pi)$ cm per sec

   ≈ 1413 cm per sec

   $\dfrac{450\pi\,\text{cm}}{1\,\text{sec}} \cdot \dfrac{1\,\text{km}}{100,000\,\text{cm}} \cdot \dfrac{3600\,\text{sec}}{1\,\text{hr}}$

   $= 50.89$ km/hr

11. $-3\left|x - \dfrac{1}{2}\right| + 5 \geq -10$

   $-3\left|x - \dfrac{1}{2}\right| \geq -15$

   $\left|x - \dfrac{1}{2}\right| \leq 5$

   $x - \dfrac{1}{2} \leq 5$ or $x - \dfrac{1}{2} \geq -5$

   $x \leq 5.5$ $x \geq -4.5$

   $\left[\dfrac{-9}{2}, \dfrac{11}{2}\right]$

13. $(0, 31), (30, 16)$

   a. $m = \dfrac{31-16}{0-30} = \dfrac{15}{-30} = \dfrac{-1}{2}$

      $y = \dfrac{-1}{2}x + 31$

   b. Every 2 years, the amount of emissions decreases by 1 millon tons.

   c. 1985 is 15 years (since 1970)

      $y = \dfrac{-1}{2}(15) + 31 = 23.5$ million tons;

      2010 is 40 years (since 1970)

      $y = \dfrac{-1}{2}(40) + 31 = 11$ million tons

Cumulative Review: Chapters 1-6

15. $f(x) = 325\cos\left(\dfrac{\pi}{6}x - \dfrac{\pi}{2}\right) + 168$

Find x when $f(x) > 330.5$,

cosine wave,

period $\dfrac{2\pi}{\left(\dfrac{\pi}{6}\right)} = 12$, amplitude 325, shifted

$-\dfrac{C}{B} = -\dfrac{-\dfrac{\pi}{2}}{\dfrac{\pi}{6}} = 3$ units to the right and 168

units up. $y = 330.5$, solutions to

$f(x) > 330.5$ occur in QI and QIII, with

solutions between these solutions.

Substitute 330.5 for $f(x)$ and isolate the

cosine function.

$330.5 = 325\cos\left(\dfrac{\pi}{6}x - \dfrac{\pi}{2}\right) + 168$

$0.5 = \cos\left(\dfrac{\pi}{6}x - \dfrac{\pi}{2}\right)$

QI, $x = \left(\cos^{-1}(0.5) + \dfrac{\pi}{2}\right)\left(\dfrac{6}{\pi}\right) = 5$

QIII, $x = \left(\pi - \left(\cos^{-1}(0.5) + \dfrac{\pi}{2}\right)\right)\left(\dfrac{6}{\pi}\right) =$

$\quad = \left(\pi - \left(\dfrac{\pi}{3} + \dfrac{\pi}{2}\right)\right)\left(\dfrac{6}{\pi}\right) = 1$

$x \in (1,5)$

17. $R(x) = (9 - 0.25x)(20 + 1x)$

$R(x) = 180 + 9x - 5x - 0.25x^2$

$R(x) = -0.25x^2 + 4x + 180$

Using a grapher, maximum occurs at $x = 8$.

8 decreases $(9 - 0.25(8)) = 7$, price of \$7.

19. $\dfrac{\cos x + 1}{\tan^2 x} = \dfrac{\cos x}{\sec x - 1}$

$\quad = \dfrac{\cos x(\sec x + 1)}{(\sec x - 1)(\sec x + 1)}$

$\quad = \dfrac{1 + \cos x}{\sec^2 x - 1}$

$\quad = \dfrac{\cos x + 1}{\tan^2 x}$

21. $\sin(2\theta) = \sin(\theta + \theta)$

$\quad = \sin\theta\cos\theta + \cos\theta\sin\theta$

$\quad = \dfrac{11}{\sqrt{202}} \cdot \dfrac{9}{\sqrt{202}} + \dfrac{9}{\sqrt{202}} \cdot \dfrac{11}{\sqrt{202}}$

$\quad = \dfrac{198}{202} = \dfrac{99}{101}$

23. a. $\dfrac{32.5 + 21.7}{2} = \dfrac{54.2}{2} = 27.1$;

$\dfrac{32.5 - 21.7}{2} = \dfrac{10.8}{2} = 5.4$;

$12 = \dfrac{2\pi}{B}$

$B = \dfrac{2\pi}{12} = \dfrac{\pi}{6}$

$5.4\sin\left(\dfrac{\pi}{6}x - \dfrac{2\pi}{3}\right) + 27.1$

b. Using a grapher, from early May until late August.

25. a. Volume of a cylinder
    b. Volume of a rectangular solid
    c. Circumference of a circle
    d. Area of a triangle

Chapter 6: Trigonometric Identities, Inverses and Equations

Connections to Calculus

1. a) $y = \dfrac{\sqrt{169+x^2}}{x}$

$x = 13\tan\theta,\ \theta \in \left(-\dfrac{\pi}{2}, \dfrac{\pi}{2}\right)$

$y = \dfrac{\sqrt{169+(13\tan\theta)^2}}{13\tan\theta}$

$y = \dfrac{\sqrt{169+169\tan^2\theta}}{13\tan\theta}$

$y = \dfrac{\sqrt{169(1+\tan^2\theta)}}{13\tan\theta}$

$y = \dfrac{\sqrt{169(\sec^2\theta)}}{13\tan\theta}$

$y = \dfrac{13\sec\theta}{13\tan\theta}$

$y = \dfrac{1}{\cos\theta} \div \dfrac{\sin\theta}{\cos\theta}$

$y = \dfrac{1}{\cos\theta} \cdot \dfrac{\cos\theta}{\sin\theta}$

$y = \dfrac{1}{\sin\theta}$

$y = \csc\theta$

b) $x = 13\tan\theta$

$\tan\theta = \dfrac{x}{13}$

$\theta = \tan^{-1}\left(\dfrac{x}{13}\right)$

$y = \csc\left(\tan^{-1}\left(\dfrac{x}{13}\right)\right)$

c) verified

3. $\dfrac{x^2}{\sqrt{16-x^2}}$; $x = 4\sin\theta$

$\dfrac{x^2}{\sqrt{16-x^2}} = \dfrac{(4\sin\theta)^2}{\sqrt{16-(4\sin\theta)^2}}$

$= \dfrac{16\sin^2\theta}{\sqrt{16-16\sin^2\theta}}$

$= \dfrac{16\sin^2\theta}{\sqrt{16\cos^2\theta}}$

$= \dfrac{16\sin^2\theta}{4\cos\theta}$

$= 4\dfrac{\sin\theta}{\cos\theta} \cdot \dfrac{1}{\cos\theta}$

$= 4\tan\theta\sec\theta$

5. $\dfrac{x}{\sqrt{9+x^2}}$; $x = 3\tan\theta$

$\dfrac{x}{\sqrt{9+x^2}} = \dfrac{3\tan\theta}{\sqrt{9+(3\tan\theta)^2}}$

$= \dfrac{3\tan\theta}{\sqrt{9+9\tan^2\theta}}$

$= \dfrac{3\tan\theta}{\sqrt{9(1+\tan^2\theta)}}$

$= \dfrac{3\tan\theta}{\sqrt{9\sec^2\theta}}$

$= \dfrac{3\tan\theta}{3\sec\theta}$

$= \dfrac{\tan\theta}{\sec\theta}$

$= \dfrac{\sin\theta}{\cos\theta} \div \dfrac{1}{\cos\theta}$

$= \dfrac{\sin\theta}{\cos\theta} \div \dfrac{\cos\theta}{1}$

$= \sin\theta$

Connections to Calculus

7. $f(x) = \dfrac{1+\cos x}{\sec x}$; $[0, 2\pi)$

$f'(x) = \dfrac{\sec x(-\sin x) - (1+\cos x)\sec x \tan x}{\sec^2 x}$

$0 = \dfrac{\sec x[(-\sin x) - (1+\cos x)\tan x]}{\sec^2 x}$

$0 = \dfrac{-\sin x - (\tan x + \cos x(\tan x))}{\sec x}$

$0 = \dfrac{-\sin x - \tan x + \cos x\left(\dfrac{\sin x}{\cos x}\right)}{\sec x}$

$0 = \dfrac{-\sin x - \tan x - \sin x}{\sec x}$

$0 = (-2\sin x - \tan x) \div \dfrac{1}{\cos x}$

$0 = (-2\sin x - \tan x)\cos x$

$0 = -2\sin x \cos x - \dfrac{\sin x}{\cos x}\cos x$

$0 = -2\sin x \cos x - \sin x$

$0 = -\sin x(2\cos x + 1)$

$-\sin x = 0$

$\sin x = 0$

$x = 0, \pi$

$2\cos x + 1 = 0$

$\cos x = -\dfrac{1}{2}$

Reference angle $= \dfrac{\pi}{3}$

Quadrants: Q2, Q3; angle $\dfrac{2\pi}{3}, \dfrac{4\pi}{3}$

$x = 0, \dfrac{2\pi}{3}, \pi, \dfrac{4\pi}{3}$

9. $f(x) = 2\sin x \cos x$;

$f'(x) = 2\sin x(-\sin x) + 2\cos x \cos x$

$0 = 2\sin x(-\sin x) + 2\cos x \cos x$

$0 = 2(\cos^2 x - \sin^2 x)$ $0 = \cos^2 x - \sin^2 x$

$0 = \cos 2x$

$2x = \dfrac{\pi}{2} + \pi k$

$x = \dfrac{\pi}{4} + \dfrac{\pi}{2}k$

$[0, 2\pi)$

$k = 0 : x = \dfrac{\pi}{4} + \dfrac{\pi}{2}(0) = \dfrac{\pi}{4}$

$k = 1 : x = \dfrac{\pi}{4} + \dfrac{\pi}{2}(1) = \dfrac{3\pi}{4}$

$k = 2 : x = \dfrac{\pi}{4} + \dfrac{\pi}{2}(2) = \dfrac{5\pi}{4}$

$k = 3 : x = \dfrac{\pi}{4} + \dfrac{\pi}{2}(3) = \dfrac{7\pi}{4}$

$x = \dfrac{\pi}{4} ; \dfrac{3\pi}{4} ; \dfrac{5\pi}{4} ; \dfrac{7\pi}{4}$

7.1 Exercises

1. ambiguous

3. I; II

5. Answers will vary.

7. $\dfrac{\sin 32°}{15} = \dfrac{\sin 18.5°}{a}$

$a \sin 32° = 15 \sin 18.5°$

$a = \dfrac{15 \sin 18.5°}{\sin 32°} \approx 8.98$

9. $\dfrac{\sin 63°}{21.9} = \dfrac{\sin C}{18.6}$

$21.9 \sin C = 18.6 \sin 63°$

$\sin C = \dfrac{18.6 \sin 63°}{21.9}$

$C = \sin^{-1}\left(\dfrac{18.6 \sin 63°}{21.9}\right) \approx 49.2°$

11. $\dfrac{\sin C}{48.5} = \dfrac{\sin 19°}{43.2}$

$43.2 \sin C = 48.5 \sin 19°$

$\sin C = \dfrac{48.5 \sin 19°}{43.2}$

$C = \sin^{-1}\left(\dfrac{48.5 \sin 19°}{43.2}\right) \approx 21.4°$

13. $\angle C = 180° - (38° + 64°) = 78°$;

$\dfrac{\sin 38°}{75} = \dfrac{\sin 64°}{b}$

$b \sin 38° = 75 \sin 64°$

$b = \dfrac{75 \sin 64°}{\sin 38°} \approx 109.5$ cm;

$\dfrac{\sin 78°}{c} = \dfrac{\sin 38°}{75}$

$c \sin 38° = 75 \sin 78°$

$c = \dfrac{75 \sin 78°}{\sin 38°} \approx 119.2$ cm

15. $\angle C = 180° - (30° + 60°) = 90°$;

$\dfrac{\sin 60°}{10\sqrt{3}} = \dfrac{\sin 30°}{a}$

$a \sin 60° = 10\sqrt{3} \sin 30°$

$a = \dfrac{10\sqrt{3} \sin 30°}{\sin 60°} = 10$ in. ;

$\dfrac{\sin 60°}{10\sqrt{3}} = \dfrac{\sin 90°}{c}$

$c \sin 60° = 10\sqrt{3} \sin 90°$

$c = \dfrac{10\sqrt{3} \sin 90°}{\sin 60°} = 20$ in.

17. Let $\angle A = 33°$, $\angle B = 102°$, $b = 19$ in.

$\angle C = 180° - (33° + 102°) = 45°$;

$\dfrac{\sin 102°}{19} = \dfrac{\sin 33°}{a}$

$a \sin 102° = 19 \sin 33°$

$a = \dfrac{19 \sin 33°}{\sin 102°} \approx 10.6$ in. ;

$\dfrac{\sin 102°}{19} = \dfrac{\sin 45°}{c}$

$c \sin 102° = 19 \sin 45°$

$c = \dfrac{19 \sin 45°}{\sin 102°} \approx 13.7$ in.

19. $\angle C = 180° - (45° + 45°) = 90°$

$\dfrac{\sin 90°}{15\sqrt{2}} = \dfrac{\sin 45°}{a}$

$a \sin 90° = 15\sqrt{2} \sin 45°$

$a = \dfrac{15\sqrt{2} \sin 45°}{\sin 90°} = 15$ mi

$b = 15$ mi

7.1 Exercises

21. $\angle A = 180° - (103.4° + 19.6°) = 57°$

$$\frac{\sin 57°}{42.7} = \frac{\sin 103.4°}{b}$$

$b\sin 57° = 42.7\sin 103.4°$

$$b = \frac{42.7\sin 103.4°}{\sin 57°} \approx 49.5 \text{ km} ;$$

$$\frac{\sin 57°}{42.7} = \frac{\sin 19.6°}{c}$$

$c\sin 57° = 42.7\sin 19.6°$

$$c = \frac{42.7\sin 19.6°}{\sin 57°} \approx 17.1 \text{ km}$$

23. Let $\angle A = 56°$, $\angle B = 112°$, $c = 0.8$ cm
$\angle C = 180° - (56° + 112°) = 12°$;

$$\frac{\sin 12°}{0.8} = \frac{\sin 56°}{a}$$

$a\sin 12° = 0.8\sin 56°$

$$a = \frac{0.8\sin 56°}{\sin 12°} \approx 3.2 \text{ cm} ;$$

$$\frac{\sin 12°}{0.8} = \frac{\sin 112°}{b}$$

$b\sin 12° = 0.8\sin 112°$

$$b = \frac{0.8\sin 112°}{\sin 12°} \approx 3.6 \text{ cm}$$

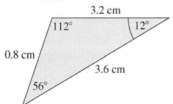

25. a.

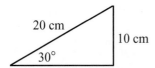

   The right triangle is a 30°-60°-90°
   triangle, so the short side is half the
   hypotenuse, or 10 cm.

   b. If side a is 8 cm, it won't reach the
   base, so no triangle is possible.

   c.

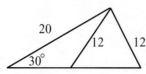

   Two triangles possible for $a = 12$.

   d.

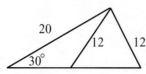

   One triangle possible for $a = 25$.

27. $\dfrac{\sin 67°}{385} = \dfrac{\sin A}{490}$; $385\sin A = 490\sin 67°$

$$\sin A = \frac{490\sin 67°}{385} = 1.17. \text{ Not possible.}$$

29. $\dfrac{\sin 30°}{12.9} = \dfrac{\sin C}{25.8}$; $12.9\sin C = 25.8\sin 30°$

$$\sin C = \frac{25.8\sin 30°}{12.9} = 1; C = 90°. \text{ (This tells}$$

us that there's only one triangle.)
$B = 180° - (30° + 90°) = 60°$
This is a 30°-60°-90° triangle, so
$b = 12.9\sqrt{3}$ mi

31. $\dfrac{\sin B}{67} = \dfrac{\sin 59°}{58}$; $58\sin B = 67\sin 59°$

$$\sin B = \frac{67\sin 59°}{58} \approx 0.9902$$

$B = \sin^{-1} 0.9902 \approx 82°$
or $B = 180° - 82° = 98°$.
For $B = 82°$:
$A = 180° - (59° + 82°) = 39°$

$$\frac{\sin 39°}{a} = \frac{\sin 59°}{58}; a\sin 59° = 58\sin 39°$$

$$a = \frac{58\sin 39°}{\sin 59°} \approx 42.6 \text{ mi} ;$$

For $B = 98°$:
$A = 180° - (98° + 59°) = 23°$

$$\frac{\sin 23°}{a} = \frac{\sin 59°}{58}; a\sin 59° = 58\sin 23°$$

$$a = \frac{58\sin 23°}{\sin 59°} \approx 26.4 \text{ mi}$$

33. $\dfrac{\sin 59°}{58} = \dfrac{\sin B}{67}$; $58\sin B = 67\sin 59°$

$$\sin B = \frac{67\sin 59°}{58} \approx 0.9902$$

$B = \sin^{-1} 0.9902 \approx 82°$ or
$B = 180° - 82° = 98°$
For $B = 82°$:
$A = 180° - (82° + 59°) = 39°$;

$$\frac{\sin 39°}{a} = \frac{\sin 59°}{58}; a\sin 59° = 58\sin 39°$$

$$a = \frac{58\sin 39°}{\sin 59°} = 42.6 \text{ ft} ;$$

For $B = 98°$:
$A = 180° - (98° + 59°) = 23°$

$$\frac{\sin 23°}{a} = \frac{\sin 59°}{58}; a\sin 59° = 58\sin 23°$$

$$a = \frac{58\sin 23°}{\sin 59°} \approx 26.4 \text{ ft}$$

35. $\dfrac{\sin 38°}{6.7} = \dfrac{\sin A}{10.9}$; $6.7 \sin A = 10.9 \sin 38°$

$\sin A = \dfrac{10.9 \sin 38°}{6.7} = 1.002$. Not possible

37. $\dfrac{\sin 62°}{2.6 \times 10^{25}} = \dfrac{\sin A}{2.9 \times 10^{25}}$

$2.6 \times 10^{25} \sin A = 2.9 \times 10^{25} \sin 62°$

$\sin A = \dfrac{2.9 \times 10^{25} \sin 62°}{2.6 \times 10^{25}} \approx 0.9848$

$A \approx 80.0°$;

$B \approx 180° - (80.0° + 62°) = 38°$

$\dfrac{\sin 62°}{2.6 \times 10^{25}} = \dfrac{\sin 38°}{b}$

$b \sin 62° = 2.6 \times 10^{25} \sin 38°$

$b = \dfrac{2.6 \times 10^{25} \sin 38°}{\sin 62°} \approx 1.8 \times 10^{25}$ mi

39. $\dfrac{\sin A}{12} = \dfrac{\sin 48°}{27}$; $27 \sin A = 12 \sin 48°$

$\sin A = \dfrac{12 \sin 48°}{27}$; $A = \sin^{-1}\left(\dfrac{12 \sin 48°}{27}\right)$

$A \approx 19.3°$;

Another possible angle: $180° - 19.3° = 160.7°$; $48° + 160.7° = 208.7° > 180°$

No second solution possible

41. $\dfrac{\sin 57°}{35.6} = \dfrac{\sin C}{40.2}$; $35.6 \sin C = 40.2 \sin 57°$

$\sin C = \dfrac{40.2 \sin 57°}{35.6}$; $C = \sin^{-1}\left(\dfrac{40.2 \sin 57°}{35.6}\right)$

$C \approx 71.3°$;

Another possible angle:
$180° - 71.3° = 108.7°$;
$108.7° + 57° = 165.7° < 180°$; Two possible solutions.

43. $\dfrac{\sin A}{280} = \dfrac{\sin 15°}{52}$; $52 \sin A = 280 \sin 15°$

$\sin A = \dfrac{280 \sin 15°}{52} = 1.39$. Not possible

45. $135° = 3(45°)$, so we get

$\sin 135° = 3 \sin 45° - 4 \sin^3 45° =$

$= 3 \cdot \dfrac{\sqrt{2}}{2} - 4\left(\dfrac{\sqrt{2}}{2}\right)^3 = \dfrac{3\sqrt{2}}{2} - 4\left(\dfrac{2\sqrt{2}}{8}\right)$

$= \dfrac{3\sqrt{2}}{2} - \dfrac{2\sqrt{2}}{2} = \dfrac{\sqrt{2}}{2}$

The reference angle for $135°$ is $45°$, and $135°$ is in QII, so $\sin 135° = \dfrac{\sqrt{2}}{2}$.

47. We are given $\theta = 20°$. Let α be the angle with vertex at Sorus and β be the angle with vertex at the Sun.

$\dfrac{\sin 20°}{51} = \dfrac{\sin \alpha}{82}$; $82 \sin 20° = 51 \sin \alpha$

$\sin \alpha = \dfrac{82 \sin 20°}{51} \approx 0.5499$

$\alpha = \sin^{-1} 0.5499 = 33.4°$ or
$\alpha = 180° - 33.4° = 146.6°$
When $\alpha = 33.4°$, the distance is the further of the two; let d_1 represent this distance.
$\beta = 180° - (33.4° + 20°) = 126.6°$

$\dfrac{\sin 126.6°}{d_1} = \dfrac{\sin 20°}{51}$

$d_1 \sin 20° = 51 \sin 126.6°$

$d_1 = \dfrac{51 \sin 126.6°}{\sin 20°} \approx 119.7$ million miles

When $\alpha = 146.6°$, the distance is the closer of the two; let d_2 represent this distance.
$\beta = 180° - (146.6° + 20°) = 13.4°$

$\dfrac{\sin 13.4°}{d_2} = \dfrac{\sin 20°}{51}$

$d_2 \sin 20° = 51 \sin 13.4°$

$d_2 = \dfrac{51 \sin 13.4°}{\sin 20°} \approx 34.6$ million miles

49. a. $\dfrac{\sin 35°}{8} = \dfrac{\sin B}{15}$; $\quad 8\sin B = 15\sin 35°$

$\sin B = \dfrac{15\sin 35°}{8} = 1.08$ Not possible

A radar with a range of 8 mi. will not detect the ship.

b. $\dfrac{\sin 35°}{12} = \dfrac{\sin B}{15}$; $\quad 12\sin B = 15\sin 35°$

$\sin B = \dfrac{15\sin 35°}{12} \approx 0.7170$

$B = \sin^{-1} 0.7170 \approx 45.8°$ or
$B = 180° - 45.8 = 134.2°$.
The closest point of detection will be when $B = 134.2°$, in which case the third angle is
$180° - (35° + 134.2°) = 10.8°$.

$\dfrac{\sin 10.8°}{d} = \dfrac{\sin 35°}{12}$

$d\sin 35° = 12\sin 10.8°$

$d = \dfrac{12\sin 10.8°}{\sin 35°} \approx 3.9$ mi

51. Segment SR is 55 km.

$\dfrac{\sin 40°}{55} = \dfrac{\sin \angle P}{80}$; $55\sin \angle P = 80\sin 40°$

$\sin \angle P = \dfrac{80\sin 40°}{55} \approx 0.9350$; $\angle P \approx 69.2°$

$\angle VRP = 180° - (40° + 69.2°) = 70.8°$

Let d_1 = the distance from V to P.

$\dfrac{\sin 70.8°}{d_1} = \dfrac{\sin 40°}{55}$; $\quad d_1 \sin 40° = 55\sin 70.8°$

$V \leftrightarrow P \quad d_1 = \dfrac{55\sin 70.8°}{\sin 40°} \approx 80.8$ km

Since $\angle P = 69.2°$,

$\angle VSR = 180° - 69.2° = 110.8°$.
Then $\angle VRS = 180° - (110.8° + 40°) = 29.2°$.
Let d_2 = the distance from V to S.

$\dfrac{\sin 29.2°}{d_2} = \dfrac{\sin 40°}{55}$; $d_2 \sin 40° = 55\sin 29.2°$

$V \leftrightarrow S \quad d_2 = \dfrac{55\sin 29.2°}{\sin 40°} \approx 41.7$ km

53. Let B be the angle at the target.

a. $\dfrac{\sin 55°}{180} = \dfrac{\sin B}{246}$

$180\sin B = 246\sin 55°$

$\sin B = \dfrac{246\sin 55°}{180} \approx 1.1$, Not possible

The arrow won't reach the target.

b. $\dfrac{\sin 55°}{a} = \dfrac{\sin 90°}{246}$

$a\sin 90° = 246\sin 55°$

$a \approx 201.5$ ft

c. We first need to find the distance $d_2 - d_1$ in the diagram below:

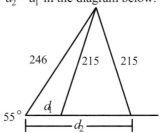

Let C = the angle at the archer.

$\dfrac{\sin 55°}{215} = \dfrac{\sin B}{246}$

$215\sin B = 246\sin 55°$

$\sin B = \dfrac{246\sin 55°}{215} \approx 0.9373$

$B = \sin^{-1} 0.9373 \approx 69.6°$ or
$B = 180° - 69.6° = 110.4°$
When $B = 69.6°$, $C = 180° - (69.6° + 55°) = 55.4°$.

$\dfrac{\sin 55°}{215} = \dfrac{\sin 55.4°}{d_2}$

$d_2 \sin 55° = 215\sin 55.4°$

$d_2 = \dfrac{215\sin 55.4°}{\sin 55°} \approx 216$ ft

When $B = 110.4°$, $C = 180° - (110.4° + 55°) = 14.6°$.

$\dfrac{\sin 55°}{215} = \dfrac{\sin 14.6°}{d_1}$

$d_1 \sin 55° = 215\sin 14.6°$

$d_1 = \dfrac{215\sin 14.6°}{\sin 55°} \approx 66$ ft

The target is in range for $216 - 66 = 150$ ft. Moving at 10 ft/sec, the target will be in range for about 15 seconds.

55. $\dfrac{\sin 26°}{8} = \dfrac{\sin A}{12}$; $8\sin A = 12\sin 26°$

$\sin A = \dfrac{12\sin 26°}{8} \approx 0.6576$

$A = \sin^{-1} 0.6576 \approx 41.1°$ or $A = 180° - 41.1° = 138.9°$. Two triangles are possible.
For $A = 41.1°$, $C_1 = 180° - (41.1° + 26°) = 112.9°$.

$\dfrac{\sin 112.9°}{c_1} = \dfrac{\sin 26°}{8}$; $c_1\sin 26° = 8\sin 112.9°$

$c_1 = \dfrac{8\sin 112.9°}{\sin 26°} \approx 16.8$ cm

Angles	Sides
$A_1 \approx 41.1°$	$a = 12\,\text{cm}$
$B = 26°$	$b = 8\,\text{cm}$
$C_1 \approx 112.9°$	$c_1 \approx 16.8\,\text{cm}$

For $A_2 = 138.9°$, $C_2 = 180° - (138.9° + 26°) = 15.1°$.

$\dfrac{\sin 15.1°}{c_2} = \dfrac{\sin 26°}{8}$; $c_2\sin 26° = 8\sin 15.1°$

$c_2 = \dfrac{8\sin 15.1°}{\sin 26°} \approx 4.8$ cm

Angles	Sides
$A_2 \approx 138.9°$	$a = 12\,\text{cm}$
$B = 26°$	$b = 8\,\text{cm}$
$C_2 = 15.1°$	$c_2 \approx 4.8\,\text{cm}$

57. From the grid, $a = 9$, $c = 5$. Consider the right triangle in the diagram as drawn: the shorter leg has length 4.

$\tan C = \dfrac{4}{9}$; $C = \tan^{-1}\left(\dfrac{4}{9}\right) \approx 24°$

Now use the Law of Sines with $a = 9$, $c = 5$, $C = 24°$.

$\dfrac{\sin 24°}{5} = \dfrac{\sin A}{9}$; $5\sin A = 9\sin 24°$

$\sin A = \dfrac{9\sin 24°}{5} \approx 0.7321$; $A \approx 47.0°$
or $A = 180° - 47.0° = 133.0°$.
For $A = 47.0°$, $B = 180° - (47.0° + 24°) = 109.0°$.

$\dfrac{\sin 109.0°}{b} = \dfrac{\sin 24°}{5}$; $b\sin 24° = 5\sin 109.0°$

$b = \dfrac{5\sin 109.0°}{\sin 24°} \approx 11.6$

Angles	Sides
$A_1 \approx 47.0°$	$a = 9\,\text{cm}$
$B_1 \approx 109.0°$	$b_1 \approx 11.6\,\text{cm}$
$C \approx 24°$	$c = 5\,\text{cm}$

For $A = 133.0°$, $B = 180° - (133.0° + 24°) = 23.0°$.

$\dfrac{\sin 23.0°}{b} = \dfrac{\sin 24°}{5}$; $b\sin 24° = 5\sin 23.0°$

$b = \dfrac{5\sin 23.0°}{\sin 24°} \approx 4.8$

Angles	Sides
$A_2 \approx 133.0°$	$a = 9\,\text{cm}$
$B_2 \approx 23.0°$	$b_2 \approx 4.8\,\text{cm}$
$C \approx 24°$	$c = 5\,\text{cm}$

59. First, find $\angle B = 180° - (32° + 53°) = 95°$.

$\dfrac{\sin 95°}{42} = \dfrac{\sin 32°}{c}$

$c\sin 95° = 42\sin 32°$

$c = \dfrac{42\sin 32°}{\sin 95°} \approx 22.3$ ft;

$\dfrac{\sin 95°}{42} = \dfrac{\sin 53°}{a}$

$a\sin 95° = 42\sin 53°$

$a = \dfrac{42\sin 53°}{\sin 95°} \approx 33.7$ ft

61. The third angle is $180° - (96° + 58°) = 26°$.
Rhymes to Tarryson:

$\dfrac{\sin 26°}{27.2} = \dfrac{\sin 96°}{\overline{RT}}$

$\overline{RT}\sin 26° = 27.2\sin 96°$

$\overline{RT} = \dfrac{27.2\sin 96°}{\sin 26°} \approx 61.7$ km;

Sexton to Tarryson:

$\dfrac{\sin 26°}{27.2} = \dfrac{\sin 58°}{\overline{ST}}$

$\overline{ST}\sin 26° = 27.2\sin 58°$

$\overline{ST} = \dfrac{27.2\sin 58°}{\sin 26°} \approx 52.6$ km

7.1 Exercises

63. The third angle is $180° - (39° + 58°) = 83°$.
The shortest side is across from the $39°$ angle, so let d be that distance.

$$\frac{\sin 83°}{5} = \frac{\sin 39°}{d}$$
$$d \sin 83° = 5 \sin 39°$$
$$d = \frac{5 \sin 39°}{\sin 83°} \approx 3.2 \text{ mi}$$

65. In the diagram provided, we need to find h.

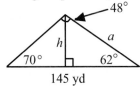

First, find a:

$$\frac{\sin 48°}{145} = \frac{\sin 70°}{a}$$
$$a \sin 48° = 145 \sin 70°$$
$$a = \frac{145 \sin 70°}{\sin 48°} \approx 183.35 \text{ yd}$$

Now use the right triangle on the right to find h:

$$\sin 62° = \frac{h}{183.35}$$
$$h = 183.35 \sin 62° \approx 161.9 \text{ yd}$$

67. Let a = the side across from the $63°$ angle and d = the base of the triangle.

$$\frac{\sin 27°}{5} = \frac{\sin 63°}{a}$$
$$a \sin 27° = 5 \sin 63°$$
$$a = \frac{5 \sin 63°}{\sin 27°} \approx 9.8 \text{ cm};$$
$$\frac{\sin 27°}{5} = \frac{\sin 90°}{d}$$
$$d \sin 27° = 5 \sin 90°$$
$$d = \frac{5 \sin 90°}{\sin 27°} \approx 11 \text{ cm}$$

The diameter of the circle is 11 cm, the base of the triangle. It is a right triangle.

69. a.

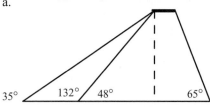

The small angle at the top of the left hand triangle is $180 - (35+132) = 13°$. Let s be the slant height of the west side, h the vertical height (dashed line).

$$\frac{\sin 13°}{1250} = \frac{\sin 35°}{s}; \quad s \sin 13° = 1250 \sin 35°$$
$$s = \frac{1250 \sin 35°}{\sin 13°} \approx 3187 \text{ m}$$

b. $\quad \sin 48° = \dfrac{h}{3187}; \quad h = 3187 \sin 48° \text{ m}$

$h \approx 2368 \text{ m}$

The east side forms a right triangle with the vertical height (2368) and a $65°$ angle.

$$\sin 65° = \frac{2368}{h_e}; \quad h_e = \frac{2368}{\sin 65°} \approx 2613$$

71. The base of the triangle is $10.2\sqrt{3}$ cm, and the hypotenuse is 20.4 cm. Using opposite over hypotenuse,

$$\sin 60° = \frac{10.2\sqrt{3}}{20.4} \text{ and } \sin 30° = \frac{10.2}{20.4}, \text{ so}$$

$$\frac{\sin 60°}{\sin 30°} = \frac{\dfrac{10.2\sqrt{3}}{20.4}}{\dfrac{10.2}{20.4}} = \sqrt{3}$$

We know that $\sin 45° = \dfrac{1}{\sqrt{2}}$ and $\sin 90° = 1$,

so $\dfrac{\sin 90°}{\sin 45°} = \dfrac{1}{\dfrac{1}{\sqrt{2}}} = \sqrt{2}$

348

73. Plug in $a = 45$, $A = 19°$, $B = 31°$

$$\frac{45+b}{45-b} = \frac{\tan\left[\frac{1}{2}(19°+31°)\right]}{\tan\left[\frac{1}{2}(19°-31°)\right]} = \frac{\tan 25°}{\tan(-6°)} \approx -4.44$$

$$\frac{45+b}{45-b} = -4.44; \quad 45+b = -4.44(45-b)$$

$$45+b = -199.8 + 4.44b$$

$$-3.44b = -244.8; \quad b \approx 71.2 \text{ cm}$$

Note that $C = 180° - (19° + 31°) = 130°$.
Plug in $a = 45$, $A = 19°$, $C = 130°$

$$\frac{45+c}{45-c} = \frac{\tan\left[\frac{1}{2}(19°+130°)\right]}{\tan\left[\frac{1}{2}(19°-130°)\right]}$$

$$= \frac{\tan 74.5°}{\tan(-55.5°)} \approx -2.48$$

$$\frac{45+c}{45-c} = -2.48; \quad 45+c = -2.48(45-c)$$

$$45+c = -111.6 + 2.48c$$

$$-1.48c = -156.6; \quad c \approx 105.9 \text{ cm}$$

75. A diagram of the first sighting:

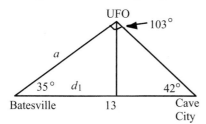

We can find side a using the law of sines, then use it to find d_1, which is the distance from the UFO's initial location to Batesville.

$$\frac{\sin 103°}{13} = \frac{\sin 42°}{a}$$

$$a \sin 103° = 13 \sin 42°$$

$$a = \frac{13 \sin 42°}{\sin 103°} \approx 8.9275 \text{ mi}$$

$$\cos 35° = \frac{d_1}{8.9275}$$

$$d_1 = 8.9275 \cos 35° \approx 7.3130 \text{ mi}$$

It will be helpful later to find the height:

$$\sin 35° = \frac{h}{8.9275}; \quad h \approx 5.1206$$

Second sighting:

$$\tan 24° = \frac{5.1206}{d_2}; \quad d_2 = \frac{5.1206}{\tan 24°} \approx 11.5011 \text{ mi}$$

The UFO's linear distance traveled is
$11.5011 - 7.3130 = 4.1881$ mi

$$\frac{4.1881 \text{ mi}}{1.2 \text{ sec}} \cdot \frac{3,600 \text{ sec}}{\text{hr}} \approx 12,564 \frac{\text{mi}}{\text{hr}}$$

77. $\tan^2 x - \sin^2 x = \tan^2 x \sin^2 x$

$$\frac{\sin^2 x}{\cos^2 x} - \sin^2 x =$$

$$\frac{\sin^2 x}{\cos^2 x} - \frac{\sin^2 x \cos^2 x}{\cos^2 x} =$$

$$\frac{\sin^2 x - \sin^2 x \cos^2 x}{\cos^2 x} =$$

$$\frac{\sin^2 x\left(1 - \cos^2 x\right)}{\cos^2 x} =$$

$$\frac{\sin^2 x \sin^2 x}{\cos^2 x} =$$

$$\frac{\sin^2 x}{\cos^2 x} \sin^2 x =$$

$$\tan^2 x \sin^2 x = \tan^2 x \sin^2 x$$

79. a. $m = \frac{2-(-3)}{4-(-5)} = \frac{5}{9}$

$$y - 2 = \frac{5}{9}(x-4); \quad y - 2 = \frac{5}{9}x - \frac{20}{9}$$

$$y = \frac{5}{9}x - \frac{2}{9}$$

b. $d = \sqrt{(2-(-3))^2 + (4-(-5))^2}$

$$= \sqrt{25+81} = \sqrt{106} \text{ units}$$

7.2 Exercises

1. cosines

3. Pythagorean

5. a. Law of Cosines Only:
   $a^2 = 37^2 + 52^2 - 2(37)(52)\cos 17°$
   $a^2 \approx 393.1; \quad a \approx 19.8$ m
   $52^2 = 37^2 + 393.1 - 2(37)(19.8)\cos C$
   $2704 = 1369 + 393.1 - 1465.2\cos C$
   $2704 = 1369 + 393.1 - 1465.2\cos C$
   $\cos C = \dfrac{941.9}{-1465.2}$
   $C = \cos^{-1}\left(\dfrac{941.9}{-1465.2}\right) \approx 130.0°$
   $B = 180 - (130 + 17) = 33.0°$
   Law of Sines:
   After we know that $a = 19.8$ m:
   $\dfrac{\sin B}{37} = \dfrac{\sin 17°}{19.8}; \quad 19.8\sin B = 37\sin 17°$
   $\sin B = \dfrac{37\sin 17°}{19.8}; \quad B = \sin^{-1}\left(\dfrac{37\sin 17°}{19.8}\right)$
   $B \approx 33.1°, \ C = 180 - (31.117) = 129.9°$

   b. The second method is simpler, Law of Sines.

7. Yes

9. No; there will be two unknowns in any of the three forms.

11. Yes

13. $a^2 = b^2 + c^2 - 2bc\cos A$
   $52.4^2 = 50^2 + 26.6^2 - 2(50)(26.6)\cos 80°$
   $2745.76 \approx 2745.7$
   $b^2 = a^2 + c^2 - 2ac\cos B$
   $50^2 = 52.4^2 + 26.6^2 - 2(52.4)(26.6)\cos 70°$
   $2500 \approx 2499.9$
   $c^2 = a^2 + b^2 - 2ab\cos C$
   $26.6^2 = 50^2 + 52.4^2 - 2(50)(52.4)\cos 30°$
   $707.6 \approx 707.8$
   With some rounding, all result in equality.

15. $4^2 = 5^2 + 6^2 - 2(5)(6)\cos B$
   $16 = 61 - 60\cos B; \ -45 = -60\cos B$
   $\cos B = \dfrac{45}{60}; \ B = \cos^{-1}\left(\dfrac{45}{60}\right) \approx 41.4°$

17. $a^2 = 9^2 + 7^2 - 2(9)(7)\cos 52° \approx 52.43$
   $a = \sqrt{52.43} \approx 7.24$

19. $10^2 = 12^2 + 15^2 - 2(12)(15)\cos A$
   $100 = 369 - 360\cos A; \ -269 = -360\cos A$
   $\cos A = \dfrac{-269}{-360}; \ A = \cos^{-1}\left(\dfrac{269}{360}\right) \approx 41.6°$

21. $c^2 = 75^2 + 32^2 - 2(75)(32)\cos 38° = 2866.55$
   $c = \sqrt{2866.55} \approx 53.5$ cm
   $\dfrac{\sin 38°}{53.5} = \dfrac{\sin B}{32}; \ 53.5\sin B = 32\sin 38°$
   $\sin B = \dfrac{32\sin 38°}{53.5}; \ B = \sin^{-1}\left(\dfrac{32\sin 38°}{53.5}\right)$
   $B \approx 21.6°; \ A = 180 - (21.6 + 38) = 120.4°$

23. $b^2 = 12.9^2 + 25.8^2 - 2(12.9)(25.8)\cos 30°$
   $b^2 \approx 255.59; b \approx 16$ mi
   $\dfrac{\sin 30°}{16} = \dfrac{\sin A}{12.9}; \ 16\sin A = 12.9\sin 30°$
   $\sin A = \dfrac{12.9\sin 30°}{16}; \ A = \sin^{-1}\left(\dfrac{12.9\sin 30°}{16}\right)$
   $A \approx 23.8°; \ C = 180 - (30 + 23.8) = 126.2°$

25. $c^2 = 538^2 + 465^2 - 2(538)(465)\cos 29°$
   $c^2 \approx 68,061.78; c \approx 260.9$ mm
   $\dfrac{\sin 29°}{260.9} = \dfrac{\sin B}{465}; \ 260.9\sin B = 465\sin 29°$
   $\sin B = \dfrac{465\sin 29°}{260.9}; \ B = \sin^{-1}\left(\dfrac{465\sin 29°}{260.9}\right)$
   $B \approx 59.8°; \ A = 180 - (59.8 + 29) = 91.2°$

27. $a^2 = b^2 + c^2 - 2bc\cos A$

$675 = 108 + 300 - 360\cos A$

$267 = -360\cos A; \quad \cos A = \dfrac{267}{-360}$

$A = \cos^{-1}\left(\dfrac{267}{-360}\right) \approx 137.9°$

$\dfrac{\sin 137.9°}{15\sqrt{3}} = \dfrac{\sin B}{6\sqrt{3}}; \quad 15\sqrt{3}\sin B = 6\sqrt{3}\sin 137.9°$

$\sin B = \dfrac{6\sqrt{3}\sin 137.9°}{15\sqrt{3}}$

$B = \sin^{-1}\left(\dfrac{6\sqrt{3}\sin 137.9°}{15\sqrt{3}}\right) \approx 15.6°$

$C = 180 - (15.6 + 137.9) = 26.5°$

29. $32.8^2 = 24.9^2 + 12.4^2 - 2(24.9)(12.4)\cos A$

$1075.84 = 773.77 - 617.52\cos A$

$302.07 = -617.52\cos A; \quad \cos A = \dfrac{302.07}{-617.52}$

$A = \cos^{-1}\left(\dfrac{302.07}{-617.52}\right) \approx 119.3°;$

$\dfrac{\sin 119.3°}{32.8} = \dfrac{\sin B}{24.9}; \quad 32.8\sin B = 24.9\sin 119.3°$

$\sin B = \dfrac{24.9\sin 119.3°}{32.8}$

$B = \sin^{-1}\left(\dfrac{24.9\sin 119.3°}{32.8}\right) \approx 41.5°$

$C = 180 - (41.5 + 119.3) = 19.2°$

31. $\left(4.1\times 10^{25}\right)^2 = \left(2.3\times 10^{25}\right)^2 + \left(2.9\times 10^{25}\right)^2$
$\qquad\qquad -2\left(2.3\times 10^{25}\right)\left(2.9\times 10^{25}\right)\cos A$

$1.7\times 10^{51} = 1.4\times 10^{51} - 1.3\times 10^{51}\cos A$

$3.0\times 10^{50} = -1.3\times 10^{51}\cos A$

$\cos A = \dfrac{3.0\times 10^{50}}{-1.3\times 10^{51}}; \quad A = \cos^{-1}\left(\dfrac{3.0\times 10^{50}}{-1.3\times 10^{51}}\right)$

$A \approx 103.3°$

$\dfrac{\sin 103.3°}{4.1\times 10^{25}} = \dfrac{\sin C}{2.9\times 10^{25}}$

$4.1\times 10^{25}\sin C = 2.9\times 10^{25}\sin 103.3°$

$\sin C = \dfrac{2.9\times 10^{25}\sin 103.3°}{4.1\times 10^{25}}$

$C = \sin^{-1}\left(\dfrac{2.9\times 10^{25}\sin 103.3°}{4.1\times 10^{25}}\right) \approx 43.5°$

$B = 180 - (43.5 + 103.3) = 33.2°$

33. $\left(12\sqrt{3}\right)^2 = 12.9^2 + 9.2^2 - 2(12.9)(9.2)\cos A$

$432 = 251.05 - 237.36\cos A$

$180.95 = -237.36\cos A; \quad \cos A = \dfrac{180.95}{-237.36}$

$A = \cos^{-1}\left(\dfrac{180.95}{-237.36}\right) \approx 139.7°$

$\dfrac{\sin 139.7°}{12\sqrt{3}} = \dfrac{\sin B}{12.9}$

$12\sqrt{3}\sin B = 12.9\sin 139.7°$

$\sin B = \dfrac{12.9\sin 139.7°}{12\sqrt{3}}$

$B = \sin^{-1}\left(\dfrac{12.9\sin 139.7°}{12\sqrt{3}}\right) \approx 23.7°$

$C = 180 - (23.7 + 139.7) = 16.6°$

35. $a^2 = b^2 + c^2 - 2bc\cos A$

$a^2 - b^2 - c^2 = -2bc\cos A$

$\dfrac{a^2 - b^2 - c^2}{-2bc} = \cos A$

$\dfrac{b^2 + c^2 - a^2}{2bc} = \cos A$

Adapting the new formula to the given triangle, where we should solve for angle C first, we get

$\cos C = \dfrac{a^2 + b^2 - c^2}{2ab}$

$\cos C = \dfrac{39^2 + 37^2 - 52^2}{2(39)(37)} \approx 0.0644$

$C = \cos^{-1} 0.0644 \approx 86.3°$

7.2 Exercises

37. $m^2 = 1435^2 + 692^2 - 2(1435)(692)\cos 99°$

$m^2 \approx 2,848774; \quad m \approx 1688$ mi

39. $198^2 = 354^2 + 423^2 - 2(354)(423)\cos P$

$39,204 = 304,245 - 299,484\cos P$

$-265,041 = -299,484\cos P$

$\cos P = \dfrac{-265,041}{-299,484}$

$P = \cos^{-1}\left(\dfrac{-265,041}{-299,484}\right) \approx 27.7°$

The heading is 27.7° north of west or a heading of 297.7°.

41. $d^2 = 1.8^2 + 2.6^2 - 2(1.8)(2.6)\cos 51°$

$d = \sqrt{10 - 9.36\cos 51°}$ mi

$= \sqrt{10 - 9.36\cos 51°}\,(5280) = 10,703.6$ ft

It cannot be constructed.

43. After 5 hours, the distances are 5(450) = 2250 miles, and 5(425) = 2125 miles. The angle between paths is 45°.

$d^2 = 2250^2 + 2125^2 - 2(2250)(2125)\cos 45°$

$d^2 \approx 2,816,416.405; \quad d \approx 1678.2$ mi

45. Call the point at the bottom left corner of the board P. Triangle PAB is a 45-45-90 triangle with legs 4, so side AB is $4\sqrt{2}$. Using the Pythagorean Theorem on triangle PBC, $BC^2 = 10^2 + 4^2$; $BC = \sqrt{116}$. Side AC is 6, so the perimeter is

$6 + 4\sqrt{2} + \sqrt{116} \approx 22.4$ cm

Again using right triangle PBC,

$\tan C = \dfrac{4}{10}; \quad C = \tan^{-1}\dfrac{4}{10} \approx 21.8°$

$\dfrac{\sin 21.8°}{4\sqrt{2}} = \dfrac{\sin B}{6}; \quad 4\sqrt{2}\sin B = 6\sin 21.8°$

$\sin B = \dfrac{6\sin 21.8°}{4\sqrt{2}}; \quad B = \sin^{-1}\left(\dfrac{6\sin 21.8°}{4\sqrt{2}}\right)$

$B \approx 23.2°; \quad A = 180 - (21.8 + 23.2) = 135°$

47. $20^2 = 12^2 + 9^2 - 2(12)(9)\cos C$

$400 = 225 - 216\cos C$

$175 = -216\cos C; \quad \cos C = \dfrac{175}{-216}$

$C = \cos^{-1}\left(\dfrac{175}{-216}\right) \approx 144.1°$

$\dfrac{\sin 144.1°}{20} = \dfrac{\sin B}{9}; \quad 20\sin B = 9\sin 144.1°$

$\sin B = \dfrac{9\sin 144.1°}{20}; \quad B = \sin^{-1}\left(\dfrac{9\sin 144.1°}{20}\right)$

$B \approx 15.3°$

$A \approx 180 - (144.1 + 15.3) = 20.6°$

49. A regular pentagon can be made from five triangles, each with an angle of $\dfrac{360}{5} = 72°$

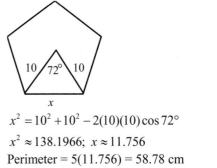

$x^2 = 10^2 + 10^2 - 2(10)(10)\cos 72°$

$x^2 \approx 138.1966; \quad x \approx 11.756$

Perimeter = 5(11.756) = 58.78 cm

51. Side AC is the hypotenuse of a right triangle with right angle at (3, 0) that has legs 3 and 4, so its length is 5. Side CB is the hypotenuse of a right triangle with right angle at (12, 0) that has legs 12 and 5 so its length is 13. The third side is the hypotenuse of a right triangle with sides 1 and 9, so its length is $\sqrt{9^2 + 1^2} = \sqrt{82}$.

$13^2 = 5^2 + \sqrt{82}^2 - 2(5)(\sqrt{82})\cos A$

$169 = 107 - 10\sqrt{82}\cos A$

$62 = -10\sqrt{82}\cos A; \quad \cos A = \dfrac{62}{-10\sqrt{82}}$

$A = \cos^{-1}\left(\dfrac{62}{-10\sqrt{82}}\right) \approx 133.2°$

$\dfrac{\sin 133.2°}{13} = \dfrac{\sin B}{5}; \quad 13\sin B = 5\sin 133.2°$

$\sin B = \dfrac{5\sin 133.2°}{13}; \quad B = \sin^{-1}\left(\dfrac{5\sin 133.2°}{13}\right)$

$B \approx 16.3°; \quad C = 180 - (133.2 + 16.3) = 30.5°$

53. Diagonal: $\sqrt{20^2 + 30^2} = \sqrt{1300} \approx 36.06$

$\tan \alpha = \dfrac{20}{30}$

$\alpha = \tan^{-1}\left(\dfrac{2}{3}\right) \approx 33.7°$;

$A = \dfrac{1}{2} bc \sin \alpha$

$A = \dfrac{1}{2}\sqrt{1300}\,(15)\sin 33.7°$

≈ 150 square feet

55. $42° + 65° + x = 180°$

$x = 73°$;

$A = \dfrac{c^2 \sin A \sin B}{2 \sin C}$

$A = \dfrac{299^2 \sin 42° \sin 65°}{2 \sin 73°} \approx 28346.7$;

a. $\dfrac{28346.7}{43560} \approx 0.65$ or 65%

b. $3{,}000{,}000(0.65) = \$1{,}950{,}000$

57. $p = \dfrac{1289 + 1063 + 922}{2} = 1637$

$A = \sqrt{1637(1637-1289)(1637-922)(1637-1063)}$

$A \approx 483{,}529$ km^2

59. The sum of the two smaller sides is 889, which is less than the longer one.

61. $53.9 = 78\cos 25° + 37\cos 117° \approx 53.9$

$a^2 = b^2 + c^2 - 2bc\cos A$

$b^2 = a^2 + c^2 - 2ac\cos B$

Substitute the right side of the first expression in for a^2 in the second:

$b^2 = \left(b^2 + c^2 - 2bc\cos A\right) + c^2 - 2ac\cos B$

$0 = 2c^2 - 2bc\cos A - 2ac\cos B$

(Divide both sides by 2c)

$0 = c - b\cos A - a\cos B$

$-c = -b\cos A - a\cos B$

$c = b\cos A + a\cos B$

63. $2\log_2 4 + 2\log_2 3 - 2\log_2 6$

$= \log_2 4^2 + \log_2 3^2 - \log_2 6^2$

$= \log_2\left(\dfrac{16 \cdot 9}{36}\right) = \log_2 4 = \log_2 2^2 = 2$

65. $y = -5,\ r = 13;\ x = \sqrt{13^2 - 5^2} = 12$, positive since cosine is positive.

$\csc x = -\dfrac{13}{5};\quad \cos x = \dfrac{12}{13};\quad \sec x = \dfrac{13}{12}$

$\tan x = -\dfrac{5}{12};\quad \cot x = -\dfrac{12}{5}$

Technology Highlight

Exercise 1: They would be equal; verified

Exercise 3: $|\mathbf{v}| = 9.5; |\mathbf{v}| = 9.5$ for all values of θ

7.3 Exercises

1. Scalar

3. Directed line

5. Answers will vary.

7.

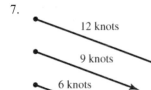

12 knots

9 knots

6 knots

9.

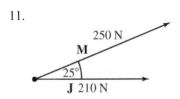

11.

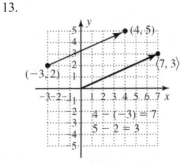

13.

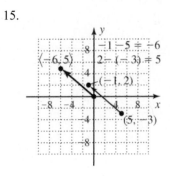

15.

17. Terminal point $= (-2+7, -3+2) = (5, -1)$;
$$|\mathbf{v}| = \sqrt{7^2 + 2^2} = \sqrt{53}$$

19. Terminal point $= (2-3, 6-5) = (-1, 1)$;
$$|\mathbf{v}| = \sqrt{(-3)^2 + (-5)^2} = \sqrt{34}$$

21. a.

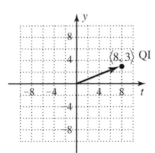

b. $|\mathbf{v}| = \sqrt{8^2 + 3^2} = \sqrt{73}$

c. $\tan\theta = \dfrac{3}{8}$; $\theta = \tan^{-1}\left(\dfrac{3}{8}\right) \approx 20.6°$

23. a.

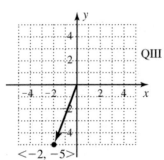

b. $|\mathbf{v}| = \sqrt{(-2)^2 + (-5)^2} = \sqrt{29}$

c. $\tan\theta = \dfrac{-5}{-2}$; $\theta = \tan^{-1}\left(\dfrac{5}{2}\right) \approx 68.2°$

25. $a = 12\cos 25° \approx 10.9$; $b = 12\sin 25° \approx 5.1$
$\langle -10.9, 5.1 \rangle$

27. $a = 140.5\cos 41° \approx 106.0$
$b = 140.5\sin 41° \approx 92.2$
$\langle 106.0, -92.2 \rangle$

29. $a = 10\cos 15° \approx 9.7$; $b = 10\sin 15° \approx 2.6$
$\langle -9.7, -2.6 \rangle$

31. a. $\mathbf{u} + \mathbf{v} = \langle 2 + (-3), 3 + 6 \rangle = \langle -1, 9 \rangle$

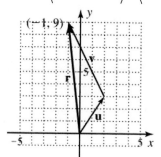

33. a. $\mathbf{u} + \mathbf{v} = \langle 7 + 1, -2 + 6 \rangle = \langle 8, 4 \rangle$

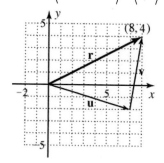

b. $\mathbf{u} - \mathbf{v} = \langle 2 - (-3), 3 - 6 \rangle = \langle 5, -3 \rangle$

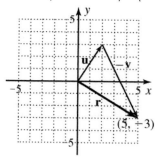

b. $\mathbf{u} - \mathbf{v} = \langle 7 - 1, -2 - 6 \rangle = \langle 6, -8 \rangle$

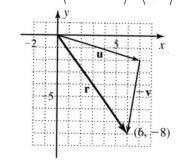

c. $2\mathbf{u} + 1.5\mathbf{v} = \langle 4, 6 \rangle + \langle -4.5, 9 \rangle = \langle -0.5, 15 \rangle$

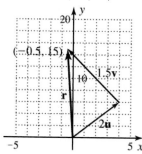

c. $2\mathbf{u} + 1.5\mathbf{v} = \langle 14, -4 \rangle + \langle 1.5, 9 \rangle = \langle 15.5, 5 \rangle$

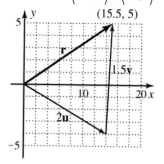

d. $\mathbf{u} - 2\mathbf{v} = \langle 2, 3 \rangle - \langle -6, 12 \rangle = \langle 8, -9 \rangle$

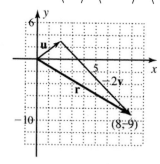

d. $\mathbf{u} - 2\mathbf{v} = \langle 7, -2 \rangle - \langle 2, 12 \rangle = \langle 5, -14 \rangle$

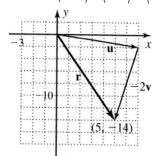

7.3 Exercises

35. a. $\mathbf{u}+\mathbf{v}=\langle -4+1,2+4\rangle =\langle -3,6\rangle$

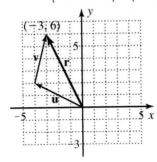

b. $\mathbf{u}-\mathbf{v}=\langle -4-1,2-4\rangle =\langle -5,-2\rangle$

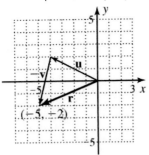

c. $2\mathbf{u}+1.5\mathbf{v}=\langle -8,4\rangle +\langle 1.5,6\rangle =\langle -6.5,10\rangle$

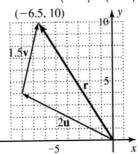

d. $\mathbf{u}-2\mathbf{v}=\langle -4,2\rangle -\langle 2,8\rangle =\langle -6,-6\rangle$

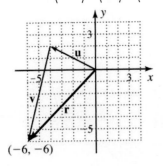

37. True

39. False $(\mathbf{c}+\mathbf{d}=\mathbf{h})$

41. True

43. $\mathbf{u}+\mathbf{v}=\langle 1+7,4+2\rangle =\langle 8,6\rangle$

$\mathbf{u}-\mathbf{v}=\langle 1-7,4-2\rangle =\langle -6,2\rangle$

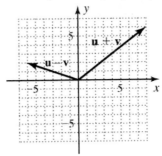

45. $\mathbf{u}+\mathbf{v}=\langle -1+(-8),-3+(-3)\rangle =\langle -9,-6\rangle$

$\mathbf{u}-\mathbf{v}=\langle -1-(-8),-3-(-3)\rangle =\langle 7,0\rangle$

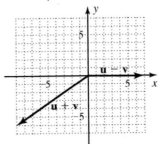

47. $\mathbf{u}+\mathbf{v}=\langle -5+2,-3+(-3)\rangle =\langle -3,-6\rangle$

$\mathbf{u}-\mathbf{v}=\langle -5-2,-3-(-3)\rangle =\langle -7,0\rangle$

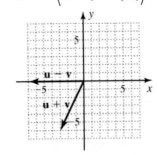

49.

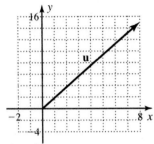

$$\langle 8,15 \rangle = 8\mathbf{i} + 15\mathbf{j}$$

$$|\mathbf{u}| = \sqrt{8^2 + 15^2} = \sqrt{289} = 17$$

51.

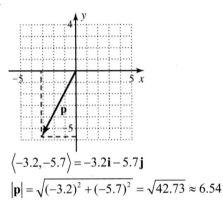

$$\langle -3.2, -5.7 \rangle = -3.2\mathbf{i} - 5.7\mathbf{j}$$

$$|\mathbf{p}| = \sqrt{(-3.2)^2 + (-5.7)^2} = \sqrt{42.73} \approx 6.54$$

53. a.

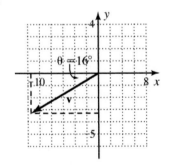

b. $a = 12\cos 16° \approx 11.5$
$b = 12\sin 16° \approx 3.3$
$\mathbf{v} = \langle -11.5, -3.3 \rangle$

c. $\mathbf{v} = -11.5\mathbf{i} - 3.3\mathbf{j}$

55. a.

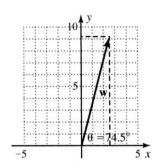

b. $a = 9.5\cos 74.5° \approx 2.5$
$b = 9.5\sin 74.5° \approx 9.2$
$\mathbf{w} = \langle 2.5, 9.2 \rangle$

c. $\mathbf{w} = 2.5\mathbf{i} + 9.2\mathbf{j}$

57. a. $\mathbf{v}_1 + \mathbf{v}_2 = (2-4)\mathbf{i} + (-3+5)\mathbf{j} = -2\mathbf{i} + 2\mathbf{j}$

$\text{Mag.} = \sqrt{(-2)^2 + 2^2} = \sqrt{8} = 2\sqrt{2}$

$\theta_r = \tan^{-1}\left(\dfrac{2}{-2}\right) = -45°$; In QII,

$\theta = 135°$

b. $\mathbf{v}_1 - \mathbf{v}_2 = (2-(-4))\mathbf{i} + (-3-5)\mathbf{j} = 6\mathbf{i} - 8\mathbf{j}$

$\text{Mag.} = \sqrt{6^2 + (-8)^2} = 10$

$\theta_r = \tan^{-1}\left(\dfrac{-8}{6}\right) \approx -53.1°$; In QIV,

$\theta = 306.9°$

c. $2\mathbf{v}_1 + 1.5\mathbf{v}_2 = 4\mathbf{i} - 6\mathbf{j} + -6\mathbf{i} + 7.5\mathbf{j}$
$= -2\mathbf{i} + 1.5\mathbf{j}$

$\text{Mag.} = \sqrt{(-2)^2 + (1.5)^2} = \sqrt{6.25} = 2.5$

$\theta_r = \tan^{-1}\left(\dfrac{1.5}{-2}\right) \approx -36.9°$; In QII,

$\theta = 143.1°$

d. $\mathbf{v}_1 - 2\mathbf{v}_2 = 2\mathbf{i} - 3\mathbf{j} - (-8\mathbf{i} + 10\mathbf{j})$
$= 10\mathbf{i} - 13\mathbf{j}$

$\text{Mag.} = \sqrt{10^2 + (-13)^2} \approx 16.4$

$\theta_r = \tan^{-1}\left(\dfrac{-13}{10}\right) \approx -52.4°$; In QIV,

$\theta = 307.6°$

7.3 Exercises

59. a. $\mathbf{v}_1 + \mathbf{v}_2 = 5\sqrt{2}\mathbf{i} + 7\mathbf{j} + (-3\sqrt{2}\mathbf{i} - 5\mathbf{j})$
$= 2\sqrt{2}\mathbf{i} + 2\mathbf{j}$
Mag. $= \sqrt{(2\sqrt{2})^2 + 2^2} \approx 3.5$
$\theta = \tan^{-1}\left(\dfrac{2}{2\sqrt{2}}\right) \approx 35.3°$

 b. $\mathbf{v}_1 - \mathbf{v}_2 = 5\sqrt{2}\mathbf{i} + 7\mathbf{j} - (-3\sqrt{2}\mathbf{i} - 5\mathbf{j})$
$= 8\sqrt{2}\mathbf{i} + 12\mathbf{j}$
Mag. $= \sqrt{(8\sqrt{2})^2 + 12^2} \approx 16.5$
$\theta = \tan^{-1}\left(\dfrac{12}{8\sqrt{2}}\right) \approx 46.7°$

 c. $2\mathbf{v}_1 + 1.5\mathbf{v}_2 = 10\sqrt{2}\mathbf{i} + 14\mathbf{j} +$
$(-4.5\sqrt{2}\mathbf{i} - 7.5\mathbf{j}) = 5.5\sqrt{2}\mathbf{i} + 6.5\mathbf{j}$
Mag. $= \sqrt{(5.5\sqrt{2})^2 + (6.5)^2} \approx 10.1$
$\theta = \tan^{-1}\left(\dfrac{6.5}{5.5\sqrt{2}}\right) \approx 39.9°$

 d. $\mathbf{v}_1 - 2\mathbf{v}_2 = 5\sqrt{2}\mathbf{i} + 7\mathbf{j} - (-6\sqrt{2}\mathbf{i} - 10\mathbf{j})$
$= 11\sqrt{2}\mathbf{i} + 17\mathbf{j}$
Mag. $= \sqrt{(11\sqrt{2})^2 + 17^2} \approx 23.0$
$\theta = \tan^{-1}\left(\dfrac{17}{11\sqrt{2}}\right) \approx 47.5°$

61. a. $\mathbf{v}_1 + \mathbf{v}_2 = 12\mathbf{i} + 4\mathbf{j} + (-4\mathbf{i}) = 8\mathbf{i} + 4\mathbf{j}$
Mag. $= \sqrt{8^2 + 4^2} \approx 8.9$
$\theta = \tan^{-1}\left(\dfrac{4}{8}\right) \approx 26.6°$

 b. $\mathbf{v}_1 - \mathbf{v}_2 = 12\mathbf{i} + 4\mathbf{j} - (-4\mathbf{i}) = 16\mathbf{i} + 4\mathbf{j}$
Mag. $= \sqrt{16^2 + 4^2} \approx 16.5$
$\theta = \tan^{-1}\left(\dfrac{4}{16}\right) \approx 14.0°$

 c. $2\mathbf{v}_1 + 1.5\mathbf{v}_2 = 24\mathbf{i} + 8\mathbf{j} + (-6\mathbf{i}) = 18\mathbf{i} + 8\mathbf{j}$
Mag. $= \sqrt{18^2 + 8^2} \approx 19.7$
$\theta = \tan^{-1}\left(\dfrac{8}{18}\right) \approx 24.0°$

 d. $\mathbf{v}_1 - 2\mathbf{v}_2 = 12\mathbf{i} + 4\mathbf{j} - (-8\mathbf{i}) = 20\mathbf{i} + 4\mathbf{j}$
Mag. $= \sqrt{20^2 + 4^2} \approx 20.4$
$\theta = \tan^{-1}\left(\dfrac{4}{20}\right) \approx 11.3°$

63. $|\mathbf{u}| = \sqrt{7^2 + 24^2} = 25$; Unit vector $=$
$\left\langle \dfrac{7}{25}, \dfrac{24}{25} \right\rangle$. Mag. $= \sqrt{\left(\dfrac{7}{25}\right)^2 + \left(\dfrac{24}{25}\right)^2} = 1$

65. $|\mathbf{p}| = \sqrt{(-20)^2 + 21^2} = 29$; $\mathbf{u} = \left\langle -\dfrac{20}{29}, \dfrac{21}{29} \right\rangle$
$|\mathbf{u}| = \sqrt{\left(-\dfrac{20}{29}\right)^2 + \left(\dfrac{21}{29}\right)^2} = 1$

67. Mag. $= \sqrt{20^2 + (-21)^2} = 29$; $\mathbf{u} = \dfrac{20}{29}\mathbf{i} - \dfrac{21}{29}\mathbf{j}$
$|\mathbf{u}| = \sqrt{\left(\dfrac{20}{29}\right)^2 + \left(-\dfrac{21}{29}\right)^2} = 1$

69. Mag. $= \sqrt{3.5^2 + 12^2} = 12.5$;
$\mathbf{u} = \dfrac{3.5}{12.5}\mathbf{i} + \dfrac{12}{12.5}\mathbf{j} = \dfrac{7}{25}\mathbf{i} + \dfrac{24}{25}\mathbf{j}$
$|\mathbf{u}| = \sqrt{\left(\dfrac{7}{25}\right)^2 + \left(\dfrac{24}{25}\right)^2} = 1$

71. $|\mathbf{v}_1| = \sqrt{13^2 + 3^2} = \sqrt{178}$;
$\mathbf{u} = \left\langle \dfrac{13}{\sqrt{178}}, \dfrac{3}{\sqrt{178}} \right\rangle$
$|\mathbf{u}| = \sqrt{\left(\dfrac{13}{\sqrt{178}}\right)^2 + \left(\dfrac{3}{\sqrt{178}}\right)^2} = 1$

73. Mag. $= \sqrt{6^2 + 11^2} = \sqrt{157}$
$\mathbf{u} = \dfrac{6}{\sqrt{157}}\mathbf{i} + \dfrac{11}{\sqrt{157}}\mathbf{j}$
$|\mathbf{u}| = \sqrt{\left(\dfrac{6}{\sqrt{157}}\right)^2 + \left(\dfrac{11}{\sqrt{157}}\right)^2} = 1$

75. $\left|\mathbf{p}\right| = \sqrt{2^2 + 7^2} = \sqrt{53}; \quad \cos 52° = \dfrac{\left|\mathbf{r}\right|}{\sqrt{53}}$

$\left|\mathbf{r}\right| = \sqrt{53}\cos 52° \approx 4.48$

$\left|\mathbf{q}\right| = \sqrt{10^2 + 4^2} = \sqrt{116} = 2\sqrt{29}$

$\mathbf{r} = 4.48\left\langle \dfrac{10}{2\sqrt{29}}, \dfrac{4}{2\sqrt{29}} \right\rangle$

$= 4.48\left\langle \dfrac{5}{\sqrt{29}}, \dfrac{2}{\sqrt{29}} \right\rangle \approx \left\langle 4.16, 1.66 \right\rangle$

77. $\left|\mathbf{p}\right| = \sqrt{4^2 + (-6)^2} = \sqrt{52}; \quad \cos 36° = \dfrac{\left|\mathbf{r}\right|}{\sqrt{52}}$

$\left|\mathbf{r}\right| = \sqrt{52}\cos 36° \approx 5.83$

$\left|\mathbf{q}\right| = \sqrt{8^2 + (-3)^2} = \sqrt{73}$

$\mathbf{r} = 5.83\left\langle \dfrac{8}{\sqrt{73}}, \dfrac{-3}{\sqrt{73}} \right\rangle \approx \left\langle 5.46, -2.05 \right\rangle$

79. $\left|\mathbf{v}\right| = \sqrt{5^2 + 9^2 + 10^2} = \sqrt{206} \approx 14.4$

81. Find the vertical component of $\mathbf{W}_2$:

$a = 700\sin 32°$
Find the angle that makes the vertical component of $\mathbf{W}_1 = 700\sin 32°$:

$700\sin 32° = 900\sin\theta; \sin\theta = \dfrac{700\sin 32°}{900}$

$\theta = \sin^{-1}\left(\dfrac{700\sin 32°}{900} \right) \approx 24.3°$

83. Horizontal: $a = 100\cos 37° \approx 79.9\,\frac{\text{ft}}{\text{sec}}$

Vertical: $b = 100\sin 37° \approx 60.2\,\frac{\text{ft}}{\text{sec}}$

85. The plane vector makes a 75° angle with the positive x–axis, and the wind vector makes a 10° angle with the positive x-axis. Find the components of each:

Plane: $a = 250\cos 75° \approx 64.7$
$\qquad b = 250\sin 75° \approx 241.5$
Wind: $a = 35\cos 10° \approx 34.5$
$\qquad b = 35\sin 10° \approx 6.1$
Resultant: $\left\langle 64.7 + 34.5, 241.5 + 6.1 \right\rangle$
$\qquad\qquad = \left\langle 99.2, 247.6 \right\rangle$

Mag. $= \sqrt{99.2^2 + 247.6^2} \approx 266.7$ mph

$\theta = \tan^{-1}\left(\dfrac{99.2}{247.6} \right) \approx 21.8°$ from positive x-axis:

Heading $= 68.2°$

87. $x = 85\cos 15° \approx 82.10$
$y = 85\sin 15° \approx 22.00$
(82.10 cm, 22.00 cm)

89. $1 \cdot \mathbf{u} = 1 \cdot \left\langle a, b \right\rangle = \left\langle 1 \cdot a, 1 \cdot b \right\rangle = \left\langle a, b \right\rangle = \mathbf{u}$

91. $\mathbf{u} - \mathbf{v} = \left\langle a, b \right\rangle - \left\langle c, d \right\rangle = \left\langle a - c, b - d \right\rangle$
$\qquad = \left\langle a + (-c), b + (-d) \right\rangle = \left\langle a, b \right\rangle + \left\langle -c, -d \right\rangle$
$\qquad = \mathbf{u} + (-\mathbf{v})$

93. $(ck)\mathbf{u} = ck\left\langle a, b \right\rangle = \left\langle cka, ckb \right\rangle = c\left\langle ka, kb \right\rangle$
$\qquad = c(k\mathbf{u}) = \left\langle cka, ckb \right\rangle = \left\langle kca, kcb \right\rangle$
$\qquad = k\left\langle ca, cb \right\rangle = k(c\mathbf{u})$

95. $\mathbf{u} + (-\mathbf{u}) = \left\langle a, b \right\rangle + \left\langle -a, -b \right\rangle = \left\langle a - a, b - b \right\rangle$
$\qquad = \left\langle 0, 0 \right\rangle$

97. $(c + k)\mathbf{u} = (c + k)\left\langle a, b \right\rangle$
$\qquad = \left\langle (c + k)a, (c + k)b \right\rangle$
$\qquad = \left\langle ca + ka, cb + kb \right\rangle = \left\langle ca, cb \right\rangle + \left\langle ka, kb \right\rangle$
$c\left\langle a, b \right\rangle + k\left\langle a, b \right\rangle = c\mathbf{u} + k\mathbf{u}$

99. Find the components of each vector, then add all horiz. and vert. components.
$\mathbf{p} = \left\langle 1, 3 \right\rangle; \quad \mathbf{r} = \left\langle 3, 3 \right\rangle; \quad \mathbf{s} = \left\langle 4, -1 \right\rangle$
$\mathbf{t} = \left\langle 2, -4 \right\rangle; \quad \mathbf{u} = \left\langle -4, -3 \right\rangle; \quad \mathbf{v} = \left\langle -6, 2 \right\rangle$
Horizontal: $1 + 3 + 4 + 2 + -4 + -6 = 0$
Vertical: $3 + 3 + -1 + -4 + -3 + 2 = 0$

101. Answers will vary. One possibility: Place the first segment at 0°; it will end at (45, 0). Place the second to reach the point (51, 39.6), and the third to reach (80, 20). In this case, the second segment has components

$\langle 6, 39.6 \rangle$ and $\theta = \tan^{-1}\left(\dfrac{39.6}{6}\right) \approx 81.4°$, while

the third segment has components

$\langle 29, -19.6 \rangle$ and $\theta = \tan^{-1}\left(\dfrac{-19.6}{29}\right) \approx -34°$.

103. a. $\ln(2(3) - 7) = \ln(-1)$ not a real number

b. $\dfrac{5}{3 - 3}$ not possible

c. $\sqrt{\dfrac{1}{3}(3) - 5} = \sqrt{-4}$ not a real number

105. $g(x) = x^3 - 7x$

$g(x) = x(x^2 - 7) = x(x + \sqrt{7})(x - \sqrt{7})$

$x = 0, x = \pm\sqrt{7}$

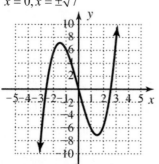

Mid-Chapter Check

1. $\dfrac{\sin A}{a} = \dfrac{\sin B}{b}$; $a\sin B = b\sin A$

$\sin B = \dfrac{b\sin A}{a}$

3. $a^2 = 207^2 + 250^2 - 2(250)(207)\cos 31°$

$a^2 = 16{,}632.2$; $a \approx 129$ m

$\dfrac{\sin 31}{129} = \dfrac{\sin B}{250}$; $129\sin B = 250\sin 31°$

$\sin B = \dfrac{250\sin 31°}{129}$; $B = \sin^{-1}\left(\dfrac{250\sin 31°}{129}\right)$

$B \approx 86.5°$; $C = 180 - (31 + 86.5) = 62.5°$

5. $\dfrac{\sin 44°}{2.1} = \dfrac{\sin C}{2.8}$; $2.1\sin C = 2.8\sin 44°$

$\sin C = \dfrac{2.8\sin 44°}{2.1}$; $C = \sin^{-1}\left(\dfrac{2.8\sin 44°}{2.1}\right)$

$C \approx 67.9°$ or $180 - 67.9 = 112.1°$
For $C = 67.9°$:
$B = 180 - (44 + 67.9) = 68.1°$
$\dfrac{\sin 44°}{2.1} = \dfrac{\sin 68.1°}{b}$
$b\sin 44° = 2.1\sin 68.1°$
$b = \dfrac{2.1\sin 68.1°}{\sin 44°} \approx 2.8$ km
For $C = 112.1°$:
$B = 180 - (112.1 + 44) = 23.9°$
$\dfrac{\sin 44°}{2.1} = \dfrac{\sin 23.9°}{b}$; $b\sin 44° = 2.1\sin 23.9°$
$b = \dfrac{2.1\sin 23.9°}{\sin 44°} \approx 1.2$ km

7. $\dfrac{75}{\sin 25°} = \dfrac{h}{\sin 20°}$; $h\sin 25° = 75\sin 20°$

$h = \dfrac{75\sin 20°}{\sin 25°} \approx 60.7$ ft

9. Adding the appropriate radii to get lengths:

13 ⟋ 16
‾‾‾‾‾‾‾
21

$21^2 = 13^2 + 16^2 - 2(13)(16)\cos\beta$
$441 = 425 - 416\cos\beta$
$16 = -416\cos\beta$; $\cos\beta = \dfrac{16}{-416}$
$\beta = \cos^{-1}\left(-\dfrac{16}{416}\right) \approx 92.2°$
$\dfrac{\sin 92.2°}{21} = \dfrac{\sin\alpha}{16}$; $21\sin\alpha = 16\sin 92.2°$
$\sin\alpha = \dfrac{16\sin 92.2°}{21}$; $\alpha = \sin^{-1}\left(\dfrac{16\sin 92.2°}{21}\right)$
$\alpha \approx 49.6°$; $\gamma = 180 - (92.2 + 49.6) = 38.2°$

Reinforcing Basic Concepts

1. $\dfrac{\sin 35°}{11.6} = \dfrac{\sin B}{20}$; $11.6\sin B = 20\sin 35°$

$\sin B = \dfrac{20\sin 35°}{11.6}$; $B = \sin^{-1}\left(\dfrac{20\sin 35°}{11.6}\right)$

$B \approx 81.5°$; $C \approx 180 - (81.5 + 35) = 63.5°$
The measurements are very close.

Chapter 7: Applications of Trigonometry

7.4 Exercises

1. equilibrium, zero

3. orthogonal

5. Answers will vary.

7. $\mathbf{F} = \mathbf{F_1} + \mathbf{F_2} = \langle -8+2, -3-5 \rangle = \langle -6, -8 \rangle$
   $-1\mathbf{F} = \langle 6, 8 \rangle$

9. $\mathbf{F} = \mathbf{F_1} + \mathbf{F_2} + \mathbf{F_3} = \langle -2+2+5, -7-7+4 \rangle$
   $\mathbf{F} = \langle 5, -10 \rangle; \quad -1\mathbf{F} = \langle -5, 10 \rangle$

11. $\mathbf{F} = \mathbf{F_1} + \mathbf{F_2} = (5+1)\mathbf{i} + (-2+10)\mathbf{j} = 6\mathbf{i} + 8\mathbf{j}$
    $-1\mathbf{F} = -6\mathbf{i} - 8\mathbf{j}$

13. $\mathbf{F} = \mathbf{F_1} + \mathbf{F_2} + \mathbf{F_3} = (2.5 - 0.3)\mathbf{i}$
    $+(4.7 + 6.9 - 12)\mathbf{j} = 2.2\mathbf{i} - 0.4\mathbf{j}$
    $-1\mathbf{F} = -2.2\mathbf{i} + 0.4\mathbf{j}$

15. $\mathbf{F_1} = \langle 10\cos 104°, 10\sin 104° \rangle = \langle -2.42, 9.70 \rangle$
    $\mathbf{F_2} = \langle 6\cos 25°, 6\sin 25° \rangle = \langle 5.44, 2.54 \rangle$
    $\mathbf{F_3} = \langle 9\cos(-20°), 9\sin(-20°) \rangle = \langle 8.46, -3.08 \rangle$
    $\mathbf{F} = \langle -2.42+5.44+8.46, 9.70+2.54-3.08 \rangle$
    $\mathbf{F} = \langle 11.48, 9.16 \rangle; \quad -1\mathbf{F} = \langle -11.48, -9.16 \rangle$

17. $\mathbf{F_1} + \mathbf{F_2} = \langle 19+5, 10+17 \rangle = \langle 24, 27 \rangle$
    $\mathbf{F_3} = \langle -24, -27 \rangle$

19. $\mathbf{F_1} = \langle 2210\cos 40°, 2210\sin 40° \rangle$
    $= \langle 1693.0, 1420.6 \rangle$
    $\mathbf{F_2} = \langle 2500\cos 130°, 2500\sin 130° \rangle$
    $= \langle -1607.0, 1915.1 \rangle$
    $\mathbf{F_1} + \mathbf{F_2} = \langle 86.0, 3335.7 \rangle$
    $\mathbf{F_3} = \langle -86, -3335.7 \rangle$
    $|\mathbf{F_3}| = \sqrt{(-86)^2 + (-3335.7)^2} \approx 3336.8;$
    $\theta_r = \tan^{-1}\left(\frac{-3335.7}{-86}\right) \approx 88.5;$ In QIII,
    268.5°

21. $\mathbf{comp_v u} = 50\cos 42° = 37.16\,\text{kg}$

23. $\mathbf{comp_v u} = 1525\cos 65° = 644.49\,\text{lbs}$

25. $\mathbf{comp_v u} = 3010\cos 30° = 2606.74\,\text{kg}$

27. $\mathbf{G}$ makes an angle of 55° with the incline (**v**)
    $\mathbf{comp_v G} = 500\cos 55° \approx 286.79\,\text{lb}$

29. Let β = the angle between $\mathbf{G}$ and the incline, and let θ = the angle of incline.
    $325\cos\beta = 225; \quad \cos\beta = \frac{225}{325}$
    $\beta = \cos^{-1}\left(\frac{225}{325}\right) \approx 46.2°$
    $\theta = 90 - 46.2 = 43.8°$

31. $W = (15\,m)(75\,N) = 1125\,\text{N-m}$

33. $R = \frac{175^2 \sin 45° \cos 45°}{16} \approx 957.0\ \text{ft}$

35. The component of force in the direction of movement is $250\cos 30°$ lb.
    $W = (300)(250\cos 30°) = 64,951.9\ \text{ft-lb}$

37. $45,000 = |\mathbf{F}|\cos 5° \cdot 100$
    $|\mathbf{F}| = \frac{45,000}{100\cos 5°} = 451.72\,\text{lb}$

39. $W = 30\cos 20° \cdot 100 \approx 2819.08\,\text{N-m}$

41. $|\mathbf{F}| = \sqrt{15^2 + 10^2} \approx 18.0$
    $\theta_f = \tan^{-1}\left(\frac{10}{15}\right) \approx 33.7°$
    $\theta_v = \tan^{-1}\left(\frac{5}{50}\right) \approx 5.7°$
    Angle between vectors: 28.0°
    $\mathbf{comp_v F} = |\mathbf{F}|\cos\theta = 18.0\cos 28.0° \approx 15.9$
    $|\mathbf{v}| = \sqrt{50^2 + 5^2} \approx 50.2$
    $W = 15.9(50.2) \approx 800\ \text{ft-lb}$

7.4 Exercises

43. $|\mathbf{F}| = \sqrt{8^2 + 2^2} \approx 8.2$

    $\theta_f = \tan^{-1}\left(\dfrac{2}{8}\right) \approx 14.0°$

    $\theta_v = \tan^{-1}\left(\dfrac{-1}{15}\right) \approx -3.8°$

    Angle between vectors: 17.8°
    $\mathbf{comp_v F} = 8.2\cos 17.8° \approx 7.8$

    $|\mathbf{v}| = \sqrt{(-1)^2 + 15^2} \approx 15.0$
    $W = 7.8(15) \approx 117 \;\; \text{ft-lb}$

45. $\mathbf{F} \cdot \mathbf{v} = \langle 15, 10 \rangle \cdot \langle 50, 5 \rangle = 15 \cdot 50 + 10 \cdot 5 = 800$
    Verified

47. $\mathbf{F} \cdot \mathbf{v} = \langle 8, 2 \rangle \cdot \langle 15, -1 \rangle = 8 \cdot 15 + 2 \cdot -1 = 118$
    Verified

49. a. $\mathbf{p} \cdot \mathbf{q} = 5 \cdot 3 + 2 \cdot 7 = 29$
    b. $|\mathbf{p}| = \sqrt{5^2 + 2^2} = \sqrt{29}$

    $|\mathbf{q}| = \sqrt{3^2 + 7^2} = \sqrt{58}$

    $\theta = \cos^{-1}\left(\dfrac{29}{\sqrt{29}\sqrt{58}}\right) = 45°$

51. a. $\mathbf{p} \cdot \mathbf{q} = (-2)(-6) + 3(-4) = 0$
    b. $\theta = \cos^{-1}\left(\dfrac{0}{|\mathbf{p}||\mathbf{q}|}\right) = \cos^{-1} 0 = 90°$

53. a. $\mathbf{p} \cdot \mathbf{q} = \left(7\sqrt{2}\right)\left(2\sqrt{2}\right) + (-3)(9) = 1$
    b. $|\mathbf{p}| = \sqrt{\left(7\sqrt{2}\right)^2 + \left(2\sqrt{2}\right)^2} = \sqrt{106}$

    $|\mathbf{q}| = \sqrt{(-3)^2 + 9^2} = \sqrt{90}$

    $\theta = \cos^{-1}\left(\dfrac{1}{\sqrt{106}\sqrt{90}}\right) \approx 89.4°$

55. $\mathbf{u} \cdot \mathbf{v} = 7(4) + (-2)(14) = 0$ Yes

57. $\mathbf{u} \cdot \mathbf{v} = (-6)(-8) + (-3)(15) = 3$ No

59. $\mathbf{u} \cdot \mathbf{v} = (-2)(9) + (-6)(-3) = 0$ Yes

61. $|\mathbf{v}| = \sqrt{7^2 + 1^2} = \sqrt{50}$

    $\mathbf{u} \cdot \mathbf{v} = 3(7) + 5(1) = 26$

    $\mathbf{comp_v u} = \dfrac{26}{\sqrt{50}} \approx 3.68$

63. $|\mathbf{v}| = \sqrt{0^2 + (-10)^2} = 10$

    $\mathbf{u} \cdot \mathbf{v} = (-7)(0) + 4(-10) = -40$

    $\mathbf{comp_v u} = \dfrac{-40}{10} = -4$

65. $|\mathbf{v}| = \sqrt{6^2 + \left(5\sqrt{3}\right)^2} \approx \sqrt{111}$

    $\mathbf{u} \cdot \mathbf{v} = \left(7\sqrt{2}\right)6 + (-3)\left(5\sqrt{3}\right) = 42\sqrt{2} - 15\sqrt{3}$

    $\mathbf{comp_v u} = \dfrac{42\sqrt{2} - 15\sqrt{3}}{\sqrt{111}} \approx 3.17$

67. a. $|\mathbf{v}| = \sqrt{8^2 + 3^2} = \sqrt{73}$

    $\mathbf{u} \cdot \mathbf{v} = 2(8) + 6(3) = 34$

    $\mathbf{proj_v u} = \left(\dfrac{34}{73}\right)\langle 8, 3 \rangle \approx \langle 3.73, 1.40 \rangle$

    b. $\mathbf{u}_1 = \langle 3.73, 1.40 \rangle; \quad \mathbf{u}_2 = \mathbf{u} - \mathbf{u}_1$
    $= \langle 2 - 3.73, 6 - 1.40 \rangle = \langle -1.73, 4.60 \rangle$

69. a. $|\mathbf{v}| = \sqrt{(-6)^2 + 1^2} = \sqrt{37}$

    $\mathbf{u} \cdot \mathbf{v} = (-2)(-6) + (-8)(1) = 4$

    $\mathbf{proj_v u} = \left(\dfrac{4}{37}\right)\langle -6, 1 \rangle \approx \langle -0.65, 0.11 \rangle$

    b. $\mathbf{u}_1 = \langle -0.65, 0.11 \rangle; \quad \mathbf{u}_2 = \mathbf{u} - \mathbf{u}_1$
    $= \langle -2 + 0.65, -8 - 0.11 \rangle = \langle -1.35, -8.11 \rangle$

71. a. $|\mathbf{v}| = \sqrt{12^2 + 2^2} = \sqrt{148}$

    $\mathbf{u} \cdot \mathbf{v} = 10(12) + 5(2) = 130$

    $\mathbf{proj_v u} = \left(\dfrac{130}{148}\right)(12\mathbf{i} + 2\mathbf{j})$

    $\approx 10.54\mathbf{i} + 1.76\mathbf{j}$

    b. $\mathbf{u}_1 = 10.54\mathbf{i} + 1.76\mathbf{j}; \quad \mathbf{u}_2 = \mathbf{u} - \mathbf{u}_1$
    $= (10 - 10.54)\mathbf{i} + (5 - 1.76)\mathbf{j}$
    $= -0.54\mathbf{i} + 3.24\mathbf{j}$

73. a. $x = (250\cos 60°)(3) = 375$ ft

$y = (250\sin 60°)(3) - 16(3)^2 \approx 505.52$ ft

b. $y = (250\sin 60°)t - 16t^2 = 250$

$-16t^2 + 216.51t - 250 = 0$
Solve using quadratic formula:
$t \approx 1.27$ sec, 12.26 sec

75. a. $x = (200\cos 45°)(3) \approx 424.26$ ft

$y = (200\sin 45°)(3) - 16(3)^2 \approx 280.26$ ft

b. $y = (200\sin 45°)t - 16t^2 = 250$

$-16t^2 + 141.42t - 250 = 0$
Solve using quadratic formula:
$t \approx 2.44$ sec, 6.40 sec

77. $y = (90\sin 65°)(1.2) - 16(1.2)^2 \approx 74.84$ ft
To find another time, set height equal to
74.84: $y = (90\sin 65°)t - 16t^2 = 74.84$

$-16t^2 + 81.57t - 74.84 = 0$
Solve using quadratic formula:
$t \approx 1.2, 3.9$; After 3.9 seconds, which is
about 2.7 seconds later.

79. $\mathbf{w} \cdot (\mathbf{u} + \mathbf{v}) = \langle e, f \rangle \cdot \langle a+c, b+d \rangle$

$= e(a+c) + f(b+d) = ea + ec + fb + fd$

$= (ea + fb) + (ec + fd)$

$= \langle e, f \rangle \cdot \langle a, b \rangle + \langle e, f \rangle \cdot \langle c, d \rangle$

$= \mathbf{w} \cdot \mathbf{u} + \mathbf{w} \cdot \mathbf{v}$

81. $\mathbf{0} \cdot \mathbf{u} = \langle 0, 0 \rangle \cdot \langle a, b \rangle = 0(a) + 0(b) = 0$

$\mathbf{u} \cdot \mathbf{0} = \langle a, b \rangle \cdot \langle 0, 0 \rangle = a(0) + b(0) = 0$

83. $\mathbf{u} \cdot \mathbf{v} = 1(5) + 5(2) = 15$

$|\mathbf{u}| = \sqrt{1^2 + 5^2} = \sqrt{26}; \quad |\mathbf{v}| = \sqrt{5^2 + 2^2} = \sqrt{29}$

$\cos\theta = \dfrac{15}{\sqrt{26}\sqrt{29}}$

$\theta = \cos^{-1}\left(\dfrac{15}{\sqrt{26}\sqrt{29}}\right) \approx 56.9°$

Slope of $1\mathbf{i} + 5\mathbf{j} = \dfrac{\Delta y}{\Delta x} = \dfrac{5}{1} = 5$

Slope of $5\mathbf{i} + 2\mathbf{j} = \dfrac{\Delta y}{\Delta x} = \dfrac{2}{5}$

$\tan\theta = \dfrac{\frac{2}{5} - 5}{1 + \frac{2}{5}\cdot 5} = \dfrac{-23/5}{3} = -\dfrac{23}{15}$

$\theta = \tan^{-1}\left(-\dfrac{23}{15}\right) = -56.9$

The angle between is 56.9°. Answers to last
part of question will vary.

85. $2.9e^{-0.25t} + 7.6 = 438$

$2.9e^{-0.25t} = 430.4$

$e^{-0.25t} = \dfrac{430.4}{2.9}$

$\ln e^{-0.25t} = \ln\left(\dfrac{430.4}{2.9}\right)$

$-0.25t = \ln\left(\dfrac{430.4}{2.9}\right)$

$t = \dfrac{\ln\left(\dfrac{430.4}{2.9}\right)}{-0.25} \approx -20$

87. $a^2 = 172^2 + 250^2 - 2(172)(250)\cos 32°$

$a \approx 138.4$ m ;

$\dfrac{\sin C}{172} = \dfrac{\sin 32°}{138.4}$

$\sin C = \dfrac{172\sin 32°}{138.4}$

$C = \sin^{-1}\left(\dfrac{172\sin 32°}{138.4}\right)$

$C \approx 41.2°$

$180° - 41.2° - 32° = 106.8°$

Angles	Sides
$A = 32°$	138.4 m
$B \approx 106.8°$	250 m
$C \approx 41.2°$	172 m

$P = 138.4 + 250 + 172 = 560.4$ m

$A = \dfrac{1}{2}(250)(172)\sin 32° \approx 11393.3$ m^2

7.5 Exercises

1. modulus; argument

3. multiply; add

5. $|z| = \sqrt{(-1)^2 + (-\sqrt{3})^2} = \sqrt{4} = 2$

$\theta_r = \tan^{-1}\left(\dfrac{-\sqrt{3}}{-1}\right) = 60°$; In QIII, $\theta = 240°$

$-1 - \sqrt{3}i = 2(\cos 240° + i\sin 240°)$

7.

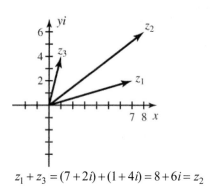

$z_1 + z_3 = (7 + 2i) + (1 + 4i) = 8 + 6i = z_2$

9.

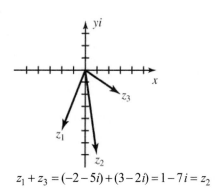

$z_1 + z_3 = (-2 - 5i) + (3 - 2i) = 1 - 7i = z_2$

11.

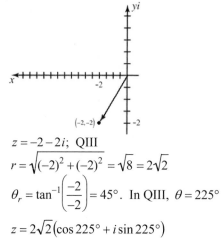

$z = -2 - 2i$; QIII

$r = \sqrt{(-2)^2 + (-2)^2} = \sqrt{8} = 2\sqrt{2}$

$\theta_r = \tan^{-1}\left(\dfrac{-2}{-2}\right) = 45°$. In QIII, $\theta = 225°$

$z = 2\sqrt{2}\left(\cos 225° + i\sin 225°\right)$

13.

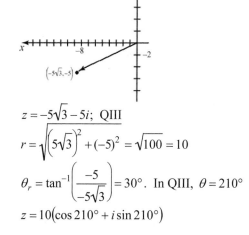

$z = -5\sqrt{3} - 5i$; QIII

$r = \sqrt{\left(5\sqrt{3}\right)^2 + (-5)^2} = \sqrt{100} = 10$

$\theta_r = \tan^{-1}\left(\dfrac{-5}{-5\sqrt{3}}\right) = 30°$. In QIII, $\theta = 210°$

$z = 10\left(\cos 210° + i\sin 210°\right)$

15.

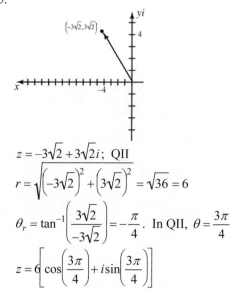

$z = -3\sqrt{2} + 3\sqrt{2}i$; QII

$r = \sqrt{\left(-3\sqrt{2}\right)^2 + \left(3\sqrt{2}\right)^2} = \sqrt{36} = 6$

$\theta_r = \tan^{-1}\left(\dfrac{3\sqrt{2}}{-3\sqrt{2}}\right) = -\dfrac{\pi}{4}$. In QII, $\theta = \dfrac{3\pi}{4}$

$z = 6\left[\cos\left(\dfrac{3\pi}{4}\right) + i\sin\left(\dfrac{3\pi}{4}\right)\right]$

17.

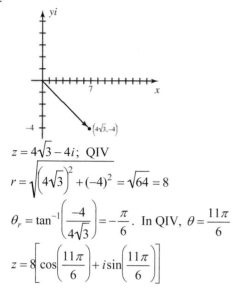

$z = 4\sqrt{3} - 4i$; QIV

$r = \sqrt{\left(4\sqrt{3}\right)^2 + (-4)^2} = \sqrt{64} = 8$

$\theta_r = \tan^{-1}\left(\dfrac{-4}{4\sqrt{3}}\right) = -\dfrac{\pi}{6}$. In QIV, $\theta = \dfrac{11\pi}{6}$

$z = 8\left[\cos\left(\dfrac{11\pi}{6}\right) + i\sin\left(\dfrac{11\pi}{6}\right)\right]$

19.

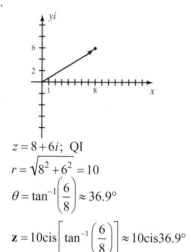

$z = 8 + 6i$; QI

$r = \sqrt{8^2 + 6^2} = 10$

$\theta = \tan^{-1}\left(\dfrac{6}{8}\right) \approx 36.9°$

$\mathbf{z} = 10\,\text{cis}\left[\tan^{-1}\left(\dfrac{6}{8}\right)\right] \approx 10\,\text{cis}\,36.9°$

21.

$z = -5 - 12i$; QIII

$r = \sqrt{(-5)^2 + (-12)^2} = 13$

$\theta_r = \tan^{-1}\left(\dfrac{-12}{-5}\right) \approx 67.4°$. QIII, $\theta = 247.4°$

$z = 13\,\text{cis}\left[180 + \tan^{-1}\left(\dfrac{12}{5}\right)\right] \approx 13\,\text{cis}\,247.4°$

23.

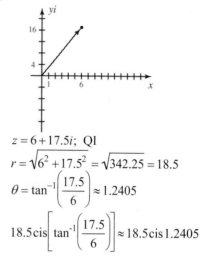

$z = 6 + 17.5i$; QI

$r = \sqrt{6^2 + 17.5^2} = \sqrt{342.25} = 18.5$

$\theta = \tan^{-1}\left(\dfrac{17.5}{6}\right) \approx 1.2405$

$18.5\,\text{cis}\left[\tan^{-1}\left(\dfrac{17.5}{6}\right)\right] \approx 18.5\,\text{cis}\,1.2405$

25.

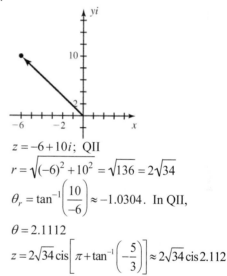

$z = -6 + 10i$; QII

$r = \sqrt{(-6)^2 + 10^2} = \sqrt{136} = 2\sqrt{34}$

$\theta_r = \tan^{-1}\left(\dfrac{10}{-6}\right) \approx -1.0304$. In QII,

$\theta = 2.1112$

$z = 2\sqrt{34}\,\text{cis}\left[\pi + \tan^{-1}\left(-\dfrac{5}{3}\right)\right] \approx 2\sqrt{34}\,\text{cis}\,2.112$

27. $r = 2, \theta = \dfrac{\pi}{4}$

$z = 2\,\text{cis}\left(\dfrac{\pi}{4}\right) = 2\left[\cos\left(\dfrac{\pi}{4}\right) + i\sin\left(\dfrac{\pi}{4}\right)\right]$

$= 2\left[\dfrac{\sqrt{2}}{2} + i\dfrac{\sqrt{2}}{2}\right] = \sqrt{2} + \sqrt{2}i$

29. $r = 4\sqrt{3}, \theta = \dfrac{\pi}{3}$

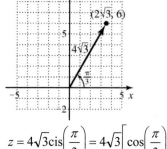

$$z = 4\sqrt{3}\,\mathrm{cis}\left(\frac{\pi}{3}\right) = 4\sqrt{3}\left[\cos\left(\frac{\pi}{3}\right) + i\sin\left(\frac{\pi}{3}\right)\right]$$

$$= 4\sqrt{3}\left[\frac{1}{2} + i\frac{\sqrt{3}}{2}\right] = 2\sqrt{3} + 6i$$

31. $r = 17, \theta = \tan^{-1}\left(\dfrac{15}{8}\right)$

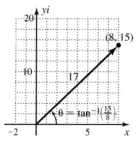

$$z = 17\,\mathrm{cis}\left[\tan^{-1}\left(\frac{15}{8}\right)\right]$$

$$= 17\left[\cos\left(\tan^{-1}\left(\frac{15}{8}\right)\right) + i\sin\left(\tan^{-1}\left(\frac{15}{8}\right)\right)\right]$$

$$= 17\left[\frac{8}{17} + i\frac{15}{17}\right] = 8 + 15i$$

33. $r = 6, \theta = \pi - \tan^{-1}\left(\dfrac{5}{\sqrt{11}}\right)$

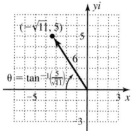

First, pretend that the "$\pi -$" is not there:

$$\cos\left[\tan^{-1}\left(\frac{5}{\sqrt{11}}\right)\right] = \frac{\sqrt{11}}{6}$$

$$\sin\left[\tan^{-1}\left(\frac{5}{\sqrt{11}}\right)\right] = \frac{5}{6}$$

Now use subtraction identities: For convenience, let $\theta = \tan^{-1}\left(\dfrac{5}{\sqrt{11}}\right)$.

$$\cos(\pi - \theta) = \cos\pi\cos\theta + \sin\pi\sin\theta$$

$$= -1\left(\frac{\sqrt{11}}{6}\right) + 0 = -\frac{\sqrt{11}}{6}$$

$$\sin(\pi - \theta) = \sin\pi\cos\theta - \cos\pi\sin\theta$$

$$= 0 - (-1)\frac{5}{6} = \frac{5}{6}$$

$$z = 6\,cis\left[\pi - \tan^{-1}\frac{5}{\sqrt{11}}\right]$$

$$= 6\left[\cos(\pi - \theta) + i\sin(\pi - \theta)\right]$$

$$= 6\left[-\frac{\sqrt{11}}{6} + i\frac{5}{6}\right] = -\sqrt{11} + 5i$$

35. $r_1 = \sqrt{(-2)^2 + 2^2} = \sqrt{8} = 2\sqrt{2}$

$\theta_{1r} = \tan^{-1}\left(\dfrac{2}{-2}\right) = -45°.$ In QII, $\theta_1 = 135°$

$r_2 = \sqrt{3^2 + (-3)^2} = \sqrt{18} = 3\sqrt{2}$

$\theta_2 = \tan^{-1}\left(\dfrac{3}{3}\right) = 45°$

$(-2 + 2i)(3 + 3i) = -6 - 6i + 6i + 6i^2$

$\qquad = -12 + 0i$

$r_1 r_2 = \left(2\sqrt{2}\right)\left(3\sqrt{2}\right) = 12\,;$

$\theta_1 + \theta_2 = 135 + 45 = 180°$

$12(\cos 180° + i\sin 180°) = -12 + 0i$

37. $r_1 = \sqrt{\sqrt{3}^2 + 1^2} = \sqrt{4} = 2$

$\theta_1 = \tan^{-1}\left(\dfrac{1}{\sqrt{3}}\right) = 30°$

$r_2 = \sqrt{1^2 + \sqrt{3}^2} = \sqrt{4} = 2$

$\theta_2 = \tan^{-1}\left(\dfrac{\sqrt{3}}{1}\right) = 60°$

$\dfrac{\sqrt{3} + i}{1 + \sqrt{3}i} \cdot \dfrac{1 - \sqrt{3}i}{1 - \sqrt{3}i} = \dfrac{\sqrt{3} - 3i + i - \sqrt{3}i^2}{1 - \sqrt{3}i + \sqrt{3}i - 3i^2}$

$\qquad = \dfrac{2\sqrt{3} - 2i}{4} = \dfrac{\sqrt{3}}{2} - \dfrac{1}{2}i\,;$

$\dfrac{r_1}{r_2} = \dfrac{2}{2} = 1\,;\quad \theta_1 - \theta_2 = 30 - 60 = -30°$

$1(\cos(-30°) + i\sin(-30°)) = \dfrac{\sqrt{3}}{2} - \dfrac{1}{2}i$

39. $r_1 r_2 = 24\,;\quad \theta_1 + \theta_2 = \dfrac{5\pi}{6} + \dfrac{\pi}{6} = \pi$

$z_1 z_2 = 24\operatorname{cis}\pi = -24 + 0i\,;$

$\dfrac{r_1}{r_2} = \dfrac{8}{3}\,;\quad \theta_1 - \theta_2 = \dfrac{5\pi}{6} - \dfrac{\pi}{6} = \dfrac{2\pi}{3}$

$\dfrac{z_1}{z_2} = \dfrac{8}{3}\operatorname{cis}\dfrac{2\pi}{3} = \dfrac{8}{3}\left(-\dfrac{1}{2} + i\dfrac{\sqrt{3}}{2}\right) = -\dfrac{4}{3} + \dfrac{4\sqrt{3}}{3}i$

41. $r_1 r_2 = \left(2\sqrt{3}\right)\left(7\sqrt{3}\right) = 42$

$\theta_1 + \theta_2 = \pi + \dfrac{5\pi}{6} = \dfrac{11\pi}{6}$

$z_1 z_2 = 42\operatorname{cis}\dfrac{11\pi}{6} = 42\left(\dfrac{\sqrt{3}}{2} - i\dfrac{1}{2}\right)$

$\qquad = 21\sqrt{3} - 21i\,;$

$\dfrac{r_1}{r_2} = \dfrac{2\sqrt{3}}{7\sqrt{3}} = \dfrac{2}{7}\,;\quad \theta_1 - \theta_2 = \pi - \dfrac{5\pi}{6} = \dfrac{\pi}{6}$

$\dfrac{z_1}{z_2} = \dfrac{2}{7}\operatorname{cis}\dfrac{\pi}{6} = \dfrac{2}{7}\left(\dfrac{\sqrt{3}}{2} + i\dfrac{1}{2}\right) = \dfrac{\sqrt{3}}{7} + \dfrac{1}{7}i$

43. $r_1 r_2 = 9(1.8) = 16.2$

$\theta_1 + \theta_2 = \dfrac{\pi}{15} + \dfrac{2\pi}{3} = \dfrac{11\pi}{15}$

$z_1 z_2 = 16.2\left[\cos\left(\dfrac{11\pi}{15}\right) + i\sin\left(\dfrac{11\pi}{15}\right)\right]$

$\qquad \approx -10.84 + 12.04i\,;$

$\dfrac{r_1}{r_2} = \dfrac{9}{1.8} = 5\,;\quad \theta_1 - \theta_2 = \dfrac{\pi}{15} - \dfrac{2\pi}{3} = -\dfrac{3\pi}{5}$

$\dfrac{z_1}{z_2} = 5\left[\cos\left(-\dfrac{3\pi}{5}\right) + i\sin\left(-\dfrac{3\pi}{5}\right)\right]$

$\qquad \approx -1.55 - 4.76i$

45. $r_1 r_2 = 40\,;\quad \theta_1 + \theta_2 = 60 + 30 = 90°$

$z_1 z_2 = 40\operatorname{cis}90° = 0 + 40i\,;$

$\dfrac{r_1}{r_2} = \dfrac{10}{4} = \dfrac{5}{2}\,;\quad \theta_1 - \theta_2 = 60 - 30 = 30°$

$\dfrac{z_1}{z_2} = \dfrac{5}{2}\operatorname{cis}30° = \dfrac{5}{2}\left(\dfrac{\sqrt{3}}{2} + i\dfrac{1}{2}\right) = \dfrac{5\sqrt{3}}{4} + \dfrac{5}{4}i$

47. $r_1 r_2 = \left(5\sqrt{2}\right)\left(2\sqrt{2}\right) = 20$

$\theta_1 + \theta_2 = 210 + 30 = 240°$

$z_1 z_2 = 20\operatorname{cis}240° = 20\left(-\dfrac{1}{2} - i\dfrac{\sqrt{3}}{2}\right)$

$\qquad = -10 - 10\sqrt{3}i\,;$

$\dfrac{r_1}{r_2} = \dfrac{5\sqrt{2}}{2\sqrt{2}} = \dfrac{5}{2}\,;\quad \theta_1 - \theta_2 = 210 - 30 = 180°$

$\dfrac{z_1}{z_2} = \dfrac{5}{2}\operatorname{cis}180° = -\dfrac{5}{2} + 0i$

7.5 Exercises

49. $r_1 r_2 = 6(1.5) = 9; \ \theta_1 + \theta_2 = 82 + 27 = 109°$

$z_1 z_2 = 9(\cos 109° + i \sin 109°) \approx -2.93 + 8.5i$;

$\dfrac{r_1}{r_2} = \dfrac{6}{1.5} = 4; \ \ \theta_1 - \theta_2 = 82 - 27 = 55°$

$\dfrac{z_1}{z_2} = 4(\cos 55° + i \sin 55°) \approx 2.29 + 3.28i$

51. Distance from u to v:

$d = \sqrt{(10-2)^2 + \left(\sqrt{3} - \sqrt{3}\right)^2} = 8$

From v to w:

$d = \sqrt{(6-10)^2 + \left(5\sqrt{3} - \sqrt{3}\right)^2} = \sqrt{16 + 48} = 8$

From w to u:

$d = \sqrt{(2-6)^2 + \left(\sqrt{3} - 5\sqrt{3}\right)^2} = \sqrt{16 + 48} = 8$

All sides have length 8.

$u^2 = \left(2 + \sqrt{3}i\right)\left(2 + \sqrt{3}i\right) = 4 + 4\sqrt{3}i + 3i^2$

$u^2 = 1 + 4\sqrt{3}i;$

$v^2 = \left(10 + \sqrt{3}i\right)\left(10 + \sqrt{3}i\right) = 100 + 20\sqrt{3}i + 3i^2$

$v^2 = 97 + 20\sqrt{3}i;$

$w^2 = \left(6 + 5\sqrt{3}i\right)\left(6 + 5\sqrt{3}i\right) = 36 + 60\sqrt{3}i + 75i^2$

$w^2 = -39 + 60\sqrt{3}i;$

$uv = \left(2 + \sqrt{3}i\right)\left(10 + \sqrt{3}i\right) = 20 + 12\sqrt{3}i + 3i^2$

$uv = 17 + 12\sqrt{3}i;$

$uw = \left(2 + \sqrt{3}i\right)\left(6 + 5\sqrt{3}i\right) = 12 + 16\sqrt{3}i + 15i^2$

$uw = -3 + 16\sqrt{3}i;$

$vw = \left(10 + \sqrt{3}i\right)\left(6 + 5\sqrt{3}i\right) = 60 + 56\sqrt{3}i + 15i^2$

$vw = 45 + 56\sqrt{3}i;$

$u^2 + v^2 + w^2 = \left(1 + 4\sqrt{3}i\right) + \left(97 + 20\sqrt{3}i\right)$

$\quad + \left(-39 + 60\sqrt{3}i\right) = 59 + 84\sqrt{3}i;$

$uv + uw + vw = \left(17 + 12\sqrt{3}i\right) + \left(-3 + 16\sqrt{3}i\right)$

$\quad + 45 + 56\sqrt{3}i = 59 + 84\sqrt{3}i$

53. a. $A = 170; \ \ V(t) = 170\sin(f(2\pi t))$

$V(t) = 170\sin(60(2\pi t)) = 170\sin(120\pi t)$

b. One cycle is 1/6 = 0.0167 seconds, so our table should go up to 0.008.

t	V	t	V
0	0	0.005	161.7
0.001	62.6	0.006	131.0
0.002	116.4	0.007	81.9
0.003	153.8	0.008	21.3
0.004	169.7		

c. The graph of V is at height 140 at about $t = 0.00257$ sec.

55. a. $Z = 15 + j(12 - 4) = 15 + 8j$ (QI)

$|Z| = \sqrt{15^2 + 8^2} = 17$

$\theta = \tan^{-1}\left(\dfrac{8}{15}\right) \approx 28.1°$

$Z = 17 \text{ cis } 28.1°$

b. $V_{RLC} = I|Z| = 3(17) = 51$ V

57. a. $Z = 7 + j(6 - 11) = 7 - 5j$ (QIV)

$|Z| = \sqrt{7^2 + (-5)^2} = \sqrt{74} \approx 8.60;$

$\theta_r = \tan^{-1}\left(\dfrac{-5}{7}\right) \approx -35.5°.$ In QIV,

$\theta = 324.5°$

$Z = 8.60 \text{ cis } 324.5°$

b. $V_{RLC} = I|Z| = 1.8(8.60) = 15.48$ V

59. a. $Z = 12 + j(5 - 0) = 12 + 5j$ (QI)

$|Z| = \sqrt{12^2 + 5^2} = 13;$

$\theta = \tan^{-1}\left(\dfrac{5}{12}\right) \approx 22.6°$

$Z = 13 \text{ cis } 22.6°$

b. $V_{RLC} = I|Z| = 1.7(13) = 22.1$ V

61. Both are in QI.

$$r_I = \sqrt{\sqrt{3}^2 + 1^2} = \sqrt{4} = 2$$

$$\theta_I = \tan^{-1}\left(\frac{1}{\sqrt{3}}\right) = 30°$$

$$r_Z = \sqrt{5^2 + 5^2} = \sqrt{50} = 5\sqrt{2}$$

$$\theta_Z = \tan^{-1}\left(\frac{5}{5}\right) = 45°$$

$$I = 2\text{cis}30°;$$

$$Z = 5\sqrt{2}\text{cis}45°$$

$$V = IZ = 2\left(5\sqrt{2}\right)\text{cis}(30° + 45°)$$

$$= 10\sqrt{2}\,\text{cis}\,75°$$

63. $r_I = \sqrt{3^2 + 2^2} = \sqrt{13}$

$$\theta_I = \tan^{-1}\left(-\frac{2}{3}\right) \approx -33.7° \text{ or } 326.3°$$

$$r_Z = \sqrt{2^2 + 3.75^2} = 4.25 = \frac{17}{4}$$

$$\theta_Z = \tan^{-1}\left(\frac{3.75}{2}\right) = 61.9°$$

$$I = \sqrt{13}\text{cis}326.3°;$$

$$Z = \frac{17}{4}\text{cis}61.9°$$

$$V = IZ = \sqrt{13}\left(\frac{17}{4}\right)\text{cis}(326.3° + 61.9°)$$

$$= \frac{17\sqrt{13}}{4}\,\text{cis}\,388.2° = \frac{17\sqrt{13}}{4}\,\text{cis}\,28.2°$$

65. $r_V = \sqrt{2^2 + 2\sqrt{3}^2} = \sqrt{16} = 4$

$$\theta_V = \tan^{-1}\left(\frac{2\sqrt{3}}{2}\right) = 60°$$

$$r_Z = \sqrt{4^2 + (-4)^2} = \sqrt{32} = 4\sqrt{2}$$

$$\theta_Z = \tan^{-1}\left(\frac{-4}{4}\right) = -45° \text{ or } 315°$$

$$V = 4\text{cis}60°;$$

$$Z = 4\sqrt{2}\text{cis}315°$$

$$I = \frac{V}{Z} = \frac{4}{4\sqrt{2}}\text{cis}(60° - (-45°))$$

$$= \frac{\sqrt{2}}{2}\,\text{cis}\,105°$$

67. $r_V = \sqrt{3^2 + (-4)^2} = \sqrt{25} = 5$

$$\theta_V = \tan^{-1}\left(\frac{-4}{3}\right) = -53.1° \text{ or } 306.9°$$

$$r_Z = \sqrt{4^2 + 7.5^2} = 8.5$$

$$\theta_Z = \tan^{-1}\left(\frac{7.5}{4}\right) = 61.9°$$

$$V = 5\text{cis}306.9°;$$

$$Z = 8.5\text{cis}61.9°$$

$$I = \frac{V}{Z} = \frac{5}{8.5}\text{cis}(306.9° - 61.9°)$$

$$= \frac{10}{17}\,\text{cis}\,245°$$

69. $r_1 = \sqrt{1^2 + 2^2} = \sqrt{5}$

$$\theta_1 = \tan^{-1}\left(\frac{2}{1}\right) \approx 63.4° \quad (\text{Q1})$$

$$r_2 = \sqrt{3^2 + 2^2} = \sqrt{13}$$

$$\theta_{2r} = \tan^{-1}\left(\frac{-2}{3}\right) \approx -33.7°. \text{ In QIV,}$$

$$\theta_2 = 326.3°$$

$$Z_1 Z_2 = \sqrt{5}\sqrt{13}\,\text{cis}\,(63.4° + 326.3°)$$

$$= \sqrt{65}\,\text{cis}\,389.7° = \sqrt{65}\,\text{cis}\,29.7°$$

$$Z_1 + Z_2 = (1 + 2j) + (3 - 2j) = 4$$

$$Z = \frac{Z_1 Z_2}{Z_1 + Z_2} = \frac{\sqrt{65}\,\text{cis}\,29.7°}{4}$$

71. $\dfrac{r_1}{r_2}\left[\cos(\alpha - \beta) + i\sin(\alpha - \beta)\right]$

$= \dfrac{r_1}{r_2}(\cos\alpha\cos\beta + \sin\alpha\sin\beta)$

$\qquad + i\dfrac{r_1}{r_2}(\sin\alpha\cos\beta + \cos\alpha\sin\beta)$

$\dfrac{\cos\alpha + i\sin\alpha}{\cos\beta + i\sin\beta} \cdot \dfrac{\cos\beta - i\sin\beta}{\cos\beta - i\sin\beta}$

$= \dfrac{\cos\alpha\cos\beta - i\cos\alpha\sin\beta + i\cos\beta\sin\alpha - i^2\sin\alpha\sin\beta}{\cos^2\beta - i^2\sin^2\beta}$

$= \dfrac{\cos\alpha\cos\beta + \sin\alpha\sin\beta}{\cos^2\beta + \sin^2\beta} + i\dfrac{\sin\alpha\cos\beta + \cos\alpha\sin\beta}{\cos^2\beta + \sin^2\beta}$

$= \dfrac{\cos\alpha\cos\beta + \sin\alpha\sin\beta}{1} + i\dfrac{\sin\alpha\cos\beta + \cos\alpha\sin\beta}{1}$

Note that $\dfrac{r_1}{r_2}$ times this expression is equal

to $\dfrac{z_1}{z_2}$, and is also equal to the expanded

version of the right side, and we're done.

73. The slope for segment 1 is $\dfrac{\Delta y}{\Delta x} = \dfrac{24}{7}$, so we

need slope $-\dfrac{7}{24}$. We can accomplish this

with $y = -7, x = 24$, or $y = 7, x = -24$. This
gives us $-24 + 7i$ and $24 - 7i$. The
magnitude of each is the same as the
magnitude of z_1, so we need to divide each
by 5 to get magnitude one-fifth as great:

$z_2 = \dfrac{24}{5} - \dfrac{7}{5}i, z_3 = -\dfrac{24}{5} + \dfrac{7}{5}i$

75. $350 = 750\sin\left(2x - \dfrac{\pi}{4}\right) - 25$

$375 = 750\sin\left(2x - \dfrac{\pi}{4}\right)$

$\dfrac{1}{2} = \sin\left(2x - \dfrac{\pi}{4}\right)$

For $x \in [0, 2\pi)$, $2x \in [0, 4\pi)$ so we need all
numbers in $[0, 4\pi)$ for which sine is ½.

$2x - \dfrac{\pi}{4} = \dfrac{\pi}{6}, \dfrac{5\pi}{6}, \dfrac{13\pi}{6}, \dfrac{17\pi}{6}$

(Add $\dfrac{\pi}{4}$ to each side)

$2x = \dfrac{\pi}{6} + \dfrac{\pi}{4}, \dfrac{5\pi}{6} + \dfrac{\pi}{4}, \dfrac{13\pi}{6} + \dfrac{\pi}{4}, \dfrac{17\pi}{6} + \dfrac{\pi}{4}$

$2x = \dfrac{5\pi}{12}, \dfrac{13\pi}{12}, \dfrac{29\pi}{12}, \dfrac{37\pi}{12}$

$x = \dfrac{5\pi}{24}, \dfrac{13\pi}{24}, \dfrac{29\pi}{24}, \dfrac{37\pi}{24}$

77.

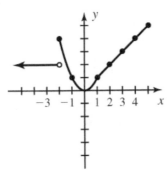

7.6 Exercises

1. $r^5[\cos(5\theta) + i\sin(5\theta)]$; DeMoivre's

3. complex

5. $z_5 = 2\text{cis}366° = 2\text{cis}6°$
$z_6 = 2\text{cis}438° = 2\text{cis}78°$
$z_7 = 2\text{cis}510° = 2\text{cis}150°$
These are equal to z_0, z_1, and z_2.

7. $r = \sqrt{3^2 + 3^2} = \sqrt{18} = 3\sqrt{2}$
$\theta = \tan^{-1}\left(\dfrac{3}{3}\right) = 45°$ (QI); $n = 4$

$(3 + 3i)^4 = (3\sqrt{2})^4[\cos(180°) + i\sin180°)]$
$= 324[-1 + 0] = -324$

9. $r = \sqrt{(-1)^2 + (\sqrt{3})^2} = \sqrt{4} = 2$; $n = 3$
$\theta_r = \tan^{-1}\left(\dfrac{\sqrt{3}}{-1}\right) = -60°$; $\theta = 120°$ (QII)

$(-1 + \sqrt{3}i)^3 = 2^3[\cos360° + i\sin360°]$
$= 8[1 + 0] = 8$

11. $r = \sqrt{\left(\dfrac{1}{2}\right)^2 + \left(-\dfrac{\sqrt{3}}{2}\right)^2} = 1$; $n = 5$
$\theta = \tan^{-1}\left(\dfrac{-\sqrt{3}/2}{1/2}\right) = -60°$ (QIV)

$\left(\dfrac{1}{2} - \dfrac{\sqrt{3}}{2}i\right)^5 = 1^5[\cos(-300°) + i\sin(-300°)]$

$= 1\left[-\dfrac{1}{2} + i\dfrac{\sqrt{3}}{2}\right] = -\dfrac{1}{2} + \dfrac{\sqrt{3}}{2}i$

13. $r = \sqrt{\left(\dfrac{\sqrt{2}}{2}\right)^2 + \left(-\dfrac{\sqrt{2}}{2}\right)^2} = 1$; $n = 6$
$\theta = \tan^{-1}\left(\dfrac{-\sqrt{2}/2}{\sqrt{2}/2}\right) = -45°$ (QIV)

$\left(\dfrac{\sqrt{2}}{2} - \dfrac{\sqrt{2}}{2}i\right)^6 = 1^6[\cos(-270°) + i\sin(-270°)]$
$= 1[0 + i] = i$

15. $r = \sqrt{(2\sqrt{3})^2 + (-2)^2} = \sqrt{16} = 4$; $n = 3$
$\theta = \tan^{-1}\left(\dfrac{-2}{2\sqrt{3}}\right) = -30°$ or $330°$ (QIV)

$(2\sqrt{3} - 2i)^3 = 4^3[\cos(-90°) + i\sin(-90°)]$
$= 64[0 - i] = -64i$

17. $r = \sqrt{\left(-\dfrac{1}{2}\right)^2 + \left(\dfrac{1}{2}\right)^2} = \sqrt{\dfrac{1}{2}} = \dfrac{\sqrt{2}}{2}$; $n = 5$
$\theta_r = \tan^{-1}(-1) = -45°$; $\theta = 135°$ in QII
$5(135°) = 675°$; θ coterminal with $315°$

$\left(-\dfrac{1}{2} + \dfrac{1}{2}i\right)^5 = \left(\dfrac{\sqrt{2}}{2}\right)^5[\cos315° + i\sin315°]$

$= \dfrac{\sqrt{2}}{8}\left[\dfrac{\sqrt{2}}{2} - i\dfrac{\sqrt{2}}{2}\right] = \dfrac{1}{8} - \dfrac{1}{8}i$

19. $r = 2$, $\theta = 90°$
$z^4 = 2^4[\cos360° + i\sin360°] = 16$
$z^3 = 2^3[\cos270° + i\sin270°] = 8[0 - i] = -8i$
$z^2 = 4i^2 = -4$
$z^4 + 3z^3 - 6z^2 + 12z - 40$
$= 16 + 3(-8i) - 6(-4) + 12(2i) - 40$
$= 16 - 24i + 24 + 24i - 40 = 0$

21. $r = \sqrt{(-3)^2 + (-3)^2} = \sqrt{18} = 3\sqrt{2}$
$\theta_r = \tan^{-1}\left(\dfrac{-3}{-3}\right) = 45°$; $\theta = 225°$ (QIII)

$4(225°) = 900°$, coterminal with $180°$
$3(225°) = 675°$, coterminal with $315°$
$2(225°) = 450°$, coterminal with $90°$
$z^4 = (3\sqrt{2})^4[\cos180° + i\sin180°]$
$= 324[-1 + 0i] = -324$
$z^3 = (3\sqrt{2})^3[\cos315° + i\sin315°]$
$= 54\sqrt{2}\left[\dfrac{\sqrt{2}}{2} - i\dfrac{\sqrt{2}}{2}\right] = 54 - 54i$
$z^2 = (3\sqrt{2})^2[\cos90° + i\sin90°]$
$= 18[0 + i] = 18i$
$z^4 + 6z^3 + 19z^2 + 6z + 18$
$= -324 + 6(54 - 54i) + 19(18i) + 6(-3 - 3i) + 18$
$= -324 + 324 - 324i + 342i - 18 - 18i + 18 = 0$
Verified

23. $r = \sqrt{\sqrt{3}^2 + (-1)^2} = 2; \quad \theta = \tan^{-1}\left(\dfrac{-1}{\sqrt{3}}\right) = -30°$

$z^5 = 2^5\left[\cos(-150°) + i\sin(-150°)\right]$

$= 32\left[-\dfrac{\sqrt{3}}{2} - i\dfrac{1}{2}\right] = -16\sqrt{3} - 16i$

$z^4 = 2^4\left[\cos(-120°) + i\sin(-120°)\right]$

$= 16\left[-\dfrac{1}{2} - i\dfrac{\sqrt{3}}{2}\right] = -8 - 8\sqrt{3}i$

$z^3 = 2^3\left[\cos(-90°) + i\sin(-90°)\right] = -8i$

$z^2 = 2^2\left[\cos(-60°) + i\sin(-60°)\right]$

$= 4\left[\dfrac{1}{2} - i\dfrac{\sqrt{3}}{2}\right] = 2 - 2\sqrt{3}i$

$z^5 + z^4 - 4z^3 - 4z^2 + 16z + 16$

$= -16\sqrt{3} - 16i - 8 - 8\sqrt{3}i - 4(-8i)$

$\qquad -4(2 - 2\sqrt{3}i) + 16(\sqrt{3} - i) + 16$

$= -16\sqrt{3} - 16i - 8 - 8\sqrt{3}i + 32i - 8$

$\qquad + 8\sqrt{3}i + 16\sqrt{3} - 16i + 16 = 0$

Verified

25. $r = \sqrt{1^2 + (2)^2} = \sqrt{5}; \quad \theta = \tan^{-1}(2)$

$z^4 = \sqrt{5}^4\left[\cos\left(4\tan^{-1}(2)\right) + i\sin\left(4\tan^{-1}(2)\right)\right]$

$= -7 - 24i$

$z^3 = \sqrt{5}^3\left[\cos\left(3\tan^{-1}(2)\right) + i\sin\left(3\tan^{-1}(2)\right)\right]$

$= -11 - 2i$

$z^2 = \sqrt{5}^2\left[\cos\left(2\tan^{-1}(2)\right) + i\sin\left(2\tan^{-1}(2)\right)\right]$

$= -3 + 4i$

$z^4 - 4z^3 + 7z^2 - 6z - 10$

$= -7 - 24i - 4(-11 - 2i) + 7(-3 + 4i)$

$\qquad - 6(1 + 2i) - 10$

$= -7 - 24i + 44 + 8i - 21 + 28i - 6 - 12i - 10$

$= 0$

Verified

27. $r = 1, \quad \theta = 0°, \quad n = 5$

$\sqrt[5]{1} = 1, \quad \dfrac{0°}{5} + \dfrac{360°k}{5} = 72°k$

$z_0 = \text{cis}\, 0° = 1$

$z_1 = \text{cis}\, 72° \approx 0.3090 + 0.9511i$

$z_2 = \text{cis}\, 144° \approx -0.8090 + 0.5878i$

$z_3 = \text{cis}\, 216° \approx -0.8090 - 0.5878i$

$z_4 = \text{cis}\, 288° \approx 0.3090 - 0.9511i$

29. $r = 243, \quad \theta = 0°, \quad n = 5$

$\sqrt[3]{243} = 3; \quad \dfrac{0°}{5} + \dfrac{360°k}{5} = 72°k$

$z_0 = 3\,\text{cis}\, 0° = 3$

$z_1 = 3\,\text{cis}\, 72° \approx 0.9271 + 2.8532i$

$z_2 = 3\,\text{cis}\, 144° \approx -2.4271 + 1.7634i$

$z_3 = 3\,\text{cis}\, 216° \approx -2.4271 - 1.7634i$

$z_4 = 3\,\text{cis}\, 288° \approx 0.9271 - 2.8532i$

31. $r = 27, \quad \theta = 270°, \quad n = 3$

$\sqrt[3]{27} = 3; \quad \dfrac{270°}{3} + \dfrac{360°k}{3} = 90° + 120°k$

$z_0 = 3\,\text{cis}\, 90° = 3[0 + i] = 3i$

$z_1 = 3\,\text{cis}\, 210° = 3\left(-\dfrac{\sqrt{3}}{2} - \dfrac{1}{2}i\right) = -\dfrac{3\sqrt{3}}{2} - \dfrac{3}{2}i$

$z_2 = 3\,\text{cis}\, 330° = 3\left(\dfrac{\sqrt{3}}{2} - \dfrac{1}{2}i\right) = \dfrac{3\sqrt{3}}{2} - \dfrac{3}{2}i$

33. $x^5 - 32 = 0; \quad x^5 = 32; \quad$ For $z = 32, r = 32,$

$\theta = 0°; \quad \sqrt[5]{32} = 2, \quad \dfrac{0°}{5} + \dfrac{360°k}{5} = 72°k$

$z_0 = 2\,\text{cis}[0°] = 2$

$z_1 = 2\,\text{cis}[72°] \approx 0.6180 + 1.9021i$

$z_2 = 2\,\text{cis}[144°] \approx -1.6180 + 1.1756i$

$z_3 = 2\,\text{cis}[216°] \approx -1.6780 - 1.1756i$

$z_4 = 2\,\text{cis}[288°] \approx 0.6180 - 1.9021i$

35. $x^3 - 27i = 0; \quad x^3 = 27i; \quad$ For $z = 27i, r = 27,$

$\theta = 90°, \quad \sqrt[3]{27} = 3,$

$\dfrac{90°}{3} + \dfrac{360°k}{3} = 30° + 120°k$

$z_0 = 3\,\text{cis}\, 30° = 3\left[\dfrac{\sqrt{3}}{2} + i\dfrac{1}{2}\right] = \dfrac{3\sqrt{3}}{2} + \dfrac{3}{2}i$

$z_1 = 3\,\text{cis}\, 150° = 3\left[-\dfrac{\sqrt{3}}{2} + i\dfrac{1}{2}\right] = -\dfrac{3\sqrt{3}}{2} + \dfrac{3}{2}i$

$z_2 = 3\,\text{cis}\, 270° = 3[0 - i] = -3i$

37. $x^5 - \sqrt{2} - \sqrt{2}i = 0$; $x^5 = \sqrt{2} + \sqrt{2}i$. For

$z = \sqrt{2} + \sqrt{2}i$, $r = 2$, $\theta = 45°$

$\sqrt[5]{2}$; $\dfrac{45°}{5} + \dfrac{360°k}{5} = 9° + 72°k$

$z_0 = \sqrt[5]{2}\,\text{cis}\,9° \approx 1.1346 + 1.1797i$

$z_1 = \sqrt[5]{2}\,\text{cis}\,81° \approx 0.1797 + 1.1346i$

$z_2 = \sqrt[5]{2}\,\text{cis}\,153° \approx -1.0235 + 0.5215i$

$z_3 = \sqrt[5]{2}\,\text{cis}\,225° \approx -0.8123 - 0.8123i$

$z_4 = \sqrt[5]{2}\,\text{cis}\,297° \approx 0.5215 - 1.0235i$

39. $x^3 - 1 = (x-1)(x^2 + x + 1) = 0$

$x - 1 = 0 \implies x = 1$

$x^2 + x + 1 = 0 \implies$

$x = \dfrac{-1 \pm \sqrt{1 - 4(1)(1)}}{2} = -\dfrac{1}{2} \pm \dfrac{\sqrt{3}}{2}i$

These are the same results as in Ex. 3.

41. $r = \sqrt{(-8)^2 + \left(8\sqrt{3}\right)^2} = \sqrt{256} = 16$; $n = 4$

$\theta_r = \tan^{-1}\left(\dfrac{8\sqrt{3}}{-8}\right) = -60°$; $\theta = 120°$ (QII)

$\sqrt[4]{16} = 2$; $\dfrac{120°}{4} + \dfrac{360°k}{4} = 30° + 90°k$

$z_0 = 2[\cos 30° + i\sin 30°] = 2\left[\dfrac{\sqrt{3}}{2} + i\dfrac{1}{2}\right]$

$\quad = \sqrt{3} + i$

$z_1 = 2[\cos 120° + i\sin 120°] = 2\left[-\dfrac{1}{2} + i\dfrac{\sqrt{3}}{2}\right]$

$\quad = -1 + \sqrt{3}i$

$z_2 = 2[\cos 210° + i\sin 210°] = 2\left[-\dfrac{\sqrt{3}}{2} - i\dfrac{1}{2}\right]$

$\quad = -\sqrt{3} - i$

$z_3 = 2[\cos 300° + i\sin 300°] = 2\left[\dfrac{1}{2} - i\dfrac{\sqrt{3}}{2}\right]$

$\quad = 1 - \sqrt{3}i$

43. $r = \sqrt{(-7)^2 + (-7)^2} = \sqrt{98} = 7\sqrt{2}$; $n = 4$

$\theta_r = \tan^{-1}\left(\dfrac{-7}{-7}\right) = 45°$; $\theta = 225°$ (QIII)

$\sqrt[4]{7\sqrt{2}} \approx 1.7738$;

$\dfrac{225°}{4} + \dfrac{360°k}{4} = 56.25° + 90°k$

$z_0 = 1.7738\,\text{cis}\,56.25° \approx 0.9855 + 1.4749i$

$z_1 = 1.7738\,\text{cis}\,146.25° \approx -1.4749 + 0.9855i$

$z_2 = 1.7738\,\text{cis}\,236.25° \approx -0.9855 - 1.4749i$

$z_3 = 1.7738\,\text{cis}\,326.25° \approx 1.4749 - 0.9855i$

45. $z^3 - 6z + 4 = 0$, $p = -6$, $q = 4$

$D = \dfrac{4(-6)^3 + 27(4)^2}{108} = \dfrac{-864 + 432}{108} = -4$

$-\dfrac{q}{2} + \sqrt{D} = -2 + 2i$; $r = \sqrt{8}$, $\theta = 135°$ (QII)

$\sqrt[3]{r} = 8^{\frac{1}{6}}$; $\dfrac{135°}{3} + \dfrac{360°k}{3} = 45° + 120°k$

$z_0 = 8^{\frac{1}{6}}\,\text{cis}\,45°$; $z_1 = 8^{1/6}\,\text{cis}\,165°$

$z_2 = 8^{\frac{1}{6}}\,\text{cis}\,285°$

$-\dfrac{q}{2} - \sqrt{D} = -2 - 2i$; $r = \sqrt{8}$, $\theta = 225°$ (QIV)

$\sqrt[3]{r} = 8^{\frac{1}{6}}$; $\dfrac{225°}{3} + \dfrac{360°k}{3} = 75° + 120°k$

$z_0 = 8^{\frac{1}{6}}\,\text{cis}\,75°$; $z_1 = 8^{\frac{1}{6}}\,\text{cis}\,195°$

$z_2 = 8^{\frac{1}{6}}\,\text{cis}\,315°$

47. We'll need four times each of the angles given to use DeMoivre's Theorem: $4(15) = 60°$; $4(105) = 420°$, coterminal with $60°$; $4(195) = 780°$, coterminal with $60°$; $4(285) = 1140°$, coterminal with $60°$. So raising all four given z's to the fourth power results in

$2^4\,\text{cis}\,60° = 16\left[\dfrac{1}{2} + i\dfrac{\sqrt{3}}{2}\right] = 8 + 8\sqrt{3}i$.

Verified

7.6 Exercises

49. a. $Z = 3 + 4j$; $r = 5$, $\theta = \tan^{-1}\left(\dfrac{4}{3}\right)$

$Z^3 = 5^3\left(\cos\left(3\tan^{-1}\dfrac{4}{3}\right) + j\sin\left(3\tan^{-1}\dfrac{4}{3}\right)\right)$

$= -117 + 44j$;

$Z^2 = 5^3\left(\cos\left(2\tan^{-1}\dfrac{4}{3}\right) + j\sin\left(2\tan^{-1}\dfrac{4}{3}\right)\right)$

$= -7 + 24j$

$3Z^2 = -21 + 72j$

b. $\dfrac{Z^3}{3Z^2} = \dfrac{-117 + 44j}{-21 + 72j} \cdot \dfrac{-21 - 72j}{-21 - 72j}$

$= \dfrac{2457 + 8424j - 924j - 3168j^2}{441 - 5184j^2}$

$= \dfrac{5625 + 7500i}{5625} = 1 + \dfrac{4}{3}j$

c. $\dfrac{Z}{3} = \dfrac{3 + 4j}{3} = 1 + \dfrac{4}{3}j$; Verified

51. Answers will vary.

53. For $1 + 2i$, $r = \sqrt{5}$, $\theta = \tan^{-1}\left(\dfrac{2}{1}\right)$. The related right triangle is:

$\sin\theta = \dfrac{2}{\sqrt{5}}$, $\cos\theta = \dfrac{1}{\sqrt{5}}$

$\sin(4\theta) = \sin(2(2\theta)) = 2\sin(2\theta)\cos(2\theta)$

$= 2\left(2\sin\theta\cos\theta\right)\left(\cos^2\theta - \sin^2\theta\right)$

$= 2\left(2\left(\dfrac{2}{\sqrt{5}}\right)\left(\dfrac{1}{\sqrt{5}}\right)\right)\left(\left(\dfrac{1}{\sqrt{5}}\right)^2 - \left(\dfrac{2}{\sqrt{5}}\right)^2\right)$

$= \dfrac{8}{5}\left(-\dfrac{3}{5}\right) = -\dfrac{24}{25}$

$\cos^2(4\theta) = 1 - \sin^2(4\theta) = 1 - \left(-\dfrac{24}{25}\right)^2$

$= 1 - \dfrac{576}{625} = \dfrac{49}{625}$; $\cos(4\theta) = -\dfrac{7}{25}$ (Note that a quick approximation on the calculator shows us that $\theta = \tan^{-1}(2) \approx 63°$, so $4\theta \approx 252°$, and is in QIII).

$z^4 = \sqrt{5}^4\left[\cos(4\theta) + i\sin(4\theta)\right] = 25\left[-\dfrac{7}{25} - i\dfrac{24}{25}\right]$

$= -7 - 24i$

55. Look at the solutions to 45, and note that

$8^{\frac{1}{4}} = \left(2^3\right)^{\frac{1}{4}} = 2^{\frac{3}{4}} = \sqrt{2}$

Add the roots from 45 whose angles add to 360°:

$8^{\frac{1}{6}}\operatorname{cis}45° + 8^{\frac{1}{6}}\operatorname{cis}315°$

$= \sqrt{2}\left[\dfrac{\sqrt{2}}{2} + i\dfrac{\sqrt{2}}{2}\right] + \sqrt{2}\left[\dfrac{\sqrt{2}}{2} - i\dfrac{\sqrt{2}}{2}\right]$

$= \dfrac{2}{2} + \dfrac{2}{2}i + \dfrac{2}{2} - \dfrac{2}{2}i = 2$;

$8^{\frac{1}{6}}\operatorname{cis}165° + 8^{\frac{1}{6}}\operatorname{cis}195°$

$\approx \left(-1.3660 + 0.3660i\right) + \left(-1.3660 - 03.660i\right)$

$= -2.7320$

$8^{\frac{1}{6}}\operatorname{cis}285° + 8^{\frac{1}{6}}\operatorname{cis}75°$

$\approx \left(0.3660 - 1.3660i\right) + \left(0.3660 + 1.3660i\right)$

$= 0.7320$;

Note: Using sum and difference identities, all three solutions can actually be found in exact form. The latter two are $-1 - \sqrt{3}$ and $-1 + \sqrt{3}$.

57. $\dfrac{\tan^2 x}{\sec x + 1} = \dfrac{\sec^2 x - 1}{\sec x + 1} = \dfrac{\left(\sec x + 1\right)\left(\sec x - 1\right)}{\sec x + 1}$

$= \sec x - 1 = \dfrac{1}{\cos x} - 1 = \dfrac{1}{\cos x} - \dfrac{\cos x}{\cos x}$

$= \dfrac{1 - \cos x}{\cos x}$

59. Goes through $(-2, 4)$ and $(3, 0)$:

$m = \dfrac{0 - 4}{3 - (-2)} = -\dfrac{4}{5}$

$y - 0 = -\dfrac{4}{5}(x - 3)$; $y = -\dfrac{4}{5}x + \dfrac{12}{5}$

Summary and Concept Review

1. $A = 180 - (21 + 123) = 36°$

$\dfrac{\sin 123°}{293} = \dfrac{\sin 21°}{b}; \quad b \sin 123° = 293 \sin 21°$

$b = \dfrac{293 \sin 21°}{\sin 123°} \approx 125.20$ cm

$\dfrac{\sin 123°}{293} = \dfrac{\sin 36°}{a}; \quad a \sin 123° = 293 \sin 36°$

$a = \dfrac{293 \sin 36°}{\sin 123°} \approx 205.35$ cm

3. The third angle is $180 - (110 + 25) = 45°$.

$\dfrac{\sin 45°}{70} = \dfrac{\sin 25°}{h}; \quad h \sin 45° = 70 \sin 25°$

$h = \dfrac{70 \sin 25°}{\sin 45°} \approx 41.84$ ft

5. $\dfrac{\sin 35°}{67} = \dfrac{\sin B}{105}; \quad 67 \sin B = 105 \sin 35°$

$\sin B = \dfrac{105 \sin 35°}{67}; \quad B = \sin^{-1}\left(\dfrac{105 \sin 35°}{67}\right)$

$B \approx 64.0°$ or $180 - 64.0 = 116.0°$

For $B_1 = 64.0°$:

$C_1 = 180 - (35 + 64.0) = 81.0°$

$\dfrac{\sin 35°}{67} = \dfrac{\sin 81.0°}{c_1}; \quad c_1 \sin 35° = 67 \sin 81.0°$

$c_1 = \dfrac{67 \sin 81.0°}{\sin 35°} \approx 115.37$ cm

For $B_2 = 116.0°$:

$C_2 = 180 - (35 + 116.0) = 29.0°$

$\dfrac{\sin 35°}{67} = \dfrac{\sin 29°}{c_2}; \quad c_2 \sin 35° = 67 \sin 29°$

$c_2 = \dfrac{67 \sin 29°}{\sin 35°} \approx 56.63$ cm

7. $81 = 369 - 360 \cos B; \quad -360 \cos B = -288$

$\cos B = \dfrac{-288}{-360}; \quad B = \cos^{-1} 0.8 \approx 36.9°$

9. Let A, B and C be the angles from largest to smallest.

$1820^2 = 1250^2 + 720^2 - 2(720)(1250) \cos C$

$3,312,400 = 2,080,900 - 1,800,000 \cos C$

$1,231,500 = -1,800,000 \cos C$

$\cos C = \dfrac{1,231,500}{-1,800,000}; \quad C \approx 133.2°$

$\dfrac{\sin B}{1250} = \dfrac{\sin 133.2°}{1820}$

$1820 \sin B = 1250 \sin 133.2°$

$\sin B = \dfrac{1250 \sin 133.2°}{1820}; \quad B \approx 30.1°$

$A = 180 - (133.2 + 30.1) = 16.7°$

11.

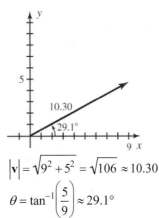

$|\mathbf{v}| = \sqrt{9^2 + 5^2} = \sqrt{106} \approx 10.30$

$\theta = \tan^{-1}\left(\dfrac{5}{9}\right) \approx 29.1°$

13. Horiz. component $= 18 \cos 52° \approx 11.08$

Vert. component $= 18 \sin 52° \approx 14.18$

15. $|\mathbf{u}| = \sqrt{7^2 + 12^2} = \sqrt{193}$

$\dfrac{7}{\sqrt{193}} \mathbf{i} + \dfrac{12}{\sqrt{193}} \mathbf{j}$

17. Karl's velocity in the direction across the stream is 3 mi/hr, so he'll make it ½ mile in 1/6 hours. The current will carry him $\dfrac{1}{6}$ mile downstream.

19. Resultant: $\langle -20 + 45, 70 + 53 \rangle = \langle 25, 123 \rangle$

Additional: $\langle -25, -123 \rangle$

21. $2(-18) + 9d = 0; \quad 9d = 36; \quad d = 4$

23. $\mathbf{W} = \mathbf{F} \cdot \mathbf{V} = 50(85) + 15(6) = 4340$ ft-lb

25. $\mathbf{F} = 75\cos 25°$
$\mathbf{W} = (75\cos 25°)(120) = 8156.77$ ft-lb

27. $r = \sqrt{(-1)^2 + (-\sqrt{3})^2} = \sqrt{4} = 2$

$\theta_r = \tan^{-1}\left(\dfrac{-\sqrt{3}}{-1}\right) = 60°; \quad \theta = 240°$ (QIII)

$z = 2[\cos 240° + i\sin 240°]$

29.

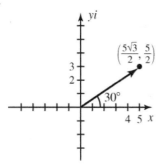

31. $I = \dfrac{V}{Z} = \dfrac{4\sqrt{3} - 4j}{1 - \sqrt{3}j} \cdot \dfrac{1 + \sqrt{3}j}{1 + \sqrt{3}j}$

$= \dfrac{4\sqrt{3} + 12j - 4j - 4\sqrt{3}j^2}{1 - 3j^2}$

$= \dfrac{8\sqrt{3} + 8j}{4} = 2\sqrt{3} + 2j$

33. $r = \sqrt{(-1)^2 + \sqrt{3}^2} = \sqrt{4} = 2$

$\theta_r = \tan^{-1}\left(\dfrac{\sqrt{3}}{-1}\right) = -60°; \quad \theta = 120°$ (QII)

$\left(-1 + \sqrt{3}i\right)^5 = 2^5[\cos(5(120°)) + i\sin(5(120°))]$
$= 32[\cos 240° + i\sin 240°]$ (240° coterminal with 600°)

$32\left[-\dfrac{1}{2} - i\dfrac{\sqrt{3}}{2}\right] = -16 - 16\sqrt{3}i$

35. $r = 125; \quad \theta = 90°; \quad \sqrt[3]{125} = 5$

$\dfrac{90°}{3} + \dfrac{360°k}{3} = 30° + 120°k$

$z_0 = 5[\cos 30° + i\sin 30°] = 5\left[\dfrac{\sqrt{3}}{2} + i\dfrac{1}{2}\right]$

$= \dfrac{5\sqrt{3}}{2} + \dfrac{5}{2}i$ or $4.3301 + 2.5i$

$z_1 = 5[\cos 150° + i\sin 250°] = 5\left[-\dfrac{\sqrt{3}}{2} - i\dfrac{1}{2}\right]$

$= -\dfrac{5\sqrt{3}}{2} + \dfrac{5}{2}i$ or $-4.3301 + 2.5i$

$z_2 = 5[\cos 270° + i\sin 270°] = 5[0 - i] = -5i$

37. $r = \sqrt{2^2 + 2^2} = \sqrt{8}; \quad \theta = \tan^{-1}(1) = 45°$
The other roots will have the same r and will be spaced 90° apart.

$2\sqrt{2}\,\text{cis}\,135° = 2\sqrt{2}\left[-\dfrac{\sqrt{2}}{2} + i\dfrac{\sqrt{2}}{2}\right] = -2 + 2i$

The remaining two must be conjugates of the two we have: $2 - 2i$ and $-2 - 2i$

39. $\dfrac{5\sqrt{3}}{2} + \dfrac{5}{2}i; \quad r = \sqrt{\left(\dfrac{5\sqrt{3}}{2}\right)^2 + \left(\dfrac{5}{2}\right)^2} = 5$

$\theta = \tan^{-1}\left(\dfrac{\frac{5}{2}}{\frac{5\sqrt{3}}{2}}\right) = 30°$

$\left(\dfrac{5\sqrt{3}}{2} + \dfrac{5}{2}i\right)^3 = 5^3\,\text{cis}\,90° = 125i$

$-\dfrac{5\sqrt{3}}{2} + \dfrac{5}{2}i; \quad r = \sqrt{\left(\dfrac{5\sqrt{3}}{2}\right)^2 + \left(\dfrac{5}{2}\right)^2} = 5$

$\theta_r = \tan^{-1}\left(\dfrac{\frac{5}{2}}{-\frac{5\sqrt{3}}{2}}\right) = -30°; \theta = 150°$ (QII)

$\left(-\dfrac{5\sqrt{3}}{2} + \dfrac{5}{2}i\right)^3 = 5^3\,\text{cis}\,450° = 125i$

$(-5i)^3 = -125i^3 = 125i$

Mixed Review

1. $A = 180 - (27 + 112) = 41°$

   $\dfrac{\sin 112°}{19} = \dfrac{\sin 27°}{b}$; $\quad b\sin 112° = 19\sin 27°$

   $b = \dfrac{19\sin 27°}{\sin 112°} \approx 9.30$ in.

   $\dfrac{\sin 112°}{19} = \dfrac{\sin 41°}{a}$; $\quad a\sin 112° = 19\sin 41°$

   $a = \dfrac{19\sin 41°}{\sin 112°} \approx 13.44$ in.

Angles	Sides
$A = 41°$	$a \approx 13.44$ in.
$B = 27°$	$b \approx 9.30$ in.
$C = 112°$	$c = 19$ in.

   $A = \dfrac{19^2 \sin 27° \sin 41°}{2\sin 112°}$

   Area ≈ 58 in.^{2}

3. $x = 21\cos 40° \approx 16.09$
   $y = 21\sin 40 \approx 13.50$

5. Missing angle $= 180 - (35 + 122) = 23°$

   $\dfrac{\sin 23°}{120} = \dfrac{\sin 35°}{h}$; $\quad h\sin 23° = 120\sin 35°$

   $h = \dfrac{120\sin 35°}{\sin 23°} \approx 176.15$ ft

7. The plane's velocity vector makes an angle
   of 60° with the pos. x-axis, so it's
   components are $x = 750\cos 60° = 375$,
   $y = 750\sin 60° \approx 649.52$. The wind vector
   is $\langle 0, 50 \rangle$. The resultant is
   $\mathbf{v} = \langle 375, 699.52 \rangle$.

   Mag. $= \sqrt{375^2 + 699.52^2} \approx 793.70$ mph;

   $\theta = \tan^{-1}\left(\dfrac{699.52}{375}\right) \approx 61.8°$. An angle of

   61.8° with the x-axis is a heading of 28.2°

9. $\dfrac{\sin 31°}{36} = \dfrac{\sin B}{24}$; $\quad 36\sin B = 24\sin 31°$

   $\sin B = \dfrac{24\sin 31°}{36}$; $\quad B = \sin^{-1}\left(\dfrac{24\sin 31°}{36}\right)$

   $B \approx 20.1°$ or $180 - 20.1 = 159.9°$
   Second one not possible since its sum with
   31° is over 180°. So $B \approx 20.1°$, and
   $C = 180 - (31 + 20.1) = 128.9°$

   $\dfrac{\sin 128.9°}{c} = \dfrac{\sin 31°}{36}$

   $c\sin 31° = 36\sin 128.9°$

   $c = \dfrac{36\sin 128.9°}{\sin 31°} \approx 54.4$ m

Angles	Sides
$A = 31°$	$a = 36$ m
$B \approx 20.1°$	$b = 24$ m
$C \approx 128.9°$	$c \approx 54.4$ m

11. $\dfrac{\sin 35°}{12} = \dfrac{\sin \theta}{20}$; $\quad 12\sin\theta = 20\sin 35°$

    $\sin\theta = \dfrac{20\sin 35°}{10} = 1.15$; No.

    To find the smallest possible angle, we need

    α so that $\dfrac{20\sin\alpha}{10} = 1$; $\quad \sin\alpha = \dfrac{1}{2}$, and the

    smallest angle is 30°.

13. a.

    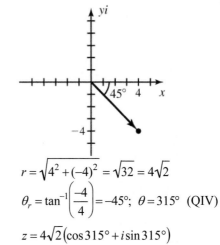

    $r = \sqrt{4^2 + (-4)^2} = \sqrt{32} = 4\sqrt{2}$

    $\theta_r = \tan^{-1}\left(\dfrac{-4}{4}\right) = -45°$; $\quad \theta = 315°$ (QIV)

    $z = 4\sqrt{2}\left(\cos 315° + i\sin 315°\right)$

    b.

    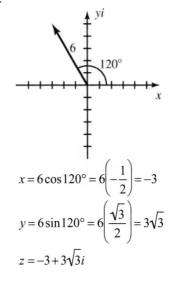

    $x = 6\cos 120° = 6\left(-\dfrac{1}{2}\right) = -3$

    $y = 6\sin 120° = 6\left(\dfrac{\sqrt{3}}{2}\right) = 3\sqrt{3}$

    $z = -3 + 3\sqrt{3}i$

15. Set vertical components equal:

$$418\sin 10° = 320\sin\theta; \quad \sin\theta = \frac{418\sin 10°}{320}$$

$$\theta = \sin^{-1}\left(\frac{418\sin 10°}{320}\right) \approx 13.1°$$

17. $\mathbf{u}\cdot\mathbf{v} = -12(19) + (-16)(-13) = -20$

$$|\mathbf{v}| = \sqrt{19^2 + (-13)^2} = \sqrt{530}$$

$$\text{comp}_\mathbf{v}\,\mathbf{u} = \frac{-20}{\sqrt{530}} \approx -0.87$$

$$\mathbf{proj}_\mathbf{v}\,\mathbf{u} = \frac{-20}{530}\langle 19,-13\rangle = -\frac{38}{53}\mathbf{i} + \frac{26}{53}\mathbf{j}$$

19. $r = \sqrt{(-2)^2 + (2\sqrt{3})^2} = 4$

$$\theta_r = \tan^{-1}\left(\frac{2\sqrt{3}}{-2}\right) = -60°; \quad \theta = 120° \text{ (QII)}$$

$$\sqrt[4]{4} = \sqrt{2}; \quad \frac{120°}{4} + \frac{360°k}{4} = 30° + 90°k$$

$$z_0 = \sqrt{2}\,\text{cis}\,30° = \sqrt{2}\left(\frac{\sqrt{3}}{2} + i\frac{1}{2}\right) = \frac{\sqrt{6}}{2} + \frac{\sqrt{2}}{2}i$$

$$z_1 = \sqrt{2}\,\text{cis}\,120° = \sqrt{2}\left(-\frac{1}{2} + i\frac{\sqrt{3}}{2}\right)$$

$$= -\frac{\sqrt{2}}{2} + \frac{\sqrt{6}}{2}i$$

$$z_2 = \sqrt{2}\,\text{cis}\,210° = \sqrt{2}\left(-\frac{\sqrt{3}}{2} - i\frac{1}{2}\right)$$

$$= -\frac{\sqrt{6}}{2} - \frac{\sqrt{2}}{2}i$$

$$z_3 = \sqrt{2}\,\text{cis}\,300° = \sqrt{2}\left(\frac{1}{2} - i\frac{\sqrt{3}}{2}\right)$$

$$= \frac{\sqrt{2}}{2} - \frac{\sqrt{6}}{2}i$$

Practice Test

1. The angle at the fire is 73°.

$$\frac{\sin 39°}{d} = \frac{\sin 73°}{10}; \quad d\sin 73° = 10\sin 39°$$

$$d = \frac{10\sin 39°}{\sin 73°} \approx 6.58 \text{ mi}$$

3. Let $b = 6$, $a = 15$, $B = 20°$

$$\frac{\sin 20°}{6} = \frac{\sin A}{15}; \quad 6\sin A = 15\sin 20°$$

$$\sin A = \frac{15\sin 20°}{6}; \quad A = \sin^{-1}\left(\frac{15\sin 20°}{6}\right)$$

$A \approx 58.8°$ or $121.2°$

For $A_1 = 58.8°$:

$$C_1 = 180 - (20 + 58.8) = 101.2°$$

$$\frac{\sin 101.2°}{c_1} = \frac{\sin 20°}{6}; \quad c_1\sin 20° = 6\sin 101.2°$$

$$c_1 = \frac{6\sin 101.2°}{\sin 20°} \approx 17.2 \text{ in}$$

For $A_2 = 121.2°$:

$$C_2 = 180 - (20 + 121.2) = 38.8°$$

$$\frac{\sin 38.8°}{c_2} = \frac{\sin 20°}{6}; \quad c_2\sin 20° = 6\sin 38.8°$$

$$c_2 = \frac{6\sin 38.8°}{\sin 20°} \approx 10.99 \text{ in}$$

Angles	Sides (in.)
$A_1 \approx 58.8°$	$a = 15$
$B = 20°$	$b = 6$
$C_1 \approx 101.2°$	$c_1 \approx 17.21$

Angles	Sides
$A_2 \approx 121.2°$	$a = 15$
$B = 20°$	$b = 6$
$C_2 \approx 38.8°$	$c_2 \approx 11.0$

5. a. $\dfrac{\sin 53°}{25} = \dfrac{\sin \theta}{35}$; $25 \sin \theta = 35 \sin 53°$

$\sin \theta = \dfrac{35 \sin 53°}{25} = 1.11$; No

   b. $\dfrac{\sin 53°}{28} = \dfrac{\sin \theta}{35}$; $28 \sin \theta = 35 \sin 53°$

$\sin \theta = \dfrac{35 \sin 53°}{25} \approx 1$; Only 1 throw

   c. With a range of 35 yd, the target is in range from the bottom left corner of the triangle, and we'll get an isosceles triangle with angles 53°, 53° and 74°.

$\dfrac{\sin 74°}{d} = \dfrac{\sin 53°}{35}$; $d \sin 53° = 35 \sin 74°$

$d = \dfrac{35 \sin 74°}{\sin 53°} \approx 42.13$ yd

$42.13 \text{ yd} \cdot \dfrac{1 \sec}{5 \text{ yd}} \approx 8.43$ sec

7. $1025^2 = 1020^2 + 977^2 - 2(1020)(977)\cos P$
$1,050,625 = 1,994,929 - 1,993,080 \cos P$
$-944,304 = -1,993,080 \cos P$

$\cos P = \dfrac{944,304}{1,993,080}$

$P = \cos^{-1}\left(\dfrac{944,304}{1,993,080}\right) \approx 61.7°$;

$\dfrac{\sin 61.7°}{1025} = \dfrac{\sin B}{1020}$

$1025 \sin B = 1020 \sin 61.7°$

$\sin B = \dfrac{1020 \sin 61.7°}{1025}$

$B = \sin^{-1}\left(\dfrac{1020 \sin 61.7°}{1025}\right) \approx 61.2°$;

$M = 180 - (61.7 + 61.2) = 57.1°$;

$p = \dfrac{1020 + 1025 + 977}{2} = 1511$

$A = \sqrt{1511(1511-1020)(1511-1025)(1511-977)}$

Area about $438,795 \text{ mi}^2$

9. Set vert. components equal:
$250 \sin 30° = 210 \sin \theta$

$\sin \theta = \dfrac{250 \sin 30°}{210}$

$\theta = \sin^{-1}\left(\dfrac{250 \sin 30°}{210}\right) \approx 36.5°$

11. $\mathbf{F}_1 = \langle 150 \cos 42°, 150 \sin 42° \rangle$
$\quad = \langle 111.47, 100.37 \rangle$

$\mathbf{F}_2 = \langle 110 \cos 113°, 110 \sin 113° \rangle$
$\quad = \langle -42.98, 101.26 \rangle$

Resultant $= \langle 68.49, 201.63 \rangle$

$\mathbf{F} = \langle -68.49, -201.63 \rangle$

$|\mathbf{F}| = \sqrt{68.49^2 + 201.63^2} \approx 212.94 \text{N}$

$\theta_r = \tan^{-1}\left(\dfrac{201.63}{68.49}\right) \approx 71.2°$; $\theta = 251.2°$

13. $y = 110 \sin 50°(2) - 16(2)^2 \approx 104.53$ ft;
$-16t^2 + 110 \sin 50° t = 104.53$
$-16t^2 + 110 \sin 50° t - 104.53 = 0$
Quadratic formula: $t \approx 1.2,\ 3.27$. It will be at that height again after 3.27 sec.

15. $|z_1| = \sqrt{(-6)^2 + 6^2} = \sqrt{72} = 6\sqrt{2}$

$|z_2| = \sqrt{4^2 + \left(-4\sqrt{3}\right)^2} = 8$

$\theta_{1r} = \tan^{-1}\left(\dfrac{6}{-6}\right) = -45°$; $\theta_1 = 135°$ (QII)

$\theta_2 = \tan^{-1}\left(\dfrac{-4\sqrt{3}}{4}\right) = -60°$ (QIV)

$z = z_1 z_2 = 6\sqrt{2}(8)\operatorname{cis}(135° + (-60°))$
$\quad = 48\sqrt{2} \operatorname{cis} 75° \approx 17.57 + 65.57i$

$|z| = \sqrt{17.57^2 + 65.57^2} \approx 67.88$

$|z_1||z_2| = 48\sqrt{2} \approx 67.88$

$\theta = \tan^{-1}\left(\dfrac{65.57}{17.57}\right) \approx 75° = \theta_1 + \theta_2$

17. $r = \sqrt{2^2 + 2\sqrt{3}^2} = \sqrt{16} = 4$

$\theta = \tan^{-1}\left(\dfrac{2\sqrt{3}}{2}\right) = 60° \quad (QI)$

$z^5 = 4^5 \operatorname{cis} 300° = 1024\left(\dfrac{1}{2} - i\dfrac{\sqrt{3}}{2}\right)$

$= 512 - 512\sqrt{3}i$

$z^3 = 4^3 \operatorname{cis} 180° = 64(-1 + 0i) = -64$

$z^2 = 4^2 \operatorname{cis} 120° = 16\left(-\dfrac{1}{2} + i\dfrac{\sqrt{3}}{2}\right)$

$= -8 + 8\sqrt{3}i$

$z^5 + 3z^3 + 64z^2 + 192$

$= 512 - 512\sqrt{3}i + 3(-64) + 64(-8 + 8\sqrt{3}i)$

$= 192$

$= 512 - 512\sqrt{3}i - 192 - 512 + 512\sqrt{3}i - 192$

$= 0$

Verified

19. $u = z^2; \quad u^2 - 6u + 58 = 0$

$u = \dfrac{6 \pm \sqrt{6^2 - 4(1)(58)}}{2} = \dfrac{6 \pm 14i}{2} = 3 \pm 7i$

So $z^2 = 3 \pm 7i$, and we need the square roots of $3 + 7i$ and $3 - 7i$. For both,

$r = \sqrt{3^2 + 7^2} = \sqrt{58}$, and $\sqrt{r} = 58^{1/4}$

For $3 + 7i$: $\theta = \tan^{-1}\dfrac{7}{3} \approx 66.8°$

$\dfrac{66.8°}{2} + \dfrac{360°k}{2} = 33.4° + 180°k$

$z_0 = 58^{1/4} \operatorname{cis} 33.4° \approx 2.3039 + 1.5192i;$

$z_1 = 58^{1/4} \operatorname{cis} 213.4° \approx -2.3039 - 1.5192i;$

For $3 - 7i$, $\theta = -66.8°$ (QIV)

$z_0 = 58^{1/4} \operatorname{cis} -33.4° \approx 2.3039 - 1.5192i;$

$z_1 = 58^{1/4} \operatorname{cis} 146.6° \approx -2.3039 + 1.5192i$

Calculator Exploration and Discovery

1. a. Approx. 50.5 ft
    b. Approx. 50.5 ft
    c. Approx. 224.54 ft
    d. Approx. 3.55 sec

3. a. 111.87 ft
    b. 132.04 ft
    c. 443.16 ft
    d. 5.75 sec
    It will clear the fence.

Strengthening Core Skills

1. Let **u** and **v** be the force vectors for the ropes, and **w** for the weight.

$\mathbf{u} = -|\mathbf{u}|\cos 25°\mathbf{i} + |\mathbf{u}|\sin 25°\mathbf{j}$

$\approx -0.9063|\mathbf{u}|\mathbf{i} + 0.4226|\mathbf{u}|\mathbf{j}$

$\mathbf{v} = |\mathbf{v}|\cos 20°\mathbf{i} + |\mathbf{v}|\sin 20°\mathbf{j}$

$\approx 0.9397|\mathbf{v}|\mathbf{i} + 0.3420|\mathbf{v}|\mathbf{j}$

$\mathbf{w} = -500\mathbf{j}$

The sum of all first components and all second components must be zero.

$\begin{cases} -0.9063|\mathbf{u}| + 0.9397|\mathbf{v}| = 0 \\ 0.4226|\mathbf{u}| + 0.3429|\mathbf{v}| - 500 = 0 \end{cases}$

Solving this system, we get

$|\mathbf{u}| = 664.46$ lb, $|\mathbf{v}| = 640.86$ lb

3. The system remains the same as in the example, except the 180 in the second equation is replaced with 200. The solution is now $x = 537.49$ lb, $y = 547.13$ lb. The rope will hold.

Cumulative Review

1. This is a 30-60-90 triangle: $\beta = 60°$, c is twice a, or 40 m, and b is $\sqrt{3}$ times a, or $20\sqrt{3}$ m.

3. $A = \pi^2 \left(R^2 - r^2 \right) = \pi^2 R^2 - \pi^2 r^2$

   $\pi^2 R^2 = A + \pi^2 r^2$; $R^2 = \dfrac{1}{\pi^2} \left(A + (\pi r)^2 \right)$

   $R = \dfrac{1}{\pi} \sqrt{A + (\pi r)^2}$

5.

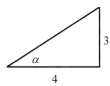

4

The third side is length 5 (Pythagorean triple), and since cosine is positive, sine is negative.

$\sin \alpha = -\dfrac{3}{5}$; $\csc \alpha = -\dfrac{5}{3}$; $\cos \alpha = \dfrac{4}{5}$

$\sec \alpha = \dfrac{5}{4}$; $\tan \alpha = -\dfrac{3}{4}$; $\cot \alpha = -\dfrac{4}{3}$

7. $x = \dfrac{-8 \pm \sqrt{64 - 4(5)(2)}}{10} = \dfrac{-8 \pm \sqrt{24}}{10}$

   $= -\dfrac{8}{10} \pm \dfrac{2\sqrt{6}}{10} = -\dfrac{4}{5} \pm \dfrac{\sqrt{6}}{5}$

9. Using a right triangle, or the Pythagorean Identity, we find that if
   $\cos 53° \approx 0.6$, $\sin 53° \approx 0.8$ and if
   $\cos 72° \approx 0.3$, $\sin 72° \approx 0.95$
   $19° = 72° - 53°$, so $\cos 19° = \cos(72° - 53°)$

   $= \cos 72° \cos 53° + \sin 72° \sin 53°$
   $\approx (0.3)(0.60) + (0.95)(0.8) = 0.94$
   Similarly, $\cos 125° = \cos(72° + 53°)$

   $= \cos 72° \cos 53° - \sin 72° \sin 53° = -0.58$

11. a. $A = \dfrac{1}{2}(1475)(2008)\sin 25.9°$

    $A = 646859.7684 \text{ ft}^2$

    $\dfrac{A}{43560} \approx 14.85 \text{ acres}$

    $(1485 \text{ acres})(\$4500 / \text{acre}) = \$66,825$

    b. Pythagorean triple 5, 12, 13.

    $d = \sqrt{7^2 + 7^2} = 7\sqrt{2}$;
    From origin to (5, 12), 13 units
    From origin to (12,5), 13 units
    From (5, 12) to (12, 5), $7\sqrt{2}$ units;

    $p = \dfrac{13 + 13 + 7\sqrt{2}}{2} = 13 + \dfrac{7\sqrt{2}}{2}$

    $A = \sqrt{\left(13 + \dfrac{7\sqrt{2}}{2}\right)\left(13 + \dfrac{7\sqrt{2}}{2} - 13\right)\left(13 + \dfrac{7\sqrt{2}}{2} - 13\right)\left(13 + \dfrac{7\sqrt{2}}{2} - 7\sqrt{2}\right)}$

    $A = \sqrt{\left(13 + \dfrac{7\sqrt{2}}{2}\right)\left(\dfrac{7\sqrt{2}}{2}\right)\left(\dfrac{7\sqrt{2}}{2}\right)\left(13 - \dfrac{7\sqrt{2}}{2}\right)}$

    $A \approx 59.5 \text{ mi}^2$

13. a. $m = \dfrac{y_2 - y_1}{x_2 - x_1}$

    b. $\left(\dfrac{x_2 + x_1}{2}, \dfrac{y_2 + y_1}{2} \right)$

    c. $x = \dfrac{-b \pm \sqrt{b^2 - 4ac}}{2a}$

    d. $d = \sqrt{(x_2 - x_1)^2 + (y_2 - y_1)^2}$

    e. $A = Pe^{rt}$

15. $a^2 = 31^2 + 52^2 - 2(31)(52)\cos 37°$
    $a^2 \approx 1090.1991$; $a \approx 33 \text{ cm}$

    $\dfrac{\sin 37°}{33} = \dfrac{\sin B}{31}$; $33 \sin B = 31 \sin 37°$

    $\sin B = \dfrac{31 \sin 37°}{33}$; $B = \sin^{-1}\left(\dfrac{31 \sin 37°}{33} \right)$

    $B \approx 34.4°$; $C = 180 - (37 + 34.4) = 108.6°$

17. Complement of 28° is 62°.
    $F = 900 \cos 62° \approx 422.5 \text{ lb}$

19.

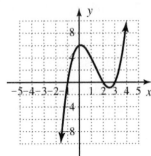

$f(x) < 0$ for $x \in (-\infty, -1)$ and $x \in (2, 3)$.

21. $r = \sqrt{1^2 + \left(-\sqrt{3}\right)^2} = \sqrt{4} = 2$

$\theta = \tan^{-1}\left(\dfrac{-\sqrt{3}}{1}\right) = -60° \quad \text{(QIV)}$

$\left(1 - \sqrt{3}\right)^8 = 2^8 \text{cis}(8(-60°)) = 256\,\text{cis}(-480°)$

$= 256\,\text{cis}\,240° = 256\left(-\dfrac{1}{2} - i\dfrac{\sqrt{3}}{2}\right)$

$= -128 - 128i\sqrt{3}$

23. $R = 0.08/12 \approx 0.00667$

$P = \dfrac{AR}{(1 + R)^{nt} - 1} = \dfrac{10{,}000\left(\frac{0.08}{12}\right)}{(1.00667)^{12t} - 1}$

$200 = \dfrac{66.67}{(1.00667)^{12t} - 1}$

$200\left(1.00667^{12t} - 1\right) = 66.67$

$200\left(1.00667^{12t}\right) - 200 = 66.67$

$200\left(1.00667^{12t}\right) = 266.67$

$1.00667^{12t} = 1.33; \quad 12t\ln 1.00667 = \ln(1.33)$

$t = \dfrac{\ln(1.33)}{12\ln 1.00667} \approx 3.6 \text{ yr}$

25. $A = 2$; $P = 2\pi$, so $B = 1$; the graph is shifted $\dfrac{\pi}{4}$ units left, so $C = \dfrac{\pi}{4}$.

Connections to Calculus

1.

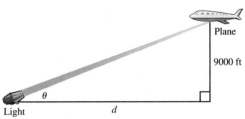

Plane

9000 ft

θ

Light d

a) $\tan\theta = \dfrac{9000}{d}$

$\theta = \tan^{-1}\left(\dfrac{9000}{d}\right)$

b) $\theta = \tan^{-1}\left(\dfrac{9000}{12850}\right)$

$\theta \approx 35°$

3.

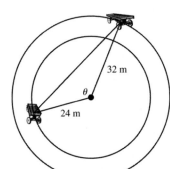

32 m

θ

24 m

a) $d^2 = 24^2 + 32^2 - 2\cdot 24\cdot 32\cos\theta$

$d = \sqrt{24^2 + 32^2 - 2\cdot 24\cdot 32\cos\theta}$

b) $d = \sqrt{24^2 + 32^2 - 2\cdot 24\cdot 32\cos 150°}$

$d \approx 54.13 \text{ m}$

c) $45^2 = 24^2 + 32^2 - 2\cdot 24\cdot 32\cos\theta$

$45^2 - 24^2 - 32^2 = -1536\cos\theta$

$425 = -1536\cos\theta$

$\dfrac{425}{-1536} = \cos\theta$

$\cos^{-1}\left(\dfrac{-425}{1536}\right) = \theta$

$\theta \approx 106.1°$

5.

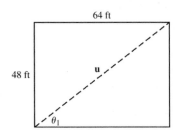

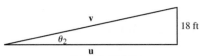

a) $|v| = \sqrt{48^2 + 64^2 + 18^2} = 82$ ft

b) $\cos\theta = \dfrac{u \cdot v}{|u||v|}$

$u = <48, 64, 0>\quad v = <48, 64, 18>$

$|u| = \sqrt{48^2 + 64^2 + 0^2} = 80$

$u \cdot v = 48 \cdot 48 + 64 \cdot 64 + 0 \cdot 18$

$= 6400$

$\cos\theta = \dfrac{6400}{80 \cdot 82} = \dfrac{40}{41}$

$\theta = \cos^{-1}\left(\dfrac{40}{41}\right)$

$\theta \approx 12.68°$

Verify:

$u = \sqrt{48^2 + 64^2} = 80$

$\tan\theta = \dfrac{18}{80}$

$\theta = \tan^{-1}\left(\dfrac{18}{80}\right)$

$\theta \approx 12.68$

7.

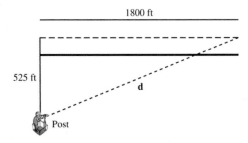

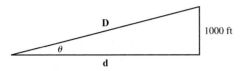

$D = <525, 1800, 1000>\quad d = <525, 1800, 0>$

a) $|D| = \sqrt{525^2 + 1800^2 + 1000^2}$

$= \sqrt{4515625} = 2125$ ft

b) $\cos\theta = \dfrac{d \cdot D}{|d||D|}$

$d \cdot D = 525 \cdot 525 + 1800 \cdot 1800 + 1000 \cdot 0 = 3515625$

$|d| = \sqrt{525^2 + 1800^2} = \sqrt{3515625} = 1875$

$\cos\theta = \dfrac{3515625}{1875 \cdot 2125} = \dfrac{15}{17}$

$\theta = \cos^{-1}\left(\dfrac{15}{17}\right)$

$\theta \approx 28.07°$

Verify:

$\sin\theta = \dfrac{1000}{2125}$

$\theta\sin^{-1}\left(\dfrac{1000}{2125}\right) \approx 28.07°$

9.

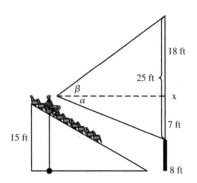

a) $\theta = \alpha + \beta$

$\tan\alpha = \dfrac{18}{x} \rightarrow \alpha = \tan^{-1}\left(\dfrac{18}{x}\right)$

$\tan\beta = \dfrac{7}{x} \rightarrow \beta = \tan^{-1}\left(\dfrac{7}{x}\right)$

$\theta = \alpha + \beta = \tan^{-1}\left(\dfrac{18}{x}\right) + \tan^{-1}\left(\dfrac{7}{x}\right)$

b) $x = 46$ ft

$\theta = \tan^{-1}\left(\dfrac{18}{46}\right) + \tan^{-1}\left(\dfrac{7}{46}\right) = 30.2°$

8.1 Technology Highlight

1. $(-1,4)$

8.1 Exercises

1. Inconsistent

3. Consistent; independent

5. Multiply the 1st equation by 6 and the 2nd equation by 10.

7. $\begin{cases} 7x - 4y = 24 \\ 4x + 3y = 15 \end{cases}$

$\begin{cases} y = \dfrac{7}{4}x - 6 \\ y = -\dfrac{4}{3}x + 5 \end{cases}$

9. $A: \quad y = x + 2$

11. $C: \quad x + 3y = -3$

13. $E: \quad y = x + 2$

$\quad\quad x + 3y = -3$

15. $\begin{cases} 3x + y = 11 \\ -5x + y = -13 \end{cases}$

$(3, 2)$

$3x + y = 11$ $\quad\quad\quad -5x + y = -13$

$3(3) + 2 = 11$ $\quad\quad\quad -5(3) + 2 = -13$

$9 + 2 = 11$ $\quad\quad\quad\quad -15 + 2 = -13$

$\quad\quad 11 = 11$ $\quad\quad\quad\quad\quad -13 = -13$

Yes

17. $\begin{cases} 8x - 24y = -17 \\ 12x + 30y = 2 \end{cases}$

$\left(-\dfrac{7}{8}, \dfrac{5}{12} \right)$

$8x - 24y = -17$

$8\left(-\dfrac{7}{8} \right) - 24\left(\dfrac{5}{12} \right) = -17$

$-7 - 10 = -17$

$-17 = -17;$

$12x + 30y = 2$

$12\left(-\dfrac{7}{8} \right) + 30\left(\dfrac{5}{12} \right) = 2$

$-\dfrac{84}{8} + \dfrac{150}{12} = 2$

$2 = 2$

Yes

19. $\begin{cases} 3x + 2y = 12 \\ \quad x - y = 9 \end{cases}$

$\begin{cases} y = -\dfrac{3}{2}x + 6 \\ y = x - 9 \end{cases}$

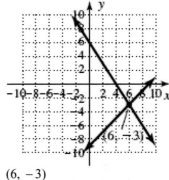

$(6, -3)$

21. $\begin{cases} 5x - 2y = 4 \\ x + 3y = -15 \end{cases}$

$\begin{cases} y = \dfrac{5}{2}x - 2 \\ y = -\dfrac{1}{3}x - 5 \end{cases}$

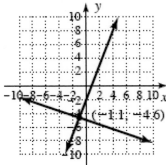

Estimate: $(-1.1, -4.6)$

23. $\begin{cases} x = 5y - 9 \\ x - 2y = -6 \end{cases}$

$\begin{aligned} x - 2y &= -6 \\ (5y - 9) - 2y &= -6 \\ 5y - 9 - 2y &= -6 \\ 3y &= 3 \\ y &= 1; \end{aligned}$

$\begin{aligned} x - 2y &= -6 \\ x - 2(1) &= -6 \\ x - 2 &= -6 \\ x &= -4 \end{aligned}$

$(-4, 1)$

25. $\begin{cases} y = \dfrac{2}{3}x - 7 \\ 3x - 2y = 19 \end{cases}$

$\begin{aligned} 3x - 2y &= 19 \\ 3x - 2\left(\dfrac{2}{3}x - 7\right) &= 19 \\ 3x - \dfrac{4}{3}x + 14 &= 19 \\ \dfrac{5}{3}x &= 5 \\ x &= 3; \end{aligned}$

$\begin{aligned} y &= \dfrac{2}{3}x - 7 \\ y &= \dfrac{2}{3}(3) - 7 \\ y &= 2 - 7 \\ y &= -5 \end{aligned}$

$(3, -5)$

27. $\begin{cases} 3x - 4y = 24 \\ 5x + y = 17 \end{cases}$

Equation 2, variable y

$\begin{cases} 3x - 4y = 24 \\ y = -5x + 17 \end{cases}$

$\begin{aligned} 3x - 4y &= 24 \\ 3x - 4(-5x + 17) &= 24 \\ 3x + 20x - 68 &= 24 \\ 23x &= 92 \\ x &= 4; \end{aligned}$

$\begin{aligned} 5x + y &= 17 \\ 5(4) + y &= 17 \\ 20 + y &= 17 \\ y &= -3 \end{aligned}$

$(4, -3)$

29. $\begin{cases} 0.7x + 2y = 5 \\ x - 1.4y = 11.4 \end{cases}$

Equation 2, variable x

$\begin{cases} 0.7x + 2y = 5 \\ x = 1.4y + 11.4 \end{cases}$

$\begin{aligned} 0.7x + 2y &= 5 \\ 0.7(1.4y + 11.4) + 2y &= 5 \\ 0.98y + 7.98 + 2y &= 5 \\ 2.98y &= -2.98 \\ y &= -1; \end{aligned}$

$\begin{aligned} x - 1.4y &= 11.4 \\ x - 1.4(-1) &= 11.4 \\ x + 1.4 &= 11.4 \\ x &= 10 \end{aligned}$

$(10, -1)$

31. $\begin{cases} 5x - 6y = 2 \\ x + 2y = 6 \end{cases}$

Equation 2, variable x

$\begin{cases} 5x - 6y = 2 \\ x = -2y + 6 \end{cases}$

$5x - 6y = 2$
$5(-2y + 6) - 6y = 2$
$-10y + 30 - 6y = 2$
$-16y = -28$
$y = \dfrac{7}{4};$

$x + 2y = 6$

$x + 2\left(\dfrac{7}{4}\right) = 6$

$x + \dfrac{7}{2} = 6$

$x = \dfrac{5}{2}$

$\left(\dfrac{5}{2}, \dfrac{7}{4}\right)$

33. $\begin{cases} 2x - 4y = 10 \\ 3x + 4y = 5 \end{cases}$

R1 + R2 = Sum
$2x - 4y = 10$
$3x + 4y = 5$
$5x = 15$
$x = 3;$
$3x + 4y = 5$
$3(3) + 4y = 5$
$9 + 4y = 5$
$4y = -4$
$y = -1$
$(3, -1)$

35. $\begin{cases} 4x - 3y = 1 \\ 3y = -5x - 19 \end{cases}$

$\begin{cases} 4x - 3y = 1 \\ 5x + 3y = -19 \end{cases}$

R1 + R2 = Sum
$4x - 3y = 1$
$5x + 3y = -19$
$9x = -18$
$x = -2;$
$3y = -5x - 19$
$3y = -5(-2) - 19$
$3y = 10 - 19$
$3y = -9$
$y = -3$
$(-2, -3)$

37. $\begin{cases} 2x = -3y + 17 \\ 4x - 5y = 12 \end{cases}$

$\begin{cases} 2x + 3y = 17 \\ 4x - 5y = 12 \end{cases}$

-2R1 + R2 = Sum
$-4x - 6y = -34$
$4x - 5y = 12$
$-11y = -22$
$y = 2;$
$2x = -3y + 17$
$2x = -3(2) + 17$
$2x = -6 + 17$
$2x = 11$
$x = \dfrac{11}{2}$

$\left(\dfrac{11}{2}, 2\right)$

39. $\begin{cases} 0.5x + 0.4y = 0.2 \\ 0.3y = 1.3 + 0.2x \end{cases}$

$\begin{cases} 0.5x + 0.4y = 0.2 \\ -0.2x + 0.3y = 1.3 \end{cases}$

$20R1 + 50\ R2 = Sum$

$10x + 8y = 4$

$-10x + 15y = 65$

$23y = 69$

$y = 3;$

$0.5x + 0.4y = 0.2$

$0.5x + 0.4(3) = 0.2$

$0.5x + 1.2 = 0.2$

$0.5x = -1$

$x = -2$

$(-2, 3)$

41. $\begin{cases} 0.32m - 0.12n = -1.44 \\ -0.24m + 0.08n = 1.04 \end{cases}$

$200\ R1 + 300\ R2 = Sum$

$64m - 24n = -288$

$-72m + 24n = 312$

$-8m = 24$

$m = -3;$

$0.32m - 0.12n = -1.44$

$0.32(-3) - 0.12n = -1.44$

$-0.96 - 0.12n = -1.44$

$-0.12n = -0.48$

$n = 4$

$(-3, 4)$

43. $\begin{cases} -\dfrac{1}{6}u + \dfrac{1}{4}v = 4 \\ \dfrac{1}{2}u - \dfrac{2}{3}v = -11 \end{cases}$

$18R1 + 6R2 = Sum$

$-3u + 4.5v = 72$

$3u - 4v = -66$

$0.5v = 6$

$v = 12;$

$-\dfrac{1}{6}u + \dfrac{1}{4}v = 4$

$-\dfrac{1}{6}u + \dfrac{1}{4}(12) = 4$

$-\dfrac{1}{6}u + 3 = 4$

$-\dfrac{1}{6}u = 1$

$u = -6$

$(-6, 12)$

45. $\begin{cases} 4x + \dfrac{3}{4}y = 14 \\ -9x + \dfrac{5}{8}y = -13 \end{cases}$

$9R1 + 4R2 = Sum$

$36x + \dfrac{27}{4}y = 126$

$-36x + \dfrac{5}{2}y = -52$

$\dfrac{37}{4}y = 74$

$y = 8;$

$4x + \dfrac{3}{4}y = 14$

$4x + \dfrac{3}{4}(8) = 14$

$4x + 6 = 14$

$4x = 8$

$x = 2$

$(2, 8);$ Consistent/independent

8.1 Exercises

47. $\begin{cases} 0.2y = 0.3x + 4 \\ 0.6x - 0.4y = -1 \end{cases}$

$\begin{cases} -0.3x + 0.2y = 4 \\ 0.6x - 0.4y = -1 \end{cases}$

2R1 + R2 = Sum
$-0.6x + 0.4y = 8$
$0.6x - 0.4y = -1$

$\qquad\qquad 0 \neq 7$

No Solution; Inconsistent

49. $\begin{cases} 6x - 22 = -y \\ 3x + \dfrac{1}{2}y = 11 \end{cases}$

$\begin{cases} 6x + y = 22 \\ 3x + \dfrac{1}{2}y = 11 \end{cases}$

R1 − 2R2 = Sum
$6x + y = 22$
$-6x - y = -22$

$\qquad\qquad 0 = 0$

$\{(x, y) | 6x + y = 22\}$

Consistent/dependent

51. $\begin{cases} -10x + 35y = -5 \\ y = 0.25x \end{cases}$

$\qquad -10x + 35y = -5$
$\qquad -10x + 35(0.25x) = -5$
$\qquad\quad -10x + 8.75x = -5$
$\qquad\qquad\quad -1.25x = -5$
$\qquad\qquad\qquad\quad x = 4;$

$y = 0.25x$
$y = 0.25(4)$
$y = 1$

(4, 1); Consistent/Independent

53. $\begin{cases} 7a + b = -25 \\ 2a - 5b = 14 \end{cases}$

5R1 + R2 = Sum
$35a + 5b = -125$

$2a - 5b = 14$
$37a = -111$
$a = -3;$

$2a - 5b = 14$
$2(-3) - 5b = 14$
$-6 - 5b = 14$
$-5b = 20$
$b = -4$

$(-3, -4)$; Consistent/Independent

55. $\begin{cases} 4a = 2 - 3b \\ 6b + 2a = 7 \end{cases}$

$\begin{cases} 4a + 3b = 2 \\ 2a + 6b = 7 \end{cases}$

R1 − 2R2 = Sum
$4a + 3b = 2$
$-4a - 12b = -14$
$\qquad\quad -9b = -12$
$\qquad\quad\;\; b = \dfrac{4}{3};$

$4a + 3b = 2$

$4a + 3\left(\dfrac{4}{3}\right) = 2$

$4a + 4 = 2$
$4a = -2$

$a = -\dfrac{1}{2}$

$\left(-\dfrac{1}{2}, \dfrac{4}{3}\right)$; Consistent/Independent

57. $\begin{cases} 2x + 4y = 6 \\ x + 12 = 4y \end{cases}$

$2x + 4y = 6$
$2x + x + 12 = 6$
$3x = -6$
$x = -2;$

$x + 12 = 4y$
$-2 + 12 = 4y$
$10 = 4y$

$\dfrac{5}{2} = y$

$\left(-2, \dfrac{5}{2}\right)$

59. $\begin{cases} 5x - 11y = 21 \\ 11y = 5 - 8x \end{cases}$

$5x - 11y = 21$
$5x - (5 - 8x) = 21$
$5x - 5 + 8x = 21$
$13x = 26$
$x = 2;$

$11y = 5 - 8x$

$11y = 5 - 8(2)$

$11y = 5 - 16$

$11y = -11$

$y = -1$

$(2, -1)$

61. $\begin{cases} (R+C)T_1 = D_1 \\ (R-C)T_2 = D_2 \end{cases}$

$\begin{cases} (R+C)\,1 = 5 \\ (R-C)\,3 = 9 \end{cases}$

$\begin{cases} R+C = 5 \\ 3R-3C = 9 \end{cases}$

$3R1 + R2 = $ Sum

$3R + 3C = 15$

$3R - 3C = 9$

$6R = 24$

$R = 4;$

$R + C = 5$

$4 + C = 5$

$C = 1$

The current was 1 mph.
He can row 4 mph in still water.

63. Let a represent the number of adult tickets.
Let c represent the number of child tickets.

$\begin{cases} 9a + 6.50c = 30495 \\ a + c = 3800 \end{cases}$

Multiply the second equation by -9.

$\begin{cases} 9a + 6.50c = 30495 \\ -9a - 9c = -34200 \end{cases}$

$-2.5c = -3705$

$c = \dfrac{-3705}{-2.5}$

$c = 1482;$

$a + 1482 = 3800$

$a = 3800 - 1482$

$a = 2318$

2318 adult tickets and 1482 child tickets
were sold.

65. Let r represent the price per gallon of
regular unleaded gasoline.
Let p represent represent the price per gallon
of premium gasoline.

$\begin{cases} 20r + 17p = 144.89 \\ p = 0.10 + r \end{cases}$

$20r + 17(0.10 + r) = 144.89$

$20r + 1.7 + 17r = 144.89$

$37r = 144.89 - 1.7$

$r = \$3.87$

$p = 0.10 + 3.87 = \$3.97$

Premium: \$3.97, Regular: \$3.87.

67. Let s represent the loan made to the science
major.
Let n represent the loan made to the nursing
student.

$\begin{cases} 0.07s + 0.06n = 635 \\ s + n = 10000 \end{cases}$

$s = 10000 - n$

$0.07(10000 - n) + 0.06n = 635$

$700 - 0.07n + 0.06n = 635$

$-0.01n = 635 - 700$

$-0.01n = -65$

$n = \dfrac{-65}{-0.01}$

$n = 6500;$

$s = 10000 - 65000 = 3500$

\$6500 was loaned to the nursing student.
\$3500 was loaned to the science major.

69. Let q represent the number of quarters.
Let d represent represent the number of
dimes.

$\begin{cases} q + d = 225 \\ 0.25q + 0.10d = 45 \end{cases}$

$q = 225 - d;$

$0.25(225 - d) + 0.10d = 45$

$56.25 - 0.25d + 0.10d = 45$

$-0.15d = 45 - 56.25$

$-0.15d = -11.25$

$d = 75;$

$q = 225 - 75 = 150$

150 quarters, 75 dimes

71. Let x represent the number of lawns serviced
each month.

(a) Total Cost: $C(x) = 75x + 4000$

Projected Revenue: $R(x) = 115x$

$\begin{cases} y = 75x + 4000 \\ y = 115x \end{cases}$

$75x + 4000 = 115x$

$4000 = 40x$

$100 = x$

100 lawns/month

(b) $115(100) = \$11,500$/month.

73. $y = 1.5x + 3$

 (a) $5.40 = 1.5x + 3$

 $2.40 = 1.5x$

 $1.6 = x$

 Supply: 1.6 billion bu;

 $y = -2.20x + 12$

 $5.40 = -2.20x + 12$

 $-6.6 = -2.20x$

 $3 = x$

 Demand: 3 billion bu.

 Yes, supply is less than demand.

 (b) $7.05 = 1.5x + 3$

 $4.05 = 1.5x$

 $2.7 = x$

 Supply: 2.7 billion bu;

 $7.05 = -2.20x + 12$

 $-4.95 = -2.20x$

 $2.25 = x$

 Demand: 2.25 billion bu.

 Yes, demand is less than supply.

 (c) $1.5x + 3 = -2.20x + 12$

 $3.70x = 9$

 $x \approx 2.43$ billion bu;

 $y = 1.5(2.43) + 3$

 $y \approx \$6.65$

75. Let c represent the speed of the current.
 Let b represent the speed of the boat in still water.

 To the drop point: $4 = (b - c)2$

 Return to the drop point: $4 = (b + c)\dfrac{1}{2}$

 $\begin{cases} 4 = 2b - 2c \\ 4 = \dfrac{1}{2}b + \dfrac{1}{2}c \end{cases}$

 4R2

 $\begin{cases} 4 = 2b - 2c \\ 4(4) = 4\left(\dfrac{1}{2}b + \dfrac{1}{2}c\right) \end{cases}$

 $\begin{cases} 4 = 2b - 2c \\ 16 = 2b + 2c \end{cases}$

 R1 + R2

 $20 = 4b$

 $5 = b$

 5 mph;

 $4 = (b - c)2$

$4 = (5 - c)2$

$4 = 10 - 2c$

$-6 = -2c$

$3 = c$

3 mph

(a) Speed of current, 3 mph

(b) Speed of boat in still water, 5 mph

77. (a) Let w represent the speed of the walkway.
 Let j represent Jason's walking speed.
 With walkway: $256 = (j + w)32$
 Opposite direction: $256 = (j - w)320$

 $\begin{cases} 256 = 32j + 32w \\ 256 = 320j - 320w \end{cases}$

 $\dfrac{R1}{32}, \dfrac{R2}{32}$

 $\begin{cases} 8 = j + w \\ 8 = 10j - 10w \end{cases}$

 $-10R1$

 $\begin{cases} -80 = -10j - 10w \\ 8 = 10j - 10w \end{cases}$

 $-72 = -20w$

 $3.6 = w$

 3.6 ft/sec

 (b) $8 = j + w$

 $8 = j + 3.6$

 $j = 4.4$ ft/sec

79. Let d represent the year the Declaration was signed.
 Let c represent the year the Civil War ended.

 $\begin{cases} c + d = 3641 \\ c - d = 89 \end{cases}$

 $2c = 3730$

 $c = \dfrac{3730}{2}$

 $c = 1865$

 $1865 + d = 3641$

 $d = 3641 - 1865$

 $d = 1776$

 The Declaration was signed in 1776.
 The Civil War ended in 1865.

81. Let x represent Tahiti's land area.
Let y represent Tonga's land area.
$$\begin{cases} x + y = 692 \\ x = 112 + y \end{cases}$$
$$112 + y + y = 692$$
$$112 + 2y = 692$$
$$2y = 580$$
$$y = 290 \text{ mi}^2;$$
$$x + y = 692$$
$$x + 290 = 692$$
$$x = 402 \text{ mi}^2;$$
Tahiti: 402 mi^2
Tonga: 290 mi^2

83. Different slopes so they cannot be the same line or parallel lines. Therefore the system is consistent and the equations are independent.

85. Let x represent the amount invested at 6%.
Let y represent the amount invested at 8.5%.
$$\begin{cases} 0.06x + 0.085y = 1250 \\ 0.085x + 0.06y = 1375 \end{cases}$$
Multiply both equations by 1000.
$$\begin{cases} 60x + 85y = 1250000 \\ 85x + 60y = 1375000 \end{cases}$$
Multiply the first equation by -85 and the second equation by 60.
$$\begin{cases} -5100x - 7225y = -106250000 \\ 5100x + 3600y = 82500000 \end{cases}$$
$$-3625y = -23750000$$
$$y = \frac{-23750000}{-3625}$$
$$y \approx 6552$$
$$60x + 85(6552) = 1250000$$
$$60x + 556920 = 1250000$$
$$60x = 1250000 - 556920$$
$$60x = 693080$$
$$x = \frac{693080}{60}$$
$$x \approx 11551$$
$6,552 invested at 8.5%.
$11,551 invested at 6%.

87. $\theta = 112°$
$$112° + 360° = 472°;$$
$$472° + 360° = 832°;$$
$$112° - 360° = -248°;$$
$$-248° - 360° = -608°$$

89. $$\frac{\sin x - \csc x}{\csc x} = -\cos^2 x$$
$$\frac{\sin x}{\csc x} - \frac{\csc x}{\csc x} = -\cos^2 x$$
$$\frac{\sin x}{\csc x} - 1 = -\cos^2 x$$
$$\sin^2 x - 1 = -\cos^2 x$$
$$-1(1 - \sin^2 x) = -\cos^2 x$$
$$-\cos^2 x = -\cos^2 x$$

Technology Highlight

1. a. Answers will vary.
   b. Answers will vary.
   c. $(-9, -6, -5)$ and $(-2, 1, 2)$ are solutions and $(6, 2, 4)$ is not a solution.

8.2 Exercises

1. Triple

3. Equivalent systems

5. $2(2)+(-5)+z=4$
$4+(-5)+z=4$
$-1+z=4$
$z=5;$
Substitute and solve for the remaining variable.

7. $x+2y+z=9$
Answers will vary.

9. $-x+y+2z=-6$
Answers will vary.

11. $\begin{cases} x+y-2z=-1 \\ 4x-y+3z=3 \\ 3x+2y-z=4 \end{cases}$
$x+y-2z=-1$
$0+3-2(2)=-1$
$3-4=-1$
$-1=-1;$
$4x-y+3z=3$
$4(0)-3+3(2)=3$
$-3+6=3$
$3=3;$
$3x+2y-z=4$
$3(0)+2(3)-2=4$
$6-2=4$
$4=4$
Yes
$x+y-2z=-1$
$-3+4-2(1)=-1$
$1-2=-1$
$-1=-1;$
$4x-y+3z=3$
$4(-3)-4+3(1)=3$
$-12-4+3=3$
$-13\neq3$
No

13. $\begin{cases} x-y-2z=-10 \\ x-z=1 \\ z=4 \end{cases}$
$x-z=1$
$x-4=1$
$x=5;$
$x-y-2z=-10$
$5-y-2(4)=-10$
$5-y-8=-10$
$-y-3=-10$
$-y=-7$
$y=7$
$(5, 7, 4)$

15. $\begin{cases} x+3y+2z=16 \\ -2y+3z=1 \\ 8y-13z=-7 \end{cases}$
$4R2+R3=Sum$
$-8y+12z=4$
$8y-13z=-7$
$-z=-3$
$z=3;$
$-2y+3z=1$
$-2y+3(3)=1$
$-2y+9=1$
$-2y=-8$
$y=4;$
$x+3y+2z=16$
$x+3(4)+2(3)=16$
$x+12+6=16$
$x+18=16$
$x=-2$
$(-2, 4, 3)$

17. $\begin{cases} 2x - y + 4z = -7 \\ x + 2y - 5z = 13 \\ y - 4z = 9 \end{cases}$

R1 + R3 = Sum

$2x - y + 4z = -7$

$y - 4z = 9$

$2x = 2$

$x = 1$

Substitute $x = 1$ into R2 then new

R2 $- 2$R3 = Sum

$1 + 2y - 5z = 13$

$2y - 5z = 12$

$-2y + 8z = -18$

$3z = -6$

$z = -2;$

$2x - y + 4z = -7$

$2(1) - y + 4(-2) = -7$

$2 - y - 8 = -7$

$-y - 6 = -7$

$-y = -1$

$y = 1$

$(1, 1, -2)$

19. $\begin{cases} -x + y + 2z = -10 \\ x + y - z = 7 \\ 2x + y + z = 5 \end{cases}$

R1 + R2 = Sum

$-x + y + 2z = -10$

$x + y - z = 7$

$2y + z = -3$

2R1 + R2 = Sum

$-2x + 2y + 4z = -20$

$2x + y + z = 5$

$3y + 5z = -15$

$\begin{cases} 2y + z = -3 \\ 3y + 5z = -15 \end{cases}$

-5R1 + R2 = Sum

$-10y - 5z = 15$

$3y + 5z = -15$

$-7y = 0$

$y = 0;$

$2y + z = -3$

$2(0) + z = -3$

$z = -3;$

$x + y - z = 7$

$x + 0 - (-3) = 7$

$x + 3 = 7$

$x = 4$

$(4, 0, -3)$

21. $\begin{cases} 3x + y - 2z = 3 \\ x - 2y + 3z = 10 \\ 4x - 8y + 5z = 5 \end{cases}$

-4R2 + R3 = Sum

$-4x + 8y - 12z = -40$

$4x - 8y + 5z = 5$

$-7z = -35$

$z = 5$

Substitute into R1 and R2

$3x + y - 2(5) = 3$

$3x + y - 10 = 3$

$3x + y = 13$

$x - 2y + 3(5) = 10$

$x - 2y + 15 = 10$

$x - 2y = -5$

$\begin{cases} 3x + y = 13 \\ x - 2y = -5 \end{cases}$

2R1 + R2 = Sum

$6x + 2y = 26$

$x - 2y = -5$

$7x = 21$

$x = 3;$

$3x + y - 2z = 3$

$3(3) + y - 2(5) = 3$

$9 + y - 10 = 3$

$y - 1 = 3$

$y = 4$

$(3, 4, 5)$

8.2 Exercises

23. $\begin{cases} 3x - y + z = 6 \\ 2x + 2y - z = 5 \\ 2x - y + z = 5 \end{cases}$

R1 + R2 = Sum

$\begin{cases} 3x - y + z = 6 \\ 2x + 2y - z = 5 \end{cases}$

$5x + y = 11;$

R2 + R3 = Sum

$\begin{cases} 2x + 2y - z = 5 \\ 2x - y + z = 5 \end{cases}$

$4x + y = 10$

$\begin{cases} 5x + y = 11 \\ 4x + y = 10 \end{cases}$

$-R1 + R2 = Sum$

$\begin{cases} -5x - y = -11 \\ 4x + y = 10 \end{cases}$

$-x = -1$

$x = 1;$

$5x + y = 11$

$5(1) + y = 11$

$y = 6;$

$3x - y + z = 6$

$3(1) - 6 + z = 6$

$3 - 6 + z = 6$

$-3 + z = 6$

$z = 9$

$(1, 6, 9)$

25. $\begin{cases} 3x + y + 2z = 3 \\ x - 2y + 3z = 1 \\ 4x - 8y + 12z = 7 \end{cases}$

2R1 + R2 = Sum

$6x + 2y + 4z = 6$

$x - 2y + 3z = 1$

$7x + 7z = 7$

8R1 + R3 = Sum

$24x + 8y + 16z = 24$

$4x - 8y + 12z = 7$

$28x + 28z = 31$

$\begin{cases} 7x + 7z = 7 \\ 28x + 28z = 31 \end{cases}$

$-4R1 + R2 = Sum$

$-28x - 28z = -28$

$28x + 28z = 31$

$0 \neq 3$

No solution; inconsistent

27. $\begin{cases} 4x + y + 3z = 8 \\ x - 2y + 3z = 2 \end{cases}$

2R1

$\begin{cases} 8x + 2y + 6z = 8 \\ x - 2y + 3z = 2 \end{cases}$

R1 + R2 = Sum

$9x + 9z = 18$

$9z = 18 - 9x$

$z = 2 - x;$

$x - 2y + 3z = 2$

$x - 2y + 3(2 - x) = 2$

$x - 2y + 6 - 3x = 2$

$-2x + 6 - 2y = 2$

$-2y = 2x - 4$

$y = -x + 2$

$y = 2 - x;$

$(x, 2 - x, 2 - x)$

Using "p" as our parameter, the solution could be written in $(p, 2 - p, 2 - p)$ parametric form.

29. $\begin{cases} 6x - 3y + 7z = 2 \\ 3x - 4y + z = 6 \end{cases}$

R1 − 2R2 = Sum

$6x - 3y + 7z = 2$

$-6x + 8y - 2z = -12$

$5y + 5z = -10$

$5y = -5z - 10$

$y = -z - 2;$

$3x - 4y + z = 6$

$3x - 4(-z - 2) + z = 6$

$3x + 4z + 8 + z = 6$

$3x + 5z = -2$

$3x = -5z - 2$

$x = -\dfrac{5}{3}z - \dfrac{2}{3}$

$\left(-\dfrac{5}{3}z - \dfrac{2}{3}, -z - 2, z \right)$

Using "p" as our parameter, the solution could be written in $\left(-\dfrac{5}{3}p - \dfrac{2}{3}, -p - 2, p \right)$ parametric form. Other solutions are possible.

31. $\begin{cases} 3x-4y+5z=5 \\ -x+2y-3z=-3 \\ 3x-2y+z=1 \end{cases}$

R2 + R3 = Sum

$-x+2y-3z=-3$

$3x-2y+z=1$

$2x-2z=-2$

$-2z=-2x-2$

$z=x+1;$

$-x+2y-3z=-3$

$-x+2y-3(x+1)=-3$

$-x+2y-3x-3=-3$

$2y-4x=0$

$2y=4x$

$y=2x;$

$(x,2x,x+1)$

Using "p" as our parameter, the solution could be written in $(p,2p,p+1)$ parametric form. Other solutions are possible.

33. $\begin{cases} x+2y-3z=1 \\ 3x+5y-8z=7 \\ x+y-2z=5 \end{cases}$

R1 − R3 = Sum

$x+2y-3z=1$

$-x-y+2z=-5$

$y-z=-4$

$y-z=-4$

$y=z-4;$

$x+y-2z=5$

$x+z-4-2z=5$

$x-z=9$

$x=z+9$

$(z+9,z-4,z)$

Using "p" as our parameter, the solution could be written in $(p+9,p-4,p)$ parametric form. Other solutions are possible.

35. $\begin{cases} -0.2x+1.2y-2.4z=-1 \\ 0.5x-3y+6z=2.5 \\ x-6y+12z=5 \end{cases}$

-2R2 + R3 = Sum

$-1x+6y-12z=-5$

$x-6y+12z=5$

$0=0$

$\{(x,y,z) | x-6y+12z=5\}$

37. $\begin{cases} x+2y-z=1 \\ x+z=3 \\ 2x-y+z=3 \end{cases}$

R1 + 2R3 = Sum

$x+2y-z=1$

$4x-2y+2z=6$

$5x+z=7$

$\begin{cases} x+z=3 \\ 5x+z=7 \end{cases}$

−R1 + R2 = Sum

$-x-z=-3$

$5x+z=7$

$4x=4$

$x=1;$

$x+z=3$

$1+z=3$

$z=2;$

$x+2y-z=1$

$1+2y-2=1$

$2y-1=1$

$2y=2$

$y=1$

$(1,1,2)$

39. $\begin{cases} 2x-5y-4z=6 \\ x-2.5y-2z=3 \\ -3x+7.5y+6z=-9 \end{cases}$

R1 − 2R2 = Sum

$2x-5y-4z=6$

$-2x+5y+4z=-6$

$0=0$

$\left\{(x,y,z) \left| x-\dfrac{5}{2}y-2z=3\right.\right\}$

41. $\begin{cases} 4x - 5y - 6z = 5 \\ 2x - 3y + 3z = 0 \\ x + 2y - 3z = 5 \end{cases}$

R1 − 2R2 = Sum
$$4x - 5y - 6z = 5$$
$$-4x + 6y - 6z = 0$$
$$y - 12z = 5$$

R1 − 4R3 = Sum
$$4x - 5y - 6z = 5$$
$$-4x - 8y + 12z = -20$$
$$-13y + 6z = -15$$

$\begin{cases} y - 12z = 5 \\ -13y + 6z = -15 \end{cases}$

R1 + 2R2 = Sum
$$y - 12z = 5$$
$$-26y + 12z = -30$$
$$-25y = -25$$
$$y = 1;$$

$$y - 12z = 5$$
$$1 - 12z = 5$$
$$-12z = 4$$
$$z = -\frac{1}{3};$$

$$x + 2y - 3z = 5$$
$$x + 2(1) - 3\left(-\frac{1}{3}\right) = 5$$
$$x + 2 + 1 = 5$$
$$x + 3 = 5$$
$$x = 2$$

$\left(2, 1, -\dfrac{1}{3}\right)$

43. $\begin{cases} 2x + 3y - 5z = 4 \\ x + y - 2z = 3 \\ x + 3y - 4z = -1 \end{cases}$

R1 − 2R2 = Sum
$$2x + 3y - 5z = 4$$
$$-2x - 2y + 4z = -6$$
$$y - z = -2$$

R1 − 2R3 = Sum
$$2x + 3y - 5z = 4$$
$$-2x - 6y + 8z = -2$$
$$-3y + 3z = 6$$
$$y - z = -2;$$

$$y - z = -2$$
$$y = z - 2;$$

$$x + y - 2z = 3$$
$$x + z - 2 - 2z = 3$$
$$x - z = 5$$
$$x = z + 5$$

$(z + 5, z - 2, z)$

Using "p" as our parameter, the solution could be written in $(p + 5, p - 2, p)$ parametric form. Other solutions are possible.

45.
$$\begin{cases} \dfrac{x}{2}+\dfrac{y}{3}-\dfrac{z}{2}=2 \\[2mm] \dfrac{2x}{3}-y-z=8 \\[2mm] \dfrac{x}{6}+2y+\dfrac{3z}{2}=6 \end{cases}$$

$3R1 + R2 = Sum$

$$\dfrac{3}{2}x+y-\dfrac{3}{2}z=6$$

$$\dfrac{2x}{3}-y-z=8$$

$$\dfrac{13}{6}x-\dfrac{5}{2}z=14$$

$2R2 + R3 = Sum$

$$\dfrac{4}{3}x-2y-2z=16$$

$$\dfrac{x}{6}+2y+\dfrac{3z}{2}=6$$

$$\dfrac{3}{2}x-\dfrac{1}{2}z=22$$

$$\begin{cases} \dfrac{13}{6}x-\dfrac{5}{2}z=14 \\[2mm] \dfrac{3}{2}x-\dfrac{1}{2}z=22 \end{cases}$$

$R1 - 5R2 = Sum$

$$\dfrac{13}{6}x-\dfrac{5}{2}z=14$$

$$-\dfrac{15}{2}x+\dfrac{5}{2}z=-110$$

$$-\dfrac{16}{3}x=-96$$

$$x=18;$$

$$\dfrac{3}{2}x-\dfrac{1}{2}z=22$$

$$\dfrac{3}{2}(18)-\dfrac{1}{2}z=22$$

$$27-\dfrac{1}{2}z=22$$

$$-\dfrac{1}{2}z=-5$$

$$z=10;$$

$$\dfrac{2}{3}x-y-z=8$$

$$\dfrac{2}{3}(18)-y-10=8$$

$$12-y-10=8$$

$$2-y=8$$

$$-y=6$$

$$y=-6$$

$$(18,\ -6,\ 10)$$

47.
$$\begin{cases} -A+3B+2C=11 \\ 2B+C=9 \\ B+2C=8 \end{cases}$$

$R2 - 2R3$

$$2B+C=9$$

$$-2B-4C=-16$$

$$-3C=-7$$

$$C=\dfrac{7}{3};$$

$$2B+C=9$$

$$2B+\dfrac{7}{3}=9$$

$$2B=\dfrac{20}{3}$$

$$B=\dfrac{10}{3};$$

$$-A+3B+2C=11$$

$$-A+3\left(\dfrac{10}{3}\right)+2\left(\dfrac{7}{3}\right)=11$$

$$-A+10+\dfrac{14}{3}=11$$

$$-A+\dfrac{44}{3}=11$$

$$-A=-\dfrac{11}{3}$$

$$A=\dfrac{11}{3}$$

$$\left(\dfrac{11}{3},\dfrac{10}{3},\dfrac{7}{3}\right)$$

8.2 Exercises

49.
$$\begin{cases} A - 2B = 5 \\ B + 3C = 7 \\ 2A - B - C = 1 \end{cases}$$

$-2R1 + R3 = \text{Sum}$

$-2A + 4B = -10$

$-2A - B - C = 1$

$3B - C = -9$

$$\begin{cases} B + 3C = 7 \\ 3B - C = -9 \end{cases}$$

$-3R1 + R2 = \text{Sum}$

$-3B - 9C = -21$

$3B - C = -9$

$-10C = -30$

$C = 3;$

$B + 3C = 7$

$B + 3(3) = 7$

$B + 9 = 7$

$B = -2;$

$A - 2B = 5$

$A - 2(-2) = 5$

$A + 4 = 5$

$A = 1$

$(1, -2, 3)$

51.
$$\begin{cases} C = 3 \\ 2A + 3C = 10 \\ 3B - 4C = -11 \end{cases}$$

$2A + 3C = 10$

$2A + 3(3) = 10$

$2A + 9 = 10$

$2A = 1$

$A = \dfrac{1}{2};$

$3B - 4C = -11$

$3B - 4(3) = -11$

$3B - 12 = -11$

$3B = 1$

$B = \dfrac{1}{3}$

$\left(\dfrac{1}{2}, \dfrac{1}{3}, 3\right)$

53. $\left| \dfrac{Ax + By + Cz - D}{\sqrt{A^2 + B^2 + C^2}} \right|$

$A = 1, B = 1, C = 1, D = 6;$

$x = 3, y = 4, z = 5;$

$\left| \dfrac{1(3) + 1(4) + 1(5) - 6}{\sqrt{1^2 + 1^2 + 1^2}} \right| = \dfrac{6}{\sqrt{3}} \approx 3.464 \text{ units}$

55. Let M represent the amount paid for the Monet.

Let P represent the amount paid for the Picasso.

Let V represent the amount paid for the Van Gogh.

$$\begin{cases} M + P + V = 7 \\ M = P + 0.8 \\ V = 2M + 0.2 \end{cases}$$

$$\begin{cases} M + P + V = 7 \\ M - P = 0.8 \\ V - 2M = 0.2 \end{cases}$$

$R1 + R2 = \text{Sum}$

$M + P + V = 7$

$M - P = 0.8$

$2M + V = 7.8$

$$\begin{cases} -2M + V = 0.2 \\ 2M + V = 7.8 \end{cases}$$

$R1 + R2 = \text{Sum}$

$-2M + V = 0.2$

$2M + V = 7.8$

$2V = 8$

$V = 4;$

$V = 2M + 0.2$

$4 = 2M + 0.2$

$3.8 = 2M$

$1.9 = M;$

$M = P + 0.8$

$1.9 = P + 0.8$

$1.1 = P$

Monet: $1,900,000

Picasso: $1,100,000

Van Gogh: $4,000,000

57. Let c represent the gestation period of a camel.
Let e represent the gestation period of an elephant.
Let r represent the gestation period of a rhinoceros.

$$\begin{cases} c + e + r = 1520 \\ r = c + 58 \\ 2c - 162 = e \end{cases}$$

$$c + e + r = 1520$$
$$c + e + c + 58 = 1520$$
$$2c + e = 1462$$

$$\begin{cases} 2c - e = 162 \\ 2c + e = 1462 \end{cases}$$

R1 + R2 = Sum
$$2c - e = 162$$
$$2c + e = 1462$$
$$4c = 1624$$
$$c = 406;$$

$$r = c + 58$$
$$r = 406 + 58$$
$$r = 464;$$

$$2c - 162 = e$$
$$2(406) - 162 = e$$
$$812 - 162 = e$$
$$650 = e$$

Camel: 406 days
Elephant: 650 days
Rhinoceros: 464 days

59. Let x represent the wingspan of the California Condor.
Let y represent the wingspan of the Wandering Albatross.
Let z represent the wingspan of the prehistoric Quetzalcoatlus.

$$\begin{cases} x + y + z = 18.6 \\ z = 5y - 2x \\ 6x = 5y \end{cases}$$

$$\begin{cases} x + y + z = 18.6 \\ 2x - 5y + z = 0 \\ 6x - 5y = 0 \end{cases}$$

$-R1$
$$\begin{cases} -x - y - z = -18.6 \\ 2x - 5y + z = 0 \\ 6x - 5y = 0 \end{cases}$$

R1 + R2 yields $x - 6y = -18.6$

Sub-system

$$\begin{cases} x - 6y = -18.6 \\ 6x - 5y = 0 \end{cases}$$

$-6R1$
$$\begin{cases} -6x + 36y = 111.6 \\ 6x - 5y = 0 \end{cases}$$

R1 + R2
$$31y = 111.6$$
$$y = 3.6;$$

Solve for x in R3
$$6x = 5y$$
$$6x = 5(3.6)$$
$$x = 3;$$

$$x + y + z = 18.6$$
$$3.6 + 3 + z = 18.6$$
$$z = 12;$$

Albatross: 3.6 m
Condor: 3.0 m
Quetzalcoatlus: 12.0 m

61. Let f represent the number of \$5 gold pieces.
Let t represent the number of \$10 gold pieces.
Let w represent the number of \$20 gold pieces.

$$\begin{cases} f + t + w = 250 \\ 5f + 10t + 20w = 1875 \\ f = 7w \end{cases}$$

$-10R1 + R2 = $ Sum
$$-10f - 10t - 10w = -2500$$
$$5f + 10t + 20w = 1875$$
$$-5f + 10w = -625$$

$$\begin{cases} -5f + 10w = -625 \\ f - 7w = 0 \end{cases}$$

R1 + 5R2 = Sum
$$-5f + 10w = -625$$
$$5f - 35w = 0$$
$$-25w = -625$$
$$w = 25;$$

$$f = 7w$$
$$f = 7(25)$$
$$f = 175;$$

$$f + t + w = 250$$
$$175 + t + 25 = 250$$
$$200 + t = 250$$
$$t = 50$$

175 \$5 gold pieces
50 \$10 gold pieces
25 \$20 gold pieces

63. $\begin{cases} A + B = 0 \\ -6A - 3B + C = 1 \\ 9A = -9 \end{cases}$

Solve for A in R3

$9A = -9$

$A = -1;$

Solve for B in R1

$A + B = 0$

$-1 + B = 0$

$B = 1;$

Solve for C in R2

$-6A - 3B + C = 1$

$-6(-1) - 3(1) + C = 1$

$6 - 3 + C = 1$

$C = -2;$

$A = -1, B = 1, C = -2;$

$\dfrac{A}{x} + \dfrac{B}{x-3} + \dfrac{C}{(x-3)^2}$

$= \dfrac{-1}{x} + \dfrac{1}{x-3} - \dfrac{2}{(x-3)^2}$

$= \dfrac{-1(x-3)^2}{x(x-3)^2} + \dfrac{1x(x-3)}{x(x-3)^2} - \dfrac{2x}{x(x-3)^2}$

$= \dfrac{-x^2 + 6x - 9 + x^2 - 3x - 2x}{x(x-3)^2}$

$= \dfrac{x-9}{x(x-3)^2}$

Verified

65. $x^2 + y^2 + Dx + Ey + F = 0$

$\begin{cases} (2)^2 + (-1)^2 + D(2) + E(-1) + F = 0 \\ (4)^2 + (-3)^2 + D(4) + E(-3) + F = 0 \\ (2)^2 + (-5)^2 + D(2) + E(-5) + F = 0 \end{cases}$

$\begin{cases} 2D - E + F = -5 \\ 4D - 3E + F = -25 \\ 2D - 5E + F = -29 \end{cases}$

R1 + (−1)R3

$4E = 24$

$E = 6;$

R1 − R2

$-2D + 2E = 20;$

$-2D + 2(6) = 20$

$-2D = 8$

$D = -4;$

$2D - E + F = -5$

$2(-4) - 6 + F = -5$

$F = 9;$

$x^2 + y^2 - 4x + 6y + 9 = 0$

67. $\mathbf{u} = \langle 1, -7 \rangle, \mathbf{v} = \left\langle -3, \dfrac{1}{2} \right\rangle$

$\mathbf{u} + 4\mathbf{v} = \left\langle 1 + 4(-3), -7 + 4\left(\dfrac{1}{2}\right) \right\rangle$

$= \langle -11, -5 \rangle;$

$3\mathbf{u} - \mathbf{v} = \left\langle 3 - (-3), 3(-7) - \left(\dfrac{1}{2}\right) \right\rangle$

$= \left\langle 6, -\dfrac{43}{2} \right\rangle;$

69. $\log(x+2) + \log(x) = \log(3)$

$\log x(x+2) = \log 3$

$x^2 + 2x - 3 = 0$

$(x+3)(x-1) = 0$

$x = -3 \text{ or } x = 1$

$x = 1 \text{ since } x = -3 \text{ will not check.}$

8.3 Exercises

1. Template

3. Repeated linear

5. Answers will vary.

7. $\dfrac{3x+2}{(x+3)(x-2)}$

$= \dfrac{A}{x+3} + \dfrac{B}{x-2}$

9. $\dfrac{2x+5}{(x-1)^2}$

$= \dfrac{A}{x-1} + \dfrac{B}{(x-1)^2}$

11. $\dfrac{3x^2 - 2x + 5}{(x-1)(x+2)(x-3)}$

$= \dfrac{A}{x-1} + \dfrac{B}{x+2} + \dfrac{C}{x-3}$

13. $\dfrac{x^2 + 5}{x(x-3)(x+1)}$

$= \dfrac{A}{x} + \dfrac{B}{x-3} + \dfrac{C}{x+1}$

15. $\dfrac{x^2 + 2x - 4}{(x-5)^3}$

$= \dfrac{A}{x-5} + \dfrac{B}{(x-5)^2} + \dfrac{C}{(x-5)^3}$

17. $\dfrac{x^2 + x - 1}{x^2(x+2)}$

$= \dfrac{A}{x} + \dfrac{B}{x^2} + \dfrac{C}{x+2}$

19. $\dfrac{x^3 + 2x - 5}{x^2(x-5)^2}$

$= \dfrac{A}{x} + \dfrac{B}{x^2} + \dfrac{C}{x-5} + \dfrac{D}{(x-5)^2}$

21. $\dfrac{2x^2 + 3}{(x-3)(x^2 + 5x + 7)}$

$= \dfrac{A}{x-3} + \dfrac{Bx+C}{x^2 + 5x + 7}$

23. $\dfrac{x^3 + 3x - 2}{(x+1)(x^2 + 2)^2}$

$= \dfrac{A}{x+1} + \dfrac{Bx+C}{(x^2 + 2)} + \dfrac{Dx+E}{(x^2 + 2)^2}$

25. $\dfrac{2x - 27}{2x^2 + x - 15} = \dfrac{2x - 27}{(2x-5)(x+3)}$

$= \dfrac{A}{2x-5} + \dfrac{B}{x+3}$

$2x - 27 = A(x+3) + B(2x-5)$

$2x - 27 = Ax + 3A + B2x - 5B$

$2x - 27 = (A + 2B)x + 3A - 5B$

$\begin{cases} A + 2B = 2 \\ 3A - 5B = -27 \end{cases}$

$\begin{bmatrix} 1 & 2 \\ 3 & -5 \end{bmatrix} \begin{bmatrix} A \\ B \end{bmatrix} = \begin{bmatrix} 2 \\ -27 \end{bmatrix}$

$\begin{bmatrix} A \\ B \end{bmatrix} = \begin{bmatrix} -4 \\ 3 \end{bmatrix}$

$\dfrac{-4}{2x-5} + \dfrac{3}{x+3}$

27. $\dfrac{8x^2 - 3x - 7}{x^3 - x} = \dfrac{8x^2 - 3x - 7}{x(x^2 - 1)}$

$= \dfrac{8x^2 - 3x - 7}{x(x-1)(x+1)} = \dfrac{A}{x} + \dfrac{B}{x+1} + \dfrac{C}{x-1}$

$8x^2 - 3x - 7 = A(x+1)(x-1) + Bx(x-1) + Cx(x+1)$

$8x^2 - 3x - 7 = Ax^2 - A + Bx^2 - Bx + Cx^2 + Cx$

$8x^2 - 3x - 7 = x^2(A + B + C) + x(-B + C) - A$

$\begin{cases} A + B + C = 8 \\ -B + C = -3 \\ -A = -7 \end{cases}$

$\begin{bmatrix} 1 & 1 & 1 \\ 0 & -1 & 1 \\ -1 & 0 & 0 \end{bmatrix} \begin{bmatrix} A \\ B \\ C \end{bmatrix} = \begin{bmatrix} 8 \\ -3 \\ -7 \end{bmatrix}$

$\begin{bmatrix} A \\ B \\ C \end{bmatrix} = \begin{bmatrix} 7 \\ 2 \\ -1 \end{bmatrix}$

$\dfrac{7}{x} + \dfrac{2}{x+1} - \dfrac{1}{x-1}$

29. $\dfrac{3x^2+7x-1}{x^3+2x^2+x}=\dfrac{3x^2+7x-1}{x\left(x^2+2x+1\right)}$

$=\dfrac{3x^2+7x-1}{x(x+1)^2}=\dfrac{A}{x}+\dfrac{B}{x+1}+\dfrac{C}{(x+1)^2}$

$3x^2+7x-1=A(x+1)^2+Bx(x+1)+Cx$

$3x^2+7x-1=Ax^2+2Ax+A+Bx^2+Bx+Cx$

$3x^2+7x-1=x^2(A+B)+x(2A+B+C)+A$

$\begin{cases} A+B=3 \\ 2A+B+C=7 \\ A=-1 \end{cases}$

$\begin{bmatrix} 1 & 1 & 0 \\ 2 & 1 & 1 \\ 1 & 0 & 0 \end{bmatrix}\begin{bmatrix} A \\ B \\ C \end{bmatrix}=\begin{bmatrix} 3 \\ 7 \\ -1 \end{bmatrix}$

$\begin{bmatrix} A \\ B \\ C \end{bmatrix}=\begin{bmatrix} -1 \\ 4 \\ 5 \end{bmatrix}$

$\dfrac{-1}{x}+\dfrac{4}{x+1}+\dfrac{5}{(x+1)^2}$

31. $\dfrac{3x^3+3x^2+3x+5}{x^4+3x^2+2}=\dfrac{3x^3+3x^2+3x+5}{\left(x^2+1\right)\left(x^2+2\right)}$

$\dfrac{3x^3+3x^2+3x+5}{\left(x^2+1\right)\left(x^2+2\right)}=\dfrac{Ax+B}{\left(x^2+1\right)}+\dfrac{Cx+D}{\left(x^2+2\right)}$

$3x^3+3x^2+3x+5$
$=(Ax+B)\left(x^2+2\right)+(Cx+D)\left(x^2+1\right)$

$3x^3+3x^2+3x+5$
$=Ax^3+2Ax+Bx^2+2B+Cx^3+Cx+Dx^2+D$

$3x^3+3x^2+3x+5$
$=x^3(A+C)+x^2(B+D)+x(2A+C)+2B+D$

$\begin{cases} A+C=2 \\ B+D=-2 \\ 2A+C=6 \\ 2B+D=0 \end{cases}$

$\begin{bmatrix} 1 & 0 & 1 & 0 \\ 0 & 1 & 0 & 1 \\ 2 & 0 & 1 & 0 \\ 0 & 2 & 0 & 1 \end{bmatrix}\begin{bmatrix} A \\ B \\ C \\ D \end{bmatrix}=\begin{bmatrix} 3 \\ 3 \\ 3 \\ 5 \end{bmatrix}$

$\begin{bmatrix} A \\ B \\ C \\ D \end{bmatrix}=\begin{bmatrix} 0 \\ 2 \\ 3 \\ 1 \end{bmatrix}$

$\dfrac{3x^3+3x^2+3x+5}{\left(x^2+1\right)\left(x^2+2\right)}=\dfrac{2}{\left(x^2+1\right)}+\dfrac{3x+1}{\left(x^2+2\right)}$

33. $\dfrac{6x^2+x+13}{x^3+2x^2+3x+6}=\dfrac{6x^2+x+13}{(x+2)\left(x^2+3\right)}$

$=\dfrac{A}{x+2}+\dfrac{Bx+C}{x^2+3}$

$6x^2+x+13=A\left(x^2+3\right)+(Bx+C)(x+2)$

$6x^2+x+13$
$=Ax^2+3A+Bx^2+2Bx+Cx+2C$

$6x^2+x+13$
$=x^2(A+B)+x(2B+C)+3A+2C$

$\begin{cases} A+B=6 \\ 2B+C=1 \\ 3A+2C=13 \end{cases}$

$\begin{bmatrix} 1 & 1 & 0 \\ 0 & 2 & 1 \\ 3 & 0 & 2 \end{bmatrix}\begin{bmatrix} A \\ B \\ C \end{bmatrix}=\begin{bmatrix} 6 \\ 1 \\ 13 \end{bmatrix}$

$\begin{bmatrix} A \\ B \\ C \end{bmatrix}=\begin{bmatrix} 5 \\ 1 \\ -1 \end{bmatrix}$

$\dfrac{5}{x+2}+\dfrac{x-1}{x^2+3}$

35. $\dfrac{x^4-3x^2-2x+1}{x^5+2x^3+x} = \dfrac{x^4-3x^2-2x+1}{x\left(x^4+2x^2+1\right)}$

$= \dfrac{x^4-3x^2-2x+1}{x\left(x^2+1\right)^2} = \dfrac{A}{x}+\dfrac{Bx+C}{x^2+1}+\dfrac{Dx+E}{\left(x^2+1\right)^2}$

x^4-3x^2-2x+1

$= A\left(x^2+1\right)^2+x\left(Bx+C\right)\left(x^2+1\right)+x\left(Dx+E\right)$

$\quad x^4-3x^2-2x+1$

$\quad = Ax^4+2Ax^2+A+Bx^4+Bx^2$

$\quad +Cx^3+Cx+Dx^2+Ex$

x^4-3x^2-2x+1

$= x^4\left(A+B\right)+Cx^3+x^2\left(2A+B+D\right)$

$\quad +x\left(C+E\right)+A$

$\begin{cases} A+B=1 \\ C=0 \\ 2A+B+D=-3 \\ C+E=-2 \\ A=1 \end{cases}$

$\begin{bmatrix} 1 & 1 & 0 & 0 & 0 \\ 0 & 0 & 1 & 0 & 0 \\ 2 & 1 & 0 & 1 & 0 \\ 0 & 0 & 1 & 0 & 1 \\ 1 & 0 & 0 & 0 & 0 \end{bmatrix}\begin{bmatrix} A \\ B \\ C \\ D \\ E \end{bmatrix} = \begin{bmatrix} 1 \\ 0 \\ -3 \\ -2 \\ 1 \end{bmatrix}$

$\begin{bmatrix} A \\ B \\ C \\ D \\ E \end{bmatrix} = \begin{bmatrix} 1 \\ 0 \\ 0 \\ -5 \\ -2 \end{bmatrix}$

$\dfrac{1}{x}-\dfrac{5x+2}{\left(x^2-1\right)^2}$

37. $\dfrac{3x^3+2x^2+7x+3}{x^4+x^3+3x^2} = \dfrac{3x^3+2x^2+7x+3}{x^2\left(x^2+x+3\right)}$

$\quad \dfrac{3x^3+2x^2+7x+3}{x^2\left(x^2+x+3\right)}$

$= \dfrac{A}{x}+\dfrac{B}{x^2}+\dfrac{Cx+D}{\left(x^2+x+3\right)}$

$3x^3+2x^2+7x+3$

$= Ax(x^2+x+3)+B(x^2+x+3)+\left(Cx+D\right)x^2$

$3x^3+2x^2+7x+3$

$= Ax^3+Ax^2+3Ax+Bx^2+Bx+3B+Cx^3+Dx^2$

$3x^3+2x^2+7x+3$

$= x^3\left(A+C\right)+x^2\left(A+B+D\right)+x\left(3A+B\right)+3B$

$\begin{cases} A+C=3 \\ A+B+D=2 \\ 3A+B=7 \\ 3B=3 \end{cases}$

$\begin{bmatrix} 1 & 0 & 1 & 0 \\ 1 & 1 & 0 & 1 \\ 3 & 1 & 0 & 0 \\ 0 & 3 & 0 & 0 \end{bmatrix}\begin{bmatrix} A \\ B \\ C \\ D \end{bmatrix} = \begin{bmatrix} 3 \\ 2 \\ 7 \\ 3 \end{bmatrix}$

$\begin{bmatrix} A \\ B \\ C \\ D \\ E \end{bmatrix} = \begin{bmatrix} 2 \\ 1 \\ 1 \\ -1 \end{bmatrix}$

$= \dfrac{2}{x}+\dfrac{1}{x^2}+\dfrac{x-1}{\left(x^2+x+3\right)}$

39. $\dfrac{3x^2+10x+4}{8-x^3} = \dfrac{3x^2+10x+4}{\left(2-x\right)\left(4+2x+x^2\right)}$

$= \dfrac{A}{2-x}+\dfrac{Bx+C}{4+2x+x^2}$

$3x^2+10x+4 = A\left(4+2x+x^2\right)+\left(Bx+C\right)\left(2-x\right)$

$3x^2+10x+4$

$= 4A+2Ax+Ax^2+2Bx-Bx^2+2C-Cx$

$3x^2+10x+4$

$= x^2\left(A-B\right)+x\left(2A+2B-C\right)+4A+2C$

$\begin{cases} A-B=3 \\ 2A+2B-C=10 \\ 4A+2C=4 \end{cases}$

$\begin{bmatrix} 1 & -1 & 0 \\ 2 & 2 & -1 \\ 4 & 0 & 2 \end{bmatrix}\begin{bmatrix} A \\ B \\ C \end{bmatrix} = \begin{bmatrix} 3 \\ 10 \\ 4 \end{bmatrix}$

$\begin{bmatrix} A \\ B \\ C \end{bmatrix} = \begin{bmatrix} 3 \\ 0 \\ -4 \end{bmatrix}$

$\dfrac{3}{2-x}-\dfrac{4}{4+2x+x^2}$

41. $\dfrac{5x+13}{(x+3)^2} = \dfrac{A}{x+3} + \dfrac{B}{(x+3)^2}$

$5x+13 = A(x+3)+B$

$5x+13 = Ax+3A+B$

$\begin{cases} A=5 \\ 3A+B=13 \end{cases}$

$3(5)+B=13, B=-2$

$\dfrac{5}{x+3} + \dfrac{-2}{(x+3)^2}$

43. $\dfrac{2x^3+x^2+5x+1}{\left(x^2+1\right)^2} = \dfrac{Ax+B}{(x^2+1)} + \dfrac{Cx+D}{(x^2+1)^2}$

$2x^3+x^2+5x+1 = (Ax+B)(x^2+1)+Cx+D$

$2x^3+x^2+5x+1 = Ax^3+Ax+Bx^2+B+Cx+D$

$2x^3+x^2+5x+1 = Ax^3+Bx^2+x(A+C)+B+D$

$\begin{cases} A=2 \\ B=1 \\ A+C=5 \\ B+D=1 \end{cases}$

$C=3, D=0$

$\dfrac{2x+1}{(x^2+1)} + \dfrac{3x}{(x^2+1)^2}$

45. $\dfrac{2x^2-4x+5}{(x-1)^3}$

$= \dfrac{A}{(x-1)} + \dfrac{B}{(x-1)^2} + \dfrac{C}{(x-1)^3}$

$2x^2-4x+5 = A(x-1)^2+B(x-1)+C$

$2x^2-4x+5 = Ax^2-2Ax+A+Bx-B+C$

$2x^2-4x+5 = Ax^2+x(-2A+B)+A-B+C$

$\begin{cases} A=2 \\ -2A+B=-4 \\ A-B+C=5 \end{cases}$

$-2A+B=-4$

$-2(2)+B=-4$

$B=0;$

$A-B+C=5$

$2-0+C=5$

$C=3;$

$\dfrac{2}{(x-1)} + \dfrac{0}{(x-1)^2} + \dfrac{3}{(x-1)^3}$

$\dfrac{2}{(x-1)} + \dfrac{3}{(x-1)^3}$

47. $\dfrac{1}{P\left(100 - \dfrac{100}{10}P\right)}$

$= \dfrac{1}{P(100-10P)}$

$= \dfrac{A}{P} + \dfrac{B}{100-10P}$

$= \dfrac{A(100-10P)+BP}{P(100-10P)}$

$1 = 100A - 10PA + BP$

$1 = 100A + P(B-10A)$

$\begin{cases} 0 = B-10A \\ 1 = 100A \end{cases}$

$1 = 100A$

$\dfrac{1}{100} = A;$

$0 = B-10A$

$0 = B - 10\left(\dfrac{1}{100}\right)$

$B = \dfrac{1}{10}$

$\dfrac{1}{P(100-10P)} = \dfrac{\frac{1}{100}}{P} + \dfrac{\frac{1}{10}}{100-10P}$

49.

$$\frac{1}{P\left(10-\frac{10}{100}P\right)}$$

$$=\frac{1}{P\left(10-\frac{1}{10}P\right)}$$

$$=\frac{A}{P}+\frac{B}{10-\frac{1}{10}P}$$

$$=\frac{A\left(10-\frac{1}{10}P\right)+BP}{P\left(10-\frac{1}{10}P\right)}$$

$$1=10A-\frac{1}{10}PA+BP$$

$$1=10A+P\left(B-\frac{1}{10}A\right)$$

$$\begin{cases}0=B-\frac{1}{10}A\\1=10A\end{cases}$$

$$1=10A$$

$$\frac{1}{10}=A;$$

$$0=B-\frac{1}{10}A$$

$$0=B-\frac{1}{10}\left(\frac{1}{10}\right)$$

$$B=\frac{1}{100}$$

$$\frac{1}{P\left(10-\frac{1}{10}P\right)}=\frac{\frac{1}{10}}{P}+\frac{\frac{1}{100}}{10-\frac{1}{10}P}$$

51. $t(x)=\dfrac{1}{x^2+x}$

$$\frac{1}{x(x+1)}=\frac{A}{x}+\frac{B}{x+1}$$

$$=\frac{A(x+1)+Bx}{x(x+1)}$$

$$1=Ax+A+Bx$$

$$1=x(A+B)+A$$

$$\begin{cases}A+B=0\\A=1\end{cases}$$

$$A+B=0$$

$$1+B=0$$

$$B=-1;$$

$$\frac{1}{x(x+1)}=\frac{1}{x}+\frac{-1}{x+1}=\frac{1}{x}-\frac{1}{x+1};$$

Given Sum: $\dfrac{1}{1\cdot2}+\dfrac{1}{2\cdot3}+\dfrac{1}{3\cdot4}+...+\dfrac{1}{49\cdot50};$

Using decomposed form:

$$=\left(1-\frac{1}{2}\right)+\left(\frac{1}{2}-\frac{1}{3}\right)+\left(\frac{1}{3}-\frac{1}{4}\right)+...+\left(\frac{1}{49}-\frac{1}{50}\right)$$

All interior terms sum to zero

$$=1-\frac{1}{50}=\frac{49}{50}$$

53. $t(x)=\dfrac{1}{(2x-1)(2x+1)}$

$$\frac{1}{(2x-1)(2x+1)}=\frac{A}{2x-1}+\frac{B}{2x+1}$$

$$=\frac{A(2x+1)+B(2x-1)}{(2x-1)(2x+1)}$$

$$1=2Ax+A+2Bx-B$$

$$1=x(2A+2B)+(A-B)$$

$$\begin{cases}2A+2B=0\\A-B=1\end{cases}$$

$$\begin{cases}2A+2B=0\\2A-2B=2\end{cases}$$

$$4A=2$$

$$A=\frac{1}{2};$$

$$A-B=1$$

$$\frac{1}{2}-B=1$$

$$B=-\frac{1}{2}$$

$$\frac{1}{(2x-1)(2x+1)}=\frac{\frac{1}{2}}{2x-1}+\frac{-\frac{1}{2}}{2x+1}$$

$$=\frac{1}{2}\left(\frac{1}{2x-1}-\frac{1}{2x+1}\right);$$

Given sum: $\dfrac{1}{1\cdot3}+\dfrac{1}{3\cdot5}+\dfrac{1}{5\cdot7}+...+\dfrac{1}{123\cdot125}$

Using decomposed form:

$$=\frac{1}{2}\left[\left(\frac{1}{1}-\frac{1}{3}\right)+\left(\frac{1}{3}-\frac{1}{5}\right)+\left(\frac{1}{5}-\frac{1}{7}\right)+...+\left(\frac{1}{123}-\frac{1}{125}\right)\right]$$

All interior terms sum to zero

$$=\frac{1}{2}\left(\frac{1}{1}-\frac{1}{125}\right)=\frac{62}{125}$$

8.3 Exercises

55. $\dfrac{\ln x + 2}{(\ln x - 2)(\ln x - 1)^2}$

$= \dfrac{A}{(\ln x - 2)} + \dfrac{B}{(\ln x - 1)} + \dfrac{C}{(\ln x - 1)^2}$

$= \dfrac{A(\ln x - 1)^2 + B(\ln x - 2)(\ln x - 1) + C(\ln x - 2)}{(\ln x - 2)(\ln x - 1)^2}$

$\ln x + 2$

$= A(\ln x)^2 - 2A\ln x + A + B(\ln x)^2$
$\qquad - 3B\ln x + 2B + C\ln x - 2C$

$= (\ln x)^2 (A + B) + \ln x(-2A - 3B + C) + A$
$\qquad + 2B - 2C$

$\begin{cases} A + B = 0 \\ -2A - 3B + C = 1 \\ A + 2B - 2C = 2 \end{cases}$

$\begin{bmatrix} 1 & 1 & 0 \\ -2 & -3 & 1 \\ 1 & 2 & -2 \end{bmatrix} \begin{bmatrix} A \\ B \\ C \end{bmatrix} = \begin{bmatrix} 0 \\ 1 \\ 2 \end{bmatrix}$

$\begin{bmatrix} A \\ B \\ C \end{bmatrix} = \begin{bmatrix} 4 \\ -4 \\ -3 \end{bmatrix}$

$\dfrac{\ln x + 2}{(\ln x - 2)(\ln x - 1)^2}$

$= \dfrac{4}{(\ln x - 2)} - \dfrac{4}{(\ln x - 1)} - \dfrac{3}{(\ln x - 1)^2}$

57. $\dfrac{x+2}{(x-1)(1-x)} = \dfrac{x+2}{(x-1)(-1)(x-1)}$

$= \dfrac{x+2}{-1(x-1)^2} = \dfrac{-x-2}{(x-1)^2}$

$= \dfrac{A}{(x-1)} + \dfrac{B}{(x-1)^2}$

$= \dfrac{A(x-1) + B}{(x-1)^2}$

$-x - 2 = Ax - A + B$

$\begin{cases} A = -1 \\ -A + B = -2 \end{cases}$

$-(-1) + B = -2$

$B = -3;$

$\dfrac{-x-2}{(x-1)^2} = \dfrac{-1}{x-1} + \dfrac{-3}{(x-1)^2}$

59.

$$\begin{array}{r} 2x - 1 \\ x^2 - x + 6 \overline{\smash{)}2x^3 - 3x^2 + 13x - 5} \\ \underline{2x^3 - 2x^2 + 12x} \\ -1x^2 + x - 5 \\ \underline{-1x^2 + x - 6} \\ 1 \end{array}$$

61. $\dfrac{\cos^3 \theta}{\sin \theta} = \cot \theta - \cos \theta \sin \theta$

$= \dfrac{\cos \theta}{\sin \theta} - \cos \theta \sin \theta$

$= \dfrac{\cos \theta}{\sin \theta} - \dfrac{\cos \theta \sin^2 \theta}{\sin \theta}$

$= \dfrac{\cos \theta - \cos \theta \sin^2 \theta}{\sin \theta}$

$= \dfrac{\cos \theta (1 - \sin^2 \theta)}{\sin \theta}$

$= \dfrac{\cos \theta (\cos^2 \theta)}{\sin \theta}$

$= \dfrac{\cos^3 \theta}{\sin \theta}$

8.4 Technology Highlight:

Exercise 1:

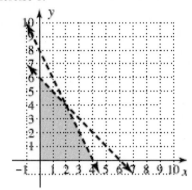

Exercise 3:

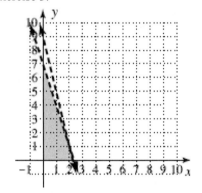

8.4 Exercises

1. Half planes

3. Solution

5. The feasible region may be bordered by three or more oblique lines, with two of them intersecting outside and away from the feasible region.

7. $2x + y > 3$

$\quad(0, 0)$ No
$$2x + y > 3$$
$$2(0) + 0 > 3$$
$$0 + 0 > 3$$
$$0 > 3;$$

$\quad(3, -5)$ No
$$2x + y > 3$$
$$2(3) + (-5) > 3$$
$$6 - 5 > 3$$
$$1 > 3;$$

$\quad(-3, -4)$ No
$$2x + y > 3$$
$$2(-3) + (-4) > 3$$
$$-6 - 4 > 3$$
$$-10 > 3;$$

$\quad(-3, 9)$ No
$$2x + y > 3$$
$$2(-3) + 9 > 3$$
$$-6 + 9 > 3$$
$$3 > 3$$

9. $4x - 2y \le -8$

$\quad(0, 0)$ No
$$4x - 2y \le -8$$
$$4(0) - 2(0) \le -8$$
$$0 - 0 \le -8$$
$$0 \le -8;$$

$\quad(-3, 5)$ Yes
$$4x - 2y \le -8$$
$$4(-3) - 2(5) \le -8$$
$$-12 - 10 \le -8$$
$$-22 \le -8;$$

$\quad(-3, -2)$ Yes
$$4x - 2y \le -8$$
$$4(-3) - 2(-2) \le -8$$
$$-12 + 4 \le -8$$
$$-8 \le -8;$$

$\quad(-1, 1)$ No
$$4x - 2y \le -8$$
$$4(-1) - 2(1) \le -8$$
$$-4 - 2 \le -8$$
$$-6 \le -8$$

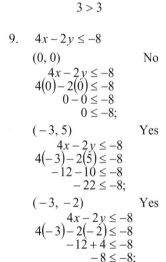

11. $x + 2y < 8$

$2y < -x + 8$

$y < -\dfrac{1}{2}x + 4$

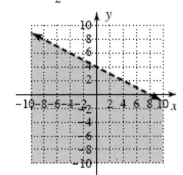

13. $2x - 3y \geq 9$

$-3y \geq -2x + 9$

$y \leq \dfrac{2}{3}x - 3$

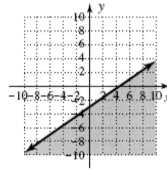

15. $\begin{cases} 5y - x \geq 10 \\ 5y + 2x \leq -5 \end{cases}$

$(-2, 1)$ No

$5y - x \geq 10$

$5(1) - (-2) \geq 10$

$5 + 2 \geq 10$

$7 \geq 10;$

$(-5, -4)$ No

$5y - x \geq 10$

$5(-4) - (-5) \geq 10$

$-20 + 5 \geq 10$

$-15 \geq 10;$

$(-6, 2)$ No

$5y - x \geq 10$

$5(2) - (-6) \geq 10$

$10 + 6 \geq 10$

$16 \geq 10;$

$5y + 2x \leq -5$

$5(2) + 2(-6) \leq -5$

$10 - 12 \leq -5$

$-2 \leq -5;$

$(-8, 2.2)$ Yes

$5y - x \geq 10$

$5(2.2) - (-8) \geq 10$

$11 + 8 \geq 10$

$19 \geq 10;$

$5y + 2x \leq -5$

$5(2.2) + 2(-8) \leq -5$

$11 - 16 \leq -5$

$-5 \leq -5$

17. $\begin{cases} x + 2y \geq 1 \\ 2x - y \leq -2 \end{cases}$

$\begin{cases} y \geq -\dfrac{1}{2}x + \dfrac{1}{2} \\ y \geq 2x + 2 \end{cases}$

Test Point: $(-1, 2)$

$x + 2y \geq 1$

$-1 + 2(2) \geq 1$

$-1 + 4 \geq 1$

$3 \geq 1;$

$2x - y \leq -2$

$2(-1) - 2 \leq -2$

$-2 - 2 \leq -2$

$-4 \leq -2$

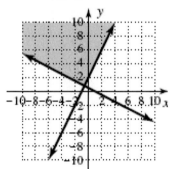

19. $\begin{cases} 3x + y > 4 \\ x > 2y \end{cases}$

$\begin{cases} y > -3x + 4 \\ y < \dfrac{1}{2}x \end{cases}$

Test Point: $(3, 0)$

$3x + y > 4$
$3(3) + 0 > 4$
$\qquad 9 > 4;$

$x > 2y$

$3 > 2(0)$

$3 > 0$

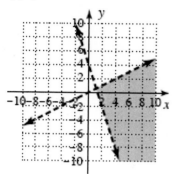

21. $\begin{cases} 2x + y < 4 \\ 2y > 3x + 6 \end{cases}$

$\begin{cases} y < -2x + 4 \\ y > \dfrac{3}{2}x + 3 \end{cases}$

Test Point: $(-3, 3)$

$2x + y < 4$
$2(-3) + 3 < 4$
$\quad -6 + 3 < 4$
$\qquad -3 < 4;$

$2y > 3x + 6$

$2(3) > 3(-3) + 6$

$\quad 6 > -9 + 6$

$\quad 6 > -3$

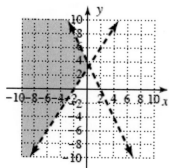

23. $\begin{cases} x > -3y - 2 \\ x + 3y \le 6 \end{cases}$

$\begin{cases} y > -\dfrac{1}{3}x - \dfrac{2}{3} \\ y \le -\dfrac{1}{3}x + 2 \end{cases}$

Test Point: $(0, 0)$

$x > -3y - 2$
$0 > -3(0) - 2$
$0 > -2;$

$x + 3y \le 6$

$0 + 3(0) \le 6$

$\qquad 0 \le 6$

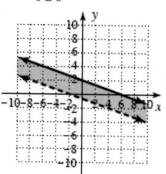

25. $\begin{cases} 5x + 4y \ge 20 \\ x - 1 \ge y \end{cases}$

$\begin{cases} y \ge -\dfrac{5}{4}x + 5 \\ y \le x - 1 \end{cases}$

Test Point: $(6, 0)$

$5x + 4y \ge 20$
$5(6) + 4(0) \ge 20$
$\qquad 30 \ge 20;$

$x - 1 \ge y$

$6 - 1 \ge 0$

$\qquad 5 \ge 0$

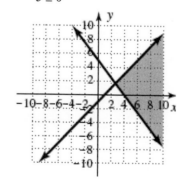

27. $\begin{cases} 0.2x > -0.3y - 1 \\ 0.3x + 0.5y \le 0.6 \end{cases}$

$\begin{cases} y > -\dfrac{2}{3}x - \dfrac{10}{3} \\ y \le -\dfrac{3}{5}x + \dfrac{6}{5} \end{cases}$

Test Point: $(0, 0)$

$0.2x > -0.3y - 1$
$0.2(0) > -0.3(0) - 1$
$0 > -1;$

$0.3x + 0.5y \le 0.6$
$0.3(0) + 0.5(0) \le 0.6$
$0 \le 0.6$

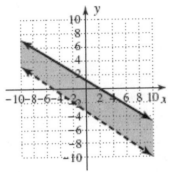

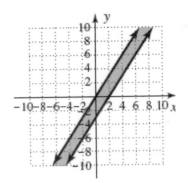

29. $\begin{cases} y \le \dfrac{3}{2}x \\ 4y \ge 6x - 12 \end{cases}$

$\begin{cases} y \le \dfrac{3}{2}x \\ y \ge \dfrac{3}{2}x - 3 \end{cases}$

Test Point: $(1, 0)$

$y \le \dfrac{3}{2}x$
$0 \le \dfrac{3}{2}(1)$
$0 \le \dfrac{3}{2};$

$4y \ge 6x - 12$
$4(0) \ge 6(1) - 12$
$0 \ge -6$

31. $\begin{cases} \dfrac{-2}{3}x + \dfrac{3}{4}y \le 1 \\ \dfrac{1}{2}x + 2y \ge 3 \end{cases}$

$\begin{cases} y \le \dfrac{8}{9}x + \dfrac{4}{3} \\ y \ge -\dfrac{1}{4}x + \dfrac{3}{2} \end{cases}$

Test Point: $(6, 4)$

$-\dfrac{2}{3}x + \dfrac{3}{4}y \le 1$

$-\dfrac{2}{3}(6) + \dfrac{3}{4}(4) \le 1$
$-4 + 3 \le 1$
$-1 \le 1;$

$\dfrac{1}{2}x + 2y \ge 3$

$\dfrac{1}{2}(6) + 2(4) \ge 3$

$3 + 8 \ge 3$
$11 \ge 3$

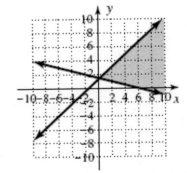

33. $\begin{cases} x - y \geq -4 \\ 2x + y \leq 4 \\ x \geq 1 \end{cases}$

$\begin{cases} y \leq x + 4 \\ y \leq -2x + 4 \\ x \geq 1 \end{cases}$

Test Point: (1.5, 0.5)

$x - y \geq -4$
$1.5 - 0.5 \geq -4$
$1 \geq -4;$

$2x + y \leq 4$
$2(1.5) + 0.5 \leq 4$
$3 + 0.5 \leq 4$
$3.5 \leq 4;$

$x \geq 1$
$1.5 \geq 1$

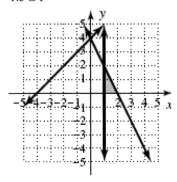

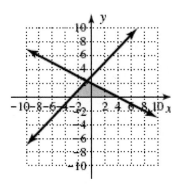

35. $\begin{cases} y \leq x + 3 \\ x + 2y \leq 4 \\ y \geq 0 \end{cases}$

$\begin{cases} y \leq x + 3 \\ y \leq -\dfrac{1}{2}x + 2 \\ y \geq 0 \end{cases}$

Test Point: (1, 1)

$y \leq x + 3$
$1 \leq 1 + 3$
$1 \leq 4;$

$x + 2y \leq 4$
$1 + 2(1) \leq 4$
$1 + 2 \leq 4$
$3 \leq 4;$

$y \geq 0$
$1 \geq 0$

37. $\begin{cases} 2x + 3y \leq 18 \\ x \geq 0 \\ y \geq 0 \end{cases}$

$\begin{cases} y \leq -\dfrac{2}{3}x + 6 \\ x \geq 0 \\ y \geq 0 \end{cases}$

Test Point: (2, 2)

$2x + 3y \leq 18$
$2(2) + 3(2) \leq 18$
$4 + 6 \leq 18$
$10 \leq 18;$

$x \geq 0$
$2 \geq 0;$

$y \geq 0$
$2 \geq 0$

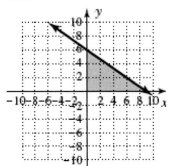

39. $\begin{cases} y - x \leq 1 \\ x + y > 3 \end{cases}$

41. $\begin{cases} y - x \leq 1 \\ x + y < 3 \\ y \geq 0 \end{cases}$

43.

Point	Objective Function $f(x,y)=12x+10y$	Result
(0, 0)	$f(0,0)=12(0)+10(0)$	0
(0, 8.5)	$f(0,8.5)=12(0)+10(8.5)$	85
(7, 0)	$f(7,0)=12(7)+10(0)$	84
(5, 3)	$f(5,3)=12(5)+10(3)$	90

Maximum value occurs at (5, 3).

45.

Point	Objective Function $f(x,y)=8x+15y$	Result
(0, 20)	$f(0,20)=8(0)+15(20)$	300
(35,0)	$f(35,0)=8(35)+15(0)$	280
(5, 15)	$f(5,15)=8(5)+15(15)$	265
(12,11)	$f(12,11)=8(12)+15(11)$	261

Minimum value occurs at (12, 11).

47. $\begin{cases} x+2y \le 6 \\ 3x+y \le 8 \\ x \ge 0 \\ y \ge 0 \end{cases}$

$\begin{cases} y \le -\frac{1}{2}x+3 \\ y \le -3x+8 \\ x \ge 0 \\ y \ge 0 \end{cases}$

Corner Point	Objective Function $f(x,y)=8x+5y$	Result
(0, 0)	$f(0,0)=8(0)+5(0)$	0
(0, 3)	$f(0,3)=8(0)+5(3)$	15
$\left(\frac{8}{3},0\right)$	$f\left(\frac{8}{3},0\right)=8\left(\frac{8}{3}\right)+5(0)$	$\frac{64}{3}$
(2, 2)	$f(2,2)=8(2)+5(2)$	26

Maximum value: (2, 2)

49. $\begin{cases} 3x+2y \ge 18 \\ 3x+4y \ge 24 \\ x \ge 0 \\ y \ge 0 \end{cases}$

$\begin{cases} y \ge -\frac{3}{2}x+9 \\ y \ge -\frac{3}{4}x+6 \\ x \ge 0 \\ y \ge 0 \end{cases}$

Corner Point	Objective Function $f(x,y)=36x+40y$	Result
(0, 9)	$f(0,9)=36(0)+40(9)$	360
(4, 3)	$f(4,3)=36(4)+40(3)$	264
(8, 0)	$f(8,0)=36(8)+40(0)$	288

Minimum value: (4, 3)

51. $\begin{cases} 20H < 200 \\ \frac{1}{2}(20)H > 50 \\ H > 0 \end{cases}$

$20H < 200$

$H < 10;$

$10H > 50$

$H > 5;$

$5 < H < 10$

53. Let J represent the amount of money given to Julius.
Let A represent the amount of money given to Anthony.
$$\begin{cases} J + A \leq 50000 \\ J \geq 20000 \\ A \leq 25000 \end{cases}$$

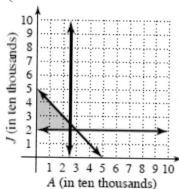

55. Let C represent the number of acres of corn.
Let S represent the number of acres of soybeans.
$$\begin{cases} C + S \leq 500 \\ 3C + 2S \leq 1300 \end{cases}$$
$$P = 900C + 800S$$
$$\begin{cases} S \leq -C + 500 \\ 2S \leq -3C + 1300 \end{cases}$$
$$\begin{cases} S \leq -C + 500 \\ S \leq \dfrac{-3}{2}C + 650 \end{cases}$$
Using a grapher, the corner points are:
$$(0, 500), \left(433\frac{1}{3}, 0\right), (300, 200)$$
$$P = 900(0) + 800(500) = 400,000;$$
$$P = 900\left(433\frac{1}{3}\right) + 800(0) = 390,000;$$
$$P = 900(300) + 800(200) = 430,000;$$
300 acres of corn, 200 acres of soybeans

57. Let x represent the number of sheet metal screws.
Let y represent the number of wood screws.
$$\begin{cases} 20x + 5y \leq 3(60)(60) \\ 15x + 15y \leq 3(60)(60) \\ 5x + 20y \leq 3(60)(60) \end{cases}$$
$$R = 0.10x + 0.12y$$
$$\begin{cases} 5y \leq -20x + 10800 \\ 15y \leq -15x + 10800 \\ 20y \leq -5x + 10800 \end{cases}$$
$$\begin{cases} y \leq -4x + 2160 \\ y \leq -1x + 720 \\ y \leq \dfrac{-1}{4}x + 540 \end{cases}$$
Using a grapher, the corner points are:
$$(0, 540), (240, 480), (480, 240), (540, 0)$$
$$R = 0.10(0) + 0.12(540) = 64.80;$$
$$R = 0.10(240) + 0.12(480) = 81.60;$$
$$R = 0.10(480) + 0.12(240) = 76.80;$$
$$R = 0.10(540) + 0.12(0) = 54;$$
240 sheet metal screws; 480 wood screws

59. Let t represent the number of ounces of traditional sandwiches.
Let d represent the number of ounces of Double-T's.
$$\begin{cases} 2t + 4d \leq 250 \\ 3t + 5d \leq 345 \\ t \geq 0 \\ d \geq 0 \end{cases}$$
$$R = 2t + 3.50d$$
$$\begin{cases} 2t \leq -4d + 250 \\ 3t \leq -5d + 345 \end{cases}$$
$$\begin{cases} t \leq -2d + 125 \\ t \leq -\dfrac{5}{3}d + 115 \end{cases}$$
Using a grapher, the pt of intersection is $(30, 65)$.
$$R = 2(65) + 3.50(30) = 235$$
65 traditionals, 30 Double-T's.

61. Let A represent the number of thousand gallons shipped from OK to CO.
    Let B represent the number of thousand gallons shipped from OK to MS.
    Let C represent the number of thousand gallons shipped from TX to CO.
    Let D represent the number of thousand gallons shipped from TX to MS.
    $\text{Cost} = 0.05A + 0.075C + 0.06B + 0.065D$

    $\begin{cases} A + C = 220 \Rightarrow C = 220 - A \\ B + D = 250 \Rightarrow D = 250 - B \end{cases}$

    Cost
    $= 0.05A + 0.075(220 - A) + 0.06B + 0.065(250 - B)$
    $= 0.05A + 16.5 - 0.75A + 0.06B + 16.25 - 0.065B$
    $= -0.25A - 0.005B + 32.75$;

    $A + B \le 320;$

    $C + D \le 240$
    but $C = 220 - A$ and $D = 250 - B$
    Thus,
    $220 - A + 250 - B \le 240$
    $-A - B \le -230$
    $A + B \ge 230;$

    $A \ge 0, B \ge 0, C \ge 0, D \ge 0$
    $220 - A \ge 0, 250 - B \ge 0$
    $A \le 220, B \le 250;$

    $\begin{cases} A + B \le 320 \\ A + B \ge 230 \\ A \le 220 \\ B \le 250 \end{cases}$

    $\begin{cases} B \le -A + 320 \\ B \ge -A + 230 \\ A \le 220 \\ B \le 250 \end{cases}$

    Using a grapher, the corner points are:
    $(220,100), (220,10), (70,250), (0,250)$

    $\text{Cost} = -0.7(220) - 0.005(100) + 32.75 = -121.75$

    $\text{Cost} = -0.7(220) - 0.005(10) + 32.75 = -121.30;$

    $\text{Cost} = -0.7(70) - 0.005(250) + 32.75 = -17.5;$

    $\text{Cost} = -0.7(0) - 0.005(250) + 32.75 = 31.5$

    $A = 220, B = 100,$

    $C = 220 - A = 0, D = 250 - B = 150;$

    220,000 gallons from OK to CO,
    100,000 gallons from OK to MS,
    0 thousand gallons from TX to CO,
    150,000 gallons from TX to MS

63. $\begin{cases} x \ge 0 \\ y \ge 0 \\ y \le 3 \\ x \le 3 \end{cases}$

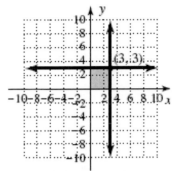

The graph is a rectangle.

Corner Point	Objective Function $f(x,y) = 4.5x + 7.2y$	Result
(0, 0)	$f(0,0) = 4.5(0) + 7.2(0)$	0
(0, 3)	$f(0,3) = 4.5(0) + 7.2(3)$	21.6
(3, 3)	$f(3,3) = 4.5(3) + 7.2(3)$	35.1
(3, 0)	$f(3,0) = 4.5(3) + 7.2(0)$	13.5

Maximum value: (3, 3)
Optimal solutions occur at vertices.

65. $\sqrt{3^2 + 4^2} = 5$;

$\cos\theta = -\dfrac{3}{5}; \csc\theta = \dfrac{5}{4}; \cot\theta = -\dfrac{3}{4}$

67. $r = \dfrac{kl}{d^2}$

$1500 = \dfrac{k(8)}{(0.004)^2}$

$0.024 = 8k$

$0.003 = k$

$r = \dfrac{0.003l}{d^2}$

$r = \dfrac{0.003(2.7)}{(0.005)^2}$

$r = \dfrac{0.0081}{0.000025}$

$r = 324 \ \Omega$

Chapter 8 Mid-Chapter Check

1. $\begin{cases} x - 3y = -2 \\ 2x + y = 3 \end{cases}$

$x = 3y - 2$

$2x + y = 3$

$2(3y - 2) + y = 3$

$6y - 4 + y = 3$

$7y = 7$

$y = 1$

$x - 3y = -2$

$x - 3(1) = -2$

$x - 3 = -2$

$x = 1$

$(1, 1)$; Consistent

3. Let x represent the amount of 40% acid.
   Let y represent the amount of 48% acid.

$\begin{cases} x + 10 = y \\ 0.40x + 0.64(10) = 0.48y \end{cases}$

$\begin{cases} x + 10 = y \\ 40x + 64(10) = 48y \end{cases}$

$40x + 640 = 48(x + 10)$

$40x + 640 = 48x + 480$

$-8x = -160$

$x = 20$

20 ounces

5. $\begin{cases} x + 2y - 3z = 3 \\ 2x + 4y - 6z = 6 \\ x - 2y + 5z = -1 \end{cases}$

The second equation is a multiple of the first equation.

7. $\begin{cases} 2x + 3y - 4z = -4 \\ x - 2y + z = 0 \\ -3x - 2y + 2z = -1 \end{cases}$

R1 + 4R2 = Sum

$2x + 3y - 4z = -4$

$4x - 8y + 4z = 0$

$6x - 5y = -4$

R1 + 2R3 = Sum

$2x + 3y - 4z = -4$

$-6x - 4y + 4z = -2$

$-4x - y = -6$

$\begin{cases} 6x - 5y = -4 \\ -4x - y = -6 \end{cases}$

R1 − 5R2 = Sum

$6x - 5y = -4$

$20x + 5y = 30$

$26x = 26$

$x = 1;$

$-4x - y = -6$

$-4(1) - y = -6$

$-4 - y = -6$

$-y = -2$

$y = 2;$

$x - 2y + z = 0$

$1 - 2(2) + z = 0$

$1 - 4 + z = 0$

$-3 + z = 0$

$z = 3$

$(1, 2, 3)$

9. Let x represent Mozart's age.
   Let y represent Morphy's age.
   Let z represent Pascal's age.

   $$\begin{cases} x + y + z = 37 \\ y = 2x - 3 \\ z = y + 3 \end{cases}$$

   $$\begin{cases} x + y + z = 37 \\ -2x + y = -3 \\ -y + z = 3 \end{cases}$$

   -R1 + R2 = Sum
   $-x - y - z = -37$
   $\underline{\quad -2x + y = -3}$
   $-3x - z = -40$
   R1 + R2 = Sum
   $x + y + z = 37$
   $\underline{\quad -y + z = 3}$
   $x + 2z = 40$

   $$\begin{cases} -3x - z = -40 \\ x + 2z = 40 \end{cases}$$

   2R1 + R2 = Sum
   $-6x - 2z = -80$
   $\underline{\quad x + 2z = 40}$
   $-5x = -40$
   $x = 8;$

   $y = 2x - 3$
   $y = 2(8) - 3$
   $y = 16 - 3$
   $y = 13;$

   $z = y + 3$
   $z = 13 + 3$
   $z = 16$

   Mozart: 8 years
   Morphy: 13 years
   Pascal: 16 years

Reinforcing Basic Concepts

1. $$\begin{cases} 15.3R + 35.7P = 211.14 \\ P = R + 0.10 \end{cases}$$
   Premium: $4.17/gal
   Regular: $4.07/gal

8.5 Technology Highlight

Exercise 1: (10, 12)

8.5 Exercises

1. Square

3. 2 by 3, 1

5. Multiply R_1 by –2 and add that result to R_2.
   This sum will be the new R_2.

7. $\begin{bmatrix} 1 & 0 \\ 2.1 & 1 \\ -3 & 5.8 \end{bmatrix}$
   $3 \times 2,\ 5.8$

9.
   $\begin{bmatrix} 1 & 0 & 4 \\ 1 & 3 & -7 \\ 5 & -1 & 2 \\ 2 & -3 & 9 \end{bmatrix}$
   $4 \times 3,\ -1$

11. $\begin{cases} x + 2y - z = 1 \\ x + z = 3 \\ 2x - y + z = 3 \end{cases}$
    $\begin{bmatrix} 1 & 2 & -1 & \vdots & 1 \\ 1 & 0 & 1 & \vdots & 3 \\ 2 & -1 & 1 & \vdots & 3 \end{bmatrix}$
    Diagonal entries 1, 0, 1

13. $\begin{bmatrix} 1 & 4 & \bigg| & 5 \\ 0 & 1 & \bigg| & \frac{1}{2} \end{bmatrix}$
    $\begin{cases} x + 4y = 5 \\ y = \dfrac{1}{2} \end{cases}$
    $x + 4y = 5$
    $x + 4\left(\dfrac{1}{2}\right) = 5$
    $x + 2 = 5$
    $x = 3$
    $\left(3, \dfrac{1}{2}\right)$

15. $\begin{bmatrix} 1 & 2 & -1 & \big| & 0 \\ 0 & 1 & 2 & \big| & 2 \\ 0 & 0 & 1 & \big| & 3 \end{bmatrix}$
    $\begin{cases} x + 2y - z = 0 \\ y + 2z = 2 \\ z = 3 \end{cases}$
    $y + 2z = 2$
    $y + 2(3) = 2$
    $y + 6 = 2$
    $y = -4;$
    $x + 2y - z = 0$
    $x + 2(-4) - (3) = 0$
    $x - 8 - 3 = 0$
    $x - 11 = 0$
    $x = 11$
    $(11,\ -4,\ 3)$

17. $\begin{bmatrix} 1 & 3 & -4 & \bigg| & 29 \\ 0 & 1 & -\frac{3}{2} & \bigg| & \frac{21}{2} \\ 0 & 0 & 1 & \bigg| & 3 \end{bmatrix}$
    $\begin{cases} x + 3y - 4z = 29 \\ y - \dfrac{3}{2}z = \dfrac{21}{2} \\ z = 3 \end{cases}$
    $y - \dfrac{3}{2}z = \dfrac{21}{2}$
    $y - \dfrac{3}{2}(3) = \dfrac{21}{2}$
    $y - \dfrac{9}{2} = \dfrac{21}{2}$
    $y = \dfrac{30}{2}$
    $y = 15;$
    $x + 3y - 4z = 29$
    $x + 3(15) - 4(3) = 29$
    $x + 45 - 12 = 29$
    $x + 33 = 29$
    $x = -4$
    $(-4,\ 15,\ 3)$

8.5 Exercises

19. $\begin{bmatrix} \frac{1}{2} & -3 & | & -1 \\ -5 & 2 & | & 4 \end{bmatrix} 2R1 \rightarrow R1$

$\begin{bmatrix} 1 & -6 & | & -2 \\ -5 & 2 & | & 4 \end{bmatrix} 5R1 + R2 \rightarrow R2$

$\begin{bmatrix} 1 & -6 & | & -2 \\ 0 & -28 & | & -6 \end{bmatrix}$

21. $\begin{bmatrix} -2 & 1 & 0 & | & 4 \\ 5 & 8 & 3 & | & -5 \\ 1 & -3 & 3 & | & 2 \end{bmatrix} R1 \leftrightarrow R3$

$\begin{bmatrix} 1 & -3 & 3 & | & 2 \\ 5 & 8 & 3 & | & -5 \\ -2 & 1 & 0 & | & 4 \end{bmatrix} -5R1 + R2 \rightarrow R2$

$\begin{bmatrix} 1 & -3 & 3 & | & 2 \\ 0 & 23 & -12 & | & -15 \\ -2 & 1 & 0 & | & 4 \end{bmatrix}$

23. $\begin{bmatrix} 3 & 1 & 1 & | & 8 \\ 6 & -1 & -1 & | & 10 \\ 4 & -2 & -3 & | & 22 \end{bmatrix} -2R1 + R2 \rightarrow R2$

$\begin{bmatrix} 3 & 1 & 1 & | & 8 \\ 0 & -3 & -3 & | & -6 \\ 4 & -2 & -3 & | & 34 \end{bmatrix} -4R1 + 3R3 \rightarrow R3$

$\begin{bmatrix} 3 & 1 & 1 & | & 8 \\ 0 & -3 & -3 & | & -6 \\ 0 & -10 & -13 & | & 34 \end{bmatrix}$

25. $\begin{bmatrix} 1 & 3 & 0 & | & 2 \\ -2 & 4 & 1 & | & 1 \\ 3 & -1 & -2 & | & 9 \end{bmatrix}$

$2R1 + R2 \rightarrow R2$

$-3R1 + R3 \rightarrow R3$

27. $\begin{bmatrix} 1 & 2 & 0 & | & 10 \\ 5 & 1 & 2 & | & 6 \\ -4 & 3 & -3 & | & 2 \end{bmatrix}$

$-5R1 + R2 \rightarrow R2$

$4R1 + R3 \rightarrow R3$

29. $\begin{cases} 0.15g - 0.35h = -0.5 \\ -0.12g + 0.25h = 0.1 \end{cases}$

$\begin{cases} 15g - 35h = -50 \\ -12g + 25h = 10 \end{cases}$

$\begin{bmatrix} 15 & -35 & | & -50 \\ -12 & 25 & | & 10 \end{bmatrix} \frac{1}{15}R1 \rightarrow R1$

$\begin{bmatrix} 1 & -\frac{7}{3} & | & -\frac{10}{3} \\ -12 & 25 & | & 10 \end{bmatrix} 12R1 + R2 \rightarrow R2$

$\begin{bmatrix} 1 & -\frac{7}{3} & | & -\frac{10}{3} \\ 0 & -3 & | & -30 \end{bmatrix} -\frac{1}{3}R2 \rightarrow R2$

$\begin{bmatrix} 1 & -\frac{7}{3} & | & -\frac{10}{3} \\ 0 & 1 & | & 10 \end{bmatrix}$

$h = 10;$

$0.15g - 0.35h = -0.5$

$0.15g - 0.35(10) = -0.5$

$0.15g - 3.5 = -0.5$

$0.15g = 3$

$g = 20$

$(20, 10)$

31. $\begin{cases} x - 2y + 2z = 7 \\ 2x + 2y - z = 5 \\ 3x - y + z = 6 \end{cases}$

$\begin{bmatrix} 1 & -2 & 2 & 7 \\ 2 & 2 & -1 & 5 \\ 3 & -1 & 1 & 6 \end{bmatrix}$ $-2R1 + R2 \rightarrow R2$

$\begin{bmatrix} 1 & -2 & 2 & 7 \\ 0 & 6 & -5 & -9 \\ 3 & -1 & 1 & 6 \end{bmatrix}$ $-3R1 + R3 \rightarrow R3$

$\begin{bmatrix} 1 & -2 & 2 & 7 \\ 0 & 6 & -5 & -9 \\ 0 & 5 & -5 & -15 \end{bmatrix}$ $-\dfrac{5}{6}R2 + R3 \rightarrow R3$

$\begin{bmatrix} 1 & -2 & 2 & 7 \\ 0 & 6 & -5 & -9 \\ 0 & 0 & -\dfrac{5}{6} & -\dfrac{15}{2} \end{bmatrix}$ $-\dfrac{6}{5}R3 \rightarrow R3$

$\begin{bmatrix} 1 & -2 & 2 & 7 \\ 0 & 6 & -5 & -9 \\ 0 & 0 & 1 & 9 \end{bmatrix}$

$z = 9$;

$6y - 5z = -9$

$6y - 5(9) = -9$

$6y - 45 = -9$

$6y = 36$

$y = 6$;

$x - 2y + 2z = 7$

$x - 2(6) + 2(9) = 7$

$x - 12 + 18 = 7$

$x + 6 = 7$

$x = 1$

$(1, 6, 9)$

33. $\begin{cases} x + 2y - z = 1 \\ x + z = 3 \\ 2x - y + z = 3 \end{cases}$

$\begin{bmatrix} 1 & 2 & -1 & 1 \\ 1 & 0 & 1 & 3 \\ 2 & -1 & 1 & 3 \end{bmatrix}$ $-R1 + R2 \rightarrow R2$

$\begin{bmatrix} 1 & 2 & -1 & 1 \\ 0 & -2 & 2 & 2 \\ 2 & -1 & 1 & 3 \end{bmatrix}$ $-2R1 + R3 \rightarrow R3$

$\begin{bmatrix} 1 & 2 & -1 & 1 \\ 0 & -2 & 2 & 2 \\ 0 & -5 & 3 & 1 \end{bmatrix}$ $-\dfrac{5}{2}R2 + R3 \rightarrow R3$

$\begin{bmatrix} 1 & 2 & -1 & 1 \\ 0 & -2 & 2 & 2 \\ 0 & 0 & -2 & -4 \end{bmatrix}$ $-\dfrac{1}{2}R3 \rightarrow R3$

$\begin{bmatrix} 1 & 2 & -1 & 1 \\ 0 & -2 & 2 & 2 \\ 0 & 0 & 1 & 2 \end{bmatrix}$

$z = 2$;

$x + z = 3$

$x + 2 = 3$

$x = 1$;

$x + 2y - z = 1$

$1 + 2y - 2 = 1$

$2y - 1 = 1$

$2y = 2$

$y = 1$

$(1, 1, 2)$

35. $\begin{cases} -x+y+2z=2 \\ x+y-z=1 \\ 2x+y+z=4 \end{cases}$

$\begin{bmatrix} -1 & 1 & 2 & | & 2 \\ 1 & 1 & -1 & | & 1 \\ 2 & 1 & 1 & | & 4 \end{bmatrix}$ $R1 \leftrightarrow R2$

$\begin{bmatrix} 1 & 1 & -1 & | & 1 \\ -1 & 1 & 2 & | & 2 \\ 2 & 1 & 1 & | & 4 \end{bmatrix}$ $R1+R2 \to R2$

$\begin{bmatrix} 1 & 1 & -1 & | & 1 \\ 0 & 2 & 1 & | & 3 \\ 2 & 1 & 1 & | & 4 \end{bmatrix}$ $-2R1+R3 \to R3$

$\begin{bmatrix} 1 & 1 & -1 & | & 1 \\ 0 & 2 & 1 & | & 3 \\ 0 & -1 & 3 & | & 2 \end{bmatrix}$ $\frac{1}{2}R2+R3 \to R3$

$\begin{bmatrix} 1 & 1 & -1 & | & 1 \\ 0 & 2 & 1 & | & 3 \\ 0 & 0 & \frac{7}{2} & | & \frac{7}{2} \end{bmatrix}$ $\frac{2}{7}R3 \to R3$

$\begin{bmatrix} 1 & 1 & -1 & | & 1 \\ 0 & 2 & 1 & | & 3 \\ 0 & 0 & 1 & | & 1 \end{bmatrix}$

$z=1;$
$2y+z=3$
$2y+1=3$
$2y=2$
$y=1;$
$x+y-z=1$
$x+1-1=1$
$x=1$
$(1,1,1)$

37. $\begin{cases} 4x-8y+8z=24 \\ 2x-6y+3z=13 \\ 3x+4y-z=-11 \end{cases}$

$\begin{bmatrix} 4 & -8 & 8 & | & 24 \\ 2 & -6 & 3 & | & 13 \\ 3 & 4 & -1 & | & -11 \end{bmatrix}$ $\frac{1}{4}R1 \to R1$

$\begin{bmatrix} 1 & -2 & 2 & | & 6 \\ 2 & -6 & 3 & | & 13 \\ 3 & 4 & -1 & | & -11 \end{bmatrix}$ $-2R1+R2 \to R2$

$\begin{bmatrix} 1 & -2 & 2 & | & 6 \\ 0 & -2 & -1 & | & 1 \\ 3 & 4 & -1 & | & -11 \end{bmatrix}$ $-3R1+R3 \to R3$

$\begin{bmatrix} 1 & -2 & 2 & | & 6 \\ 0 & -2 & -1 & | & 1 \\ 0 & 10 & -7 & | & -29 \end{bmatrix}$ $5R2+R3 \to R3$

$\begin{bmatrix} 1 & -2 & 2 & | & 6 \\ 0 & -2 & -1 & | & 1 \\ 0 & 0 & -12 & | & -24 \end{bmatrix}$ $-\frac{1}{12}R3 \to R3$

$\begin{bmatrix} 1 & -2 & 2 & | & 6 \\ 0 & -2 & -1 & | & 1 \\ 0 & 0 & 1 & | & 2 \end{bmatrix}$

$z=2;$
$-2y-z=1$
$-2y-2=1$
$-2y=3$
$y=-\frac{3}{2};$
$3x+4y-z=-11$
$3x+4\left(-\frac{3}{2}\right)-2=-11$
$3x-6-2=-11$
$3x-8=-11$
$3x=-3$
$x=-1$
$\left(-1, \frac{-3}{2}, 2\right)$

39. $\begin{cases} x+3y+5z=20 \\ 2x+3y+4z=16 \\ x+2y+3z=12 \end{cases}$

$\begin{bmatrix} 1 & 3 & 5 & | & 20 \\ 2 & 3 & 4 & | & 16 \\ 1 & 2 & 3 & | & 12 \end{bmatrix}$ $R2 - R1 \to R2$

$\begin{bmatrix} 1 & 3 & 5 & | & 20 \\ 1 & 0 & -1 & | & -4 \\ 1 & 2 & 3 & | & 12 \end{bmatrix}$ $R3 - R1 \to R3$

$\begin{bmatrix} 1 & 3 & 5 & | & 20 \\ 1 & 0 & -1 & | & -4 \\ 0 & -1 & -2 & | & -8 \end{bmatrix}$ $R2 \leftrightarrow R3$

$\begin{bmatrix} 1 & 3 & 5 & | & 20 \\ 0 & -1 & -2 & | & -8 \\ 1 & 0 & -1 & | & -4 \end{bmatrix}$ $3R2 + R1 \to R1$

$\begin{bmatrix} 1 & 0 & -1 & | & -4 \\ 0 & -1 & -2 & | & -8 \\ 1 & 0 & -1 & | & -4 \end{bmatrix}$ $-R1 + R3 \to R3$

$\begin{bmatrix} 1 & 0 & -1 & | & -4 \\ 0 & -1 & -2 & | & -8 \\ 0 & 0 & 0 & | & 0 \end{bmatrix}$

$x - z = -4$

$\quad x = z - 4;$

$-y - 2z = -8$

$-y = 2z - 8$

$y = -2z + 8;$

Linear dependence; $(p-4, -2p+8, p)$

41. $\begin{cases} 3x - 4y + 2z = -2 \\ \dfrac{3}{2}x - 2y + z = -1 \\ -6x + 8y - 4z = 4 \end{cases}$

$\begin{bmatrix} 3 & -4 & 2 & | & -2 \\ \dfrac{3}{2} & -2 & 1 & | & -1 \\ -6 & 8 & -4 & | & 4 \end{bmatrix}$ $\dfrac{1}{3}R1 \to R1$

$\begin{bmatrix} 1 & -\dfrac{4}{3} & \dfrac{2}{3} & | & -\dfrac{2}{3} \\ \dfrac{3}{2} & -2 & 1 & | & -1 \\ -6 & 8 & -4 & | & 4 \end{bmatrix}$ $-\dfrac{3}{2}R1 + R2 \to R2$

$\begin{bmatrix} 1 & -\dfrac{4}{3} & \dfrac{2}{3} & | & -\dfrac{2}{3} \\ 0 & 0 & 0 & | & 0 \\ -6 & 8 & -4 & | & 4 \end{bmatrix}$ $6R1 + R3 \to R3$

$\begin{bmatrix} 1 & -\dfrac{4}{3} & \dfrac{2}{3} & | & -\dfrac{2}{3} \\ 0 & 0 & 0 & | & 0 \\ 0 & 0 & 0 & | & 0 \end{bmatrix}$ $3R1 \to R1$

$\begin{bmatrix} 3 & -4 & 2 & | & -2 \\ 0 & 0 & 0 & | & 0 \\ 0 & 0 & 0 & | & 0 \end{bmatrix}$

Coincident dependence;

$\{(x, y, z) | 3x - 4y + 2z = -2\}$

43. $\begin{cases} 2x - y + 3z = 1 \\ 2y + 6z = 2 \\ x - \dfrac{1}{2}y + \dfrac{3}{2}z = 5 \end{cases}$

In terms of z:

$\begin{bmatrix} 2 & -1 & 3 & | & 1 \\ 0 & 2 & 6 & | & 2 \\ 1 & -\dfrac{1}{2} & \dfrac{3}{2} & | & 5 \end{bmatrix}$ $R1 \leftrightarrow R3$

$\begin{bmatrix} 1 & -\dfrac{1}{2} & \dfrac{3}{2} & | & 5 \\ 0 & 2 & 6 & | & 2 \\ 2 & -1 & 3 & | & 1 \end{bmatrix}$ $-2R1 + R3 \to R3$

$\begin{bmatrix} 1 & -\dfrac{1}{2} & \dfrac{3}{2} & | & 5 \\ 0 & 2 & 6 & | & 2 \\ 0 & 0 & 0 & | & -9 \end{bmatrix}$

$0 \neq -9$

No solution

45. $\begin{cases} -2x+4y-3z = 4 \\ 5x-6y+7z = -12 \\ x+2y+z = -4 \end{cases}$

In terms of z:

$\begin{bmatrix} -2 & 4 & -3 & | & 4 \\ 5 & -6 & 7 & | & -12 \\ 1 & 2 & 1 & | & -4 \end{bmatrix} R1 \leftrightarrow R3$

$\begin{bmatrix} 1 & 2 & 1 & | & -4 \\ 5 & -6 & 7 & | & -12 \\ -2 & 4 & -3 & | & 4 \end{bmatrix} -5R1+R2 \leftrightarrow R2$

$\begin{bmatrix} 1 & 2 & 1 & | & -4 \\ 0 & -16 & 2 & | & 8 \\ -2 & 4 & -3 & | & 4 \end{bmatrix} 2R1+R3 \rightarrow R3$

$\begin{bmatrix} 1 & 2 & 1 & | & -4 \\ 0 & -16 & 2 & | & 8 \\ 0 & 8 & -1 & | & -4 \end{bmatrix} -\frac{1}{16}R2 \rightarrow R2$

$\begin{bmatrix} 1 & 2 & 1 & | & -4 \\ 0 & 1 & -\frac{1}{8} & | & -\frac{1}{2} \\ 0 & 8 & -1 & | & -4 \end{bmatrix} -8R2+R3 \rightarrow R3$

$\begin{bmatrix} 1 & 2 & 1 & | & -4 \\ 0 & 1 & -\frac{1}{8} & | & -\frac{1}{2} \\ 0 & 0 & 0 & | & 0 \end{bmatrix} -2R2+R1 \rightarrow R1$

$\begin{bmatrix} 1 & 0 & \frac{5}{4} & | & -3 \\ 0 & 1 & -\frac{1}{8} & | & -\frac{1}{2} \\ 0 & 0 & 0 & | & 0 \end{bmatrix}$

$x + \frac{5}{4}z = -3$

$\qquad x = -\frac{5}{4}z - 3;$

$y - \frac{1}{8}z = -\frac{1}{2}$

$\qquad y = \frac{1}{8}z - \frac{1}{2}$

$\left(-\frac{5}{4}p - 3, \frac{1}{8}p - \frac{1}{2}, p \right)$

47.

$A = \pm\frac{1}{2}\left(x_1 y_2 - x_2 y_1 + x_2 y_3 - x_3 y_2 + x_3 y_1 - x_1 y_3 \right)$

$(6, -2), (-5, 4), (-1, 7)$

$A = \pm\frac{1}{2}\left(6(4) - (-5)(-2) + (-5)(7) - (-1)(4) + (-1)(-2) - 6(7) \right)$

$= \pm\frac{1}{2}\left(24 - 10 - 35 + 4 + 2 - 42 \right)$

$= \pm\frac{1}{2}(-57)$

$= 28.5$

28.5 units^2

49. Let x represent the Heat's score.
Let y represent the Maverick's score.

$\begin{cases} x - y = 3 \\ x + y = 187 \end{cases}$

$\begin{bmatrix} 1 & -1 & | & 3 \\ 1 & 1 & | & 187 \end{bmatrix} -R1+R2 \rightarrow R2$

$\begin{bmatrix} 1 & -1 & | & 1 \\ 0 & 2 & | & 184 \end{bmatrix} \frac{1}{2}R2 \rightarrow R2$

$\begin{bmatrix} 1 & -1 & | & 1 \\ 0 & 1 & | & 92 \end{bmatrix}$

$y = 92 ;$

$x - y = 3$

$x - 92 = 3$

$\qquad x = 95$

Heat: 95, Mavericks: 92

51. Let x represent Poe's book.
Let y represent Baum's book.
Let z represent Wouk's book.

$$\begin{cases} x+y+z=100000 \\ x+2z=y \\ z=2x \end{cases}$$

$$\begin{cases} x+y+z=100000 \\ x-y+2z=0 \\ -2x+z=0 \end{cases}$$

$$\begin{bmatrix} 1 & 1 & 1 & | & 100000 \\ 1 & -1 & 2 & | & 0 \\ -2 & 0 & 1 & | & 0 \end{bmatrix} \quad -R1+R2 \rightarrow R2$$

$$\begin{bmatrix} 1 & 1 & 1 & | & 100000 \\ 0 & -2 & 1 & | & -100000 \\ -2 & 0 & 1 & | & 0 \end{bmatrix} \quad 2R1+R3 \rightarrow R3$$

$$\begin{bmatrix} 1 & 1 & 1 & | & 100000 \\ 0 & -2 & 1 & | & -100000 \\ 0 & 2 & 3 & | & 200000 \end{bmatrix} \quad -\frac{1}{2}R2 \rightarrow R2$$

$$\begin{bmatrix} 1 & 1 & 1 & | & 100000 \\ 0 & 1 & -\frac{1}{2} & | & 50000 \\ 0 & 2 & 3 & | & 200000 \end{bmatrix} \quad -2R2+R3 \rightarrow R3$$

$$\begin{bmatrix} 1 & 1 & 1 & | & 100000 \\ 0 & 1 & -\frac{1}{2} & | & 50000 \\ 0 & 0 & 4 & | & 100000 \end{bmatrix} \quad \frac{1}{4}R3 \rightarrow R3$$

$$\begin{bmatrix} 1 & 1 & 1 & | & 100000 \\ 0 & 1 & -\frac{1}{2} & | & 50000 \\ 0 & 0 & 1 & | & 25000 \end{bmatrix}$$

$z=25000$;

$$y-\frac{1}{2}z=50000$$

$$y-\frac{1}{2}(25000)=50000$$

$$y-12500=50000$$

$$y=62500;$$

$$x+y+z=100000$$

$$x+62500+25000=100000$$

$$x+87500=100000$$

$$x=12500$$

Poe: \$12,500
Baum: \$62,500
Wouk: \$25,000

53. Let A represent the measure of angle A.
Let B represent the measure of angle B.
Let C represent the measure of angle C.

$$\begin{cases} A+B+C=180 \\ A+C=3B \\ C=2B+10 \end{cases}$$

$$\begin{cases} A+B+C=180 \\ -A+3B-C=0 \\ -2B+C=10 \end{cases}$$

$$\begin{bmatrix} 1 & 1 & 1 & | & 180 \\ -1 & 3 & -1 & | & 0 \\ 0 & -2 & 1 & | & 10 \end{bmatrix} \quad R1+R2 \rightarrow R2$$

$$\begin{bmatrix} 1 & 1 & 1 & | & 180 \\ 0 & 4 & 0 & | & 180 \\ 0 & -2 & 1 & | & 10 \end{bmatrix} \quad \frac{1}{2}R2+R3 \rightarrow R3$$

$$\begin{bmatrix} 1 & 1 & 1 & | & 180 \\ 0 & 4 & 0 & | & 180 \\ 0 & 0 & 1 & | & 100 \end{bmatrix}$$

$C=100$;

$$4B=180$$

$$B=45;$$

$$A+B+C=180$$

$$A+45+100=180$$

$$A+145=180$$

$$A=35$$

$$A=35°, B=45°, C=100°$$

8.5 Exercises

55. Let x represent the amount of money invested in the 4% savings fund.

Let y represent the amount of money invested in the 7% money market.

Let z represent the amount of money invested in the 8% government bonds.

$$\begin{cases} x+y+z=2.5 \\ 0.04x+0.07y+0.08z=0.178 \\ z=2y+0.3 \end{cases}$$

$$\begin{cases} x+y+z=2.5 \\ 40x+70y+80z=178 \\ -20y+10z=3 \end{cases}$$

$$\begin{bmatrix} 1 & 1 & 1 & 2.5 \\ 40 & 70 & 80 & 178 \\ 0 & -20 & 10 & 3 \end{bmatrix} \quad -40R1+R2 \rightarrow R2$$

$$\begin{bmatrix} 1 & 1 & 1 & 2.5 \\ 0 & 30 & 40 & 78 \\ 0 & -20 & 10 & 3 \end{bmatrix} \quad \frac{1}{30}R2 \rightarrow R2$$

$$\begin{bmatrix} 1 & 1 & 1 & 2.5 \\ 0 & 1 & \frac{4}{3} & 2.6 \\ 0 & -20 & 10 & 3 \end{bmatrix} \quad 20R2+R3 \rightarrow R3$$

$$\begin{bmatrix} 1 & 1 & 1 & 2.5 \\ 0 & 1 & \frac{4}{3} & 2.6 \\ 0 & 0 & \frac{110}{3} & 55 \end{bmatrix} \quad \frac{3}{110}R3 \rightarrow R3$$

$$\begin{bmatrix} 1 & 1 & 1 & 2.5 \\ 0 & 1 & \frac{4}{3} & 2.6 \\ 0 & 0 & 1 & 1.5 \end{bmatrix}$$

$z=1.5$;

$$y+\frac{4}{3}z=2.6$$

$$y+\frac{4}{3}(1.5)=2.6$$

$$y+2=2.6$$

$$y=0.6;$$

$$x+y+z=2.5$$

$$x+0.6+1.5=2.5$$

$$x+2.1=2.5$$

$$x=0.4$$

$0.4 million at 4%

$0.6 million at 7%

$1.5 million at 8%

57. $\begin{cases} x+y=180-71 \\ x-59=y \end{cases}$

$\begin{cases} x+y=109 \\ x-y=59 \end{cases}$

$$\begin{bmatrix} 1 & 1 & 109 \\ 1 & -1 & 59 \end{bmatrix} \quad -R1+R2 \rightarrow R2$$

$$\begin{bmatrix} 1 & 1 & 109 \\ 0 & -2 & -50 \end{bmatrix} \quad -\frac{1}{2}R2 \rightarrow R2$$

$$\begin{bmatrix} 1 & 1 & 109 \\ 0 & 1 & 25 \end{bmatrix}$$

$y=25$;

$$x+y=109$$

$$x+25=109$$

$$x=84$$

$x=84°; y=25°$

59. a. $z_1=-1-3i$ to trig form

$$r=\sqrt{(-1)^2+(-3)^2}=\sqrt{1+9}=\sqrt{10}$$

$$\theta=\tan^{-1}\left(\frac{3}{1}\right)+\pi$$

$$z_1=\sqrt{10}\ \text{cis}\left[\pi+\tan^{-1}(3)\right]$$

b. $z_2=5\text{cis}\left(\frac{2\pi}{3}\right)$

$$z_2=5\left[\cos\left(\frac{2\pi}{3}\right)+i\sin\left(\frac{2\pi}{3}\right)\right]$$

$$z_2=5\left[-\frac{1}{2}+i\left(\frac{\sqrt{3}}{2}\right)\right]$$

$$z_2=-\frac{5}{2}+\frac{5\sqrt{3}}{2}i$$

61. $C(t)=15\ln(t+1)$

$$30=15\ln(t+1)$$

$$2=\ln(t+1)$$

$$e^2=e^{\ln(t+1)}$$

$$e^2=t+1$$

$$e^2-1=t$$

$$t\approx6.39$$

$C>30,000$ in the year 2011.

8.6 Exercises

1. $a_{ij} = b_{ij}$

3. Scalar

5. Answers will vary.

7. $\begin{bmatrix} 1 & -3 \\ 5 & -7 \end{bmatrix}$

2×2, $a_{12} = -3$, $a_{21} = 5$

9. $\begin{bmatrix} 2 & -3 & 0.5 \\ 0 & 5 & 6 \end{bmatrix}$

2×3, $a_{12} = -3$, $a_{23} = 6$, $a_{22} = 5$

11. $\begin{bmatrix} -2 & 1 & -7 \\ 0 & 8 & 1 \\ 5 & -1 & 4 \end{bmatrix}$

3×3, $a_{12} = 1$, $a_{23} = 1$, $a_{31} = 5$

13. $\begin{bmatrix} \sqrt{1} & \sqrt{4} & \sqrt{8} \\ \sqrt{16} & \sqrt{32} & \sqrt{64} \end{bmatrix} = \begin{bmatrix} 1 & 2 & 2\sqrt{2} \\ 4 & 4\sqrt{2} & 8 \end{bmatrix}$

True.

15. $\begin{bmatrix} -2 & 3 & a \\ 2b & -5 & 4 \\ 0 & -9 & 3c \end{bmatrix} = \begin{bmatrix} c & 3 & -4 \\ 6 & -5 & -a \\ 0 & -3b & -6 \end{bmatrix}$

Conditional, $c = -2$, $a = -4$, $b = 3$

17. $A + H$

$= \begin{bmatrix} 2 & 3 \\ 5 & 8 \end{bmatrix} + \begin{bmatrix} 8 & -3 \\ -5 & 2 \end{bmatrix}$

$= \begin{bmatrix} 2+8 & 3+(-3) \\ 5+(-5) & 8+2 \end{bmatrix}$

$= \begin{bmatrix} 10 & 0 \\ 0 & 10 \end{bmatrix}$

19. $F + H$

$= \begin{bmatrix} 6 & -3 & 9 \\ 12 & 0 & -6 \end{bmatrix} + \begin{bmatrix} 8 & -3 \\ -5 & 2 \end{bmatrix}$

Not possible, different order.

21. $3H - 2A$

$= 3 \begin{bmatrix} 8 & -3 \\ -5 & 2 \end{bmatrix} - 2 \begin{bmatrix} 2 & 3 \\ 5 & 8 \end{bmatrix}$

$= \begin{bmatrix} 24 & -9 \\ -15 & 6 \end{bmatrix} - \begin{bmatrix} 4 & 6 \\ 10 & 16 \end{bmatrix}$

$= \begin{bmatrix} 24-4 & -9-6 \\ -15-10 & 6-16 \end{bmatrix}$

$= \begin{bmatrix} 20 & -15 \\ -25 & -10 \end{bmatrix}$

23. $\frac{1}{2}E - 3D$

$= \frac{1}{2} \begin{bmatrix} 1 & -2 & 0 \\ 0 & -1 & 2 \\ 4 & 3 & -6 \end{bmatrix} - 3 \begin{bmatrix} 1 & 0 & 0 \\ 0 & 1 & 0 \\ 0 & 0 & 1 \end{bmatrix}$

$= \begin{bmatrix} \frac{1}{2} & -1 & 0 \\ 0 & -\frac{1}{2} & 1 \\ 2 & \frac{3}{2} & -3 \end{bmatrix} - \begin{bmatrix} 3 & 0 & 0 \\ 0 & 3 & 0 \\ 0 & 0 & 3 \end{bmatrix}$

$= \begin{bmatrix} \frac{1}{2}-3 & -1-0 & 0-0 \\ 0-0 & -\frac{1}{2}-3 & 1-0 \\ 2-0 & \frac{3}{2}-0 & -3-3 \end{bmatrix}$

$= \begin{bmatrix} \frac{-5}{2} & -1 & 0 \\ 0 & \frac{-7}{2} & 1 \\ 2 & \frac{3}{2} & -6 \end{bmatrix}$

25. ED

$= \begin{bmatrix} 1 & -2 & 0 \\ 0 & -1 & 2 \\ 4 & 3 & -6 \end{bmatrix} \begin{bmatrix} 1 & 0 & 0 \\ 0 & 1 & 0 \\ 0 & 0 & 1 \end{bmatrix}$

$= \begin{bmatrix} 1+0+0 & 0+(-2)+0 & 0+0+0 \\ 0+0+0 & 0+(-1)+0 & 0+0+2 \\ 4+0+0 & 0+3+0 & 0+0+(-6) \end{bmatrix}$

$= \begin{bmatrix} 1 & -2 & 0 \\ 0 & -1 & 2 \\ 4 & 3 & -6 \end{bmatrix}$

27. *AH*

$$= \begin{bmatrix} 2 & 3 \\ 5 & 8 \end{bmatrix} \begin{bmatrix} 8 & -3 \\ -5 & 2 \end{bmatrix}$$

$$= \begin{bmatrix} 2(8)+3(-5) & 2(-3)+3(2) \\ 5(8)+8(-5) & 5(-3)+8(2) \end{bmatrix}$$

$$= \begin{bmatrix} 16-15 & -6+6 \\ 40-40 & -15+16 \end{bmatrix}$$

$$= \begin{bmatrix} 1 & 0 \\ 0 & 1 \end{bmatrix}$$

29. *FD*

$$= \begin{bmatrix} 6 & -3 & 9 \\ 12 & 0 & -6 \end{bmatrix} \begin{bmatrix} 1 & 0 & 0 \\ 0 & 1 & 0 \\ 0 & 0 & 1 \end{bmatrix}$$

$$= \begin{bmatrix} 6+0+0 & 0+(-3)+0 & 0+0+9 \\ 12+0+0 & 0+0+0 & 0+0+(-6) \end{bmatrix}$$

$$= \begin{bmatrix} 6 & -3 & 9 \\ 12 & 0 & -6 \end{bmatrix}$$

31. *HF*

$$= \begin{bmatrix} 8 & -3 \\ -5 & 2 \end{bmatrix} \begin{bmatrix} 6 & -3 & 9 \\ 12 & 0 & -6 \end{bmatrix}$$

$$= \begin{bmatrix} 8(6)+(-3)(12) & 8(-3)+(-3)(0) & 8(9)+(-3)(-6) \\ -5(6)+2(12) & -5(-3)+2(0) & -5(9)+2(-6) \end{bmatrix}$$

$$= \begin{bmatrix} 48-36 & -24+0 & 72+18 \\ -30+24 & 15+0 & -45-12 \end{bmatrix}$$

$$= \begin{bmatrix} 12 & -24 & 90 \\ -6 & 15 & -57 \end{bmatrix}$$

33. *H²*

$$= \begin{bmatrix} 8 & -3 \\ -5 & 2 \end{bmatrix} \begin{bmatrix} 8 & -3 \\ -5 & 2 \end{bmatrix}$$

$$= \begin{bmatrix} 8(8)+(-3)(-5) & 8(-3)+(-3)(2) \\ -5(8)+2(-5) & -5(-3)+2(2) \end{bmatrix}$$

$$= \begin{bmatrix} 64+15 & -24-6 \\ -40-10 & 15+4 \end{bmatrix}$$

$$= \begin{bmatrix} 79 & -30 \\ -50 & 19 \end{bmatrix}$$

35. *FE*

$$= \begin{bmatrix} 6 & -3 & 9 \\ 12 & 0 & -6 \end{bmatrix} \begin{bmatrix} 1 & -2 & 0 \\ 0 & -1 & 2 \\ 4 & 3 & -6 \end{bmatrix}$$

$$= \begin{bmatrix} 6+0+36 & -12+3+27 & 0+-6-54 \\ 12+0-24 & -24+0-18 & 0+0+36 \end{bmatrix}$$

$$= \begin{bmatrix} 42 & 18 & -60 \\ -12 & -42 & 36 \end{bmatrix}$$

37. *C + H*

$$= \begin{bmatrix} \dfrac{\sqrt{3}}{2} & \dfrac{\sqrt{3}}{3} \\ \sqrt{3} & 2\sqrt{3} \end{bmatrix} + \begin{bmatrix} -\dfrac{3}{19} & \dfrac{4}{57} \\ \dfrac{1}{19} & \dfrac{5}{57} \end{bmatrix}$$

$$= \begin{bmatrix} \dfrac{\sqrt{3}}{2}+\left(-\dfrac{3}{19}\right) & \dfrac{\sqrt{3}}{3}+\dfrac{4}{57} \\ \sqrt{3}+\dfrac{1}{19} & 2\sqrt{3}+\dfrac{5}{57} \end{bmatrix}$$

$$\approx \begin{bmatrix} 0.71 & 0.65 \\ 1.78 & 3.55 \end{bmatrix}$$

39. *E + G*

$$= \begin{bmatrix} 1 & -2 & 0 \\ 0 & -1 & 2 \\ 4 & 3 & -6 \end{bmatrix} + \begin{bmatrix} 0 & \dfrac{3}{4} & \dfrac{1}{4} \\ -\dfrac{1}{2} & \dfrac{3}{8} & \dfrac{1}{8} \\ -\dfrac{1}{4} & \dfrac{11}{16} & \dfrac{1}{16} \end{bmatrix}$$

$$= \begin{bmatrix} 1+0 & -2+\dfrac{3}{4} & 0+\dfrac{1}{4} \\ 0+\left(-\dfrac{1}{2}\right) & -1+\dfrac{3}{8} & 2+\dfrac{1}{8} \\ 4+\left(-\dfrac{1}{4}\right) & 3+\dfrac{11}{16} & -6+\dfrac{1}{16} \end{bmatrix}$$

$$\approx \begin{bmatrix} 1 & -1.25 & 0.25 \\ -0.5 & -0.63 & 2.13 \\ 3.75 & 3.69 & -5.94 \end{bmatrix}$$

41. AH

$$= \begin{bmatrix} -5 & 4 \\ 3 & 9 \end{bmatrix} \begin{bmatrix} -\dfrac{3}{19} & \dfrac{4}{57} \\ \dfrac{1}{19} & \dfrac{5}{57} \end{bmatrix}$$

$$= \begin{bmatrix} -5\left(-\dfrac{3}{19}\right)+4\left(\dfrac{1}{19}\right) & -5\left(\dfrac{4}{57}\right)+4\left(\dfrac{5}{57}\right) \\ 3\left(-\dfrac{3}{19}\right)+9\left(\dfrac{1}{19}\right) & 3\left(\dfrac{4}{57}\right)+9\left(\dfrac{5}{57}\right) \end{bmatrix}$$

$$= \begin{bmatrix} 1 & 0 \\ 0 & 1 \end{bmatrix}$$

43. EG

$$= \begin{bmatrix} 1 & -2 & 0 \\ 0 & -1 & 2 \\ 4 & 3 & -6 \end{bmatrix} \begin{bmatrix} 0 & \dfrac{3}{4} & \dfrac{1}{4} \\ -\dfrac{1}{2} & \dfrac{3}{8} & \dfrac{1}{8} \\ -\dfrac{1}{4} & \dfrac{11}{16} & \dfrac{1}{16} \end{bmatrix}$$

$$= \begin{bmatrix} 0+1+0 & \dfrac{3}{4}-\dfrac{3}{4}+0 & \dfrac{1}{4}-\dfrac{1}{4}+0 \\ 0+\dfrac{1}{2}-\dfrac{1}{2} & 0-\dfrac{3}{8}+\dfrac{11}{8} & 0-\dfrac{1}{8}+\dfrac{1}{8} \\ 0-\dfrac{3}{2}+\dfrac{3}{2} & 3+\dfrac{9}{8}-\dfrac{33}{8} & 1+\dfrac{3}{8}-\dfrac{3}{8} \end{bmatrix}$$

$$= \begin{bmatrix} 1 & 0 & 0 \\ 0 & 1 & 0 \\ 0 & 0 & 1 \end{bmatrix}$$

45. HB

$$= \begin{bmatrix} -\dfrac{3}{19} & \dfrac{4}{57} \\ \dfrac{1}{19} & \dfrac{5}{57} \end{bmatrix} \begin{bmatrix} 1 & 0 \\ 0 & 1 \end{bmatrix}$$

$$= \begin{bmatrix} -\dfrac{3}{19} & \dfrac{4}{57} \\ \dfrac{1}{19} & \dfrac{5}{57} \end{bmatrix}$$

47. DG

$$= \begin{bmatrix} 1 & 0 & 0 \\ 0 & 1 & 0 \\ 0 & 0 & 1 \end{bmatrix} \begin{bmatrix} 0 & \dfrac{3}{4} & \dfrac{1}{4} \\ -\dfrac{1}{2} & \dfrac{3}{8} & \dfrac{1}{8} \\ -\dfrac{1}{4} & \dfrac{11}{16} & \dfrac{1}{16} \end{bmatrix}$$

$$= \begin{bmatrix} 0 & \dfrac{3}{4} & \dfrac{1}{4} \\ \dfrac{-1}{2} & \dfrac{3}{8} & \dfrac{1}{8} \\ \dfrac{-1}{4} & \dfrac{11}{16} & \dfrac{1}{16} \end{bmatrix}$$

49. C^2

$$= \begin{bmatrix} \dfrac{\sqrt{3}}{2} & \dfrac{\sqrt{3}}{3} \\ \sqrt{3} & 2\sqrt{3} \end{bmatrix} \begin{bmatrix} \dfrac{\sqrt{3}}{2} & \dfrac{\sqrt{3}}{3} \\ \sqrt{3} & 2\sqrt{3} \end{bmatrix}$$

$$= \begin{bmatrix} \dfrac{3}{4}+1 & \dfrac{1}{2}+2 \\ \dfrac{3}{2}+6 & 1+12 \end{bmatrix}$$

$$= \begin{bmatrix} 1.75 & 2.5 \\ 7.5 & 13 \end{bmatrix}$$

51. FG

$$= \begin{bmatrix} -0.52 & 0.002 & 1.032 \\ 1.021 & -1.27 & 0.019 \end{bmatrix} \begin{bmatrix} 0 & \dfrac{3}{4} & \dfrac{1}{4} \\ -\dfrac{1}{2} & \dfrac{3}{8} & \dfrac{1}{8} \\ -\dfrac{1}{4} & \dfrac{11}{16} & \dfrac{1}{16} \end{bmatrix}$$

$$\approx \begin{bmatrix} -0.26 & 0.32 & -0.07 \\ 0.63 & 0.30 & 0.10 \end{bmatrix}$$

8.6 Exercises

53. (a) $AB \neq BA$

$$AB = \begin{bmatrix} -1 & 3 & 5 \\ 2 & 7 & -1 \\ 4 & 0 & 6 \end{bmatrix} \begin{bmatrix} 0.3 & -0.4 & 1.2 \\ -2.5 & 2 & 0.9 \\ 1 & -0.5 & 0.2 \end{bmatrix}$$

$$= \begin{bmatrix} -2.8 & 3.9 & 2.5 \\ -17.9 & 13.7 & 8.5 \\ 7.2 & -4.6 & 6 \end{bmatrix}$$

$$BA = \begin{bmatrix} 0.3 & -0.4 & 1.2 \\ -2.5 & 2 & 0.9 \\ 1 & -0.5 & 0.2 \end{bmatrix} \begin{bmatrix} -1 & 3 & 5 \\ 2 & 7 & -1 \\ 4 & 0 & 6 \end{bmatrix}$$

$$= \begin{bmatrix} 3.7 & -1.9 & 9.1 \\ 10.1 & 6.5 & -9.1 \\ -1.2 & -0.5 & 6.7 \end{bmatrix}$$

$AB \neq BA$; Verified

(b) $AC \neq CA$

$$AC = \begin{bmatrix} -1 & 3 & 5 \\ 2 & 7 & -1 \\ 4 & 0 & 6 \end{bmatrix} \begin{bmatrix} 45 & -1 & 3 \\ -6 & 10 & -15 \\ 21 & -28 & 36 \end{bmatrix}$$

$$= \begin{bmatrix} 42 & -109 & 132 \\ 27 & 96 & -135 \\ 306 & -172 & 228 \end{bmatrix}$$

$$CA = \begin{bmatrix} 45 & -1 & 3 \\ -6 & 10 & -15 \\ 21 & -28 & 36 \end{bmatrix} \begin{bmatrix} -1 & 3 & 5 \\ 2 & 7 & -1 \\ 4 & 0 & 6 \end{bmatrix}$$

$$= \begin{bmatrix} -35 & 128 & 244 \\ -34 & 52 & -130 \\ 67 & -133 & 349 \end{bmatrix}$$

$AC \neq CA$; Verified

(c) $BC \neq CB$

BC

$$= \begin{bmatrix} 0.3 & -0.4 & 1.2 \\ -2.5 & 2 & 0.9 \\ 1 & -0.5 & 0.2 \end{bmatrix} \begin{bmatrix} 45 & -1 & 3 \\ -6 & 10 & -15 \\ 21 & -28 & 36 \end{bmatrix}$$

$$= \begin{bmatrix} 41.1 & -37.9 & 50.1 \\ -105.6 & -2.7 & -5.1 \\ 52.2 & -11.6 & 17.7 \end{bmatrix}$$

CB

$$= \begin{bmatrix} 45 & -1 & 3 \\ -6 & 10 & -15 \\ 21 & -28 & 36 \end{bmatrix} \begin{bmatrix} 0.3 & -0.4 & 1.2 \\ -2.5 & 2 & 0.9 \\ 1 & -0.5 & 0.2 \end{bmatrix}$$

$$= \begin{bmatrix} 19 & -21.5 & 53.7 \\ -41.8 & 29.9 & -1.2 \\ 112.3 & -82.4 & 7.2 \end{bmatrix}$$

$BC \neq CB$; Verified

55. $(B+C)A = BA + CA$

$$(B+C)A = \begin{bmatrix} 45.3 & -1.4 & 4.2 \\ -8.5 & 12 & -14.1 \\ 22 & -28.5 & 36.2 \end{bmatrix} A$$

$$= \begin{bmatrix} -31.3 & 126.1 & 253.1 \\ -23.9 & 58.5 & -139.1 \\ 65.8 & -133.5 & 355.7 \end{bmatrix}$$

$BA + CA$

$$= \begin{bmatrix} 3.7 & -1.9 & 9.1 \\ 10.1 & 6.5 & -9.1 \\ -1.2 & -0.5 & 6.7 \end{bmatrix} + \begin{bmatrix} -35 & 128 & 244 \\ -34 & 52 & -130 \\ 67 & -133 & 349 \end{bmatrix}$$

$$= \begin{bmatrix} -31.3 & 26.1 & 253.1 \\ -23.9 & 58.5 & -139.1 \\ 65.8 & -133.5 & 355.7 \end{bmatrix}$$

$(B+C)A = BA + CA$; Verified

57. $$\begin{bmatrix} 2 & 2 \\ 4.35 & 0 \end{bmatrix} \cdot \begin{bmatrix} 6.374 \\ 4.35 \end{bmatrix}$$

$$= \begin{bmatrix} 2(6.374) + 2(4.35) \\ 4.35(6.374) + 0(4.35) \end{bmatrix}$$

$$= \begin{bmatrix} 21.448 \\ 27.7269 \end{bmatrix}$$

$P = 2l + 2w$

$P = 2(6.374) + 2(4.35)$

$P = 12.748 + 8.7$

$P = 21.448$;

$A = lw$

$A = 6.374(4.35)$

$A = 27.7269$

$P = 21.448 \, \text{cm}, \quad A = 27.7269 \, \text{cm}^2$

59. a.

$$V \rightarrow \begin{array}{c} S \\ D \\ P \end{array}\begin{bmatrix} T & S \\ 3820 & 1960 \\ 2460 & 1240 \\ 1540 & 920 \end{bmatrix}$$

$$M \rightarrow \begin{array}{c} S \\ D \\ P \end{array}\begin{bmatrix} T & S \\ 4220 & 2960 \\ 2960 & 3240 \\ 1640 & 820 \end{bmatrix}$$

b. $\begin{bmatrix} 4220 & 2960 \\ 2960 & 3240 \\ 1640 & 820 \end{bmatrix} - \begin{bmatrix} 3820 & 1960 \\ 2460 & 1240 \\ 1540 & 920 \end{bmatrix}$

$= \begin{bmatrix} 4220-3820 & 2960-1960 \\ 2960-2460 & 3240-1240 \\ 1640-1540 & 820-920 \end{bmatrix}$

$= \begin{bmatrix} 400 & 1000 \\ 500 & 2000 \\ 100 & -100 \end{bmatrix}$

$400 + 1000 + 500 + 2000 + 100 - 100$
$= 3900$
3,900 more by Minsk

c. $V \rightarrow 1.04 \begin{bmatrix} 3820 & 1960 \\ 2460 & 1240 \\ 1540 & 920 \end{bmatrix}$

$= \begin{bmatrix} 3972.8 & 2038.4 \\ 2558.4 & 1289.6 \\ 1601.6 & 956.8 \end{bmatrix}$;

$M \rightarrow 1.04 \begin{bmatrix} 4220 & 2960 \\ 2960 & 3240 \\ 1640 & 820 \end{bmatrix}$

$= \begin{bmatrix} 4388.8 & 3078.4 \\ 3078.4 & 3369.6 \\ 1705.6 & 852.8 \end{bmatrix}$

d. $\begin{bmatrix} 3972.8 & 2038.4 \\ 2558.4 & 1289.6 \\ 1601.6 & 956.8 \end{bmatrix} + \begin{bmatrix} 4388.8 & 3078.4 \\ 3078.4 & 3369.6 \\ 1705.6 & 852.8 \end{bmatrix}$

$= \begin{bmatrix} 8361.6 & 5116.8 \\ 5636.8 & 4659.2 \\ 3307.2 & 1809.6 \end{bmatrix}$

61. $\begin{bmatrix} 1500 & 500 & 2500 \end{bmatrix}\begin{bmatrix} 9 & 6 & 5 & 4 \\ 7 & 5 & 7 & 6 \\ 2 & 3 & 5 & 2 \end{bmatrix}$

$= \begin{bmatrix} 22000 & 19000 & 23500 & 14000 \end{bmatrix}$

Total profit for north: \$22,000.
Total profit for south: \$19,000.
Total profit for east: \$23,500.
Total profit for west: \$14,000.

63. a. $10(8) + 8(1.5) + 18(0.9) = \$108.20$
    b. $8(7.5) + 12(1.75) + 20(1) = \$101.00$

c. $\begin{bmatrix} 8 & 12 & 20 \\ 10 & 8 & 18 \end{bmatrix}\begin{bmatrix} 8 & 7.5 & 10 \\ 1.5 & 1.75 & 2 \\ 0.9 & 1 & 0.75 \end{bmatrix}$

$= \begin{array}{c} \text{Science} \\ \text{Math} \end{array}\begin{bmatrix} 100 & 101 & 119 \\ 108.2 & 107 & 129.5 \end{bmatrix}$

1st row, total cost for science from each restaurant.
2nd row, Total cost for math from each restaurant.

65. $\begin{bmatrix} 25 & 18 & 21 \\ 22 & 19 & 18 \end{bmatrix}\begin{bmatrix} 0.6 & 0.1 & 0.3 \\ 0.5 & 0.2 & 0.3 \\ 0.4 & 0.2 & 0.4 \end{bmatrix}$

$= \begin{bmatrix} 32.4 & 10.3 & 21.3 \\ 29.9 & 9.6 & 19.5 \end{bmatrix}$

a. Approximately 10 females
b. Approximately 20 males
c. The approximate number of females expected to join the writing club

67. 1st, 3rd rows entries double, 2nd row entries double and increase by 1 in a_{21}, a_{23} positions, and a_{22} stays a 1.

$\begin{bmatrix} 2^{n-1} & 0 & 2^{n-1} \\ 2^n - 1 & 1 & 2^n - 1 \\ 2^{n-1} & 0 & 2^{n-1} \end{bmatrix}$

69. $\begin{bmatrix} 2 & 1 \\ -3 & -2 \end{bmatrix} \cdot \begin{bmatrix} a & b \\ c & d \end{bmatrix} = \begin{bmatrix} 1 & 0 \\ 0 & 1 \end{bmatrix}$

$\begin{cases} 2a + c = 1 \\ -3a - 2c = 0 \end{cases}$ $\begin{cases} 2b + d = 0 \\ -3b - 2d = 1 \end{cases}$

$2R1 + R2 = Sum$ $2R1 + R2 = Sum$

$\quad 4a + 2c = 2$ $\quad 4b + 2d = 0$

$\quad -3a - 2c = 0$ $\quad -3b - 2d = 1$

$\quad a = 2$ $\quad b = 1$

$\quad 2a + c = 1$ $\quad 2b + d = 0$

$\quad 2(2) + c = 1$ $\quad 2(1) + d = 0$

$\quad\quad 4 + c = 1$ $\quad\quad 2 + d = 0$

$\quad\quad\quad c = -3$ $\quad\quad\quad d = -2$

$a = 2, b = 1, c = -3, d = -2$

71. $\cos\left(\cos^{-1} 0.3211\right) = 0.3211$

73. $\dfrac{x^3 - 9x + 10}{x - 2}$

$\begin{array}{r|rrrr} 2 & 1 & 0 & -9 & 10 \\ & & 2 & 4 & -10 \\ \hline & 1 & 2 & -5 & \underline{|0} \end{array}$

$= x^2 + 2x - 5$

8.7 Exercises

1. Main diagonal; zeroes

3. Identity

5. Answers will vary.

7. $A = \begin{bmatrix} 2 & 5 \\ -3 & -7 \end{bmatrix} \cdot \begin{bmatrix} a & b \\ c & d \end{bmatrix} = \begin{bmatrix} 2 & 5 \\ -3 & -7 \end{bmatrix}$

$\begin{bmatrix} 2a + 5c & 2b + 5d \\ -3a - 7c & -3b - 7d \end{bmatrix} = \begin{bmatrix} 2 & 5 \\ -3 & -7 \end{bmatrix}$

$\begin{cases} 2a + 5c = 2 \\ -3a - 7c = -3 \end{cases}$

$3R1 + 2R2$

$\begin{cases} 6a + 15c = 6 \\ -6a - 14c = -6 \end{cases}$

$\quad\quad\quad\quad c = 0 \; ;$

$2a + 5c = 2$

$2a + 5(0) = 2$

$\quad\quad 2a = 2$

$\quad\quad\quad a = 1$

$\begin{cases} 2b + 5d = 5 \\ -3b - 7d = -7 \end{cases}$

$3R1 + 2R2$

$\begin{cases} 6b + 15d = 15 \\ -6b - 14d = -14 \end{cases}$

$\quad\quad\quad\quad d = 1 \; ;$

$2b + 5d = 5$

$2b + 5(1) = 5$

$\quad 2b + 5 = 5$

$\quad\quad 2b = 0$

$\quad\quad\quad b = 0$

$\begin{bmatrix} a & b \\ c & d \end{bmatrix} = \begin{bmatrix} 1 & 0 \\ 0 & 1 \end{bmatrix}$

9. $A = \begin{bmatrix} 0.4 & 0.6 \\ 0.3 & 0.2 \end{bmatrix} \cdot \begin{bmatrix} a & b \\ c & d \end{bmatrix} = \begin{bmatrix} 0.4 & 0.6 \\ 0.3 & 0.2 \end{bmatrix}$

$\begin{bmatrix} 0.4a + 0.6c & 0.4b + 0.6d \\ 0.3a + 0.2c & 0.3b + 0.2d \end{bmatrix} = \begin{bmatrix} 0.4 & 0.6 \\ 0.3 & 0.2 \end{bmatrix}$

$\begin{cases} 0.4a + 0.6c = 0.4 \\ 0.3a + 0.2c = 0.3 \end{cases}$

$30R1 + (-40)R2$

$\begin{cases} 12a + 18c = 12 \\ -12a - 8c = -12 \end{cases}$

$\quad\quad\quad\quad 10c = 0$

$\quad\quad\quad\quad\quad c = 0 \; ;$

$0.4a + 0.6c = 0.4$

$0.4a + 0.6(0) = 0.4$

$\quad\quad 0.4a = 0.4$

$\quad\quad\quad a = 1 \; ;$

$\begin{cases} 0.4b + 0.6d = 0.6 \\ 0.3b + 0.2d = 0.2 \end{cases}$

$30R1 + (-40)R2$

$\begin{cases} 12b + 18d = 18 \\ -12b - 8d = -8 \end{cases}$

$\quad\quad\quad\quad 10d = 10$

$\quad\quad\quad\quad\quad d = 1 \; ;$

$0.4b + 0.6d = 0.6$

$0.4b + 0.6(1) = 0.6$

$\quad 0.4b + 0.6 = 0.6$

$\quad\quad 0.4b = 0$

$\quad\quad\quad b = 0 \; ;$

$\begin{bmatrix} a & b \\ c & d \end{bmatrix} = \begin{bmatrix} 1 & 0 \\ 0 & 1 \end{bmatrix}$

11. $\begin{bmatrix} -3 & 8 \\ -4 & 10 \end{bmatrix} \cdot \begin{bmatrix} 1 & 0 \\ 0 & 1 \end{bmatrix}$

$= \begin{bmatrix} -3(1)+8(0) & -3(0)+8(1) \\ -4(1)+10(0) & -4(0)+10(1) \end{bmatrix}$

$= \begin{bmatrix} -3 & 8 \\ -4 & 10 \end{bmatrix}$;

$\begin{bmatrix} 1 & 0 \\ 0 & 1 \end{bmatrix} \cdot \begin{bmatrix} -3 & 8 \\ -4 & 10 \end{bmatrix}$

$= \begin{bmatrix} 1(-3)+0(-4) & 1(8)+0(10) \\ 0(-3)+1(-4) & 0(8)+1(10) \end{bmatrix}$

$= \begin{bmatrix} -3 & 8 \\ -4 & 10 \end{bmatrix}$

$AI = IA = A$

13. $\begin{bmatrix} -4 & 1 & 6 \\ 9 & 5 & 3 \\ 0 & -2 & 1 \end{bmatrix} \cdot \begin{bmatrix} 1 & 0 & 0 \\ 0 & 1 & 0 \\ 0 & 0 & 1 \end{bmatrix}$

$= \begin{bmatrix} -4 & 1 & 6 \\ 9 & 5 & 3 \\ 0 & -2 & 1 \end{bmatrix}$;

$\begin{bmatrix} 1 & 0 & 0 \\ 0 & 1 & 0 \\ 0 & 0 & 1 \end{bmatrix} \cdot \begin{bmatrix} -4 & 1 & 6 \\ 9 & 5 & 3 \\ 0 & -2 & 1 \end{bmatrix}$

$= \begin{bmatrix} -4 & 1 & 6 \\ 9 & 5 & 3 \\ 0 & -2 & 1 \end{bmatrix}$

$AI = IA = A$

15. $\begin{bmatrix} 5 & -4 \\ 2 & 2 \end{bmatrix}$

$A^{-1} = \frac{1}{ad-bc} \begin{bmatrix} d & -b \\ -c & a \end{bmatrix}$

$A^{-1} = \frac{1}{5(2)-(-4)(2)} \begin{bmatrix} 2 & 4 \\ -2 & 5 \end{bmatrix}$

$A^{-1} = \frac{1}{18} \begin{bmatrix} 2 & 4 \\ -2 & 5 \end{bmatrix}$

$A^{-1} = \begin{bmatrix} \frac{1}{9} & \frac{2}{9} \\ \frac{-1}{9} & \frac{5}{18} \end{bmatrix}$

17. $\begin{bmatrix} 1 & -3 \\ 4 & -10 \end{bmatrix}$

$A^{-1} = \frac{1}{ad-bc} \begin{bmatrix} d & -b \\ -c & a \end{bmatrix}$

$A^{-1} = \frac{1}{1(-10)-(-3)(4)} \begin{bmatrix} -10 & 3 \\ -4 & 1 \end{bmatrix}$

$A^{-1} = \frac{1}{2} \begin{bmatrix} -10 & 3 \\ -4 & 1 \end{bmatrix}$

$A^{-1} = \begin{bmatrix} -5 & \frac{3}{2} \\ -2 & \frac{1}{2} \end{bmatrix}$

19. $A = \begin{bmatrix} 1 & 5 \\ -2 & -9 \end{bmatrix}$ $B = \begin{bmatrix} -9 & -5 \\ 2 & 1 \end{bmatrix}$

$AB = \begin{bmatrix} 1 & 5 \\ -2 & -9 \end{bmatrix} \begin{bmatrix} -9 & -5 \\ 2 & 1 \end{bmatrix}$

$= \begin{bmatrix} 1(-9)+5(2) & 1(-5)+5(1) \\ -2(-9)-9(2) & -2(-5)-9(1) \end{bmatrix}$

$= \begin{bmatrix} 1 & 0 \\ 0 & 1 \end{bmatrix}$;

$BA = \begin{bmatrix} -9 & -5 \\ 2 & 1 \end{bmatrix} \begin{bmatrix} 1 & 5 \\ -2 & -9 \end{bmatrix}$

$= \begin{bmatrix} -9(1)-5(-2) & -9(5)-5(-9) \\ 2(1)+1(-2) & 2(5)+1(-9) \end{bmatrix}$

$= \begin{bmatrix} 1 & 0 \\ 0 & 1 \end{bmatrix}$

$AB = BA = I$

21. $A = \begin{bmatrix} 4 & -5 \\ 0 & 2 \end{bmatrix}$ $B = \begin{bmatrix} \dfrac{1}{4} & \dfrac{5}{8} \\ 0 & \dfrac{1}{2} \end{bmatrix}$

$AB = \begin{bmatrix} 4 & -5 \\ 0 & 2 \end{bmatrix} \begin{bmatrix} \dfrac{1}{4} & \dfrac{5}{8} \\ 0 & \dfrac{1}{2} \end{bmatrix}$

$= \begin{bmatrix} 4\left(\dfrac{1}{4}\right) - 5(0) & 4\left(\dfrac{5}{8}\right) - 5\left(\dfrac{1}{2}\right) \\ 0\left(\dfrac{1}{4}\right) + 2(0) & 0\left(\dfrac{5}{8}\right) + 2\left(\dfrac{1}{2}\right) \end{bmatrix}$

$= \begin{bmatrix} 1 & 0 \\ 0 & 1 \end{bmatrix};$

$BA = \begin{bmatrix} \dfrac{1}{4} & \dfrac{5}{8} \\ 0 & \dfrac{1}{2} \end{bmatrix} \begin{bmatrix} 4 & -5 \\ 0 & 2 \end{bmatrix}$

$= \begin{bmatrix} \dfrac{1}{4}(4) + \dfrac{5}{8}(0) & \dfrac{1}{4}(-5) + \dfrac{5}{8}(2) \\ 0(4) + \dfrac{1}{2}(0) & 0(-5) + \dfrac{1}{2}(2) \end{bmatrix}$

$= \begin{bmatrix} 1 & 0 \\ 0 & 1 \end{bmatrix}$

$AB = BA = I$

23. $A = \begin{bmatrix} -2 & 3 & 1 \\ 5 & 2 & 4 \\ 2 & 0 & -1 \end{bmatrix}$

$A^{-1} = B = \begin{bmatrix} -\dfrac{2}{39} & \dfrac{1}{13} & \dfrac{10}{39} \\ \dfrac{1}{3} & 0 & \dfrac{1}{3} \\ -\dfrac{4}{39} & \dfrac{2}{13} & -\dfrac{19}{39} \end{bmatrix};$

$AB = \begin{bmatrix} -2 & 3 & 1 \\ 5 & 2 & 4 \\ 2 & 0 & -1 \end{bmatrix} \begin{bmatrix} -\dfrac{2}{39} & \dfrac{1}{13} & \dfrac{10}{39} \\ \dfrac{1}{3} & 0 & \dfrac{1}{3} \\ -\dfrac{4}{39} & \dfrac{2}{13} & -\dfrac{19}{39} \end{bmatrix}$

$= \begin{bmatrix} 1 & 0 & 0 \\ 0 & 1 & 0 \\ 0 & 0 & 1 \end{bmatrix}$

$BA = \begin{bmatrix} -\dfrac{2}{39} & \dfrac{1}{13} & \dfrac{10}{39} \\ \dfrac{1}{3} & 0 & \dfrac{1}{3} \\ -\dfrac{4}{39} & \dfrac{2}{13} & -\dfrac{19}{39} \end{bmatrix} \begin{bmatrix} -2 & 3 & 1 \\ 5 & 2 & 4 \\ 2 & 0 & -1 \end{bmatrix}$

$= \begin{bmatrix} 1 & 0 & 0 \\ 0 & 1 & 0 \\ 0 & 0 & 1 \end{bmatrix}$

$AB = BA = I$

25. $A = \begin{bmatrix} -7 & 5 & -3 \\ 1 & 9 & 0 \\ 2 & -2 & -5 \end{bmatrix}$

$A^{-1} = B = \begin{bmatrix} -\dfrac{9}{80} & \dfrac{31}{400} & \dfrac{27}{400} \\ \dfrac{1}{80} & \dfrac{41}{400} & -\dfrac{3}{400} \\ -\dfrac{1}{20} & -\dfrac{1}{100} & -\dfrac{17}{100} \end{bmatrix};$

$AB = \begin{bmatrix} -7 & 5 & -3 \\ 1 & 9 & 0 \\ 2 & -2 & -5 \end{bmatrix} \begin{bmatrix} -\dfrac{9}{80} & \dfrac{31}{400} & \dfrac{27}{400} \\ \dfrac{1}{80} & \dfrac{41}{400} & -\dfrac{3}{400} \\ -\dfrac{1}{20} & -\dfrac{1}{100} & -\dfrac{17}{100} \end{bmatrix}$

$= \begin{bmatrix} 1 & 0 & 0 \\ 0 & 1 & 0 \\ 0 & 0 & 1 \end{bmatrix};$

$BA = \begin{bmatrix} -\dfrac{9}{80} & \dfrac{31}{400} & \dfrac{27}{400} \\ \dfrac{1}{80} & \dfrac{41}{400} & -\dfrac{3}{400} \\ -\dfrac{1}{20} & -\dfrac{1}{100} & -\dfrac{17}{100} \end{bmatrix} \begin{bmatrix} -7 & 5 & -3 \\ 1 & 9 & 0 \\ 2 & -2 & -5 \end{bmatrix}$

$= \begin{bmatrix} 1 & 0 & 0 \\ 0 & 1 & 0 \\ 0 & 0 & 1 \end{bmatrix}$

$AB = BA = I$

27. $\begin{cases} 2x - 3y = 9 \\ -5x + 7y = 8 \end{cases}$

$\begin{bmatrix} 2 & -3 \\ -5 & 7 \end{bmatrix}\begin{bmatrix} x \\ y \end{bmatrix} = \begin{bmatrix} 9 \\ 8 \end{bmatrix}$

29. $\begin{cases} x + 2y - z = 1 \\ x + z = 3 \\ 2x - y + z = 3 \end{cases}$

$\begin{bmatrix} 1 & 2 & -1 \\ 1 & 0 & 1 \\ 2 & -1 & 1 \end{bmatrix}\begin{bmatrix} x \\ y \\ z \end{bmatrix} = \begin{bmatrix} 1 \\ 3 \\ 3 \end{bmatrix}$

31. $\begin{cases} -2w + x - 4y + 5 = -3 \\ 2w - 5x + y - 3z = 4 \\ -3w + x + 6y + z = 1 \\ w + 4x - 5y + z = -9 \end{cases}$

$\begin{bmatrix} -2 & 1 & -4 & 5 \\ 2 & -5 & 1 & -3 \\ -3 & 1 & 6 & 1 \\ 1 & 4 & -5 & 1 \end{bmatrix}\begin{bmatrix} w \\ x \\ y \\ z \end{bmatrix} = \begin{bmatrix} -3 \\ 4 \\ 1 \\ -9 \end{bmatrix}$

33. $\begin{cases} 0.05x - 3.2y = -15.8 \\ 0.02x + 2.4y = 12.08 \end{cases}$

$\begin{bmatrix} 0.05 & -3.2 \\ 0.02 & 2.4 \end{bmatrix}\begin{bmatrix} x \\ y \end{bmatrix} = \begin{bmatrix} -15.8 \\ 12.08 \end{bmatrix}$

Using the grapher,

$A^{-1} = \begin{bmatrix} \dfrac{300}{23} & \dfrac{400}{23} \\ -\dfrac{5}{46} & \dfrac{25}{92} \end{bmatrix}$;

$A^{-1}\left(\begin{bmatrix} 0.05 & -3.2 \\ 0.02 & 2.4 \end{bmatrix}\begin{bmatrix} x \\ y \end{bmatrix}\right) = A^{-1}\begin{bmatrix} -15.8 \\ 12.08 \end{bmatrix}$

$\left(A^{-1}\begin{bmatrix} 0.05 & -3.2 \\ 0.02 & 2.4 \end{bmatrix}\right)\begin{bmatrix} x \\ y \end{bmatrix} = A^{-1}\begin{bmatrix} -15.8 \\ 12.08 \end{bmatrix}$

$\begin{bmatrix} 1 & 0 \\ 0 & 1 \end{bmatrix}\begin{bmatrix} x \\ y \end{bmatrix} = A^{-1}\begin{bmatrix} -15.8 \\ 12.08 \end{bmatrix}$

$\begin{bmatrix} x \\ y \end{bmatrix} = \begin{bmatrix} 4 \\ 5 \end{bmatrix}$

$(4, 5)$

35. $\begin{cases} -\dfrac{1}{6}u + \dfrac{1}{4}v = 1 \\ \dfrac{1}{2}u - \dfrac{2}{3}v = -2 \end{cases}$

$\begin{bmatrix} -\dfrac{1}{6} & \dfrac{1}{4} \\ \dfrac{1}{2} & -\dfrac{2}{3} \end{bmatrix}\begin{bmatrix} u \\ v \end{bmatrix} = \begin{bmatrix} 1 \\ -2 \end{bmatrix}$

$A^{-1} = \begin{bmatrix} 48 & 18 \\ 36 & 12 \end{bmatrix}$;

$\begin{bmatrix} 48 & 18 \\ 36 & 12 \end{bmatrix}\left(\begin{bmatrix} -\dfrac{1}{6} & \dfrac{1}{4} \\ \dfrac{1}{2} & -\dfrac{2}{3} \end{bmatrix}\begin{bmatrix} u \\ v \end{bmatrix}\right) = \begin{bmatrix} 48 & 18 \\ 36 & 12 \end{bmatrix}\begin{bmatrix} 1 \\ -2 \end{bmatrix}$

$\left(\begin{bmatrix} 48 & 18 \\ 36 & 12 \end{bmatrix}\begin{bmatrix} -\dfrac{1}{6} & \dfrac{1}{4} \\ \dfrac{1}{2} & -\dfrac{2}{3} \end{bmatrix}\right)\begin{bmatrix} u \\ v \end{bmatrix} = \begin{bmatrix} 48 & 18 \\ 36 & 12 \end{bmatrix}\begin{bmatrix} 1 \\ -2 \end{bmatrix}$

$\begin{bmatrix} 1 & 0 \\ 0 & 1 \end{bmatrix}\begin{bmatrix} u \\ v \end{bmatrix} = \begin{bmatrix} 48 & 18 \\ 36 & 12 \end{bmatrix}\begin{bmatrix} 1 \\ -2 \end{bmatrix}$

$\begin{bmatrix} u \\ v \end{bmatrix} = \begin{bmatrix} 12 \\ 12 \end{bmatrix}$

$(12, 12)$

37. $\begin{cases} \dfrac{-1}{8}a + \dfrac{3}{5}b = \dfrac{5}{6} \\ \dfrac{5}{16}a - \dfrac{3}{2}b = \dfrac{-4}{5} \end{cases}$

$\begin{bmatrix} -\dfrac{1}{8} & \dfrac{3}{5} \\ \dfrac{5}{16} & -\dfrac{3}{2} \end{bmatrix}\begin{bmatrix} a \\ b \end{bmatrix} = \begin{bmatrix} \dfrac{5}{6} \\ \dfrac{4}{5} \end{bmatrix}$

No Solution; matrix is singular

8.7 Exercises

39. $\begin{cases} 0.2x - 1.6y + 2z = -1.9 \\ -0.4x - y + 0.6z = -1 \\ 0.8x + 3.2y - 0.4z = 0.2 \end{cases}$

$$\begin{bmatrix} 0.2 & -1.6 & 2 \\ -0.4 & -1 & 0.6 \\ 0.8 & 3.2 & -0.4 \end{bmatrix} \begin{bmatrix} x \\ y \\ z \end{bmatrix} = \begin{bmatrix} -1.9 \\ -1 \\ 0.2 \end{bmatrix}$$

Using the grapher,

$$A^{-1} = -\begin{bmatrix} \dfrac{95}{111} & -\dfrac{120}{37} & -\dfrac{65}{111} \\ \dfrac{20}{111} & \dfrac{35}{37} & \dfrac{115}{222} \\ \dfrac{10}{37} & \dfrac{40}{37} & \dfrac{35}{74} \end{bmatrix};$$

$$A^{-1}\left(\begin{bmatrix} 0.2 & -1.6 & 2 \\ -0.4 & -1 & 0.6 \\ 0.8 & 3.2 & -0.4 \end{bmatrix} \begin{bmatrix} x \\ y \\ z \end{bmatrix}\right) = A^{-1}\begin{bmatrix} -1.9 \\ -1 \\ 0.2 \end{bmatrix}$$

$$\left(A^{-1}\begin{bmatrix} 0.2 & -1.6 & 2 \\ -0.4 & -1 & 0.6 \\ 0.8 & 3.2 & -0.4 \end{bmatrix}\right) \begin{bmatrix} x \\ y \\ z \end{bmatrix} = A^{-1}\begin{bmatrix} -1.9 \\ -1 \\ 0.2 \end{bmatrix}$$

$$\begin{bmatrix} 1 & 0 & 0 \\ 0 & 1 & 0 \\ 0 & 0 & 1 \end{bmatrix} \begin{bmatrix} x \\ y \\ z \end{bmatrix} = A^{-1}\begin{bmatrix} -1.9 \\ -1 \\ 0.2 \end{bmatrix}$$

$$\begin{bmatrix} x \\ y \\ z \end{bmatrix} = \begin{bmatrix} 1.5 \\ -0.5 \\ -1.5 \end{bmatrix}$$

$(1.5, -0.5, -1.5)$

41. $\begin{cases} x - 2y + 2z = 6 \\ 2x - 1.5y + 1.8z = 2.8 \\ \dfrac{-2}{3}x + \dfrac{1}{2}y - \dfrac{3}{5}z = -\dfrac{11}{30} \end{cases}$

$$\begin{bmatrix} 1 & -2 & 2 \\ 2 & -1.5 & 1.8 \\ -\dfrac{2}{3} & \dfrac{1}{2} & -\dfrac{3}{5} \end{bmatrix} \begin{bmatrix} x \\ y \\ z \end{bmatrix} = \begin{bmatrix} 6 \\ 2.8 \\ -\dfrac{11}{30} \end{bmatrix}$$

Singular; no solution

43. $\begin{cases} -2w + 3x - 4y + 5z = -3 \\ 0.2w - 2.6x + y - 0.4z = 2.4 \\ -3w + 3.2x + 2.8y + z = 6.1 \\ 1.6w + 4x - 5y + 2.6z = -9.8 \end{cases}$

$$\begin{bmatrix} -2 & 3 & -4 & 5 \\ 0.2 & -2.6 & 1 & -0.4 \\ -3 & 3.2 & 2.8 & 1 \\ 1.6 & 4 & -5 & 2.6 \end{bmatrix} \begin{bmatrix} w \\ x \\ y \\ z \end{bmatrix} = \begin{bmatrix} -3 \\ 2.4 \\ 6.1 \\ -9.8 \end{bmatrix}$$

Using the grapher,

$$A^{-1} = \begin{bmatrix} -0.35859 & 1.15741 & 0.36978 & 0.72543 \\ -0.10811 & -0.13514 & 0.13514 & 0.13514 \\ -0.23646 & 0.933507 & 0.44725 & 0.42633 \\ -0.06774 & 1.290852 & 0.42463 & 0.55015 \end{bmatrix}$$

$$A^{-1}\left(\begin{bmatrix} -2 & 3 & -4 & 5 \\ 0.2 & -2.6 & 1 & -0.4 \\ -3 & 3.2 & 2.8 & 1 \\ 1.6 & 4 & -5 & 2.6 \end{bmatrix} \begin{bmatrix} w \\ x \\ y \\ z \end{bmatrix}\right) = A^{-1}\begin{bmatrix} -3 \\ 2.4 \\ 6.1 \\ -9.8 \end{bmatrix}$$

$$\left(A^{-1}\begin{bmatrix} -2 & 3 & -4 & 5 \\ 0.2 & -2.6 & 1 & -0.4 \\ -3 & 3.2 & 2.8 & 1 \\ 1.6 & 4 & -5 & 2.6 \end{bmatrix}\right) \begin{bmatrix} w \\ x \\ y \\ z \end{bmatrix} = A^{-1}\begin{bmatrix} -3 \\ 2.4 \\ 6.1 \\ -9.8 \end{bmatrix}$$

$$\begin{bmatrix} 1 & 0 & 0 & 0 \\ 0 & 1 & 0 & 0 \\ 0 & 0 & 1 & 0 \\ 0 & 0 & 0 & 1 \end{bmatrix} \begin{bmatrix} w \\ x \\ y \\ z \end{bmatrix} = A^{-1}\begin{bmatrix} -3 \\ 2.4 \\ 6.1 \\ -9.8 \end{bmatrix}$$

$$\begin{bmatrix} w \\ x \\ y \\ z \end{bmatrix} = \begin{bmatrix} -1 \\ -0.5 \\ 1.5 \\ 0.5 \end{bmatrix}$$

$(-1, -0.5, 1.5, 0.5)$

45. $\begin{bmatrix} 4 & -7 \\ 3 & -5 \end{bmatrix}$

    $\det A = 4(-5) - 3(-7) = -20 + 21 = 1$

    yes

47. $\begin{bmatrix} 1.2 & -0.8 \\ 0.3 & -0.2 \end{bmatrix}$

    $\det A = 1.2(-0.2) - (0.3)(-0.8)$

    $= -0.24 + 0.24 = 0$

    no

49. $A = \begin{bmatrix} 1 & 0 & -2 \\ 0 & -1 & -1 \\ 2 & 1 & -4 \end{bmatrix}$

    $\det A$

    $= 1\begin{vmatrix} -1 & -1 \\ 1 & -4 \end{vmatrix} - 0\begin{vmatrix} 0 & -1 \\ 2 & -4 \end{vmatrix} + (-2)\begin{vmatrix} 0 & -1 \\ 2 & 1 \end{vmatrix}$

    $= 1(4+1) - 0 - 2(0+2)$

    $= 5 - 0 - 4$

    $= 1$

51. $C = \begin{bmatrix} -2 & 3 & 4 \\ 0 & 6 & 2 \\ 1 & -1.5 & -2 \end{bmatrix}$

    $\det C$

    $= 0\begin{vmatrix} 3 & 4 \\ -1.5 & -2 \end{vmatrix} + 6\begin{vmatrix} -2 & 4 \\ 1 & -2 \end{vmatrix} - 2\begin{vmatrix} -2 & 3 \\ 1 & -1.5 \end{vmatrix}$

    $= 0 + 6(4-4) - 2(3-3)$

    $= 0$

    Singular Matrix

53. $A = \begin{bmatrix} 1 & 0 & 3 & -4 \\ 2 & 5 & 0 & 1 \\ 8 & 15 & 6 & -5 \\ 0 & 8 & -4 & 1 \end{bmatrix}$

    $\det A = 0$

    Singular

55. $\begin{bmatrix} 2 & -3 & 1 \\ 4 & -1 & 5 \\ 1 & 0 & -2 \end{bmatrix}$

    $\begin{bmatrix} 2 & -3 & 1 \\ 4 & -1 & 5 \\ 1 & 0 & -2 \end{bmatrix}\begin{matrix} 2 & -3 \\ 4 & -1 \\ 1 & 0 \end{matrix}$

    $2(-1)(-2) = 4,$

    $(-3)(5)(1) = -15,$

    $1(4)(0) = 0,$

    $4 + (-15) + 0 = -11;$

    $(1)(-1)(1) = -1,$

    $(0)(5)(2) = 0,$

    $(-2)(4)(-3) = 24,$

    $-1 + 0 + 24 = -23;$

    $-11 - 23 = -34$

57. $\begin{bmatrix} 1 & -1 & 2 \\ 3 & -2 & 4 \\ 4 & 3 & 1 \end{bmatrix}$

    $\begin{bmatrix} 1 & -1 & 2 \\ 3 & -2 & 4 \\ 4 & 3 & 1 \end{bmatrix}\begin{matrix} 1 & -1 \\ 3 & -2 \\ 4 & 3 \end{matrix}$

    $1(-2)(1) = -2,$

    $(-1)(4)(4) = -16,$

    $2(3)(3) = 18,$

    $-2 + (-16) + 18 = 0;$

    $(4)(-2)(2) = -16,$

    $(3)(4)(1) = 12,$

    $(1)(3)(-1) = -3,$

    $-16 + 12 + (-3) = -7;$

    $0 - (-7) = 7$

59. $\begin{cases} x - 2y + 2z = 7 \\ 2x + 2y - z = 5 \\ 3x - y + z = 6 \end{cases}$

(1) $\begin{bmatrix} 1 & -2 & 2 \\ 2 & 2 & -1 \\ 3 & -1 & 1 \end{bmatrix} \begin{bmatrix} x \\ y \\ z \end{bmatrix} = \begin{bmatrix} 7 \\ 5 \\ 6 \end{bmatrix}$

(2) $\qquad \det A$

$= 1 \begin{vmatrix} 2 & -1 \\ -1 & 1 \end{vmatrix} - (-2) \begin{vmatrix} 2 & -1 \\ 3 & 1 \end{vmatrix} + 2 \begin{vmatrix} 2 & 2 \\ 3 & -1 \end{vmatrix}$

$= 1(2-1) + 2(2+3) + 2(-2-6)$

$= 1 + 10 - 16$

$= -5$

(3) $A^{-1} = \begin{bmatrix} -0.2 & 0 & 0.4 \\ 1 & 1 & -1 \\ 1.6 & 1 & -1.2 \end{bmatrix}$

$$X = A^{-1}B$$

$X = \begin{bmatrix} -0.2 & 0 & 0.4 \\ 1 & 1 & -1 \\ 1.6 & 1 & -1.2 \end{bmatrix} \begin{bmatrix} 7 \\ 5 \\ 6 \end{bmatrix}$

$X = \begin{bmatrix} 1 \\ 6 \\ 9 \end{bmatrix}$

$(1, 6, 9)$

61. $\begin{cases} x - 3y + 4z = -1 \\ 4x - y + 5z = 7 \\ 3x + 2y + z = -3 \end{cases}$

(1) $\begin{bmatrix} 1 & -3 & 4 \\ 4 & -1 & 5 \\ 3 & 2 & 1 \end{bmatrix} \begin{bmatrix} x \\ y \\ z \end{bmatrix} = \begin{bmatrix} -1 \\ 7 \\ -3 \end{bmatrix}$

(2) $\det A$

$= 1 \begin{vmatrix} -1 & 5 \\ 2 & 1 \end{vmatrix} - (-3) \begin{vmatrix} 4 & 5 \\ 3 & 1 \end{vmatrix} + 4 \begin{vmatrix} 4 & -1 \\ 3 & 2 \end{vmatrix}$

$= 1(-1-10) + 3(4-15) + 4(8+3)$

$= -11 - 33 + 44$

$= 0$

Singular

63. $A = \begin{bmatrix} 3 & -5 \\ 2 & 1 \end{bmatrix}$

$A^{-1} = \dfrac{1}{ad - bc} \begin{bmatrix} d & -b \\ -c & a \end{bmatrix}$

$A^{-1} = \dfrac{1}{3(1) - (-5)(2)} \begin{bmatrix} 1 & 5 \\ -2 & 3 \end{bmatrix}$

$A^{-1} = \dfrac{1}{13} \begin{bmatrix} 1 & 5 \\ -2 & 3 \end{bmatrix}$

$A^{-1} = \begin{bmatrix} \dfrac{1}{13} & \dfrac{5}{13} \\ \dfrac{-2}{13} & \dfrac{3}{13} \end{bmatrix}$;

$AA^{-1} = \begin{bmatrix} 3 & -5 \\ 2 & 1 \end{bmatrix} \begin{bmatrix} \dfrac{1}{13} & \dfrac{5}{13} \\ \dfrac{-2}{13} & \dfrac{3}{13} \end{bmatrix} = \begin{bmatrix} 1 & 0 \\ 0 & 1 \end{bmatrix}$

$A^{-1}A = \begin{bmatrix} \dfrac{1}{13} & \dfrac{5}{13} \\ \dfrac{-2}{13} & \dfrac{3}{13} \end{bmatrix} \begin{bmatrix} 3 & -5 \\ 2 & 1 \end{bmatrix} = \begin{bmatrix} 1 & 0 \\ 0 & 1 \end{bmatrix}$

$AA^{-1} = A^{-1}A = I$

65. $C = \begin{bmatrix} 0.3 & -0.4 \\ -0.6 & 0.8 \end{bmatrix}$

$C^{-1} = \dfrac{1}{ad - bc} \begin{bmatrix} d & -b \\ -c & a \end{bmatrix}$

$C^{-1} = \dfrac{1}{0.3(0.8) - (-0.4)(-0.6)} \begin{bmatrix} 0.8 & 0.4 \\ 0.6 & 0.3 \end{bmatrix}$

$C^{-1} = \dfrac{1}{0} \begin{bmatrix} 0.8 & 0.4 \\ 0.6 & 0.3 \end{bmatrix}$

Singular

67. Let B represent the number of behemoth slushies sold.

Let G represent the number of gargantuan slushies sold.

Let M represent the number of mammoth slushies sold.

Let J represent the number of jumbo slushies sold.

$$\begin{cases} 2.59B + 2.29G + 1.99M + 1.59J = 402.29 \\ 60B + 48G + 36M + 24J = 7884 \\ B + G + M + J = 191 \\ B = J + 1 \end{cases}$$

$$\begin{cases} 2.59B + 2.29G + 1.99M + 1.59J = 402.29 \\ 60B + 48G + 36M + 24J = 7884 \\ B + G + M + J = 191 \\ B - J = 1 \end{cases}$$

$$\begin{bmatrix} 2.59 & 2.29 & 1.99 & 1.59 \\ 60 & 48 & 36 & 24 \\ 1 & 1 & 1 & 1 \\ 1 & 0 & 0 & -1 \end{bmatrix} \begin{bmatrix} B \\ G \\ M \\ J \end{bmatrix} = \begin{bmatrix} 402.29 \\ 7884 \\ 191 \\ 1 \end{bmatrix}$$

$$A^{-1} = \begin{bmatrix} -10 & \frac{1}{4} & \frac{109}{10} & 1 \\ 10 & -\frac{1}{6} & -\frac{139}{10} & -2 \\ 10 & -\frac{1}{3} & -\frac{69}{10} & 1 \\ -10 & \frac{1}{4} & \frac{109}{10} & 0 \end{bmatrix};$$

$$X = A^{-1}B$$

$$X = \begin{bmatrix} -10 & \frac{1}{4} & \frac{109}{10} & 1 \\ 10 & -\frac{1}{6} & -\frac{139}{10} & -2 \\ 10 & -\frac{1}{3} & -\frac{69}{10} & 1 \\ -10 & \frac{1}{4} & \frac{109}{10} & 0 \end{bmatrix} \begin{bmatrix} 402.29 \\ 7884 \\ 191 \\ 1 \end{bmatrix}$$

$$X = \begin{bmatrix} 31 \\ 52 \\ 78 \\ 30 \end{bmatrix}$$

31 behemoth
52 gargantuan
78 mammoth
30 jumbo

69. Let J represent the playing time of *Jumpin' Jack Flash*.

Let T represent the playing time of *Tumbling Dice*.

Let W represent the playing time of *Wild Horses*.

Let Y represent the playing time of You Can't Always Get What You Want.

$$\begin{cases} J + T + W + Y = 20.75 \\ J + T = Y \\ W = J + 2 \\ Y = 2T \end{cases}$$

$$\begin{cases} J + T + W + Y = 20.75 \\ J + T - Y = 0 \\ -J + W = 2 \\ -2T + Y = 0 \end{cases}$$

$$\begin{bmatrix} 1 & 1 & 1 & 1 \\ 1 & 1 & 0 & -1 \\ -1 & 0 & 1 & 0 \\ 0 & -2 & 0 & 1 \end{bmatrix} \begin{bmatrix} J \\ T \\ W \\ Y \end{bmatrix} = \begin{bmatrix} 20.75 \\ 0 \\ 2 \\ 0 \end{bmatrix}$$

$$A^{-1} = \begin{bmatrix} 0.2 & 0.6 & -0.2 & 0.4 \\ 0.2 & -0.4 & -0.2 & -0.6 \\ 0.2 & 0.6 & 0.8 & 0.4 \\ 0.4 & -0.8 & -0.4 & -0.2 \end{bmatrix};$$

$$X = A^{-1}B$$

$$X = \begin{bmatrix} 0.2 & 0.6 & -0.2 & 0.4 \\ 0.2 & -0.4 & -0.2 & -0.6 \\ 0.2 & 0.6 & 0.8 & 0.4 \\ 0.4 & -0.8 & -0.4 & -0.2 \end{bmatrix} \begin{bmatrix} 20.75 \\ 0 \\ 2 \\ 0 \end{bmatrix}$$

$$X = \begin{bmatrix} 3.75 \\ 3.75 \\ 5.75 \\ 7.5 \end{bmatrix}$$

Jumpin' Jack Flash: 3.75 min
Tumbling Dice: 3.75 min
Wild Horses: 5.75 min
You Can't Always Get What You Want: 7.5 min

71. Let A represent the number of Clock A manufactured.
Let B represent the number of Clock B manufactured.
Let C represent the number of Clock C manufactured.
Let D represent the number of Clock D manufactured.

$$\begin{cases} 2.2A + 2.5B + 2.75C + 3D = 262 \\ 1.2A + 1.4B + 1.8C + 2D = 160 \\ 0.2A + 0.25B + 0.3C + 0.5D = 29 \\ 0.5A + 0.55B + 0.75C + D = 68 \end{cases}$$

$$\begin{bmatrix} 2.2 & 2.5 & 2.75 & 3 \\ 1.2 & 1.4 & 1.8 & 2 \\ 0.2 & 0.25 & 0.3 & 0.5 \\ 0.5 & 0.55 & 0.75 & 1 \end{bmatrix} \begin{bmatrix} A \\ B \\ C \\ D \end{bmatrix} = \begin{bmatrix} 262 \\ 160 \\ 29 \\ 68 \end{bmatrix}$$

$$A^{-1} = \begin{bmatrix} \dfrac{30}{11} & -\dfrac{205}{22} & -\dfrac{210}{11} & 20 \\ 0 & 5 & 20 & -20 \\ -\dfrac{20}{11} & \dfrac{50}{11} & -\dfrac{80}{11} & 0 \\ 0 & -\dfrac{3}{2} & 4 & 2 \end{bmatrix};$$

$$X = A^{-1}B$$

$$X = \begin{bmatrix} \dfrac{30}{11} & -\dfrac{205}{22} & -\dfrac{210}{11} & 20 \\ 0 & 5 & 20 & -20 \\ -\dfrac{20}{11} & \dfrac{50}{11} & -\dfrac{80}{11} & 0 \\ 0 & -\dfrac{3}{2} & 4 & 2 \end{bmatrix} \begin{bmatrix} 262 \\ 160 \\ 29 \\ 68 \end{bmatrix}$$

$$X = \begin{bmatrix} 30 \\ 20 \\ 40 \\ 12 \end{bmatrix}$$

30 of clock A
20 of clock B
40 of clock C
12 of clock D

73.

$$\begin{cases} p_1 = \dfrac{70 + 64 + p_2 + p_3}{4} \\ p_2 = \dfrac{80 + 64 + p_4 + p_1}{4} \\ p_3 = \dfrac{70 + 96 + p_1 + p_4}{4} \\ p_4 = \dfrac{96 + 80 + p_2 + p_3}{4} \end{cases}$$

$$\begin{cases} 4p_1 - p_2 - p_3 = 134 \\ -p_1 + 4p_2 - p_4 = 144 \\ -p_1 + 4p_3 - p_4 = 166 \\ -p_2 - p_3 + 4p_4 = 176 \end{cases}$$

$$\begin{bmatrix} 4 & -1 & -1 & 0 \\ -1 & 4 & 0 & -1 \\ -1 & 0 & 4 & -1 \\ 0 & -1 & -1 & 4 \end{bmatrix} \begin{bmatrix} p_1 \\ p_2 \\ p_3 \\ p_4 \end{bmatrix} = \begin{bmatrix} 134 \\ 144 \\ 166 \\ 176 \end{bmatrix}$$

$$A^{-1} = \begin{bmatrix} \dfrac{7}{24} & \dfrac{1}{12} & \dfrac{1}{12} & \dfrac{1}{24} \\ \dfrac{1}{12} & \dfrac{7}{24} & \dfrac{1}{24} & \dfrac{1}{12} \\ \dfrac{1}{12} & \dfrac{1}{24} & \dfrac{7}{24} & \dfrac{1}{12} \\ \dfrac{1}{24} & \dfrac{1}{12} & \dfrac{1}{12} & \dfrac{7}{24} \end{bmatrix}$$

$$A^{-1} \begin{bmatrix} 4 & -1 & -1 & 0 \\ -1 & 4 & 0 & -1 \\ -1 & 0 & 4 & -1 \\ 0 & -1 & -1 & 4 \end{bmatrix} \begin{bmatrix} p_1 \\ p_2 \\ p_3 \\ p_4 \end{bmatrix} = A^{-1} \begin{bmatrix} 134 \\ 144 \\ 166 \\ 176 \end{bmatrix}$$

$$\begin{bmatrix} 1 & 0 & 0 & 0 \\ 0 & 1 & 0 & 0 \\ 0 & 0 & 1 & 0 \\ 0 & 0 & 0 & 1 \end{bmatrix} \begin{bmatrix} p_1 \\ p_2 \\ p_3 \\ p_4 \end{bmatrix} = \begin{bmatrix} 72.25 \\ 74.75 \\ 80.25 \\ 82.75 \end{bmatrix}$$

$p_1 = 72.25°, p_2 = 74.75°, p_3 = 80.25°, p_4 = 82.75°$

75. $ax^3 + bx^2 + cx + d = 0$

$(-4, -6), (-1, 0), (1, -16)$ and $(3, 8)$

$a(-4)^3 + b(-4)^2 + c(-4) + d = -6$
$-64a + 16b - 4c + d = -6;$

$a(-1)^3 + b(-1)^2 + c(-1) + d = 0$
$-a + b - c + d = 0;$

$a(1)^3 + b(1)^2 + c(1) + d = -16$
$a + b + c + d = -16;$

$a(3)^3 + b(3)^2 + c(3) + d = 8$
$27a + 9b + 3c + d = 8;$

$$\begin{cases} -64a + 16b - 4c + d = -6 \\ -a + b - c + d = 0 \\ a + b + c + d = -16 \\ 27a + 9b + 3c + d = 8 \end{cases}$$

$$\begin{bmatrix} -64 & 16 & -4 & 1 \\ -1 & 1 & -1 & 1 \\ 1 & 1 & 1 & 1 \\ 27 & 9 & 3 & 1 \end{bmatrix} \begin{bmatrix} a \\ b \\ c \\ d \end{bmatrix} = \begin{bmatrix} -6 \\ 0 \\ -16 \\ 8 \end{bmatrix}$$

$$A^{-1} = \begin{bmatrix} -\dfrac{1}{105} & \dfrac{1}{24} & -\dfrac{1}{20} & \dfrac{1}{56} \\ \dfrac{1}{35} & 0 & -\dfrac{1}{10} & \dfrac{1}{14} \\ \dfrac{1}{105} & -\dfrac{13}{24} & \dfrac{11}{20} & -\dfrac{1}{56} \\ -\dfrac{1}{35} & \dfrac{1}{2} & \dfrac{3}{5} & -\dfrac{1}{14} \end{bmatrix};$$

$X = A^{-1}B$

$$X = \begin{bmatrix} -\dfrac{1}{105} & \dfrac{1}{24} & -\dfrac{1}{20} & \dfrac{1}{56} \\ \dfrac{1}{35} & 0 & -\dfrac{1}{10} & \dfrac{1}{14} \\ \dfrac{1}{105} & -\dfrac{13}{24} & \dfrac{11}{20} & -\dfrac{1}{56} \\ -\dfrac{1}{35} & \dfrac{1}{2} & \dfrac{3}{5} & -\dfrac{1}{14} \end{bmatrix} \begin{bmatrix} -6 \\ 0 \\ -16 \\ 8 \end{bmatrix}$$

$$X = \begin{bmatrix} 1 \\ 2 \\ -9 \\ -10 \end{bmatrix}$$

$y = x^3 + 2x^2 - 9x - 10$

77. Let x represent the number of ounces of food for Food I.

Let y represent the number of ounces of food for Food II.

Let z represent the number of ounces of food for Food III.

$$\begin{cases} 2x + 4y + 3z = 20 \\ 4x + 2y + 5z = 30 \\ 5x + 6y + 7z = 44 \end{cases}$$

$$\begin{bmatrix} 2 & 4 & 3 \\ 4 & 2 & 5 \\ 5 & 6 & 7 \end{bmatrix} \begin{bmatrix} x \\ y \\ z \end{bmatrix} = \begin{bmatrix} 20 \\ 30 \\ 44 \end{bmatrix}$$

$$A^{-1} = \begin{bmatrix} 8 & 5 & -7 \\ 1.5 & 0.5 & -1 \\ -7 & -4 & 6 \end{bmatrix};$$

$X = A^{-1}B$

$$X = \begin{bmatrix} 8 & 5 & -7 \\ 1.5 & 0.5 & -1 \\ -7 & -4 & 6 \end{bmatrix} \begin{bmatrix} 20 \\ 30 \\ 44 \end{bmatrix}$$

$$X = \begin{bmatrix} 2 \\ 1 \\ 4 \end{bmatrix}$$

2 oz food I
1 oz food II
4 oz food III

79. Answers will vary.

81.

a. $\begin{bmatrix} 1 & -2 & 3 \\ -4 & 5 & -6 \\ 2 & 5 & 3 \end{bmatrix}$ $4R1 + R2 \rightarrow R2$

$\begin{bmatrix} 1 & -2 & 3 \\ 0 & -3 & 6 \\ 2 & 5 & 3 \end{bmatrix}$ $-2R1 + R3 \rightarrow R3$

$\begin{bmatrix} 1 & -2 & 3 \\ 0 & -3 & 6 \\ 0 & 9 & -3 \end{bmatrix}$ $3R2 + R3 \rightarrow R3$

$\begin{bmatrix} 1 & -2 & 3 \\ 0 & -3 & 6 \\ 0 & 0 & 15 \end{bmatrix}$ $-\dfrac{2}{3}R2 + R1 \rightarrow R1$

$\begin{bmatrix} 1 & 0 & -1 \\ 0 & -3 & 6 \\ 0 & 0 & 15 \end{bmatrix}$ $\dfrac{1}{15}R3 + R1 \rightarrow R1$

$\begin{bmatrix} 1 & 0 & 0 \\ 0 & -3 & 6 \\ 0 & 0 & 15 \end{bmatrix}$ $-\dfrac{2}{5}R3 + R2 \rightarrow R2$

$\begin{bmatrix} 1 & 0 & 0 \\ 0 & -3 & 0 \\ 0 & 0 & 15 \end{bmatrix}$

$(1)(-3)(15) = -45$

b. $\begin{bmatrix} 2 & 5 & -1 \\ -2 & -3 & 4 \\ 4 & 6 & 5 \end{bmatrix}$ $R1 + R2 \rightarrow R2$

$\begin{bmatrix} 2 & 5 & -1 \\ 0 & 2 & 3 \\ 4 & 6 & 5 \end{bmatrix}$ $-2R1 + R3 \rightarrow R3$

$\begin{bmatrix} 2 & 5 & -1 \\ 0 & 2 & 3 \\ 0 & -4 & 7 \end{bmatrix}$ $2R2 + R3 \rightarrow R3$

$\begin{bmatrix} 2 & 5 & -1 \\ 0 & 2 & 3 \\ 0 & 0 & 13 \end{bmatrix}$ $-\dfrac{5}{2}R2 + R1 \rightarrow R1$

$\begin{bmatrix} 2 & 0 & -\dfrac{17}{2} \\ 0 & 2 & 3 \\ 0 & 0 & 13 \end{bmatrix}$ $\dfrac{17}{26}R3 + R1 \rightarrow R1$

$\begin{bmatrix} 2 & 0 & 0 \\ 0 & 2 & 3 \\ 0 & 0 & 13 \end{bmatrix}$ $-\dfrac{3}{13}R3 + R2 \rightarrow R2$

$\begin{bmatrix} 2 & 0 & 0 \\ 0 & 2 & 0 \\ 0 & 0 & 13 \end{bmatrix}$

$(2)(2)(13) = 52$

c. $\begin{bmatrix} -2 & 4 & 1 \\ 5 & 7 & -2 \\ 3 & -8 & -1 \end{bmatrix}$ $R1 + R3 \to R1$

$\begin{bmatrix} 1 & -4 & 0 \\ 5 & 7 & -2 \\ 3 & -8 & -1 \end{bmatrix}$ $-5R1 + R2 \to R2$

$\begin{bmatrix} 1 & -4 & 0 \\ 0 & 27 & -2 \\ 3 & -8 & -1 \end{bmatrix}$ $-3R1 + R3 \to R3$

$\begin{bmatrix} 1 & -4 & 0 \\ 0 & 27 & -2 \\ 0 & 4 & -1 \end{bmatrix}$ $-2R3 + R2 \to R2$

$\begin{bmatrix} 1 & -4 & 0 \\ 0 & 19 & 0 \\ 0 & 4 & -1 \end{bmatrix}$ $-\dfrac{4}{19}R2 + R3 \to R3$

$\begin{bmatrix} 1 & -4 & 0 \\ 0 & 19 & 0 \\ 0 & 0 & -1 \end{bmatrix}$ $\dfrac{4}{19}R2 + R1 \to R1$

$\begin{bmatrix} 1 & 0 & 0 \\ 0 & 19 & 0 \\ 0 & 0 & -1 \end{bmatrix}$

$(1)(19)(-1) = -19$

d. $\begin{bmatrix} 3 & -1 & 4 \\ 0 & -2 & 6 \\ -2 & 1 & -3 \end{bmatrix}$ $R1 + R3 \to R1$

$\begin{bmatrix} 1 & 0 & 1 \\ 0 & -2 & 6 \\ -2 & 1 & -3 \end{bmatrix}$ $2R1 + R3 \to R3$

$\begin{bmatrix} 1 & 0 & 1 \\ 0 & -2 & 6 \\ 0 & 1 & -1 \end{bmatrix}$ $\dfrac{1}{2}R2 + R3 \to R3$

$\begin{bmatrix} 1 & 0 & 1 \\ 0 & -2 & 6 \\ 0 & 0 & 2 \end{bmatrix}$ $-3R3 + R2 \to R2$

$\begin{bmatrix} 1 & 0 & 1 \\ 0 & -2 & 0 \\ 0 & 0 & 2 \end{bmatrix}$ $-\dfrac{1}{2}R3 + R1 \to R1$

$\begin{bmatrix} 1 & 0 & 0 \\ 0 & -2 & 0 \\ 0 & 0 & 2 \end{bmatrix}$

$(1)(-2)(2) = -4$

83. $y = -125\cos(3t)$

Amplitude $= |-125| = 125$

Period $= \dfrac{2\pi}{3}$

85. $-3|2x+5| - 7 \le -19$

$-3|2x+5| \le -12$

$|2x+5| \ge 4$

$2x + 5 \ge 4$ or $2x + 5 \le -4$

$2x \ge -1$ or $2x \le -9$

$x \ge -\dfrac{1}{2}$ or $x \le -\dfrac{9}{2}$

$x \in \left(-\infty, -\dfrac{9}{2}\right] \cup \left[-\dfrac{1}{2}, \infty\right)$

8.8 Exercises

1. $a_{11}a_{22} - a_{21}a_{12}$

3. Constant

5. Answers will vary.

7. $\begin{cases} 2x + 5y = 7 \\ -3x + 4y = 1 \end{cases}$

$D = \begin{vmatrix} 2 & 5 \\ -3 & 4 \end{vmatrix}$; $D_x = \begin{vmatrix} 7 & 5 \\ 1 & 4 \end{vmatrix}$; $D_y = \begin{vmatrix} 2 & 7 \\ -3 & 1 \end{vmatrix}$

9. $\begin{cases} 4x + y = -11 \\ 3x - 5y = -60 \end{cases}$

$D = \begin{vmatrix} 4 & 1 \\ 3 & -5 \end{vmatrix} = -20 - 3 = -23$;

$D_x = \begin{vmatrix} -11 & 1 \\ -60 & -5 \end{vmatrix} = 55 + 60 = 115$;

$D_y = \begin{vmatrix} 4 & -11 \\ 3 & -60 \end{vmatrix} = -240 + 33 = -207$;

$x = \dfrac{D_x}{D} = \dfrac{115}{-23} = -5$;

$y = \dfrac{D_y}{D} = \dfrac{-207}{-23} = 9$

$(-5, 9)$

11. $\begin{cases} \dfrac{x}{8} + \dfrac{y}{4} = 1 \\ \dfrac{y}{5} = \dfrac{x}{2} + 6 \end{cases}$

$\begin{cases} \dfrac{x}{8} + \dfrac{y}{4} = 1 \\ -\dfrac{x}{2} + \dfrac{y}{5} = 6 \end{cases}$

$D = \begin{vmatrix} \dfrac{1}{8} & \dfrac{1}{4} \\ -\dfrac{1}{2} & \dfrac{1}{5} \end{vmatrix} = \dfrac{1}{40} + \dfrac{1}{8} = \dfrac{3}{20}$;

$D_x = \begin{vmatrix} 1 & \dfrac{1}{4} \\ 6 & \dfrac{1}{5} \end{vmatrix} = \dfrac{1}{5} - \dfrac{3}{2} = -\dfrac{13}{10}$;

$D_y = \begin{vmatrix} \dfrac{1}{8} & 1 \\ -\dfrac{1}{2} & 6 \end{vmatrix} = \dfrac{3}{4} + \dfrac{1}{2} = \dfrac{5}{4}$;

$x = \dfrac{D_x}{D} = \dfrac{-\dfrac{13}{10}}{\dfrac{3}{20}} = -\dfrac{260}{30} = -\dfrac{26}{3}$;

$y = \dfrac{D_y}{D} = \dfrac{\dfrac{5}{4}}{\dfrac{3}{20}} = \dfrac{100}{12} = \dfrac{25}{3}$

$\left(\dfrac{-26}{3}, \dfrac{25}{3} \right)$

13. $\begin{cases} 0.6x - 0.3y = 8 \\ 0.8x - 0.4y = -3 \end{cases}$

$D = \begin{vmatrix} 0.6 & -0.3 \\ 0.8 & -0.4 \end{vmatrix} = -0.24 + 0.24 = 0$

No Solution; determinant cannot be zero.

15. a.
$$\begin{cases} 4x - y + 2z = -5 \\ -3x + 2y - z = 8 \\ x - 5y + 3z = -3 \end{cases}$$

$$D = \begin{vmatrix} 4 & -1 & 2 \\ -3 & 2 & -1 \\ 1 & -5 & 3 \end{vmatrix}$$

$$D = 4\begin{vmatrix} 2 & -1 \\ -5 & 3 \end{vmatrix} + 1\begin{vmatrix} -3 & -1 \\ 1 & 3 \end{vmatrix} + 2\begin{vmatrix} -3 & 2 \\ 1 & -5 \end{vmatrix}$$

$$D = 4(1) + 1(-8) + 2(13) = 22; \text{ solutions possible}$$

$$D_x = \begin{vmatrix} -5 & -1 & 2 \\ 8 & 2 & -1 \\ -3 & -5 & 3 \end{vmatrix};$$

$$D_y = \begin{vmatrix} 4 & -5 & 2 \\ -3 & 8 & -1 \\ 1 & -3 & 3 \end{vmatrix};$$

$$D_z = \begin{vmatrix} 4 & -1 & -5 \\ -3 & 2 & 8 \\ 1 & -5 & -3 \end{vmatrix}$$

b.
$$\begin{cases} 4x - y + 2z = -5 \\ -3x + 2y - z = 8 \\ x + y + z = -3 \end{cases}$$

$$D = \begin{vmatrix} 4 & -1 & 2 \\ -3 & 2 & -1 \\ 1 & 1 & 1 \end{vmatrix}$$

$$D = 4\begin{vmatrix} 2 & -1 \\ 1 & 1 \end{vmatrix} + 1\begin{vmatrix} -3 & -1 \\ 1 & 1 \end{vmatrix} + 2\begin{vmatrix} -3 & 2 \\ 1 & 1 \end{vmatrix}$$

$$D = 4(3) + 1(-2) + 2(-5) = 0$$

$D = 0$; Cramer's Rule cannot be used. The sum of the first two rows of D gives row 3.

17.
$$\begin{cases} x + 2y + 5z = 10 \\ 3x + 4y - z = 10 \\ x - y - z = -2 \end{cases}$$

$$D = \begin{vmatrix} 1 & 2 & 5 \\ 3 & 4 & -1 \\ 1 & -1 & -1 \end{vmatrix} = -36;$$

$$D_x = \begin{vmatrix} 10 & 2 & 5 \\ 10 & 4 & -1 \\ -2 & -1 & -1 \end{vmatrix} = -36;$$

$$D_y = \begin{vmatrix} 1 & 10 & 5 \\ 3 & 10 & -1 \\ 1 & -2 & -1 \end{vmatrix} = -72;$$

$$D_z = \begin{vmatrix} 1 & 2 & 10 \\ 3 & 4 & 10 \\ 1 & -1 & -2 \end{vmatrix} = -36;$$

$$x = \frac{D_x}{D} = \frac{-36}{-36} = 1;$$

$$y = \frac{D_y}{D} = \frac{-72}{-36} = 2;$$

$$z = \frac{D_z}{D} = \frac{-36}{-36} = 1$$

$$(1, 2, 1)$$

19.
$$\begin{cases} y + 2z = 1 \\ 4x - 5y + 8z = -8 \\ 8x - 9z = 9 \end{cases}$$

$$D = \begin{vmatrix} 0 & 1 & 2 \\ 4 & -5 & 8 \\ 8 & 0 & -9 \end{vmatrix} = 180;$$

$$D_x = \begin{vmatrix} 1 & 1 & 2 \\ -8 & -5 & 8 \\ 9 & 0 & -9 \end{vmatrix} = 135;$$

$$D_y = \begin{vmatrix} 0 & 1 & 2 \\ 4 & -8 & 8 \\ 8 & 9 & -9 \end{vmatrix} = 300;$$

$$D_z = \begin{vmatrix} 0 & 1 & 1 \\ 4 & -5 & -8 \\ 8 & 0 & 9 \end{vmatrix} = -60;$$

$$x = \frac{D_x}{D} = \frac{135}{180} = \frac{3}{4};$$

$$y = \frac{D_y}{D} = \frac{300}{180} = \frac{5}{3};$$

$$z = \frac{D_z}{D} = \frac{-60}{180} = -\frac{1}{3}$$

$$\left(\frac{3}{4}, \frac{5}{3}, -\frac{1}{3}\right)$$

21.
$$\begin{cases} w + 2x - 3y = -8 \\ x - 3y + 5z = -22 \\ 4w - 5x = 5 \\ -y + 3z = -11 \end{cases}$$

$$D = \begin{vmatrix} 1 & 2 & -3 & 0 \\ 0 & 1 & -3 & 5 \\ 4 & -5 & 0 & 0 \\ 0 & 0 & -1 & 3 \end{vmatrix} = -16\;;$$

$$D_w = \begin{vmatrix} -8 & 2 & -3 & 0 \\ -22 & 1 & -3 & 5 \\ 5 & -5 & 0 & 0 \\ -11 & 0 & -1 & 3 \end{vmatrix} = 0\;;$$

$$D_x = \begin{vmatrix} 1 & -8 & -3 & 0 \\ 0 & -22 & -3 & 5 \\ 4 & 5 & 0 & 0 \\ 0 & -11 & -1 & 3 \end{vmatrix} = 16\;;$$

$$D_y = \begin{vmatrix} 1 & 2 & -8 & 0 \\ 0 & 1 & -22 & 5 \\ 4 & -5 & 5 & 0 \\ 0 & 0 & -11 & 3 \end{vmatrix} = -32\;;$$

$$D_z = \begin{vmatrix} 1 & 2 & -3 & -8 \\ 0 & 1 & -3 & -22 \\ 4 & -5 & 0 & 5 \\ 0 & 0 & -1 & -11 \end{vmatrix} = 48\;;$$

$$w = \frac{D_w}{D} = \frac{0}{-16} = 0\;;$$

$$x = \frac{D_x}{D} = \frac{16}{-16} = -1\;;$$

$$y = \frac{D_y}{D} = \frac{-32}{-16} = 2\;;$$

$$z = \frac{D_z}{D} = \frac{48}{-16} = -3$$

$$(0, -1, 2, -3)$$

23. $A = \begin{vmatrix} L & r^2 \\ -\dfrac{\pi}{2} & W \end{vmatrix}$

$$A = \begin{vmatrix} 20 & 8^2 \\ -\dfrac{\pi}{2} & 16 \end{vmatrix} = 320 - 64\left(-\dfrac{\pi}{2}\right)$$

$$= 320 + 32\pi \approx 420.5 \, \text{in}^2$$

25. $(2,1), (3,7), (5,3)$

$$A = \frac{\begin{vmatrix} x_1 & y_1 & 1 \\ x_2 & y_2 & 1 \\ x_3 & y_3 & 1 \end{vmatrix}}{2}$$

$$A = \frac{\begin{vmatrix} 2 & 1 & 1 \\ 3 & 7 & 1 \\ 5 & 3 & 1 \end{vmatrix}}{2} = \left|\frac{-16}{2}\right| = 8 \, \text{cm}^2$$

27. $(-4,2), (-6,-1), (3,-1), (5,2)$

$$A = \frac{\begin{vmatrix} x_1 & y_1 & 1 \\ x_2 & y_2 & 1 \\ x_3 & y_3 & 1 \end{vmatrix}}{2}$$

For triangle use $(-4, 2)$, $(-6, -1)$ and $(5, 2)$.

$$A = \frac{\begin{vmatrix} -4 & 2 & 1 \\ -6 & -1 & 1 \\ 5 & 2 & 1 \end{vmatrix}}{2} = \left|\frac{27}{2}\right| = \frac{27}{2}$$

$$2\left(\frac{27}{2}\right) = 27 \, \text{ft}^2$$

29. $h = 6m$; vertices $(3,5), (-4,2), (-1,6)$

$$A = \frac{\begin{vmatrix} 3 & 5 & 1 \\ -4 & 2 & 1 \\ -1 & 6 & 1 \end{vmatrix}}{2} = \left|\frac{-19}{2}\right| = 9.5$$

$$V = \frac{1}{3} Bh$$

$$V = \frac{1}{3}(9.5)(6)$$

$$V = 19 \, \text{m}^3$$

31. $(3, 0), (9, 0), (6, 4), h = 8$

$$A = \frac{\begin{vmatrix} 3 & 0 & 1 \\ 9 & 0 & 1 \\ 6 & 4 & 1 \end{vmatrix}}{2} = \left|\frac{24}{2}\right| = 12$$

$$V = Bh$$

$$V = 12(8) = 96 \, \text{in}^3$$

33. $(1,5),(-2,-1),(4,11)$

$$|A| = \begin{vmatrix} 1 & 5 & 1 \\ -2 & -1 & 1 \\ 4 & 11 & 1 \end{vmatrix} = 0$$

Yes

35. $(-2.5, 5.2),(1.2, -5.6),(2.2, -8.5)$

$$|A| = \begin{vmatrix} -2.5 & 5.2 & 1 \\ 1.2 & -5.6 & 1 \\ 2.2 & -8.5 & 1 \end{vmatrix} = 0.07$$

No

37. $2x - 3y = 7$;

$(2, -1), (-1.3, -3.2), (-3.1, -4.4)$

$(2,-1)$ Yes

$2(2) - 3(-1) = 7$

$4 + 3 = 7$

$7 = 7$;

$(-1.3, -3.2)$ Yes

$2(-1.3) - 3(-3.2) = 7$

$-2.6 + 9.6 = 7$

$7 = 7$;

$(-3.1, -4.4)$ Yes

$$|A| = \begin{vmatrix} 2 & -1 & 1 \\ -1.3 & -3.2 & 1 \\ -3.1 & -4.4 & 1 \end{vmatrix} = 0$$

39. $\dfrac{x^4 - x^2 - 2x + 1}{x^5 - 2x^3 + x} = \dfrac{x^4 - x^2 - 2x + 1}{x(x^4 - 2x^2 + 1)}$

$= \dfrac{x^4 - x^2 - 2x + 1}{x(x^2 - 1)^2} = \dfrac{A}{x} + \dfrac{Bx + C}{x^2 - 1} + \dfrac{Dx + E}{(x^2 - 1)^2}$

$x^4 - x^2 - 2x + 1$

$= A(x^2 - 1)^2 + x(Bx + C)(x^2 - 1) + x(Dx + E)$

$x^4 - x^2 - 2x + 1$

$= Ax^4 - 2Ax^2 + A + Bx^4 - Bx^2$

$+ Cx^3 - Cx + Dx^2 + Ex$

$x^4 - x^2 - 2x + 1$

$= x^4(A + B) + Cx^3 + x^2(-2A - B + D)$

$+ x(-C + E) + A$

$$\begin{cases} A + B = 1 \\ C = 0 \\ -2A - B + D = -1 \\ -C + E = -2 \\ A = 1 \end{cases}$$

$$\begin{bmatrix} 1 & 1 & 0 & 0 & 0 \\ 0 & 0 & 1 & 0 & 0 \\ -2 & -1 & 0 & 1 & 0 \\ 0 & 0 & -1 & 0 & 1 \\ 1 & 0 & 0 & 0 & 0 \end{bmatrix} \begin{bmatrix} A \\ B \\ C \\ D \\ E \end{bmatrix} = \begin{bmatrix} 1 \\ 0 \\ -1 \\ -2 \\ 1 \end{bmatrix}$$

$$A^{-1} = \begin{bmatrix} 0 & 0 & 0 & 0 & 1 \\ 1 & 0 & 0 & 0 & -1 \\ 0 & 1 & 0 & 0 & 0 \\ 1 & 0 & 1 & 0 & 1 \\ 0 & 1 & 0 & 1 & 0 \end{bmatrix}$$

$X = A^{-1}B$

$$X = \begin{bmatrix} 1 \\ 0 \\ 0 \\ 1 \\ -2 \end{bmatrix}$$

$\dfrac{1}{x} + \dfrac{x - 2}{(x^2 - 1)^2}$

41. $\dfrac{x^3 - 17x^2 + 76x - 98}{\left(x^2 - 6x + 9\right)\left(x^2 - 2x - 3\right)}$

$\dfrac{x^3 - 17x^2 + 76x - 98}{(x-3)(x-3)(x-3)(x+1)}$

$= \dfrac{x^3 - 17x^2 + 76x - 98}{(x-3)^3 (x+1)}$

$= \dfrac{A}{x+1} + \dfrac{B}{(x-3)} + \dfrac{C}{(x-3)^2} + \dfrac{D}{(x-3)^3}$

$x^3 - 17x^2 + 76x - 98$
$= A(x-3)^3 + B(x+1)(x-3)^2$
$+ C(x+1)(x-3) + D(x+1)$
$= Ax^3 - 9Ax^2 + 27Ax - 27A$
$+ Bx^3 - 5Bx^2 + 3Bx + 9B + Cx^2$
$- 2Cx - 3C + Dx + D;$
$= x^3(A+B) + x^2(-9A - 5B + C)$
$+ x(27A + 3B - 2C + D)$
$+ (-27A + 9B - 3C + D);$

$\begin{cases} A + B = 1 \\ -9A - 5B + C = -17 \\ 27A + 3B - 2C + D = 76 \\ -27A + 9B - 3C + D = -98 \end{cases}$

$\begin{bmatrix} 1 & 1 & 0 & 0 \\ -9 & -5 & 1 & 0 \\ 27 & 3 & -2 & 1 \\ -27 & 9 & -3 & 1 \end{bmatrix} \begin{bmatrix} A \\ B \\ C \\ D \end{bmatrix} = \begin{bmatrix} 1 \\ -17 \\ 76 \\ -98 \end{bmatrix}$

$A^{-1} = \begin{bmatrix} \dfrac{1}{64} & \dfrac{-1}{64} & \dfrac{1}{64} & \dfrac{-1}{64} \\ \dfrac{63}{64} & \dfrac{1}{64} & \dfrac{-1}{64} & \dfrac{1}{64} \\ \dfrac{81}{16} & \dfrac{15}{16} & \dfrac{1}{16} & \dfrac{-1}{16} \\ \dfrac{27}{4} & \dfrac{9}{4} & \dfrac{3}{4} & \dfrac{1}{4} \end{bmatrix}$

$X = A^{-1}B$

$X = \begin{bmatrix} 3 \\ -2 \\ 0 \\ 1 \end{bmatrix}$

$= \dfrac{3}{x+1} + \dfrac{-2}{(x-3)} + \dfrac{0}{(x-3)^2} + \dfrac{1}{(x-3)^3}$

$= \dfrac{3}{x+1} - \dfrac{2}{(x-3)} + \dfrac{1}{(x-3)^3}$

43. Let x represent the rate of the \$15000 investment.
Let y represent the rate of the \$25000 investment.

$\begin{cases} 15000x + 25000y = 2900 \\ 25000x + 15000y = 2700 \end{cases}$

$D = \begin{vmatrix} 15000 & 25000 \\ 25000 & 15000 \end{vmatrix} = -400000000;$

$D_x = \begin{vmatrix} 2900 & 25000 \\ 2700 & 15000 \end{vmatrix} = -24000000;$

$D_y = \begin{vmatrix} 15000 & 2900 \\ 25000 & 2700 \end{vmatrix} = -32000000;$

$x = \dfrac{D_x}{D} = \dfrac{-24000000}{-400000000} = 0.06;$

$y = \dfrac{D_y}{D} = \dfrac{-32000000}{-400000000} = 0.08$

\$15000 invested at 6%
\$25000 invested at 8%

45. Let A represent the cost per pound of apples.
Let K represent the cost per pound of kiwi.
Let P represent the cost per pound of pears.

$\begin{cases} 2A + 2K + 10P = 3.26 \\ 3A + 2K + 7P = 2.98 \\ 2A + 3K + 6P = 2.89 \end{cases}$

$D = \begin{vmatrix} 2 & 2 & 10 \\ 3 & 2 & 7 \\ 2 & 3 & 6 \end{vmatrix} = 24;$

$D_x = \begin{vmatrix} 3.26 & 2 & 10 \\ 2.98 & 2 & 7 \\ 2.89 & 3 & 6 \end{vmatrix} = 6.96;$

$D_y = \begin{vmatrix} 2 & 3.26 & 10 \\ 3 & 2.98 & 7 \\ 2 & 2.89 & 6 \end{vmatrix} = 9.36;$

$D_z = \begin{vmatrix} 2 & 2 & 3.26 \\ 3 & 2 & 2.98 \\ 2 & 3 & 2.89 \end{vmatrix} = 4.56;$

$x = \dfrac{D_x}{D} = \dfrac{6.96}{24} = 0.29;$

$y = \dfrac{D_y}{D} = \dfrac{9.36}{24} = 0.39;$

$z = \dfrac{D_z}{D} = \dfrac{4.56}{24} = 0.19$

Apples, 29 cents/lb
Kiwi, 39 cents/lb
Pears, 19 cents/lb

47. Let x represent the number of lb of \$1.90 coffee.
Let y represent the number of lb of \$2.25 coffee.
Let z represent the number of lb of \$3.50 coffee.

$$\begin{cases} x+y+z=24 \\ x=z+4 \\ 1.90x+2.25y+3.50z=58 \end{cases}$$

$$D=\begin{vmatrix} 1 & 1 & 1 \\ 1 & 0 & -1 \\ 190 & 225 & 350 \end{vmatrix}=-90;$$

$$D_x=\begin{vmatrix} 24 & 1 & 1 \\ 4 & 0 & -1 \\ 5800 & 225 & 350 \end{vmatrix}=-900;$$

$$D_y=\begin{vmatrix} 1 & 24 & 1 \\ 1 & 4 & -1 \\ 190 & 5800 & 350 \end{vmatrix}=-720;$$

$$D_z=\begin{vmatrix} 1 & 1 & 24 \\ 1 & 0 & 4 \\ 190 & 225 & 5800 \end{vmatrix}=-540;$$

$$x=\frac{D_x}{D}=\frac{-900}{-90}=10;$$

$$y=\frac{D_y}{D}=\frac{-720}{-90}=8;$$

$$z=\frac{D_z}{D}=\frac{-540}{-90}=6$$

10 lbs at \$1.90, 8 lbs at \$2.25, 6 lbs at \$3.50

49. $$\begin{cases} x+3y+5z=6 \\ 2x-4y+6z=14 \\ 9x-6y+3z=3 \end{cases}$$

(1) $-2R1+R2$

$$\begin{cases} -2x-6y-10z=-12 \\ 2x-4y+6z=14 \end{cases}$$
$$-10y-4z=2$$

$-9R1+R3.$

$$\begin{cases} -9x-27y-45z=-54 \\ 9x-6y+3z=3 \end{cases}$$
$$-33y-42z=-51$$

$$\begin{cases} -10y-4z=2 \\ -33y-42z=-51 \end{cases}$$

$-21R1+2R2$

$$\begin{cases} 210y+84z=-42 \\ -66y-84z=-102 \end{cases}$$
$$144y=-144$$
$$y=-1;$$

$$210y+84z=-42$$
$$210(-1)+84z=-42$$
$$-210+84z=-42$$
$$84z=168$$
$$z=2;$$
$$x+3y+5z=6$$
$$x+3(-1)+5(2)=6$$
$$x-3+10=6$$
$$x=-1$$
$$(-1,-1,2)$$

(2) $$\begin{bmatrix} 1 & 3 & 5 & 6 \\ 2 & -4 & 6 & 14 \\ 9 & -6 & 3 & 3 \end{bmatrix} \quad -2R1+R2 \to R2$$

$$\begin{bmatrix} 1 & 3 & 5 & 6 \\ 0 & -10 & -4 & 2 \\ 9 & -6 & 3 & 3 \end{bmatrix} \quad -9R1+R3 \to R3$$

$$\begin{bmatrix} 1 & 3 & 5 & 6 \\ 0 & -10 & -4 & 2 \\ 0 & -33 & -42 & -51 \end{bmatrix} \quad -\frac{1}{10}R2 \to R2$$

$$\begin{bmatrix} 1 & 3 & 5 & 6 \\ 0 & 1 & \frac{2}{5} & -\frac{1}{5} \\ 0 & -33 & -42 & -51 \end{bmatrix} \quad 33R2+R3 \to R3$$

$$\begin{bmatrix} 1 & 3 & 5 & 6 \\ 0 & 1 & \frac{2}{5} & -\frac{1}{5} \\ 0 & 0 & -\frac{144}{5} & -\frac{288}{5} \end{bmatrix} \quad -\frac{5}{144}R3 \to R3$$

$$\begin{bmatrix} 1 & 3 & 5 & 6 \\ 0 & 1 & \frac{2}{5} & -\frac{1}{5} \\ 0 & 0 & 1 & 2 \end{bmatrix}$$

$$z=2;$$
$$y+\frac{2}{5}z=-\frac{1}{5}$$
$$y+\frac{2}{5}(2)=-\frac{1}{5}$$
$$y+\frac{4}{5}=-\frac{1}{5}$$
$$y=-1;$$
$$x+3y+5z=6$$
$$x+3(-1)+5(2)=6$$
$$x-3+10=6$$
$$x+7=6$$
$$x=-1$$
$$(-1,-1,2)$$

8.8 Exercises

(3) $\begin{cases} x+3y+5z=6 \\ 2x-4y+6z=14 \\ 9x-6y+3z=3 \end{cases}$

$D = \begin{bmatrix} 1 & 3 & 5 \\ 2 & -4 & 6 \\ 9 & -6 & 3 \end{bmatrix} = 288\,;$

$D_x = \begin{bmatrix} 6 & 3 & 5 \\ 14 & -4 & 6 \\ 3 & -6 & 3 \end{bmatrix} = -288\,;$

$D_y = \begin{bmatrix} 1 & 6 & 5 \\ 2 & 14 & 6 \\ 9 & 3 & 3 \end{bmatrix} = -288\,;$

$D_z = \begin{bmatrix} 1 & 3 & 6 \\ 2 & -4 & 14 \\ 9 & -6 & 3 \end{bmatrix} = 576\,;$

$x = \dfrac{D_x}{D} = \dfrac{-288}{288} = -1\,;$

$y = \dfrac{D_y}{D} = \dfrac{-288}{288} = -1\,;$

$z = \dfrac{D_z}{D} = \dfrac{576}{288} = 2$

$(-1,\,-1,\,2)$

(4) $\begin{bmatrix} 1 & 3 & 5 \\ 2 & -4 & 6 \\ 9 & -6 & 3 \end{bmatrix}\begin{bmatrix} x \\ y \\ z \end{bmatrix} = \begin{bmatrix} 6 \\ 14 \\ 3 \end{bmatrix}$

$A^{-1} = \begin{bmatrix} \dfrac{1}{12} & \dfrac{-13}{96} & \dfrac{19}{144} \\ \dfrac{1}{6} & \dfrac{-7}{48} & \dfrac{1}{72} \\ \dfrac{1}{12} & \dfrac{11}{96} & \dfrac{-5}{144} \end{bmatrix}$

$X = A^{-1}B$

$X = \begin{bmatrix} -1 \\ -1 \\ 2 \end{bmatrix}$

$(-1,\,-1,\,2)$
Answers will vary.

51. $x^2 + y^2 + Dx + Ey + F = 0$

$(-1, 7), (2, 8)$ and $(5, -1)$

$x^2 + y^2 + Dx + Ey + F = 0$
$(-1)^2 + (7)^2 + D(-1) + E(7) + F = 0$
$1 + 49 - D + 7E + F = 0$
$-D + 7E + F = -50\,;$

$x^2 + y^2 + Dx + Ey + F = 0$
$(2)^2 + (8)^2 + D(2) + E(8) + F = 0$
$4 + 64 + 2D + 8E + F = 0$
$2D + 8E + F = -68\,;$

$x^2 + y^2 + Dx + Ey + F = 0$
$(5)^2 + (-1)^2 + D(5) + E(-1) + F = 0$
$25 + 1 + 5D - E + F = 0$
$5D - E + F = -26\,;$

$\begin{cases} -D + 7E + F = -50 \\ 2D + 8E + F = -68 \\ 5D - E + F = -26 \end{cases}$

$\begin{bmatrix} -1 & 7 & 1 \\ 2 & 8 & 1 \\ 5 & -1 & 1 \end{bmatrix}\begin{bmatrix} D \\ E \\ F \end{bmatrix} = \begin{bmatrix} -50 \\ -68 \\ -26 \end{bmatrix}$

$A^{-1} = \begin{bmatrix} -\dfrac{3}{10} & \dfrac{4}{15} & \dfrac{1}{30} \\ \dfrac{-1}{10} & \dfrac{1}{5} & \dfrac{-1}{10} \\ \dfrac{7}{5} & \dfrac{-17}{15} & \dfrac{11}{15} \end{bmatrix}$

$X = A^{-1}B$

$X = \begin{bmatrix} -4 \\ -6 \\ -12 \end{bmatrix}$

$D = -4,\;\; E = -6,\;\; F = -12$
$x^2 + y^2 - 4x - 6y - 12 = 0$

53. $f(x) = x^3 - 2x^2 - 7x + 6$

End behavior: down to the left, up to the right
y – intercept: $(0, 6)$
x – intercepts: $(-2.26, 0), (0.76, 0), (3.51, 0)$

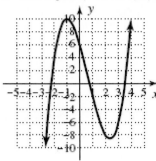

55. $\dfrac{\sin 49.0}{8.7} = \dfrac{\sin B}{11.2}$

$11.2\left(\dfrac{\sin 49.0}{8.7}\right) = \sin B$

$\angle B = \sin^{-1}\left(\dfrac{11.2 \sin 49.0}{8.7}\right)$

$\angle B \approx 76.3°$;

$\angle C \approx 180° - 76.3° - 49.0° = 54.7°$;

$\dfrac{\sin 49.0}{8.7} = \dfrac{\sin 54.7°}{c}$

$c = 8.7\left(\dfrac{\sin 54.7°}{\sin 49.0°}\right)$

$c \approx 9.4$ in.

Summary and Concept Review

1. $\begin{cases} 3x - 2y = 4 \\ -x + 3y = 8 \end{cases}$

$\begin{cases} y = \dfrac{3}{2}x - 2 \\ y = \dfrac{1}{3}x + \dfrac{8}{3} \end{cases}$

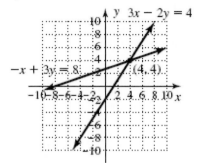

3. $\begin{cases} 2x + y = 2 \\ x - 2y = 4 \end{cases}$

$\begin{cases} y = -2x + 2 \\ y = \dfrac{1}{2}x - 2 \end{cases}$

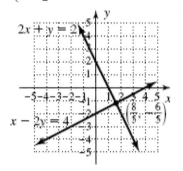

5. $\begin{cases} x + y = 4 \\ 0.4x + 0.3y = 1.7 \end{cases}$

$x = 4 - y$

$0.4x + 0.3y = 1.7$
$0.4(4 - y) + 0.3y = 1.7$
$1.6 - 0.4y + 0.3y = 1.7$
$-0.1y = 0.1$
$y = -1;$

$x + y = 4$

$x + (-1) = 4$

$x - 1 = 4$

$x = 5$

$(5, -1);$ consistent

7. $\begin{cases} 2x - 4y = 10 \\ 3x + 4y = 5 \end{cases}$

R1 + R2 = Sum

$2x - 4y = 10$

$3x + 4y = 5$

$5x = 15$

$x = 3;$

$2x - 4y = 10$

$2(3) - 4y = 10$

$6 - 4y = 10$

$-4y = 4$

$y = -1$

$(3, -1);$ consistent

9. $\begin{cases} 2x = 3y + 6 \\ 2.4x + 3.6y = 6 \end{cases}$

$\begin{cases} 2x - 3y = 6 \\ 24x + 36y = 60 \end{cases}$

12R1 + R2 = Sum

$24x - 36y = 72$

$24x + 36y = 60$

$48x = 132$

$x = \dfrac{11}{4};$

$2x = 3y + 6$

$2\left(\dfrac{11}{4}\right) = 3y + 6$

$\dfrac{11}{2} = 3y + 6$

$-\dfrac{1}{2} = 3y$

$-\dfrac{1}{6} = y$

$\left(\dfrac{11}{4}, -\dfrac{1}{6}\right);$ consistent

11. $\begin{cases} x + y - 2z = -1 \\ 4x - y + 3z = 3 \\ 3x + 2y - z = 4 \end{cases}$

R1 + R2 = Sum

$x + y - 2z = -1$

$4x - y + 3z = 3$

$5x + z = 2$

-2R1 + R3 = Sum

$-2x - 2y + 4z = 2$

$3x + 2y - z = 4$

$x + 3z = 6$

$\begin{cases} 5x + z = 2 \\ x + 3z = 6 \end{cases}$

-3R1 + R2 = Sum

$-15x - 3z = -6$

$x + 3z = 6$

$-14x = 0$

$x = 0;$

$x + 3z = 6$

$0 + 3z = 6$

$3z = 6$

$z = 2;$

$x + y - 2z = -1$

$0 + y - 2(2) = -1$

$y - 4 = -1$

$y = 3$

$(0, 3, 2)$

13. $\begin{cases} 3x + y + 2z = 3 \\ x - 2y + 3z = 1 \\ 4x - 8y + 12z = 7 \end{cases}$

-4R2 + R3 = Sum

$-4x + 8y - 12z = -4$

$4x - 8y + 12z = 7$

$0 \neq 3$

No solution; inconsistent

15. $y = ax^2 + bx + c$

$\begin{cases} -9 = a(0)^2 + b(0) + c \\ 7 = a(2)^2 + b(2) + c \\ 15 = a(6)^2 + b(6) + c \end{cases}$

$\begin{cases} c = -9 \\ 4a + 2b + c = 7 \\ 36a + 6b + c = 15 \end{cases}$

$\begin{cases} 4a + 2b - 9 = 7 \\ 36a + 6b - 9 = 15 \end{cases}$

$\begin{cases} 4a + 2b = 16 \\ 36a + 6b = 24 \end{cases}$

-3R1 + R2 = Sum

$24a = -24$

$a = -1;$

$4(-1) + 2b = 16$

$2b = 20$

$b = 10;$

$y = -1x^2 + 10x - 9$

17. $\dfrac{16x+1}{2x^2-5x-3} = \dfrac{16x+1}{(x-3)(2x+1)}$

$= \dfrac{A}{x-3} + \dfrac{B}{2x+1}$

$16x+1 = A(2x+1) + B(x-3)$

$16x+1 = 2Ax + A + Bx - 3B$

$16x+1 = (2A+B)x + A - 3B$

$\begin{cases} 2A+B=16 \\ A-3B=1 \end{cases}$

$\begin{bmatrix} 2 & 1 \\ 1 & -3 \end{bmatrix}\begin{bmatrix} A \\ B \end{bmatrix} = \begin{bmatrix} 16 \\ 1 \end{bmatrix}$

$A^{-1} = \begin{bmatrix} \dfrac{3}{7} & \dfrac{1}{7} \\ \dfrac{1}{7} & \dfrac{-2}{7} \end{bmatrix}$

$X = A^{-1}B$

$X = \begin{bmatrix} 7 \\ 2 \end{bmatrix}$

$\dfrac{7}{x-3} + \dfrac{2}{2x+1}$

19. $\dfrac{-2x^2-3x-19}{x^3-5x^2+3x-15} = \dfrac{-2x^2-3x-19}{(x-5)(x^2+3)}$

$= \dfrac{A}{x-5} + \dfrac{Bx+C}{x^2+3}$

$-2x^2-3x-19 = A(x^2+3) + (Bx+C)(x-5)$

$-2x^2-3x-19 = Ax^2 + 3A + Bx^2 - 5Bx + Cx - 5C$

$-2x^2-3x-19$

$= x^2(A+B) + x(-5B+C) + 3A - 5C$

$\begin{cases} A+B=-2 \\ -5B+C=-3 \\ 3A-5C=-19 \end{cases}$

$\begin{bmatrix} 1 & 1 & 0 \\ 0 & -5 & 1 \\ 3 & 0 & -5 \end{bmatrix}\begin{bmatrix} A \\ B \\ C \end{bmatrix} = \begin{bmatrix} -2 \\ -3 \\ -19 \end{bmatrix}$

$A^{-1} = \begin{bmatrix} \dfrac{25}{28} & \dfrac{5}{28} & \dfrac{1}{28} \\ \dfrac{3}{28} & \dfrac{-5}{28} & \dfrac{-1}{28} \\ \dfrac{15}{28} & \dfrac{3}{28} & \dfrac{-5}{28} \end{bmatrix}$

$X = A^{-1}B$

$X = \begin{bmatrix} -3 \\ 1 \\ 2 \end{bmatrix}$

$= \dfrac{-3}{x-5} + \dfrac{x+2}{x^2+3}$

21. $\dfrac{6x^2+2x+7}{x^3-1} = \dfrac{6x^2+2x+7}{(x-1)(x^2+x+1)}$

$= \dfrac{A}{x-1} + \dfrac{Bx+C}{x^2+x+1}$

$6x^2+2x+7 = A(x^2+x+1) + (Bx+C)(x-1)$

$6x^2+2x+7 = Ax^2 + Ax + A + Bx^2 - Bx + Cx - C$

$6x^2+2x+7$

$= x^2(A+B) + x(A-B+C) + A - C$

$\begin{cases} A+B=6 \\ A-B+C=2 \\ A-C=7 \end{cases}$

$\begin{bmatrix} 1 & 1 & 0 \\ 1 & -1 & 1 \\ 1 & 0 & -1 \end{bmatrix}\begin{bmatrix} A \\ B \\ C \end{bmatrix} = \begin{bmatrix} 6 \\ 2 \\ 7 \end{bmatrix}$

$A^{-1} = \begin{bmatrix} \dfrac{1}{3} & \dfrac{1}{3} & \dfrac{1}{3} \\ \dfrac{2}{3} & \dfrac{-1}{3} & \dfrac{-1}{3} \\ \dfrac{1}{3} & \dfrac{1}{3} & \dfrac{-2}{3} \end{bmatrix}$

$X = A^{-1}B$

$X = \begin{bmatrix} 5 \\ 1 \\ -2 \end{bmatrix}$

$\dfrac{5}{x-1} + \dfrac{x-2}{x^2+x+1}$

Summary and Concept Review

23. $\dfrac{-x^2-15x+22}{x^3+3x^2-9x+5}=\dfrac{-x^2-15x+22}{(x+5)(x-1)^2}$

$\dfrac{-x^2-15x+22}{(x+5)(x-1)^2}=\dfrac{A}{x+5}+\dfrac{B}{x-1}+\dfrac{C}{(x-1)^2}$

$-x^2-15x+22=A(x-1)^2+B(x+5)(x-1)+C(x+5)$

$-x^2-15x+22$

$=Ax^2-2Ax+A+Bx^2+4Bx-5B+Cx+5C$

$-x^2-15x+22$

$=x^2(A+B)+x(-2A+4B+C)+A-5B+5C$

$\begin{cases}A+B=-1\\-2A+4B+C=-15\\A-5B+5C=22\end{cases}$

$\begin{bmatrix}1&1&0\\-2&4&1\\1&-5&5\end{bmatrix}\begin{bmatrix}A\\B\\C\end{bmatrix}=\begin{bmatrix}-1\\-15\\22\end{bmatrix}$

$A^{-1}=\begin{bmatrix}\dfrac{25}{36}&\dfrac{-5}{36}&\dfrac{1}{36}\\[2mm]\dfrac{11}{36}&\dfrac{5}{36}&-\dfrac{1}{36}\\[2mm]\dfrac{1}{6}&\dfrac{1}{6}&\dfrac{1}{6}\end{bmatrix}$

$X=A^{-1}B$

$X=\begin{bmatrix}2\\-3\\1\end{bmatrix}$

$\dfrac{-x^2-15x+22}{(x+5)(x-1)^2}=\dfrac{2}{x+5}-\dfrac{3}{x-1}+\dfrac{1}{(x-1)^2}$

25. $\begin{cases}-x-y>-2\\-x+y<-4\end{cases}$

$\begin{cases}y<-x+2\\y<x-4\end{cases}$

Test point: $(2,\,-5)$

$-x-y>-2$
$-2-(-5)>-2$
$-2+5>-2$
$3>-2;$

$-x+y<-4$
$-2+(-5)<-4$
$-7<-4$

27. $\begin{cases}x+2y\geq1\\2x-y\leq-2\end{cases}$

$\begin{cases}y\geq-\dfrac{1}{2}x+\dfrac{1}{2}\\y\geq2x+2\end{cases}$

Test point: $(0,4)$
$x+2y\geq1$
$0+2(4)\geq1$
$8\geq1;$

$2x-y\leq-2$
$2(0)-4\leq-2$
$-4\leq-2$

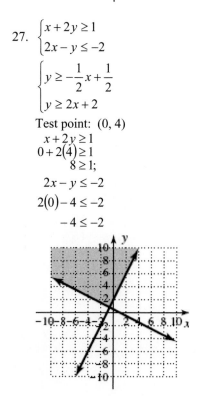

452

29. $\begin{cases} 3m + 2e \le 1000 \\ 2m + 1e \le 525 \\ m \ge 0 \\ e \ge 0 \end{cases}$

$\begin{cases} m \le \dfrac{-2e + 1000}{3} \\ m \le \dfrac{-1e + 525}{2} \\ m \ge 0 \\ e \ge 0 \end{cases}$

Corner Point	Maximize $P(e, m) = 85m + 50e$	Result
(0, 0)	$P(e, m) = 85(0) + 50(0)$	0
(0,267.5)	$P(e, m) = 85(267.5) + 50(0)$	22737.50
(425,50)	$P(e, m) = 85(50) + 50(425)$	25,500
(500,0)	$P(e, m) = 85(0) + 50(500)$	25,000

Maximize profits with 50 milk cows, 425 egg – laying chickens.

31. $\begin{cases} x - 2y + 2z = 7 \\ 2x + 2y - z = 5 \\ 3x - y + z = 6 \end{cases}$

$\begin{bmatrix} 1 & -2 & 2 & | & 7 \\ 2 & 2 & -1 & | & 5 \\ 3 & -1 & 1 & | & 6 \end{bmatrix} \quad -2R1 + R2 \to R2$

$\begin{bmatrix} 1 & -2 & 2 & | & 7 \\ 0 & 6 & -5 & | & -9 \\ 3 & -1 & 1 & | & 6 \end{bmatrix} \quad -3R1 + R3 \to R3$

$\begin{bmatrix} 1 & -2 & 2 & | & 7 \\ 0 & 6 & -5 & | & -9 \\ 0 & 5 & -5 & | & -15 \end{bmatrix} \quad \dfrac{1}{6}R2 \to R2$

$\begin{bmatrix} 1 & -2 & 2 & | & 7 \\ 0 & 1 & -\dfrac{5}{6} & | & -\dfrac{3}{2} \\ 0 & 5 & -5 & | & -15 \end{bmatrix} \quad -5R2 + R3 \to R3$

$\begin{bmatrix} 1 & -2 & 2 & | & 7 \\ 0 & 1 & -\dfrac{5}{6} & | & -\dfrac{3}{2} \\ 0 & 0 & -\dfrac{5}{6} & | & -\dfrac{15}{2} \end{bmatrix} \quad -\dfrac{6}{5}R3 \to R3$

$\begin{bmatrix} 1 & -2 & 2 & | & 7 \\ 0 & 1 & -\dfrac{5}{6} & | & -\dfrac{3}{2} \\ 0 & 0 & 1 & | & 9 \end{bmatrix}$

$z = 9$;

$y - \dfrac{5}{6}z = -\dfrac{3}{2}$

$y - \dfrac{5}{6}(9) = -\dfrac{3}{2}$

$y - \dfrac{15}{2} = -\dfrac{3}{2}$

$y = -\dfrac{3}{2} + \dfrac{15}{2}$

$y = 6;$

$x - 2y + 2z = 7$

$x - 2(6) + 2(9) = 7$

$x - 12 + 18 = 7$

$x + 6 = 7$

$x = 1$

$(1, 6, 9)$

33. $A + B$

$= \begin{bmatrix} \dfrac{-1}{4} & \dfrac{-3}{4} \\ \dfrac{-1}{8} & \dfrac{-7}{8} \end{bmatrix} + \begin{bmatrix} -7 & 6 \\ 1 & -2 \end{bmatrix}$

$= \begin{bmatrix} \dfrac{-1}{4} + (-7) & \dfrac{-3}{4} + 6 \\ \dfrac{-1}{8} + 1 & \dfrac{-7}{8} + (-2) \end{bmatrix}$

$= \begin{bmatrix} -7.25 & 5.25 \\ 0.875 & -2.875 \end{bmatrix}$

35. $C - B$

$= \begin{bmatrix} -1 & 3 & 4 \\ 5 & -2 & 0 \\ 6 & -3 & 2 \end{bmatrix} - \begin{bmatrix} -7 & 6 \\ 1 & -2 \end{bmatrix}$

Not possible

37. BA

$$= \begin{bmatrix} -7 & 6 \\ 1 & -2 \end{bmatrix} \begin{bmatrix} \dfrac{-1}{4} & \dfrac{-3}{4} \\ \dfrac{-1}{8} & \dfrac{-7}{8} \end{bmatrix}$$

$$= \begin{bmatrix} -7\left(\dfrac{-1}{4}\right)+6\left(\dfrac{-1}{8}\right) & -7\left(\dfrac{-3}{4}\right)+6\left(\dfrac{-7}{8}\right) \\ 1\left(\dfrac{-1}{4}\right)+(-2)\left(\dfrac{-1}{8}\right) & 1\left(\dfrac{-3}{4}\right)+(-2)\left(\dfrac{-7}{8}\right) \end{bmatrix}$$

$$= \begin{bmatrix} 1 & 0 \\ 0 & 1 \end{bmatrix}$$

39. $D - C$

$$= \begin{bmatrix} 2 & -3 & 0 \\ 0.5 & 1 & -1 \\ 4 & 0.1 & 5 \end{bmatrix} - \begin{bmatrix} -1 & 3 & 4 \\ 5 & -2 & 0 \\ 6 & -3 & 2 \end{bmatrix}$$

$$= \begin{bmatrix} 2-(-1) & -3-3 & 0-4 \\ 0.5-5 & 1-(-2) & -1-0 \\ 4-6 & 0.1-(-3) & 5-2 \end{bmatrix}$$

$$= \begin{bmatrix} 3 & -6 & -4 \\ -4.5 & 3 & -1 \\ -2 & 3.1 & 3 \end{bmatrix}$$

41. $-4D$

$$= -4 \begin{bmatrix} 2 & -3 & 0 \\ 0.5 & 1 & -1 \\ 4 & 0.1 & 5 \end{bmatrix}$$

$$= \begin{bmatrix} -4(2) & -4(-3) & -4(0) \\ -4(0.5) & -4(1) & -4(-1) \\ -4(4) & -4(0.1) & -4(5) \end{bmatrix}$$

$$= \begin{bmatrix} -8 & 12 & 0 \\ -2 & -4 & 4 \\ -16 & -0.4 & -20 \end{bmatrix}$$

43. $A = \begin{bmatrix} 1 & 0 \\ 0 & 1 \end{bmatrix}$

$|A| = 1(1) - 0(0) = 1;$

$B = \begin{bmatrix} 0.2 & 0.2 \\ -0.6 & 0.4 \end{bmatrix}$

$|B| = 0.2(0.4) - 0.2(-0.6) = 0.2;$

$C = \begin{bmatrix} 2 & -1 \\ 3 & 1 \end{bmatrix}$

$|C| = 2(1) - (-1)(3) = 5;$

$D = \begin{bmatrix} 10 & -6 \\ -15 & 9 \end{bmatrix}$

$|D| = 10(9) - (-6)(-15) = 0$

Matrix D is singular

45. $BC = \begin{bmatrix} 0.2 & 0.2 \\ -0.6 & 0.4 \end{bmatrix} \begin{bmatrix} 2 & -1 \\ 3 & 1 \end{bmatrix} = \begin{bmatrix} 1 & 0 \\ 0 & 1 \end{bmatrix}$

$CB = \begin{bmatrix} 2 & -1 \\ 3 & 1 \end{bmatrix} \begin{bmatrix} 0.2 & 0.2 \\ -0.6 & 0.4 \end{bmatrix} = \begin{bmatrix} 1 & 0 \\ 0 & 1 \end{bmatrix}$

It is the inverse of B.

47. $GF = \begin{bmatrix} 1 & 0 & 0 \\ 0 & 1 & 0 \\ 0 & 0 & 1 \end{bmatrix} \begin{bmatrix} 1 & -1 & 1 \\ 0 & 1 & 0 \\ -2 & 1 & -1 \end{bmatrix}$

$$= \begin{bmatrix} 1 & -1 & 1 \\ 0 & 1 & 0 \\ -2 & 1 & -1 \end{bmatrix};$$

$FG = \begin{bmatrix} 1 & -1 & 1 \\ 0 & 1 & 0 \\ -2 & 1 & -1 \end{bmatrix} \begin{bmatrix} 1 & 0 & 0 \\ 0 & 1 & 0 \\ 0 & 0 & 1 \end{bmatrix}$

$$= \begin{bmatrix} 1 & -1 & 1 \\ 0 & 1 & 0 \\ -2 & 1 & -1 \end{bmatrix}$$

It is an identity matrix.

49. $EH \neq HE$

$$EH = \begin{bmatrix} 1 & -2 & 3 \\ -2 & 1 & -5 \\ -1 & -1 & -2 \end{bmatrix}\begin{bmatrix} -1 & 0 & -1 \\ 0 & 1 & 0 \\ 2 & 1 & 1 \end{bmatrix}$$

$$= \begin{bmatrix} 5 & 1 & 2 \\ -8 & -4 & -3 \\ -3 & -3 & -1 \end{bmatrix}$$

$$HE = \begin{bmatrix} -1 & 0 & -1 \\ 0 & 1 & 0 \\ 2 & 1 & 1 \end{bmatrix}\begin{bmatrix} 1 & -2 & 3 \\ -2 & 1 & -5 \\ -1 & -1 & -2 \end{bmatrix}$$

$$= \begin{bmatrix} 0 & 3 & -1 \\ -2 & 1 & -5 \\ -1 & -4 & -1 \end{bmatrix}$$

$EH \neq HE$; verified

51. $\begin{cases} 0.5x - 2.2y + 3z = -8 \\ -0.6x - y + 2z = -7.2 \\ x + 1.5y - 0.2z = 2.6 \end{cases}$

$$\begin{bmatrix} 0.5 & -2.2 & 3 & | & -8 \\ -0.6 & -1 & 2 & | & -7.2 \\ 1 & 1.5 & -0.2 & | & 2.6 \end{bmatrix}$$

$(2, 0, -3)$

53. $\begin{cases} 2x + y = -2 \\ -x + y + 5z = 12 \\ 3x - 2y + z = -8 \end{cases}$

$$D = \begin{vmatrix} 2 & 1 & 0 \\ -1 & 1 & 5 \\ 3 & -2 & 1 \end{vmatrix} = 38 \, ;$$

$$D_x = \begin{vmatrix} -2 & 1 & 0 \\ 12 & 1 & 5 \\ -8 & -2 & 1 \end{vmatrix} = -74 \, ;$$

$$D_y = \begin{vmatrix} 2 & -2 & 0 \\ -1 & 12 & 5 \\ 3 & -8 & 1 \end{vmatrix} = 72 \, ;$$

$$D_z = \begin{vmatrix} 2 & 1 & -2 \\ -1 & 1 & 12 \\ 3 & -2 & -8 \end{vmatrix} = 62 \, ;$$

$$x = \frac{D_x}{D} = \frac{-74}{38} = -\frac{37}{19} \, ;$$

$$y = \frac{D_y}{D} = \frac{72}{38} = \frac{36}{19} \, ;$$

$$z = \frac{D_z}{D} = \frac{62}{38} = \frac{31}{19}$$

$$\left(\frac{-37}{19}, \frac{36}{19}, \frac{31}{19} \right)$$

55. $(6, 1), (-1, -6)$ and $(-6, 2)$

$$A = \frac{\begin{vmatrix} \begin{vmatrix} x_1 & y_1 & 1 \\ x_2 & y_2 & 1 \\ x_3 & y_3 & 1 \end{vmatrix} \end{vmatrix}}{2}$$

$$A = \frac{\begin{vmatrix} \begin{vmatrix} 6 & 1 & 1 \\ -1 & -6 & 1 \\ -6 & 2 & 1 \end{vmatrix} \end{vmatrix}}{2} = \begin{vmatrix} \frac{-91}{2} \end{vmatrix} = \frac{91}{2} \, \text{cm}^2$$

Mixed Review

1. a. $\begin{cases} -3x + 5y = 10 \\ 6x + 20 = 10y \end{cases}$

$\begin{cases} y = \dfrac{3}{5}x + 2 \\ y = \dfrac{3}{5}x + 2 \end{cases}$

Consistent/dependent

b. $\begin{cases} 4x - 3y = 9 \\ -2x + 5y = -10 \end{cases}$

$\begin{cases} y = \dfrac{4}{3}x - 3 \\ y = \dfrac{2}{5}x - 2 \end{cases}$

Consistent/independent

c. $\begin{cases} x - 3y = 9 \\ -6y + 2x = 10 \end{cases}$

$\begin{cases} y = \dfrac{1}{3}x - 3 \\ y = \dfrac{1}{3}x - \dfrac{5}{3} \end{cases}$

Inconsistent

3. $\begin{cases} 2x + 3y = 5 \\ -x + 5y = 17 \end{cases}$

$x = 5y - 17$

$2x + 3y = 5$

$2(5y - 17) + 3y = 5$

$10y - 34 + 3y = 5$

$13y = 39$

$y = 3;$

$-x + 5y = 17$

$-x + 5(3) = 17$

$-x + 15 = 17$

$-x = 2$

$x = -2$

$(-2, 3)$

5. $\begin{cases} x + 2y - 3z = -4 \\ -3x + 4y + z = 1 \\ 2x - 6y + z = 1 \end{cases}$

R1 + 3R2 = Sum

$x + 2y - 3z = -4$

$-9x + 12y + 3z = 3$

$-8x + 14y = -1$

R1 + 3R3 = Sum

$x + 2y - 3z = -4$

$6x - 18y + 3z = 3$

$7x - 16y = -1$

$\begin{cases} -8x + 14y = -1 \\ 7x - 16y = -1 \end{cases}$

7R1 + 8R2 = Sum

$-56x + 98y = -7$

$56x - 128y = -8$

$-30y = -15$

$y = \dfrac{1}{2};$

$-8x + 14\left(\dfrac{1}{2}\right) = -1$

$-8x + 7 = -1$

$-8x = -8$

$x = 1;$

$x + 2y - 3z = -4$

$1 + 2\left(\dfrac{1}{2}\right) - 3z = -4$

$1 + 1 - 3z = -4$

$2 - 3z = -4$

$-3z = -6$

$z = 2$

$\left(1, \dfrac{1}{2}, 2\right)$

7. $\begin{cases} \dfrac{1}{2}x + \dfrac{2}{3}y = 3 \\ \dfrac{-2}{5}x - \dfrac{1}{4}y = 1 \end{cases}$

$\begin{bmatrix} \dfrac{1}{2} & \dfrac{2}{3} & \Big| & 3 \\ \dfrac{-2}{5} & \dfrac{-1}{4} & \Big| & 1 \end{bmatrix} \quad 2R1 \to R1$

$\begin{bmatrix} 1 & \dfrac{4}{3} & \Big| & 6 \\ \dfrac{-2}{5} & \dfrac{-1}{4} & \Big| & 1 \end{bmatrix} \quad \dfrac{2}{5}R1 + R2 \to R2$

$\begin{bmatrix} 1 & \dfrac{4}{3} & \Big| & 6 \\ 0 & \dfrac{17}{60} & \Big| & \dfrac{17}{5} \end{bmatrix} \quad \dfrac{60}{17}R2 \to R2$

$\begin{bmatrix} 1 & \dfrac{4}{3} & \Big| & 6 \\ 0 & 1 & \Big| & 12 \end{bmatrix}$

$y = 12$;

$x + \dfrac{4}{3}y = 6$

$x + \dfrac{4}{3}(12) = 6$

$x + 16 = 6$

$x = -10$

$(-10, 12)$

9. $\dfrac{13-x}{x^2 - x - 6} = \dfrac{13-x}{(x+2)(x-3)}$

$= \dfrac{A}{x+2} + \dfrac{B}{x-3}$

$13 - x = A(x-3) + B(x+2)$

$13 - x = Ax - 3A + Bx + 2B$

$13 - x = x(A+B) - 3A + 2B$

$\begin{cases} A + B = -1 \\ -3A + 2B = 13 \end{cases}$

$\begin{bmatrix} 1 & 1 \\ -3 & 2 \end{bmatrix} \begin{bmatrix} A \\ B \end{bmatrix} = \begin{bmatrix} -1 \\ 13 \end{bmatrix}$

$A^{-1} = \begin{bmatrix} \dfrac{2}{5} & \dfrac{-1}{5} \\ \dfrac{3}{5} & \dfrac{1}{5} \end{bmatrix}$

$X = A^{-1}B$

$X = \begin{bmatrix} -3 \\ 2 \end{bmatrix}$

$\dfrac{13-x}{(x+2)(x-3)} = \dfrac{-3}{x+2} + \dfrac{2}{x-3}$

11. a. $-2[A][C]$

$= -2 \begin{bmatrix} 2 & -1 \\ 0 & 3 \end{bmatrix} \begin{bmatrix} 1 & -4 & 2 \\ -2 & 0 & -1 \end{bmatrix}$

$= -2 \begin{bmatrix} 4 & -8 & 5 \\ -6 & 0 & -3 \end{bmatrix}$

$= \begin{bmatrix} -8 & 16 & -10 \\ 12 & 0 & 6 \end{bmatrix}$

b. $[C][D]$

$= \begin{bmatrix} 1 & -4 & 2 \\ -2 & 0 & -1 \end{bmatrix} \begin{bmatrix} 3 & 0 & 1 \\ -1 & 2 & 0 \\ 1 & 1 & -4 \end{bmatrix}$

$= \begin{bmatrix} 9 & -6 & -7 \\ -7 & -1 & 2 \end{bmatrix}$

13. $\begin{cases} -x - 2z = 5 \\ 2y + z = -4 \\ -x + 2y = 3 \end{cases}$

$\begin{bmatrix} -1 & 0 & -2 & \Big| & 5 \\ 0 & 2 & 1 & \Big| & -4 \\ -1 & 2 & 0 & \Big| & 3 \end{bmatrix} \quad -R1 \to R1$

$\begin{bmatrix} 1 & 0 & 2 & \Big| & -5 \\ 0 & 2 & 1 & \Big| & -4 \\ -1 & 2 & 0 & \Big| & 3 \end{bmatrix} \quad R1 + R3 \to R3$

$\begin{bmatrix} 1 & 0 & 2 & \Big| & -5 \\ 0 & 2 & 1 & \Big| & -4 \\ 0 & 2 & 2 & \Big| & -2 \end{bmatrix} \quad \dfrac{1}{2}R2 \to R2$

$\begin{bmatrix} 1 & 0 & 2 & \Big| & -5 \\ 0 & 1 & \dfrac{1}{2} & \Big| & -2 \\ 0 & 2 & 2 & \Big| & -2 \end{bmatrix} \quad -2R2 + R3 \to R3$

$\begin{bmatrix} 1 & 0 & 2 & \Big| & -5 \\ 0 & 1 & \dfrac{1}{2} & \Big| & -2 \\ 0 & 0 & 1 & \Big| & 2 \end{bmatrix} \quad -\dfrac{1}{2}R3 + R2 \to R2$

$z = 2$;

$y + \dfrac{1}{2}z = -2$

$y + \dfrac{1}{2}(2) = -2$

$y + 1 = -2$

$y = -3$;

$x + 2z = -5$

$x + 2(2) = -5$

$x + 4 = -5$

$x = -9$

$(-9, -3, 2)$

15. $\begin{cases} -x+5y-2z=1 \\ 2x+3y-z=3 \\ 3x-y+3z=-2 \end{cases}$

$$D = \begin{vmatrix} -1 & 5 & -2 \\ 2 & 3 & -1 \\ 3 & -1 & 3 \end{vmatrix} = -31 \,;$$

$$D_x = \begin{vmatrix} 1 & 5 & -2 \\ 3 & 3 & -1 \\ -2 & -1 & 3 \end{vmatrix} = -33 \,;$$

$$D_y = \begin{vmatrix} -1 & 1 & -2 \\ 2 & 3 & -1 \\ 3 & -2 & 3 \end{vmatrix} = 10 \,;$$

$$D_z = \begin{vmatrix} -1 & 5 & 1 \\ 2 & 3 & 3 \\ 3 & -1 & -2 \end{vmatrix} = 57 \,;$$

$$x = \frac{D_x}{D} = \frac{-33}{-31} = \frac{33}{31} \,;$$

$$y = \frac{D_y}{D} = \frac{10}{-31} = \frac{-10}{31} \,;$$

$$z = \frac{D_z}{D} = \frac{57}{-31} = \frac{-57}{31}$$

$$\left(\frac{33}{31}, \frac{-10}{31}, \frac{-57}{31} \right)$$

17. $\begin{cases} 4x+2y \le 14 \\ 2x+3y \le 15 \\ y \ge 0 \\ x \ge 0 \end{cases}$

$\begin{cases} y \le -2x+7 \\ y \le -\dfrac{2}{3}x+5 \\ y \ge 0 \\ x \ge 0 \end{cases}$

19. Let u represent the number of unicyles.
Let b represent the number of bicycles.
Let t represent the number of tricycles.

$\begin{cases} u+b+t=21 \\ u+2b+3t=40 \\ b=2t-1 \end{cases}$

$-R1 + R2 = \text{Sum}$

$-u-b-t=-21$
$\underline{u+2b+3t=40}$
$\qquad\qquad b+2t=19$

$\begin{cases} b=2t-1 \\ b+2t=19 \end{cases}$

$b+2t=19$
$2t-1+2t=19$
$\qquad 4t=20$
$\qquad\quad t=5;$

$b=2t-1$
$b=2(5)-1$
$b=10-1$
$b=9;$
$u+b+t=21$
$u+9+5=21$
$\quad u+14=21$
$\qquad\quad u=7$

7 unicycles
9 bicycles
5 tricycles

Practice Test

1. $\begin{cases} 3x + 2y = 12 \\ -x + 4y = 10 \end{cases}$

$\begin{cases} y = -\dfrac{3}{2}x + 6 \\ y = \dfrac{1}{4}x + \dfrac{5}{2} \end{cases}$

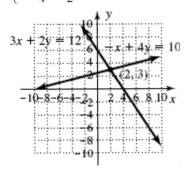

3. $\begin{cases} 5x + 8y = 1 \\ 3x + 7y = 5 \end{cases}$

$-3R1 + 5R2 = \text{Sum}$

$-15x - 24y = -3$

$15x + 35y = 25$

$11y = 22$

$y = 2;$

$3x + 7y = 5$

$3x + 7(2) = 5$

$3x + 14 = 5$

$3x = -9$

$x = -3$

$(-3, 2)$

5. a. $A - B$

$= \begin{bmatrix} -3 & -2 \\ 5 & 4 \end{bmatrix} - \begin{bmatrix} 3 & 3 \\ -3 & -5 \end{bmatrix}$

$= \begin{bmatrix} -3-3 & -2-3 \\ 5-(-3) & 4-(-5) \end{bmatrix}$

$= \begin{bmatrix} -6 & -5 \\ 8 & 9 \end{bmatrix}$

b. $\dfrac{2}{5}B = \dfrac{2}{5}\begin{bmatrix} 3 & 3 \\ -3 & -5 \end{bmatrix} = \begin{bmatrix} 1.2 & 1.2 \\ -1.2 & -2 \end{bmatrix}$

c. AB

$= \begin{bmatrix} -3 & -2 \\ 5 & 4 \end{bmatrix}\begin{bmatrix} 3 & 3 \\ -3 & -5 \end{bmatrix}$

$= \begin{bmatrix} -3 & 1 \\ 3 & -5 \end{bmatrix}$

d. A^{-1}

$= \begin{bmatrix} -2 & -1 \\ 2.5 & 1.5 \end{bmatrix}$

e. $|A| = -3(4) - (-2)(5) = -12 + 10 = -2$

7. $\begin{cases} 4x - 5y - 6z = 5 \\ 2x - 3y + 3z = 0 \\ x + 2y - 3z = 5 \end{cases}$

$\begin{bmatrix} 4 & -5 & -6 & | & 5 \\ 2 & -3 & 3 & | & 0 \\ 1 & 2 & -3 & | & 5 \end{bmatrix} R1 \leftrightarrow R3$

$\begin{bmatrix} 1 & 2 & -3 & | & 5 \\ 2 & -3 & 3 & | & 0 \\ 4 & -5 & -6 & | & 5 \end{bmatrix} -2R1 + R2 \rightarrow R2$

$\begin{bmatrix} 1 & 2 & -3 & | & 5 \\ 0 & -7 & 9 & | & -10 \\ 4 & -5 & -6 & | & 5 \end{bmatrix} -4R1 + R3 \rightarrow R3$

$\begin{bmatrix} 1 & 2 & -3 & | & 5 \\ 0 & -7 & 9 & | & -10 \\ 0 & -13 & 6 & | & -15 \end{bmatrix} -\frac{1}{7} R2 \leftrightarrow R2$

$\begin{bmatrix} 1 & 2 & -3 & | & 5 \\ 0 & 1 & -\frac{9}{7} & | & \frac{10}{7} \\ 0 & -13 & 6 & | & -15 \end{bmatrix} 13R2 + R3 \rightarrow R3$

$\begin{bmatrix} 1 & 2 & -3 & | & 5 \\ 0 & 1 & -\frac{9}{7} & | & \frac{10}{7} \\ 0 & 0 & -\frac{75}{7} & | & \frac{25}{7} \end{bmatrix} -\frac{7}{75} R3 \rightarrow R3$

$\begin{bmatrix} 1 & 2 & -3 & | & 5 \\ 0 & 1 & -\frac{9}{7} & | & \frac{10}{7} \\ 0 & 0 & 1 & | & -\frac{1}{3} \end{bmatrix}$

$z = -\frac{1}{3};$

$y - \frac{9}{7}z = \frac{10}{7}$

$y - \frac{9}{7}\left(\frac{-1}{3}\right) = \frac{10}{7}$

$y + \frac{3}{7} = \frac{10}{7}$

$y = \frac{7}{7}$

$y = 1;$

$x + 2y - 3z = 5$

$x + 2(1) - 3\left(\frac{-1}{3}\right) = 5$

$x + 2 + 1 = 5$

$x + 3 = 5$

$x = 2$

$\left(2, 1, -\frac{1}{3}\right)$

9. $\begin{cases} 2x - 5y = 11 \\ 4x + 7y = 4 \end{cases}$

$\begin{bmatrix} 2 & -5 & | & 11 \\ 4 & 7 & | & 4 \end{bmatrix}$

$A = \begin{bmatrix} 2 & -5 \\ 4 & 7 \end{bmatrix}; \quad B = \begin{bmatrix} 11 \\ 4 \end{bmatrix}$

$A^{-1} = \begin{bmatrix} \frac{7}{34} & \frac{5}{34} \\ \frac{-2}{17} & \frac{1}{17} \end{bmatrix}$

$A^{-1}B = \begin{bmatrix} \frac{97}{34} \\ \frac{-18}{17} \end{bmatrix}$

$\left(\frac{97}{34}, -\frac{18}{17}\right)$

11. Let l represent the length of the paper.
Let w represent the width of the paper.

$\begin{cases} 2l + 2w = 114.3 \\ l = 2w - 7.62 \end{cases}$

$2l + 2w = 114.3$

$2(2w - 7.62) + 2w = 114.3$

$4w - 15.24 + 2w = 114.3$

$6w = 129.54$

$w = 21.59;$

$l = 2w - 7.62$

$l = 2(21.59) - 7.62$

$l = 43.18 - 7.62$

$l = 35.56$

21.59 cm by 35.56 cm

13.
$$\begin{cases} 2C + 3B + P = 1.39 \\ 3C + 2B + 2P = 1.73 \\ C + 4B + 3P = 1.92 \end{cases}$$

$-2R1 + R2 = \text{Sum}$
$-4C - 6B - 2P = -2.78$

$\quad 3C + 2B + 2P = 1.73$
$\quad -C - 4B = -1.05$
$-3R1 + R3 = \text{Sum}$
$-6C - 9B - 3P = -4.17$

$\quad C + 4B + 3P = 1.92$
$-5C - 5B = -2.25$

$$\begin{cases} -C - 4B = -1.05 \\ -5C - 5B = -2.25 \end{cases}$$

$-5R1 + R2 = \text{Sum}$
$5C + 20B = 5.25$

$-5C - 5B = -2.25$

$\quad 15B = 3$

$\quad B = 0.20;$

$\quad -C - 4B = -1.05$

$-C - 4(0.20) = -1.05$

$\quad -C - 0.80 = -1.05$

$\quad -C = -0.25$

$\quad C = 0.25;$

$\quad 2C + 3B + P = 1.39$

$2(0.25) + 3(0.2) + P = 1.39$

$0.50 + 0.60 + P = 1.39$

$1.10 + P = 1.39$

$\quad P = 0.29$

Corn: 25¢
Beans: 20¢
Peas: 29¢

15. $h(t) = at^2 + bt + c$

$$\begin{cases} 128 = a(1)^2 + b(1) + c \\ 80 = a(2)^2 + b(2) + c \\ 44 = a(2.5)^2 + b(2.5) + c \end{cases}$$

$$\begin{cases} a + b + c = 128 \\ 4a + 2b + c = 80 \\ 6.25a + 2.5b + c = 44 \end{cases}$$

$R1 - R2 = \text{Sum}$
$-3a - b = 48$
$R1 - R3 = \text{Sum}$
$-5.25a - 1.5b = 84$

$$\begin{cases} -3a - b = 48 \\ -5.25a - 1.5b = 84 \end{cases}$$

$-1.5R1 + R2 = \text{Sum}$
$-0.75a = 12$

$a = -16;$

$-3a - b = 48$

$-3(-16) - b = 48$

$-b = 0$

$b = 0;$

$a + b + c = 128$

$-16 + 0 + c = 128$

$c = 144;$

$h(t) = -16t^2 + 144$

17. $\begin{cases} x - y \le 2 \\ x + 2y \ge 8 \end{cases}$

$\begin{cases} y \ge x - 2 \\ y \ge -\dfrac{1}{2}x + 4 \end{cases}$

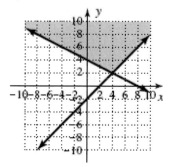

19. $P(x, y) = 4.25x + 5y$

$\begin{cases} x + y \le 50 \\ 2x + 3y \le 120 \end{cases}$

Corner Point	Objective Function $P(x, y) = 4.25x + 5y$	Result
(0, 0)	$P(0,0) = 4.25(0) + 5(0)$	0
(0, 40)	$P(0,40) = 4.25(0) + 5(40)$	200
(30, 20)	$P(30,20) = 4.25(30) + 5(20)$	227.5
(50, 0)	$P(50,0) = 4.25(50) + 5(0)$	212.50

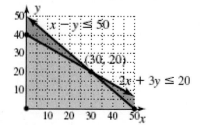

30 plain, 20 deluxe

Calculator Exploration and Discovery

Exercise 1: Answers will vary.

Exercise 3:

$\begin{cases} 2x + 2y = 15 \\ x + y = 6 \\ x + 4y = 9 \\ x, y \ge 0 \end{cases}$

(5, 1)

Strengthening Core Skills

Exercise 1:

$\begin{cases} 2x + y = -2 \\ -x + 3y - 2z = -15 \\ 3x - y + 2z = 9 \end{cases}$

$A = \begin{bmatrix} 2 & 1 & 0 \\ -1 & 3 & -2 \\ 3 & -1 & 2 \end{bmatrix};$

$A^{-1} = \begin{bmatrix} 1 & -0.5 & -0.5 \\ -1 & 1 & 1 \\ -2 & 1.25 & 1.75 \end{bmatrix}; \quad B = \begin{bmatrix} -2 \\ -15 \\ 9 \end{bmatrix}$

$A^{-1}B = \begin{bmatrix} 1 \\ -4 \\ 1 \end{bmatrix}$

(1, −4, 1)

Cumulative Review: Chapters 1-8

1. a) $9x^2 - 12x = -4$
   $9x^2 - 12x + 4 = 0$
   $(3x-2)(3x-2) = 0$
   $3x = 2$
   $x = \dfrac{2}{3}$, multiplicity 2

   b) $x^2 - 7x = 0$
   $x(x-7) = 0$
   $x = 0 \quad x - 7 = 0$
   $x = 7$

   c) $3x^3 - 15x^2 + 6x = 30$
   $3x^3 - 15x^2 + 6x - 30 = 0$
   $3x^2(x-5) + 6(x-5) = 0$
   $(3x^2 + 6)(x-5) = 0$
   $3x^2 + 6 = 0 \quad x - 5 = 0$
   $3x^2 - 6 \qquad x = 5$
   $x^2 = -2$
   $x = \pm\sqrt{-2}$
   $x = \pm i\sqrt{2}$

   d) $x^3 = 4x + 3x^2$
   $x^3 - 3x^2 - 4x = 0$
   $x(x^2 - 3x - 4) = 0$
   $x(x-4)(x+1) = 0$
   $x = 0 \quad x - 4 = 0 \quad x + 1 = 0$
   $x = 0 \qquad x = 4 \qquad x = -1$

3. $A = \pi^2\left(R^2 - r^2\right) = \pi^2 R^2 - \pi^2 r^2$
   $\pi^2 R^2 = A + \pi^2 r^2; \; R^2 = \dfrac{1}{\pi^2}\left(A + (\pi r)^2\right)$
   $R = \pm\dfrac{1}{\pi}\sqrt{A + (\pi r)^2}$

5.

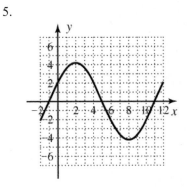

7. a. $(a+bi) + (a-bi) = 2a$ which has no imaginary part, and is real.

   b. $(a+bi)(a-bi) = a^2 + abi - abi - bi^2$
   $= a^2 + b^2$ which is also real.

9. $3x^2 - 72x + 427 = 0$
   $3\left(x^2 - 24x\right) = -427$
   $3\left(x^2 - 24x + 144\right) = -427 + 3(144)$
   $3(x-12)^2 = 5; \; (x-12)^2 = \dfrac{5}{3}$
   $x - 12 = \pm\dfrac{\sqrt{5}}{\sqrt{3}} = \pm\dfrac{\sqrt{15}}{3}$
   $x = 12 \pm \dfrac{\sqrt{15}}{3}$

11. $\left(\dfrac{\sqrt{3}}{4}, y\right), \; y > 0$

$$\left(\dfrac{\sqrt{3}}{4}\right)^2 + y^2 = 1$$

$$y^2 = 1 - \dfrac{3}{16}$$

$$y^2 = \dfrac{13}{16}$$

$$y^2 = \dfrac{\sqrt{13}}{4}$$

$$\sin\theta = \dfrac{\sqrt{13}}{4}$$

$$\cos\theta = \dfrac{\sqrt{3}}{4}$$

$$\tan\theta = \dfrac{\frac{\sqrt{13}}{4}}{\frac{\sqrt{3}}{4}} = \dfrac{\sqrt{13}}{\sqrt{3}} \cdot \dfrac{\sqrt{3}}{\sqrt{3}} = \dfrac{\sqrt{39}}{3}$$

13. a. $\quad m = \dfrac{y_2 - y_1}{x_2 - x_1}$

   b. $\quad \left(\dfrac{x_2 + x_1}{2}, \dfrac{y_2 + y_1}{2}\right)$

   c. $\quad x = \dfrac{-b \pm \sqrt{b^2 - 4ac}}{2a}$

   d. $\quad d = \sqrt{(x_2 - x_1)^2 + (y_2 - y_1)^2}$

   e. $\quad A = Pe^{rt}$

15. $-1\langle -5+8, 12+6\rangle = \langle -3, -18\rangle$

17. a. Sides: $x, 11, \sqrt{121+x^2}$

   $\dfrac{\sqrt{121+x^2}}{11}$

   b. Sides: $x, 3, \sqrt{9+x^2}$

   $\dfrac{x}{\sqrt{9+x^2}}$

19.

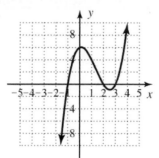

$f(x) < 0$ for $x \in (-\infty, -1)$ and $x \in (2,3)$.

21. $r = \sqrt{1^2 + \left(-\sqrt{3}\right)^2} = \sqrt{4} = 2$

$\theta = \tan^{-1}\left(\dfrac{-\sqrt{3}}{1}\right) = -60° \;\; (\text{QIV})$

$\left(1 - \sqrt{3}\right)^8 = 2^8 \text{cis}(8(-60°)) = 256\,\text{cis}(-480°)$

$= 256\,\text{cis}\,240° = 256\left(-\dfrac{1}{2} - i\dfrac{\sqrt{3}}{2}\right)$

$= -128 - 128i\sqrt{3}$

23. $R = 0.08/12 \approx 0.00667$

$P = \dfrac{AR}{(1+R)^{nt} - 1} = \dfrac{10{,}000\left(\frac{0.08}{12}\right)}{(1.00667)^{12t} - 1}$

$200 = \dfrac{66.67}{(1.00667)^{12t} - 1}$

$200\left(1.00667^{12t} - 1\right) = 66.67$

$200\left(1.00667^{12t}\right) - 200 = 66.67$

$200\left(1.00667^{12t}\right) = 266.67$

$1.00667^{12t} = 1.33; \quad 12t \ln 1.00667 = \ln(1.33)$

$t = \dfrac{\ln(1.33)}{12 \ln 1.00667} \approx 3.6 \text{ yr}$

25. $A = 2; P = 2\pi$, so $B = 1$; the graph is shifted $\dfrac{\pi}{4}$ units left, so $C = \dfrac{\pi}{4}$.

Chapter 8: Systems of Equations and Inequalities

Connections to Calculus

1.

a)
$$\frac{k}{(x+a)(x+b)} = \frac{A}{x+a} + \frac{B}{x+b}$$

$$\frac{k}{(x+a)(x+b)} = \frac{A(x+b) + B(x+a)}{(x+a)(x+b)}$$

$$k = Ax + Ab + Bx + Ba$$

$$k = (A+B)x + (Ab + Ba)$$

b)
$$\begin{cases} A+B=0 \\ Ab+Ba=k \end{cases}$$

$$A+B=0$$
$$B=-A$$
$$Ab+(-A)a=k$$
$$A(b-a)=k$$

$$\begin{cases} A+B=0 \\ A(b-a)=k \end{cases}$$

c)
$$A(b-a)=k$$

$$A = \frac{k}{(b-a)}$$

$$\frac{16}{(x+3)(x-5)}$$

$$k=16; \; a=3; \; b=-5$$

$$A = \frac{16}{-5-3} = -2$$

$$B = -A = -(-2) = 2$$

$$\frac{16}{(x+3)(x-5)} = \frac{-2}{x+3} + \frac{2}{x+5}$$

3.
$$\frac{6}{x^2-9} \Rightarrow 2k = 6$$

$$k=3$$

$$k^2 = 3^2 = 9$$

$$\frac{6}{x^2-9} = \frac{2(3)}{(x+3)(x-3)}$$

$$= \frac{A}{(x+3)} + \frac{B}{(x-3)}$$

$$= \frac{-1}{(x+3)} + \frac{1}{(x-3)}$$

5.
$$\frac{-22}{x^2+3x-28} = \frac{-22}{(x+7)(x-4)}$$

$$= \frac{A}{x+7} + \frac{B}{x-4}$$

$$k=-22; \; a=7; \; b=-4$$

$$A = \frac{k}{b-a} = \frac{-22}{-4-7} = \frac{-22}{-11} = 2$$

$$B = -A = -2$$

$$\frac{-22}{x^2+3x-28} = \frac{2}{x+7} + \frac{-2}{x-4}$$

7.
$$\frac{x+11}{x^2+13x+40} = \frac{x+11}{(x+8)(x+5)}$$

$$= \frac{A}{x+8} + \frac{B}{x+5}$$

$$k=11; \; a=8; \; b=5$$

$$A = \frac{k-a}{b-a} = \frac{11-8}{5-8} = \frac{3}{-3} = -1$$

$$B = 1 - A = 1 - (-1) = 2$$

$$\frac{x+11}{x^2+13x+40} = \frac{-1}{x+8} + \frac{2}{x+5}$$

9. $\mathbf{u} = \langle 2,8 \rangle$ and $\mathbf{v} = \langle 15,3 \rangle$

$$\text{abs}\left(\begin{vmatrix} 2 & 8 \\ 15 & 3 \end{vmatrix}\right) = |2 \cdot 3 - 8 \cdot 15|$$

$$= |6 - 120|$$

$$= |-114|$$

$$= 114 \text{ units}^2$$

Verify:

$$\mathbf{w} = \langle 2,8 \rangle + \langle 15,3 \rangle$$

$$= \langle 2+15, \ 8+3 \rangle$$

$$= \langle 17,11 \rangle$$

$$\mathbf{u} \square \mathbf{w} = 2 \square 17 + 8 \square 11 = 122$$

$$|\mathbf{w}| = \sqrt{17^2 + 11^2}$$

$$= \sqrt{289 + 121}$$

$$= \sqrt{410}$$

$$|\mathbf{u}| = \sqrt{2^2 + 8^2} = \sqrt{68} = 2\sqrt{17}$$

$$\cos\theta = \frac{\mathbf{u} \square \mathbf{w}}{|\mathbf{u}||\mathbf{w}|} = \frac{122}{2\sqrt{17}\sqrt{410}}$$

$$\cos\theta = \frac{61}{\sqrt{17}\sqrt{410}}$$

$$\theta = \cos^{-1}\left(\frac{61}{\sqrt{17}\sqrt{410}}\right) \approx 43.06°$$

A = area of $\square$ = 57 units^2

area of parallelogram $2A = 2(57) = 114$ units^2

11. $6\mathbf{i} + 3\mathbf{j}$ and $6\mathbf{i} - 3\mathbf{j}$

$\langle 6,3 \rangle$; and $\langle 6,-3 \rangle$

$$\text{abs}\left(\begin{vmatrix} 6 & 3 \\ 6 & -3 \end{vmatrix}\right) = |6(-3) - 6(3)| = |-36|$$

$$= 36 \text{ units}^2$$

Verify:

$$\mathbf{w} = \langle 6,3 \rangle + \langle 6,-3 \rangle$$

$$= \langle 6+6, \ 3+(-3) \rangle$$

$$= \langle 12,0 \rangle$$

$$\mathbf{u} \square \mathbf{w} = 6 \square 2 + 3 \square 0 = 72$$

$$|\mathbf{w}| = \sqrt{12^2 + 0^2} = \sqrt{144} = 12$$

$$|\mathbf{u}| = \sqrt{6^2 + 3^2} = \sqrt{45} = 3\sqrt{5}$$

$$\cos\theta = \frac{\mathbf{u} \square \mathbf{w}}{|\mathbf{u}||\mathbf{w}|} = \frac{72}{(3\sqrt{5})(12)} = \frac{2}{\sqrt{5}}$$

$$\theta = \cos^{-1}\left(\frac{2}{\sqrt{5}}\right) \approx 26.57°$$

area of $\square = \frac{1}{2}\left(3\sqrt{5}(12)\right)\sin 26.57° = 18$ units^2

area of parallelogram $= 2(18) = 36$ units^2

13. $\langle 5,0,0 \rangle$; $\langle 0,6,0 \rangle$; $\langle 0,0,8 \rangle$

$$\text{Volume} = \text{abs}\left(\begin{vmatrix} 5 & 0 & 0 \\ 0 & 6 & 0 \\ 0 & 0 & 8 \end{vmatrix}\right)$$

$$= 5\begin{vmatrix} 6 & 0 \\ 0 & 8 \end{vmatrix} - 0\begin{vmatrix} 0 & 0 \\ 0 & 8 \end{vmatrix} + 0\begin{vmatrix} 0 & 6 \\ 0 & 0 \end{vmatrix}$$

$$= 5(6 \cdot 8 - 0 \cdot 0) - 0 + 0$$

$$= 5(48) = 240 \text{ units}^2$$

$$V = LWH$$

$$= 5 \cdot 6 \cdot 8 = 240 \text{ units}^2$$

Chapter 9: Analytical Geometry

9.1 Exercises

1. Geometry, algebra

3. Perpendicular

5. Point, intersecting

7. Hypotenuse is from $P_2(1,2)$ to $P_3(-5,-6)$

   Midpoint $\left(\dfrac{1+(-5)}{2}, \dfrac{2+(-6)}{2}\right) = (-2,-2)$;

   Distance: $(-2,-2),(-5,2)$

   $\sqrt{\left(-2-(-5)\right)^2 + \left(-2-2\right)^2}$

   $= \sqrt{9+16} = 5$;

   Distance: $(-5,-6),(-2,-2)$

   $\sqrt{\left(-5-(-2)\right)^2 + \left(-6-(-2)\right)^2}$

   $= \sqrt{9+16} = 5$;

   Distance: $(1,2),(-2,-2)$

   $\sqrt{\left(1-(-2)\right)^2 + \left(2-(-2)\right)^2}$

   $= \sqrt{9+16} = 5$;

   Verified.

9. Hypotenuse is from $P_1(-2,1)$ to $P_2(6,-5)$

   Midpoint $\left(\dfrac{-2+6}{2}, \dfrac{1+(-5)}{2}\right) = (2,-2)$;

   Distance: $(2,-7),(2,-2)$

   $\sqrt{(2-2)^2 + \left(-2-(-7)\right)^2}$

   $= \sqrt{0+25} = 5$;

   Distance: $(-2,1),(2,-2)$

   $\sqrt{\left(2-(-2)\right)^2 + \left(-2-1\right)^2}$

   $= \sqrt{16+9} = 5$;

   Distance: $(6,-5),(2,-2)$

   $\sqrt{(2-6)^2 + \left(-2-(-5)\right)^2}$

   $= \sqrt{16+9} = 5$;

   Verified.

11. Hypotenuse is from $P_3(3,3)$ to $P_1(10,-21)$

    Midpoint $\left(\dfrac{3+10}{2}, \dfrac{3+(-21)}{2}\right) = \left(\dfrac{13}{2},-9\right)$;

    Distance: $(3,3),\left(\dfrac{13}{2},-9\right)$

    $\sqrt{\left(\dfrac{13}{2}-3\right)^2 + \left(-9-3\right)^2}$

    $= \sqrt{\left(\dfrac{7}{2}\right)^2 + 144} = \sqrt{\dfrac{625}{4}} = \dfrac{25}{2}$;

    Distance: $(-6,-9),\left(\dfrac{13}{2},-9\right)$

    $\sqrt{\left(\dfrac{13}{2}-(-6)\right)^2 + \left(-9-(-9)\right)^2}$

    $= \sqrt{\left(\dfrac{25}{2}\right)^2 + 0} = \dfrac{25}{2}$;

    Distance: $(10,-21),\left(\dfrac{13}{2},-9\right)$

    $\sqrt{\left(\dfrac{13}{2}-10\right)^2 + \left(-9-(-21)\right)^2}$

    $= \sqrt{\left(\dfrac{-7}{2}\right)^2 + 144} = \sqrt{\dfrac{625}{4}} = \dfrac{25}{2}$;

    Verified.

13. Center $(-2,-2)$, $d = 2(5) = 10$;

    $\left(x-(-2)\right)^2 + \left(y-(-2)\right)^2 = \left(\dfrac{10}{2}\right)^2$

    $(x+2)^2 + (y+2)^2 = 5^2$

15. Center $(2,-2)$, $d = 2(5) = 10$;

    $(x-2)^2 + \left(y-(-2)\right)^2 = \left(\dfrac{10}{2}\right)^2$

    $(x-2)^2 + (y+2)^2 = 5^2$

17. Center $\left(\dfrac{13}{2},-9\right)$, $d = 2\left(\dfrac{25}{2}\right) = 25$;

    $\left(x-\dfrac{13}{2}\right)^2 + \left(y-(-9)\right)^2 = \left(\dfrac{25}{2}\right)^2$

    $\left(x-\dfrac{13}{2}\right)^2 + (y+9)^2 = \left(\dfrac{25}{2}\right)^2$

9.1 Exercises

19. (a) $A(2,3)$ to $B(7,15)$

$$\sqrt{(7-2)^2+(15-3)^2}$$
$$=\sqrt{5^2+12^2}=13;$$

$A(2,3)$ to $C(-10,8)$

$$\sqrt{(-10-2)^2+(8-3)^2}$$
$$=\sqrt{(-12)^2+5^2}=13;$$

$A(2,3)$ to $D(9,14)$

$$\sqrt{(9-2)^2+(14-3)^2}$$
$$=\sqrt{7^2+11^2}=\sqrt{170};$$

$A(2,3)$ to $E(-3,-9)$

$$\sqrt{(-3-2)^2+(-9-3)^2}$$
$$=\sqrt{(-5)^2+(-12)^2}=13;$$

$A(2,3)$ to $F\left(5,4+3\sqrt{10}\right)$

$$\sqrt{(5-2)^2+\left(4+3\sqrt{10}-3\right)^2}$$
$$=\sqrt{3^2+\left(1+3\sqrt{10}\right)^2}$$
$$=\sqrt{9+1+6\sqrt{10}+3(10)}$$
$$=\sqrt{40+6\sqrt{10}};$$

$A(2,3)$ to $G(2-2\sqrt{30},10)$

$$\sqrt{\left(2-2\sqrt{30}-2\right)^2+(10-3)^2}$$
$$=\sqrt{\left(-2\sqrt{30}\right)^2+(7)^2}=\sqrt{169}=13;$$

Points of equal distance from A are: B, C, E, and G. Distance is 13.

(b) To find other points, pick any x or y value and find the other.
Pick $x=14$.
$A(2,3)$ to $(14,y)$

$$13=\sqrt{(14-2)^2+(y-3)^2}$$
$$13^2=12^2+(y-3)^2$$
$$169=144+y^2-6y+9$$
$$0=y^2-6y-16$$
$$0=(y-8)(y+2)$$
$$y-8=0 \text{ or } y+2=0$$
$$y=8 \qquad \text{or } y=-2;$$

$(14,8);(14,-2)$ are both the same distance away.
Pick $x=13$.

$A(2,3)$ to $(13,y)$

$$13=\sqrt{(13-2)^2+(y-3)^2}$$
$$13^2=11^2+(y-3)^2$$
$$169=121+y^2-6y+9$$
$$0=y^2-6y-39$$
$$y=\frac{-(-6)\pm\sqrt{(-6)^2-4(1)(-39)}}{2(1)}$$
$$=\frac{6\pm\sqrt{36+156}}{2}=\frac{6\pm\sqrt{192}}{2}$$
$$=\frac{6\pm8\sqrt{3}}{2}=3\pm4\sqrt{3};$$

$\left(13,3+4\sqrt{3}\right);\left(13,3-4\sqrt{3}\right)$ are both the same distance away.

21. $d=\left|\dfrac{Ax_1+By_1+C}{\sqrt{A^2+B^2}}\right|$

$P(-6,2)$ to $y=-\dfrac{1}{2}x+3$

$$y=-\frac{1}{2}x+3$$
$$2y=-x+6$$
$$x+2y-6=0;$$

$$d=\left|\frac{(1)(-6)+2(2)-6}{\sqrt{1^2+2^2}}\right|=\left|\frac{8}{\sqrt{5}}\right|$$
$$d=\frac{8}{\sqrt{5}}\cdot\frac{\sqrt{5}}{\sqrt{5}}=\frac{8\sqrt{5}}{5}$$

$Q(6,4)$ to $y=-\dfrac{1}{2}x+3$

$$x+2y-6=0;$$
$$d=\left|\frac{(1)(6)+2(4)-6}{\sqrt{1^2+2^2}}\right|=\left|\frac{8}{\sqrt{5}}\right|$$
$$d=\frac{8}{\sqrt{5}}\cdot\frac{\sqrt{5}}{\sqrt{5}}=\frac{8\sqrt{5}}{5}$$

Verified.

23. a. $A(0, 1)$ and $y = -1$, $y + 1 = 0$

$A(0,1)$ to $B(-6,9)$

$\sqrt{(-6-0)^2 + (9-1)^2}$

$= \sqrt{36+64} = 10;$

$B(-6,9)$ to $y+1=0$

$d = \left| \dfrac{0+9+1}{\sqrt{0^2+1^2}} \right| = 10;$

$A(0,1)$ to $C(4,4)$

$\sqrt{(4-0)^2 + (4-1)^2}$

$= \sqrt{16+9} = 5;$

$C(4,4)$ to $y+1=0$

$d = \left| \dfrac{0+4+1}{\sqrt{0^2+1^2}} \right| = 5;$

$A(0,1)$ to $D(-2\sqrt{2},6)$

$\sqrt{(-2\sqrt{2}-0)^2 + (6-1)^2}$

$= \sqrt{8+25} = \sqrt{33};$

$D(-2\sqrt{2},6)$ to $y+1=0$

$d = \left| \dfrac{0+6+1}{\sqrt{0^2+1^2}} \right| = 7;$

$A(0,1)$ to $E(4\sqrt{2},8)$

$\sqrt{(4\sqrt{2}-0)^2 + (8-1)^2}$

$= \sqrt{32+49} = 9;$

$E(4\sqrt{2},8)$ to $y+1=0$

$d = \left| \dfrac{0+8+1}{\sqrt{0^2+1^2}} \right| = 9;$

Points B, C, and E

b. Answers will vary.

Pick $y = 5$

$A(0,1)$ to $(x,5)$

$\sqrt{(x-0)^2 + (5-1)^2}$

$= \sqrt{x^2+16};$

$(x,5)$ to $y+1=0$

$d = \left| \dfrac{0+5+1}{\sqrt{0^2+1^2}} \right| = 6;$

$6 = \sqrt{x^2+16}$

$36 = x^2 + 16$

$20 = x^2$

$x = \pm\sqrt{20}$

$x = \pm 2\sqrt{5}$

$(2\sqrt{5},5), (-2\sqrt{5},5)$

25. $(0, -4)$ and $y = 4$, $y - 4 = 0$

$A(4,-1)$ to $(0,-4)$

$\sqrt{(0-4)^2 + (-4-(-1))^2}$

$= \sqrt{16+9} = 5;$

$A(4,-1)$ to $y-4=0$

$d = \left| \dfrac{0+(-1)-4}{\sqrt{0^2+1^2}} \right| = 5;$ verified

$B\left(10, -\dfrac{25}{4}\right)$ to $(0,-4)$

$\sqrt{(0-10)^2 + \left(-4-\dfrac{-25}{4}\right)^2}$

$= \sqrt{100+\dfrac{81}{16}} = \dfrac{41}{4};$

$B\left(10, \dfrac{25}{4}\right)$ to $y-4=0$

$d = \left| \dfrac{0+\dfrac{-25}{4}-4}{\sqrt{0^2+1^2}} \right| = \dfrac{41}{4};$ verified

$C\left(4\sqrt{2},-2\right)$ to $(0,-4)$

$d = \sqrt{(0-4\sqrt{2})^2 + (-4-(-2))^2}$

$= \sqrt{32+4} = 6;$

$C\left(4\sqrt{2},-2\right)$ to $y-4=0$

$d = \left| \dfrac{0-2-4}{\sqrt{0^2+1^2}} \right| = 6;$ verified

$D\left(8\sqrt{5},-20\right)$ to $(0,-4)$

$d = \sqrt{(0-8\sqrt{5})^2 + (-4-(-20))^2}$

$= \sqrt{320+256} = 24;$

$D\left(8\sqrt{5},-20\right)$ to $y-4=0$

$d = \left| \dfrac{0-20-4}{\sqrt{0^2+1^2}} \right| = 24;$ verified

9.1 Exercises

27. $(0, -4), y = 4$

$(x, 4)$ to (x, y) $\quad$ $(0, -4)$ to (x, y)

$$\sqrt{(x-x)^2 + (y-4)^2} = \sqrt{(x-0)^2 + (y-(-4))^2}$$

$$\sqrt{(y-4)^2} = \sqrt{x^2 + (y+4)^2}$$

$$y - 4 = \sqrt{x^2 + y^2 + 8y + 16}$$

$$y^2 - 8y + 16 = x^2 + y^2 + 8y + 16$$

$$-16y = x^2$$

$$y = -\frac{1}{16}x^2$$

29. Focus $(0, -2)$, directrix $y = -8$

$(0, -2)$ to (x, y) and $(x, -8)$ to (x, y)

$$\sqrt{(x-0)^2 + (y-(-2))^2} = \frac{1}{2}\sqrt{(x-x)^2 + (y-(-8))^2}$$

$$\sqrt{x^2 + (y+2)^2} = \frac{1}{2}\sqrt{(y+8)^2}$$

$$4\left(x^2 + (y+2)^2\right) = (y+8)^2$$

$$4x^2 + 4y^2 + 16y + 16 = y^2 + 16y + 64$$

$$4x^2 + 3y^2 = 48$$

31. $4x^2 + 3y^2 = 48, (-3, 2)$

$$4(-3)^2 + 3(2)^2 = 48$$

$$36 + 12 = 48$$

$$48 = 48 \text{ verified}$$

$$\left(\sqrt{12}, 0\right)$$

$$4\left(\sqrt{12}\right)^2 + 3(0)^2 = 48$$

$$48 + 0 = 48$$

$$48 = 48 \text{ verified}$$

$$d_1 = \sqrt{(0-(-3))^2 + (2-2)^2} = 3;$$

$$d_2 = \sqrt{(0-(-3))^2 + (-2-2)^2} = 5;$$

$$d_3 = \sqrt{\left(\sqrt{12}-0\right)^2 + (0-2)^2}$$

$$= \sqrt{12+4} = 4;$$

$$d_4 = \sqrt{\left(\sqrt{12}-0\right)^2 + (0-(-2))^2}$$

$$\sqrt{12+4} = 4;$$

$$d_1 + d_2 = d_3 + d_4$$

$$3 + 5 = 4 + 4$$

$$8 = 8 \text{ verified}$$

33. $x = \frac{1}{2}, (2, 0)$

$(2, 0)$ to (x, y) and $\left(\frac{1}{2}, y\right)$ to (x, y)

$$\sqrt{(x-2)^2 + (y-0)^2} = 2\sqrt{\left(x-\frac{1}{2}\right)^2 + (y-y)^2}$$

$$\sqrt{x^2 - 4x + 4 + y^2} = 2\sqrt{x^2 - x + \frac{1}{4} + 0}$$

$$x^2 - 4x + 4 + y^2 = 4\left(x^2 - x + \frac{1}{4}\right)$$

$$x^2 - 4x + 4 + y^2 = 4x^2 - 4x + 1$$

$$-3x^2 + y^2 = -3$$

$$3x^2 - y^2 = 3$$

35. $A(-8, 2), B(-2, -6), C(4, 0)$

a. orthocenter, point where altitudes meet

$$m_{\overline{AB}} = \frac{-6-2}{-2-(-8)} = \frac{-4}{3}$$

Perpendicular to $\overline{AB}: m = \frac{3}{4}$

$$m_{\overline{BC}} = \frac{0-(-6)}{4-(-2)} = 1$$

Perpendicular to $\overline{BC}: m = -1$

$$m_{\overline{CA}} = \frac{2-0}{-8-4} = -\frac{1}{6}$$

Perpendicular to $\overline{CA}: m = 6$

Equation of line from A to $\overline{BC}$

$$y - 2 = -1(x-(-8))$$

$$y - 2 = -x - 8$$

$$y = -x - 6 \text{ (Eq. 1)}$$

Equation of line from B to $\overline{CA}$

$$y - (-6) = 6(x-(-2))$$

$$y + 6 = 6x + 12$$

$$y = 6x + 6 \text{ (Eq. 2)}$$

Set Eq. 1 = Eq. 2

$$-x - 6 = 6x + 6$$

$$-7x = 12$$

$$x = -\frac{12}{7};$$

$$y = -\left(\frac{-12}{7}\right) - 6 = -\frac{30}{7}$$

$$\left(\frac{-12}{7}, \frac{-30}{7}\right)$$

b. Centroid, point where medians meet. Medians of triangle intersect in point that is two-thirds of the distance from each vertex to the midpoint of the opposite side.

$A(-8,2), B(-2,-6), C(4,0)$

Midpoint

$\overline{AB}$ $\left(\dfrac{-8+(-2)}{2}, \dfrac{2+(-6)}{2} \right) = (-5,-2)$

Midpoint $\overline{BC}$ $\left(\dfrac{-2+4}{2}, \dfrac{-6+0}{2} \right) = (1,-3)$

Midpoint $\overline{CA}$ $\left(\dfrac{4+-8}{2}, \dfrac{0+2}{2} \right) = (-2,1)$

B to midpoint of $\overline{CA}$ $(-2,-6),(-2,1)$

$\dfrac{2}{3}\sqrt{(-2-(-2))^2 + (1-(-6))^2}$

$= \dfrac{2}{3} \cdot 7 = \dfrac{14}{3}$

Using $B(-2, -6)$ to midpoint of $\overline{CA}$ $(-2,1)$,

Centroid at $(-2, y)$.

Because this is a vertical distance from

B, $y = -6 + \dfrac{14}{3} = -\dfrac{4}{3}$

$\left(-2, -\dfrac{4}{3} \right)$

37. $A(-2,0), B(2,0), C(-2,3), D\left(2\sqrt{2}, \sqrt{6}\right)$

$AC = \sqrt{(-2-(-2))^2 + (3-0)^2}$

$\qquad = \sqrt{0+9} = 3;$

$BC = \sqrt{(-2-2)^2 + (3-0)^2}$

$\qquad = \sqrt{16+9} = 5;$

$AD = \sqrt{\left(2\sqrt{2}-(-2)\right)^2 + \left(\sqrt{6}-0\right)^2}$

$\qquad = \sqrt{8\sqrt{2}+12+6} = \sqrt{8\sqrt{2}+18};$

$BD = \sqrt{\left(2\sqrt{2}-2\right)^2 + \left(\sqrt{6}-0\right)^2}$

$\qquad = \sqrt{12-8\sqrt{2}+6} = \sqrt{18-8\sqrt{2}};$

$AC+BC = 3+5 = 8;$

$AD+BD$

$\qquad = \sqrt{8\sqrt{2}+18} + \sqrt{18-8\sqrt{2}} = 8$

Verified, both add to 8.

39. $-225 = 600 + 825\sin\left(x+\dfrac{\pi}{6}\right)$

$-825 = 825\sin\left(x+\dfrac{\pi}{6}\right)$

$\dfrac{-825}{825} = \sin\left(x+\dfrac{\pi}{6}\right)$

$\sin\left(x+\dfrac{\pi}{6}\right) = -1$

$x + \dfrac{\pi}{6} = \dfrac{3\pi}{2} + 2\pi n$

$x = \dfrac{4\pi}{3} + 2\pi n$

$x = \dfrac{4\pi}{3}$

41. $h(x) = \dfrac{x^2-9}{x^2-4} = \dfrac{(x+3)(x-3)}{(x+2)(x-2)}$

HA: $y = 1$

x-intercepts $(-3, 0), (3, 0)$

y-intercept: $\left(0, \dfrac{9}{4} \right)$

VA: $x = -2, x = 2$

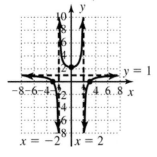

9.2 Exercises

1. $c^2 = \left| a^2 - b^2 \right|$

3. $2a, 2b$

5. Answers will vary.

7. $(x-0)^2 + (y-0)^2 = 7^2$
   $x^2 + y^2 = 49$

9. $(x-5)^2 + (y-0)^2 = \left(\sqrt{3}\right)^2$
   $(x-5)^2 + y^2 = 3$

11. Diameter endpoints: $(4, 9)$ and $(-2, 1)$
    Center = Midpoint
    $$\left(\frac{4+(-2)}{2}, \frac{9+1}{2} \right) = (1,5)$$
    Radius: $\sqrt{(4-1)^2 + (9-5)^2} = \sqrt{9+16} = 5$
    $(x-1)^2 + (y-5)^2 = (5)^2$
    $(x-1)^2 + (y-5)^2 = 25$

13. $x^2 + y^2 - 12x - 10y + 52 = 0$
    $x^2 - 12x + y^2 - 10y = -52$
    $x^2 - 12x + 36 + y^2 - 10y + 25 = -52 + 36 + 25$
    $(x-6)^2 + (y-5)^2 = 9$
    Center: $(6, 5)$, radius: $\sqrt{9} = 3$

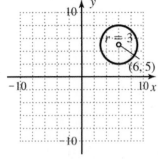

15. $x^2 + y^2 - 4x + 10y + 4 = 0$
    $x^2 - 4x + y^2 + 10y = -4$
    $x^2 - 4x + 4 + y^2 + 10y + 25 = -4 + 4 + 25$
    $(x-2)^2 + (y+5)^2 = 25$
    Center: $(2, -5)$, radius: $\sqrt{25} = 5$

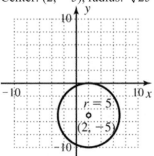

17. $x^2 + y^2 + 6x - 5 = 0$
    $x^2 + 6x + y^2 = 5$
    $x^2 + 6x + 9 + y^2 = 5 + 9$
    $(x+3)^2 + y^2 = 14$
    Center: $(-3, 0)$, radius: $\sqrt{14} \approx 3.7$

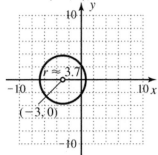

19. $\dfrac{(x-1)^2}{9} + \dfrac{(y-2)^2}{16} = 1$
    Center: $(1,2)$, $a = 3$, $b = 4$

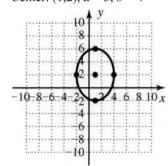

21. $\dfrac{(x-2)^2}{25} + \dfrac{(y+3)^2}{4} = 1$

Center: $(2, -3)$, $a = 5$, $b = 2$

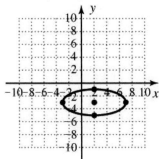

23. $\dfrac{(x+1)^2}{16} + \dfrac{(y+2)^2}{9} = 1$

Center $(-1, -2)$, $a = 4$, $b = 3$

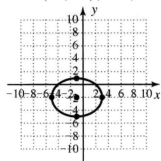

25. $x^2 + 4y^2 = 16$

   a. $\dfrac{x^2}{16} + \dfrac{y^2}{4} = 1$

   Center: $(0,0)$, $a = 4$, $b = 2$

   b. Vertices : $(-4,0), (4,0)$

   Endpts of minor axis : $(0,-2), (0,2)$

   c.

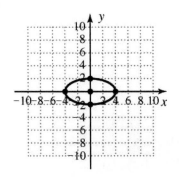

27. $16x^2 + 9y^2 = 144$

   a. $\dfrac{x^2}{9} + \dfrac{y^2}{16} = 1$

   Center: $(0,0)$, $a = 3$, $b = 4$

   b. Vertices : $(0,-4), (0,4)$

   Endpts of minor axis : $(-3,0), (3,0)$

   c.

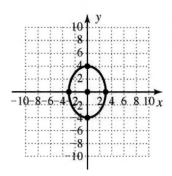

29. $2x^2 + 5y^2 = 10$

   a. $\dfrac{x^2}{5} + \dfrac{y^2}{2} = 1$

   Center: $(0,0)$, $a = \sqrt{5}$, $b = \sqrt{2}$

   b. Vertices : $(-\sqrt{5},0), (\sqrt{5},0)$

   Endpts of minor axis : $(0,-\sqrt{2}), (0,\sqrt{2})$

   c.

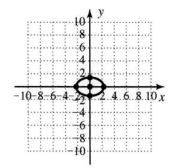

31. $(x+1)^2 + 4(y-2)^2 = 16$

$$\frac{(x+1)^2}{16} + \frac{(y-2)^2}{4} = 1$$

$$\frac{(x+1)^2}{4^2} + \frac{(y-2)^2}{2^2} = 1$$

Ellipse

Center: $(-1,2)$, $a = 4, b = 2$

Vertices: $(-5,2),(3,2)$

Endpts of minor axis: $(-1,0),(-1,4)$

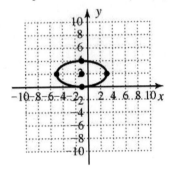

33. $2(x-2)^2 + 2(y+4)^2 = 18$

$(x-2)^2 + (y+4)^2 = 9$

Circle

Center: $(2, -4)$, Radius: 3

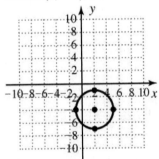

35. $4(x-1)^2 + 9(y-4)^2 = 36$

$$\frac{(x-1)^2}{9} + \frac{(y-4)^2}{4} = 1$$

$$\frac{(x-1)^2}{3^2} + \frac{(y-4)^2}{2^2} = 1$$

Ellipse

Center: $(1,4)$, $a = 3, b = 2$

Vertices: $(4,4),(-2,4)$

Endpts of minor axis: $(1,2),(1,6)$

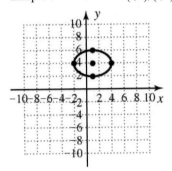

37. $4x^2 + y^2 + 6y + 5 = 0$

$4x^2 + y^2 + 6y = -5$

$4x^2 + y^2 + 6y + 9 = -5 + 9$

$4(x)^2 + (y+3)^2 = 4$

$$x^2 + \frac{(y+3)^2}{4} = 1$$

Center: $(0,-3)$

Vertices $(0,-3-2),(0,-3+2)$;

$(0,-5),(0,-1)$

Endpts of minor axis: $(0-1,-3),(0+1,-3)$;

$(-1,-3),(1,-3)$

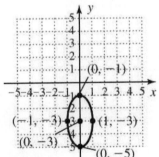

39. $x^2 + 4y^2 - 8y + 4x - 8 = 0$

$x^2 + 4x + 4y^2 - 8y = 8$

$x^2 + 4x + 4(y^2 - 2y) = 8$

$x^2 + 4x + 4 + 4(y^2 - 2y + 1) = 8 + 4 + 4$

$(x+2)^2 + 4(y-1)^2 = 16$

$\dfrac{(x+2)^2}{16} + \dfrac{(y-1)^2}{4} = 1$

$a = 4, \; b = 2$

Center: $(-2,1)$

Vertices: $(-2-4,1), (-2+4,1)$;

$\quad (-6,1), (2,1)$

Endpts of minor axis: $(-2,1-2), (-2,1+2)$;

$\quad (-2,-1), (-2,3)$

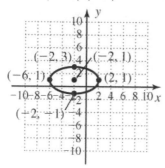

41. $5x^2 + 2y^2 + 20y - 30x + 75 = 0$

$5x^2 - 30x + 2y^2 + 20y = -75$

$5(x^2 - 6x) + 2(y^2 + 10y) = -75$

$5(x^2 - 6x + 9) + 2(y^2 + 10y + 25) = -75 + 45 + 50$

$5(x-3)^2 + 2(y+5)^2 = 20$

$\dfrac{(x-3)^2}{4} + \dfrac{(y+5)^2}{10} = 1$

$a = 2, \; b = \sqrt{10}$

Center: $(3,-5)$

Vertices: $\left(3,-5-\sqrt{10}\right), \left(3,-5+\sqrt{10}\right)$

Endpts of minor axis: $(3-2,-5), (3+2,-5)$

$\quad (1,-5), (5,-5)$

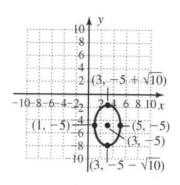

43. $2x^2 + 5y^2 - 12x + 20y - 12 = 0$

$2x^2 - 12x + 5y^2 + 20y = 12$

$2(x^2 - 6x) + 5(y^2 + 4y) = 12$

$2(x^2 - 6x + 9) + 5(y^2 + 4y + 4) = 12 + 18 + 20$

$2(x-3)^2 + 5(y+2)^2 = 50$

$\dfrac{(x-3)^2}{25} + \dfrac{(y+2)^2}{10} = 1$

$a = 5, \; b = \sqrt{10}$

Center: $(3,-2)$

Vertices $(3-5,-2), (3+5,-2)$;

$\quad (-2,-2), (8,-2)$

Endpts of minor axis:

$\quad \left(3,-2-\sqrt{10}\right), \left(3,-2+\sqrt{10}\right)$

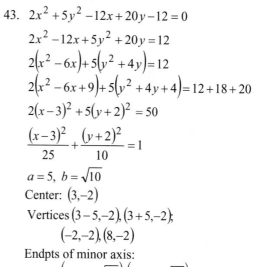

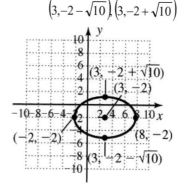

45. $c = 6, b = 8$;

$$a^2 - 8^2 = 6^2$$
$$a^2 = 100$$
$$a = 10;$$
$$2a = 20$$

47. $c = 8, b = 6$;

$$a^2 - 6^2 = 8^2$$
$$a^2 = 100$$
$$a = 10;$$
$$2a = 20$$

49. $4x^2 + 25y^2 - 16x - 50y - 59 = 0$

$$4(x^2 - 4x + 4) + 25(y^2 - 2y + 1) = 59 + 16 + 25$$
$$4(x - 2)^2 + 25(y - 1)^2 = 100$$
$$\frac{(x-2)^2}{25} + \frac{(y-1)^2}{4} = 1;$$

$a = 5, \ b = 2$
$$a^2 - b^2 = c^2$$
$$25 - 4 = c^2$$
$$21 = c^2$$
$$c = \sqrt{21};$$

   a.   Center: $(2,1)$

   b.   Vertices: $(2-5,1)$ and $(2+5,1)$

       $(-3,1)$ and $(7,1)$

   c.   Foci: $(2-\sqrt{21},1)$ and $(2+\sqrt{21},1)$

   d.   Endpoint of minor axis:

       $(2,1+2)$ and $(2,1-2)$

       $(2,3)$ and $(2,-1)$

   e.

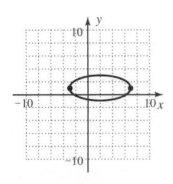

51. $25x^2 + 16y^2 - 200x + 96y + 144 = 0$

$$25(x^2 - 8x + 16) + 16(y^2 + 6y + 9) = -144 + 400 + 144$$
$$25(x-4)^2 + 16(y+3)^2 = 400$$
$$\frac{(x-4)^2}{16} + \frac{(y+3)^2}{25} = 1;$$

$a = 4, \ b = 5$
$$c^2 = 25 - 16 = 9$$
$$c = 3;$$

   a.   Center: $(4,-3)$

   b.   Vertices: $(4,-3+5)$ and $(4,-3-5)$

       $(4,2)$ and $(4,-8)$

   c.   Foci: $(4,-3+3)$ and $(4,-3-3)$

       $(4,0)$ and $(4,-6)$

   d.   Endpoint of minor axis:

       $(4-4,-3)$ and $(4+4,-3)$

       $(0,-3)$ and $(8,-3)$

   e.

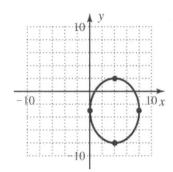

53. $6x^2 + 24x + 9y^2 + 36y + 6 = 0$

$6(x^2 + 4x + 4) + 9(y^2 + 4y + 4) = -6 + 24 + 36$

$6(x+2)^2 + 9(y+2)^2 = 54$

$\dfrac{(x+2)^2}{9} + \dfrac{(y+2)^2}{6} = 54;$

$a = 3, b = \sqrt{6}$

$c^2 = 9 - 6 = 3$

$c = \sqrt{3};$

a. Center: $(-2,-2)$

b. Vertices:

$(-2-3,-2)$ and $(-2+3,-2)$

$(-5,-2)$ and $(1,-2)$

c. Foci: $\left(-2-\sqrt{3},-2\right)$ and $\left(-2+\sqrt{3},-2\right)$

d. Endpoint of minor axis:

$\left(-2,-2+\sqrt{6}\right)$ and $\left(-2,-2-\sqrt{6}\right)$

e.

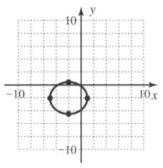

55. Vertices at $(-6,0)$ and $(6,0)$;

Foci at $(-4,0)$ and $(4,0), \leftrightarrow$;

$V: (-6,0), (6,0), a = 6$

$F: (-4,0), (4,0), c = 4$

$4^2 = 6^2 - b^2$

$b^2 = 36 - 16 = 20$

$b = \sqrt{20};$

Center: $(0,0)$

$\dfrac{x^2}{36} + \dfrac{y^2}{20} = 1$

57. Foci at $(3,-6)$ and $(3,2), \updownarrow$;

Length of minor axis: 6 units

$c = \dfrac{1}{2}\sqrt{(3-3)^2 + (2-(-6))^2} = \dfrac{1}{2}\sqrt{64} = 4;$

$2a = 6, a = 3;$

$4^2 = b^2 - 3^2$

$b^2 = 25$

$b = 5;$

Center: $\left(\dfrac{3+3}{2}, \dfrac{-6+2}{2}\right) = (3,-2);$

$\dfrac{(x-3)^2}{9} + \dfrac{(y+2)^2}{25} = 1$

59. Center: $(0, 0)$

$a = 4, b = 3$

$c^2 = 4^2 - 3^2$

$c = \pm\sqrt{7};$

$\dfrac{x^2}{4^2} + \dfrac{y^2}{3^2} = 1$

$\dfrac{x^2}{16} + \dfrac{y^2}{9} = 1$

Foci: $\left(-\sqrt{7},0\right), \left(\sqrt{7},0\right)$

61. Center: $(-3, -1)$

$a = 2, b = 4$

$c^2 = 4^2 - 2^2$

$c = \pm\sqrt{12} = \pm2\sqrt{3};$

$\dfrac{(x-(-3))^2}{2^2} + \dfrac{(y-(-1))^2}{4^2} = 1$

$\dfrac{(x+3)^2}{4} + \dfrac{(y+1)^2}{16} = 1$

Foci: $\left(-3,-1+2\sqrt{3}\right), \left(-3,-1-2\sqrt{3}\right)$

63. $A = \pi ab;$

$16x^2 + 9y^2 = 144$

$\dfrac{x^2}{9} + \dfrac{y^2}{16} = 1;$

$A = \pi(3)(4) = 12\pi$ units^2

9.2 Exercises

65. $a = 4, b = 3$

$c^2 = a^2 - b^2 = 16 - 9 = 7$

$c = \sqrt{7}$;

Spines are $\sqrt{7}$ or ≈ 2.65 ft from center.

The height of the spine occurs at $\left(\sqrt{7}, y\right)$ on

the hyperbola $\dfrac{x^2}{16} + \dfrac{y^2}{9} = 1$

$\dfrac{\left(\sqrt{7}\right)^2}{16} + \dfrac{y^2}{9} = 1$

$\dfrac{7}{16} + \dfrac{y^2}{9} = 1$

$\dfrac{y^2}{9} = 1 - \dfrac{7}{16}$

$y^2 = 9\left(\dfrac{9}{16}\right)$

$y = \dfrac{9}{4} = 2.25$

Height of the spine: 2.25 ft.

67. $a = 12, b = 8$;

$c = \sqrt{12^2 - 8^2}$

$c = \sqrt{80} = 4\sqrt{5} \approx 8.9$ ft

8.9 ft from center

$2\left(4\sqrt{5}\right) \approx 17.9$ ft apart

69. $\dfrac{x^2}{15^2} + \dfrac{y^2}{8^2} = 1$

$\dfrac{9^2}{15^2} + \dfrac{y^2}{8^2} = 1$

$\dfrac{y^2}{8^2} = 0.64$

$y^2 = 40.96$

$y = 6.4$ ft

71. $a = \dfrac{72}{2} = 36; b = \dfrac{70.5}{2} = 35.25$

$\dfrac{x^2}{36^2} + \dfrac{y^2}{(35.25)^2} = 1$

73. $P = 2\pi\sqrt{\dfrac{a^2 + b^2}{2}}$

Aphelion (max)

$c - (-a) = 156$ million miles

Perihelion

$a - c = 128$

$\begin{cases} a + c = 156 \\ a - c = 128 \end{cases}$

$2a = 284$

Semi major $a = 142$ million miles;

$142 - c = 128$

$c = 142 - 128 = 14$;

$14^2 = 142^2 - b^2$

$b^2 = 142^2 - 14^2 = 19968$

$b \approx 141$;

Semi minor ≈ 141 million miles;

$P = 2\pi\sqrt{\dfrac{142^2 + 141^2}{2}}$

$= 2\pi\sqrt{20022.5}$ million miles;

$\dfrac{889.076 \text{ million miles}}{1.296 \text{ million miles/day}} \approx 686$ days

75. $4x^2 + 9y^2 = 900$

$\dfrac{x^2}{15^2} + \dfrac{y^2}{10^2} = 1$;

$9x^2 + 25y^2 = 900$

$\dfrac{x^2}{10^2} + \dfrac{y^2}{6^2} = 1$;

$A = \pi(15)(10) - \pi(10)(6) = 90\pi$

$= 9{,}000\pi \text{ yd}^2$

77. $\dfrac{x^2}{81} - \dfrac{y^2}{36} = 1$

$a = 9, b = 6$;

$L = \dfrac{2m^2}{n}$

$L = \dfrac{2(6)^2}{9} = 8 \text{ units}$;

$81 - 36 = c^2$

$c = \sqrt{45} = 3\sqrt{5}$;

$\dfrac{\left(\sqrt{45}\right)^2}{81} - \dfrac{y^2}{36} = 1$

$\dfrac{45}{81} - \dfrac{y^2}{36} = 1$

$-\dfrac{y^2}{36} = \dfrac{4}{9}$

$y^2 = 16$

$y = \pm 4$

Verified

$\left(3\sqrt{5}, 4\right), \left(3\sqrt{5}, -4\right),$

$\left(-3\sqrt{5}, 4\right), \left(-3\sqrt{5}, -4\right)$

79. Ellipse $\dfrac{x^2}{a^2} + \dfrac{y^2}{b^2} = 1$;

$c^2 = a^2 - b^2$;

$\dfrac{a^2 - b^2}{a^2} + \dfrac{y^2}{b^2} = 1$

$b^2a^2 - b^2b^2 + a^2y^2 = a^2b^2$

$a^2y^2 - b^2b^2 = -a^2b^2 + a^2b^2$

$a^2y^2 = b^4$

$y^2 = \dfrac{b^4}{a^2}$

$y = \dfrac{b^2}{a}$;

Focal Chord is $2y$ so $L = \dfrac{2b^2}{a}$.

Verified.

81. a. $R = \dfrac{kL}{d^2}$

b. $240 = \dfrac{k(2)}{0.005^2}$

$k = 0.003$

c. $R = \dfrac{0.003(3)}{0.006^2} = 250 \ \Omega$

83. $a^2 = 250^2 + 30^2 - 2(250)(30)(cos110°)$

$a^2 \approx 68530.30215$

$a^2 = \sqrt{68530.30215} \approx 261.8 \text{ mph}$

$\dfrac{\sin x}{30} = \dfrac{\sin 110}{261.8}$

$\sin x = \dfrac{30 \sin 110}{261.8} \approx 0.10768$

$x = \sin^{-1}(0.10768)$

$x \approx 6.2$

Heading $20° + 6.2° \approx 26.2°$

261.8 mph

Technology Highlight

1. $25y^2 - 4x^2 = 100$

Vertices: $(0, 2),\ (0, -2)$

When $x = 4, y = \pm 2.5612497$

9.3 Exercises

1.   transverse

3.   midway

5.   Answers will vary.

7. $\dfrac{x^2}{9} - \dfrac{y^2}{4} = 1$

Center: (0,0)

Vertices: $(-3,0),\ (3,0)$

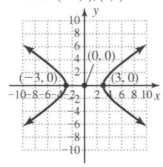

9. $\dfrac{x^2}{4} - \dfrac{y^2}{9} = 1$

Center: (0,0)

Vertices: $(-2,0),\ (2,0)$

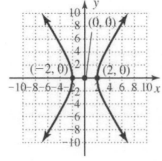

11. $\dfrac{x^2}{49} - \dfrac{y^2}{16} = 1$

Center: (0,0)

Vertices: $(-7,0),\ (7,0)$

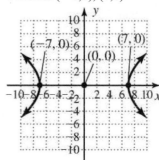

13. $\dfrac{x^2}{36} - \dfrac{y^2}{16} = 1$

Center: (0,0)

Vertices: $(-6,0),\ (6,0)$

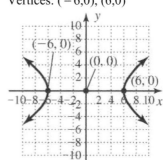

15. $\dfrac{y^2}{9} - \dfrac{x^2}{1} = 1$

Center: (0,0)

Vertices: $(0,\ -3),\ (0,3)$

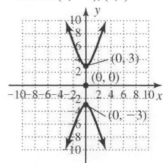

Chapter 9: Analytical Geometry

17. $\dfrac{y^2}{12} - \dfrac{x^2}{4} = 1$

Center: (0,0)

Vertices: $\left(0, -2\sqrt{3}\right), \left(0, 2\sqrt{3}\right)$

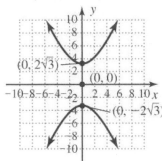

19. $\dfrac{y^2}{9} - \dfrac{x^2}{9} = 1$

Center: (0,0)

Vertices: $(0, -3), (0,3)$

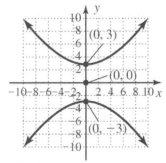

21. $\dfrac{y^2}{36} - \dfrac{x^2}{25} = 1$

Center: (0,0)

Vertices: $(0, -6), (0,6)$

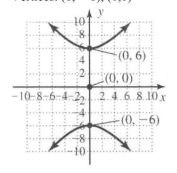

23. Vertices: $(-4, -2), (2, -2)$

Transverse Axis: $y = -2$

$\dfrac{-4+2}{2} = -1$

Center: $(-1, -2)$

Conjugate Axis: $x = -1$

25. Vertices: $(4,1), (4, -3)$

Transverse Axis: $x = 4$

$\dfrac{1+-3}{2} = -1$

Center: $(4, -1)$

Conjugate Axis: $y = -1$

27. $\dfrac{(y+1)^2}{4} - \dfrac{x^2}{25} = 1$

Center: $(0, -1)$

$a = 5, \ b = 2$

Vertices: $\left(0, -1-2\right), \left(0, -1+2\right)$;

$\left(0, -3\right), \left(0, 1\right)$

Transverse axis: $x = 0$

Conjugate axis: $y = -1$

Asymptotes: Slope $= \pm\dfrac{2}{5}$

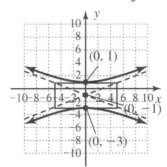

29. $\dfrac{(x-3)^2}{36} - \dfrac{(y+2)^2}{49} = 1$

Center: $(3, -2)$

$a = 6,\ b = 7$

Vertices: $(3-6,-2),(3+6,-2)$;

$(-3,-2),(9,-2)$

Transverse axis: $y = -2$

Conjugate axis: $x = 3$

Asymptotes: Slope $= \pm\dfrac{7}{6}$

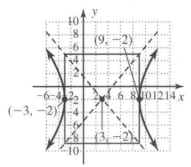

31. $\dfrac{(y+1)^2}{7} - \dfrac{(x+5)^2}{9} = 1$

Center: $(-5, -1)$

$a = 3,\ b = \sqrt{7}$

Vertices: $\left(-5,-1+\sqrt{7}\right),\left(-5,-1-\sqrt{7}\right)$

Transverse axis: $x = -5$

Conjugate axis: $y = -1$

Asymptotes: Slope $= \pm\dfrac{\sqrt{7}}{3}$

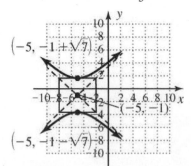

33. $(x-2)^2 - 4(y+1)^2 = 16$

$\dfrac{(x-2)^2}{16} - \dfrac{(y+1)^2}{4} = 1$

Center: $(2, -1)$

$a = 4,\ b = 2$

Vertices: $(2-4,-1),(2+4,-1)$;

$(-2,-1),(6,-1)$

Transverse axis: $y = -1$

Conjugate axis: $x = 2$

Asymptotes: Slope $= \pm\dfrac{1}{2}$

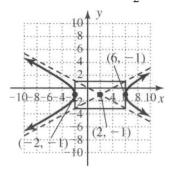

35. $2(y+3)^2 - 5(x-1)^2 = 50$

$\dfrac{(y+3)^2}{25} - \dfrac{(x-1)^2}{10} = 1$

Center: $(1, -3)$

$a = \sqrt{10},\ b = 5$

Vertices: $(1,-3+5),(1,-3-5)$;

$(1,2),(1,-8)$

Transverse axis: $x = 1$

Conjugate axis: $y = -3$

Asymptotes: Slope $= \pm\dfrac{5}{\sqrt{10}} = \pm\dfrac{\sqrt{10}}{2}$

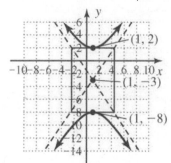

37. $12(x-4)^2 - 5(y-3)^2 = 60$

$$\frac{(x-4)^2}{5} - \frac{(y-3)^2}{12} = 1$$

Center: $(4,3)$

$a = \sqrt{5},\ b = \sqrt{12} = 2\sqrt{3}$

Vertices: $\left(4+\sqrt{5},3\right), \left(4-\sqrt{5},3\right)$

Transverse axis: $y = 3$

Conjugate axis: $x = 4$

Asymptotes: Slope $= \pm\dfrac{2\sqrt{3}}{\sqrt{5}}$

$$= \pm\frac{2\sqrt{3}}{\sqrt{5}} \cdot \frac{\sqrt{5}}{\sqrt{5}} = \pm\frac{2\sqrt{15}}{5}$$

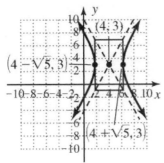

39. $16x^2 - 9y^2 = 144$

$$\frac{x^2}{9} - \frac{y^2}{16} = 1$$

Center: $(0,0)$

$a = 3,\ b = 4$

Vertices: $(0+3,0),(0-3,0)$;

$(3,0),(-3,0)$

Transverse axis: $y = 0$

Conjugate axis: $x = 0$

Asymptotes: Slope $= \pm\dfrac{4}{3}$

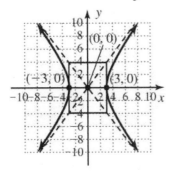

41. $9y^2 - 4x^2 = 36$

$$\frac{y^2}{4} - \frac{x^2}{9} = 1$$

Center: $(0,0)$

$a = 3,\ b = 2$

Vertices: $(0,0-2),(0,0+2)$;

$(0,-2),(0,2)$

Transverse axis: $x = 0$

Conjugate axis: $y = 0$

Asymptotes: Slope $= \pm\dfrac{2}{3}$

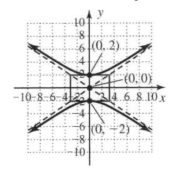

43. $12x^2 - 9y^2 = 72$

$$\frac{x^2}{6} - \frac{y^2}{8} = 1$$

Center: $(0,0)$

$a = \sqrt{6},\ b = \sqrt{8} = 2\sqrt{2}$

Vertices: $\left(\sqrt{6},0\right),\left(-\sqrt{6},0\right)$

Transverse axis: $y = 0$

Conjugate axis: $x = 0$

Asymptotes: Slope $= \pm\dfrac{2\sqrt{2}}{\sqrt{6}}$

$$= \pm\frac{2\sqrt{2}}{\sqrt{6}} \cdot \frac{\sqrt{6}}{\sqrt{6}} = \pm\frac{2\sqrt{3}}{3}$$

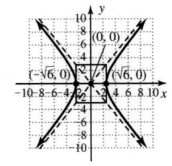

9.3 Exercises

45. $4x^2 - y^2 + 40x - 4y + 60 = 0$

$4x^2 + 40x - y^2 - 4y = -60$

$4(x^2 + 10x) - (y^2 + 4y) = -60$

$4(x^2 + 10x + 25) - (y^2 + 4y + 4) = -60 + 100 - 4$

$4(x+5)^2 - (y+2)^2 = 36$

$\dfrac{(x+5)^2}{9} - \dfrac{(y+2)^2}{36} = 1$

Center: $(-5, -2)$

$a = 3,\ b = 6$

Vertices: $(-5-3,-2), (-5+3,-2);$

$(-8,-2), (-2,-2)$

Transverse axis: $y = -2$

Conjugate axis: $x = -5$

Asymptotes: Slope $= \pm\dfrac{6}{3} = \pm 2$

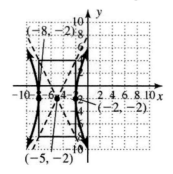

47. $x^2 - 4y^2 - 24y - 4x - 36 = 0$

$x^2 - 4x - 4y^2 - 24y = 36$

$(x^2 - 4x) - 4(y^2 + 6y) = 36$

$(x^2 - 4x + 4) - 4(y^2 + 6y + 9) = 36 + 4 - 36$

$(x-2)^2 - 4(y+3)^2 = 4$

$\dfrac{(x-2)^2}{4} - \dfrac{(y+3)^2}{1} = 1$

Center: $(2, -3)$

$a = 2,\ b = 1$

Vertices: $(2-2,-3), (2+2,-3);$

$(0,-3), (4,-3)$

Transverse axis: $y = -3$

Conjugate axis: $x = 2$

Asymptotes: Slope $= \pm\dfrac{1}{2}$

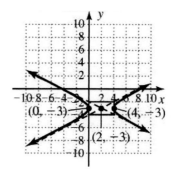

49. $-4x^2 - 4y^2 = -24$

$x^2 + y^2 = 6$

The equation contains a sum of second degree terms with equal coefficients. The equation represents a circle.

51. $x^2 + y^2 = 2x + 4y + 4$

The equation contains a sum of second degree terms with equal coefficients. The equation represents a circle.

53. $2x^2 - 4y^2 = 8$

The equation contains a difference of second degree terms. The equation represents a hyperbola.

55. $x^2 + 5 = 2y^2$

$x^2 - 2y^2 = -5$

$2y^2 - x^2 = 5$

The equation contains a difference of second degree terms. The equation represents a hyperbola.

57. $2x^2 = -2y^2 + x + 20$

$2x^2 - x + 2y^2 = 20$

The equation contains a sum of second degree terms with equal coefficients. The equation represents a circle.

59. $16x^2 + 5y^2 - 3x + 4y = 538$

The equation contains a sum of second degree terms with unequal coefficients. The equation represents an ellipse.

61. $\sqrt{(-5-5)^2 + (0-2.25)^2}$

$\qquad -\sqrt{(5-5)^2 + (0-2.25)^2} = 2a$

$\sqrt{(-5-5)^2 + (0-2.25)^2} - 2.25 = 2a$

$\qquad \sqrt{100 + 5.0625} - 2.25 = 2a$

$\qquad\qquad\qquad 8 = 2a$

$a = 4, c = 5$

$25 = 4^2 + b^2$

$9 = b^2$

$3 = b;$

$2b = 2(3) = 6;$

Dimensions: 8 x 6

63. $\sqrt{(-0-6)^2 + (-10-7.5)^2} - \sqrt{(0-6)^2 + (10-7.5)^2} = 2b$

$\qquad \sqrt{36 + (-17.5)^2} - \sqrt{36 + (2.5)^2} = 2b$

$\qquad\qquad \sqrt{342.25} - \sqrt{42.25} = 2b$

$\qquad\qquad\qquad\qquad 12 = 2b;$

$\quad b = 6, c = 10$

$100 = a^2 + 36$

$a^2 = 64$

$\quad a = 8;$

$2a = 2(8) = 16;$

Dimensions: 16 x 12

65. $4x^2 - 9y^2 - 24x + 72y - 144 = 0, \leftrightarrow$

$4(x^2 - 6x + 9) - 9(y^2 - 8y + 16) = 144 - 144 + 36$

$\qquad 4(x-3)^2 - 9(y-4)^2 = 36$

$\qquad \dfrac{(x-3)^2}{9} - \dfrac{(y-4)^2}{4} = 1;$

$a = 3, b = 2;$

$c^2 = 9 + 4 = 13$

$c = \sqrt{13};$

a. Center: $(3,4)$

b. Vertices: $(3-3,4)$ and $(3+3,4)$

$\qquad (0,4)$ and $(6,4)$

c. Foci: $(3-\sqrt{13},4)$ and $(3+\sqrt{13},4)$

d. $2a = 6, 2b = 4;$

e. Asymptotes: Slope $= \pm\dfrac{2}{3};$

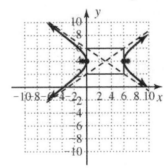

9.3 Exercises

67. $16x^2 - 4y^2 + 24y - 100 = 0, \leftrightarrow$

$$16(x^2) - 4(y^2 - 6y + 9) = 100 - 36$$

$$16(x^2) - 4(y - 3)^2 = 64$$

$$\frac{x^2}{4} - \frac{(y-3)^2}{16} = 1;$$

$a = 2, b = 4;$

$c^2 = 4 + 16$

$c^2 = 20$

$c = 2\sqrt{5};$

a. Center: $(0,3)$

b. Vertices: $(0 - 2, 3)$ and $(0 + 2, 3)$
   $(-2, 3)$ and $(2, 3)$

c. Foci: $\left(-2\sqrt{5}, 3\right)$ and $\left(2\sqrt{5}, 3\right)$

d. $2a = 4, 2b = 8;$

e. Asymptotes: Slope $= \pm 2$

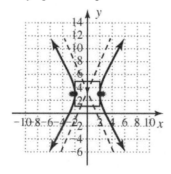

69. $9x^2 - 3y^2 - 54x - 12y + 33 = 0, \leftrightarrow$

$$9(x^2 - 6x + 9) - 3(y^2 + 4y + 4) = -33 + 81 - 12$$

$$9(x-3)^2 - 3(y+2)^2 = 36$$

$$\frac{(x-3)^2}{4} - \frac{(y+2)^2}{12} = 1;$$

$a = 2, b = \sqrt{12} = 2\sqrt{3}$

$c^2 = 4 + 12 = 16$

$c = 4;$

a. Center: $(3, -2)$

b. Vertices: $(3 - 2, -2)$ and $(3 + 2, -2)$
   $(1, -2)$ and $(5, -2)$

c. Foci: $(-1, -2), (7, -2)$

d. $2a = 4, 2b = 4\sqrt{3};$

e. Asymptotes: Slope $= \pm\sqrt{3}$

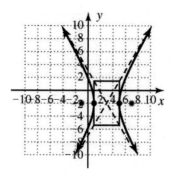

71. Vertices: $(-6, 0)$ and $(6, 0)$

Foci: $(-8, 0)$ and $(8, 0)$

$8^2 = 6^2 + b^2$

$64 = 36 + b^2$

$b^2 = 28;$

$$\frac{x^2}{36} - \frac{y^2}{28} = 1$$

73. Foci: $\left(-2, -3\sqrt{2}\right)$ and $\left(-2, 3\sqrt{2}\right)$

Length of conjugate axis: 6 units

$2a = 6$

$a = 3;$

Center: $\left(\dfrac{-2 + (-2)}{2}, \dfrac{-3\sqrt{2} + 3\sqrt{2}}{2}\right) = (-2, 0)$

$c = 3\sqrt{2};$

$\left(3\sqrt{2}\right)^2 = 3^2 + b^2$

$18 - 9 = b^2$

$b = 3;$

$$\frac{(y-0)^2}{9} - \frac{(x-(-2))^2}{9} = 1$$

$$\frac{y^2}{9} - \frac{(x+2)^2}{9} = 1$$

75. Center $(0, 0)$

$$\frac{(x-0)^2}{2^2} - \frac{(y-0)^2}{3^2} = 1$$

$$\frac{x^2}{4} - \frac{y^2}{9} = 1;$$

$c^2 = 4 + 9 = 13$

$c = \sqrt{13};$

Foci: $\left(-\sqrt{13}, 0\right), \left(\sqrt{13}, 0\right)$

77. Center $(2, 1)$

$c = 3;$

$3^2 = a^2 + 2^2$

$9 - 4 = a^2$

$a = \sqrt{5};$

$4 \times 2\sqrt{5}$

$\dfrac{(y-1)^2}{2^2} - \dfrac{(x-2)^2}{(\sqrt{5})^2} = 1$

$\dfrac{(y-1)^2}{4} - \dfrac{(x-2)^2}{5} = 1;$

79. $y = \sqrt{\dfrac{36 - 4x^2}{-9}}$

a. $y = \sqrt{\dfrac{-4(-9 + x^2)}{-9}}$

$y = \sqrt{\dfrac{4(x^2 - 9)}{9}}$

$y = \dfrac{2}{3}\sqrt{x^2 - 9}$

b. $x^2 - 9 \geq 0$

$(x + 3)(x - 3) \geq 0$

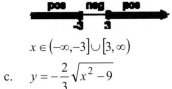

$x \in (-\infty, -3] \cup [3, \infty)$

c. $y = -\dfrac{2}{3}\sqrt{x^2 - 9}$

81. $25y^2 - 1600x^2 = 40000$

$\dfrac{y^2}{1600} - \dfrac{x^2}{25} = 1$

$\dfrac{y^2}{40^2} - \dfrac{x^2}{5^2} = 1$

40 yards

83. $1600x^2 - 400(y - 50)^2 = 640000$

$\dfrac{x^2}{400} - \dfrac{(y-50)^2}{1600} = 1$

$\dfrac{x^2}{20^2} - \dfrac{(y-50)^2}{40^2} = 1$

$20 + 20 = 40$ feet

85. 0.4 milliseconds to closer;

0.5 milliseconds to farther;

300 km/millisecond;

$0.5(300) = 150$ km; $0.4(300) = 120$ km;

$\sqrt{150^2} - \sqrt{120^2} = 2a$

$150 - 120 = 2a$

$30 = 2a$

$a = 15;$

$c = 50;$

$50^2 = 15^2 + b^2$

$2500 = 225 + b^2$

$b^2 = 2275;$

$\dfrac{x^2}{225} - \dfrac{y^2}{2275} = 1;$

$\dfrac{x^2}{225} - \dfrac{(60)^2}{2275} = 1$

$\dfrac{x^2}{225} = 1 + \dfrac{(60)^2}{2275}$

$x^2 = \dfrac{52875}{91}$

$x = \pm 24.1;$

$(24.1, 60) \text{ or } (-24.1, 60)$

87. a. $4x^2 - 32x - y^2 + 4y + 60 = 0$

$4(x^2 - 8x) - (y^2 - 4y) = -60$

$4(x^2 - 8x + 16) - (y^2 - 4y + 4) = -60 + 64 - 4$

$4(x - 4)^2 - (y - 2)^2 = 0$

$\dfrac{(x-4)^2}{\frac{1}{4}} - (y - 2)^2 = 0$

b. $x^2 - 4x + 5y^2 - 40y + 84 = 0$

$(x^2 - 4x) + 5(y^2 - 8y) = -84$

$(x^2 - 4x + 4) + 5(y^2 - 8y + 16) = -84 + 4 + 80$

$(x - 2)^2 + 5(y - 4)^2 = 0$

$(x - 2)^2 + \dfrac{(y-4)^2}{\frac{1}{5}} = 0$

Both equal 0.

89. a. $(x-5)^2 - (y+4)^2 = 57$

$$\frac{(x-5)^2}{57} - \frac{(y+4)^2}{57} = 1$$

Area of central rectangle:

$$A = \left(2\sqrt{57}\right)\left(2\sqrt{57}\right) = 228$$

b. $(x-5)^2 + (y+4)^2 = 57$

$$A = \pi\left(\sqrt{57}\right)^2 = 57\pi \approx 179.07$$

c. $9(x-5)^2 + 10(y+4)^2 = 570$

$$\frac{(x-5)^2}{\frac{190}{3}} + \frac{(y+4)^2}{57} = 1$$

$$A = \pi\left(\sqrt{\frac{190}{3}}\right)\left(\sqrt{57}\right) \approx 188.76$$

Choice a

91. $9(x-2)^2 - 25(y-3)^2 = 225$

$$\frac{(x-2)^2}{25} - \frac{(y-3)^2}{9} = 1;$$

$$c^2 = 25 + 9 = 34;$$

$$c^2 = a^2 - b^2$$

$$34 = a^2 - 9$$

$$a^2 = 43;$$

$$\frac{(x-2)^2}{43} + \frac{(y-3)^2}{9} = 1$$

93. $\cos 65° = \dfrac{x}{700}$

$$x = 700\cos 65°$$

$$x \approx 295.8 < 350$$

Yes

95. a. $x^4 + 4 = 0$

$$\left(1 + i\sqrt{2}\right)^4 = 0$$

$$\left(1 + i\sqrt{2}\right)\left(1 + i\sqrt{2}\right)\left(1 + i\sqrt{2}\right)\left(1 + i\sqrt{2}\right) = 0$$

$$\left(1 + 2i\sqrt{2} + 2i^2\right)\left(1 + 2i\sqrt{2} + 2i^2\right) = 0$$

$$\left(-1 + 2i\sqrt{2}\right)\left(-1 + 2i\sqrt{2}\right) = 0$$

$$1 - 4i\sqrt{2} + 4i^2(2) = 0$$

$$-7 - 4i\sqrt{2} \neq 0$$

b. $x^3 - 6x^2 + 11x - 12 = 0$

$$\left(1 + i\sqrt{2}\right)^3 - 6\left(1 + i\sqrt{2}\right)^2 + 11\left(1 + i\sqrt{2}\right) - 12 = 0$$

$$\left(1 + i\sqrt{2}\right)\left(1 + i\sqrt{2}\right)\left(1 + i\sqrt{2}\right) - 6\left(1 + i\sqrt{2}\right)\left(1 + i\sqrt{2}\right) + 11\left(1 + i\sqrt{2}\right) - 12 = 0$$

$$\left(1 + i\sqrt{2}\right)\left(1 + 2i\sqrt{2} + 2i^2\right) - 6\left(1 + 2i\sqrt{2} + 2i^2\right) + 11 + 11i\sqrt{2} - 12 = 0$$

$$\left(1 + i\sqrt{2}\right)\left(-1 + 2i\sqrt{2}\right) - 6 - 12i\sqrt{2} + 12 + 11 + 11i\sqrt{2} - 12 = 0$$

$$-1 + i\sqrt{2} + 2i^2(2) - 6 - 12i\sqrt{2} + 12 + 11 + 11i\sqrt{2} - 12 = 0$$

$$-1 + i\sqrt{2} - 4 - 6 - 12i\sqrt{2} + 12 + 11 + 11i\sqrt{2} - 12 = 0$$

$$0 = 0$$

c. $x^2 - 2x + 3 = 0$

$$\left(1 + i\sqrt{2}\right)^2 - 2\left(1 + i\sqrt{2}\right) + 3 = 0$$

$$1 + 2i\sqrt{2} + 2i^2 - 2 - 2i\sqrt{2} + 3 = 0$$

$$1 + 2i\sqrt{2} - 2 - 2 - 2i\sqrt{2} + 3 = 0$$

$$0 = 0$$

b and c

9.4 Exercises

1. Horizontal, right, $a < 0$

3. $(p,0)$, $x = -p$

5. Answers will vary.

7. $y = x^2 - 2x - 3$;
$0 = (x-3)(x+1)$
x-intercepts: $(-1,0), (3,0)$;
$y = (0)^2 - 2(0) - 3 = -3$
y-intercept: $(0,-3)$;
$x = \dfrac{-(-2)}{2(1)} = 1$
$y = (1)^2 - 2(1) - 3 = -4$
Vertex: $(1,-4)$;
Domain: $x \in (-\infty, \infty)$
Range: $y \in [-4, \infty)$
$y = (x^2 - 2x + 1) - 3 - 1$
$y = (x-1)^2 - 4$

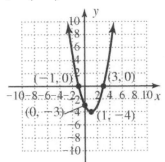

9. $y = 2x^2 - 8x - 10$
$0 = 2(x-5)(x+1)$
x-intercepts: $(-1,0), (5,0)$;
$y = 2(0)^2 - 8(0) - 10 = -10$
y-intercept: $(0,-10)$;
$x = \dfrac{-(-8)}{2(2)} = 2$
$y = 2(2)^2 - 8(2) - 10 = -18$
Vertex: $(2,-18)$;
Domain: $x \in (-\infty, \infty)$
Range: $y \in [-18, \infty)$
$y = 2(x^2 - 4x + 4) - 10 - 8$
$y = 2(x-2)^2 - 18$

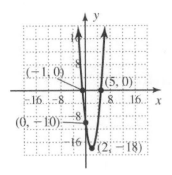

11. $y = 2x^2 + 5x - 7$;
$0 = (2x+7)(x-1)$
x-intercepts: $(-3.5,0), (1,0)$;
$y = 2(0)^2 + 5(0) - 7 = -7$
y-intercept: $(0,-7)$;
$x = \dfrac{-(5)}{2(2)} = -1.25$
$y = 2(-1.25)^2 + 5(-1.25) - 7 = -10.125$
Vertex: $(-1.25, -10.125)$;
Domain: $x \in (-\infty, \infty)$
Range: $y \in [-10.125, \infty)$
$y = 2\left(x^2 + \dfrac{5}{2}x + \dfrac{25}{16}\right) - 7 - \dfrac{25}{8}$
$y = 2\left(x + \dfrac{5}{4}\right)^2 - \dfrac{81}{8}$

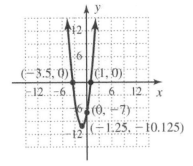

13. $x = y^2 - 2y - 3$

$x = (0)^2 - 2(0) - 3 = -3$

x-intercept: $(-3, 0)$;

$0 = (y - 3)(y + 1)$

y-intercepts: $(0, 3), (0, -1)$;

$y = \dfrac{-(-2)}{2(1)} = 1$

$x = (1)^2 - 2(1) - 3 = -4$

Vertex: $(-4, 1)$;

Domain: $x \in [-4, \infty)$

Range: $y \in (-\infty, \infty)$

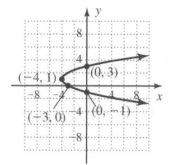

17. $x = -y^2 + 8y - 16$

$x = -(0)^2 + 8(0) - 16 = -16$

x-intercept: $(-16, 0)$;

$0 = (-y + 4)(y - 4)$

y-intercept: $(0, 4)$;

$y = \dfrac{-(8)}{2(-1)} = 4$

$x = -(4)^2 + 8(4) - 16 = 0$

Vertex: $(0, 4)$;

Domain: $x \in (-\infty, 0]$

Range: $y \in (-\infty, \infty)$

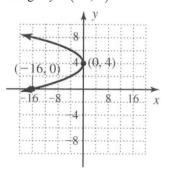

15. $x = -y^2 + 6y + 7$

$x = -(0)^2 + 6(0) + 7 = 7$

x-intercept: $(7, 0)$;

$0 = (-y + 7)(y + 1)$

y-intercepts: $(0, 7), (0, -1)$;

$y = \dfrac{-(6)}{2(-1)} = 3$

$x = -(3)^2 + 6(3) + 7 = 16$

Vertex: $(16, 3)$;

Domain: $x \in (-\infty, 16]$

Range: $y \in (-\infty, \infty)$

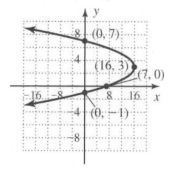

19. $x = y^2 - 6y$

$x = (y^2 - 6y + 9) - 9$

$x = (y - 3)^2 - 9$

Vertex: $(-9, 3)$;

$x = (0)^2 - 6(0) = 0$

x-intercept: $(0, 0)$;

$0 = y(y - 6)$

y-intercepts: $(0, 0), (0, 6)$;

Domain: $x \in [-9, \infty)$

Range: $y \in (-\infty, \infty)$

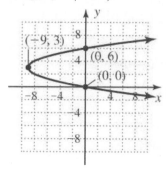

21. $x = y^2 - 4$

Vertex: $(-4, 0)$;

$x = (0)^2 - 4 = -4$

x-intercept: $(-4, 0)$;

$0 = (y + 2)(y - 2)$

y-intercepts: $(0, 2), (0, -2)$;

Domain: $x \in [-4, \infty)$

Range: $y \in (-\infty, \infty)$

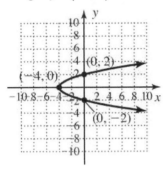

23. $x = -y^2 + 2y - 1$

$x = -(y^2 - 2y) - 1$

$x = -(y^2 - 2y + 1) - 1 + 1$

$x = -(y - 1)^2 + 0$

Vertex: $(0, 1)$;

$x = -(0)^2 + 2(0) - 1 = -1$

x-intercept: $(-1, 0)$;

$0 = (-y + 1)(y - 1)$

y-intercept: $(0, 1)$;

Domain: $x \in (-\infty, 0]$

Range: $y \in (-\infty, \infty)$

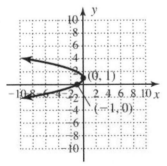

25. $x = y^2 + y - 6$

$x = \left(y^2 + y + \dfrac{1}{4} \right) - 6 - \dfrac{1}{4}$

$x = \left(y + \dfrac{1}{2} \right)^2 - \dfrac{25}{4}$

Vertex: $(-6.25, -0.5)$;

$x = (0)^2 + (0) - 6 = -6$

x-intercept: $(-6, 0)$;

$0 = (y + 3)(y - 2)$

y-intercepts: $(0, 2), (0, -3)$;

Domain: $x \in [-6.25, \infty)$

Range: $y \in (-\infty, \infty)$

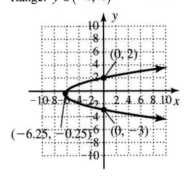

27. $x = y^2 - 10y + 4$

$x = (y^2 - 10y + 25) + 4 - 25$

$x = (y-5)^2 - 21$

Vertex: $(-21, 5)$;

$x = (0)^2 - 10(0) + 4 = 4$

x-intercept: $(4, 0)$;

$a = 1, b = -10, c = 4$

$y = \dfrac{-(-10) \pm \sqrt{(-10)^2 - 4(1)(4)}}{2(1)}$

$y = \dfrac{10 \pm \sqrt{84}}{2} = \dfrac{10 \pm 2\sqrt{21}}{2} = 5 \pm \sqrt{21}$

y-intercepts: $(0, 5 - \sqrt{21}), (0, 5 + \sqrt{21})$;

Domain: $x \in [-21, \infty)$

Range: $y \in (-\infty, \infty)$

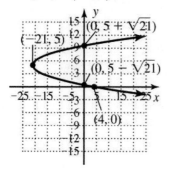

29. $x = 3 - 8y - 2y^2$

$x = -2y^2 - 8y + 3$

$x = -2(y^2 + 4y + 4) + 3 + 8$

$x = -2(y+2)^2 + 11$

Vertex: $(11, -2)$;

$x = 3 - 8(0) - 2(0)^2 = 3$

x-intercept: $(3, 0)$;

$a = -2, b = -8, c = 3$

$y = \dfrac{-(-8) \pm \sqrt{(-8)^2 - 4(-2)(3)}}{2(-2)}$

$y = \dfrac{8 \pm \sqrt{88}}{-4} = \dfrac{8 \pm 2\sqrt{22}}{-4} = \dfrac{-4 \pm \sqrt{22}}{2}$

y-intercepts: $\left(0, \dfrac{-4 + \sqrt{22}}{2}\right), \left(0, \dfrac{-4 - \sqrt{22}}{2}\right)$;

Domain: $x \in (-\infty, 11]$

Range: $y \in (-\infty, \infty)$

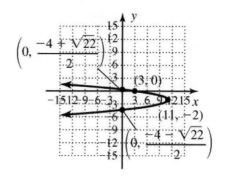

31. $y = (x-2)^2 + 3$

Vertex: $(2, 3)$;

$0 \neq (x-2)^2 + 3$

x-intercept: None;

$y = (0-2)^2 + 3 = 7$

y-intercept: $(0, 7)$;

Domain: $x \in (-\infty, \infty)$

Range: $y \in [3, \infty)$

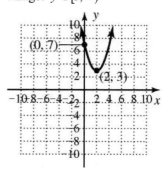

33. $x = (y-3)^2 + 2$

Vertex: $(2,3)$;

$x = (0-3)^2 + 2 = 11$

x-intercept: $(11,0)$;

$0 \neq (y-3)^2 + 2$

y-intercept: None ;

Domain: $x \in [2, \infty)$

Range: $y \in (-\infty, \infty)$

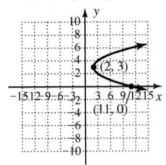

35. $x = 2(y-3)^2 + 1$

Vertex: $(1,3)$;

$x = 2(0-3)^2 + 1 = 19$

x-intercept: $(19,0)$;

$0 \neq 2(y-3)^2 + 1$

y-intercept: None;

Domain: $x \in [1, \infty)$

Range: $y \in (-\infty, \infty)$

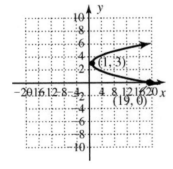

37. $x^2 = 8y$

Vertex: $(0,0)$

$8 = 4p$

$2 = p$

Focus: $(0,2)$

Length of Focal Chord: 8

Directrix: $y = -2$

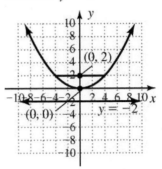

39. $x^2 = -24y$

Vertex: $(0,0)$

$-24 = 4p$

$-6 = p$

Focus: $(0,-6)$

Length of Focal Chord: 24

Directrix: $y = 6$

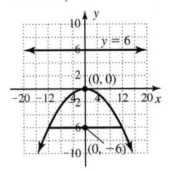

9.4 Exercises

41. $x^2 = 6y$

Vertex: $(0,0)$

$6 = 4p$

$\dfrac{3}{2} = p$

Focus: $\left(0, \dfrac{3}{2}\right)$

Length of Focal Chord: 6

Directrix: $y = \dfrac{-3}{2}$

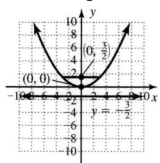

45. $y^2 = 18x$

Vertex: $(0,0)$

$18 = 4p$

$\dfrac{9}{2} = p$

Focus: $\left(\dfrac{9}{2}, 0\right)$

Length of Focal Chord: 18

Directrix: $x = -\dfrac{9}{2}$

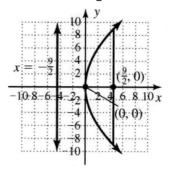

43. $y^2 = -4x$

Vertex: $(0,0)$

$-4 = 4p$

$-1 = p$

Focus: $(-1, 0)$

Length of Focal Chord: 4

Directrix: $x = 1$

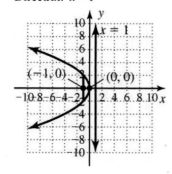

47. $y^2 = -10x$

Vertex: $(0,0)$

$-10 = 4p$

$-\dfrac{5}{2} = p$

Focus: $\left(-\dfrac{5}{2}, 0\right)$

Length of Focal Chord: 10

Directrix: $x = \dfrac{5}{2}$

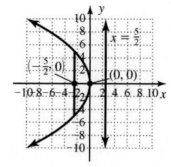

49. $x^2 - 8x - 8y + 16 = 0$

$x^2 - 8x + 16 = 8y$

$(x-4)^2 = 8y$

Vertex: $(4,0)$

$8 = 4p$

$2 = p$

Focus: $(4, 0+2) = (4,2)$

Length of Focal Chord: 8

Directrix: $y = -2$

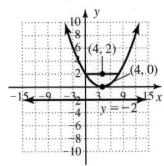

51. $x^2 - 14x - 24y + 1 = 0$

$x^2 - 14x = 24y - 1$

$x^2 - 14x + 49 = 24y - 1 + 49$

$(x-7)^2 = 24y + 48$

$(x-7)^2 = 24(y+2)$

Vertex: $(7,-2)$

$24 = 4p$

$6 = p$

Focus: $(7, -2+6) = (7,4)$

Length of Focal Chord: 24

Directrix: $y = -2 - 6 = -8$

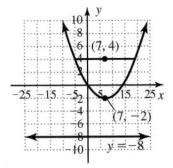

53. $3x^2 - 24x - 12y + 12 = 0$

$3x^2 - 24x = 12y - 12$

$3(x^2 - 8x + 16) = 12y - 12 + 48$

$3(x-4)^2 = 12y + 36$

$3(x-4)^2 = 12(y+3)$

$(x-4)^2 = 4(y+3)$

Vertex: $(4,-3)$

$4 = 4p$

$1 = p$

Focus: $(4, -3+1) = (4,-2)$

Length of Focal Chord: 4

Directrix: $y = -3 - 1 = -4$

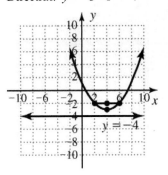

55. $y^2 - 12y - 20x + 36 = 0$

$y^2 - 12y + 36 = 20x$

$(y-6)^2 = 20x$

Vertex: $(0,6)$

$4p = 20$

$p = 5$

Focus: $(0+5, 6) = (5,6)$

Length of Focal Chord: 20

Directrix: $x = 0 - 5 = -5$

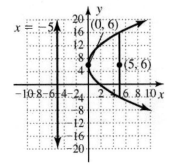

9.4 Exercises

57. $y^2 - 6y + 4x + 1 = 0$

$y^2 - 6y = -4x - 1$

$y^2 - 6y + 9 = -4x - 1 + 9$

$(y-3)^2 = -4(x-2)$

Vertex: $(2,3)$

$4p = -4$

$p = -1$

Focus: $(2-1,3) = (1,3)$

Length of Focal Chord: 4

Directrix: $x = 2 - (-1) = 3$

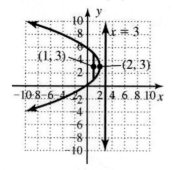

59. $2y^2 - 20y + 8x + 2 = 0$

$y^2 - 10y + 4x + 1 = 0$

$y^2 - 10y = -4x - 1$

$y^2 - 10y + 25 = -4x - 1 + 25$

$(y-5)^2 = -4(x-6)$

Vertex: $(6,5)$

$4p = -4$

$p = -1$

Focus: $(6-1,5) = (5,5)$

Length of Focal Chord: 4

Directrix: $x = 6 - (-1) = 7$

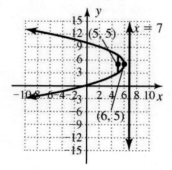

61. Focus: $(0,2)$; Directrix: $y = -2$

Opens up, Vertex $(0,0)$

$p = 2, 4p = 8;$

$(x-h)^2 = 4p(y-k)$

$x^2 = 8y$

63. Focus: $(4,0)$; Directrix: $x = -4$

Opens right, Vertex $(0,0)$

$p = 4, 4p = 16;$

$(y-k)^2 = 4p(x-h)$

$y^2 = 16x$

65. Focus: $(0,-5)$; Directrix: $y = 5$

Opens down, Vertex $(0,0)$

$p = -5, 4p = -20;$

$(x-h)^2 = 4p(y-k)$

$x^2 = -20y$

67. Vertex: $(2,-2)$; Focus: $(-1,-2)$

Opens left,

$p = -1 - 2 = -3, 4p = -12;$

$(y-k)^2 = 4p(x-h)$

$(y--2)^2 = -12(x-2)$

$(y+2)^2 = -12(x-2)$

69. Vertex: $(4,-7)$; Focus: $(4,-4)$

Opens up,

$p = -4 - -7 = 3, 4p = 12;$

$(x-h)^2 = 4p(y-k)$

$(x-4)^2 = 12(y+7)$

71. Directrix: $y = 0$; Focus: $(3,4)$

Opens up, Vertex $(3,2)$
$p = 2, 4p = 8;$

$(x-h)^2 = 4p(y-k)$

$(x-3)^2 = 8(y-2)$

73. Directrix: $x = -3$;

Endpoints of focal chord: $(1,4)$ and $(1,-4)$

Length of focal chord: 8
Opens right, Vertex $(-1,0)$, Focus: $(1,0)$

$(y-k)^2 = 4p(x-h)$

$y^2 = 8(x--1)$

$y^2 = 8(x+1)$

75. Vertex $(-2,2)$, Focus: $(-4,2)$

Opens left, $p = -2, 4p = -8;$

Directrix: $x = 0$;
Endpoints of focal chord:
$(-4,4)$ and $(-4,0)$

$(y-k)^2 = 4p(x-h)$

$(y-2)^2 = -8(x+2)$

77. $A = \dfrac{2}{3}ab$

$A = \dfrac{2}{3}(3)(8) = 16 \text{ units}^2$

79. $25x = 16y^2$

$\dfrac{25}{16}x = y^2$

Vertex: $(0,0)$

$4p = \dfrac{25}{16}$

$p = \dfrac{25}{64}$

Focus: $\left(\dfrac{25}{64}, 0\right)$

Length of Focal Chord: $\dfrac{25}{16}$

Directrix: $x = -\dfrac{25}{64}$

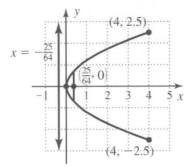

81. $y^2 = 54x$

$4p = 54$

$p = 13.5$

Focus: $(13.5, 0)$;

36 inch diameter, $y = 18$

$(18)^2 = 54x$

$x = 6$

Parabolic receiver: 6 inches

83. $x^2 = 167y$

$4p = 167$
$p = 41.75$

Focus: $(0, 41.75)$
100 feet diameter, $x = 50$
$(50)^2 = 167y$
$y \approx 14.97$
Parabolic receiver: 14.97 ft

85. 10 cm diameter: $y = 5$

$5^2 = 5x$
$5 = x;$
$4p = 5$
$p = 1.25$
The bulb will be placed a length of 1.25 cm from the vertex.
$y^2 = 5x$ or $x^2 = 5y$

87. $y = 2x^2 - 8x$

$y = 2(x^2 - 4x)$
$y + 8 = 2(x^2 - 4x + 4)$
$y + 8 = 2(x-2)^2$
$(x-2)^2 = \frac{1}{2}(y+8)$
Vertex: $(2, -8)$
$4p = \frac{1}{2}$
$p = \frac{1}{8}$

89. $A = \frac{4}{3}T$;

Area of the triangle: $A = \left|\dfrac{\det(T)}{2}\right| = \left|\dfrac{\|T\|}{2}\right|$

where $T = \begin{bmatrix} -3 & 5 & 1 \\ -6 & -3 & 1 \\ 0 & 4 & 1 \end{bmatrix}$

$A = \left|\dfrac{\det(T)}{2}\right| = \left|\dfrac{\|27\|}{2}\right| = 13.5$;

Area of oblique parabolic segment:
$A = \frac{4}{3}(13.5) = 18$ units^2

91. $x^6 - 64 = 0$

$(x^3 + 8)(x^3 - 8) = 0$
$(x+2)(x^2 - 2x + 4)(x-2)(x^2 + 2x + 4) = 0$
$x = 2, -2$
$x^2 - 2x + 4 = 0$ or $x^2 + 2x + 4 = 0$
$a = 1, b = -2, c = 4$ or $a = 1, b = 2, c = 4$
$x = \dfrac{-(-2) \pm \sqrt{(-2)^2 - 4(1)(4)}}{2(1)} = 1 \pm \sqrt{3}\, i$
or $x = \dfrac{-(2) \pm \sqrt{(2)^2 - 4(1)(4)}}{2(1)} = -1 \pm \sqrt{3}\, i$;
$x = -2, x = 2, x = 1 \pm \sqrt{3}\, i, x = -1 \pm \sqrt{3}\, i$

93. $9 \to -3$ $\qquad$ $15 \to 3$

$0 \to 60$ and $300 \to 360$
$60 + 60 = 120$ days

Mid-Chapter Check

1. $(x-4)^2 + (y+3)^2 = 9$

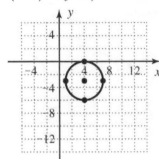

3. $\dfrac{(x-2)^2}{16} + \dfrac{(y+3)^2}{1} = 1$

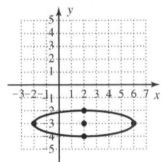

5. $\dfrac{(x+3)^2}{9} - \dfrac{(y-4)^2}{4} = 1$

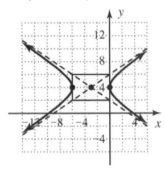

7. a. $\dfrac{(x+3)^2}{4} + \dfrac{(y-1)^2}{16} = 1$
    Center: $(-3, 1)$
    $a = 2, \ b = 4$
    $D: x \in [-3-2, -3+2]$
    $D: x \in [-5, -1]$;
    $R: y \in [1-4, 1+4]$
    $R: y \in [-3, 5]$

    b. $(x-3)^2 + (y-2)^2 = 16$
    Center: $(3,2), \ r = 4$
    $D: x \in [3-4, 3+4]$
    $D: x \in [-1, 7]$;
    $R: y \in [2-4, 2+4]$
    $R: y \in [-2, 6]$

    c. $y = (x-3)^2 - 4$
    $y + 4 = (x-3)^2$
    Vertex: $(3, -4)$
    $D: x \in (-\infty, \infty)$
    $R: y \in [-4, \infty)$

9. Vertices: $(-4, 0), (4, 0)$
    Center: $\left(\dfrac{-4+4}{2}, \dfrac{0+0}{2} \right) = (0,0)$;

    $a = 4, c = \dfrac{4\sqrt{3}}{2} = 2\sqrt{3}$

    $\left(2\sqrt{3} \right)^2 = 4^2 - b^2$

    $12 = 16 - b^2$

    $b^2 = 4$

    $b = 2$;

    $\dfrac{(x)^2}{16} + \dfrac{(y)^2}{4} = 1$

Chapter 9-Reinforcing Basic Concepts

1. $100x^2 - 400x - 18y^2 - 108y + 230 = 0$

$$100(x^2 - 4x) - 18(y^2 + 6y) = -230$$

$$100(x^2 - 4x + 4) - 18(y^2 + 6y + 9) = -230 + 400 - 162$$

$$100(x - 2)^2 - 18(y + 3)^2 = 8$$

$$\frac{25(x - 2)^2}{2} - \frac{9(y + 3)^2}{4} = 1$$

$$\frac{(x - 2)^2}{\left(\frac{\sqrt{2}}{5}\right)^2} - \frac{(y + 3)^2}{\left(\frac{2}{3}\right)^2} = 1$$

$$a = \frac{\sqrt{2}}{5}, b = \frac{2}{3}$$

3. $\dfrac{4(x + 3)^2}{49} + \dfrac{25(y - 1)^2}{36} = 1$

$$\frac{(x + 3)^2}{\frac{49}{4}} + \frac{(y - 1)^2}{\frac{36}{25}} = 1$$

$$\frac{(x + 3)^2}{\left(\frac{7}{2}\right)^2} + \frac{(y - 1)^2}{\left(\frac{6}{5}\right)^2} = 1$$

$$a = \frac{7}{2}, b = \frac{6}{5}$$

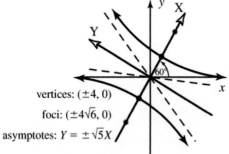

vertices: $(\pm 4, 0)$

foci: $(\pm 4\sqrt{6}, 0)$

asymptotes: $Y = \pm\sqrt{5}X$

9.5 Exercises

1.

a. Circle and Line
3 or 4 not possible

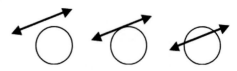

b. Parabola and Line
3 or 4 not possible

c. Circle and Parabola

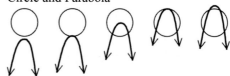

d. Circle and Hyperbola

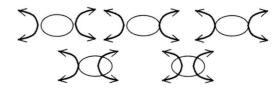

e. Hyperbola and Ellipse

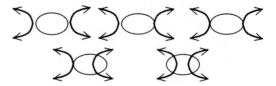

f. Circle and Ellipse

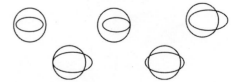

Chapter 9: Analytical Geometry

3. Region, solutions

5. Answers will vary.

7. $\begin{cases} x^2 + y = 6 & \text{Parabola} \\ x + y = 4 & \text{Line} \end{cases}$

   Solutions: $(-1,5), (2,2)$

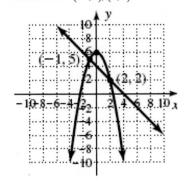

9. $\begin{cases} y - x^2 = -1 & \text{Parabola} \\ 4x^2 + y^2 = 100 & \text{Ellipse} \end{cases}$

   Solutions: $(-3,8), (3,8)$

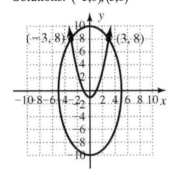

11. $\begin{cases} x^2 - y^2 = 9 & \text{Hyperbola} \\ x^2 + y^2 = 41 & \text{Circle} \end{cases}$

    Solutions: $(-5,4), (5,4), (-5,-4), (5,-4)$

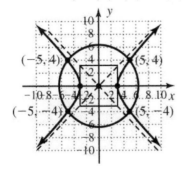

13. $\begin{cases} x^2 + y^2 = 25 \\ y - x = 1 \end{cases}$

    2nd equation : $y = x + 1$

    $x^2 + (x+1)^2 = 25$
    $x^2 + x^2 + 2x + 1 = 25$
    $2x^2 + 2x - 24 = 0$
    $x^2 + x - 12 = 0$
    $(x+4)(x-3) = 0$

    $x = -4, x = 3$;
    $y = -4 + 1 = -3;$
    $y = 3 + 1 = 4$

    Solutions: $(-4,-3), (3,4)$

15. $\begin{cases} x^2 + 4y^2 = 25 \\ x + 2y = 7 \end{cases}$

    $x = 7 - 2y$

    $(7-2y)^2 + 4y^2 = 25$
    $49 - 28y + 4y^2 + 4y^2 = 25$
    $8y^2 - 28y + 24 = 0$
    $2y^2 - 7y + 6 = 0$
    $(2y-3)(y-2) = 0$

    $y = \dfrac{3}{2}, y = 2$;

    $x = 7 - 2\left(\dfrac{3}{2}\right) = 4$;

    $x = 7 - 2(2) = 3$

    Solutions: $\left(4, \dfrac{3}{2}\right), (3,2)$

17. $\begin{cases} x^2 + y = 13 \\ 9x^2 - y^2 = 81 \end{cases}$

    $x^2 = 13 - y$

    $9(13 - y) - y^2 = 81$
    $117 - 9y - y^2 = 81$
    $y^2 + 9y - 36 = 0$
    $(y-3)(y+12) = 0$
    $y = 3$ or $y = -12$;

    $x^2 = 13 - 3$ or $x^2 = 13 - (-12)$
    $x^2 = 10$ or $x^2 = 25$
    $x = \pm\sqrt{10}$ or $x = \pm 5$

    Solutions: $\left(\sqrt{10}, 3\right), \left(-\sqrt{10}, 3\right),$
    $(5,-12), (-5,-12)$

19. $\begin{cases} x^2 + y^2 = 25 \\ 2x^2 - 3y^2 = 5 \end{cases}$

$\begin{cases} 3x^2 + 3y^2 = 75 \\ 2x^2 - 3y^2 = 5 \end{cases}$

$5x^2 = 80$
$x^2 = 16$
$x = \pm 4;$
$(4)^2 + y^2 = 25$
$y^2 = 9$
$y = \pm 3;$
Solutions: $(4,3),(4,-3),(-4,3),(-4,-3)$

21. $\begin{cases} x^2 - y = 4 \\ x^2 - y^2 = 16 \end{cases}$

$\begin{cases} x^2 - y = 4 \\ -x^2 + y^2 = -16 \end{cases}$

$y^2 - y = -12$
$y^2 - y + 12 = 0$
$a = 1, b = -1, c = 12$

$y = \dfrac{-(-1) \pm \sqrt{(-1)^2 - 4(1)(12)}}{2(1)}$

$y = \dfrac{1 \pm \sqrt{-47}}{2}$

No real solution.

23. $\begin{cases} 5x^2 - 2y^2 = 75 \\ 2x^2 + 3y^2 = 125 \end{cases}$

$\begin{cases} 15x^2 - 6y^2 = 225 \\ 4x^2 + 6y^2 = 250 \end{cases}$

$19x^2 = 475$
$x^2 = 25$
$x = \pm 5$
$5(\pm 5)^2 - 2y^2 = 75$
$-2y^2 = -50$
$y^2 = 25$
$y = \pm 5$
Solutions: $(5,-5),(5,5),(-5,5),(-5,-5)$

25. $\begin{cases} y - 5 = \log x \\ y = 6 - \log(x-3) \end{cases}$

$\begin{cases} y = \log x + 5 \\ y = 6 - \log(x-3) \end{cases}$

$\log x + 5 = 6 - \log(x-3)$
$\log x + \log(x-3) = 1$
$\log x(x-3) = 1$
$10 = x^2 - 3x$
$0 = x^2 - 3x - 10$
$0 = (x-5)(x+2)$
$x = 5 \text{ or } x = -2$
$x = -2$ is extraneous
$y = \log 5 + 5;$
Solution: $(5, \log 5 + 5)$

27. $\begin{cases} y = \ln(x^2) + 1 \\ y - 1 = \ln(x+12) \end{cases}$
Substitute in Equation 2
$y = \ln(x^2) + 1$
$\ln(x^2) + 1 - 1 = \ln(x+12)$
$\ln(x^2) = \ln(x+12)$
$x^2 = x + 12$
$x^2 - x - 12 = 0$
$(x-4)(x+3) = 0$
$x = 4 \text{ or } x = -3;$
If $x = 4, y = \ln(4^2) + 1$
$y = \ln 16 + 1;$
If $x = -3, y = \ln\left((-3)^2\right) + 1$
$y = \ln 9 + 1;$
Solutions: $(-3, \ln 9 + 1), (4, \ln 16 + 1)$

29. $\begin{cases} y - 9 = e^{2x} \\ 3 = y - 7e^x \end{cases}$

$\begin{cases} y - 9 = e^{2x} \\ -y + 3 = -7e^x \end{cases}$

R1 + R2

$-6 = e^{2x} - 7e^x$

$0 = e^{2x} - 7e^x + 6$

$(e^x - 6)(e^x - 1) = 0$

$e^x = 6$ or $e^x = 1$

$\ln e^x = \ln 6$ or $\ln e^x = \ln 1$

$x = \ln 6$ or $x = 0$

If $x = \ln 6$, $y - 9 = e^{2x}$

$y - 9 = e^{2\ln 6}$

$y - 9 = e^{\ln 36}$

$y - 9 = 36$

$y = 45$;

If $x = 0$, $y - 9 = e^{2(0)}$

$y - 9 = e^0$

$y - 9 = 1$

$y = 10$;

Solutions: $(0, 10), (\ln 6, 45)$

31. $\begin{cases} y = 4^{x+3} \\ y - 2^{x^2 + 3x} = 0 \end{cases}$

$4^{x+3} = 2^{x^2 + 3x}$

$2^{2x+6} = 2^{x^2 + 3x}$

$2x + 6 = x^2 + 3x$

$0 = x^2 + x - 6$

$(x+3)(x-2) = 0$

$x = -3$ or $x = 2$;

$y = 4^{-3+3} = 1$;

$y = 4^{2+3} = 1024$;

Solutions: $(-3, 1), (2, 1024)$

33. $\begin{cases} x^3 - y = 2x \\ y - 5x = -6 \end{cases}$

$y = 5x - 6$

$x^3 - 5x + 6 = 2x$

$x^3 - 7x + 6 = 0$

Possible rational roots: $\dfrac{\pm 1, \pm 6, \pm 2, \pm 3}{\pm 1}$

$\{\pm 1, \pm 6, \pm 2, \pm 3\}$;

$(x+3)(x-2)(x-1) = 0$

$x = -3$ or $x = 2$ or $x = 1$;

$y = 5(-3) - 6 = -21$;

$y = 5(2) - 6 = 4$;

$y = 5(1) - 6 = -1$;

Solutions: $(-3, -21), (2, 4), (1, -1)$

35. $\begin{cases} x^2 - 6x = y - 4 \\ y - 2x = -8 \end{cases}$

Solve for y in Equation 2.

$y = 2x - 8$, Substitute in Equation 1.

$x^2 - 6x = 2x - 8 - 4$

$x^2 - 8x + 12 = 0$

$(x-6)(x-2) = 0$

$x - 6 = 0$ or $x - 2 = 0$

$x = 6$ or $x = 2$;

If $x = 6$, $y - 2x = -8$

$y - 2(6) = -8$

$y = 4$;

If $x = 2$, $y - 2x = -8$

$y - 2(2) = -8$

$y = -4$;

Solutions: $(2, -4), (6, 4)$

37. $\begin{cases} x^2 + y^2 = 34 \\ y^2 + (x-3)^2 = 25 \end{cases}$

Solve for y in Equation 1.

$y^2 = -x^2 + 34$

$y = \pm\sqrt{-x^2 + 34}$;

Solve for y in Equation 2.

$y^2 + (x-3)^2 = 25$

$y^2 = -(x-3)^2 + 25$

$y = \pm\sqrt{-(x-3)^2 + 25}$;

Using a graphing calculator:

Solutions: $(3, 5), (3, -5)$

39. $\begin{cases} y = 2^x - 3 \\ y + 2x^2 = 9 \end{cases}$

$y1 = 2^x - 3;$

$y2 = -2x^2 + 9;$

Solutions: $(-2.43, -2.81), (2, 1)$

41. $\begin{cases} y = \dfrac{1}{(x-3)^2} + 2 \\ (x-3)^2 + y^2 = 10 \end{cases}$

$y1 = \dfrac{1}{(x-3)^2} + 2;$

$y^2 = -(x-3)^2 + 10$

$y2 = \pm\sqrt{-(x-3)^2 + 10}\;;$

Solutions:
$(0.72, 2.19), (2, 3), (4, 3), (5.28, 2.19)$

43. $\begin{cases} y - x^2 \geq 1 & \text{parabola} \\ x + y \leq 3 & \text{line} \end{cases}$

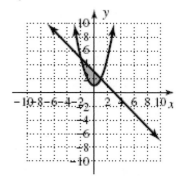

45. $\begin{cases} x^2 + y^2 > 16 & \text{circle} \\ x^2 + y^2 \leq 64 & \text{circle} \end{cases}$

Inequality 1, circle with center (0,0), radius 4.
Inequality 2, circle with center (0,0), radius 8.

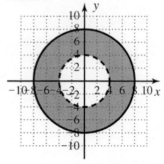

47. $\begin{cases} y - x^2 \leq -16 & \text{parabola} \\ y^2 + x^2 < 9 & \text{circle} \end{cases}$

$y \leq x^2 - 16$
$x^2 + y^2 < 9$
No solution.

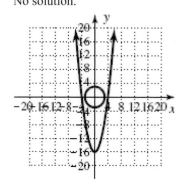

49. $\begin{cases} y^2 + x^2 \leq 25 & \text{circle} \\ |x| - 1 > -y & \text{absolute value} \end{cases}$

Inequality 1, circle with center (0,0), radius 5.
Inequality 2, absolute value with vertex (0,1).

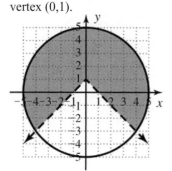

51. $h = b\sqrt{1 - \left(\dfrac{d}{a}\right)^2}$;

    $a = 50, \ b = 30$

    If $d = 20$ ft

    $h = 30\sqrt{1 - \left(\dfrac{20}{50}\right)^2}$

    $h \approx 27.5$ ft;

    If $d = 30$ ft

    $h = 30\sqrt{1 - \left(\dfrac{30}{50}\right)^2}$

    $h = 24$ ft;

    If $d = 40$ ft

    $h = 30\sqrt{1 - \left(\dfrac{40}{50}\right)^2}$

    $h = 18$ ft

53. $C(x) = 2.5x^2 - 120x + 3500$

    $R(x) = -2x^2 + 180x - 500$

    $C(x) = R(x)$

    $2.5x^2 - 120x + 3500 = -2x^2 + 180x - 500$

    $4.5x^2 - 300x + 4000 = 0$

    Using a graphing calculator,
    $x \approx 18.426$ or $x \approx 48.241$
    The company breaks even if either 18,400 or 48,200 cars are sold.

55. a. $8P^2 - 8P - 4D = 12$

    $-4D = -8P^2 + 8P + 12$

    $D = 2P^2 - 2P - 3$
    minimum: \$1.83 (when $D = 0$)

    b. $\begin{cases} 10P^2 + 6D = 144 \\ 8P^2 - 8P - 4D = 12 \end{cases}$

    $10P^2 + 6D = 144$

    $\dfrac{144 - 10P^2}{6} = D$;

    $8P^2 - 8P - 4D = 12$

    $8P^2 - 8P - 4\left(\dfrac{144 - 10P^2}{6}\right) = 12$

    $2P^2 - 2P - \left(\dfrac{144 - 10P^2}{6}\right) = 3$

    $12P^2 - 12P - 144 + 10P^2 = 18$

    $22P^2 - 12P - 162 = 0$

    $11P^2 - 6P - 81 = 0$

    $(11P + 27)(P - 3) = 0$

    $P = -\dfrac{27}{11}$ or $P = \$3$;

    $10(3)^2 + 6D = 144$

    $6D = 54$

    $D = 9$
    90,000 gallons

57. $85 = l\text{w}$

    $37 = 2l + 2\text{w}$;

    $\dfrac{85}{l} = w$

    $37 = 2l + 2\left(\dfrac{85}{l}\right)$

    $37l = 2l^2 + 170$

    $0 = 2l^2 - 37l + 170$

    $0 = (2l - 17)(l - 10)$

    $l = \dfrac{17}{2}, l = 10$;

    $w = \dfrac{85}{\left(\dfrac{17}{2}\right)} = 10$;

    $w = \dfrac{85}{(10)} = 8.5$;

    $8.5 \text{ m} \times 10 \text{ m}$

9.5 Exercises

59. Area : 45km^2

Diagonal : $\sqrt{106}$ km

$45 = lw$

$l = \dfrac{45}{w}$

$l^2 + w^2 = 106$

$\left(\dfrac{45}{w}\right)^2 + w^2 = 106$

$2025 + w^4 = 106w^2$

$w^4 - 106w^2 + 2025 = 0$

$\left(w^2 - 25\right)\left(w^2 - 81\right) = 0$

$w = \pm5, \pm9$

5 km, 9 km

61. Surface Area = 928 ft^2

Edges = 164 ft

$4w + 4l + 4w = 164$

$4l + 8w = 164$

$4l = 164 - 8w$

$l = 41 - 2w;$

$928 = 4lw + 2w^2$

$928 = 4(41 - 2w)w + 2w^2$

$928 = 164w - 8w^2 + 2w^2$

$6w^2 - 164w + 928 = 0$

$3w^2 - 82w + 464 = 0$

$(3w - 58)(w - 8) = 0$

$w = \dfrac{58}{3}$ or $w = 8;$

$w = 8, \ l = 41 - 2(8) = 25$

8 ft x 8 ft x 25 ft

63. Answers will vary.

65. Height : 18 inches

Surface Area : 4806 in^2

$4806 = lw + 2(18l) + 2(18w)$

$4806 = lw + 36l + 36w$

$4806 - 36l = lw + 36w$

$4806 - 36l = w(l + 36)$

$\dfrac{4806 - 36l}{l + 36} = w;$

$108(231) = 18(l)w$

$108(231) = 18l\left(\dfrac{4806 - 36l}{l + 36}\right)$

$24948(l + 36) = 18l(4806 - 36l)$

$24948l + 898128 = 86508l - 648l^2$

$648l^2 - 61560l + 898128 = 0$

$l^2 - 95l + 1386 = 0$

$(l - 18)(l - 77) = 0$

18 in. x 18 in. x 77 in.

67. $\dfrac{\sin 25°}{L} = \dfrac{\sin 43°}{100}$

$\dfrac{100 \sin 25°}{\sin 43°} = L$

$62 \approx L$

The length is approximately 62 ft.

69. a.   Let 2001 be year 0.

$m = \dfrac{\Delta \text{value}}{\Delta \text{time}} = \dfrac{4500 - 3300}{0 - 3} = -400$

The computer depreciates by $400 a year.

b. $y = -400x + 4500$

c. $y = -400(7) + 4500 = \$1700$

d. $700 = -400x + 4500$

$-3800 = -400x$

$9.5 = x$

9.5 years

9.6 Exercises

1. Polar

3. II, IV

5. To plot the point (r, θ) start at the origin or pole and move $|r|$ units out along the polar axis. Then move counterclockwise an angle measure of θ. You should be r units straight out from the pole in a direction of θ from the positive polar axis. If r is negative, final resting place for the point (r, θ) will be $180°$ from θ.

7. $\left(4, \dfrac{\pi}{2}\right)$

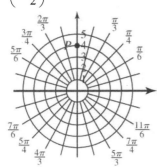

9. $\left(2, \dfrac{5\pi}{4}\right)$

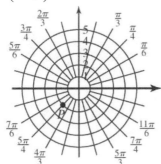

11. $\left(-5, \dfrac{5\pi}{6}\right)$

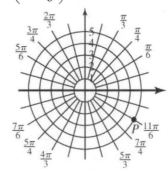

13. $\left(-3, -\dfrac{2\pi}{3}\right)$

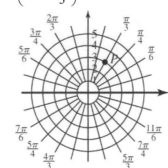

15. $P(0,4) \rightarrow \left(4, \dfrac{\pi}{2}\right)$

17. $(4, 4)$

$r = \sqrt{4^2 + 4^2} = \sqrt{16 + 16} = \sqrt{32} = 4\sqrt{2}$;

$\theta = \tan^{-1}\left(\dfrac{4}{4}\right) = \dfrac{\pi}{4}$;

$P(4,4) \rightarrow P\left(4\sqrt{2}, \dfrac{\pi}{4}\right)$

19. $\left(-4, 4\sqrt{3}\right)$

$r = \sqrt{(-4)^2 + \left(4\sqrt{3}\right)^2} = \sqrt{16 + 48} = \sqrt{64} = 8$;

$\theta_r = \tan^{-1}\left(\dfrac{4\sqrt{3}}{-4}\right) = \dfrac{-\pi}{3}$

$\theta = \pi - \dfrac{\pi}{3} = \dfrac{2\pi}{3}$;

$P\left(-4, \ 4\sqrt{3}\right) \rightarrow P\left(8, \ \dfrac{2\pi}{3}\right)$

9.6 Exercises

21. $(-4, 4)$

$r = \sqrt{(-4)^2 + 4^2} = \sqrt{32} = 4\sqrt{2}$;

$\theta_r = \tan^{-1}\left(\dfrac{4}{-4}\right) = \dfrac{-\pi}{4}$;

$\theta = \pi - \dfrac{\pi}{4} = \dfrac{3\pi}{4}$;

$P(-4, 4) \rightarrow P\left(4\sqrt{2}, \dfrac{3\pi}{4}\right)$

23. Original Point: $\left(3\sqrt{2}, \dfrac{3\pi}{4}\right)$

$\left(3\sqrt{2}, \dfrac{3\pi}{4} - 2\pi\right) \rightarrow \left(3\sqrt{2}, \dfrac{-5\pi}{4}\right)$

$\left(-3\sqrt{2}, \dfrac{3\pi}{4} + \pi\right) \rightarrow \left(-3\sqrt{2}, \dfrac{7\pi}{4}\right)$

$\left(-3\sqrt{2}, \dfrac{3\pi}{4} - \pi\right) \rightarrow \left(-3\sqrt{2}, \dfrac{-\pi}{4}\right)$

$\left(3\sqrt{2}, \dfrac{3\pi}{4} + 2\pi\right) \rightarrow \left(3\sqrt{2}, \dfrac{11\pi}{4}\right)$

25. Original Point: $\left(-2, \dfrac{11\pi}{6}\right)$

$\left(2, \dfrac{11\pi}{6} - \pi\right) \rightarrow \left(2, \dfrac{5\pi}{6}\right)$

$\left(2, \dfrac{11\pi}{6} - 3\pi\right) \rightarrow \left(2, \dfrac{-7\pi}{6}\right)$

$\left(-2, \dfrac{11\pi}{6} - 2\pi\right) \rightarrow \left(-2, \dfrac{-\pi}{6}\right)$

$\left(2, \dfrac{11\pi}{6} + \pi\right) \rightarrow \left(2, \dfrac{17\pi}{6}\right)$

27. C

29. C

31. D

33. B

35. D

37. $(-8, 0)$

$r = \sqrt{(-8)^2 + 0^2} = \sqrt{64} = 8$;

$\theta_r = 0°$, $\theta = 180°$

$P(-8, 0) \rightarrow P(8, \pi)$ or $P(8, 180°)$

39. $(4, 4)$

$r = \sqrt{4^2 + 4^2} = \sqrt{32} = 4\sqrt{2}$;

$\theta = \tan^{-1}\left(\dfrac{4}{4}\right) = \dfrac{\pi}{4}$;

$P(4, 4) \rightarrow P\left(4\sqrt{2}, \dfrac{\pi}{4}\right)$ or $P\left(4\sqrt{2}, 45°\right)$

41. $\left(5\sqrt{2}, 5\sqrt{2}\right)$

$r = \sqrt{\left(5\sqrt{2}\right)^2 + \left(5\sqrt{2}\right)^2} = \sqrt{100} = 10$;

$\theta = \tan^{-1}\left(\dfrac{5\sqrt{2}}{5\sqrt{2}}\right) = \dfrac{\pi}{4}$;

$P\left(5\sqrt{2}, 5\sqrt{2}\right) \rightarrow P\left(10, \dfrac{\pi}{4}\right)$ or $P(10, 45°)$

43. $(-5, -12)$

$r = \sqrt{(-5)^2 + (-12)^2} = \sqrt{169} = 13$;

$\theta_r = \tan^{-1}\left(\dfrac{-12}{-5}\right) = 67.4°$;

$\theta = 67.4 + 180 = 247.4°$
or $1.176 + \pi = 4.3176$;

$P(-5, -12) \rightarrow P(13, 247.4°)$ or $P(13, 4.3176)$

45. $(8, \; 45°)$

$x = 8\cos 45° = 8\left(\dfrac{\sqrt{2}}{2}\right) = 4\sqrt{2}$;

$y = 8\sin 45° = 8\left(\dfrac{\sqrt{2}}{2}\right) = 4\sqrt{2}$;

$\left(4\sqrt{2}, 4\sqrt{2}\right)$

47. $\left(4, \dfrac{3\pi}{4}\right)$

$x = 4\cos \dfrac{3\pi}{4} = 4\left(\dfrac{-\sqrt{2}}{2}\right) = -2\sqrt{2}$;

$y = 4\sin \dfrac{3\pi}{4} = 4\left(\dfrac{\sqrt{2}}{2}\right) = 2\sqrt{2}$;

$\left(-2\sqrt{2}, 2\sqrt{2}\right)$

49. $\left(-2, \dfrac{7\pi}{6}\right)$

$x = -2\cos\left(\dfrac{7\pi}{6}\right) = -2\left(\dfrac{-\sqrt{3}}{2}\right) = \sqrt{3}$;

$y = -2\sin\left(\dfrac{7\pi}{6}\right) = -2\left(\dfrac{-1}{2}\right) = 1$

$\left(\sqrt{3}, 1\right)$

51. $(-5, -135°)$

$x = -5\cos(-135°) = -5\left(\dfrac{-\sqrt{2}}{2}\right) = \dfrac{5\sqrt{2}}{2}$;

$y = -5\sin(-135°) = -5\left(\dfrac{-\sqrt{2}}{2}\right) = \dfrac{5\sqrt{2}}{2}$;

$\left(\dfrac{5\sqrt{2}}{2}, \dfrac{5\sqrt{2}}{2}\right)$

53. $r = 5$
Circle, Center: $(0,0)$

Cycle	r-value analysis	Location of graph
0 to $\dfrac{\pi}{2}$	$\lvert r \rvert$ constant at 5	QI $(r=5)$
$\dfrac{\pi}{2}$ to π	$\lvert r \rvert$ constant at 5	QII $(r=5)$
π to $\dfrac{3\pi}{2}$	$\lvert r \rvert$ constant at 5	QIII $(r=5)$
$\dfrac{3\pi}{2}$ to 2π	$\lvert r \rvert$ constant at 5	QIV $(r=5)$

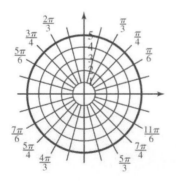

55. $\theta = \dfrac{\pi}{6}$

Straight line, points of the form $\left(r, \dfrac{\pi}{6}\right)$, $\dfrac{\pi}{6}$ constant, r varies

Cycle	r-value analysis	Location of graph
At $\dfrac{\pi}{6}$	$\lvert r \rvert$ increases from 0 to ∞	QI $(r>0)$
At $\dfrac{7\pi}{6}$	$\lvert r \rvert$ increases from 0 to $-\infty$	QI $(r>0)$

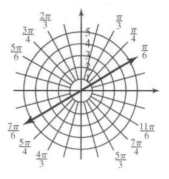

57. $r = 4\cos\theta$
Circle, Center: $(2,0)$
Closed figure limited to QI and QIV

Cycle	r-value analysis	Location of graph
0 to $\dfrac{\pi}{4}$	$\lvert r \rvert$ decreases from 4 to 2	QI $(r>0)$
$\dfrac{\pi}{4}$ to $\dfrac{\pi}{2}$	$\lvert r \rvert$ decreases from 2 to 0	QII $(r<0)$
$\dfrac{3\pi}{2}$ to $\dfrac{3\pi}{4}$	$\lvert r \rvert$ increases from 0 to 2	QIII $(r<0)$
$\dfrac{3\pi}{4}$ to 2π	$\lvert r \rvert$ increases from 2 to 4	QIV $(r>0)$

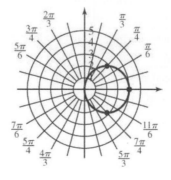

59. $r = 3 + 3\sin\theta$

Cardioid, symmetric about $\theta = \dfrac{\pi}{2}$

Cycle	r-value analysis	Location of graph
0 to $\dfrac{\pi}{2}$	$\lvert r \rvert$ increases from 3 to 6	QI $(r > 0)$
$\dfrac{\pi}{2}$ to π	$\lvert r \rvert$ decreases from 6 to 3	QII $(r > 0)$
π to $\dfrac{3\pi}{2}$	$\lvert r \rvert$ decreases from 3 to 0	QIII $(r > 0)$
$\dfrac{3\pi}{2}$ to 2π	$\lvert r \rvert$ increases from 0 to 3	QIV $(r > 0)$

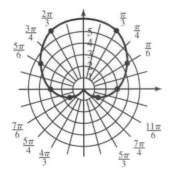

61. $r = 2 - 4\sin\theta$

Limacon, symmetric about $\theta = \dfrac{\pi}{2}$

Cycle	r-value analysis	Location of graph
0 to $\dfrac{\pi}{6}$	$\lvert r \rvert$ decreases from 2 to 0	QI $(r > 0)$
$\dfrac{\pi}{6}$ to $\dfrac{\pi}{2}$	$\lvert r \rvert$ increases from 0 to 2	QIII $(r < 0)$
$\dfrac{\pi}{2}$ to $\dfrac{2\pi}{3}$	$\lvert r \rvert$ decreases from 2 to 0	QIV $(r < 0)$
$\dfrac{2\pi}{3}$ to π	$\lvert r \rvert$ increases from 0 to 2	QII $(r > 0)$
π to $\dfrac{3\pi}{2}$	$\lvert r \rvert$ increases from 2 to 6	QIII $(r > 0)$
$\dfrac{3\pi}{2}$ to 2π	$\lvert r \rvert$ decreases from 6 to 2	QIV $(r > 0)$

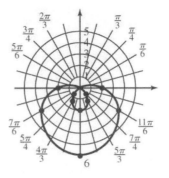

63. $r = 5\cos(2\theta)$

Four-petal rose, symmetric about $\theta = 0$

Cycle	r-value analysis	Location of graph
0 to $\dfrac{\pi}{4}$	$\lvert r \rvert$ decreases from 5 to 0	QI $(r > 0)$
$\dfrac{\pi}{4}$ to $\dfrac{\pi}{2}$	$\lvert r \rvert$ increases from 0 to 5	QIII $(r < 0)$
$\dfrac{\pi}{2}$ to $\dfrac{3\pi}{4}$	$\lvert r \rvert$ decreases from 5 to 0	QIV $(r < 0)$
$\dfrac{3\pi}{4}$ to π	$\lvert r \rvert$ increases from 0 to 5	QII $(r > 0)$
π to $\dfrac{5\pi}{4}$	$\lvert r \rvert$ decreases from 5 to 0	QIII $(r > 0)$
$\dfrac{5\pi}{4}$ to $\dfrac{3\pi}{2}$	$\lvert r \rvert$ increases from 0 to 5	QI $(r < 0)$
$\dfrac{3\pi}{2}$ to $\dfrac{7\pi}{4}$	$\lvert r \rvert$ decreases from 5 to 0	QII $(r < 0)$
$\dfrac{7\pi}{4}$ to 2π	$\lvert r \rvert$ increases from 0 to 5	QIV $(r > 0)$

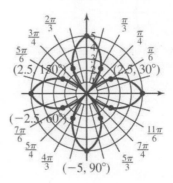

65. $r = 4\sin(2\theta)$

Four-petal rose, symmetric about $\theta = \dfrac{\pi}{2}$

Cycle	r-value analysis	Location of graph
0 to $\dfrac{\pi}{4}$	$\lvert r \rvert$ increases from 0 to 4	QI $(r > 0)$
$\dfrac{\pi}{4}$ to $\dfrac{\pi}{2}$	$\lvert r \rvert$ decreases from 4 to 0	QIII $(r > 0)$
$\dfrac{\pi}{2}$ to $\dfrac{3\pi}{4}$	$\lvert r \rvert$ increases from 0 to 4	QIV $(r < 0)$
$\dfrac{3\pi}{4}$ to π	$\lvert r \rvert$ decreases from 4 to 0	QII $(r < 0)$
π to $\dfrac{5\pi}{4}$	$\lvert r \rvert$ increases from 0 to 4	QIII $(r < 0)$
$\dfrac{5\pi}{4}$ to $\dfrac{3\pi}{2}$	$\lvert r \rvert$ decreases from 4 to 0	QI $(r < 0)$
$\dfrac{3\pi}{2}$ to $\dfrac{7\pi}{4}$	$\lvert r \rvert$ increases from 0 to 4	QII $(r > 0)$
$\dfrac{7\pi}{4}$ to 2π	$\lvert r \rvert$ decreases from 4 to 0	QIV $(r > 0)$

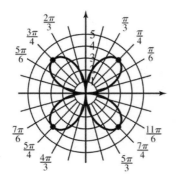

67. $r^2 = 9\sin(2\theta)$

Lemniscate, symmetric about $\theta = \dfrac{\pi}{4}$

Closed image in QI & QIII

Cycle	r-value analysis	Location of graph
0 to $\dfrac{\pi}{4}$	$\lvert r \rvert$ increases from 0 to 3	QI $(r > 0)$
$\dfrac{\pi}{4}$ to $\dfrac{\pi}{2}$	$\lvert r \rvert$ decreases from 3 to 0	QI $(r > 0)$
π to $\dfrac{5\pi}{4}$	$\lvert r \rvert$ increases from 0 to 3	QIII $(r > 0)$
$\dfrac{5\pi}{4}$ to $\dfrac{3\pi}{2}$	$\lvert r \rvert$ decreases from 3 to 0	QIII $(r > 0)$

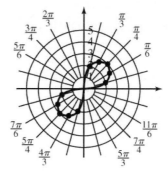

9.6 Exercises

69. $r = 4\sin\left(\dfrac{\theta}{2}\right)$

Symmetric about $\theta = \dfrac{\pi}{2}$ and $\theta = 0$

Cycle	r-value analysis	Location of graph
0 to $\dfrac{\pi}{2}$	$\lvert r\rvert$ increases from 0 to $2\sqrt{2}$	QI $(r > 0)$
$\dfrac{\pi}{2}$ to π	$\lvert r\rvert$ increases from 4 to $2\sqrt{2}$	QII $(r > 0)$
π to $\dfrac{3\pi}{2}$	$\lvert r\rvert$ decreases from 4 to $2\sqrt{2}$	QIII $(r > 0)$
$\dfrac{3\pi}{2}$ to 2π	$\lvert r\rvert$ decreases from $2\sqrt{2}$ to 0	QIV $(r > 0)$
2π to $\dfrac{5\pi}{2}$	$\lvert r\rvert$ increases from 0 to $2\sqrt{2}$	QIII $(r < 0)$
$\dfrac{5\pi}{2}$ to 3π	$\lvert r\rvert$ increases from $2\sqrt{2}$ to 4	QIV $(r < 0)$
3π to $\dfrac{7\pi}{2}$	$\lvert r\rvert$ decreases from 4 to $2\sqrt{2}$	QI $(r < 0)$
$\dfrac{7\pi}{2}$ to 4π	$\lvert r\rvert$ decreases from $2\sqrt{2}$ to 0	QII $(r < 0)$

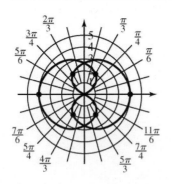

71. $r = 4\sqrt{1 - \sin^2\theta}$, a hippopede

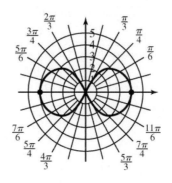

73. $r = 2\cos\theta\cot\theta$, a cissoid

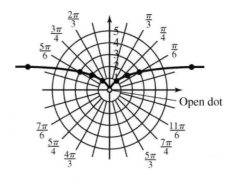

Open dot

75. $r = 8\sin\theta\cos^2\theta$, a bifoliate

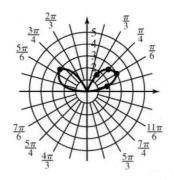

77. $M =$

$$\left(\frac{8\cos 30° + 6\cos 45°}{2}, \frac{8\sin 30° + 6\sin 45°}{2} \right)$$

$$= \left(\frac{8\left(\frac{\sqrt{3}}{2}\right) + 6\left(\frac{\sqrt{2}}{2}\right)}{2}, \frac{8\left(\frac{1}{2}\right) + 6\left(\frac{\sqrt{2}}{2}\right)}{2} \right)$$

$$= \left(\frac{4\sqrt{3} + 3\sqrt{2}}{2}, \frac{4 + 3\sqrt{2}}{2} \right)$$

$(6, 45°) \rightarrow \left(3\sqrt{2}, 3\sqrt{2}\right)$

$x = 6\cos 45° = 6\left(\frac{\sqrt{2}}{2}\right) = 3\sqrt{2}$;

$y = 6\sin 45° = 6\left(\frac{\sqrt{2}}{2}\right) = 3\sqrt{2}$;

$(8, 30°) \rightarrow \left(4\sqrt{3}, 4\right)$

$x = 8\cos 30° = 8\left(\frac{\sqrt{3}}{2}\right) = 4\sqrt{3}$;

$y = 8\sin 30° = 8\left(\frac{1}{2}\right) = 4$;

$$M = \left(\frac{x_1 + x_2}{2} \right), \left(\frac{y_1 + y_2}{2} \right)$$

$$= \left(\frac{3\sqrt{2} + 4\sqrt{3}}{2}, \frac{3\sqrt{2} + 4}{2} \right), \text{ yes}$$

79. $r = 4 + 4\cos\theta$

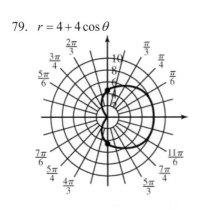

81. $r = 4\cos(5\theta)$

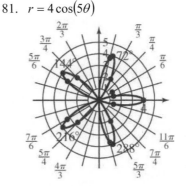

83. $r^2 = 16\cos(2\theta)$

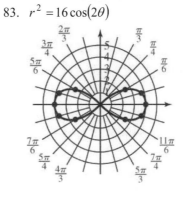

85. $r = 4\sin\theta$

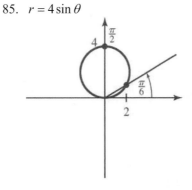

87. a: This is a circle through $\left(6, 0°\right)$ symmetric about the polar axis.

89. g: This is a circle through $\left(6, \frac{\pi}{2}\right)$ symmetric about $\theta = \frac{\pi}{2}$.

91. f: This is a limacon symmetric about $\theta = \frac{\pi}{2}$ with an inner loop. Thus $a < b$.

9.6 Exercises

93. b: This is a cardoid symmetric about $\theta = \dfrac{\pi}{2}$

through $\left(6, \dfrac{3\pi}{2}\right)$.

95. $(7200, 45°)$ and $(0, 90°)$

$a = 7200$

$r^2 = 7200^2 \sin(2\theta)$

97. 5 blades; $r = 15$ mm

$r = 15\cos(5\theta)$ or $r = 15\sin(5\theta)$

99. $r = a\theta; r = \dfrac{1}{2}\theta$

π, π, π , Answers will vary.

101. Consider $r = a\sqrt{\cos(2\theta)}$ and

$r = -a\sqrt{\cos(2\theta)}$; both satisfy

$r^2 = a^2\cos(2\theta)$. Thus, (r, θ) and $(-r, \theta)$

will both be on the curve. The same is true

with $r = a\sqrt{\sin(2\theta)}$ and $r = -a\sqrt{\sin(2\theta)}$.

103. $A = \pi r^2$; $r = 6$

$A = \pi(6)^2$

$A = 36\pi$

Area $= (0.25)36\pi = 9\pi$ units^2

105. $r = \dfrac{6}{2 + 4\sin\theta}$

$2r + 4r\sin\theta = 6$

$r + 2r\sin\theta = 3$

$\sqrt{x^2 + y^2} + 2y = 3$

$\left(\sqrt{x^2 + y^2}\right)^2 = (3 - 2y)$

$x^2 + y^2 = 9 - 12y + 4y^2$

$3y^2 - x^2 - 12y + 9 = 0$

107. $20 = 5 - 30\sin\left(2t - \dfrac{\pi}{6}\right)$

$\dfrac{15}{-30} = \dfrac{-30}{-30}\sin\left(2t - \dfrac{\pi}{6}\right)$

$\dfrac{-1}{2} = \sin\left(2t - \dfrac{\pi}{6}\right)$

Since $\sin^{-1}\left(\dfrac{1}{2}\right) = \dfrac{\pi}{6}$ and because r is

negative $\theta \in \text{QIII} \& \theta \in \text{QIV}$,

$\theta = \dfrac{7\pi}{6}$ or $\theta = \dfrac{11\pi}{6}$

$2t - \dfrac{\pi}{6} = \dfrac{7\pi}{6} + 2\pi k$

$2t = \dfrac{8\pi}{6} + 2\pi k$

$t = \dfrac{2\pi}{3} + \pi k$;

If $k = 0, t = \dfrac{2\pi}{3}$; If $k = 1, t = \dfrac{2\pi}{3} + \pi = \dfrac{5\pi}{3}$;

$2t - \dfrac{\pi}{6} = \dfrac{11\pi}{6} + 2\pi k$

$2t = \dfrac{12\pi}{6} + 2\pi k$

$t = \dfrac{2\pi}{2} + \pi k = \pi + \pi k$;

If $k = -1, t = \pi - \pi = 0$;

If $k = 0, t = \pi + 0 = \pi$;

$x \in \left\{0, \dfrac{2\pi}{3}, \pi, \dfrac{5\pi}{3}\right\}$

109. $D : x \in [-5, 2) \cup (2, 5]$

$R : y \in [-3, 2) \cup \{4\}$

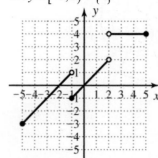

Technology Highlight

1. Verified.

9.7 Exercises

1. Rotation of axes; $\dfrac{B}{A-C}$

3. Invariants

5. If you use rotation of axes, the equation can be written more simply lacking the xy-term. The simpler form allows one to look at the equation and understand what type of graph it represents and important features of the graph.

7. Hyperbola, $xy = -4$

Asymptotes: $y = \pm \dfrac{b}{a}x$

Because rotation is $45°$ and $\tan 45° = 1$

$\dfrac{b}{a} = 1$

$2^2 + (-2)^2 = a^2$

$4 + 4 = a^2$

$8 = a^2$

$a = \sqrt{8} = 2\sqrt{2}$;

$a = 2\sqrt{2}$ so $b = 2\sqrt{2}$

$\dfrac{Y^2}{\left(2\sqrt{2}\right)^2} - \dfrac{X^2}{\left(2\sqrt{2}\right)^2} = 1$

$\dfrac{Y^2}{8} - \dfrac{X^2}{8} = 1$

9. $\left(6\sqrt{2}, 6\right)$

$X = x\cos\beta + y\sin\beta$

$X = 6\sqrt{2}\cos 45° + 6\sin 45°$

$X = 6\sqrt{2}\left(\dfrac{1}{\sqrt{2}}\right) + 6\left(\dfrac{1}{\sqrt{2}}\right) = 6 + \dfrac{6}{\sqrt{2}}$

$X = 6 + \dfrac{6}{\sqrt{2}} \cdot \dfrac{\sqrt{2}}{\sqrt{2}} = 6 + \dfrac{6\sqrt{2}}{2} = 6 + 3\sqrt{2}$;

$Y = -x\sin\beta + y\cos\beta$

$Y = -6\sqrt{2}\sin 45° + 6\cos 45°$

$Y = -6\sqrt{2}\left(\dfrac{1}{\sqrt{2}}\right) + 6\left(\dfrac{1}{\sqrt{2}}\right)$

$Y = -6 + \dfrac{6}{\sqrt{2}} = -6 + \dfrac{6\sqrt{2}}{2} = -6 + 3\sqrt{2}$;

$X = 6 + 3\sqrt{2}$; $Y = -6 + 3\sqrt{2}$

11. $(0, 5)$

$X = x\cos\beta + y\sin\beta$

$X = 0\cos 45° + 5\sin 45°$

$X = 0 + 5\left(\dfrac{1}{\sqrt{2}}\right) = \dfrac{5\sqrt{2}}{2}$;

$Y = -x\sin\beta + y\cos\beta$

$Y = 0\sin 45° + 5\cos 45°$

$Y = 5\left(\dfrac{1}{\sqrt{2}}\right) = \dfrac{5\sqrt{2}}{2}$;

$X = \dfrac{5\sqrt{2}}{2}$; $Y = \dfrac{5\sqrt{2}}{2}$

13. $\beta = 30°$; $(X, Y) = \left(2, 2\sqrt{3}\right)$

$x = X\cos\beta - Y\sin\beta$

$x = 2\cos 30° - 2\sqrt{3}\sin 30°$

$x = 2\left(\dfrac{\sqrt{3}}{2}\right) - 2\sqrt{3}\left(\dfrac{1}{2}\right) = \sqrt{3} - \sqrt{3} = 0$;

$y = X\sin\beta + Y\cos\beta$

$y = 2\sin 30° + 2\sqrt{3}\cos 30°$

$y = 2\left(\dfrac{1}{2}\right) + 2\sqrt{3}\left(\dfrac{\sqrt{3}}{2}\right) = 1 + 3 = 4$;

$x = 0, y = 4$

15. $\beta = 30°$; $(X, Y) = (3, 4)$

$x = X\cos\beta - Y\sin\beta$

$x = 3\cos 30° - 4\sin 30°$

$x = 3\left(\dfrac{\sqrt{3}}{2}\right) - 4\left(\dfrac{1}{2}\right) = \dfrac{3\sqrt{3}}{2} - 2$;

$y = X\sin\beta + Y\cos\beta$

$y = 3\sin 30° + 4\cos 30°$

$y = 3\left(\dfrac{1}{2}\right) + 4\left(\dfrac{\sqrt{3}}{2}\right) = \dfrac{3}{2} + 2\sqrt{3}$;

$x = \dfrac{3\sqrt{3}}{2} - 2$; $y = \dfrac{3}{2} + 2\sqrt{3}$

17. $X^2 - Y^2 = 9$; $60°$

$\cos 60° = \dfrac{1}{2}$; $\sin 60° = \dfrac{\sqrt{3}}{2}$

$(x\cos\beta + y\sin\beta)^2$

$\qquad -(y\cos\beta - x\sin\beta)^2 = 9$

$\left(\dfrac{1}{2}x + \dfrac{\sqrt{3}}{2}y\right)^2 - \left(\dfrac{1}{2}y - \dfrac{\sqrt{3}}{2}x\right)^2 = 9$

$\left(\dfrac{1}{4}x^2 + \dfrac{\sqrt{3}}{2}xy + \dfrac{3}{4}y^2\right)$

$\qquad -\left(\dfrac{1}{4}y^2 - \dfrac{\sqrt{3}}{2}xy + \dfrac{3}{4}x^2\right) = 9$

$\dfrac{1}{4}x^2 + \dfrac{\sqrt{3}}{2}xy + \dfrac{3}{4}y^2 - \dfrac{1}{4}y^2$

$\qquad + \dfrac{\sqrt{3}}{2}xy - \dfrac{3}{4}x^2 = 9$

$\dfrac{-x^2}{2} + xy\sqrt{3} + \dfrac{y^2}{2} = 9$

19. $3x^2 + 2xy + 3y^2 = 9$; $45°$

$\beta = 45°$; $\cos 45° = \sin 45° = \dfrac{\sqrt{2}}{2}$

$x = \dfrac{\sqrt{2}}{2}X - \dfrac{\sqrt{2}}{2}Y$; $\quad y = \dfrac{\sqrt{2}}{2}X + \dfrac{\sqrt{2}}{2}Y$;

$3\left(\dfrac{\sqrt{2}}{2}X - \dfrac{\sqrt{2}}{2}Y\right)^2 + 2\left(\dfrac{\sqrt{2}}{2}X - \dfrac{\sqrt{2}}{2}Y\right)$

$\cdot\left(\dfrac{\sqrt{2}}{2}X + \dfrac{\sqrt{2}}{2}Y\right) + 3\left(\dfrac{\sqrt{2}}{2}X + \dfrac{\sqrt{2}}{2}Y\right)^2 = 9$

$3\left(\dfrac{1}{2}X^2 - XY + \dfrac{1}{2}Y^2\right) + 2\left(\dfrac{1}{2}X^2 - \dfrac{1}{2}Y^2\right)$

$\qquad + 3\left(\dfrac{1}{2}X^2 + XY + \dfrac{1}{2}Y^2\right) = 9$

$\dfrac{3}{2}X^2 - 3XY + \dfrac{3}{2}Y^2 + X^2 - Y^2$

$\qquad + \dfrac{3}{2}X^2 + 3XY + \dfrac{3}{2}Y^2 = 9$

$4X^2 + 2Y^2 = 9$

21. (a) $x^2 + 4xy + y^2 - 2 = 0$

$A = 1, C = 1, B = 4$

$\tan(2\beta) = \dfrac{B}{A - C}$

$\tan(2\beta) = \dfrac{4}{1 - 1}$ Undefined

$2\beta = \dfrac{\pi}{2}$; $\beta = \dfrac{\pi}{4}$

$\sin\dfrac{\pi}{4} = \cos\dfrac{\pi}{4} = \dfrac{\sqrt{2}}{2}$

$x = \dfrac{\sqrt{2}}{2}X - \dfrac{\sqrt{2}}{2}Y$; $y = \dfrac{\sqrt{2}}{2}X + \dfrac{\sqrt{2}}{2}Y$

$\left(\dfrac{\sqrt{2}}{2}X - \dfrac{\sqrt{2}}{2}Y\right)^2 + 4\left(\dfrac{\sqrt{2}}{2}X - \dfrac{\sqrt{2}}{2}Y\right)$

$\cdot\left(\dfrac{\sqrt{2}}{2}X + \dfrac{\sqrt{2}}{2}Y\right) + \left(\dfrac{\sqrt{2}}{2}X + \dfrac{\sqrt{2}}{2}Y\right)^2 = 2$

$\left(\dfrac{1}{2}X^2 - XY + \dfrac{1}{2}Y^2\right) + 4\left(\dfrac{1}{2}X^2 - \dfrac{1}{2}Y^2\right)$

$\qquad + \left(\dfrac{1}{2}X^2 + XY + \dfrac{1}{2}Y^2\right) = 2$

$3X^2 - Y^2 = 2$

(b)

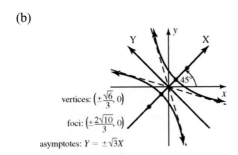

vertices: $\left(\pm\frac{\sqrt{6}}{3}, 0\right)$

foci: $\left(\pm\frac{2\sqrt{10}}{3}, 0\right)$

asymptotes: $Y = \pm\sqrt{3}X$

23. (a) $5x^2 + 6xy + 5y^2 = 16$

$A = 5, B = 6, C = 5$

$\tan(2\beta) = \dfrac{6}{5-5} = $ Undefined

$2\beta = \dfrac{\pi}{2}$; $\beta = \dfrac{\pi}{4}$

$\sin\dfrac{\pi}{4} = \cos\dfrac{\pi}{4} = \dfrac{\sqrt{2}}{2}$;

$x = \dfrac{\sqrt{2}}{2}X - \dfrac{\sqrt{2}}{2}Y$; $y = \dfrac{\sqrt{2}}{2}X + \dfrac{\sqrt{2}}{2}Y$;

$5\left(\dfrac{\sqrt{2}}{2}X - \dfrac{\sqrt{2}}{2}Y\right)^2 + 6\left(\dfrac{\sqrt{2}}{2}X - \dfrac{\sqrt{2}}{2}Y\right)$

$\cdot\left(\dfrac{\sqrt{2}}{2}X + \dfrac{\sqrt{2}}{2}Y\right) + 5\left(\dfrac{\sqrt{2}}{2}X + \dfrac{\sqrt{2}}{2}Y\right)^2 = 16$

$5\left(\dfrac{1}{2}X^2 - XY + \dfrac{1}{2}Y^2\right) + 6\left(\dfrac{1}{2}X^2 - \dfrac{1}{2}Y^2\right)$

$+ 5\left(\dfrac{1}{2}X^2 + XY + \dfrac{1}{2}Y^2\right) = 16$

$\dfrac{5}{2}X^2 - 5XY + \dfrac{5}{2}Y^2 + 3X^2 - 3Y^2$

$+\dfrac{5}{2}X^2 + 5XY + \dfrac{5}{2}Y^2 = 16$

$8X^2 + 2Y^2 = 16$

$4X^2 + Y^2 = 8$

(b)

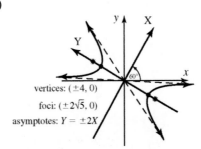

vertices: $(0, \pm 2\sqrt{2})$

foci: $(0, \pm\sqrt{6})$

minor
axis endpoints: $(\pm\sqrt{6}, 0)$

25. (a) $x^2 + 10\sqrt{3}xy + 11y^2 = -64$

$A = 1, B = 10\sqrt{3}, C = 11$

$\tan(2\beta) = \dfrac{10\sqrt{3}}{1-11} = -\sqrt{3}$

$2\beta = 120°$; $\beta = 60°$

$\sin 60° = \dfrac{\sqrt{3}}{2}$; $\cos 60° = \dfrac{1}{2}$;

$x = \left(\dfrac{1}{2}X - \dfrac{\sqrt{3}}{2}Y\right)$; $y = \left(\dfrac{\sqrt{3}}{2}X + \dfrac{1}{2}Y\right)$

$\left(\dfrac{1}{2}X - \dfrac{\sqrt{3}}{2}Y\right)^2 + 10\sqrt{3}\left(\dfrac{1}{2}X - \dfrac{\sqrt{3}}{2}Y\right)$

$\cdot\left(\dfrac{\sqrt{3}}{2}X + \dfrac{1}{2}Y\right) + 11\left(\dfrac{\sqrt{3}}{2}X + \dfrac{1}{2}Y\right)^2 = -64$

$\dfrac{1}{4}X^2 - \dfrac{\sqrt{3}}{2}XY + \dfrac{3}{4}Y^2$

$+ 10\sqrt{3}\left(\dfrac{\sqrt{3}}{4}X^2 - \dfrac{1}{2}XY - \dfrac{\sqrt{3}}{4}Y^2\right)$

$+ 11\left(\dfrac{3}{4}X^2 + \dfrac{\sqrt{3}}{2}XY + \dfrac{1}{4}Y^2\right) = -64$

$\dfrac{1}{4}X^2 - \dfrac{\sqrt{3}}{2}XY + \dfrac{3}{4}Y^2 + \dfrac{30}{4}X^2$

$- \dfrac{10\sqrt{3}}{2}XY^2 - \dfrac{30}{4}Y^2 + \dfrac{33}{4}X^2$

$+ \dfrac{11\sqrt{3}}{2}XY + \dfrac{11}{4}Y^2 = -64$

$16X^2 - 4Y^2 = -64$

$Y^2 - 4X^2 = 16$

(b)

vertices: $(\pm 4, 0)$

foci: $(\pm 2\sqrt{5}, 0)$

asymptotes: $Y = \pm 2X$

27. (a) $3x^2 - 2\sqrt{3}xy + y^2 - 8x - 8\sqrt{3}y = 0$

$A = 3, B = -2\sqrt{3}, C = 1$

$\tan(2\beta) = \dfrac{-2\sqrt{3}}{3-1} = -\sqrt{3}$

$2\beta = 120°; \quad \beta = 60°$

$\sin 60° = \dfrac{\sqrt{3}}{2}; \cos 60° = \dfrac{1}{2}$

$x = \dfrac{1}{2}X - \dfrac{\sqrt{3}}{2}Y; \quad y = \dfrac{\sqrt{3}}{2}X + \dfrac{1}{2}Y;$

$3\left(\dfrac{1}{2}X - \dfrac{\sqrt{3}}{2}Y\right)^2$

$-2\sqrt{3}\left(\dfrac{1}{2}X - \dfrac{\sqrt{3}}{2}Y\right)\cdot\left(\dfrac{\sqrt{3}}{2}X + \dfrac{1}{2}Y\right)$

$+\left(\dfrac{\sqrt{3}}{2}X + \dfrac{1}{2}Y\right)^2 - 8\left(\dfrac{1}{2}X - \dfrac{\sqrt{3}}{2}Y\right)$

$-8\sqrt{3}\left(\dfrac{\sqrt{3}}{2}X + \dfrac{1}{2}Y\right) = 0$

$3\left(\dfrac{1}{4}X^2 - \dfrac{\sqrt{3}}{2}XY + \dfrac{3}{4}Y^2\right)$

$-2\sqrt{3}\left(\dfrac{\sqrt{3}}{4}X^2 - \dfrac{1}{2}XY - \dfrac{\sqrt{3}}{4}Y^2\right)$

$+\left(\dfrac{3}{4}X^2 + \dfrac{\sqrt{3}}{2}XY + \dfrac{1}{4}Y^2\right)$

$-8\left(\dfrac{1}{2}X - \dfrac{\sqrt{3}}{2}Y\right)$

$-8\sqrt{3}\left(\dfrac{\sqrt{3}}{2}X + \dfrac{1}{2}Y\right) = 0$

$\dfrac{3}{4}X^2 - \dfrac{3\sqrt{3}}{2}XY + \dfrac{9}{4}Y^2 - \dfrac{6}{4}X^2$

$+\dfrac{2\sqrt{3}}{2}XY + \dfrac{6}{4}Y^2 + \dfrac{3}{4}X^2 + \dfrac{\sqrt{3}}{2}XY$

$+\dfrac{1}{4}Y^2 - \dfrac{8}{2}X + \dfrac{8\sqrt{3}}{2}Y$

$-\dfrac{24}{2}X - \dfrac{8\sqrt{3}}{2}Y = 0$

$\dfrac{16Y^2}{4} - 16X = 0$

$4Y^2 - 16X = 0$

$Y^2 - 4X = 0$

(b)

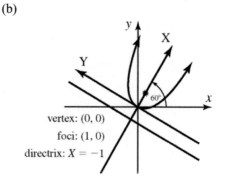

vertex: $(0, 0)$

foci: $(1, 0)$

directrix: $X = -1$

29. (a) $13x^2 - 6\sqrt{3}xy + 7y^2 - 100 = 0$

$A = 13, B = -6\sqrt{3}, C = 7$

$\tan(2\beta) = \dfrac{-6\sqrt{3}}{13-7} = -\sqrt{3}$

$2\beta = 120°; \quad \beta = 60°$

$\sin 60° = \dfrac{\sqrt{3}}{2}; \cos 60° = \dfrac{1}{2};$

$x = \dfrac{1}{2}X - \dfrac{\sqrt{3}}{2}Y; \quad y = \dfrac{\sqrt{3}}{2}X + \dfrac{1}{2}Y;$

$13\left(\dfrac{1}{2}X - \dfrac{\sqrt{3}}{2}Y\right)^2 - 6\sqrt{3}\left(\dfrac{1}{2}X - \dfrac{\sqrt{3}}{2}Y\right)$

$\cdot\left(\dfrac{\sqrt{3}}{2}X + \dfrac{1}{2}Y\right) + 7\left(\dfrac{\sqrt{3}}{2}X + \dfrac{1}{2}Y\right)^2 = 100$

$13\left(\dfrac{1}{4}X^2 - \dfrac{\sqrt{3}}{2}XY + \dfrac{3}{4}Y^2\right)$

$-6\sqrt{3}\left(\dfrac{\sqrt{3}}{4}X^2 - \dfrac{1}{2}XY + \dfrac{\sqrt{3}}{4}Y^2\right)$

$+7\left(\dfrac{3}{4}X^2 + \dfrac{\sqrt{3}}{2}XY + \dfrac{1}{4}Y^2\right) = 100$

$\dfrac{13}{4}X^2 - \dfrac{13\sqrt{3}}{2}XY + \dfrac{39}{4}Y^2 - \dfrac{18}{4}X^2$

$+\dfrac{6\sqrt{3}}{2}XY + \dfrac{18}{4}Y^2 + \dfrac{21}{4}X^2$

$+\dfrac{7\sqrt{3}}{2}XY + \dfrac{7}{4}Y^2 = 100$

$\dfrac{16}{4}X^2 + \dfrac{64}{4}Y^2 = 100$

$4X^2 + 16Y = 100$

$X^2 + 4Y = 25$

Chapter 9: Analytical Geometry

(b)

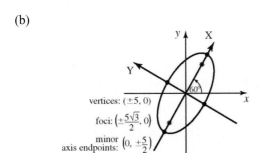

vertices: $(\pm 5, 0)$

foci: $\left(\pm \frac{5\sqrt{3}}{2}, 0\right)$

minor axis endpoints: $\left(0, \pm \frac{5}{2}\right)$

31. $12x^2 + 24xy + 5y^2 - 40x - 30y = 25$

$A = 12, B = 24, C = 5$

$B^2 - 4AC = 24^2 - 4(12)(5) = 336$

$336 > 0$; Hyperbola;

$\tan(2\beta) = \dfrac{B}{A-C} = \dfrac{24}{12-5} = \dfrac{24}{7}$;

$\sqrt{24^2 + 7^2} = 25$; $\cos(2\beta) = \dfrac{7}{25}$;

$\cos\beta = \sqrt{\dfrac{1+\cos(2\beta)}{2}} = \sqrt{\dfrac{1+\dfrac{7}{25}}{2}} = \dfrac{4}{5}$;

$\sin\beta = \sqrt{\dfrac{1-\cos(2\beta)}{2}} = \sqrt{\dfrac{1-\dfrac{7}{25}}{2}} = \dfrac{3}{5}$

33. $x^2 - 2xy + y^2 - 5 = 0$

(a) $A = 1, B = -2, C = 1$

$B^2 - 4AC = (-2)^2 - 4(1)(1) = 0$

Parabola

(b) $\tan 2\beta = \dfrac{2}{1-1} = $ undefined

$2\beta = \dfrac{\pi}{2}, \beta = \dfrac{\pi}{4}$; $\beta = 45°$

$\sin 45° = \dfrac{\sqrt{2}}{2}$; $\cos 45° = \dfrac{\sqrt{2}}{2}$;

$x^2 = \left(\dfrac{\sqrt{2}}{2}X - \dfrac{\sqrt{2}}{2}Y\right)^2 = \dfrac{X^2}{2} - XY + \dfrac{Y^2}{2}$;

$-2xy = -2\left(\dfrac{X\sqrt{2}}{2} - \dfrac{Y\sqrt{2}}{2}\right)\cdot\left(\dfrac{X\sqrt{2}}{2} + \dfrac{Y\sqrt{2}}{2}\right)$

$= -2\left(\dfrac{X^2}{2} - \dfrac{Y^2}{2}\right) = -X^2 + Y^2$;

$y^2 = \left(\dfrac{\sqrt{2}}{2}X + \dfrac{\sqrt{2}}{2}Y\right)^2 = \dfrac{X^2}{2} + XY + \dfrac{Y^2}{2}$;

$\dfrac{X^2}{2} - XY + \dfrac{Y^2}{2} - X^2 + Y^2$

$+ \dfrac{X^2}{2} + XY + \dfrac{Y^2}{2} = 5$

$\dfrac{4Y^2}{2} = 5$

$2Y^2 = 5$

c. Invariants: $F = f = -5$;

$A + C = a + c \Rightarrow 1+1 = 0+2$;

$B^2 - 4AC = b^2 - 4ac$

$(-2)^2 - 4\cdot 1\cdot 1 = (0)^2 - 4\cdot 0\cdot 2$

$0 = 0$

35. $3x^2 + \sqrt{3}xy + 4y^2 + 4x = 1$

(a) $A = 3, B = \sqrt{3}, C = 4$

$B^2 - 4AC = \left(\sqrt{3}\right)^2 - 4(3)(4) = -45$

Circle or Ellipse

(b) $\tan(2\beta) = \dfrac{\sqrt{3}}{3-4} = \dfrac{-\sqrt{3}}{1}$

$2\beta = 120°; \quad \beta = 60°$

$\sin 60° = \dfrac{\sqrt{3}}{2}, \cos 60° = \dfrac{1}{2};$

$y = \dfrac{\sqrt{3}}{2}X + \dfrac{1}{2}Y; \ x = \dfrac{1}{2}X - \dfrac{\sqrt{3}}{2}Y;$

$3\left(\dfrac{1}{2}X - \dfrac{\sqrt{3}}{2}Y\right)^2 + \sqrt{3}\left(\dfrac{1}{2}X - \dfrac{\sqrt{3}}{2}Y\right)$

$\cdot\left(\dfrac{\sqrt{3}}{2}X + \dfrac{1}{2}Y\right) + 4\left(\dfrac{\sqrt{3}}{2}X + \dfrac{1}{2}Y\right)^2$

$+ 4\left(\dfrac{1}{2}X - \dfrac{\sqrt{3}}{2}Y\right) = 1$

$3\left(\dfrac{1}{4}X^2 - \dfrac{\sqrt{3}}{2}XY + \dfrac{3}{4}Y^2\right)$

$+ \sqrt{3}\left(\dfrac{\sqrt{3}}{4}X^2 - \dfrac{1}{2}XY - \dfrac{\sqrt{3}}{4}Y^2\right)$

$+ 4\left(\dfrac{3}{4}X^2 + \dfrac{\sqrt{3}}{2}XY + \dfrac{1}{4}Y^2\right)$

$+ 2X - 2\sqrt{3}Y = 1$

$\dfrac{3}{4}X^2 - \dfrac{3\sqrt{3}}{2}XY + \dfrac{9}{4}Y^2 + \dfrac{3}{4}X^2$

$- \dfrac{\sqrt{3}}{2}XY - \dfrac{3}{4}Y^2 + \dfrac{12X^2}{4} + \dfrac{4\sqrt{3}}{2}XY$

$+ \dfrac{4}{4}Y^2 + 2X - 2\sqrt{3}Y = 1$

$\dfrac{18}{4}X^2 + \dfrac{10}{4}Y^2 + 2X - 2\sqrt{3}Y = 1$

$\dfrac{9}{2}X^2 + \dfrac{5}{2}Y^2 + 2x - 2\sqrt{3}Y = 1$

(c) Invariants: $F = f = -1;$

$A + C = a + c \Rightarrow \dfrac{9}{2} + \dfrac{5}{2} = 3 + 4 = 7;$

$B^2 - 4AC = b^2 - 4ac$

$(0)^2 - 4\cdot\dfrac{9}{2}\cdot\dfrac{5}{2} = \left(\sqrt{3}\right)^2 - 4\cdot 3\cdot 4$

$-45 = -45$

37. f

39. g

41. h

43. $r = \dfrac{4}{2 + 2\sin\theta} = \dfrac{2}{1 + 1\sin\theta}$

$e = 1$; Parabola

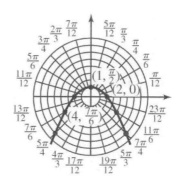

45. $r = \dfrac{12}{6 - 3\sin\theta} = \dfrac{2}{1 - \dfrac{1}{2}\sin\theta}$

$e = \dfrac{1}{2}; \ 0 < \dfrac{1}{2} < 1$; Ellipse

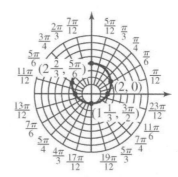

47. $r = \dfrac{6}{2 + 4\cos\theta} = \dfrac{3}{1 + 2\cos\theta}$

$e = 2; \ 2 > 1$; Hyperbola

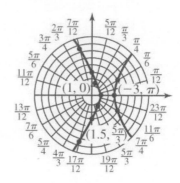

Chapter 9: Analytical Geometry

49. $r = \dfrac{5}{5 + 4\cos\theta} = \dfrac{1}{1 + 0.8\cos\theta}$

$e = 0.8$; $0 < 0.8 < 1$; Ellipse

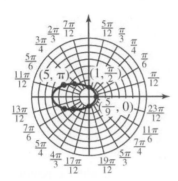

51. Ellipse, $e = 0.8$

Directrix to focus: $d = 4$

$r = \dfrac{de}{1 \pm e\cos\theta}$

$= \dfrac{4(0.8)}{1 - 0.8\cos\theta}$

$= \dfrac{3.2}{1 - 0.8\cos\theta}$

Answers may vary.

$de = 3.2$ and $e = 0.8$

53. Parabola, vertex at $(2, \pi)$

$e = 1$ at $(2, \pi)$;

$2 = \dfrac{d}{1 - \cos\pi}$

$2 = \dfrac{d}{1 - (-1)}$

$2 = \dfrac{d}{2}$

$d = 4$;

$r = \dfrac{de}{1 \pm \cos\theta} = \dfrac{4}{1 - \cos\theta}$

55. Hyperbola, $e = 1.5$, vertex at $\left(3, \dfrac{\pi}{2}\right)$

$3 = \dfrac{d(1.5)}{1 + 1.5\sin\dfrac{\pi}{2}}$

$3 = \dfrac{d(1.5)}{2.5}$

$7.5 = d(1.5)$

$r = \dfrac{7.5}{1 + 1.5\sin\theta}$

57. $r = \dfrac{C}{A\cos\theta + B\sin\theta}$; $2x + 3y = 12$

$A = 2, B = 3, C = 12$

(a) $r = \dfrac{12}{2\cos\theta + 3\sin\theta}$

(b)

$-\dfrac{r\left(\dfrac{\pi}{2}\right)}{r(0)} = -\dfrac{\dfrac{12}{2\cos\left(\dfrac{\pi}{2}\right) + 3\sin\left(\dfrac{\pi}{2}\right)}}{\dfrac{12}{2\cos(0) + 3\sin(0)}} = \dfrac{-2}{3}$

59. $L = 2\pi\sqrt{0.5\left(a^2 + b^2\right)}$

Jupiter: aphelion: 507; perihelion: 460

Length of major axes:

$2a = (507 + 460)$

$2a = 967$

$a = 483.5$

$a = c + \text{perihelion}$

$483.5 = c + 460$

$23.5 = c$

Eccentricity: $e = \dfrac{c}{a}$

$e = \dfrac{23.5}{483.5}$

$e \approx 0.0486$;

Saturn: aphelion: 941; perihelion: 840

Length of major axes:

$2a = 941 + 840$

$2a = 1781$

$a = 890.5$

$a = c + \text{perihelion}$

$890.5 = c + 840$

$50.5 = c$

Eccentricity: $e = \dfrac{c}{a}$

$e = \dfrac{50.5}{890.5}$

$e \approx 0.0567$

9.7 Exercises

61. Pluto: semi major axis $= 3647; e = 0.244$
    Perihelion $a = (1 - e)$
    $= 3647(1 - 0.244)$
    $= 3647(0.756)$
    ≈ 2757.1 million miles

63. $L = 2\pi\sqrt{0.5(a^2 + b^2)}$
    Jupiter: aphelion: 507; perihelion: 460
      Length of major axes:
      $2a = (507 + 460)$
      $2a = 967$
      $a = 483.5$
      $a = c + \text{perihelion}$
      $483.5 = c + 460$
      $23.5 = c$

    Eccentricity: $e = \dfrac{c}{a}$

    $e = \dfrac{23.5}{483.5}$
    $e \approx 0.04836$;
    Saturn: aphelion: 941; perihelion: 840
      Length of major axes:
      $2a = 941 + 840$
      $2a = 1781$
      $a = 890.5$
      $a = c + \text{perihelion}$
      $890.5 = c + 840$
      $50.5 = c$

    Eccentricity: $e = \dfrac{c}{a}$

    $e = \dfrac{50.5}{890.5}$
    $e \approx 0.0567$
    $L = 2\pi\sqrt{0.5(a^2 + b^2)}$
    Uranus: aphelion: 1866; perihelion: 1703
      Length of major axes:
      $2a = 1866 + 1703$
      $2a = 3569$
      $a = 1784.5$
      $a = c + \text{perihelion}$
      $1784.5 = c + 1703$
      $81.5 = c$

    Eccentricity: $e = \dfrac{c}{a}$

    $e = \dfrac{81.5}{1784.5}$
    $e \approx 0.0457$;

Neptune: aphelion: 2824; perihelion: 2762
  Length of major axes:
  $2a = 2824 + 2762$
  $2a = 5586$
  $a = 2793$
  $a = c + \text{perihelion}$
  $2793 = c + 2762$
  $31 = c$

  Eccentricity: $e = \dfrac{c}{a}$

  $e = \dfrac{31}{2793}$
  $e \approx 0.0111$
Saturn has the greatest orbital eccentricity:
  $e \approx 0.0567$

65. Jupiter: major axes $2a = 460 + 507$
    $2a = 967$
    $a = 483.5$;
    $a = c + \text{perihelion}$
    $483.5 = c + 460$
    $23.5 = c$;
    $e = \dfrac{23.5}{483.5} \approx 0.0486$
    Polar equation:
    $r = \dfrac{a(1 - e^2)}{1 - e\cos\theta}$
    $r = \dfrac{483.5(1 - 0.0486^2)}{1 - 0.0486\cos\theta}$
    $r \approx \dfrac{482.36}{1 - 0.0486\cos\theta}$

67. Uranus:
    Length of major axes:
    $2a = 1866 + 1703$
    $2a = 3569$
    $a = 1784.5$;
    $a = c + \text{perihelion}$
    $1784.5 = c + 1703$
    $81.5 = c$;

    Eccentricity: $e = \dfrac{c}{a}$

    $e = \dfrac{81.5}{1784.5} \approx 0.0457$;
    Polar equation:
    $r = \dfrac{a(1 - e^2)}{1 - e\cos\theta}$
    $r = \dfrac{1784.5(1 - 0.0457^2)}{1 - 0.0457\cos\theta} \approx \dfrac{1780.77}{1 - 0.0457\cos\theta}$

Chapter 9: Analytical Geometry

69. Jupiter: $r \approx \dfrac{482.36}{1 - 0.0486 \cos \theta}$

$r \approx \dfrac{482.36}{1 - 0.0486 \cos \left(\dfrac{\pi}{2} \right)} \approx 482.3$

Saturn: $r \approx \dfrac{887.64}{1 - 0.0567 \cos \theta}$

$r \approx \dfrac{887.64}{1 - 0.0567 \cos \left(\dfrac{\pi}{2} \right)} \approx 887.6$

Uranus: $r \approx \dfrac{1780.77}{1 - 0.0457 \cos \theta}$

$r \approx \dfrac{1780.77}{1 - 0.0457 \cos \left(\dfrac{\pi}{2} \right)} \approx 1780.7$

Neptune: $r \approx \dfrac{2792.66}{1 - 0.0111 \cos \theta}$

$r \approx \dfrac{2792.66}{1 - 0.0111 \cos \left(\dfrac{\pi}{2} \right)} \approx 2792.6$

Distance from Jupiter to Saturn:
  $887.6 - 482.3 = 405.3$ million mi
Distance from Jupiter to Uranus:
  $1780.7 - 482.3 = 1298.4$ million mi
Distance from Jupiter to Neptune:
  $2792.6 - 482.3 = 2310.3$ million mi
Distance from Saturn to Uranus:
  $1780.7 - 887.6 = 893.1$ million mi
Distance from Saturn to Neptune:
  $2792.6 - 887.6 = 1905.0$ million mi
Distance from Uranus to Neptune:
  $2792.6 - 1780.7 = 1011.9$ million mi

71. $L = 4$ ft, $w = 4(0.618) = 2.472$ ft

$a = 2, b = \dfrac{2.472}{2} = 1.236$

$c^2 = a^2 - b^2$

$c^2 = 2^2 - 1.236^2$

$c^2 = 2.472304$

$c \approx 1.57236$;

$e = \dfrac{c}{a} = \dfrac{1.57236}{2} = 0.7862$

$de = 2(1 - 0.7862)^2 = 0.7638$

$r = \dfrac{de}{1 \pm e \cos \theta}$

$r = \dfrac{0.7638}{1 \pm 0.7862 \cos \theta}$

73. $L = 1.5$m, $w = 1.5(0.618) = 0.927$m

$a = \dfrac{1.5}{2} = 0.75, b = \dfrac{0.927}{2} = 0.4635$

$c^2 = a^2 - b^2$

$c^2 = 0.75^2 - 0.4635^2$

$c \approx 0.58963$;

$e = \dfrac{c}{a} = \dfrac{0.58963}{0.75} = 0.7862$;

$de = 0.75 \left(1 - 0.7862^2 \right) = 0.2864$

$r = \dfrac{de}{1 \pm e \cos \theta}$

$r = \dfrac{0.2864}{1 \pm 0.7862 \cos \theta}$

75. $A = \pi a b$; Price: \$75; Cost $= 75 \pi a b$
   a) $L = 4, a = 2, b = 1.236$
       Cost $= 75 \pi (2)(1.236) = \$582.45$
   b) $L = 3.5, a = 1.75, b = 1.0815$
       Cost $= 75 \pi (1.75)(1.0815) = \$445.94$
   c) $L = 1.5, a = 0.75, b = 0.4635$
       Cost $= 807 \pi (0.75)(0.4635) = \$881.32$
   d) $L = 0.5, a = 0.25, b = 0.1545$
       Cost $= 807 \pi (0.25)(0.1545) = \$97.92$

77. $d = 3, e = 1$

$r = \dfrac{3}{1 - \cos \theta}$

79. $\begin{cases} X = x \cos \beta + y \sin \beta \\ Y = y \cos \beta - x \sin \beta \end{cases}$

Solve for y:
Multiply 1st equation by $\sin \beta$:

$X \sin \beta = x \sin \beta \cos \beta + y \sin^2 \beta$

Multiply 2nd equation by $\cos \beta$:

$Y \cos \beta = y \cos^2 \beta - x \sin \beta \cos \beta$

Equation 1 + Equation 2:

$X \sin \beta + Y \cos \beta = y \sin^2 \beta + y \cos^2 \beta$

$X \sin \beta + Y \cos \beta = y \left(\sin^2 \beta + \cos^2 \beta \right)$

$X \sin \beta + Y \cos \beta = y$

9.7 Exercises

81. $a \rightarrow A\cos^2\beta + B\sin\beta\cos\beta + C\sin^2\beta$

$b \rightarrow -2A\sin\beta\cos\beta + B(\cos^2\beta - \sin^2\beta)$
$\quad +2C\sin\beta\cos\beta$

$c \rightarrow A\sin^2\beta - B\sin\beta\cos\beta + C\cos^2\beta$

$f \rightarrow F$

a. $b^2 - 4ac = B^2 - 4AC$

$(-2A\sin\beta\cos\beta + B(\cos^2\beta - \sin^2\beta)$
$\quad + 2C\sin\beta\cos\beta)^2 - 4(A\cos^2\beta$
$\quad + B\sin\beta\cos\beta + C\sin^2\beta)$
$\quad \cdot (A\sin^2\beta - B\sin\beta\cos\beta$
$\quad + C\cos^2\beta) =$

$4A^2\sin^2\beta\cos^2\beta - 2AB\sin\beta\cos\beta$
$\quad \cdot (\cos^2\beta - \sin^2\beta)$
$\quad - 4AC\sin^2\beta\cos^2\beta$
$\quad - 2AB\sin\beta\cos\beta(\cos^2\beta - \sin^2\beta)$
$\quad + B^2(\cos^2\beta - \sin^2\beta)^2$
$\quad + 2BC\sin\beta\cos\beta(\cos^2\beta - \sin^2\beta)$
$\quad - 4AC\sin^2\beta\cos^2\beta$
$\quad + 2BC\sin\beta\cos\beta(\cos^2\beta - \sin^2\beta)$
$\quad + 4C^2\sin^2\beta\cos^2\beta$
$\quad - 4(A^2\sin^2\beta\cos^2\beta$
$\quad - AB\sin\beta\cos^3\beta + AC\cos^4\beta$
$\quad + AB\sin^3\beta\cos\beta - B^2\sin^2\beta\cos^2\beta$
$\quad + BC\sin\beta\cos^3\beta + AC\sin^4\beta$
$\quad - BC\sin^3\beta\cos\beta$
$\quad + C^2\sin^2\beta\cos^2\beta) =$

$4A^2\sin^2\beta\cos^2\beta - 2AB\sin\beta\cos\beta$
$\quad \cdot (\cos^2\beta - \sin^2\beta)$
$\quad - 4AC\sin^2\beta\cos^2\beta$
$\quad - 2AB\sin\beta\cos\beta(\cos^2\beta - \sin^2\beta)$
$\quad + B^2(\cos^2\beta - \sin^2\beta)^2$
$\quad + 2BC\sin\beta\cos\beta(\cos^2\beta - \sin^2\beta)$
$\quad - 4AC\sin^2\beta\cos^2\beta$
$\quad + 2BC\sin\beta\cos\beta(\cos^2\beta - \sin^2\beta)$
$\quad + 4C^2\sin^2\beta\cos^2\beta$
$\quad - 4A^2\sin^2\beta\cos^2\beta$
$\quad + 4AB\sin\beta\cos^3\beta - 4AC\cos^4\beta$
$\quad - 4AB\sin^3\beta\cos\beta$

$\quad + 4B^2\sin^2\beta\cos^2\beta$
$\quad - 4BC\sin\beta\cos^3\beta - 4AC\sin^4\beta$
$\quad + 4BC\sin^3\beta\cos\beta$
$\quad - 4C^2\sin^2\beta\cos^2\beta =$

$B^2\Big[(\cos^2\beta - \sin^2\beta)^2$
$\quad + 4\sin^2\beta\cos^2\beta\Big]$
$\quad - 4AC\Big[\sin^2\beta\cos^2\beta$
$\quad + \sin^2\beta\cos^2\beta$
$\quad + \cos^4\beta + \sin^4\beta\Big]$
$\quad - 2AB\Big[\sin\beta\cos\beta(\cos^2\beta - \sin^2\beta)$
$\quad + \sin\beta\cos\beta(\cos^2\beta - \sin^2\beta)$
$\quad - 2\sin\beta\cos^3\beta - 2\sin^3\beta\cos\beta\Big]$
$\quad + 2BC\Big[\sin\beta\cos\beta(\cos^2\beta - \sin^2\beta)$
$\quad + \sin\beta\cos\beta(\cos^2\beta - \sin^2\beta)$
$\quad - 2\sin\beta\cos^3\beta + 2\sin^3\beta\cos\beta\Big] =$

$B^2\Big[(\cos^2\beta - \sin^2\beta)^2 + 4\sin^2\beta\cos^2\beta\Big]$
$\quad - 4AC\Big[\sin^4\beta + 2\sin^2\beta + \cos^4\beta\Big]$
$\quad - 2AB\Big[2\sin\beta\cos^3\beta + 2\sin^3\beta\cos\beta$
$\quad - 2\sin\beta\cos^3\beta - 2\sin^3\beta\cos\beta\Big]$
$\quad + 2BC\Big[2\sin\beta\cos^3\beta$
$\quad + 2\sin^3\beta\cos\beta - 2\sin\beta\cos^3\beta$
$\quad - 2\sin^3\beta\cos\beta\Big] =$

$B^2\Big[\cos^4\beta - 2\cos^2\beta\sin^2\beta + \sin^4\beta$
$\quad + 4\sin^2\beta\cos^2\beta\Big]$
$\quad - 4AC\Big[(\sin^2\beta + \cos^2\beta)^2\Big] =$

$B^2\Big[\cos^4\beta + 2\cos^2\beta\sin^2\beta + \sin^4\beta\Big]$
$\quad - 4AC[1] =$

$B^2\Big[(\cos^2\beta + \sin^2\beta)^2\Big] - 4AC =$

$B^2 - 4AC =$

524

b. $a + c = A + C$

$A\cos^2\beta + B\sin\beta\cos\beta + C\sin^2\beta$
$\quad + A\sin^2\beta - B\sin\beta\cos\beta$
$\quad + C\cos^2\beta =$
$A\left(\cos^2\beta + \sin^2\beta\right) + C\left(\sin^2\beta + \cos^2\beta\right) =$
$A + C = A + C$

c. $f = F$. The rotation formulas affect only those terms having xy-variables. The constant term is not affected.

83. Equation of hyperbola with center at pole:
$r = 2R\cos(\theta - \beta)$;
$0^2 + 0^2 - 6\sqrt{2}(0) - 6\sqrt{2}(0) = 0$
$0 = 0$; verified
$x^2 + y^2 - 6\sqrt{2}x - 6\sqrt{2}y = 0$
$x^2 - 6\sqrt{2}x + 18 + y^2 - 6\sqrt{2}y + 18$
$\quad = 0 + 18 + 18$
$\left(x - 3\sqrt{2}\right)^2 + \left(y - 3\sqrt{2}\right)^2 = 36$
Center $\left(3\sqrt{2}, 3\sqrt{2}\right)$
Radius $= R = \sqrt{36} = 6$
$r = 2 \cdot 6\cos(\theta - \beta)$; $\beta = \dfrac{\pi}{4}$;
$\cos\left(\theta - \dfrac{\pi}{4}\right) = \cos\theta\cos\dfrac{\pi}{4} + \sin\theta\sin\dfrac{\pi}{4}$
$= \dfrac{\sqrt{2}}{2}(\cos\theta + \sin\theta)$;
$r = 2R\cos(\theta - \beta)$
$r = 2(6)\left[\dfrac{\sqrt{2}}{2}(\cos\theta + \sin\theta)\right]$
$r = 12\left[\dfrac{\sqrt{2}}{2}(\cos\theta + \sin\theta)\right]$
$r = 6\sqrt{2}(\cos\theta + \sin\theta)$

85. $25x^2 + 840xy - 16y^2 = 400$

$\tan(2\beta) = \dfrac{B}{A - C} = \dfrac{840}{81}$;

$\sqrt{840^2 + 41^2} = 841$; $\cos(2\beta) = \dfrac{41}{841}$;

$\cos\beta = \sqrt{\dfrac{1 + \cos(2\beta)}{2}} = \sqrt{\dfrac{1 + \dfrac{41}{841}}{2}} = \dfrac{21}{29}$;

$\sin\beta = \sqrt{\dfrac{1 - \cos(2\beta)}{2}} = \sqrt{\dfrac{1 - \dfrac{41}{841}}{2}} = \dfrac{20}{29}$;

$x = \dfrac{21}{29}X - \dfrac{20}{29}Y$; $y = \dfrac{20}{29}X + \dfrac{21}{29}Y$;

$25\left(\dfrac{21}{29}X - \dfrac{20}{29}Y\right)^2$

$\quad + 840\left(\dfrac{21}{29}X - \dfrac{20}{29}Y\right)\left(\dfrac{20}{29}X + \dfrac{21}{29}Y\right)$

$\quad - 16\left(\dfrac{20}{29}X + \dfrac{21}{29}Y\right)^2 = 400$

$25\left(\dfrac{441}{841}X^2 - \dfrac{840}{841}XY + \dfrac{400}{841}Y^2\right)$

$\quad + 840\left(\dfrac{420}{841}X^2 + \dfrac{441}{841}XY\right.$

$\quad\left. - \dfrac{400}{841}XY - \dfrac{420}{841}Y^2\right)$

$\quad - 16\left(\dfrac{400}{841}X^2 + \dfrac{840}{841}XY + \dfrac{441}{841}Y^2\right)$

$= 400$

$\dfrac{11025}{841}X^2 - \dfrac{25 \cdot 840}{841}XY + \dfrac{10000}{841}Y^2$

$\quad + \dfrac{352800}{841}X^2 + \dfrac{840(41)}{841}XY$

$\quad - \dfrac{352800}{841}Y^2 - \dfrac{6400}{841}X^2$

$\quad - \dfrac{16 \cdot 840}{841}XY + \dfrac{7056}{841}Y^2 = 400$

$\dfrac{35712}{841}X^2 - 416Y^2 - 400 = 0$

$425X^2 - 416Y^2 = 400$

87. $(0,0),(2\sqrt{3},2),(2\sqrt{3}-2,2+2\sqrt{3}),(-2,2\sqrt{3})$

$$\begin{bmatrix} \cos(-30°) & -\sin(-30°) \\ \sin(-30°) & \cos(-30°) \end{bmatrix} \cdot \begin{bmatrix} 0 \\ 0 \end{bmatrix} = \begin{bmatrix} 0 \\ 0 \end{bmatrix}$$

$$\begin{bmatrix} \cos(-30°) & -\sin(-30°) \\ \sin(-30°) & \cos(-30°) \end{bmatrix} \cdot \begin{bmatrix} 2\sqrt{3} \\ 2 \end{bmatrix} = \begin{bmatrix} 4 \\ 0 \end{bmatrix}$$

$$\begin{bmatrix} \cos(-30°) & -\sin(-30°) \\ \sin(-30°) & \cos(-30°) \end{bmatrix} \cdot \begin{bmatrix} 2\sqrt{3}-2 \\ 2+2\sqrt{3} \end{bmatrix} = \begin{bmatrix} 4 \\ 4 \end{bmatrix}$$

$$\begin{bmatrix} \cos(-30°) & -\sin(-30°) \\ \sin(-30°) & \cos(-30°) \end{bmatrix} \cdot \begin{bmatrix} -2 \\ 2\sqrt{3} \end{bmatrix} = \begin{bmatrix} 0 \\ 4 \end{bmatrix}$$

$(0,0),(4,0),\ (4,4)$ and $(0,4)$

89. $21.7 = 77.5e^{-0.0052x} - 44.95$

$66.65 = 77.5e^{-0.0052x}$

$\dfrac{66.65}{77.5} = e^{-0.0052x}$

$\ln\left(\dfrac{66.65}{77.5}\right) = \ln e^{-0.0052x}$

$\ln\left(\dfrac{66.65}{77.5}\right) = -0.0052x$

$x = \dfrac{\ln\left(\dfrac{66.65}{77.5}\right)}{-0.0052}$

$x \approx 29.0$

91. $360° - 325° = 35°\,;\,35° + 90° = 125°\,;$

$360° - 10° = 350°\,;$

Vector coordinates: $\langle 12\cos 125°, 12\sin 125°\rangle,$

$\langle 5\cos 350°, 5\sin 350°\rangle$

Resultant:

$\langle 12\cos 125° + 5\cos 350°, 12\sin 125° + 5\sin 350°\rangle$

$= \langle -1.9589, 8.9616\rangle\,;$

$\sqrt{(-1.9589)^2 + (8.9616)^2} \approx 9.2\,;$

$\tan^{-1}\left(-\dfrac{8.9616}{1.9589}\right) \approx -77.7\,;$

$360° - 77.7° = 282.3°$ in QIV or

$180° - 77.7° = 102.3°$ in QII

Heading: $360° - (102.3° - 90°) = 347.7°$

Magnitude 9.2, heading $347.7°$

Technology Highlights

1. Verified.

9.8 Exercises

1. Parameter

3. Direction

5. Answers will vary.

7. $x = t + 2; t \in [-3,3]$

$y = t^2 - 1$

a) A parabola with vertex at $(2, -1)$.

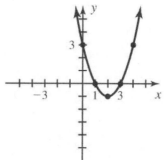

b) $x - 2 = t\,;$

$y = (x - 2)^2 - 1$

$y = x^2 - 4x + 4 - 1$

$y = x^2 - 4x + 3$

9. $x = (2-t)^2$; $t \in [0,5]$

$y = (t-3)^2$

a) A parabola

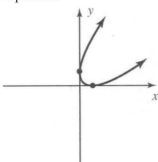

b) $\pm\sqrt{x} = 2-t$

$t = 2 \pm \sqrt{x}$;

$y = \left(2 \pm \sqrt{x} - 3\right)^2$

$y = \left(\pm\sqrt{x} - 1\right)^2$

$y = x \pm 2\sqrt{x} + 1$

11. $x = \dfrac{5}{t}$; $t \neq 0$; $t \in [-3.5, 3.5]$

$y = t^2$

a) Power function with $p = -2$

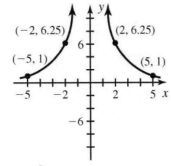

b) $x = \dfrac{5}{t}$

$xt = 5$

$t = \dfrac{5}{x}$;

$y = \left(\dfrac{5}{x}\right)^2$

$y = \dfrac{25}{x^2}$, $x \neq 0$

13. $x = 4\cos t$, $t \in [0, 2\pi)$

$y = 3\sin t$

a) Ellipse

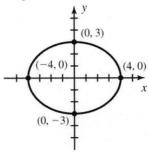

b) $x^2 = (4\cos t)^2$; $y^2 = (3\sin t)^2$

$\begin{cases} x^2 = 16\cos^2 t \\ y^2 = 9\sin^2 t \end{cases}$

$\begin{cases} \dfrac{x^2}{16} = \cos^2 t \\ \dfrac{y^2}{9} = \sin^2 t \end{cases}$

$\dfrac{x^2}{16} + \dfrac{y^2}{9} = \cos^2 t + \sin^2 t$

$\dfrac{x^2}{16} + \dfrac{y^2}{9} = 1$

15. $x = 4\sin(2t)$; $t \in [0, 2\pi)$

$y = 6\cos t$

a) Lissajous figure

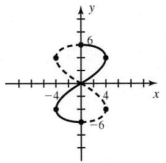

b) $x = 4\sin 2t$

$\dfrac{x}{4} = \sin 2t$;

$\sin^{-1}\left(\dfrac{x}{4}\right) = 2t$

$\dfrac{1}{2}\sin^{-1}\left(\dfrac{x}{4}\right) = t$

$y = 6\cos t$

$y = 6\cos\left[\dfrac{1}{2}\sin^{-1}\left(\dfrac{x}{4}\right)\right]$

9.8 Exercises

17. $\begin{cases} x = \dfrac{-3}{\tan t}, \ t \in (0, \pi) \\ y = 5\sin(2t) \end{cases}$

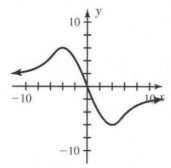

19. $y = 3x - 2$

   i) $x = t$
        $y = 3t - 2$

   ii) $x = \dfrac{1}{3}t$

        $y = 3\left(\dfrac{1}{3}t\right) - 2$

        $y = t - 2$

   iii) $x = \cos t$
        $y = 3\cos t - 2$

21. $y = (x + 3)^2 + 1$

   i) $x = t$
        $y = (t + 3)^2 + 1$

   ii) $x = t - 3$
        $y = [(t - 3) + 3]^2 + 1$
        $y = t^2 + 1$

   iii) $x = \tan t - 3$
        $\tan t = x + 3$
        $x = \tan t - 3$
        $y = \tan^2 t + 1 = \sec^2 t$

        $t \notin \left\{ \left(k + \dfrac{1}{2}\right)\pi, k \in Z \right\}$

        $t \notin \left\{ \dfrac{(2k + 1)\pi}{2}, k \in Z \right\}$

23. $y = \tan^2(x - 2) + 1$

   1) $x = t;$
        $y = \tan^2(t - 2) + 1$

        $t \notin \left\{ \pi k + \dfrac{\pi}{2} + 2, k \in Z \right\}$

   2) $t = x - 2$
        $x = t + 2;$
        $y = \tan^2(t + 2 - 2) + 1$
        $y = \tan^2 t + 1$
        $y = \sec^2 t$

        $t \notin \left\{ \left(k + \dfrac{1}{2}\right)\pi, k \in Z \right\}$

   3) $x = \tan^{-1} t + 2$
        $y = \tan^2\left(\tan^{-1} t + 2 - 2\right) + 1$
        $y = \tan^2\left(\tan^{-1} t\right) + 1$
        $y = t^2 + 1$
        $t \in R$

25. $y = 4(x - 3)^2 + 1$

   1) $x = t$
        $y = 4(t - 3)^2 + 1$

   2) $x = t + 3$
        $y = 4t^2 + 1$

   3) $x = \dfrac{1}{2}\tan t + 3$

        $y = \sec^2 t$

Using grapher, verified.

528

27. $x = 8\cos t + 2\cos(4t)$;

    $y = 8\sin t - 2\sin(4t)$;

   a)   Hypocycloid (5-cusp)

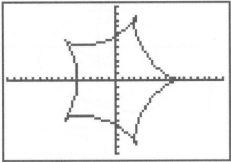

   b) x-intercepts:

      $t = 0, x = 10, y = 0$ and

      $t = \pi, x = -6, y = 0$;

      y-intercepts:

      $t \approx 1.757, x = 0, y \approx 6.5$

      $t \approx 4.527, x = 0, y \approx -6.5$

      Min x-value: -8.1

      Max x-value: 10

      Min y-value: -9.5

      Max y-value: 9.5

29. $\begin{cases} x = \dfrac{2}{\tan t} \\ y = 8\sin t \cos t \end{cases}$

   a)   Serpentine curve

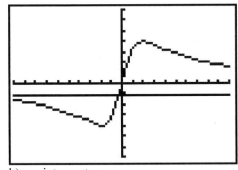

   b) x-intercept: none

      y-intercept: none

      Min x-value: none

      Max x-value: none

      Min y-value: -4

      Max y-value: 4

31. $\begin{cases} x = 2(\cos t + t\sin t) \\ y = 2(\sin t - t\cos t) \end{cases}$

   a)   Involute of a circle

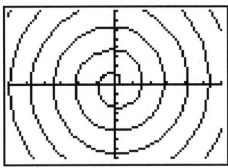

   b) x-intercepts:

      $t = 0, x = 2, y = 0$

      $t \approx 4.493, x = -9.2, y = 0$

      Infinitely many others.

      y-intercepts:

      $t \approx 2.79, x = 0, y = 5.9$

      $t \approx 6.12, x = 0, y \approx -12.4$

      Infinitely many others.

      No minimum or maximum values for

      x or y.

33. $\begin{cases} x = 3t - \sin t \\ y = 3 - \cos t \end{cases}$

   a)   Curtate cycloid

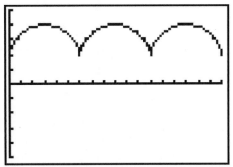

   b) x-intercept: none

      y-intercept: $t = 0, x = 0, y = 2$

      Min x-value: none

      Max x-value: none

      Min y-value: 2

      Max y-value: 4

35. $x = 2[3\cos t - \cos(3t)]$
    $y = 2[3\sin t - \sin(3t)]$

a) Nephroid

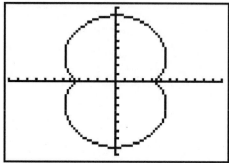

b) x-intercepts: $t = 0, (4,0)$

$t = \pi, (-4, 0)$

y-intercepts: $t = \dfrac{\pi}{2}, (0,8)$

$t = \dfrac{3\pi}{2}, (0, -8)$

Min x-value: ≈ -5.657
Max x-value: ≈ 5.657
Min y-value: -8
Max y-value: 8

37. $x = 6\sin(2t)$
    $y = 8\cos(t)$

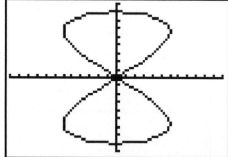

Box to frame curve: width 12, length 16.
Over interval $[0, 2\pi]$ graph crosses itself 2 times.

39. $x = 5\sin(7t)$
    $y = 7\cos(4t)$

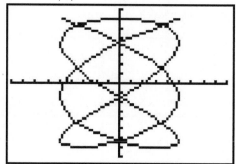

Box to frame curve: width 10, length 14.
Over interval $[0, 2\pi]$ graph crosses itself 9 times.

41. $x = 10\sin(1.5t)$
    $y = 10\cos(2.5t)$

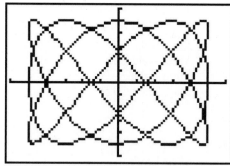

Box to frame curve: width 20, length 20.
Over interval $[0, 4\pi]$ graph crosses itself 23 times.

43. $\begin{cases} x = \dfrac{a}{\tan t} \\ y = b\sin t \cos t \end{cases}$

The maximum value as graph swells is at

$(x, y) = \left(a, \dfrac{b}{2} \right)$.

The minimum value as graph dips to the

valley is at $(x, y) = \left(-a, \dfrac{-b}{2} \right)$.

45. $x(t) = 2kt, \; y(t) = \dfrac{2k}{1+t^2}$

    a)   The curve is approaching $y = 2$ as t

       approaches $\dfrac{3\pi}{2}$, but $\cot\left(\dfrac{3\pi}{2}\right)$ is

       undefined, and the trig form seems to

       indicate a "hole" at $t = \dfrac{3\pi}{2}$, $x = 0$, $y = 2$.

       The algebraic form does not have this
       problem and shows a maximum defined
       at $t = 0$, $x = 0$, $y = 2$.

    b)   As $|t| \to \infty$, $y(t) \to 0$

    c)   The maximum value occurs at $(0, 2k)$.

47. $x = 75t \cos 36$

    $y = 75t \sin 36 - 16t^2$

    a)   At $t = 1.24$
       $x = 75.239$
       $y = 30.062$
       Yes, goes through hoop.

    b)   At $t = 2.7552$
       $x = 167.1779$
       $y = 0$
       Yes.

    c) $168 - 167.18 = 0.82$ ft

49. $x = 80t \cos 29°$

    $y = 80t \sin 29° - 16t^2$;

    $150 = 80t \cos 29°$;
    $2.144 \approx t$;

    $y = 80(2.144)\sin 29° - 16(2.144)^2$
    $y = 9.5769 < 10$

    Will not clear goal post. The kick is short.

51. $x = 6\cos t$; $y = 2\sin t$; $t = 2$ to 3

t	x	y
2	-2.50	1.82
2.09	-2.98	1.74
2.225	-3.65	1.59
3	-5.94	0.28

Left and downward

53. $\begin{cases} x - 5y + z = 3 \\ 5x + y - 7z = -9 \\ 2x + 3y - 4z = -6 \end{cases}$

$7R_1 + R_2$

$\begin{cases} 7x - 35y + 7z = 21 \\ 5x + y - 7z = -9 \end{cases}$

$12x - 34y = 12$ (equation 4)

$4R_1 + R_3$

$\begin{cases} 4x - 20y + 4z = 12 \\ 2x + 3y - 4z = -6 \end{cases}$

$6x - 17y = 6$ (equation 5)

$-2R_5, + R_4$

$-12x + 34y = -12$

$12x - 34y = 12$

$0 = 0$

dependent equations

$\dfrac{34y}{34} = \dfrac{12x}{34} - \dfrac{12}{34}$

$y = \dfrac{6}{17}x - \dfrac{6}{17}$;

$x - 5\left(\dfrac{6}{17}x - \dfrac{6}{17}\right) + z = 3$

$x - \dfrac{30x}{17} + \dfrac{30}{17} + z = 3$

$\dfrac{-13x}{17} + z = \dfrac{21}{17}$

$z = \dfrac{13}{17}x + \dfrac{21}{17}$;

$\left(x, \dfrac{6}{17}x - \dfrac{6}{17}, \dfrac{13}{17}x + \dfrac{21}{17}\right)$

$\left(t, \dfrac{6}{17}t - \dfrac{6}{17}, \dfrac{13}{17}t + \dfrac{21}{17}\right)$

9.8 Exercises

55. $\begin{cases} x + y - 5z = -4 \\ 2y - 3z = -1 \\ x - 3y + z = -3 \end{cases}$

$-R_1 + R_3$

$\begin{cases} -x - y + 5z = 4 \\ x - 3y + z = -3 \end{cases}$

$\overline{-4y + 6z = 1}$ (equation 4)

$2R_2 + R_4$

$\begin{cases} 4y - 6z = -2 \\ -4y + 6z = 1 \end{cases}$

$\overline{0 \neq -1}$

Inconsistent; no solution.

57.

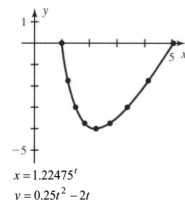

$x = 1.22475^t$

$y = 0.25t^2 - 2t$

The parametric equations fit the data very well.

59. Graphing both functions on the same screen, we note that although the paths intersect at approximately (5.3, 3.7) and (4.1, 1.5), the first particle arrives at these locations for T = 5.00 and T = 4.90 respectively, while the second particle arrives at T = 2.17 and T = 3.40. No, the particles would not collide.

61. $(1 - 0.20)x = 39.96$

$0.80x = 39.96$

$x = \dfrac{39.96}{.8} = 49.95$

$49.95 - 39.96 = 9.99$

$\dfrac{9.99}{39.96} = 0.25$ or 25%

63. $f(x) = x^3 + 2x^2 - 5x - 6$

End behavior: left falls $(\downarrow)$

Right rises $(\uparrow)$

y-intercept $(0, -6)$

$f(0) = 0^3 + 2(0)^2 - 5(0) - 6 = -6$

x-intercepts $(2, 0)(-3, 0)(-1, 0)$

$\begin{array}{r|rrr} 2 & 1 & 2 & -5 & -6 \\ & & 2 & 8 & 6 \\ \hline & 1 & 4 & 3 & \underline{0} \end{array}$

$x^2 + 4x + 3$

$(x + 3)(x + 1)$

$x = -3 \quad x = -1$

Mid-interval points

$x = -4$

$f(x) = (x - 2)(x + 3)(x + 1)$

$f(-4) = (-4 - 2)(-4 + 3)(-4 + 1)$

$= (-6)(-1)(-3) = -18$

$x = -2$

$f(-2) = (-2 - 2)(-2 + 3)(-2 + 1)$

$= (-4)(1)(-1) = 6$

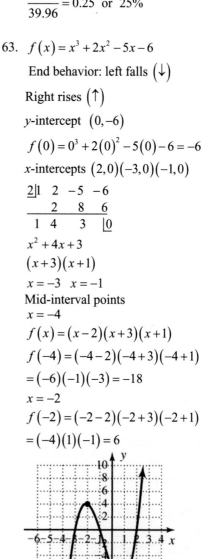

Summary and Concept Review

1. $(-3,-4),(-5,4),(3,6),(5,-2)$

    $(-3,-4),(-5,4)$

    $m = \dfrac{4-(-4)}{-5-(-3)} = -4$;

    $d = \sqrt{(-3-(-5))^2 + (-4-4)^2}$
    $= \sqrt{4+64} = \sqrt{68} = 2\sqrt{17}$;

    $(-5,4),(3,6)$

    $m = \dfrac{6-4}{3-(-5)} = \dfrac{1}{4}$;

    $d = \sqrt{(3--5)^2 + (6-4)^2}$
    $= \sqrt{64+4} = \sqrt{68} = 2\sqrt{17}$;

    $(3,6),(5,-2)$

    $m = \dfrac{-2-6}{5-3} = -4$;

    $d = \sqrt{(5-3)^2 + (-2-6)^2}$
    $= \sqrt{4+64} = \sqrt{68} = 2\sqrt{17}$;

    $(5,-2),(-3,-4)$

    $m = \dfrac{-4-(-2)}{-3-5} = \dfrac{1}{4}$;

    $d = \sqrt{(-3-5)^2 + (-4-(-2))^2}$
    $= \sqrt{64+4} = \sqrt{68} = 2\sqrt{17}$;

    Figure is a square, all sides are equal in length and consecutive segments are perpendicular.
    Verified.

3. $(-3,6)$ to $(6,-9)$ and $(-5,-2)$ to $(5,4)$

    Midpoint $(-5,-2)$ to $(5,4)$

    $\left(\dfrac{-5+5}{2}, \dfrac{-2+4}{2}\right) = (0,1)$;

    Slope $(-5,-2)$ to $(5,4)$

    $m = \dfrac{4-(-2)}{5-(-5)} = \dfrac{3}{5}$;

    Slope $(-3,6)$ to $(6,-9)$

    $m = \dfrac{-9-6}{6-(-3)} = -\dfrac{5}{3}$;

    The two segments are perpendicular.
    The equation of the line through $(-3, 6)$ and $(6, -9)$:

    $y - 6 = -\dfrac{5}{3}\left(x-(-3)\right)$

    $y - 6 = -\dfrac{5}{3}x - 5$

    $y = -\dfrac{5}{3}x + 1$;

    At $x = 0$, $y = 1$
    Line containing $(-3, 6)$ and $(6, -9)$
    contains $(0, 1)$.
    $(-3, 6)$ to $(6, -9)$ is perpendicular bisector
    to $(-5, -2)$ to $(-5, -4)$.

5. $x^2 + y^2 = 16$

    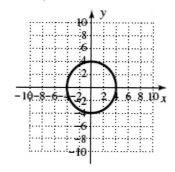

7. $9x^2 + y^2 - 18x - 27 = 0$

$9x^2 - 18x + y^2 = 27$

$9(x^2 - 2x) + y^2 = 27$

$9(x^2 - 2x + 1) + y^2 = 27 + 9$

$9(x-1)^2 + y^2 = 36$

$\dfrac{(x-1)^2}{4} + \dfrac{y^2}{36} = 1$

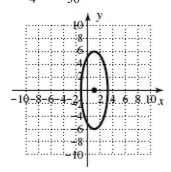

9. $\dfrac{(x+3)^2}{16} + \dfrac{(y-2)^2}{9} = 1$

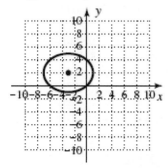

11. a. Vertices: $(-13,0),(13,0)$

Foci: $(-12,0),(12,0)$

Center: $\left(\dfrac{-13+13}{2}, \dfrac{0+0}{2}\right) = (0,0)$

$a = 13, c = 12$

$12^2 = 13^2 - b^2$

$144 = 169 - b^2$

$b = 5;$

$\dfrac{(x-0)^2}{13^2} + \dfrac{(y-0)^2}{5^2} = 1$

$\dfrac{x^2}{169} + \dfrac{y^2}{25} = 1$

b. Foci: $(0,-16),(0,16)$

Major axis: $2a = 40$, $a = 20$, $c = 16$

Center: $\left(\dfrac{0+0}{2}, \dfrac{-16+16}{2}\right) = (0,0)$

$16^2 = 20^2 - b^2$

$256 = 400 - b^2$

$b = 12;$

$\dfrac{(x-0)^2}{20^2} + \dfrac{(y-0)^2}{12^2} = 1$

$\dfrac{x^2}{400} + \dfrac{y^2}{144} = 1$

13 $4y^2 - 25x^2 = 100$

$\dfrac{y^2}{25} - \dfrac{x^2}{4} = 1$

$a = 2,\ b = 5$

Hyperbola

Center: $(0,0)$

Vertices: $(0,5), (0,-5)$

$a^2 + b^2 = c^2$

$25 + 4 = c^2$

$\sqrt{29} = c$

Foci: $\left(0, \sqrt{29}\right), \left(0, -\sqrt{29}\right)$

Asymptotes: $y = \pm\dfrac{5}{2}x$

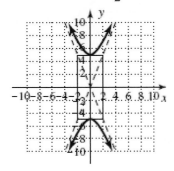

15. $\dfrac{(x+2)^2}{9} - \dfrac{(y-1)^2}{4} = 1$

Hyperbola

Center: $(-2,1)$

$a = 3,\ b = 2$

Vertices: $(-2-3,1), (-2+3,1)$

$(-5,1), (1,1)$;

$a^2 + b^2 = c^2$

$9 + 4 = c^2$

$\sqrt{13} = c$

Foci: $\left(-2-\sqrt{13},1\right), \left(-2+\sqrt{13},1\right)$

Asymptotes: $y - 1 = \pm\dfrac{2}{3}(x+2)$

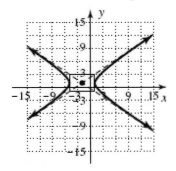

17. $x^2 - 4y^2 - 12x - 8y + 16 = 0$

$x^2 - 12x - 4y^2 - 8y = -16$

$\left(x^2 - 12x\right) - 4\left(y^2 + 2y\right) = -16$

$\left(x^2 - 12x + 36\right) - 4\left(y^2 + 2y + 1\right) = -16 + 36 - 4$

$(x-6)^2 - 4(y+1)^2 = 16$

$\dfrac{(x-6)^2}{16} - \dfrac{(y+1)^2}{4} = 1$

Hyperbola

Center: $(6,-1)$

$a = 4,\ b = 2$

Vertices: $(6+4,-1), (6-4,-1)$

$(10,-1), (2,-1)$;

$a^2 + b^2 = c^2$

$16 + 4 = c^2$

$2\sqrt{5} = c$

Foci: $\left(6+2\sqrt{5},-1\right), \left(6-2\sqrt{5},-1\right)$

Asymptotes: $y + 1 = \pm\dfrac{1}{2}(x-6)$

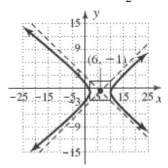

Summary and Concept Review

19. a.   Vertices: $(-15,0), (15,0)$

Foci: $(-17,0), (17,0)$

$a^2 + b^2 = c^2$

$225 + b^2 = 289$

$b^2 = 64$

$\dfrac{x^2}{225} - \dfrac{y^2}{64} = 1$

b.   Foci: $(0,-5), (0,5)$

Vertical Length of central rectangle: 8 units

$a^2 + b^2 = c^2$

$4^2 + b^2 = 5^2$

$b^2 = 9$

$\dfrac{y^2}{16} - \dfrac{x^2}{9} = 1$

21. $x = y^2 - 4$

Parabola

x-intercept: (-4,0)

y-intercepts: (0,2), (0,-2)

Vertex: $(-4,0)$

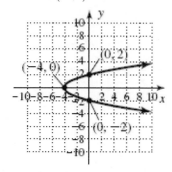

23. $x^2 = -20y$

$4p = -20$

$p = -5$

Parabola

Vertex: $(0,0)$

Focus: $(0,-5)$

Length of Focal Chord: 20

Directrix: $y = 5$

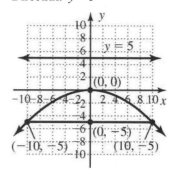

25. $\begin{cases} x^2 + y^2 = 25 & \text{Circle} \\ y - x = -1 & \text{Line} \end{cases}$

2nd equation: $y = x - 1$

$x^2 + (x-1)^2 = 25$

$x^2 + x^2 - 2x + 1 = 25$

$2x^2 - 2x - 24 = 0$

$x^2 - x - 12 = 0$

$(x-4)(x+3) = 0$

$x = 4, x = -3 \, ;$

$y = 4 - 1 = 3 \, ;$

$y = -3 - 1 = -4$

Solutions: $(4,3), (-3,-4)$

27. $\begin{cases} -x^2 + y = -1 \text{ Parabola} \\ x^2 + y^2 = 7 \quad \text{Circle} \end{cases}$

R1 + R2

$y^2 + y = 6$

$y^2 + y - 6 = 0$

$(y+3)(y-2) = 0$

$y = -3$ or $y = 2$;

If $y = -3, -x^2 + y = -1$

$-x^2 - 3 = -1$

$-x^2 = 2$

$x^2 = -2$ not real;

If $y = 2, -x^2 + y = -1$

$-x^2 + 2 = -1$

$-x^2 = -3$

$x^2 = 3$

$x = \pm\sqrt{3}$;

Solutions: $\left(\sqrt{3},2\right),\left(-\sqrt{3},2\right)$

29. $\begin{cases} y \le x^2 - 2 \\ x^2 + y^2 \le 16 \end{cases}$

Inequality 1 is a parabola with vertex $(0, -2)$.
Inequality 2 is a circle with center $(0,0)$, radius 4.

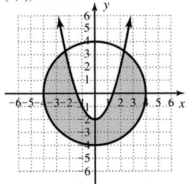

31. $r = 5\sin\theta$

Circle, Center $\left(0, \dfrac{5}{2}\right)$

Symmetric about $\theta = \dfrac{\pi}{2}$

Cycle	r-value analysis	Location of graph
0 to $\dfrac{\pi}{2}$	$\|r\|$ increases from 0 to 5	QI $(r > 0)$
$\dfrac{\pi}{2}$ to π	$\|r\|$ decreases from 5 to 0	QI $(r > 0)$
π to $\dfrac{3\pi}{2}$	$\|r\|$ increases from 0 to 5	QII $(r < 0)$
$\dfrac{3\pi}{2}$ to 2π	$\|r\|$ decreases from 5 to 0	QII $(r < 0)$

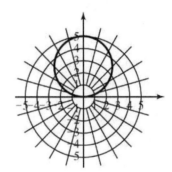

33. $r = 2 + 4\cos\theta$

Limacon, Symmetric about $\theta = 0$

Cycle	r-value analysis	Location of graph
0 to $\dfrac{\pi}{2}$	$\lvert r \rvert$ decreases from 6 to 2	QI $(r > 0)$
$\dfrac{\pi}{2}$ to $\dfrac{2\pi}{3}$	$\lvert r \rvert$ decreases from 2 to 0	QII $(r > 0)$
$\dfrac{2\pi}{3}$ to π	$\lvert r \rvert$ increases from 0 to 2	QIV $(r < 0)$
π to $\dfrac{4\pi}{3}$	$\lvert r \rvert$ decreases from 2 to 0	QII $(r < 0)$
$\dfrac{4\pi}{3}$ to $\dfrac{3\pi}{2}$	$\lvert r \rvert$ increases from 0 to 2	QIII $(r > 0)$
$\dfrac{3\pi}{2}$ to 2π	$\lvert r \rvert$ increases from 2 to 6	QIV $(r > 0)$

35. $2x^2 - 4xy + 2y^2 - 8\sqrt{2}\,y - 24 = 0$

$\theta = 45°$, $\tan(2\beta) = \dfrac{-4}{2-2} = $ undefined;

$2\beta = \dfrac{\pi}{2}$, $\beta = \dfrac{\pi}{4}$;

$x = \dfrac{\sqrt{2}}{2}X - \dfrac{\sqrt{2}}{2}Y$; $y = \dfrac{\sqrt{2}}{2}X + \dfrac{\sqrt{2}}{2}Y$;

$2\left(\dfrac{\sqrt{2}}{2}X - \dfrac{\sqrt{2}}{2}Y\right)^2$

$-4\left(\dfrac{\sqrt{2}}{2}X - \dfrac{\sqrt{2}}{2}Y\right)\left(\dfrac{\sqrt{2}}{2}X + \dfrac{\sqrt{2}}{2}Y\right)$

$+2\left(\dfrac{\sqrt{2}}{2}X + \dfrac{\sqrt{2}}{2}Y\right)^2$

$-8\sqrt{2}\left(\dfrac{\sqrt{2}}{2}X + \dfrac{\sqrt{2}}{2}Y\right) = 24$

$2\left(\dfrac{1}{2}X^2 - XY + \dfrac{1}{2}Y^2\right)$

$-4\left(\dfrac{1}{2}X^2 - \dfrac{1}{2}Y^2\right) + 2\left(\dfrac{1}{2}X^2 + XY + \dfrac{1}{2}Y^2\right)$

$\qquad -\dfrac{16}{2}X - \dfrac{16}{2}X = 24$

$X^2 - 2XY + Y^2 - 2X^2 + 2Y^2 + X^2$
$\qquad + 2XY + Y^2 - 8X - 8Y = 24$

$4Y^2 - 8X - 8Y = 24$

$Y^2 - 2X - 2Y = 6$

$Y^2 - 2Y = 2X + 6$

$Y^2 - 2Y + 1 = 2X + 6 + 1$

$(Y - 1)^2 = 2X + 7$

$(Y - 1)^2 = 2\left(X + \dfrac{7}{2}\right)$

vertex: $\left(-\dfrac{7}{2}, 1\right)$

foci: $(-3, 1)$

y-intercepts: $(0, \sqrt{7})$ and $(0, -\sqrt{7} + 1)$

37. $r = \dfrac{9}{3 - 2\cos\theta} = \dfrac{3}{1 - \dfrac{2}{3}\cos\theta}$

$e = \dfrac{2}{3}$, Ellipse

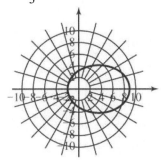

39. $r = \dfrac{4}{3 + 3\sin\theta} = \dfrac{\dfrac{4}{3}}{1 + \sin\theta}$

$e = 1$, Parabola

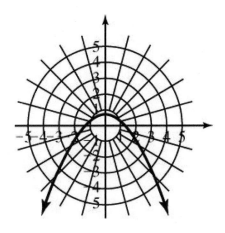

41. $x = t - 4$; $t \in [-3,3]$; $y = -2t^2 + 3$

$t = x + 4$;

$y = -2(x + 4)^2 + 3$

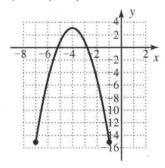

43. $x = -3\sin t$; $t \in [0, 2\pi)$

$y = 4\cos t$

$\dfrac{x}{-3} = \sin t \rightarrow \dfrac{x^2}{9} = \sin^2 t$

$\dfrac{y}{4} = \cos t \rightarrow \dfrac{y^2}{16} = \cos^2 t$

$\dfrac{x^2}{9} + \dfrac{y^2}{16} = \sin^2 t + \cos^2 t$

$\dfrac{x^2}{9} + \dfrac{y^2}{16} = 1$

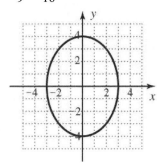

45. $x = 4\sin(5t)$; $y = 8\cos t$

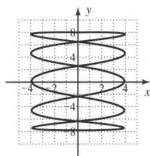

$x \in [-4, 4]$; $y \in [-8, 8]$

Mixed Review

1. $9x^2 + 9y^2 = 54$
   $x^2 + y^2 = 6$
   Circle
   Center: $(0,0)$, Radius: $r = \sqrt{6}$

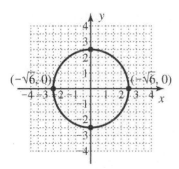

3. $9y^2 - 25x^2 = 225$
   $\dfrac{y^2}{25} - \dfrac{x^2}{9} = 1$
   Hyperbola
   Center: $(0,0)$
   Vertices: $(0,5),(0,-5)$
   $a = 3,\ b = 5$;
   $a^2 + b^2 = c^2$
   $25 + 9 = c^2$
   $\sqrt{34} = c$
   Foci: $\left(0,\sqrt{34}\right),\left(0,-\sqrt{34}\right)$
   CA: $(3,0),(-3,0)$
   Asymptotes: $y = \pm\dfrac{5}{3}x$
   $L = \dfrac{2 \cdot 9}{5} = 3\dfrac{3}{5} = 3.6$

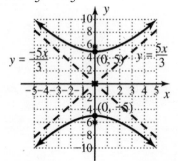

5. $4(x-1)^2 - 36(y+2)^2 = 144$
   $\dfrac{(x-1)^2}{36} - \dfrac{(y+2)^2}{4} = 1$
   Hyperbola
   Center: $(1,-2)$
   $a = 6,\ b = 2$
   Vertices: $(-5,-2),(7,-2)$
   $a^2 + b^2 = c^2$
   $36 + 4 = c^2$
   $2\sqrt{10} = c$
   Foci: $\left(1-2\sqrt{10},-2\right),\left(1+2\sqrt{10},-2\right)$
   CA: $(1,-2+2),(1,-2-2)$
   $\quad\ (1,0),(1,-4)$
   Asymptotes: $y + 2 = \pm\dfrac{1}{3}(x-1)$
   $L = \dfrac{2 \cdot 4}{6} = \dfrac{4}{3}$

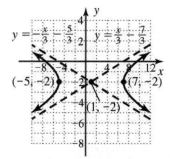

7. $y = -2x^2 - 10x + 15$

$y = -2(x^2 + 5x) + 15$

$y = -2\left(x^2 + 5x + \dfrac{25}{4}\right) + 15 + \dfrac{25}{2}$

$y = -2\left(x + \dfrac{5}{2}\right)^2 + \dfrac{55}{2}$;

$y - \dfrac{55}{2} = -2\left(x + \dfrac{5}{2}\right)^2$

$-\dfrac{1}{2}\left(y - \dfrac{55}{2}\right) = \left(x + \dfrac{5}{2}\right)^2$;

$4p = -\dfrac{1}{2}$

$p = -\dfrac{1}{8}$

Parabola

Vertex: $\left(-\dfrac{5}{2}, \dfrac{55}{2}\right)$

Focus: $\left(-\dfrac{5}{2}, \dfrac{55}{2} - \dfrac{1}{8}\right) = \left(-\dfrac{5}{2}, \dfrac{209}{8}\right)$

Directrix: $y = \dfrac{55}{2} + \dfrac{1}{8} = \dfrac{211}{8}$

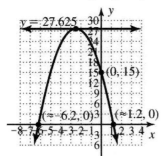

9. $x = y^2 + 2y + 3$

$x = (y^2 + 2y + 1) + 3 - 1$

$x = (y + 1)^2 + 2$;

$(x - 2) = (y + 1)^2$;

$4p = 1$

$p = \dfrac{1}{4}$

Parabola

Vertex: $(2, 1)$

Focus: $\left(2 + \dfrac{1}{4}, 1\right) = \left(\dfrac{9}{4}, 1\right)$

Directrix: $x = 2 - \dfrac{1}{4} = \dfrac{7}{4}$

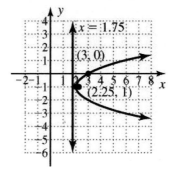

11. $x^2 - 8x - 8y + 16 = 0$

$x^2 - 8x + 16 = 8y$

$(x - 4)^2 = 8y$;

$4p = 8$

$p = 2$

Parabola

Vertex: $(4, 0)$

Focus: $(4, 0 + 2) = (4, 2)$

Directrix: $y = 0 - 2 = -2$

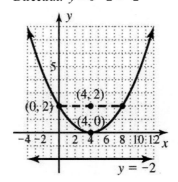

Summary and Concept Review

13. $4x^2 - 25y^2 - 24x + 150y - 289 = 0$

$4x^2 - 24x - 25y^2 + 150y = 289$

$4(x^2 - 6x) - 25(y^2 - 6y) = 289$

$4(x^2 - 6x + 9) - 25(y^2 - 6y + 9) = 289 + 36 - 225$

$4(x-3)^2 - 25(y-3)^2 = 100$

$\dfrac{(x-3)^2}{25} - \dfrac{(y-3)^2}{4} = 1$

Hyperbola

Center: $(3,3)$

Vertices: $(3+5,3), (3-5,3)$

$(8,3), (-2,3)$

$a^2 + b^2 = c^2$

$25 + 4 = c^2$

$\sqrt{29} = c$

Foci: $\left(3+\sqrt{29},3\right), \left(3-\sqrt{29},3\right)$

CA: $(3,3+2), (3,3-2)$

$(3,5), (3,1)$

Asymptotes: $y - 3 = \pm\dfrac{2}{5}(x-3)$

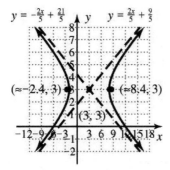

15. $49(x+2)^2 + (y-3)^2 = 49$

$\dfrac{(x+2)^2}{1} + \dfrac{(y-3)^2}{49} = 1$

Ellipse

Center: $(-2,3)$

Vertices: $(-2, 3+7), (-2, 3-7)$

$(-2,10), (-2,-4)$

$a^2 = b^2 + c^2$

$49 = 1 + c^2$

$4\sqrt{3} = c$

Foci: $\left(-2, 3 - 4\sqrt{3}\right), \left(-2, 3 + 4\sqrt{3}\right)$

CA: $(-2+1,3), (-2-1,3)$

$(-1,3), (-3,3)$

$L = \dfrac{2 \cdot 1}{7} = \dfrac{2}{7}$

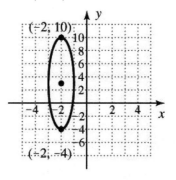

17. $x = (t-2)^2 \; ; \; y = (t-4)^2$

Parabola

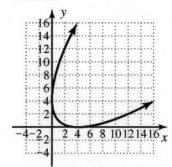

542

19. a. $\begin{cases} 4x^2 - y^2 = -9 \\ x^2 + 3y^2 = 79 \end{cases}$

Multiply first equation by 3.

$\begin{cases} 12x^2 - 3y^2 = -27 \\ x^2 + 3y^2 = 79 \end{cases}$

$13x^2 = 52$

$x^2 = 4$

$x = \pm 2$;

$4(\pm 2)^2 + 9 = y^2$

$25 = y^2$

$y = \pm 5$;

$(2, 5), (2, -5), (-2, 5), (-2, -5)$

b. $\begin{cases} 4x^2 + 9y^2 = 36 \\ x^2 + 3y = 6 \end{cases}$

Multiply the 2nd equation by -4.

$\begin{cases} 4x^2 + 9y^2 = 36 \\ -4x^2 - 12y = -24 \end{cases}$

$9y^2 - 12y = 12$

$3y^2 - 4y - 4 = 0$

$(3y + 2)(y - 2) = 0$

$y = -\dfrac{2}{3}; y = 2$;

$x^2 + 3y = 6$

$x^2 + 3\left(\dfrac{-2}{3}\right) = 6$

$x^2 = 8$

$x = \pm 2\sqrt{2}$;

$x^2 + 3y = 6$

$x^2 + 3(2) = 6$

$x^2 = 0$

$x = 0$;

$(0, 2), \left(2\sqrt{2}, \dfrac{-2}{3}\right), \left(-2\sqrt{2}, \dfrac{-2}{3}\right)$

21. $2a = 100; a = 50$

$2b = 60; b = 30$

$x = 50\cos\theta; x = 50\cos t$

$y = 30\sin\theta; y = 30\sin t$

23. a) $r = \dfrac{84}{100 + 70\cos\theta} = \dfrac{0.84}{1 + 0.7\cos\theta}$

$e = \dfrac{c}{a} = \dfrac{70}{100}$; Elliptic

$a(1 - e^2) = 0.84$

$a(1 - 0.7^2) = 0.84$

$a = \dfrac{0.84}{1 - 0.7^2} = 1.64716$;

$e \cdot a = c$

$0.7(1.64716) = c$

$1.153012 = c$

$a - c = 1.64716 - 1.153012 = 0.494148$

perihelion = 0.494 million mi

b) $r = \dfrac{31}{5 - 5\sin\theta} = \dfrac{\dfrac{31}{5}}{1 - \sin\theta}$

$e = \dfrac{5}{5} = 1$; Parabolic

$d = \dfrac{31}{5} = 6.2$

Perihelion = 3.1 million mi

25. $\dfrac{X^2}{80^2} - \dfrac{Y^2}{400^2} = 1$; $\beta = 45°$

$X = x\cos\beta + y\sin\beta$

$Y = -x\sin\beta + y\cos\beta$

$\dfrac{(x\cos\beta + y\sin\beta)^2}{80^2} - \dfrac{(-x\sin\beta + y\cos\beta)^2}{400^2} = 1$

$\dfrac{(x\cos45° + y\sin45°)^2}{80^2} - \dfrac{(-x\sin45° + y\cos45°)^2}{400^2} = 1$

$\dfrac{\left(x\left(\dfrac{\sqrt{2}}{2}\right) + y\left(\dfrac{\sqrt{2}}{2}\right)\right)^2}{80^2} - \dfrac{\left(-x\sin\left(\dfrac{\sqrt{2}}{2}\right) + y\left(\dfrac{\sqrt{2}}{2}\right)\right)^2}{400^2} = 1$

$\dfrac{\dfrac{1}{2}x^2 + xy + \dfrac{1}{2}y^2}{80^2} - \dfrac{\dfrac{1}{2}x^2 - xy + \dfrac{1}{2}y^2}{400^2} = 1$

$25\left(\dfrac{1}{2}x^2 + xy + \dfrac{1}{2}y^2\right)$

$\qquad -\dfrac{1}{2}x^2 + xy - \dfrac{1}{2}y^2 = 400^2$

$\dfrac{25}{2}x^2 + 25xy + \dfrac{25}{2}y^2 - \dfrac{1}{2}x^2 + xy - \dfrac{1}{2}y^2 = 400^2$

$\dfrac{24}{2}x^2 + 26xy + \dfrac{24}{2}y^2 = 160,000$

$12x^2 + 26xy + 12y^2 = 160,000$

Practice Test

Practice Test

1. Circle (c)

3. Hyperbola (b)

5. $x^2 + y^2 - 4x + 10y + 20 = 0$

    $x^2 - 4x + y^2 + 10y = -20$

    $(x^2 - 4x + 4) + (y^2 + 10y + 25) = -20 + 4 + 25$

    $(x - 2)^2 + (y + 5)^2 = 9$

    Center: (2,-5), $r = 3$

    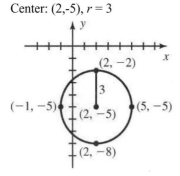

7. $r = \dfrac{10}{5 - 4\cos\theta} = \dfrac{2}{1 - \dfrac{4}{5}\cos\theta}$

    $e = \dfrac{4}{5}$

    Ellipse;

    If $\theta = 0, r = 10$;If $\theta = \pi, r = \dfrac{10}{9}$;

    If $\theta = \dfrac{\pi}{2}, r = 2$; If $\theta = \dfrac{3\pi}{2}, r = 2$

    Major vertices: $\left(\dfrac{-10}{9}, 0\right), (10, 0)$

    Center $(x, y) = \left(\dfrac{40}{9}, 0\right)$

    Minor vertices: $\left(\dfrac{4}{9}, \dfrac{10}{3}\right), \left(\dfrac{4}{9}, \dfrac{-10}{3}\right)$

    Foci: $(0, 0), \left(\dfrac{80}{9}, 0\right)$

    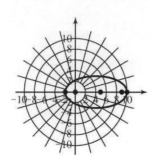

9. $\dfrac{(y + 3)^2}{9} - \dfrac{(x - 2)^2}{16} = 1$

    Hyperbola

    Center: $(2, -3)$

    Vertices:

    $(2, -3 + 3) = (2, 0)$

    $(2, -3 - 3) = (2, -6)$

    Foci:

    $(2, -3 + 5) = (2, 2)$

    $(2, -3 - 5) = (2, -8)$

    Asymptotes:

    $(y + 3) = \pm \dfrac{3}{4}(x - 2)$

    $y = -\dfrac{3}{4}x - \dfrac{3}{2}$; $y = \dfrac{3}{4}x - \dfrac{9}{2}$

    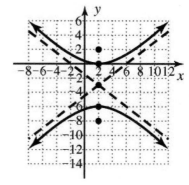

11. $80x^2 + 120xy + 45y^2 - 100y - 44 = 0$

    $B^2 - 4AC = 120^2 - 4(80)45 = 0$; Parabola

    $\tan(2\beta) = \dfrac{120}{80 - 45} = \dfrac{24}{7}$; $2\beta = \tan^{-1}\left(\dfrac{24}{7}\right)$

    $2\beta = 73.74°$; $\beta = 36.87°$;

    $\tan(2\beta) = \dfrac{24}{7}$;

    $\sqrt{24^2 + 7^2} = 25$; $\cos(2\beta) = \dfrac{7}{25}$;

    $\cos\beta = \sqrt{\dfrac{1 + \dfrac{7}{25}}{2}} = \sqrt{\dfrac{32}{25} \cdot \dfrac{1}{2}} = \sqrt{\dfrac{16}{25}} = \dfrac{4}{5}$;

    $\sqrt{5^2 - 4^2} = 3$;

    $\cos\beta = \dfrac{4}{5}$; $\sin\beta = \dfrac{3}{5}$

Chapter 9: Analytical Geometry

13. $r = 3 + 3\cos\theta$

Cardioid

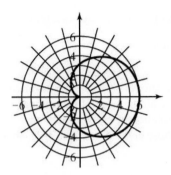

15. $r = 6\sin(2\theta)$

Four-petal rose

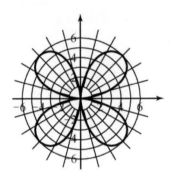

17. $x = (t-3)^2 + 1$; $y = t + 2$; $y - 2 = t$

$x = (y - 2 - 3)^2 + 1$

$x = (y - 5)^2 + 1$

Parabola, opens right

19. a. $\begin{cases} 4x^2 - y^2 = 16 & \text{Hyperbola} \\ y - x = 2 & \text{Line} \end{cases}$

2nd equation: $y = x + 2$

$4x^2 - (x+2)^2 = 16$

$4x^2 - x^2 - 4x - 4 = 16$

$3x^2 - 4x - 20 = 0$

$(3x - 10)(x + 2) = 0$

$x = \dfrac{10}{3}, x = -2$;

$y = \dfrac{10}{3} + 2 = \dfrac{16}{3}$;

$y = -2 + 2 = 0$

Solutions: $\left(\dfrac{10}{3}, \dfrac{16}{3}\right), (-2, 0)$

b. $\begin{cases} 4y^2 - x^2 = 4 & \text{Hyperbola} \\ x^2 + y^2 = 4 & \text{Circle} \end{cases}$

$5y^2 = 8$

$y^2 = \dfrac{8}{5}$

$y = \pm\sqrt{\dfrac{8}{5}}$;

$4\left(\pm\dfrac{2\sqrt{10}}{5}\right)^2 - x^2 = 4$

$-x^2 = 4 - 4\left(\dfrac{8}{5}\right)$

$x^2 = \dfrac{12}{5}$

$x = \pm\sqrt{\dfrac{12}{5}}$;

Solutions: $\left(\sqrt{\dfrac{12}{5}}, \sqrt{\dfrac{8}{5}}\right), \left(\sqrt{\dfrac{12}{5}}, -\sqrt{\dfrac{8}{5}}\right),$

$\left(-\sqrt{\dfrac{12}{5}}, \sqrt{\dfrac{8}{5}}\right), \left(-\sqrt{\dfrac{12}{5}}, -\sqrt{\dfrac{8}{5}}\right),$

21. $x = v_0 t \cos \theta$

$x = 80t \cos 28°$;

$y = v_0 t \sin \theta - 16t^2$

$y = 80t \sin 28° - 16t^2$;

At $x = 165$, $y \approx 0.43$

The ball is 0.43 ft above the ground at $x = 165$ feet, and will likely go into the goal.

23. $y = (x-1)^2 - 4$;

$D : x \in (-\infty, \infty)$

$R : y \in [-4, \infty)$

Focus: $\left(1, -\dfrac{15}{4}\right)$

25. $\dfrac{(x+2)^2}{9} + \dfrac{(y-1)^2}{25} = 1$

Domain: $x \in [-5, 1]$

Range: $y \in [-4, 6]$

Calculator Exploration & Discovery

1. A rotation of π or $-\pi$ causes a reflection about the line $\theta = \dfrac{\pi}{2}$.

3. If the numerator is changed to a negative, the radii are all opposite of the original, thus there is a rotation of $\pm\pi$. The graph overlaps those created by a reflection about $\theta = \dfrac{\pi}{2}$.

5. A rotation around $\dfrac{2\pi}{5}$ causes Galois to intersect both Agnesi and Erdös.

Strengthening Core Skills

1. #31 from 10.4

$12x^2 + 24xy + 5y^2 - 40x - 30y = 25$

$A = 12, B = 24, C = 5$

$B^2 - 4AC = 24^2 - 4(12)(5) = 336$

Hyperbola

$\tan(2\beta) = \dfrac{B}{A-C} = \dfrac{24}{12-5} = \dfrac{24}{7};$

$\cos(2\beta) = \dfrac{7}{25};$

$\cos\beta = \sqrt{\dfrac{1+\cos(2\beta)}{2}} = \sqrt{\dfrac{1+\dfrac{7}{25}}{2}}$

$= \sqrt{\dfrac{\dfrac{32}{25}}{\dfrac{2}{1}}} = \sqrt{\dfrac{16}{25}} = \dfrac{4}{5};$

$\sin\beta = \sqrt{\dfrac{1-\cos(2\beta)}{2}} = \sqrt{\dfrac{1-\dfrac{7}{25}}{2}}$

$= \sqrt{\dfrac{\dfrac{18}{25}}{\dfrac{2}{1}}} = \sqrt{\dfrac{9}{25}} = \dfrac{3}{5};$

$x = X\cos\beta - Y\sin\beta$

$x = \dfrac{4}{5}X - \dfrac{3}{5}Y = \dfrac{4X-3Y}{5};$

$y = X\sin\beta - +Y\cos\beta$

$y = \dfrac{3}{5}X + \dfrac{4}{5}Y = \dfrac{3X+4Y}{5};$

$x^2 = \left(\dfrac{4X-3Y}{5}\right)^2 = \dfrac{16X^2 - 24XY + 9Y^2}{25};$

$y^2 = \left(\dfrac{3X+4Y}{5}\right)^2 = \dfrac{9X^2 + 24XY + 16Y^2}{25};$

$xy = \left(\dfrac{4X-3Y}{5}\right)\left(\dfrac{3X+4Y}{5}\right)$

$xy = \dfrac{12X^2 + 16XY - 9XY - 12Y^2}{25}$

$xy = \dfrac{12X^2 + 7XY - 12Y^2}{25};$

$25 = 12x^2 + 24xy + 5y^2 - 40x - 30y$

$25 = 12\left(\dfrac{16X^2 - 24XY + 9Y^2}{25}\right)$

$+ 24\left(\dfrac{12X^2 + 7XY - 12Y^2}{25}\right)$

$+ 5\left(\dfrac{9X^2 + 24XY + 16Y^2}{25}\right)$

$- 40\left(\dfrac{4X-3Y}{5}\right) - 3\left(\dfrac{3X+4Y}{5}\right)$

$625 = 192X^2 - 288XY + 108Y^2 + 288X^2$

$+ 168XY - 288Y^2 + 456X^2 + 120XY$

$+ 80Y^2 - 800X + 600Y - 45X - 60Y$

$625 = 525X^2 - 100Y^2 - 845X + 540Y$

$1 = \dfrac{21X^2}{25} - \dfrac{4Y^2}{25} - \dfrac{169X}{125} + \dfrac{108Y}{125}$

#32 from 10.4

$25x^2 + 840xy - 16y^2 - 400 = 0$

$A = 25, B = 840, C = -16$

$B^2 - 4AC = 840^2 - 4(25)(-16) = 707200$

$707,200 > 0;$ Hyperbola;

$\tan(2\beta) = \dfrac{B}{A-C} = \dfrac{840}{25+16} = \dfrac{840}{41};$

$\sqrt{840^2 + 41^2} = 841;\ \cos(2\beta) = \dfrac{41}{841};$

$\cos\beta = \sqrt{\dfrac{1+\cos 2\beta}{2}} = \sqrt{\dfrac{1+\dfrac{41}{841}}{2}} = \dfrac{21}{29};$

$\sin\beta = \sqrt{\dfrac{1-\cos(2\beta)}{2}} = \sqrt{\dfrac{1-\dfrac{41}{841}}{2}} = \dfrac{20}{29};$

$x = X\cos\beta - Y\sin\beta$

$x = \dfrac{21}{29}X - \dfrac{20}{29}Y = \dfrac{21X-20Y}{29};$

$y = X\sin\beta - +Y\cos\beta$

$y = \dfrac{20}{29}X + \dfrac{21}{29}Y = \dfrac{20X+21Y}{29};$

$$x^2 = \left(\frac{21X - 20Y}{29}\right)^2 = \frac{441X^2 - 840XY + 400Y^2}{841};$$

$$y^2 = \left(\frac{20X + 21Y}{29}\right)^2 = \frac{400X^2 + 840XY + 441Y^2}{841};$$

$$xy = \left(\frac{21X - 20Y}{29}\right)\left(\frac{20X + 21Y}{29}\right)$$

$$xy = \frac{420X^2 + 441XY - 400XY - 420Y^2}{841}$$

$$xy = \frac{420X^2 + 41XY - 420Y^2}{841};$$

$$400 = 25x^2 + 840xy - 16y^2$$

$$400 = 25\left(\frac{441X^2 - 840XY + 400Y^2}{841}\right)$$

$$+840\left(\frac{420X^2 + 41XY - 420Y^2}{841}\right)$$

$$-16\left(\frac{400X^2 + 840XY + 441Y^2}{841}\right)$$

$$336400 = 11025X^2 - 21000XY + 10000Y^2$$

$$+352800X^2 + 34440XY - 352800Y^2$$

$$-6400X^2 - 13440XY - 7056Y^2$$

$$3364000 = 357425X^2 - 349856Y^2$$

$$1 = \frac{17X^2}{16} - \frac{26Y^2}{25}$$

Yes, the calculations are much "cleaner" with the right triangle definitions.

Cumulative Review

1. $\sqrt{x+2} + 2 = \sqrt{3x+4}$

$\left(\sqrt{x+2} + 2\right)^2 = \left(\sqrt{3x+4}\right)^2$

$x + 2 + 4\sqrt{x+2} + 4 = 3x + 4$

$4\sqrt{x+2} = 2x - 2$

$\left(4\sqrt{x+2}\right)^2 = (2x - 2)^2$

$16(x+2) = 4x^2 - 8x + 4$

$16x + 32 = 4x^2 - 8x + 4$

$0 = 4x^2 - 24x - 28$

$0 = 4\left(x^2 - 6x - 7\right)$

$(x-7)(x+1) = 0$

$x - 7 = 0 \ \text{ or } \ x + 1 = 0$

$x = 7 \quad \text{ or } \ x = -1$

$x = 7, x = -1$ is extraneous.

3. $4 \cdot 2^{x+1} = \dfrac{1}{8}$

$2^2 \cdot 2^{x+1} = 2^{-3}$

$2^{x+3} = 2^{-3}$

$x + 3 = -3$

$x = -6$

5. $\log_3 81 = x$

$3^x = 81$

$3^x = 3^4$

$x = 4$

7. $-6 \tan x = 2\sqrt{3}$

$\dfrac{-6}{-6} \tan x = \dfrac{2\sqrt{3}}{-6}$

$\tan x = \dfrac{-\sqrt{3}}{3}$

$x = \dfrac{5\pi}{6} + \pi k, \text{k} \in Z$

9. $\dfrac{\sin 27°}{18} = \dfrac{\sin x}{35}$

$\dfrac{35\sin 27°}{18} = \sin x$

$0.882759305 \approx \sin x$

$\sin^{-1}(0.882759305) = x$

$x \approx 61.98° + 360°k, k \in Z$ (QI)

$x \approx 180° - 61.98° = 118.02°$ (QII)

$x \approx 118.02° + 360°k, k \in Z$

11. $P = \dfrac{Kd}{S}$

$18 = \dfrac{K850}{1000}$

$\dfrac{18000}{850} = k$

$k = \dfrac{360}{17}$

$P = \dfrac{\left(\dfrac{360}{17}\right)d}{S}$

$P = \dfrac{\left(\dfrac{360}{17}\right)1400}{1200}$

$P = 24.7$ peso/kg

13. $x^2 = 540^2 + 850^2 - 2(540)(850)\cos 110°$

$x^2 = 1328074.492$

$x = 1152.4$ yards wide

15. $y = \sqrt{x-3} + 1$

Node: (3,1)

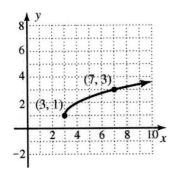

17. $h(x) = \dfrac{x-2}{x^2 - 9}$

Horizontal asymptote: $y = 0$
(deg num < deg den)

$h(x) = \dfrac{x-2}{(x+3)(x-3)}$

Vertical asymptotes: $x = 3$ or $x = -3$;

$h(0) = \dfrac{0-2}{0^2 - 9} = \dfrac{2}{9} \approx 0.22$

y-intercept: (0,0.22);

$0 = \dfrac{x-2}{x^2 - 9}$

x-intercept: (2,0);

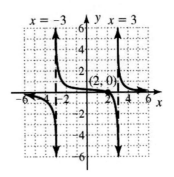

19. $f(x) = \log_2(x+1)$

$f(0) = \log_2(0+1) = 0$

y-intercept: (0,0);

$x + 1 = 0$

Vertical asymptote: $x = -1$

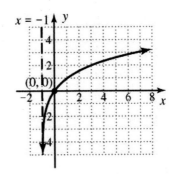

21. $4(x-1)^2 - 36(y+2)^2 = 144$

$$\frac{(x-1)^2}{36} - \frac{(y+2)^2}{4} = 1$$

Hyperbola

Center: $(1,-2)$

$a = 6, \ b = 2$

Vertices: $(1-6,-2),(1+6,-2)$

$\quad (-5,-2),(7,-2)$;

$a^2 + b^2 = c^2$

$36 + 4 = c^2$

$2\sqrt{10} = c$

Foci: $(1-2\sqrt{10},-2),(1+2\sqrt{10},-2)$

CA: $(1,-2+2),(1,-2-2)$

$\quad (1,0),(1,-4)$

Asymptotes: $y + 2 = \pm\frac{1}{3}(x-1)$

$$y + 2 = \frac{1}{3}x - \frac{1}{3}; \ y + 2 = -\frac{1}{3}x + \frac{1}{3}$$

$$y = \frac{1}{3}x - \frac{7}{3}; \ y = -\frac{1}{3}x - \frac{5}{3}$$

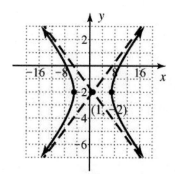

23. $r = 4\cos(2\theta)$

Four-petal rose

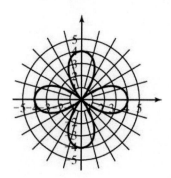

25. $\cos\theta = \dfrac{\mathbf{u}\cdot\mathbf{v}}{\|\mathbf{u}\|\cdot\|\mathbf{v}\|} = \dfrac{-4(3)+5(7)}{\sqrt{(-4)^2+5^2}\sqrt{3^2+7^2}}$

$\cos\theta = \dfrac{23}{\sqrt{41\cdot58}}$

$\theta \approx 61.9°$

27. $\begin{cases} x^2 + y^2 = 25 \\ 64x^2 + 12y^2 = 768 \end{cases}$

$\quad\begin{cases} -12x^2 - 12y^2 = -300 \\ 64x^2 + 12y^2 = 768 \end{cases}$

$\quad 52x^2 = 468$

$\quad x^2 = 9$

$\quad x = \pm 3$;

$\quad (\pm 3)^2 + y^2 = 25$

$\quad y^2 = 16$

$\quad y = \pm 4$;

$\quad (3,4),(3,-4),(-3,4),(-3,-4)$

29. $y = \dfrac{3x^3 - 2x^2 + x - 3}{x^4 + x^2}$

$$\frac{3x^3 - 2x^2 + x - 3}{x^2(x^2+1)} = \frac{A}{x} + \frac{B}{x^2} + \frac{Cx+D}{x^2+1}$$

$3x^3 - 2x^2 + x - 3 = Ax(x^2+1) + B(x^2+1) + (Cx+D)x^2$

$\quad = Ax^3 + Ax + Bx^2 + B + Cx^3 + Dx^2$

$\quad = (A+C)x^3 + (B+D)x^2 + Ax + B$

$3 = A+C \quad B+D=-2 \quad A=1 \quad B=-3$

$3 = 1+C \quad -3+D=-2$

$C=2 \qquad D=1$

$$= \frac{1}{x} + \frac{3}{x^2} + \frac{2x+1}{x^2+1}$$

30. $R(w) = -w^4 + 25w^3 - 200w^2 + 560w - 234$

$\begin{array}{r|rrrrr} 5 & -1 & 25 & -200 & 560 & -234 \\ & & -5 & 100 & -500 & 300 \\ \hline & -1 & 20 & -100 & 60 & \underline{|66} \end{array}$

$66 million

Connections to Calculus

1. $r = 1 + \cos\theta$;

   Slope of the tangent line:

a) $\dfrac{\cos\theta + \sin\theta(-\sin\theta) + \cos\theta\cos\theta}{-\sin\theta + 2\cos\theta(-\sin\theta)}$

$$= \frac{\cos\theta - \sin^2\theta + \cos^2\theta}{-\sin\theta - 2\sin\theta\cos\theta}$$

$$= \frac{\cos^2\theta - \sin^2\theta + \cos\theta}{-(2\sin\theta\cos\theta + \sin\theta)}$$

$$= -\frac{\cos 2\theta + \cos\theta}{\sin 2\theta + \sin\theta}$$

b) Slope of the tangent line:

   at $\theta = \dfrac{\pi}{2}$

$$= -\frac{\cos 2\left(\dfrac{\pi}{2}\right) + \cos\dfrac{\pi}{2}}{\sin 2\left(\dfrac{\pi}{2}\right) + \sin\dfrac{\pi}{2}}$$

$$= -\frac{\cos\pi + 0}{\sin\pi + 1}$$

$$= -\frac{-1+0}{0+1} = 1$$

   at $\theta = \dfrac{11\pi}{6}$

$$= -\frac{\cos 2\left(\dfrac{11\pi}{6}\right) + \cos\dfrac{11\pi}{6}}{\sin 2\left(\dfrac{11\pi}{6}\right) + \sin\dfrac{11\pi}{6}}$$

$$= -\frac{\cos\left(\dfrac{11\pi}{3}\right) + \dfrac{\sqrt{3}}{2}}{\sin\left(\dfrac{11\pi}{3}\right) + \left(-\dfrac{1}{2}\right)}$$

$$= -\frac{\cos\left(\dfrac{5\pi}{3}\right) + \dfrac{\sqrt{3}}{2}}{\sin\left(\dfrac{5\pi}{3}\right) + \left(-\dfrac{1}{2}\right)}$$

$$= -\frac{\dfrac{1}{2} + \dfrac{\sqrt{3}}{2}}{-\dfrac{\sqrt{3}}{2} - \dfrac{1}{2}} = -\frac{\dfrac{1+\sqrt{3}}{2}}{\dfrac{-\sqrt{3}-1}{2}} = 1$$

c) Tangent line (horizontal):

$$0 = \cos 2\theta + \cos\theta$$
$$0 = 2\cos^2\theta - 1 + \cos\theta$$
$$= 2\cos^2\theta + \cos\theta - 1$$
$$= (2\cos\theta - 1)(\cos\theta + 1)$$

$2\cos\theta - 1 = 0 \qquad \cos\theta + 1 = 0$

$\cos\theta = \dfrac{1}{2} \qquad\qquad \cos\theta = -1$

$\theta = \dfrac{\pi}{3}, \dfrac{5\pi}{3} \qquad\qquad \theta = \pi$

$\theta \neq \pi$ because π makes both numerator and denominator 0.

$\theta = \dfrac{\pi}{3}, \dfrac{5\pi}{3}$

d) Tangent line (vertical):

$$0 = \sin 2\theta + \sin\theta$$
$$0 = 2\sin\theta\cos\theta + \sin\theta$$
$$= \sin\theta(2\cos\theta + 1)$$

$\sin\theta = 0 \quad\text{or}\quad 2\cos\theta + 1 = 0$

$\theta = 0, \pi \qquad\qquad \cos\theta = -\dfrac{1}{2}$

$\theta = \dfrac{2\pi}{3}, \dfrac{4\pi}{3}$

$\theta \neq \pi$ because π makes both numerator and denominator 0.

$\theta = 0, \dfrac{2\pi}{3}, \dfrac{4\pi}{3}$

3. $r = \cos\theta - \sin\theta$

a) slope of the tangent line:

$$= \frac{\cos\theta\cos\theta + \sin\theta(-\sin\theta) - 2\sin\theta\cos\theta}{2\cos\theta(-\sin\theta) - [\sin\theta(-\sin\theta) + \cos\theta\cos\theta]}$$

$$= \frac{\cos^2\theta - \sin^2\theta - \sin 2\theta}{-2\sin\theta\cos\theta - \left[\cos^2\theta - \sin^2\theta\right]}$$

$$= -\frac{\cos 2\theta - \sin 2\theta}{\sin 2\theta + \cos 2\theta}$$

$$= \frac{\sin 2\theta - \cos 2\theta}{\sin 2\theta + \cos 2\theta}$$

b) slope at $\theta = \dfrac{\pi}{4}$

$$= \frac{\sin 2\left(\dfrac{\pi}{4}\right) - \cos 2\left(\dfrac{\pi}{4}\right)}{\sin 2\left(\dfrac{\pi}{4}\right) + \cos 2\left(\dfrac{\pi}{4}\right)}$$

$$= \frac{\sin\left(\dfrac{\pi}{2}\right) - \cos\left(\dfrac{\pi}{2}\right)}{\sin\left(\dfrac{\pi}{2}\right) + \cos\left(\dfrac{\pi}{2}\right)} = \frac{1-0}{1+0} = 1$$

slope at $\theta = \dfrac{3\pi}{4}$

$$= \frac{\sin 2\left(\dfrac{3\pi}{4}\right) - \cos 2\left(\dfrac{3\pi}{4}\right)}{\sin 2\left(\dfrac{3\pi}{4}\right) + \cos 2\left(\dfrac{3\pi}{4}\right)}$$

$$= \frac{\sin\left(\dfrac{3\pi}{2}\right) - \cos\left(\dfrac{3\pi}{2}\right)}{\sin\left(\dfrac{3\pi}{2}\right) + \cos\left(\dfrac{3\pi}{2}\right)} = \frac{-1-0}{-1+0} = 1$$

c) Tangent line (horizontal) over $\left[0, 2\pi\right)$:

$0 = \sin 2\theta - \cos 2\theta$

$\sin 2\theta = \cos 2\theta$

$\dfrac{\sin 2\theta}{\cos 2\theta} = 1$

$\tan 2\theta = 1$

$2\theta = \dfrac{\pi}{4} + \pi k$

$\theta = \dfrac{\pi}{8} + \dfrac{\pi}{2}k$

$\theta = \dfrac{\pi}{8}, \dfrac{5\pi}{8}, \dfrac{9\pi}{8}, \dfrac{13\pi}{8}$

d) Tangent line (vertical):

$0 = \sin 2\theta + \cos 2\theta$

$\sin 2\theta = -\cos 2\theta$

$\dfrac{\sin 2\theta}{\cos 2\theta} = -1$

$\tan 2\theta = -1$

$2\theta = \dfrac{3\pi}{4} + \pi k$

$\theta = \dfrac{3\pi}{8} + \dfrac{\pi}{2}k$

$\theta = \dfrac{3\pi}{8}, \dfrac{7\pi}{8}, \dfrac{11\pi}{8}, \dfrac{15\pi}{8}$

5. $\begin{cases} r = \dfrac{3}{2} \\ r = 1 + \cos^2\theta \end{cases}$

$\dfrac{3}{2} = 1 + \cos^2\theta$

$\dfrac{1}{2} = \cos^2\theta$

$\cos\theta = \pm\sqrt{\dfrac{1}{2}}$

$\cos\theta = \pm\dfrac{\sqrt{2}}{2}$

$\theta = \dfrac{\pi}{4}, \dfrac{3\pi}{4}, \dfrac{5\pi}{4}, \dfrac{7\pi}{4}$

Verified on calculator.

7. $\begin{cases} r = \sin(2\theta) \text{ four-leaf rose} \\ r = \cos\theta \text{ circle} \end{cases}$

$\sin 2\theta = \cos\theta$

$2\sin\theta\cos\theta = \cos\theta$

$2\sin\theta\cos\theta - \cos\theta = 0$

$\cos\theta(2\sin\theta - 1) = 0$

$\cos\theta = 0 \qquad 2\sin\theta - 1 = 0$

$\theta = \dfrac{\pi}{2}, \dfrac{3\pi}{2} \qquad\qquad \sin\theta = \dfrac{1}{2}$

$\qquad\qquad\qquad\qquad \theta = \dfrac{\pi}{6}, \dfrac{5\pi}{6}$

Verified on calculator.

9. $\begin{cases} r = 4\cos\theta \\ r = 8\sin\theta\cos\theta \end{cases}$

$4\cos\theta = 8\sin\theta\cos\theta$

$\cos\theta = 2\sin\theta\cos\theta$

$2\sin\theta\cos\theta - \cos\theta = 0$

$\cos\theta(2\sin\theta - 1) = 0$

$\cos\theta = 0 \qquad 2\sin\theta - 1 = 0$

$\theta = \dfrac{\pi}{2}, \dfrac{3\pi}{2} \qquad \theta = \dfrac{\pi}{6}, \dfrac{5\pi}{6}$

$\theta = \dfrac{\pi}{2} \qquad r = 4\cos\dfrac{\pi}{2}$

$\qquad\qquad r = 0$

$\theta = \dfrac{3\pi}{2} \qquad r = 4\cos\dfrac{3\pi}{2} = 0$

$\theta = \dfrac{\pi}{6} \qquad r = 4\cos\dfrac{\pi}{6} = 4\left(\dfrac{\sqrt{3}}{2}\right) = 2\sqrt{3}$

$\theta = \dfrac{5\pi}{6} \quad r = 4\cos\dfrac{5\pi}{6} = 4\left(-\dfrac{\sqrt{3}}{2}\right) = -2\sqrt{3}$

$(r, \theta): \left(0, \dfrac{\pi}{2}\right)\left(2\sqrt{3}, \dfrac{\pi}{6}\right)\left(-2\sqrt{3}, \dfrac{5\pi}{6}\right)$

$(x, y): \left(3, \sqrt{3}\right), \left(3, -\sqrt{3}\right), (0, 0)$

$x = 0\cos\dfrac{\pi}{2} \qquad x = 2\sqrt{3}\cos\dfrac{\pi}{6} = 2\sqrt{3}\left(\dfrac{\sqrt{3}}{2}\right)$

$r = -2\sqrt{3}\cos\dfrac{5\pi}{6} = -2\sqrt{3}\left(-\dfrac{\sqrt{3}}{2}\right)$

$y = 0\sin\dfrac{\pi}{2} \qquad y = 2\sqrt{3}\sin\dfrac{\pi}{6} = 2\sqrt{3}\left(\dfrac{1}{2}\right)$

$r = -2\sqrt{3}\sin\dfrac{5\pi}{6} = -2\sqrt{3}\left(\dfrac{1}{2}\right)$

10.1 Exercises

10.1 Technology Highlight

Exercise 1: $a_n = \dfrac{1}{3^n}$

First 10 terms:

$\dfrac{1}{3}, \dfrac{1}{9}, \dfrac{1}{27}, \dfrac{1}{81}, \dfrac{1}{243}, \dfrac{1}{729}, \dfrac{1}{2187},$

$\dfrac{1}{6561}, \dfrac{1}{19683}, \dfrac{1}{59049}$

Sum of first 10 terms: sum $\to 0.5$

Exercise 3: $a_n = \dfrac{1}{(2n-1)(2n+1)}$

First 10 terms:

$\dfrac{1}{3}, \dfrac{1}{15}, \dfrac{1}{35}, \dfrac{1}{63}, \dfrac{1}{99}, \dfrac{1}{143}, \dfrac{1}{195}, \dfrac{1}{255}, \dfrac{1}{323}, \dfrac{1}{399}$

Sum of first 10 terms: sum $\to 0.5$

10.1 Exercises

1. Pattern; order

3. Increasing

5. Formula defining the sequence uses the preceding term. Answers will vary.

7. $a_n = 2n - 1$

$a_1 = 2(1) - 1 = 2 - 1 = 1$;
$a_2 = 2(2) - 1 = 4 - 1 = 3$;
$a_3 = 2(3) - 1 = 6 - 1 = 5$;
$a_4 = 2(4) - 1 = 8 - 1 = 7$;
$a_8 = 2(8) - 1 = 16 - 1 = 15$;
$a_{12} = 2(12) - 1 = 24 - 1 = 23$;
$1, 3, 5, 7; a_8 = 15; a_{12} = 23$

9. $a_n = 3n^2 - 3$

$a_1 = 3(1)^2 - 3 = 3(1) - 3 = 3 - 3 = 0$;
$a_2 = 3(2)^2 - 3 = 3(4) - 3 = 12 - 3 = 9$;
$a_3 = 3(3)^2 - 3 = 3(9) - 3 = 27 - 3 = 24$;
$a_4 = 3(4)^2 - 3 = 3(16) - 3 = 48 - 3 = 45$;
$a_8 = 3(8)^2 - 3 = 3(64) - 3 = 192 - 3 = 189$;
$a_{12} = 3(12)^2 - 3 = 3(144) - 3 = 432 - 3 = 429$;
$0, 9, 24, 45; a_8 = 189; a_{12} = 429$

11. $a_n = (-1)^n n$

$a_1 = (-1)^1(1) = -1(1) = -1$;
$a_2 = (-1)^2(2) = 1(2) = 2$;
$a_3 = (-1)^3(3) = -1(3) = -3$;
$a_4 = (-1)^4(4) = 1(4) = 4$;
$a_8 = (-1)^8(8) = 1(8) = 8$;
$a_{12} = (-1)^{12}(12) = 1(12) = 12$;
$-1, 2, -3, 4; a_8 = 8; a_{12} = 12$

13. $a_n = \dfrac{n}{n+1}$

$a_1 = \dfrac{1}{1+1} = \dfrac{1}{2}$;

$a_2 = \dfrac{2}{2+1} = \dfrac{2}{3}$;

$a_3 = \dfrac{3}{3+1} = \dfrac{3}{4}$;

$a_4 = \dfrac{4}{4+1} = \dfrac{4}{5}$;

$a_8 = \dfrac{8}{8+1} = \dfrac{8}{9}$;

$a_{12} = \dfrac{12}{12+1} = \dfrac{12}{13}$;

$\dfrac{1}{2}, \dfrac{2}{3}, \dfrac{3}{4}, \dfrac{4}{5}; a_8 = \dfrac{8}{9}; a_{12} = \dfrac{12}{13}$

15. $a_n = \left(\dfrac{1}{2}\right)^n$

$a_1 = \left(\dfrac{1}{2}\right)^1 = \dfrac{1}{2}$;

$a_2 = \left(\dfrac{1}{2}\right)^2 = \dfrac{1}{4}$;

$a_3 = \left(\dfrac{1}{2}\right)^3 = \dfrac{1}{8}$;

$a_4 = \left(\dfrac{1}{2}\right)^4 = \dfrac{1}{16}$;

$a_8 = \left(\dfrac{1}{2}\right)^8 = \dfrac{1}{256}$;

$a_{12} = \left(\dfrac{1}{2}\right)^{12} = \dfrac{1}{4096}$;

$\dfrac{1}{2}, \dfrac{1}{4}, \dfrac{1}{8}, \dfrac{1}{16}; a_8 = \dfrac{1}{256}; a_{12} = \dfrac{1}{4096}$

17. $a_n = \dfrac{1}{n}$

$a_1 = \dfrac{1}{1} = 1$;

$a_2 = \dfrac{1}{2}$;

$a_3 = \dfrac{1}{3}$;

$a_4 = \dfrac{1}{4}$;

$a_8 = \dfrac{1}{8}$;

$a_{12} = \dfrac{1}{12}$;

$1, \dfrac{1}{2}, \dfrac{1}{3}, \dfrac{1}{4}$; $a_8 = \dfrac{1}{8}$; $a_{12} = \dfrac{1}{12}$

19. $a_n = \dfrac{(-1)^n}{n(n+1)}$

$a_1 = \dfrac{(-1)^1}{1(1+1)} = \dfrac{-1}{2}$;

$a_2 = \dfrac{(-1)^2}{2(2+1)} = \dfrac{1}{2(3)} = \dfrac{1}{6}$;

$a_3 = \dfrac{(-1)^3}{3(3+1)} = \dfrac{-1}{3(4)} = \dfrac{-1}{12}$;

$a_4 = \dfrac{(-1)^4}{4(4+1)} = \dfrac{1}{4(5)} = \dfrac{1}{20}$;

$a_8 = \dfrac{(-1)^8}{8(8+1)} = \dfrac{1}{8(9)} = \dfrac{1}{72}$;

$a_{12} = \dfrac{(-1)^{12}}{12(12+1)} = \dfrac{1}{12(13)} = \dfrac{1}{156}$;

$\dfrac{-1}{2}, \dfrac{1}{6}, \dfrac{-1}{12}, \dfrac{1}{20}$; $a_8 = \dfrac{1}{72}$; $a_{12} = \dfrac{1}{156}$

21. $a_n = (-1)^n 2^n$

$a_1 = (-1)^1 2^1 = -1(2) = -2$;

$a_2 = (-1)^2 2^2 = 1(4) = 4$;

$a_3 = (-1)^3 2^3 = -1(8) = -8$;

$a_4 = (-1)^4 2^4 = 1(16) = 16$;

$a_8 = (-1)^8 2^8 = 1(256) = 256$;

$a_{12} = (-1)^{12} 2^{12} = 1(4096) = 4096$;

-2, 4, -8, 16; $a_8 = 256$; $a_{12} = 4096$

23. $a_n = n^2 - 2$

$a_9 = (9)^2 - 2 = 81 - 2 = 79$

25. $a_n = \dfrac{(-1)^{n+1}}{n}$

$a_5 = \dfrac{(-1)^{5+1}}{5} = \dfrac{(-1)^6}{5} = \dfrac{1}{5}$

27. $a_n = 2\left(\dfrac{1}{2}\right)^{n-1}$

$a_7 = 2\left(\dfrac{1}{2}\right)^{7-1} = 2\left(\dfrac{1}{2}\right)^6 = 2\left(\dfrac{1}{64}\right) = \dfrac{1}{32}$

29. $a_n = \left(1 + \dfrac{1}{n}\right)^n$

$a_{10} = \left(1 + \dfrac{1}{10}\right)^{10} = \left(\dfrac{11}{10}\right)^{10}$

31. $a_n = \dfrac{1}{(n)(2n+1)}$

$a_4 = \dfrac{1}{(4)[2(4)+1]} = \dfrac{1}{4(9)} = \dfrac{1}{36}$

33. $\begin{cases} a_1 = 2 \\ a_n = 5a_{n-1} - 3 \end{cases}$

$a_2 = 5a_{2-1} - 3 = 5a_1 - 3$
$= 5(2) - 3 = 10 - 3 = 7$;
$a_3 = 5a_{3-1} - 3 = 5a_2 - 3$
$= 5(7) - 3 = 35 - 3 = 32$;
$a_4 = 5a_{4-1} - 3 = 5a_3 - 3$
$= 5(32) - 3 = 160 - 3 = 157$;
$a_5 = 5a_{5-1} - 3 = 5a_4 - 3$
$= 5(157) - 3 = 785 - 3 = 782$;
$2, 7, 32, 157, 782$

10.1 Exercises

35. $\begin{cases} a_1 = -1 \\ a_n = (a_{n-1})^2 + 3 \end{cases}$

$a_2 = (a_{2-1})^2 + 3 = (a_1)^2 + 3$
$= (-1)^2 + 3 = 1 + 3 = 4;$
$a_3 = (a_{3-1})^2 + 3 = (a_2)^2 + 3$
$= (4)^2 + 3 = 16 + 3 = 19;$
$a_4 = (a_{4-1})^2 + 3 = (a_3)^2 + 3$
$= (19)^2 + 3 = 361 + 3 = 364;$
$a_5 = (a_{5-1})^2 + 3 = (a_4)^2 + 3$
$= (364)^2 + 3 = 132496 + 3 = 132499;$
$-1, 4, 19, 364, 132499$

37. $\begin{cases} c_1 = 64, c_2 = 32 \\ c_n = \dfrac{c_{n-2} - c_{n-1}}{2} \end{cases}$

$c_3 = \dfrac{c_{3-2} - c_{3-1}}{2} = \dfrac{c_1 - c_2}{2}$
$= \dfrac{64 - 32}{2} = \dfrac{32}{2} = 16;$
$c_4 = \dfrac{c_{4-2} - c_{4-1}}{2} = \dfrac{c_2 - c_3}{2}$
$= \dfrac{32 - 16}{2} = \dfrac{16}{2} = 8;$
$c_5 = \dfrac{c_{5-2} - c_{5-1}}{2} = \dfrac{c_3 - c_4}{2}$
$= \dfrac{16 - 8}{2} = \dfrac{8}{2} = 4;$
$64, 32, 16, 8, 4$

39. $\dfrac{8!}{5!} = \dfrac{8 \cdot 7 \cdot 6 \cdot 5!}{5!} = 8 \cdot 7 \cdot 6 = 336$

41. $\dfrac{9!}{7! \, 2!} = \dfrac{9 \cdot 8 \cdot 7!}{7! \, 2!} = \dfrac{9 \cdot 8}{2 \cdot 1} = \dfrac{72}{2} = 36$

43. $\dfrac{8!}{2! \, 6!} = \dfrac{8 \cdot 7 \cdot 6!}{2! \, 6!} = \dfrac{8 \cdot 7}{2 \cdot 1} = \dfrac{56}{2} = 28$

45. $a_n = \dfrac{n!}{(n+1)!}$

$a_1 = \dfrac{1!}{(1+1)!} = \dfrac{1!}{2!} = \dfrac{1!}{2 \cdot 1!} = \dfrac{1}{2}$
$a_2 = \dfrac{2!}{(2+1)!} = \dfrac{2!}{3!} = \dfrac{2!}{3 \cdot 2!} = \dfrac{1}{3}$
$a_3 = \dfrac{3!}{(3+1)!} = \dfrac{3!}{4!} = \dfrac{3!}{4 \cdot 3!} = \dfrac{1}{4}$
$a_4 = \dfrac{4!}{(4+1)!} = \dfrac{4!}{5!} = \dfrac{4!}{5 \cdot 4!} = \dfrac{1}{5}$
$\dfrac{1}{2}, \dfrac{1}{3}, \dfrac{1}{4}, \dfrac{1}{5}$

47. $a_n = \dfrac{(n+1)!}{(3n)!}$

$a_1 = \dfrac{(1+1)!}{(3 \cdot 1)!} = \dfrac{2!}{3!} = \dfrac{2!}{3 \cdot 2!} = \dfrac{1}{3}$
$a_2 = \dfrac{(2+1)!}{(3 \cdot 2)!} = \dfrac{3!}{6!} = \dfrac{3!}{6 \cdot 5 \cdot 4 \cdot 3!} = \dfrac{1}{120}$
$a_3 = \dfrac{(3+1)!}{(3 \cdot 3)!} = \dfrac{4!}{9!} = \dfrac{4!}{9 \cdot 8 \cdot 7 \cdot 6 \cdot 5 \cdot 4!} = \dfrac{1}{15120}$
$a_4 = \dfrac{(4+1)!}{(3 \cdot 4)!} = \dfrac{5!}{12!}$
$= \dfrac{5!}{12 \cdot 11 \cdot 10 \cdot 9 \cdot 8 \cdot 7 \cdot 6 \cdot 5!} = \dfrac{1}{3991680}$
$\dfrac{1}{3}, \dfrac{1}{120}, \dfrac{1}{15120}, \dfrac{1}{3991680}$

49. $a_n = \dfrac{n^n}{n!}$

$a_1 = \dfrac{1^1}{1!} = \dfrac{1}{1} = 1$
$a_2 = \dfrac{2^2}{2!} = \dfrac{4}{2!} = \dfrac{4}{2 \cdot 1} = \dfrac{4}{2} = 2$
$a_3 = \dfrac{3^3}{3!} = \dfrac{27}{3 \cdot 2 \cdot 1} = \dfrac{27}{6} = \dfrac{9}{2}$
$a_4 = \dfrac{4^4}{4!} = \dfrac{256}{4 \cdot 3 \cdot 2 \cdot 1} = \dfrac{256}{24} = \dfrac{32}{3}$
$1, 2, \dfrac{9}{2}, \dfrac{32}{3}$

51. $a_n = n$

$S_5 = a_1 + a_2 + a_3 + a_4 + a_5$

$S_5 = 1 + 2 + 3 + 4 + 5$

$S_5 = 15$

53. $a_n = 2n - 1$

$S_8 = a_1 + a_2 + a_3 + a_4 + a_5 + a_6 + a_7 + a_8$

$S_8 = 1 + 3 + 5 + 7 + 9 + 11 + 13 + 15$

$S_8 = 64$

55. $a_n = \dfrac{1}{n}$

$S_5 = a_1 + a_2 + a_3 + a_4 + a_5$

$S_5 = 1 + \dfrac{1}{2} + \dfrac{1}{3} + \dfrac{1}{4} + \dfrac{1}{5}$

$S_5 = \dfrac{137}{60}$

57. $\displaystyle\sum_{i=1}^{4} (3i - 5)$

$= (3(1) - 5) + (3(2) - 5) + (3(3) - 5) + (3(4) - 5)$

$= -2 + 1 + 4 + 7 = 10$

59. $\displaystyle\sum_{k=1}^{5} \left(2k^2 - 3\right)$

$= \left(2(1)^2 - 3\right) + \left(2(2)^2 - 3\right) + \left(2(3)^2 - 3\right) +$

$\left(2(4)^2 - 3\right) + \left(2(5)^2 - 3\right)$

$= (2 - 3) + (8 - 3) + (18 - 3) + (32 - 3) + (50 - 3)$

$= -1 + 5 + 15 + 29 + 47 = 95$

61. $\displaystyle\sum_{k=1}^{7} (-1)^k k$

$= (-1)^1(1) + (-1)^2(2) + (-1)^3(3) + (-1)^4(4) +$

$(-1)^5(5) + (-1)^6(6) + (-1)^7(7)$

$= -1 + 2 - 3 + 4 - 5 + 6 - 7 = -4$

63. $\displaystyle\sum_{i=1}^{4} \dfrac{i^2}{2}$

$= \dfrac{1^2}{2} + \dfrac{2^2}{2} + \dfrac{3^2}{2} + \dfrac{4^2}{2}$

$= \dfrac{1}{2} + \dfrac{4}{2} + \dfrac{9}{2} + \dfrac{16}{2} = 15$

65. $\displaystyle\sum_{j=3}^{7} 2j$

$= 2(3) + 2(4) + 2(5) + 2(6) + 2(7)$

$= 6 + 8 + 10 + 12 + 14 = 50$

67. $\displaystyle\sum_{k=3}^{8} \dfrac{(-1)^k}{k(k-2)}$

$= \dfrac{(-1)^3}{3(3-2)} + \dfrac{(-1)^4}{4(4-2)} + \dfrac{(-1)^5}{5(5-2)} + \dfrac{(-1)^6}{6(6-2)} +$

$\dfrac{(-1)^7}{7(7-2)} + \dfrac{(-1)^8}{8(8-2)}$

$= \dfrac{-1}{3} + \dfrac{1}{8} - \dfrac{1}{15} + \dfrac{1}{24} - \dfrac{1}{35} + \dfrac{1}{48}$

$= \dfrac{-27}{112}$

69. $4 + 8 + 12 + 16 + 20$

$\displaystyle\sum_{n=1}^{5} (4n)$

71. $-1 + 4 - 9 + 16 - 25 + 36$

$\displaystyle\sum_{n=1}^{6} (-1)^n n^2$

73. $a^n = n + 3; \ S_5$

$\displaystyle\sum_{n=1}^{5} (n + 3)$

75. $a^n = \dfrac{n^2}{3}; \ $ third partial sum

$\displaystyle\sum_{n=1}^{3} \dfrac{n^2}{3}$

77. $a^n = \dfrac{n}{2^n}; \ $ sum for $n = 3$ to 7

$\displaystyle\sum_{n=3}^{7} \dfrac{n}{2^n}$

79. $\displaystyle\sum_{i=1}^{5}(4i-5)=\sum_{i=1}^{5}4i+\sum_{i=1}^{5}(-5)$

$\displaystyle=4\sum_{i=1}^{5}i+\sum_{i=1}^{5}(-5)$

$=4(15)+(-5)(5)=35$

81. $\displaystyle\sum_{k=1}^{4}(3k^2+k)=\sum_{k=1}^{4}3k^2+\sum_{k=1}^{4}k$

$\displaystyle=3\sum_{k=1}^{4}k^2+\sum_{k=1}^{4}k$

$=3(30)+10=100$

83. $a_n=3n-2:S_n=\dfrac{n(3n-1)}{2}$

$a_n=3n-2=1,4,7,10...,(3n-2),...$

$S_5=\dfrac{5(3(5)-1)}{2}=\dfrac{5(14)}{2}=\dfrac{70}{2}=35$;

$a_1=3(1)-2=1$;

$a_2=3(2)-2=4$;

$a_3=3(3)-2=7$;

$a_4=3(4)-2=10$;

$a_5=3(5)-2=13$;

$1+4+7+10+13=35$

85. $a_n=(0.8)^{n-1}(6000)$

$a_1=(0.8)^{1-1}(6000)=6000$;

$a_2=(0.8)^{2-1}(6000)=0.8(6000)=4800$;

$a_3=(0.8)^{3-1}(6000)=(0.8)^2(6000)=3840$;

$a_4=(0.8)^{4-1}(6000)=(0.8)^3(6000)=3072$;

$a_5=(0.8)^{5-1}(6000)=(0.8)^4(6000)=2457.60$;

$a_6=(0.8)^{6-1}(6000)=(0.8)^5(6000)=1966.08$

6000; 4800; 3840; 3072; 2457.60;
1966.08

87. $5.20, 5.70, 6.20, 6.70, 7.20$
$8(7.20)(240)=\$13,824$

89. $b_0=1500$; $b_n=1.05b_{n-1}+100$

$b_1=1.05b_{1-1}+100=1.05(1500)+100=1675$;

$b_2=1.05b_{2-1}+100=1.05b_1+100$
$=1.05(1675)+100=1858.75$;

$b_3=1.05b_{3-1}+100=1.05b_2+100$
$=1.05(1858.75)+100=2051.69$;

$b_4=1.05b_{4-1}+100=1.05b_3+100$
$=1.05(2051.69)+100=2254.27$;

$b_5=1.05b_{5-1}+100=1.05b_4+100$
$=1.05(2254.27)+100=2466.98$;

$b_6=1.05b_{6-1}+100=1.05b_6+100$
$=1.05(2466.98)+100=2690.33$

Approximately 2690

91. $\displaystyle\sum_{i=1}^{n}(a_i\pm b_i)=\sum_{i=1}^{n}a_i\pm\sum_{i=1}^{n}b_i$

$\displaystyle\sum_{i=1}^{n}(a_i\pm b_i)$

$=[a_1\pm b_1]+[a_2\pm b_2]+...+[a_n\pm b_n]$

$=[a_1+a_2+...+a_n]\pm[b_1+b_2+...+b_n]$

$\displaystyle=\sum_{i=1}^{n}a_i\pm\sum_{i=1}^{n}b_i$

Verified

93. $\displaystyle\sum_{k=1}^{n}\frac{1}{3^{k}}$

$\displaystyle\sum_{k=1}^{4}\frac{1}{3^{k}}$

$S_4 = a_1 + a_2 + a_3 + a_4$

$= \dfrac{1}{3} + \dfrac{1}{9} + \dfrac{1}{27} + \dfrac{1}{81} = \dfrac{40}{81}$

$\displaystyle\sum_{k=1}^{8}\frac{1}{3^{k}}$

$S_8 = S_4 + a_5 + a_6 + a_7 + a_8$

$= \dfrac{40}{81} + \dfrac{1}{243} + \dfrac{1}{729} + \dfrac{1}{2187} + \dfrac{1}{6561} \approx 0.5\ ;$

$\displaystyle\sum_{k=1}^{12}\frac{1}{3^{k}}$

$S_{12} = S_8 + a_9 + a_{10} + a_{11} + a_{12}$

$= 0.5 + \dfrac{1}{19,683} + \dfrac{1}{59,049} + \dfrac{1}{177,147} + \dfrac{1}{531,441}$

$\approx \dfrac{1}{2}$

Approaches $\dfrac{1}{2}$

95. $\csc x \sin\left(\dfrac{\pi}{2} - x\right) = -1$

$\csc x (\cos x) = -1$

$\dfrac{1}{\sin x} \cdot \dfrac{\cos x}{1} = -1$

$\dfrac{\cos x}{\sin x} = -1$

$\cot x = -1$

$\theta_R = \dfrac{\pi}{4}$

$x = \left\{ \dfrac{3\pi}{4}, \dfrac{7\pi}{4} \right\}$

97. $a = 0.4\text{m} \quad b = 0.5\text{m} \quad c = 0.3\text{m}$

$(0.5)^2 = (0.4)^2 + (0.3)^2 - 2(0.4)(0.3)\cos B$

$\dfrac{0.5^2 - 0.4^2 - 0.3^2}{-2(0.4)(0.3)} = \cos B$

$0 = \cos B$

$\angle B = 90°$

$\dfrac{\sin A}{0.4} = \dfrac{\sin 90°}{0.5}$

$\sin A = \dfrac{0.4 \sin 90°}{0.5}$

$\sin A = 0.8$

$\angle A = \sin^{-1}(0.8) \approx 53.1°$

$\angle C = 180° - 90° - 53.1° = 36.9°$

$\angle A \approx 53.1° \quad \angle B = 90° \quad \angle C \approx 36.9°$

10.2 Exercises

1. Common difference.

3. $\dfrac{n(a_1 + a_n)}{2}$; n^{th}

5. Answers will vary.

7. $-5, -2, 1, 4, 7, 10, \ldots$
   $-2 - (-5) = -2 + 5 = 3$;
   $1 - (-2) = 1 + 2 = 3$;
   $4 - 1 = 3$;
   $7 - 4 = 3$;
   $10 - 7 = 3$;
   Arithmetic; $d = 3$

9. $-0.5, 3, 5.5, 8, 10.5, \ldots$
   $3 - (-0.5) = 3 + 0.5 = 3.5$;
   $5.5 - 3 = 2.5$;
   $8 - 5.5 = 2.5$;
   $10.5 - 8 = 2.5$;
   Arithmetic; $d = 2.5$

11. $2, 3, 5, 7, 11, 13, 17, \ldots$
    $3 - 2 = 1$;
    $5 - 3 = 2$;
    Not arithmetic; all prime.

13. $\dfrac{1}{24}, \dfrac{1}{12}, \dfrac{1}{8}, \dfrac{1}{6}, \dfrac{5}{24}, \ldots$
    $\dfrac{1}{12} - \dfrac{1}{24} = \dfrac{2}{24} - \dfrac{1}{24} = \dfrac{1}{24}$;
    $\dfrac{1}{8} - \dfrac{1}{12} = \dfrac{3}{24} - \dfrac{2}{24} = \dfrac{1}{24}$;
    $\dfrac{1}{6} - \dfrac{1}{8} = \dfrac{4}{24} - \dfrac{3}{24} = \dfrac{1}{24}$;
    $\dfrac{5}{24} - \dfrac{1}{6} = \dfrac{5}{24} - \dfrac{4}{24} = \dfrac{1}{24}$;
    Arithmetic; $d = \dfrac{1}{24}$

15. $1, 2, 4, 9, 16, 25, 36, \ldots$
    $2 - 1 = 1$;
    $4 - 2 = 2$;
    Not arithmetic; $a_n = n^2$

17. $\pi, \dfrac{5\pi}{6}, \dfrac{2\pi}{3}, \dfrac{\pi}{2}, \dfrac{\pi}{3}, \dfrac{\pi}{6}, \ldots$
    $\dfrac{5\pi}{6} - \pi = \dfrac{5\pi}{6} - \dfrac{6\pi}{6} = \dfrac{-\pi}{6}$;
    $\dfrac{2\pi}{3} - \dfrac{5\pi}{6} = \dfrac{4\pi}{6} - \dfrac{5\pi}{6} = \dfrac{-\pi}{6}$;
    $\dfrac{\pi}{2} - \dfrac{2\pi}{3} = \dfrac{3\pi}{6} - \dfrac{4\pi}{6} = \dfrac{-\pi}{6}$;
    $\dfrac{\pi}{3} - \dfrac{\pi}{2} = \dfrac{2\pi}{6} - \dfrac{3\pi}{6} = \dfrac{-\pi}{6}$;
    $\dfrac{\pi}{6} - \dfrac{\pi}{3} = \dfrac{\pi}{6} - \dfrac{2\pi}{6} = \dfrac{-\pi}{6}$;
    Arithmetic; $d = \dfrac{-\pi}{6}$

19. $a_1 = 2, d = 3$
    $2 + 3 = 5$;
    $5 + 3 = 8$;
    $8 + 3 = 11$;
    2, 5, 8, 11

21. $a_1 = 7, d = -2$
    $7 - 2 = 5$;
    $5 - 2 = 3$;
    $3 - 2 = 1$;
    7, 5, 3, 1

23. $a_1 = 0.3, d = 0.03$
    $0.3 + 0.03 = 0.33$;
    $0.33 + 0.03 = 0.36$;
    $0.36 + 0.03 = 0.39$;
    0.3, 0.33, 0.36, 0.39

25. $a_1 = \dfrac{3}{2}, d = \dfrac{1}{2}$
    $\dfrac{3}{2} + \dfrac{1}{2} = \dfrac{4}{2} = 2$;
    $2 + \dfrac{1}{2} = \dfrac{4}{2} + \dfrac{1}{2} = \dfrac{5}{2}$;
    $\dfrac{5}{2} + \dfrac{1}{2} = \dfrac{6}{2} = 3$;
    $\dfrac{3}{2}, 2, \dfrac{5}{2}, 3$

27. $a_1 = \dfrac{3}{4}, d = -\dfrac{1}{8}$

$\dfrac{3}{4} - \dfrac{1}{8} = \dfrac{6}{8} - \dfrac{1}{8} = \dfrac{5}{8}$;

$\dfrac{5}{8} - \dfrac{1}{8} = \dfrac{4}{8} = \dfrac{1}{2}$;

$\dfrac{1}{2} - \dfrac{1}{8} = \dfrac{4}{8} - \dfrac{1}{8} = \dfrac{3}{8}$;

$\dfrac{3}{4}, \dfrac{5}{8}, \dfrac{1}{2}, \dfrac{3}{8}$

29. $a_1 = -2, d = -3$

$-2 - 3 = -5$;

$-5 - 3 = -8$;

$-8 - 3 = -11$;

$-2, -5, -8, -11$

31. $2, 7, 12, 17, \ldots$

$a_1 = 2, d = 5$

$a_n = a_1 + (n-1)d$

$a_n = 2 + (n-1)5$

$a_n = 2 + 5n - 5$

$a_n = 5n - 3$;

$a_6 = 5(6) - 3 = 30 - 3 = 27$;

$a_{10} = 5(10) - 3 = 50 - 3 = 47$;

$a_{12} = 5(12) - 3 = 60 - 3 = 57$

33. $\$5.10, \$5.25, \$5.40, \ldots$

$a_1 = 5.10, d = 0.15$

$a_n = a_1 + (n-1)d$

$a_n = 5.10 + (n-1)(0.15)$

$a_n = 5 + 0.15n - 0.15$

$a_n = 0.15n + 4.95$;

$a_6 = 0.15(6) + 4.95 = 0.90 + 4.95 = 5.85$;

$a_{10} = 0.15(10) + 4.95 = 1.50 + 4.95 = 6.45$

$a_{12} = 0.15(12) + 4.95 = 1.80 + 4.95 = 6.75$

$\$5.85, \$6.45, \$6.75$

35. $\dfrac{3}{2}, \dfrac{9}{4}, 3, \dfrac{15}{4}, \ldots$

$a_1 = \dfrac{3}{2}, d = \dfrac{3}{4}$

$a_n = a_1 + (n-1)d$

$a_n = \dfrac{3}{2} + (n-1)\left(\dfrac{3}{4}\right)$

$a_n = \dfrac{3}{2} + \dfrac{3}{4}n - \dfrac{3}{4}$

$a_n = \dfrac{3}{4}n + \dfrac{3}{4}$;

$a_6 = \dfrac{3}{4}(6) + \dfrac{3}{4} = \dfrac{9}{2} + \dfrac{3}{4} = \dfrac{18}{4} + \dfrac{3}{4} = \dfrac{21}{4}$;

$a_{10} = \dfrac{3}{4}(10) + \dfrac{3}{4} = \dfrac{15}{2} + \dfrac{3}{4} = \dfrac{30}{4} + \dfrac{3}{4} = \dfrac{33}{4}$;

$a_{12} = \dfrac{3}{4}(12) + \dfrac{3}{4} = 9 + \dfrac{3}{4} = \dfrac{36}{4} + \dfrac{3}{4} = \dfrac{39}{4}$

37. $a_1 = 5, d = 4$; Find a_{15}

$a_n = a_1 + (n-1)d$

$a_n = 5 + (n-1)4$

$a_n = 5 + 4n - 4$

$a_n = 4n + 1$;

$a_{15} = 4(15) + 1 = 60 + 1 = 61$

39. $a_1 = \dfrac{3}{2}, d = -\dfrac{1}{12}$; Find a_7

$a_n = a_1 + (n-1)d$

$a_n = \dfrac{3}{2} + (n-1)\left(-\dfrac{1}{12}\right)$

$a_n = \dfrac{3}{2} - \dfrac{1}{12}n + \dfrac{1}{12}$

$a_n = \dfrac{18}{12} - \dfrac{1}{12}n + \dfrac{1}{12}$

$a_n = -\dfrac{1}{12}n + \dfrac{19}{12}$;

$a_7 = -\dfrac{1}{12}(7) + \dfrac{19}{12} = \dfrac{-7}{12} + \dfrac{19}{12} = \dfrac{12}{12} = 1$

41. $a_1 = -0.025, d = 0.05$; Find a_{50}

$a_n = a_1 + (n-1)d$

$a_n = -0.025 + (n-1)(0.05)$

$a_n = -0.025 + 0.05n - 0.05$

$a_n = 0.05n - 0.075$;

$a_{50} = 0.05(50) - 0.075 = 2.5 - 0.075 = 2.425$

43. $a_1 = 2, a_n = -22, d = -3$

$a_n = a_1 + (n-1)d$

$-22 = 2 + (n-1)(-3)$

$-22 = 2 - 3n + 3$

$-22 = -3n + 5$

$-27 = -3n$

$9 = n$

45. $a_1 = 0.4, a_n = 10.9, d = 0.25$

$a_n = a_1 + (n-1)d$

$10.9 = 0.4 + (n-1)(0.25)$

$10.9 = 0.4 + 0.25n - 0.25$

$10.9 = 0.15 + 0.25n$

$10.75 = 0.25n$

$43 = n$

47. $-3, -0.5, 2, 4.5, 7, ..., 47$

$a_1 = -3, a_n = 47, d = 2.5$

$a_n = a_1 + (n-1)d$

$47 = -3 + (n-1)(2.5)$

$47 = -3 + 2.5n - 2.5$

$47 = -5.5 + 2.5n$

$52.5 = 2.5n$

$21 = n$

49. $\dfrac{1}{12}, \dfrac{1}{8}, \dfrac{1}{6}, \dfrac{5}{24}, \dfrac{1}{4}, ..., \dfrac{9}{8}$

$a_1 = \dfrac{1}{12}, a_n = \dfrac{9}{8}, d = \dfrac{1}{24}$

$a_n = a_1 + (n-1)d$

$\dfrac{9}{8} = \dfrac{1}{12} + (n-1)\left(\dfrac{1}{24}\right)$

$\dfrac{9}{8} = \dfrac{1}{12} + \dfrac{1}{24}n - \dfrac{1}{24}$

$\dfrac{9}{8} = \dfrac{1}{24} + \dfrac{1}{24}n$

$\dfrac{13}{12} = \dfrac{1}{24}n$

$26 = n$

51. $a_3 = 7, a_7 = 19$

$a_7 = a_3 + 4d$

$19 = 7 + 4d$

$12 = 4d$

$3 = d;$

$a_7 = a_1 + 6d$

$19 = a_1 + 6(3)$

$19 = a_1 + 18$

$1 = a_1;$

$d = 3, a_1 = 1$

53. $a_2 = 1.025, a_{26} = 10.025$

$a_{26} = a_2 + 24d$

$10.025 = 1.025 + 24d$

$9 = 24d$

$0.375 = d;$

$a_{26} = a_1 + 25d$

$10.025 = a_1 + 25(0.375)$

$10.025 = a_1 + 9.375$

$0.65 = a_1;$

$d = 0.375, a_1 = 0.65$

55. $a_{10} = \dfrac{13}{18}, a_{24} = \dfrac{27}{2}$

$a_{24} = a_{10} + 14d$

$\dfrac{27}{2} = \dfrac{13}{18} + 14d$

$\dfrac{115}{9} = 14d$

$\dfrac{115}{126} = d;$

$a_{24} = a_1 + 23d$

$\dfrac{27}{2} = a_1 + 23\left(\dfrac{115}{126}\right)$

$\dfrac{27}{2} = a_1 + \dfrac{2645}{126}$

$\dfrac{-472}{63} = a_1;$

$d = \dfrac{115}{126}, a_1 = \dfrac{-472}{63}$

57. $\displaystyle\sum_{n=1}^{30}(3n-4)$

    Initial terms: 1, 2, 5, ...
    $a_1 = -1; d = 3; n = 30$
    $a_{30} = a_1 + 29d$
    $a_{30} = -1 + 29(3) = -1 + 87 = 86$;

    $S_{30} = n\left(\dfrac{a_1 + a_n}{2}\right)$

    $S_{30} = 30\left(\dfrac{-1 + 86}{2}\right)$

    $S_{30} = 30\left(\dfrac{85}{2}\right)$

    $S_{30} = 1275$

59. $\displaystyle\sum_{n=1}^{37}\left(\dfrac{3}{4}n + 2\right)$

    Initial terms: $\dfrac{11}{4}, \dfrac{7}{2}, \dfrac{17}{4}, ...$

    $a_1 = \dfrac{11}{4}; d = \dfrac{3}{4}; n = 37$

    $a_{37} = a_1 + 36d$

    $a_{37} = \dfrac{11}{4} + 36\left(\dfrac{3}{4}\right) = \dfrac{11}{4} + 27 = \dfrac{119}{4}$;

    $S_{37} = n\left(\dfrac{a_1 + a_{37}}{2}\right)$

    $S_{37} = 37\left(\dfrac{\dfrac{11}{4} + \dfrac{119}{4}}{2}\right)$

    $S_{37} = 37\left(\dfrac{\dfrac{65}{2}}{2}\right)$

    $S_{37} = 601.25$

61. $\displaystyle\sum_{n=4}^{15}(3 - 5n)$

    Initial terms: ...,-17, -22, -27, ...
    $a_4 = -17; d = -5; n = 12$
    $a_{15} = a_4 + 11d$
    $a_{15} = -17 + 11(-5) = -17 - 55 = -72$;

    $S_{12} = n\left(\dfrac{a_4 + a_{15}}{2}\right)$

$S_{12} = 12\left(\dfrac{-17 + (-72)}{2}\right)$

$S_{12} = 12\left(\dfrac{-89}{2}\right)$

$S_{12} = -534$

−534

63. $-12 + (-9.5) + (-7) + (-4.5) + ...$

    Find S_{15} ; $a_1 = -12$; $d = 2.5$; $n = 15$

    $S_n = \dfrac{n}{2}[2a_1 + (n-1)d]$

    $S_{15} = \dfrac{15}{2}[2(-12) + (15-1)(2.5)]$

    $S_{15} = \dfrac{15}{2}[-24 + 14(2.5)]$

    $S_{15} = \dfrac{15}{2}[-24 + 35]$

    $S_{15} = \dfrac{15}{2}[11]$

    $S_{15} = 82.5$

65. $0.003 + 0.173 + 0.343 + 0.513 + ...$

    Find S_{30} ; $a_1 = 0.003$; $d = 0.17$; $n = 30$

    $S_n = \dfrac{n}{2}[2a_1 + (n-1)d]$

    $S_{30} = \dfrac{30}{2}[2(0.003) + (30-1)(0.17)]$
    $S_{30} = 15[0.006 + (29)(0.17)]$
    $S_{30} = 15[0.006 + 4.93]$
    $S_{30} = 15[4.936]$
    $S_{30} = 74.04$

67. $\sqrt{2} + 2\sqrt{2} + 3\sqrt{2} + 4\sqrt{2} + ...$

    Find S_{20} ; $a_1 = \sqrt{2}$; $d = \sqrt{2}$; $n = 20$

    $S_n = \dfrac{n}{2}[2a_1 + (n-1)d]$

    $S_{20} = \dfrac{20}{2}[2\sqrt{2} + (20-1)\sqrt{2}]$

    $S_{20} = 10(2\sqrt{2} + 19\sqrt{2})$

    $S_{20} = 10(21\sqrt{2})$

    $S_{20} = 210\sqrt{2}$

69. $S_n = \dfrac{n(n+1)}{2}$

$1 + 2 + 3 + 4 + 5 + 6 = 21$;

$S_6 = \dfrac{6(6+1)}{2} = \dfrac{6(7)}{2} = \dfrac{42}{2} = 21$;

$S_{75} = \dfrac{75(75+1)}{2} = \dfrac{75(76)}{2} = \dfrac{5700}{2} = 2850$

71. $a_1 = 33$; $d = -3$; $a_n = 0$

$a_n = a_1 + (n-1)d$

$0 = 33 + (n-1)(-3)$

$0 = 33 - 3n + 3$

$0 = 36 - 3n$

$-36 = -3n$

$12 = n$

12 half-hours after 5 P.M. is 11 P.M.

73. (a) $a_1 = 10$; $d = \dfrac{-3}{4}$; $n = 7$

$a_n = a_1 + (n-1)d$

$a_7 = 10 + (7-1)\left(\dfrac{-3}{4}\right)$

$a_7 = 10 + 6\left(\dfrac{-3}{4}\right)$

$a_7 = 10 - \dfrac{9}{2}$

$a_7 = 5.5$;

5.5 inches

(b) $S_n = \dfrac{n(a_1 + a_n)}{2}$

$S_7 = \dfrac{7(10 + 5.5)}{2} = \dfrac{7(15.5)}{2} = 54.25$;

54.25 inches

75. $a_1 = 100$; $d = 20$; $a_n = 2500$

$a_n = a_1 + (n-1)d$

$a_7 = 100 + (7-1)(20) = 100 + 6(20) = 220$;

$220;

$a_{12} = 100 + (12-1)(20) = 100 + 11(20) = 320$;

$S_n = \dfrac{n(a_1 + a_n)}{2}$

$S_{12} = \dfrac{12(100 + 320)}{2} = \dfrac{12(420)}{2} = 2520$

$2520; yes

77. (a) $19, 11.8, 4.6, -2.6, -9.8, -17, -24.2$

1st difference:

$11.8 - 19 = -7.2$

$4.6 - 11.8 = -7.2$

$-2.6 - 4.6 = -7.2$

$-9.8 - (-2.6) = -7.2$

$-17 - (-9.8) = -7.2$

$-24.2 - (-17) = -7.2$

Linear function.

(b) $-10.31, -10.94, -11.99, -13.46, -15.35...$

1st difference:

$-10.94 + 10.31 = -0.63$

$-11.99 + 10.94 = -1.05$

$-13.46 + 11.99 = -1.47$

$-15.35 + 13.46 = -1.89$;

-0.63, -1.05, -1.47, -1.89

2nd difference:

$-1.05 + 0.63 = -0.42$

$-1.47 + 1.05 = -0.42$

$-1.89 + 1.47 = -0.42$

Quadratic.

79. $f(t) = 7\sin\left(\dfrac{\pi}{3}t - \dfrac{\pi}{6}\right) + 10$

$= 7\sin\left(\dfrac{\pi}{3}\left(t - \dfrac{1}{2}\right)\right) + 10$

$A: |7| = 7$

$P: \dfrac{2\pi}{\left(\dfrac{\pi}{3}\right)} = 2\pi \cdot \dfrac{3}{\pi} = 6$

HS: Right $\dfrac{1}{2}$ unit

VS: up 10 units

$PI: \dfrac{1}{2} \le t \le 6 + \dfrac{1}{2}$

$\dfrac{1}{2} \le t \le \dfrac{13}{2}$

81. $(0, 972), (5, 1217)$

$m = \dfrac{1217 - 972}{5 - 0} = \dfrac{245}{5} = 49$

$y - y_1 = m(x - x_1)$

$y - 972 = 49(x - 0)$

$y - 972 = 49x$

$y = 49x + 972$;

$f(x) = 49x + 972$;

$f(8) = 49(8) + 972 = 392 + 972 = 1364$

Chapter 10: Additional Topics in Algebra

10.3 Exercises

1. Multiplying.

3. $a_1 r^{n-1}$

5. Answers will vary.

7. $4, 8, 16, 32, \ldots$

$\dfrac{8}{4} = 2; \ \dfrac{16}{8} = 2; \ \dfrac{32}{16} = 2$

$r = 2$

9. $3, -6, 12, -24, 48, \ldots$

$\dfrac{-6}{3} = -2; \ \dfrac{12}{-6} = -2; \ \dfrac{-24}{12} = -2; \ \dfrac{48}{-24} = -2$

$r = -2$

11. $2, 5, 10, 17, 26, \ldots$

$\dfrac{5}{2}; \ \dfrac{10}{5} = 2; \ \dfrac{17}{10}; \ \dfrac{26}{17}; \ $ not geometric

$a_n = n^2 + 1$

13. $3, 0.3, 0.03, 0.003, \ldots$

$\dfrac{0.3}{3} = 0.1; \ \dfrac{0.03}{0.3} = 0.1; \ \dfrac{0.003}{0.03} = 0.1$

$r = 0.1$

15. $-1, 3, -12, 60, -360, \ldots$

$\dfrac{3}{-1} = -3; \ \dfrac{-12}{3} = -4; \ \dfrac{60}{-12} = -5; \ \dfrac{-360}{60} = -6$

Not geometric; ratio of terms decreases by 1.

17. $25, 10, 4, \dfrac{8}{5}, \ldots$

$\dfrac{10}{25} = \dfrac{2}{5}; \ \dfrac{4}{10} = \dfrac{2}{5}; \ \dfrac{\frac{8}{5}}{4} = \dfrac{8}{20} = \dfrac{2}{5}$

$r = \dfrac{2}{5}$

19. $\dfrac{1}{2}, \dfrac{1}{4}, \dfrac{1}{8}, \dfrac{1}{16}, \ldots$

$\dfrac{\frac{1}{4}}{\frac{1}{2}} = \dfrac{1}{2}; \ \dfrac{\frac{1}{8}}{\frac{1}{4}} = \dfrac{1}{2}; \ \dfrac{\frac{1}{16}}{\frac{1}{8}} = \dfrac{1}{2}$

$r = \dfrac{1}{2}$

21. $3, \dfrac{12}{x}, \dfrac{48}{x^2}, \dfrac{192}{x^3}, \ldots$

$\dfrac{\frac{12}{x}}{3} = \dfrac{12}{3x} = \dfrac{4}{x}; \ \dfrac{\frac{48}{x^2}}{\frac{12}{x}} = \dfrac{48x}{12x^2} = \dfrac{4}{x};$

$\dfrac{\frac{192}{x^3}}{\frac{48}{x^2}} = \dfrac{192x^2}{48x^3} = \dfrac{4}{x}$

$r = \dfrac{4}{x}$

23. $240, 120, 40, 10, 2, \ldots$

$\dfrac{120}{240} = \dfrac{1}{2}; \ \dfrac{40}{120} = \dfrac{1}{3}; \ \dfrac{10}{40} = \dfrac{1}{4}; \ \dfrac{2}{10} = \dfrac{1}{5}$

Not geometric $a_n = \dfrac{240}{n!}$

25. $a_1 = 5, r = 2$

$a_2 = 5 \cdot 2 = 10;$

$a_3 = 10 \cdot 2 = 20;$

$a_4 = 20 \cdot 2 = 40;$

$5, 10, 20, 40$

27. $a_1 = -6, r = -\dfrac{1}{2}$

$a_2 = -6\left(\dfrac{-1}{2}\right) = 3;$

$a_3 = 3\left(\dfrac{-1}{2}\right) = \dfrac{-3}{2};$

$a_4 = \dfrac{-3}{2}\left(\dfrac{-1}{2}\right) = \dfrac{3}{4};$

$-6, 3, \dfrac{-3}{2}, \dfrac{3}{4}$

29. $a_1 = 4, r = \sqrt{3}$

$a_2 = 4\sqrt{3};$

$a_3 = 4\sqrt{3}\left(\sqrt{3}\right) = 12;$

$a_4 = 12\sqrt{3};$

$4, 4\sqrt{3}, 12, 12\sqrt{3}$

31. $a_1 = 0.1, r = 0.1$

$a_2 = 0.1(0.1) = 0.01$;

$a_3 = 0.01(0.1) = 0.001$;

$a_4 = 0.001(0.1) = 0.0001$;

$0.1, 0.01, 0.001, 0.0001$

33. $a_1 = -24, \ r = \dfrac{1}{2}$; find a_7

$a_n = a_1 r^{n-1}$

$a_7 = -24\left(\dfrac{1}{2}\right)^{7-1} = -24\left(\dfrac{1}{2}\right)^6$

$= -24\left(\dfrac{1}{64}\right) = -\dfrac{3}{8}$

35. $a_1 = -\dfrac{1}{20}, \ r = -5$; find a_4

$a_n = a_1 r^{n-1}$

$a_4 = -\dfrac{1}{20}(-5)^{4-1} = -\dfrac{1}{20}(-5)^3$

$= -\dfrac{1}{20}(-125) = \dfrac{25}{4}$

37 $a_1 = 2, r = \sqrt{2}$; find a_7

$a_n = a_1 r^{n-1}$

$a_7 = 2\left(\sqrt{2}\right)^{7-1} = 2\left(\sqrt{2}\right)^6 = 2(8) = 16$

39. $\dfrac{1}{27}, -\dfrac{1}{9}, \dfrac{1}{3}, -1, 3, \ldots$

$r = \dfrac{-\dfrac{1}{9}}{\dfrac{1}{27}} = \dfrac{-27}{9} = -3$

$a_1 = \dfrac{1}{27}; \ r = -3$

$a_n = \dfrac{1}{27}(-3)^{n-1}$

$a_6 = \dfrac{1}{27}(-3)^{6-1} = \dfrac{1}{27}(-3)^5 = \dfrac{1}{27}(-243) = -9$;

$a_{10} = \dfrac{1}{27}(-3)^{10-1} = \dfrac{1}{27}(-3)^9$

$= \dfrac{1}{27}(-19683) = -729$;

$a_{12} = \dfrac{1}{27}(-3)^{12-1} = \dfrac{1}{27}(-3)^{11}$

$= \dfrac{1}{27}(-177147) = -6561$

41. $729, 243, 81, 27, 9, \ldots$

$r = \dfrac{243}{729} = \dfrac{1}{3}$

$a_1 = 729; \ r = \dfrac{1}{3}$

$a_n = 729\left(\dfrac{1}{3}\right)^{n-1}$

$a_6 = 729\left(\dfrac{1}{3}\right)^{6-1} = 729\left(\dfrac{1}{3}\right)^5 = 729\left(\dfrac{1}{243}\right) = 3$;

$a_{10} = 729\left(\dfrac{1}{3}\right)^{10-1} = 729\left(\dfrac{1}{3}\right)^9$

$= 729\left(\dfrac{1}{19683}\right) = \dfrac{1}{27}$;

$a_{12} = 729\left(\dfrac{1}{3}\right)^{12-1} = 729\left(\dfrac{1}{3}\right)^{11}$

$= 729\left(\dfrac{1}{177147}\right) = \dfrac{1}{243}$

43. $\dfrac{1}{2}, \dfrac{\sqrt{2}}{2}, 1, \sqrt{2}, 2, \ldots$

$r = \dfrac{\dfrac{\sqrt{2}}{2}}{\dfrac{1}{2}} = \dfrac{2\sqrt{2}}{2} = \sqrt{2}$

$a_1 = \dfrac{1}{2}; \ r = \sqrt{2}$

$a_n = \dfrac{1}{2}\left(\sqrt{2}\right)^{n-1}$

$a_6 = \dfrac{1}{2}\left(\sqrt{2}\right)^{6-1} = \dfrac{1}{2}\left(\sqrt{2}\right)^5 = \dfrac{1}{2}\left(4\sqrt{2}\right) = 2\sqrt{2}$;

$a_{10} = \dfrac{1}{2}\left(\sqrt{2}\right)^{10-1} = \dfrac{1}{2}\left(\sqrt{2}\right)^9 = \dfrac{1}{2}\left(16\sqrt{2}\right) = 8\sqrt{2}$;

$a_{12} = \dfrac{1}{2}\left(\sqrt{2}\right)^{12-1} = \dfrac{1}{2}\left(\sqrt{2}\right)^{11} = \dfrac{1}{2}\left(32\sqrt{2}\right) = 16\sqrt{2}$

45. $0.2, 0.08, 0.032, 0.0128,...$

$r = \dfrac{0.08}{0.2} = 0.4$

$a_1 = 0.2;\ \ r = 0.4$

$a_n = 0.2(0.4)^{n-1}$

$a_6 = 0.2(0.4)^{6-1} = 0.2(0.4)^5$
$\quad = 0.2(0.01024) = 0.002048;$

$a_{10} = 0.2(0.4)^{10-1} = 0.2(0.4)^9$
$\quad = 0.2(0.000261244) = 0.0000524288;$

$a_{12} = 0.2(0.4)^{12-1} = 0.2(0.4)^{11}$
$\quad = 0.2(0.00004194304)$
$\quad = 0.000008388608$

47. $a_1 = 9,\ a_n = 729,\ r = 3$

$a_n = a_1 r^{n-1}$

$729 = 9(3)^{n-1}$

$81 = 3^{n-1}$

$3^4 = 3^{n-1}$

$4 = n - 1$

$5 = n$

49. $a_1 = 16, a_n = \dfrac{1}{64},\ r = \dfrac{1}{2}$

$a_n = a_1 r^{n-1}$

$\dfrac{1}{64} = 16\left(\dfrac{1}{2}\right)^{n-1}$

$\dfrac{1}{1024} = \left(\dfrac{1}{2}\right)^{n-1}$

$\left(\dfrac{1}{2}\right)^{10} = \left(\dfrac{1}{2}\right)^{n-1}$

$10 = n - 1$

$11 = n$

51. $a_1 = -1, a_n = -1296,\ r = \sqrt{6}$

$a_n = a_1 r^{n-1}$

$-1296 = -1\left(\sqrt{6}\right)^{n-1}$

$1296 = \left(\sqrt{6}\right)^{n-1}$

$\left(\sqrt{6}\right)^8 = \left(\sqrt{6}\right)^{n-1}$

$8 = n - 1$

$9 = n$

53. $2, -6, 18, -54, ..., -4374$

$r = \dfrac{-6}{2} = -3;\ \ a_1 = 2;\ \ a_n = -4374$

$a_n = a_1 r^{n-1}$

$-4374 = 2(-3)^{n-1}$

$-2187 = (-3)^{n-1}$

$(-3)^7 = (-3)^{n-1}$

$7 = n - 1$

$8 = n$

55. $64, 32\sqrt{2}, 32, 16\sqrt{2}, ..., 1$

$r = \dfrac{32\sqrt{2}}{64} = \dfrac{\sqrt{2}}{2};\ \ a_1 = 64;\ \ a_n = 1$

$a_n = a_1 r^{n-1}$

$1 = 64\left(\dfrac{\sqrt{2}}{2}\right)^{n-1}$

$\dfrac{1}{64} = \left(\dfrac{\sqrt{2}}{2}\right)^{n-1}$

$\left(\dfrac{\sqrt{2}}{2}\right)^{12} = \left(\dfrac{\sqrt{2}}{2}\right)^{n-1}$

$12 = n - 1$

$13 = n$

57. $\dfrac{3}{8}, -\dfrac{3}{4}, \dfrac{3}{2}, -3, ..., 96$

$r = \dfrac{-\dfrac{3}{4}}{\dfrac{3}{8}} = \dfrac{-24}{12} = -2;\ \ a_1 = \dfrac{3}{8};\ \ a_n = 96$

$a_n = a_1 r^{n-1}$

$96 = \dfrac{3}{8}(-2)^{n-1}$

$256 = (-2)^{n-1}$

$(-2)^8 = (-2)^{n-1}$

$8 = n - 1$

$9 = n$

59. $a_3 = 324, \, a_7 = 64$

$a_7 = a_3 \cdot r^4$

$64 = 324r^4$

$\dfrac{64}{324} = r^4$

$\sqrt[4]{\dfrac{16}{81}} = r$

$\dfrac{2}{3} = r;$

$a_7 = a_1 \cdot \left(\dfrac{2}{3}\right)^6$

$64 = a_1\left(\dfrac{2}{3}\right)^6$

$64 = a_1\left(\dfrac{64}{729}\right)$

$729 = a_1$

$r = \dfrac{2}{3}, \, a_1 = 729$

61. $a_4 = \dfrac{4}{9}, \, a_8 = \dfrac{9}{4}$

$a_8 = a_4 \cdot r^4$

$\dfrac{9}{4} = \dfrac{4}{9}(r)^4$

$\dfrac{81}{16} = r^4$

$\sqrt[4]{\dfrac{81}{16}} = r$

$\dfrac{3}{2} = r;$

$a_8 = a_1\left(\dfrac{3}{2}\right)^7$

$\dfrac{9}{4} = a_1\left(\dfrac{3}{2}\right)^7$

$\dfrac{9}{4} = a_1\left(\dfrac{2187}{128}\right)$

$\dfrac{32}{243} = a_1$

$r = \dfrac{3}{2}, \, a_1 = \dfrac{32}{243}$

63. $a_4 = \dfrac{32}{3}, \, a_8 = 54$

$a_8 = a_4 \cdot r^4$

$54 = \left(\dfrac{32}{3}\right)r^4$

$\dfrac{81}{16} = r^4$

$\sqrt[4]{\dfrac{81}{16}} = r$

$\dfrac{3}{2} = r;$

$a_8 = a_1\left(\dfrac{3}{2}\right)^7$

$54 = a_1\left(\dfrac{3}{2}\right)^7$

$54 = a_1\left(\dfrac{2187}{128}\right)$

$\dfrac{256}{81} = a_1$

$r = \dfrac{3}{2}, \, a_1 = \dfrac{256}{81}$

65. $a_1 = 8, r = -2$, find S_{12}

$S_n = \dfrac{a_1(1-r^n)}{1-r}$

$S_{12} = \dfrac{8(1-(-2)^{12})}{1+3} = \dfrac{8(1-4096)}{1+2}$

$= \dfrac{8(-4095)}{3} = -10{,}920$

67. $a_1 = 96, r = \dfrac{1}{3}$, find S_5

$S_n = \dfrac{a_1(1-r^n)}{1-r}$

$S_5 = \dfrac{96\left(1-\left(\dfrac{1}{3}\right)^5\right)}{1-\dfrac{1}{3}} = \dfrac{96\left(1-\dfrac{1}{243}\right)}{\dfrac{2}{3}}$

$= \dfrac{96\left(\dfrac{242}{243}\right)}{\dfrac{2}{3}} = \dfrac{3872}{27} \approx 143.41$

69. $a_1 = 8, r = \dfrac{3}{2}$, find S_7

$$S_n = \frac{a_1\left(1 - r^n\right)}{1 - r}$$

$$S_7 = \frac{8\left(1 - \left(\dfrac{3}{2}\right)^7\right)}{1 - \dfrac{3}{2}} = \frac{8\left(1 - \dfrac{2187}{128}\right)}{-\dfrac{1}{2}}$$

$$= \frac{8\left(\dfrac{-2059}{128}\right)}{-\dfrac{1}{2}} = \frac{2059}{8} = 257.375$$

71. $2 + 6 + 18 + \ldots$; find S_6

$$a_1 = 2; \quad r = \frac{6}{2} = 3$$

$$S_n = \frac{a_1\left(1 - r^n\right)}{1 - r}$$

$$S_6 = \frac{2\left(1 - 3^6\right)}{1 - 3} = \frac{2\left(1 - 729\right)}{-2} = \frac{2\left(-728\right)}{-2} = 728$$

73. $16 - 8 + 4 - \ldots$; find S_8

$$a_1 = 16; \quad r = \frac{-8}{16} = \frac{-1}{2}$$

$$S_n = \frac{a_1\left(1 - r^n\right)}{1 - r}$$

$$S_8 = \frac{16\left(1 - \left(\dfrac{-1}{2}\right)^8\right)}{1 - \left(-\dfrac{1}{2}\right)} = \frac{16\left(1 - \dfrac{1}{256}\right)}{\dfrac{3}{2}}$$

$$= \frac{16\left(\dfrac{255}{256}\right)}{\dfrac{3}{2}} = \frac{85}{8} = 10.625$$

75. $\dfrac{4}{3} + \dfrac{2}{9} + \dfrac{1}{27} + \ldots$; find S_9

$$a_1 = \frac{4}{3}; \quad r = \frac{1}{6}$$

$$S_n = \frac{a_1\left(1 - r^n\right)}{1 - r}$$

$$S_9 = \frac{\dfrac{4}{3}\left(1 - \left(\dfrac{1}{6}\right)^9\right)}{1 - \dfrac{1}{6}} = \frac{\dfrac{4}{3}\left(1 - \dfrac{1}{10077696}\right)}{\dfrac{5}{6}}$$

$$= \frac{\dfrac{4}{3}\left(\dfrac{10077695}{10077696}\right)}{\dfrac{5}{6}} \approx 1.60$$

77. $\displaystyle\sum_{j=1}^{5} 4^j$

Initial terms: 4, 16, 64, ….

$$a_1 = 4; \quad r = \frac{16}{4} = 4; \quad n = 5$$

$$S_n = \frac{a_1\left(1 - r^n\right)}{1 - r}$$

$$S_5 = \frac{4\left(1 - 4^5\right)}{1 - 4} = \frac{4\left(1 - 1024\right)}{-3}$$

$$= \frac{4\left(-1023\right)}{-3} = 1,364$$

10.3 Exercises

79. $\displaystyle\sum_{k=1}^{8} 5\left(\frac{2}{3}\right)^{k-1}$

$5\left(\frac{2}{3}\right)^{0} = 5$;

$5\left(\frac{2}{3}\right)^{1} = \frac{10}{3}$;

$5\left(\frac{2}{3}\right)^{2} = \frac{20}{9}$;

$5\left(\frac{2}{3}\right)^{3} = \frac{40}{27}$

Initial terms: $5, \dfrac{10}{3}, \dfrac{20}{9}, \dfrac{40}{27}, \dots$

$a_1 = 5; \quad r = \dfrac{\frac{20}{9}}{\frac{10}{3}} = \dfrac{2}{3}; \quad n = 8$

$S_n = \dfrac{a_1\left(1 - r^n\right)}{1 - r}$

$S_8 = \dfrac{5\left(1 - \left(\frac{2}{3}\right)^8\right)}{1 - \frac{2}{3}} = \dfrac{5\left(1 - \frac{256}{6561}\right)}{\frac{1}{3}}$

$= \dfrac{5\left(\frac{6305}{6561}\right)}{\frac{1}{3}} = \dfrac{31525}{2187} \approx 14.41$

81. $\displaystyle\sum_{i=4}^{10} 9\left(-\frac{1}{2}\right)^{i-1}$

$9\left(\frac{-1}{2}\right)^{3} = \frac{-9}{8}$;

$9\left(\frac{-1}{2}\right)^{4} = \frac{9}{16}$;

$9\left(\frac{-1}{2}\right)^{5} = \frac{-9}{32}$;

$9\left(\frac{-1}{2}\right)^{6} = \frac{9}{64}$

Initial terms: $\dfrac{-9}{8}, \dfrac{9}{16}, \dfrac{-9}{32}, \dfrac{9}{64}, \dots$

$a_1 = \dfrac{-9}{8}; \quad r = \dfrac{\frac{-9}{32}}{\frac{9}{16}} = \dfrac{-1}{2}; \quad n = 7$

$S_n = \dfrac{a_1\left(1 - r^n\right)}{1 - r}$

$S_7 = \dfrac{\frac{-9}{8}\left(1 - \left(\frac{-1}{2}\right)^7\right)}{1 - \left(\frac{-1}{2}\right)} = \dfrac{\frac{-9}{8}\left(1 + \frac{1}{128}\right)}{\frac{3}{2}}$

$= \dfrac{\frac{-9}{8}\left(\frac{129}{128}\right)}{\frac{3}{2}} = \dfrac{-387}{512} \approx -0.76$

83. $a_2 = -5, a_5 = \dfrac{1}{25}$, find S_5

Find r :

$a_5 = a_2 r^3$

$\dfrac{1}{25} = -5r^3$

$\dfrac{-1}{125} = r^3$

$\sqrt[3]{\dfrac{-1}{125}} = r$

$\dfrac{-1}{5} = r;$

Find a_1 :

$a_5 = a_1 r^4$

$\dfrac{1}{25} = a_1 \left(\dfrac{-1}{5}\right)^4$

$\dfrac{1}{25} = a_1 \left(\dfrac{1}{625}\right)$

$25 = a_1;$

$a_1 = 25; \quad r = \dfrac{-1}{5}$

$S_n = \dfrac{a_1\left(1 - r^n\right)}{1 - r}$

$S_5 = \dfrac{25\left(1 - \left(\dfrac{-1}{5}\right)^5\right)}{1 - \left(\dfrac{-1}{5}\right)} = \dfrac{25\left(1 + \dfrac{1}{3125}\right)}{\dfrac{6}{5}}$

$= \dfrac{25\left(\dfrac{3126}{3125}\right)}{\dfrac{6}{5}} = \dfrac{521}{25}$

85. $a_3 = \dfrac{4}{9}, a_7 = \dfrac{9}{64}$, find S_6

Find r :

$a_7 = a_3 r^4$

$\dfrac{9}{64} = \dfrac{4}{9} r^4$

$\dfrac{81}{256} = r^4$

$\sqrt[4]{\dfrac{81}{256}} = r$

$\dfrac{3}{4} = r;$

Find a_1 :

$a_7 = a_1 r^6$

$\dfrac{9}{64} = a_1 \left(\dfrac{3}{4}\right)^6$

$\dfrac{9}{64} = a_1 \left(\dfrac{729}{4096}\right)$

$\dfrac{64}{81} = a_1;$

$a_1 = \dfrac{64}{81}; \quad r = \dfrac{3}{4}$

$S_n = \dfrac{a_1\left(1 - r^n\right)}{1 - r}$

$S_6 = \dfrac{\dfrac{64}{81}\left(1 - \left(\dfrac{3}{4}\right)^6\right)}{1 - \dfrac{3}{4}} = \dfrac{\dfrac{64}{81}\left(1 - \dfrac{729}{4096}\right)}{\dfrac{1}{4}}$

$= \dfrac{\dfrac{64}{81}\left(\dfrac{3367}{4096}\right)}{\dfrac{1}{4}} = \dfrac{3367}{1296}$

87. $a_3 = 2\sqrt{2}, a_6 = 8,$ find S_7

Find r :

$a_6 = a_3 r^3$

$8 = 2\sqrt{2}(r)^3$

$\dfrac{4}{\sqrt{2}} = r^3$

$2\sqrt{2} = r^3$

$2 \cdot 2^{\frac{1}{2}} = r^3$

$2^{\frac{3}{2}} = r^3$

$\left(2^{\frac{3}{2}}\right)^{\frac{1}{3}} = r$

$\sqrt{2} = r;$

Find a_1 :

$a_6 = a_1 r^5$

$8 = a_1 \left(\sqrt{2}\right)^5$

$8 = a_1 \left(2^{\frac{5}{2}}\right)$

$\dfrac{2^3}{2^{\frac{5}{2}}} = a_1$

$2^{3-\frac{5}{2}} = a_1$

$\sqrt{2} = a_1$

$a_1 = \sqrt{2}; \ r = \sqrt{2}$

$S_n = \dfrac{a_1\left(1 - r^n\right)}{1 - r}$

$S_7 = \dfrac{\sqrt{2}\left(1 - \left(\sqrt{2}\right)^7\right)}{1 - \sqrt{2}} = \dfrac{\sqrt{2}\left(1 - \left(\sqrt{2}\right)^6\left(\sqrt{2}\right)\right)}{1 - \sqrt{2}}$

$= \dfrac{\sqrt{2}\left(1 - 8\left(\sqrt{2}\right)\right)}{1 - \sqrt{2}} = \dfrac{\sqrt{2} - 16}{1 - \sqrt{2}}$

$= \dfrac{\sqrt{2} - 16}{1 - \sqrt{2}} \cdot \dfrac{1 + \sqrt{2}}{1 + \sqrt{2}}$

$= \dfrac{\sqrt{2} + 2 - 16 - 16\sqrt{2}}{1 - 2}$

$= \dfrac{-14 - 15\sqrt{2}}{-1} = 14 + 15\sqrt{2}$

89. $3 + 6 + 12 + 24 + \ldots$

$\dfrac{6}{3} = 2;$ No

91. $9 + 3 + 1 + \ldots$

$\dfrac{3}{9} = \dfrac{1}{3}; \ \left|\dfrac{1}{3}\right| < 1$

$a_1 = 9; \ r = \dfrac{1}{3}$

$S_\infty = \dfrac{a_1}{1 - r}$

$S_\infty = \dfrac{9}{1 - \dfrac{1}{3}} = \dfrac{9}{\dfrac{2}{3}} = \dfrac{27}{2}$

93. $25 + 10 + 4 + \dfrac{8}{5} + \ldots$

$\dfrac{10}{25} = \dfrac{2}{5}; \ \left|\dfrac{2}{5}\right| < 1$

$a_1 = 25; \ r = \dfrac{2}{5}$

$S_\infty = \dfrac{a_1}{1 - r}$

$S_\infty = \dfrac{25}{1 - \dfrac{2}{5}} = \dfrac{25}{\dfrac{3}{5}} = \dfrac{125}{3}$

95. $6+3+\dfrac{3}{2}+\dfrac{3}{4}+...$

$\dfrac{3}{6}=\dfrac{1}{2};\ \left|\dfrac{1}{2}\right|<1$

$a_1=6;\ r=\dfrac{1}{2}$

$S_\infty=\dfrac{a_1}{1-r}$

$S_\infty=\dfrac{6}{1-\dfrac{1}{2}}=\dfrac{6}{\dfrac{1}{2}}=12$

97. $6-3+\dfrac{3}{2}-\dfrac{3}{4}+...$

$\dfrac{-3}{6}=\dfrac{-1}{2};\ \left|\dfrac{-1}{2}\right|<1$

$a_1=6;\ r=\dfrac{-1}{2}$

$S_\infty=\dfrac{a_1}{1-r}$

$S_\infty=\dfrac{6}{1-\dfrac{-1}{2}}=\dfrac{6}{\dfrac{3}{2}}=\dfrac{12}{3}=4$

99. $0.3+0.03+0.003+...$

$\dfrac{0.03}{0.3}=\dfrac{1}{10};\ \left|\dfrac{1}{10}\right|<1$

$a_1=0.3;\ r=\dfrac{1}{10}$

$S_\infty=\dfrac{a_1}{1-r}$

$S_\infty=\dfrac{0.3}{1-\dfrac{1}{10}}=\dfrac{0.3}{\dfrac{9}{10}}=\dfrac{3}{9}=\dfrac{1}{3}$

101. $\displaystyle\sum_{k=1}^{\infty}\dfrac{3}{4}\left(\dfrac{2}{3}\right)^k$

$\dfrac{3}{4}\left(\dfrac{2}{3}\right)^1=\dfrac{1}{2};$

$\dfrac{3}{4}\left(\dfrac{2}{3}\right)^2=\dfrac{3}{4}\left(\dfrac{4}{9}\right)=\dfrac{1}{3};$

$\dfrac{3}{4}\left(\dfrac{2}{3}\right)^3=\dfrac{3}{4}\left(\dfrac{8}{27}\right)=\dfrac{2}{9};$

Initial terms: $\dfrac{1}{2}+\dfrac{1}{3}+\dfrac{2}{9}+....$

$\dfrac{\dfrac{1}{3}}{\dfrac{1}{2}}=\dfrac{2}{3};\ \left|\dfrac{2}{3}\right|<1$

$a_1=\dfrac{1}{2};\ r=\dfrac{2}{3}$

$S_\infty=\dfrac{a_1}{1-r}$

$S_\infty=\dfrac{\dfrac{1}{2}}{1-\dfrac{2}{3}}=\dfrac{\dfrac{1}{2}}{\dfrac{1}{3}}=\dfrac{3}{2}$

103. $\displaystyle\sum_{j=1}^{\infty}9\left(-\dfrac{2}{3}\right)^j$

$9\left(\dfrac{-2}{3}\right)^1=-6;$

$9\left(\dfrac{-2}{3}\right)^2=9\left(\dfrac{4}{9}\right)=4;$

$9\left(\dfrac{-2}{3}\right)^3=9\left(\dfrac{-8}{27}\right)=\dfrac{-8}{3}$

Initial terms: $-6+4-\dfrac{8}{3}+....$

$\dfrac{4}{-6}=\dfrac{2}{-3};\ \left|\dfrac{-2}{3}\right|<1$

$a_1=-6;\ r=\dfrac{-2}{3}$

$S_\infty=\dfrac{a_1}{1-r}$

$S_\infty=\dfrac{-6}{1-\dfrac{-2}{3}}=\dfrac{-6}{\dfrac{5}{3}}=\dfrac{-18}{5}$

10.3 Exercises

105. $S_n = \dfrac{n^2(n+1)^2}{4}$; $1^3 + 2^3 + 3^3 + \ldots 8^3$

$S_8 = \dfrac{8^2(8+1)^2}{4} = \dfrac{64(9)^2}{4} = \dfrac{64(81)}{4} = 1296$

$1^3 + 2^3 + 3^3 + 4^3 + 5^3 + 6^3 + 7^3 + 8^3$

$= 1 + 8 + 27 + 64 + 125 + 216 + 343 + 512$

$= 1296$

107. $a_1 = 24$; $r = 0.8$; $n = 7$

Initial terms: 24, 19.2, 15.36, ...

$a_n = a_1(r)^{n-1}$

$a_7 = 24(0.8)^6 = 24(0.262144) \approx 6.3$;

$S_\infty = \dfrac{a_1}{1-r}$

$S_\infty = \dfrac{24}{1-0.8} = \dfrac{24}{0.2} = 120$

about 6.3 ft; 120 ft

109. $a_1 = 46000$; $r = 0.20$; $n = 4$

$a_n = a_1(1-r)^n$

$a_4 = 46000(1-0.2)^4$

$a_4 = 46000(0.8)^4$

$a_4 = 46000(0.4096)$

$a_4 = 18841.60$;

$5000 = 46000(1-0.2)^n$

$0.1086956522 = 0.8^n$

$\ln 0.1086956522 = n \ln 0.8$

$\dfrac{\ln 0.1086956522}{\ln 0.8} = n$

$10 \approx n$;

about \$18841.60; 10 years

111. $a_0 = 160$; $a_1 = 160(0.97) = 155.2$;

$r = 0.03$; $n = 8$

$a_n = a_1(1-r)^{n-1}$

$a_8 = 155.2(1-0.03)^{8-1}$

$a_8 = 155.2(0.97)^7$

$a_8 \approx 125.4$;

$118 = 155.2(0.97)^{n-1}$

$\dfrac{118}{155.2} = 0.97^{n-1}$

$\ln \dfrac{118}{155.2} = (n-1)\ln 0.97$

$\dfrac{\ln \dfrac{118}{155.2}}{\ln 0.97} = n-1$

$9 \approx n-1$

$10 \approx n$

about 125.4 gpm; about 10 months

113. $a_1 = 277$; $r = 0.023$; $n = 10$

$a_n = a_1(1+r)^n$

$a_{10} = 277(1+0.023)^{10}$

$a_{10} = 277(1.023)^{10}$

$a_{10} \approx 347.7$

about 347.7 million

115. $a_1 = 50$; $r = 2$; $n = 10$

10 half-hours in 5 hours

$a_n = a_1(r)^n$

$a_{10} = 50(2)^{10} = 50(1024) = 51200$;

$204800 = 50(2)^n$

$4096 = 2^n$

$(2)^{12} = 2^n$

$12 = n$

51,200 bacteria; 12 half hours later or 6 hours

117. $a_1 = \frac{4}{5}(2) = 1.6; \quad r = \frac{4}{5}; \quad n = 7$

$a_n = a_1(r)^{n-1}$

$a_7 = 1.6\left(\frac{4}{5}\right)^6 = 1.6\left(\frac{4096}{15625}\right) \approx 0.42 ;$

$S_\infty = \frac{a_1}{1-r}$

$S_\infty = \frac{1.6}{1-\frac{4}{5}} = \frac{1.6}{\frac{1}{5}} = 8$ up

Down: $S = \frac{2}{1-\frac{4}{5}} = \frac{2}{\frac{1}{5}} = 10$

approximately 0.42 m; 18 m

119. $a_1 = 462; \quad r = \frac{2}{5}; \quad n = 5$

$a_n = a_1(1-r)^n$

$a_5 = 462\left(1-\frac{2}{5}\right)^5$

$a_5 = 462\left(\frac{3}{5}\right)^5$

$a_5 = 462\left(\frac{243}{3125}\right)$

$a_5 = 35.9$

$12.9 = 462\left(\frac{3}{5}\right)^n$

$0.0279 = \left(\frac{3}{5}\right)^n$

$\ln 0.0279 = n \ln 0.6$

$\frac{\ln 0.0279}{\ln 0.6} = n$

35.9 in^3; about 7 strokes

121. $a_0 = 40000; \quad d = 1750; \quad r = 0.96$

$40000 + 1750n = 40000(1.04)^n$

$Y_1 = Y_2$

Using a grapher: about 6 years

123. $S_n = \sum_{k=1}^{n} \log(k)$

$S_n = \log n!$

125. $f(x) = x^2 + 5x + 9$

$x = \frac{-5 \pm \sqrt{(5)^2 - 4(1)(9)}}{2(1)}$

$x = \frac{-5 \pm \sqrt{25 - 36}}{2}$

$x = \frac{-5 \pm \sqrt{-11}}{2}$

$x = \frac{-5}{2} \pm \frac{\sqrt{11}}{2}i$

127. $h(x) = \frac{x^2}{x-1}$

Vertical asymptote: $x = 1$
Horizontal asymptote: none
(deg num > deg den)
Oblique asymptote: $y = x$

$$\begin{array}{r} x \\ x-1 \overline{) x^2 } \\ -\underline{(x^2 - x)} \\ x \end{array}$$

y-intercept: (0,0)

$h(0) = \frac{0^2}{0-1} = 0$

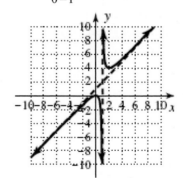

575

10.4 Exercises

1. Finite; universally

3. Induction hypothesis

5. Answers will vary.

7. $a_n = 10n - 6$
    $a_4 = 10(4) - 6 = 40 - 6 = 34$;
    $a_5 = 10(5) - 6 = 50 - 6 = 44$;
    $a_k = 10k - 6$;
    $a_{k+1} = 10(k+1) - 6 = 10k + 10 - 6 = 10k + 4$

9. $a_n = n$
    $a_4 = 4$;
    $a_5 = 5$;
    $a_k = k$;
    $a_{k+1} = k + 1$

11. $a_n = 2^{n-1}$
    $a_4 = 2^{4-1} = 2^3 = 8$;
    $a_5 = 2^{5-1} = 2^4 = 16$;
    $a_k = 2^{k-1}$;
    $a_{k+1} = 2^{k+1-1} = 2^k$

13. $S_n = n(5n - 1)$
    $S_4 = 4(5(4) - 1) = 4(20 - 1) = 4(19) = 76$;
    $S_5 = 5(5(5) - 1) = 5(25 - 1) = 5(24) = 120$;
    $S_k = k(5k - 1)$;
    $S_{k+1} = (k+1)(5(k+1) - 1) = (k+1)(5k + 5 - 1)$
    $= (k+1)(5k + 4)$

15. $S_n = \dfrac{n(n+1)}{2}$
    $S_4 = \dfrac{4(4+1)}{2} = \dfrac{4(5)}{2} = 10$;
    $S_5 = \dfrac{5(5+1)}{2} = \dfrac{5(6)}{2} = 15$;
    $S_k = \dfrac{k(k+1)}{2}$;
    $S_{k+1} = \dfrac{(k+1)(k+1+1)}{2} = \dfrac{(k+1)(k+2)}{2}$

17. $S_n = 2^n - 1$
    $S_4 = 2^4 - 1 = 16 - 1 = 15$;
    $S_5 = 2^5 - 1 = 32 - 1 = 31$;
    $S_k = 2^k - 1$;
    $S_{k+1} = 2^{k+1} - 1$

19. $a_n = 10n - 6$; $S_n = n(5n - 1)$
    $S_4 = 4(5(4) - 1) = 4(20 - 1) = 4(19) = 76$;
    $a_5 = 10(5) - 6 = 50 - 6 = 44$;
    $S_5 = 5(5(5) - 1) = 5(25 - 1) = 5(24) = 120$;
    $S_4 + a_5 = S_5$
    $76 + 44 = 120$
    $\quad 120 = 120$
    Verified

21. $a_n = n$; $S_n = \dfrac{n(n+1)}{2}$
    $S_4 = \dfrac{4(4+1)}{2} = \dfrac{4(5)}{2} = 10$;
    $a_5 = 5$;
    $S_5 = \dfrac{5(5+1)}{2} = \dfrac{5(6)}{2} = 15$;
    $S_4 + a_5 = S_5$
    $10 + 5 = 15$
    $\quad 15 = 15$
    Verified

23. $a_n = 2^{n-1}$; $S_n = 2^n - 1$
    $S_4 = 2^4 - 1 = 16 - 1 = 15$;
    $a_5 = 2^{5-1} = 2^4 = 16$;
    $S_5 = 2^5 - 1 = 32 - 1 = 31$;
    $S_4 + a_5 = S_5$
    $15 + 16 = 31$
    $\quad 31 = 31$
    Verified

25. $a_n = n^3$; $S_n = (1+2+3+4+...+n)^2$

    a.   for $n = 1$, $1 = (1)^2$;

        for $n = 5$,

        $1+8+27+64+125 = (1+2+3+4+5)^2$

        $225 == (15)^2$;

        for $n = 9$,

        $1+8+27+64+125+216+343+512+729$

        $= (1+2+3+4+5+6+7+8+9)^2$

        $2025 = (45)^2$

        $2025 = 2025$

    b.   The needed components are:

        $a_n = n^3$; $a_k = k^3$; $a_{k+1} = (k+1)^3$

        $S_k = (1+2+3+4+...+k)^2$;

        $S_{k+1} = (1+2+3+4+...+k+(k+1))^2$

    1.   Show S_n is true for $n = 1$.

        $S_1 = (1)^2 = 1$

        Verified

    2.   Assume S_k is true:

        $1+8+27+...+k^3 = (1+2+3+4+...+k)^2$

        and use it to show the truth of S_{k+1}

        follows. That is:

        $1+8+27+...+k^3 + (k+1)^3$

        $= (1+2+3+...+k+(k+1))^2$

        $S_k + a_{k+1} = S_{k+1}$

        Working with the left hand side:

        $1+8+27+...+k^3 + (k+1)^3$

        $= 1+8+27+...+k^3 + (k+1)^2(k+1)$

        $= 1+8+27+...+k^3 + k(k+1)^2 + (k+1)^2$

        $= 1+8+27+...+k^3 + k(k+1)(k+1) + (k+1)^2$

        $= 1+8+27+...+k^3$

        $+ 2\dfrac{k(k+1)}{2}(k+1) + (k+1)^2$

        $= (1+2+3+...+k)^2$

        $+ 2(1+2+3+...+k)(k+1) + (k+1)^2$

        Factoring as a trinomial:

        $= ((1+2+3+...+k)+(k+1))^2$

        Since the truth of S_{k+1} follows from

        S_k, the formula is true for all n.

27. $2+4+6+8+10+...+2n$;

    The needed components are:

    $a_n = 2n$; $a_k = 2k$; $a_{k+1} = 2(k+1)$

    $S_n = n(n+1)$; $S_k = k(k+1)$;

    $S_{k+1} = (k+1)(k+1+1) = (k+1)(k+2)$

    1.   Show S_n is true for $n = 1$.

        $S_1 = 1(1+1) = 1(2) = 2$

        Verified

    2.   Assume S_k is true:

        $2+4+6+8+10+...+2k = k(k+1)$

        and use it to show the truth of S_{k+1}

        follows. That is:

        $2+4+6+...+2k+2(k+1) = (k+1)(k+2)$

        $S_k + a_{k+1} = S_{k+1}$

        Working with the left hand side:

        $2+4+6+...+2k+2(k+1)$

        $= k(k+1)+2(k+1)$

        $= k^2 + k + 2k + 2$

        $= k^2 + 3k + 2$

        $= (k+1)(k+2)$

        $= S_{k+1}$

    Since the truth of S_{k+1} follows from

    S_k, the formula is true for all n.

29. $5 + 10 + 15 + 20 + 25 + \ldots + 5n$

The needed components are:

$a_n = 5n$; $a_k = 5k$; $a_{k+1} = 5(k+1)$;

$S_n = \dfrac{5n(n+1)}{2}$; $S_k = \dfrac{5k(k+1)}{2}$;

$S_{k+1} = \dfrac{5(k+1)(k+1+1)}{2} = \dfrac{5(k+1)(k+2)}{2}$

1. Show S_n is true for $n = 1$.

$S_1 = \dfrac{5(1)(1+1)}{2} = \dfrac{5(2)}{2} = 5$

Verified

2. Assume S_k is true:

$5 + 10 + 15 + \ldots + 5k = \dfrac{5k(k+1)}{2}$

and use it to show the truth of S_{k+1} follows. That is:

$5 + 10 + 15 + \ldots + 5k + 5(k+1)$

$= \dfrac{5(k+1)(k+1+1)}{2}$

$S_k + a_{k+1} = S_{k+1}$

Working with the left hand side:

$5 + 10 + 15 + \ldots + 5k + 5(k+1)$

$= \dfrac{5k(k+1)}{2} + 5(k+1)$

$= \dfrac{5k(k+1) + 10(k+1)}{2}$

$= \dfrac{(k+1)(5k+10)}{2}$

$= \dfrac{5(k+1)(k+2)}{2}$

$= S_{k+1}$

Since the truth of S_{k+1} follows from S_k, the formula is true for all n.

31. $5 + 9 + 13 + 17 + \ldots + 4n + 1$

The needed components are:

$a_n = 4n + 1$; $a_k = 4k + 1$;

$a_{k+1} = 4(k+1) + 1 = 4k + 4 + 1 = 4k + 5$;

$S_n = n(2n+3)$; $S_k = k(2k+3)$;

$S_{k+1} = (k+1)(2(k+1)+3) = (k+1)(2k+5)$

1. Show S_n is true for $n = 1$.

$S_1 = 1(2(1)+3) = 5$

Verified

2. Assume S_k is true:

$5 + 9 + 13 + 17 + \ldots + 4k + 1 = k(2k+3)$

and use it to show the truth of S_{k+1} follows. That is:

$5 + 9 + 13 + 17 + \ldots + 4k + 1 + 4(k+1) + 1$

$= (k+1)(2(k+1)+3)$

$S_k + a_{k+1} = S_{k+1}$

Working with the left hand side:

$5 + 9 + 13 + 17 + \ldots + 4k + 1 + 4k + 5$

$= k(2k+3) + 4k + 5$

$= 2k^2 + 3k + 4k + 5 = 2k^2 + 7k + 5$

$= (k+1)(2k+5) = S_{k+1}$

Since the truth of S_{k+1} follows from S_k, the formula is true for all n.

33. $3 + 9 + 27 + 81 + 243 + \ldots + 3^n$

The needed components are:

$a_n = 3^n$; $a_k = 3^k$; $a_{k+1} = 3^{k+1}$;

$S_n = \dfrac{3(3^n - 1)}{2}$; $S_k = \dfrac{3(3^k - 1)}{2}$;

$S_{k+1} = \dfrac{3(3^{k+1} - 1)}{2}$

1. Show S_n is true for $n = 1$.

$$S_1 = \frac{3(3^1 - 1)}{2} = \frac{3(3 - 1)}{2} = \frac{3(2)}{2} = 3$$

Verified

2. Assume S_k is true:

$$3 + 9 + 27 + \ldots + 3^k = \frac{3(3^k - 1)}{2}$$

and use it to show the truth of S_{k+1} follows. That is:

$$3 + 9 + 27 + \ldots + 3^k + 3^{k+1} = \frac{3(3^{k+1} - 1)}{2}$$

$S_k + a_{k+1} = S_{k+1}$

Working with the left hand side:

$3 + 9 + 27 + \ldots + 3^k + 3^{k+1}$

$= \dfrac{3(3^k - 1)}{2} + 3^{k+1}$

$= \dfrac{3(3^k - 1) + 2(3^{k+1})}{2}$

$= \dfrac{3^{k+1} - 3 + 2(3^{k+1})}{2}$

$= \dfrac{3(3^{k+1}) - 3}{2}$

$= \dfrac{3(3^{k+1} - 1)}{2}$

$= S_{k+1}$

Since the truth of S_{k+1} follows from S_k, the formula is true for all n.

35. $2 + 4 + 8 + 16 + 32 + 64 + \ldots + 2^n$

The needed components are:

$a_n = 2^n$; $a_k = 2^k$; $a_{k+1} = 2^{k+1}$;

$S_n = 2^{n+1} - 2$; $S_k = 2^{k+1} - 2$;

$S_{k+1} = 2^{k+1+1} - 2 = 2^{k+2} - 2$

1. Show S_n is true for $n = 1$.

$S_n = 2^{n+1} - 2$

$S_1 = 2^{1+1} - 2 = 2^2 - 2 = 4 - 2 = 2$

Verified

2. Assume S_k is true:

$2 + 4 + 8 + \ldots + 2^k = 2^{k+1} - 2$

and use it to show the truth of S_{k+1} follows. That is:

$2 + 4 + 8 + \ldots + 2^k + 2^{k+1} = 2^{k+1} - 2$

$S_k + a_{k+1} = S_{k+1}$

Working with the left hand side:

$2 + 4 + 8 + \ldots + 2^k + 2^{k+1}$

$= 2^{k+1} - 2 + 2^{k+1}$

$= 2(2^{k+1}) - 2$

$= 2^{k+2} - 2$

$= S_{k+1}$

Since the truth of S_{k+1} follows from S_k, the formula is true for all n.

37. $\dfrac{1}{1(3)}+\dfrac{1}{3(5)}+\dfrac{1}{5(7)}+...+\dfrac{1}{(2n-1)(2n+1)}$

The needed components are:

$a_n=\dfrac{1}{(2n-1)(2n+1)}$; $a_k=\dfrac{1}{(2k-1)(2k+1)}$;

$a_{k+1}=\dfrac{1}{(2(k+1)-1)(2(k+1)+1)}=\dfrac{1}{(2k+1)(2k+3)}$;

$S_n=\dfrac{n}{2n+1}$; $S_k=\dfrac{k}{2k+1}$;

$S_{k+1}=\dfrac{k+1}{2(k+1)+1}=\dfrac{k+1}{2k+3}$

1. Show S_n is true for $n=1$.

$S_n=\dfrac{n}{2n+1}$

$S_1=\dfrac{1}{2(1)+1}=\dfrac{1}{2+1}=\dfrac{1}{3}$

Verified

2. Assume S_k is true:

$\dfrac{1}{3}+\dfrac{1}{15}+\dfrac{1}{35}+...+\dfrac{1}{(2k-1)(2k+1)}=\dfrac{k}{2k+1}$

and use it to show the truth of S_{k+1} follows. That is:

$\dfrac{1}{3}+\dfrac{1}{15}+\dfrac{1}{35}+...+\dfrac{1}{(2k-1)(2k+1)}$

$+\dfrac{1}{(2(k+1)-1)(2(k+1)+1)}=\dfrac{k+1}{2(k+1)+1}$

$S_k+a_{k+1}=S_{k+1}$

Working with the left hand side:

$\dfrac{1}{3}+\dfrac{1}{15}+\dfrac{1}{35}+...+\dfrac{1}{(2k-1)(2k+1)}+\dfrac{1}{(2k+1)(2k+3)}$

$=\dfrac{k}{2k+1}+\dfrac{1}{(2k+1)(2k+3)}$

$=\dfrac{k(2k+3)+1}{(2k+1)(2k+3)}$

$=\dfrac{2k^2+3k+1}{(2k+1)(2k+3)}$

$=\dfrac{(2k+1)(k+1)}{(2k+1)(2k+3)}$

$=\dfrac{k+1}{2k+3}$

$=S_{k+1}$

Since the truth of S_{k+1} follows from S_k, the formula is true for all n.

39. $S_n:3^n\geq 2n+1$

$S_k:3^k\geq 2k+1$

$S_{k+1}:3^{k+1}\geq 2(k+1)+1$

1. Show S_n is true for $n=1$.

$S_1:$

$3^1\geq 2(1)+1$

$3\geq 2+1$

$3\geq 3$

Verified

2. Assume $S_k:3^k\geq 2k+1$ is true and use it to show the truth of S_{k+1} follows. That is: $3^{k+1}\geq 2k+3$.

Working with the left hand side:

$\begin{aligned}3^{k+1}&=3(3^k)\\&\geq 3(2k+1)\\&\geq 6k+3\end{aligned}$

Since k is a positive integer,

$6k+3\geq 2k+3$

Showing $S_{k+1}:3^{k+1}\geq 2k+3$

Verified

41. $S_n:3\cdot 4^{n-1}\leq 4^n-1$

$S_k:3\cdot 4^{k-1}\leq 4^k-1$

$S_{k+1}:3\cdot 4^k\leq 4^{k+1}-1$

1. Show S_n is true for $n=1$.

$S_1:$

$3\cdot 4^{1-1}\leq 4^1-1$

$3\cdot 4^0\leq 4-1$

$3\cdot 1\leq 3$

$3\leq 3$

Verified

2. Assume $S_k:3\cdot 4^{k-1}\leq 4^k-1$ is true. and use it to show the truth of S_{k+1} follows. That is: $3\cdot 4^k\leq 4^{k+1}-1$.

Working with the left hand side:

$\begin{aligned}3\cdot 4^k&=3\cdot 4(4^{k-1})\\&=4\cdot 3(4^{k-1})\\&\leq 4(4^k-1)\\&\leq 4^{k+1}-4\end{aligned}$

Since k is a positive integer,

$4^{k+1}-4\leq 4^{k+1}-1$

Showing that $3\cdot 4^k\leq 4^{k+1}-1$

43. $n^2 - 7n$ is divisible by 2

    1. Show S_n is true for $n = 1$.

$$S_n : n^2 - 7n = 2m$$

$$S_1 :$$

$$(1)^2 - 7(1) = 2m$$
$$1 - 7 = 2m$$
$$-6 = 2m$$

    Verified

    2. Assume $S_k : k^2 - 7k = 2m$ for $m \in Z$.
and use it to show the truth of S_{k+1}
follows. That is:
$(k+1)^2 - 7(k+1) = 2p$ for $p \in Z$.
Working with the left hand side:

$$= (k+1)^2 - 7(k+1)$$
$$= k^2 + 2k + 1 - 7k - 7$$
$$= k^2 - 7k + 2k - 6$$
$$= 2m + 2k - 6$$
$$= 2(m + k - 3)$$

    is divisible by 2.

45. $n^3 + 3n^2 + 2n$ is divisible by 3

    1. Show S_n is true for $n = 1$.

$$S_n : n^3 + 3n^2 + 2n = 3m$$

$$S_1 :$$

$$(1)^3 + 3(1)^2 + 2(1) = 3m$$
$$1 + 3 + 2 = 3m$$
$$6 = 3m$$
$$2 = m$$

    Verified

    2. Assume $S_k : k^3 + 3k^2 + 2k = 3m$ for
$m \in Z$ and use it to show the truth of
S_{k+1} follows.
That is:
$S_{k+1} : (k+1)^3 + 3(k+1)^2 + 2(k+1) = 3p$
for $p \in Z$.
Working with the left hand side:

$(k+1)^3 + 3(k+1)^2 + 2(k+1)$ is true.

$$= k^3 + 3k^2 + 3k + 1 + 3(k^2 + 2k + 1) + 2k + 2$$
$$= k^3 + 3k^2 + 2k + 3 + 3(k^2 + 2k + 1) + 3k + 3$$
$$= k^3 + 3k^2 + 2k + 3(k^2 + 2k + 1) + 3(k + 1)$$
$$= 3m + 3(k^2 + 2k + 1) + 3(k + 1)$$

    is divisible by 3.

47. $6^n - 1$ is divisible by 5

    1. Show S_n is true for $n = 1$.

$$S_n : 6^n - 1 = 5m$$

$$S_1 :$$

$$6^1 - 1 = 5m$$
$$6 - 1 = 5m$$
$$5 = 5m$$
$$1 = m$$

    Verified

    2. Assume $S_k : 6^k - 1 = 5m$ for $m \in Z$ and
use it to show the truth of S_{k+1} follows.
That is: $S_{k+1} : 6^{k+1} - 1 = 5p$ for $p \in Z$.
Working with the left hand side:

$$= 6^k - 1$$
$$= 6(6^k) - 1$$
$$= 6(5m + 1) - 1$$
$$= 30m + 6 - 1$$
$$= 30m + 5$$
$$= 5(6m + 1)$$

    is divisible by 5,
    Verified

10.4 Exercises

49. $\dfrac{x^n - 1}{x - 1} = \left(1 + x + x^2 + x^3 + \ldots + x^{n-1}\right)$

The needed components are:

$a_k = x^{k-1}$, $S_k = \dfrac{x^k - 1}{x - 1}$, $S_{k+1} = \dfrac{x^{k+1} - 1}{x - 1}$

1. Show S_n is true for $n = 1$.

$S_k = \dfrac{x^k - 1}{x - 1}$

$S_1 = \dfrac{x^1 - 1}{x - 1} = 1$

Verified

2. Assume S_k is true:

$1 + x + x^2 + x^3 + \ldots + x^{k-1}$

$= \dfrac{x^k - 1}{x - 1}$

and use it to show the truth of S_{k+1} follows. That is:

$1 + x + x^2 + x^3 + \ldots + x^{k-1} + x^{k+1-1}$

$= \dfrac{x^{k+1} - 1}{x - 1}$

$S_k + a_{k+1} = S_{k+1}$

Working with the left hand side:

$1 + x + x^2 + x^3 + \ldots + x^{k-1} + x^{k+1-1}$

$= \dfrac{x^k - 1}{x - 1} + x^k$

$= \dfrac{x^k - 1}{x - 1} + \dfrac{x^k(x - 1)}{x - 1}$

$= \dfrac{x^k - 1 + x^k(x - 1)}{x - 1}$

$= \dfrac{x^k - 1 + x^{k+1} - x^k}{x - 1}$

$= \dfrac{x^{k+1} - 1}{x - 1}$

$= S_{k+1}$

Since the truth of S_{k+1} follows from S_k, the formula is true for all n.

51. $(\sin\theta + \cos\theta)^2 + (\sin\theta - \cos\theta)^2 = 2$

$\sin^2\theta + 2\sin\theta\cos\theta + \cos^2\theta + \sin^2\theta - 2\sin\theta\cos\theta + \cos^2\theta = 2$

$\sin^2\theta + \cos^2\theta + \sin^2\theta + \cos^2\theta = 2$

$1 + 1 = 2$

$2 = 2$

52. Domain: $(-\infty, \infty)$

Range: $[-2, \infty)$

53. Center: (4, 3); Point on circle: (1, 7);

$d = \sqrt{(4-1)^2 + (3-7)^2} = 5$, $r = 5$

$(y - 3)^2 + (x - 4)^2 = 25$

54. $p = \langle \sqrt{3}, -1 \rangle$ $q = \langle 1, 1 \rangle$

a) $p \cdot q = \sqrt{3}(1) + (-1)(1)$

$= \sqrt{3} - 1$

b) $\|p\| = \sqrt{(\sqrt{3})^2 + (-1)} = \sqrt{3 + 1} = \sqrt{4} = 2$

$\|q\| = \sqrt{1^2 + 1^2} = \sqrt{2}$

$\cos\theta = \dfrac{p \cdot q}{\|p\| \cdot \|q\|} = \dfrac{\sqrt{3} - 1}{2(\sqrt{2})} = 0.2588$

$\theta = \cos^{-1}(0.2588)$

$\theta \approx 75°$

Mid-Chapter Check

1. $a_n = 7n - 4$

   $a_1 = 7(1) - 4 = 7 - 4 = 3$;

   $a_2 = 7(2) - 4 = 14 - 4 = 10$;

   $a_3 = 7(3) - 4 = 21 - 4 = 17$;

   $a_9 = 7(9) - 4 = 63 - 4 = 59$

3. $a_n = (-1)^n (2n - 1)$

   $a_1 = (-1)^1 (2(1) - 1) = -1(2 - 1) = -1(1) = -1$;

   $a_2 = (-1)^2 (2(2) - 1) = 1(4 - 1) = 1(3) = 3$;

   $a_3 = (-1)^3 (2(3) - 1) = -1(6 - 1) = -1(5) = -5$;

   $a_9 = (-1)^9 (2(9) - 1) = -1(18 - 1) = -17$

5. $1 + 4 + 7 + 10 + 13 + 16$

   $\displaystyle\sum_{n=1}^{6} (3k - 2)$

7. $a_n = a_1 r^{n-1}$; e

   nth term formula for a geometric series

9. $a_n = a_1 + (n-1)d$; b

   nth term formula for an arithmetic series

11. (a) $2, 5, 8, 11, \ldots$

    $a_1 = 2$; $d = 5 - 2 = 3$

    $a_n = 2 + (n-1)3$

    $a_n = 2 + 3n - 3$

    $a_n = 3n - 1$

    (b) $\dfrac{3}{2}, \dfrac{9}{4}, 3, \dfrac{15}{4}, \ldots$

    $a_1 = \dfrac{3}{2}$; $d = \dfrac{9}{4} - \dfrac{3}{2} = \dfrac{3}{4}$

    $a_n = \dfrac{3}{2} + (n-1)\dfrac{3}{4}$

    $a_n = \dfrac{3}{2} + \dfrac{3}{4}n - \dfrac{3}{4}$

    $a_n = \dfrac{3}{4}n + \dfrac{3}{4}$

13. $\dfrac{1}{2} + \dfrac{3}{2} + \dfrac{5}{2} + \dfrac{7}{2} + \ldots + \dfrac{31}{2}$

    $a_1 = \dfrac{1}{2}$; $d = 1$

    $\dfrac{31}{2} = \dfrac{1}{2} + (n-1)(1)$

    $\dfrac{31}{2} = \dfrac{1}{2} + n - 1$

    $\dfrac{31}{2} = n - \dfrac{1}{2}$

    $16 = n$;

    $S_{16} = \dfrac{16\left(\dfrac{1}{2} + \dfrac{31}{2}\right)}{2} = \dfrac{16(16)}{2} = 128$

15. $a_3 = -81$; $a_7 = -1$

$a_7 = a_3 \cdot r^4$

$-1 = -81r^4$

$\dfrac{1}{81} = r^4$

$\dfrac{1}{3} = r$;

$a_7 = a_1 r^6$

$-1 = a_1 \left(\dfrac{1}{3}\right)^6$

$-1 = \dfrac{1}{729} a_1$

$-729 = a_1$;

$S_{10} = \dfrac{-729\left(1 - \left(\dfrac{1}{3}\right)^{10}\right)}{1 - \dfrac{1}{3}}$

$= \dfrac{-729\left(1 - \dfrac{1}{59049}\right)}{\dfrac{2}{3}}$

$= \dfrac{-729\left(\dfrac{59048}{59049}\right)}{\dfrac{2}{3}}$

$= \dfrac{-29524}{27}$

17. $\dfrac{1}{54} + \dfrac{1}{18} + \dfrac{1}{6} + \dots + \dfrac{81}{2}$

$a_1 = \dfrac{1}{54}$; $\quad r = \dfrac{\dfrac{1}{18}}{\dfrac{1}{54}} = 3$

$\dfrac{81}{2} = \dfrac{1}{54}(3)^{n-1}$

$2187 = 3^{(n-1)}$

$3^7 = 3^{(n-1)}$

$7 = n - 1$

$n = 8$;

$S_8 = \dfrac{\dfrac{1}{54}\left(1 - 3^8\right)}{1 - 3} = \dfrac{\dfrac{1}{54}(1 - 6561)}{-2}$

$= \dfrac{\dfrac{1}{54}(-6560)}{-2} = \dfrac{6560}{108} = \dfrac{1640}{27}$

19. $60, 59, 58, \dots, 10$

$a_1 = 60$; $\quad d = -1$

$10 = 60 + (n-1)(-1)$

$10 = 60 - n + 1$

$10 = 61 - n$

$-51 = -n$

$51 = n$;

$S_{51} = \dfrac{51(60+10)}{2} = \dfrac{51(70)}{2} = 1785$

Reinforcing Basic Concepts

1. $a_n = 0.125x^3 - 2.5x^2 + 12x$

$\$71,500$

10.5 Technology Highlight

Exercise 1:

   (a) $_9C_2 = 36$

   (b) $_9C_3 = 84$

   (c) $_9C_4 = 126$

   (d) $_9C_5 = 126$

Exercise 3: $_6C_3 = 20$

10.5 Exercises

1. Experiment, well defined.

3. Distinguishable.

5. Answers will vary.

7. (a)

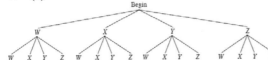

   (b) $WW, WX, WY, WZ,$
       $XW, XX, XY, XZ,$
       $YW, YX, YY, YZ,$
       ZW, ZX, ZY, ZZ

9. 32

11. $25^3 = 15,625$

13. $26 \cdot 26 \cdot 4 \cdot 10 \cdot 10 \cdot 10 = 2,704,000$

15. (a) $9^5 = 59,049$
    (b) $9 \cdot 8 \cdot 7 \cdot 6 \cdot 5 = 15,120$

17. $4 \cdot 6 \cdot 5 \cdot 3 = 360$
    360 if double vegetables are not allowed,
    $4 \cdot 6 \cdot 6 \cdot 3 = 432$
    432 if double vegetables are allowed.

19. (a) $5 \cdot 4 \cdot 3 \cdot 2 = 120$
    (b) $5^4 = 625$
    (c) $2 \cdot 3 \cdot 2 \cdot 1 = 12$

21. $4 \cdot 3 \cdot 2 \cdot 1 = 24$

23. $2 \cdot 2 \cdot 1 \cdot 1 = 4$

25. $5 \cdot 4 \cdot 3 \cdot 2 \cdot 1 = 120$

27. $1 \cdot 3 \cdot 2 \cdot 1 = 6$

29. $_{10}P_3 = 10 \cdot 9 \cdot 8 = 720$;

$$_nP_r = \frac{n!}{(n-r)!}$$

$$_{10}P_3 = \frac{10!}{(10-3)!} = \frac{10 \cdot 9 \cdot 8 \cdot 7!}{7!} = 720$$

31 $_9P_4 = 9 \cdot 8 \cdot 7 \cdot 6 = 3024$;

$$_nP_r = \frac{n!}{(n-r)!}$$

$$_9P_4 = \frac{9!}{(9-4)!} = \frac{9 \cdot 8 \cdot 7 \cdot 6 \cdot 5!}{5!} = 3024$$

33. $_8P_7 = 8 \cdot 7 \cdot 6 \cdot 5 \cdot 4 \cdot 3 \cdot 2 = 40320$

$$_nP_r = \frac{n!}{(n-r)!}$$

$$_8P_7 = \frac{8!}{(8-7)!}$$

$$= \frac{8 \cdot 7 \cdot 6 \cdot 5 \cdot 4 \cdot 3 \cdot 2 \cdot 1!}{1!} = 40320$$

35. T, R and A
    $_3P_3 = 3 \cdot 2 \cdot 1 = 6$
    TRA, TAR, RTA, RAT, ART, ATR
    3 actual words

37. $_{10}P_2 = 10 \cdot 9 = 90$

39. $_8P_3 = 8 \cdot 7 \cdot 6 = 336$

41. (a) $_6P_6 = 6 \cdot 5 \cdot 4 \cdot 3 \cdot 2 \cdot 1 = 720$
    (b) $_6P_3 = 6 \cdot 5 \cdot 4 = 120$
    (c) $_4P_4 = \cdot 4 \cdot 3 \cdot 2 \cdot 1 = 24$

43. $\dfrac{_nP_n}{p!} = \dfrac{_6P_6}{2!} = \dfrac{6 \cdot 5 \cdot 4 \cdot 3 \cdot 2 \cdot 1}{2 \cdot 1} = 360$

45. $\dfrac{_nP_n}{p!q!} = \dfrac{_6P_6}{2!3!} = \dfrac{6 \cdot 5 \cdot 4 \cdot 3 \cdot 2 \cdot 1}{2 \cdot 1 \cdot 3 \cdot 2 \cdot 1} = \dfrac{120}{2} = 60$

10.5 Exercises

47. $\dfrac{_nP_n}{p!\,q!} = \dfrac{_6P_6}{2!3!} = \dfrac{6\cdot5\cdot4\cdot3\cdot2\cdot1}{2\cdot1\cdot3\cdot2\cdot1} = \dfrac{120}{2} = 60$

49. Logic

$\dfrac{_nP_n}{p!} = \dfrac{_5P_5}{1!} = \dfrac{5\cdot4\cdot3\cdot2\cdot1}{1} = 120$

51. Lotto

$\dfrac{_nP_n}{p!\,q!} = \dfrac{_5P_5}{2!2!} = \dfrac{5\cdot4\cdot3\cdot2\cdot1}{2\cdot1\cdot2\cdot1} = \dfrac{60}{2} = 30$

53. A, A, A, N, N, B

$\dfrac{_nP_n}{p!\,q!} = \dfrac{_6P_6}{3!2!} = \dfrac{6\cdot5\cdot4\cdot3\cdot2\cdot1}{3\cdot2\cdot1\cdot2\cdot1} = \dfrac{120}{2} = 60$

BANANA

55. $_9C_4$

(a) $_nC_r = \dfrac{_nP_r}{r!}$

$_9C_4 = \dfrac{_9P_4}{4!} = \dfrac{9\cdot8\cdot7\cdot6}{4\cdot3\cdot2\cdot1} = \dfrac{3024}{24} = 126$;

(b) $_nC_r = \dfrac{n!}{r!\,(n-r)!}$

$_9C_4 = \dfrac{9!}{4!\,(9-4)!} = \dfrac{9\cdot8\cdot7\cdot6\cdot5!}{4\cdot3\cdot2\cdot1\cdot5!}$

$= \dfrac{3024}{24} = 126$

57. $_8C_5$

(a) $_nC_r = \dfrac{_nP_r}{r!}$

$_8C_5 = \dfrac{_8P_5}{5!} = \dfrac{8\cdot7\cdot6\cdot5\cdot4}{5\cdot4\cdot3\cdot2\cdot1}$

$= \dfrac{6720}{120} = 56$;

(b) $_nC_r = \dfrac{n!}{r!\,(n-r)!}$

$_8C_5 = \dfrac{8!}{5!\,(8-5)!} = \dfrac{8\cdot7\cdot6\cdot5!}{5!\cdot3\cdot2\cdot1}$

$= \dfrac{336}{6} = 56$

59. $_6C_6$

(a) $_nC_r = \dfrac{_nP_r}{r!}$

$_6C_6 = \dfrac{_6P_6}{6!} = \dfrac{6\cdot5\cdot4\cdot3\cdot2\cdot1}{6\cdot5\cdot4\cdot3\cdot2\cdot1} = 1$;

(b) $_nC_r = \dfrac{n!}{r!\,(n-r)!}$

$_6C_6 = \dfrac{6!}{6!\,(6-6)!} = \dfrac{6!}{6!} = 1$

61. $_9C_4, _9C_5$

$_9C_4 = \dfrac{9!}{4!\,(9-4)!} = \dfrac{9!}{4!5!} = \dfrac{9\cdot8\cdot7\cdot6\cdot5!}{4\cdot3\cdot2\cdot1\cdot5!}$

$_9C_4 = \dfrac{3024}{24} = 126$;

$_9C_5 = \dfrac{9!}{5!\,(9-5)!} = \dfrac{9!}{5!4!} = \dfrac{9\cdot8\cdot7\cdot6\cdot5!}{5!\cdot4\cdot3\cdot2\cdot1}$

$_9C_5 = \dfrac{3024}{24} = 126$

Verified

63. $_8C_5, _8C_3$

$_8C_5 = \dfrac{8!}{5!\,(8-5)!} = \dfrac{8!}{5!3!}$

$= \dfrac{8\cdot7\cdot6\cdot5!}{5!\cdot3\cdot2\cdot1} = \dfrac{336}{6} = 56$;

$_8C_3 = \dfrac{8!}{3!\,(8-3)!} = \dfrac{8!}{3!5!}$

$= \dfrac{8\cdot7\cdot6\cdot5!}{3\cdot2\cdot1\cdot5!} = \dfrac{336}{6} = 56$

Verified

65. $_{12}C_4 = \dfrac{12!}{4!\,(12-4)!} = \dfrac{12!}{12!8!} = \dfrac{12\cdot11\cdot10\cdot9\cdot8!}{4\cdot3\cdot2\cdot1\cdot8!}$

$= \dfrac{11880}{24} = 495$

67. $_{14}C_3 = \dfrac{14!}{3!\,(14-3)!} = \dfrac{14!}{3!\,11!} = \dfrac{14\cdot13\cdot12\cdot11!}{3\cdot2\cdot1\cdot11!}$

$= \dfrac{2184}{6} = 364$

69. $_{10}C_5 = \dfrac{10!}{5!\,(10-5)!} = \dfrac{10!}{5!5!} = \dfrac{10\cdot9\cdot8\cdot7\cdot6\cdot5!}{5\cdot4\cdot3\cdot2\cdot1\cdot5!}$

$= \dfrac{30240}{120} = 252$

71. $8! = 40,320$

73. $_nP_r = \dfrac{n!}{(n-r)!}$

$_8P_3 = \dfrac{8!}{(8-3)!} = \dfrac{8!}{5!} = \dfrac{8 \cdot 7 \cdot 6 \cdot 5!}{5!} = 336$

75. $_{20}C_5 = \dfrac{20!}{5!(20-5)!} = \dfrac{20!}{5!\,15!}$

$= \dfrac{20 \cdot 19 \cdot 18 \cdot 17 \cdot 16 \cdot 15!}{5 \cdot 4 \cdot 3 \cdot 2 \cdot 1 \cdot 15!} = \dfrac{1860480}{120} = 15,504$

77. $_8C_4 = \dfrac{8!}{4!(8-4)!} = \dfrac{8 \cdot 7 \cdot 6 \cdot 5 \cdot 4!}{4 \cdot 3 \cdot 2 \cdot 1 \cdot 4!}$

$= \dfrac{1680}{24} = 70$

79. (a) $7! = 5,040$;

$n! \approx \sqrt{2\pi} \cdot \left(n^{n+0.5}\right) \cdot e^{-n}$

$7! \approx \sqrt{2\pi} \cdot \left(7^{7+0.5}\right) \cdot e^{-7}$

$7! \approx \sqrt{2\pi} \cdot \left(7^{7.5}\right)\left(e^{-7}\right)$

$7! \approx 4980.395832$;

$\dfrac{5040 - 4980}{5040} = 0.0119 \approx 1.2\%$

(b) $10! = 3,628,800$;

$10! \approx \sqrt{2\pi} \cdot \left(10^{10+0.5}\right) \cdot e^{-10}$

$10! \approx \sqrt{2\pi} \cdot \left(10^{10.5}\right) \cdot e^{-10}$

$10! \approx 3598695.619$;

$\dfrac{3,628,800 - 3,598,696}{3,628,800} \approx 0.83\%$

81. $6^5 = 7776$

83. $9 \cdot 6 \cdot 6 = 324$

85. $8 \cdot 10 \cdot 10 = 800$

87. Exchanges: $8 \cdot 10 \cdot 10 = 800$
Area Codes: $8 \cdot 10 \cdot 10 = 800 - 16 = 784$
Final digits: $10 \cdot 10 \cdot 10 \cdot 10 = 10000$
$784 \cdot 800 \cdot 10000 = 6,272,000,000$

89. $9 \cdot 10 \cdot 10 \cdot 24 \cdot 24 = 518,400$

91. $9 \cdot 9 \cdot 8 \cdot 24 \cdot 23 = 357,696$

93. Five

$_8P_5 = \dfrac{8!}{(8-5)!} = \dfrac{8 \cdot 7 \cdot 6 \cdot 5 \cdot 4 \cdot 3!}{3!} = 6,720$

95. One

$_8P_1 = \dfrac{8!}{(8-1)!} = \dfrac{8 \cdot 7!}{7!} = 8$

97. $7 \cdot 2 \cdot 6! = 10,080$

7 ways they can sit side by side;
2 ways they can sit together, teacher 1 on the
left and teacher 2 on the right, or teacher 2
on the left and teacher 1 on the right;
the students can be seated randomly.

99. $1 \cdot 7! = 1 \cdot 7 \cdot 6 \cdot 5 \cdot 4 \cdot 3 \cdot 2 \cdot 1 = 5,040$

101. $2 \cdot 2 \cdot 6 \cdot 5 \cdot 4 \cdot 3 \cdot 2 \cdot 1 = 2880$

103. $_{15}C_6 = \dfrac{15!}{6!(15-6)!}$

$= \dfrac{15 \cdot 14 \cdot 13 \cdot 12 \cdot 11 \cdot 10 \cdot 9!}{6 \cdot 5 \cdot 4 \cdot 3 \cdot 2 \cdot 1 \cdot 9!}$

$_{15}C_6 = \dfrac{3603600}{720} = 5005$

105. $_{10}P_3 = \dfrac{10 \cdot 9 \cdot 8 \cdot 7!}{7!} = 720$

107. $26 \cdot 25 \cdot 9 \cdot 9 = 52,650$; no

109.(a) $_{10}C_3 \cdot {}_7C_2 = {}_{10}C_2 \cdot {}_8C_5$

$$\frac{10!}{7!3!} \cdot \frac{7!}{5!2!} = \frac{10!}{2!3!5!}$$

$$\frac{10!}{2!8!} \cdot \frac{8!}{5!3!} = \frac{10!}{2!3!5!}$$

(b) $_9C_3 \cdot {}_6C_2 = {}_9C_2 \cdot {}_7C_4$

$$\frac{9!}{3!6!} \cdot \frac{6!}{2!4!} = \frac{9!}{2!3!4!}$$

$$\frac{9!}{2!7!} \cdot \frac{7!}{4!3!} = \frac{9!}{2!3!4!}$$

(c) $_{11}C_4 \cdot {}_7C_5 = {}_{11}C_5 \cdot {}_6C_4$

$$\frac{11!}{4!7!} \cdot \frac{7!}{5!2!} = \frac{11!}{2!4!5!}$$

$$\frac{11!}{5!6!} \cdot \frac{6!}{4!2!} = \frac{11!}{2!4!5!}$$

(d) $_8C_3 \cdot {}_5C_2 = {}_8C_2 \cdot {}_6C_3$

$$\frac{8!}{3!5!} \cdot \frac{5!}{2!3!} = \frac{8!}{2!3!3!}$$

$$\frac{8!}{2!6!} \cdot \frac{6!}{3!3!} = \frac{8!}{2!3!3!}$$

111. $\begin{cases} 2x+y<6 \\ x+2y<6 \\ x \geq 0 \\ y \geq 0 \end{cases}$

$\begin{cases} y<-2x+6 \\ y<-\dfrac{1}{2}x+6 \\ x \geq 0 \\ y \geq 0 \end{cases}$

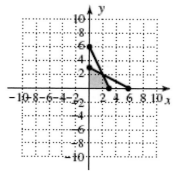

113. $\cos(2\alpha)\cos(3\alpha) - \sin(2\alpha)\sin(3\alpha)$

$= \cos(2\alpha + 3\alpha)$

$= \cos(5\alpha)$

10.6 Technology Highlight

Exercise 1: The "middle values" are repeated for $n=7$ and $n=5$ but not $n=6$ because of the symmetry of $_nC_r$.

Exercise 2: $\left({}_{10}C_4 \right)\left({}_8C_3 \right) = 11{,}760$

10.6 Exercises

1. $P(E) = \dfrac{n(E)}{n(S)}$

3. $0 \le P(E) \le 1$

   $P(S) = 1$, and $P(\sim S) = 0$

5. Answers will vary.

7. $S = \{HH, HT, TH, TT\}$; $\dfrac{1}{4}$

9. $S = \{$coach of Patriots, Cougars, Angels, Sharks, Eagles, Stars$\}$; $\dfrac{1}{6}$

11. $S = \{$nine index cards 1-9$\}$

    $P(E) = \dfrac{4}{9}$

13. $S = \{52 \text{ cards}\}$

    a. drawing a Jack: $\dfrac{4}{52} = \dfrac{1}{13}$

    b. drawing a spade: $\dfrac{13}{52} = \dfrac{1}{4}$

    c. drawing a black card: $\dfrac{26}{52} = \dfrac{1}{2}$

    d. drawing a red three: $\dfrac{2}{52} = \dfrac{1}{26}$

15. $S = \{$three males, five females$\}$

    $P(E_1) = \dfrac{1}{8}$;

    $P(E_2) = \dfrac{5}{8}$;

    $P(E_3) = \dfrac{6}{8} = \dfrac{3}{4}$

17. $S = \{$Spinner 1-4$\}$

    a. P(green): $\dfrac{3}{4}$

    b. P(less than 5): $\dfrac{4}{4} = 1$

    c. P(2): $\dfrac{1}{4}$

    d. P(prime number): $\dfrac{2}{4} = \dfrac{1}{2}$

    (1 is not considered a prime number)

19. $P(\sim C) = 1 - P(C) = 1 - \dfrac{13}{52} = \dfrac{39}{52} = \dfrac{3}{4}$

21. $10! = 3,628,800$

    $P(\sim 2) = 1 - P(2) = 1 - \dfrac{1}{7} = \dfrac{6}{7}$

23. $P(\sim F) = 1 - P(F) = 1 - 0.009 = 0.991$

25. a. $P(\text{Sum} < 4) = \dfrac{3}{36} = \dfrac{1}{12}$

    b. $P(\sim \text{Sum} < 11) = 1 - P(\text{Sum} > 10) = 1 - \dfrac{3}{36}$

    $= \dfrac{33}{36} = \dfrac{11}{12}$

    c. $P(\sim \text{Sum}9) = 1 - P(9) = 1 - \dfrac{4}{36} = \dfrac{32}{36} = \dfrac{8}{9}$

    d. $P(\sim D) = 1 - P(D) = 1 - \dfrac{6}{36} = \dfrac{30}{36} = \dfrac{5}{6}$

27. $n(E) = {}_6C_3 \cdot {}_4C_2 ; n(S) = {}_{10}C_5$

    $P(E) = \dfrac{{}_6C_3 \cdot {}_4C_2}{{}_{10}C_5} = \dfrac{120}{252} = \dfrac{10}{21}$

29. $n(E) = {}_9C_6 \cdot {}_5C_3 ; n(S) = {}_{14}C_9$

    $P(E) = \dfrac{{}_9C_6 \cdot {}_5C_3}{{}_{14}C_9} = \dfrac{840}{2002} = \dfrac{60}{143}$

10.6 Exercises

31. a. $P(\text{all red}) = \dfrac{_{26}C_5}{_{52}C_5}$

    $= \dfrac{65780}{2598960} = \dfrac{253}{9996} \approx 0.025$

    b. $P(\text{all numbered}) = \dfrac{_{36}C_5}{_{52}C_5}$

    $= \dfrac{376992}{2598960} = \dfrac{66}{455} \approx 0.145$

    b; $0.145 - 0.025 = 0.12$
    about 12 %

33. a. exactly two vegetarians

    $P(E) = \dfrac{_9C_2 \cdot _{15}C_4}{_{24}C_6} = \dfrac{49140}{134596} = 0.3651$

    b. exactly four non-vegetarians

    $P(E) = \dfrac{_{15}C_4 \cdot _9C_2}{_{24}C_6} = \dfrac{49140}{134596} = 0.3651$

    c. at least three vegetarians
    $1 - \left(P(0\text{veg}) + P(1\text{veg}) + P(2\text{veg}) \right)$

    $= 1 - \left(\dfrac{_9C_0 \cdot _{15}C_6 + _9C_1 \cdot _{15}C_5 + _9C_2 \cdot _{15}C_4}{_{24}C_6} \right)$

    $= 1 - \left(\dfrac{81172}{134596} \right) = 0.3969$

35. $P(E_1) = 0.7; P(E_2) = 0.5; P(E_1 \cap E_2) = 0.3$
    $P(E_1 \cup E_2) = P(E_1) + P(E_2) - P(E_1 \cap E_2)$
    $P(E_1 \cup E_2) = 0.7 + 0.5 - 0.3 = 0.9$

37. $P(E_1) = \dfrac{3}{8}; P(E_2) = \dfrac{3}{4}; P(E_1 \cup E_2) = \dfrac{15}{18}$

    $P(E_1 \cup E_2) = P(E_1) + P(E_2) - P(E_1 \cap E_2)$

    $\dfrac{15}{18} = \dfrac{3}{8} + \dfrac{3}{4} - P(E_1 \cap E_2)$

    $\dfrac{15}{18} = \dfrac{9}{8} - P(E_1 \cap E_2)$

    $-\dfrac{7}{24} = -P(E_1 \cap E_2)$

    $\dfrac{7}{24} = P(E_1 \cap E_2)$

39. $P(E_1 \cup E_2) = 0.72; P(E_2) = 0.56;$
    $P(E_1 \cap E_2) = 0.43$
    $P(E_1 \cup E_2) = P(E_1) + P(E_2) - P(E_1 \cap E_2)$
    $0.72 = P(E_1) + 0.56 - 0.43$
    $0.72 = P(E_1) + 0.13$
    $0.59 = P(E_1)$

41. a. $P(\text{multiple of 3 and odd}) = \dfrac{6}{36} = \dfrac{1}{6}$

    b. $P(\text{sum} > 5 \text{ and a } 3) = \dfrac{7}{36}$

    c. $P(\text{even and} > 9) = \dfrac{4}{36} = \dfrac{1}{9}$

    d. $P(\text{odd and} < 10) = \dfrac{16}{36} = \dfrac{4}{9}$

43. a. $P(\text{woman and sergeant}) = \dfrac{4}{50} = \dfrac{2}{25}$

    b. $P(\text{man and private}) = \dfrac{9}{50}$

    c. $P(\text{private and sergeant}) = \dfrac{0}{50} = 0$

    d. $P(\text{woman and officer}) = \dfrac{4}{50} = \dfrac{2}{25}$

    e. $P(\text{person in military}) = \dfrac{50}{50} = 1$

45. $\dfrac{9 \cdot 5 \cdot 10 + 9 \cdot 5 \cdot 10 - 9 \cdot 5 \cdot 5}{9 \cdot 10 \cdot 10} = \dfrac{3}{4}$

47. A number greater than 4000 can have digits
    4, 5, 6, 7, 8, or 9 in the first position.
    (6 choices)
    Multiples of 5 must end in 0 or 5.
    (2 choices)

    $\dfrac{6 \cdot 10 \cdot 10 \cdot 10}{9 \cdot 10 \cdot 10 \cdot 10} + \dfrac{9 \cdot 10 \cdot 10 \cdot 2}{9 \cdot 10 \cdot 10 \cdot 10} - \dfrac{6 \cdot 10 \cdot 10 \cdot 2}{9 \cdot 10 \cdot 10 \cdot 10}$

    $= \dfrac{6}{9} + \dfrac{1}{5} - \dfrac{6}{45} = \dfrac{11}{15}$

Chapter 10: Additional Topics in Algebra

49. a. E_1 = boxcars; E_2 = snake eyes
$$P(E_1 \cup E_2) = P(E_1) + P(E_2)$$
$$P(E_1 \cup E_2) = \frac{1}{36} + \frac{1}{36} = \frac{2}{36} = \frac{1}{18}$$

b. E_1 = sum of 7; E_2 = sum of 11
$$P(E_1 \cup E_2) = P(E_1) + P(E_2)$$
$$P(E_1 \cup E_2) = \frac{6}{36} + \frac{2}{36} = \frac{8}{36} = \frac{2}{9}$$

c. E_1 = even numbered sum;
E_2 = prime sum
$$P(E_1 \cup E_2)$$
$$= P(E_1) + P(E_2) - P(E_1 \cap E_2)$$
$$P(E_1 \cup E_2) = \frac{18}{36} + \frac{15}{36} - \frac{1}{36} = \frac{32}{36} = \frac{8}{9}$$

d. E_1 = odd numbered sum;
E_2 = multiple of four
$$P(E_1 \cup E_2) = P(E_1) + P(E_2)$$
$$P(E_1 \cup E_2) = \frac{18}{36} + \frac{9}{36} = \frac{27}{36} = \frac{3}{4}$$

e. E_1 = a sum of 15; E_2 = multiple of 12
$$P(E_1 \cup E_2) = P(E_1) + P(E_2)$$
$$P(E_1 \cup E_2) = \frac{0}{36} + \frac{1}{36} = \frac{1}{36}$$

f. E = prime number sum
$$P(E) = \frac{15}{36} = \frac{5}{12}$$

51. $P(n) = \left(\frac{1}{4}\right)^n$

a. $P(\text{spins a 2}) = \left(\frac{1}{4}\right)^1 = \frac{1}{4}$

b. $P(\text{all 4 spin a 2}) = \left(\frac{1}{4}\right)^4 = \frac{1}{256}$

c. Answers will vary.

53. a. $P(x \geq 2) = 0.25 + 0.08 = 0.33$

b. $P(x < 2) = 0.07 + 0.28 + 0.32 = 0.67$

c. $P(x \leq 4)$
$= 0.08 + 0.25 + 0.32 + 0.28 + 0.07 = 1$

d. $P(x > 4) = 0$

e. $P(x < 2 \text{ or } x > 4) = 0.67 + 0 = 0.67$

f. $P(x \geq 3) = 0.08$

55. Total = 200

a. $P(\text{Isosceles}) = \dfrac{\frac{1}{2}(200)}{200} = \dfrac{1}{2}$

b. $P(\text{Right triangle}) = \dfrac{\frac{1}{2}(200)}{200} = \dfrac{1}{2}$

c. $5^2 + h^2 = 10^2$
$h^2 = 75$
$h = 5\sqrt{3};$
$$P(\text{Equilateral}) = \dfrac{\frac{1}{2}(10)5\sqrt{3}}{200} = \dfrac{\sqrt{3}}{8}$$
≈ 0.2165

57. a. $P(x \geq 4) = \dfrac{\pi(6)^2}{\pi(8)^2} = \dfrac{36}{64} = \dfrac{9}{16}$

b. $P(x \geq 6) = \dfrac{\pi(4)^2}{\pi(8)^2} = \dfrac{16}{64} = \dfrac{1}{4}$

c. $P(\text{exactly 8}) = \dfrac{\pi(2)^2}{\pi(8)^2} = \dfrac{4}{64} = \dfrac{1}{16}$

d. $P(x = 4) = \dfrac{\pi(6)^2 - \pi(4)^2}{\pi(8)^2} = \dfrac{20\pi}{64\pi} = \dfrac{5}{16}$

59. n = 13, 3R, 6B, 4W

   a. $P(\text{red, blue}) = \dfrac{3}{13} \cdot \dfrac{6}{12} = \dfrac{3}{26}$

   b. $P(\text{blue, red}) = \dfrac{6}{13} \cdot \dfrac{3}{12} = \dfrac{3}{26}$

   c. $P(\text{white, white}) = \dfrac{4}{13} \cdot \dfrac{3}{12} = \dfrac{1}{13}$

   d. $P(\text{blue, not red}) = \dfrac{6}{13} \cdot \dfrac{9}{12} = \dfrac{9}{26}$

   e. $P(\text{white, not blue}) = \dfrac{4}{13} \cdot \dfrac{6}{12} = \dfrac{2}{13}$

   f. $P(\text{not red, not blue})$

$$= \dfrac{6}{13} \cdot \dfrac{3}{12} + \dfrac{6}{13} \cdot \dfrac{4}{12} + \dfrac{4}{13} \cdot \dfrac{3}{12} + \dfrac{4}{13} \cdot \dfrac{3}{12}$$

$$= \dfrac{3}{26} + \dfrac{4}{26} + \dfrac{2}{26} + \dfrac{2}{26} = \dfrac{11}{26}$$

61.   Let C represent correct.
Let W represent wrong.

   a. $P(\text{Grade} \geq 80\%) = \left(\dfrac{1}{2}\right)^3 = \dfrac{1}{8}$

With 3 questions, the only grade greater than or equal to 80% would be a 100% since 2 out of three questions correct only give 67%.

Possible outcomes: CCC, CCW, CWW, CWC, WCC, WCW, WWC, WWW

   b. $P(\text{Grade} \geq 80\%) = \left(\dfrac{1}{2}\right)^4 = \dfrac{1}{16}$

With 4 questions, the only grade greater than or equal to 80% would be a 100% since 3 out of four questions correct only give 75%.

Possible outcomes: CCCC, CCCW, CCWC, CCWW, CWCC, CWCW, CWWC, CWWW, WCCW, WCCC, WCWC, WCWW, WWCC, WWCW, WWWC, WWWW.

   c. $P(\text{Grade} \geq 80\%) = \dfrac{1}{32} + \dfrac{5}{32} = \dfrac{6}{32} = \dfrac{3}{16}$

With 5 questions, the only grade greater than 80% would be a 100% but 4 out of five questions would give 80%. Thus, we need 5 or 4 correct answers.

Possible outcomes: CCCCC, CCCCW, CCCWC, CCWW, CCWCC, CCWCW, CCWWC, CCWWW, CWCCC, CWCCW, CWCWC, CWCWW, CWWCC, CWWCW, CWWWC, CWWWW, WCCCC, WCCCW, WCCWC, WCWW, WCWCC, WCWCW, WCWWC, WCWWW, WWCCC, WWCCW, WWCWC, WWCWW, WWWCC, WWWCW, WWWWC, WWWWW.

63.   a. $P(\text{career and opposed}) = \dfrac{47}{100}$

   b. $P(\text{medical and supported}) = \dfrac{8}{100} = \dfrac{2}{25}$

   c. $P(\text{military and opposed}) = \dfrac{3}{100}$

   d. $P(\text{legal or business and opposed})$

$$= \dfrac{18}{100} = \dfrac{9}{50}$$

   e. $P(\text{academic or medical and supported})$

$$= \dfrac{11}{100}$$

65. a. $\dfrac{_6C_4 \cdot _5C_4}{_{15}C_8} = \dfrac{5}{429}$

    b. $\dfrac{_4C_3 \cdot _6C_5}{_{15}C_8} = \dfrac{8}{2145}$

67. $\dfrac{8!}{2!3!} = \dfrac{8 \cdot 7 \cdot 6 \cdot 5 \cdot 4 \cdot 3!}{2 \cdot 3!} = 3360$;

    $P(\text{parallel}) = \dfrac{1}{3360}$

69. $\left(\dfrac{1}{2}\right)^x$ where $x = $ number of flips

    $P(\text{exactly 20 heads}) = \left(\dfrac{1}{2}\right)^{20} = \dfrac{1}{1048576}$

    P(winning the lottery) = will vary; but
    P(exactly 20 heads) > P(winning the lottery).

71. $\csc\theta = 3 \quad \cos\theta < 0 \rightarrow \theta \in \text{QII}$

    $1^2 + b^2 = 3^2$

    $b^2 = 9 - 1$

    $b^2 = 8$

    $b = 2\sqrt{2}$

    $\sin\theta = \dfrac{1}{3}$

    $\cos\theta = \dfrac{-2\sqrt{2}}{3}$

    $\tan\theta = -\dfrac{1}{2\sqrt{2}}$

    $\cot\theta = -2\sqrt{2}$

    $\sec\theta = -\dfrac{3}{2\sqrt{2}}$

73. $\cos\theta = \dfrac{-21}{29} \quad \theta \in \text{QII}$

    $a^2 + (-21)^2 = 29^2$

    $a^2 = 841 - 441$

    $a^2 = 400$

    $a = 20$

    $\cos(2\theta) = \cos^2\theta - \sin^2\theta = \left(\dfrac{-21}{29}\right)^2 - \left(\dfrac{20}{29}\right)^2$

    $= \dfrac{441}{841} - \dfrac{400}{841} = \dfrac{41}{841}$

    $\sin(2\theta) = 2\sin\theta\cos\theta = 2\left(\dfrac{20}{29}\right)\left(\dfrac{-21}{29}\right)$

    $= \dfrac{-840}{841}$

    $\tan(2\theta) = \dfrac{2\tan\theta}{1 - \tan^2\theta}$

    $= \dfrac{2\left(\dfrac{20}{-21}\right)}{1 - \left(\dfrac{-20}{21}\right)^2}$

    $= \dfrac{\dfrac{40}{-21}}{1 - \dfrac{400}{441}}$

    $= \dfrac{-40}{21} \div \left(\dfrac{441 - 400}{441}\right)$

    $= \dfrac{-40}{21} \cdot \dfrac{441}{41} = \dfrac{-840}{41}$

10.7 Exercises

1. One

3. $\left(a+\left(-2b\right)\right)^5$

5. Answers will vary.

7. $(x+y)^5$

$x^5+5x^4y+10x^3y^2+10x^2y^3+5xy^4+y^5$

9. $(2x+3)^4$

$=(2x)^4+4(2x)^3(3)+6(2x)^2(3)^2$
$\quad+4(2x)(3)^3+3^4$
$=16x^4+4\left(24x^3\right)+6\left(36x^2\right)+4(54x)+81$
$=16x^4+96x^3+216x^2+216x+81$

11. $(1-2i)^5$

$=1^5+5(1)^4(-2i)+10(1)^3(-2i)^2$
$\quad+10(1)^2(-2i)^3+5(1)(-2i)^4+(-2i)^5$
$=1-10i-40+80i+80-32i$
$=41+38i$

13. $\binom{n}{r}=\dfrac{n!}{r!(n-r)!}$

$\binom{7}{4}=\dfrac{7!}{4!(7-4)!}=\dfrac{7\cdot6\cdot5\cdot4!}{4!\cdot3\cdot2\cdot1}=\dfrac{210}{6}=35$

15. $\binom{n}{r}=\dfrac{n!}{r!(n-r)!}$

$\binom{5}{3}=\dfrac{5!}{3!(5-3)!}=\dfrac{5\cdot4\cdot3!}{3!2\cdot1}=\dfrac{20}{2}=10$

17. $\binom{n}{r}=\dfrac{n!}{r!(n-r)!}$

$\binom{20}{17}=\dfrac{20!}{17!(20-17)!}=\dfrac{20\cdot19\cdot18\cdot17!}{17!3\cdot2\cdot1}$

$=\dfrac{6840}{6}=1140$

19. $\binom{n}{r}=\dfrac{n!}{r!(n-r)!}$

$\binom{40}{3}=\dfrac{40!}{3!(40-3)!}=\dfrac{40\cdot39\cdot38\cdot37!}{3\cdot2\cdot1\cdot37!}$

$=\dfrac{59280}{6}=9880$

21. $\binom{n}{r}=\dfrac{n!}{r!(n-r)!}$

$\binom{6}{0}=\dfrac{6!}{0!(6-0)!}=\dfrac{6!}{6!}=1$

23. $\binom{n}{r}=\dfrac{n!}{r!(n-r)!}$

$\binom{15}{15}=\dfrac{15!}{15!(15-15)!}=\dfrac{15!}{15!(0)!}=1$

25. $(c+d)^5$; $a=c$; $b=d$; $n=5$

$(c+d)^5$

$=\binom{5}{0}c^5d^0+\binom{5}{1}c^4d+\binom{5}{2}c^3d^2+\binom{5}{3}c^2d^3$
$\qquad+\binom{5}{4}cd^4+\binom{5}{5}c^0d^5$

$=1c^5+5c^4d+\dfrac{5!}{2!3!}c^3d^2+\dfrac{5!}{3!2!}c^2d^3$
$\qquad+\dfrac{5!}{4!1!}cd^4+1d^5$

$=c^5+5c^4d+10c^3d^2+10c^2d^3+5cd^4+d^5$

27. $(a-b)^6$; $a=a$; $b=b$; $n=6$

$(a-b)^6$

$=\binom{6}{0}a^6b^0-\binom{6}{1}a^5b^1+\binom{6}{2}a^4b^2-\binom{6}{3}a^3b^3$
$\qquad+\binom{6}{4}a^2b^4-\binom{6}{5}ab^5+\binom{6}{6}a^0b^6$

$=1a^6b^0-6a^5b+\dfrac{6!}{2!4!}a^4b^2-\dfrac{6!}{3!3!}a^3b^3$
$\qquad+\dfrac{6!}{2!4!}a^2b^4-\dfrac{6!}{1!5!}ab^5+1a^0b^6$

$=a^6-6a^5b+15a^4b^2-20a^3b^3$
$\qquad+15a^2b^4-6ab^5+b^6$

29. $(2x-3)^4$; $a=2x$; $b=3$; $n=4$

$$=\binom{4}{0}(2x)^4(3)^0-\binom{4}{1}(2x)^3(3)^1+\binom{4}{2}(2x)^2(3)^2$$
$$-\binom{4}{3}(2x)^1(3)^3+\binom{4}{4}(2x)^0(3)^4$$
$$=1(16x^4)-4(8x^3)(3)+\frac{4!}{2!2!}(4x^2)(9)$$
$$-\frac{4!}{3!1!}(2x)(27)+1(1)(81)$$
$$=16x^4-96x^3+6(36x^2)-4(54x)+81$$
$$=16x^4-96x^3+216x^2-216x+81$$

31. $(1-2i)^3$; $a=1$; $b=2i$; $n=3$

$(1-2i)^3$

$$=\binom{3}{0}1^3(2i)^0-\binom{3}{1}1^2(2i)+\binom{3}{2}1(2i)^2-\binom{3}{3}1^0(2i)^3$$
$$=1(1)-3(1)(2i)+\frac{3!}{2!1!}(4i^2)-(8i^3)$$
$$=1-6i+3(-4)+8i$$
$$=1-6i-12+8i$$
$$=-11+2i$$

33. $(x+2y)^9$; $a=x$; $b=2y$; $n=9$

$$=\binom{9}{0}x^9(2y)^0+\binom{9}{1}x^8(2y)^1+\binom{9}{2}x^7(2y)^2$$
$$=1(x^9)(1)+9x^8(2y)+\frac{9!}{2!7!}x^7(4y^2)$$
$$=x^9+18x^8y+36x^7(4y^2)$$
$$=x^9+18x^8y+144x^7y^2$$

35. $\left(v^2-\frac{1}{2}w\right)^{12}$; $a=v^2$; $b=\frac{1}{2}w$; $n=12$

$$=\binom{12}{0}(v^2)^{12}\left(\frac{1}{2}w\right)^0-\binom{12}{1}(v^2)^{11}\left(\frac{1}{2}w\right)^1$$
$$+\binom{12}{2}(v^2)^{10}\left(\frac{1}{2}w\right)^2$$
$$=v^{24}-12\left(v^{22}\left(\frac{1}{2}w\right)\right)+\frac{12!}{2!10!}v^{20}\left(\frac{1}{4}w^2\right)$$
$$=v^{24}-6v^{22}w+66v^{20}\left(\frac{1}{4}w^2\right)$$
$$=v^{24}-6v^{22}w+\frac{33}{2}v^{20}w^2$$

37. $(x+y)^7$; 4th term

$a=x$; $b=y$; $n=7$; $r=3$

$$\binom{n}{r}a^{n-r}b^r$$
$$\binom{7}{3}(x)^{7-3}y^3=\frac{7!}{3!4!}x^4y^3=35x^4y^3$$

39. $(p-2)^8$; 7th term

$a=p$; $b=2$; $n=8$; $r=6$

$$\binom{n}{r}a^{n-r}b^r$$
$$\binom{8}{6}p^{8-6}(2)^6=\frac{8!}{6!2!}p^2(64)$$
$$=28p^2(64)=1792p^2$$

41. $(2x+y)^{12}$; 11th term

$a=2x$; $b=y$; $n=12$; $r=10$

$$\binom{n}{r}a^{n-r}b^r$$
$$\binom{12}{10}(2x)^{12-10}y^{10}=\frac{12!}{10!2!}(2x)^2y^{10}$$
$$=66(4x^2)y^{10}=264x^2y^{10}$$

43. $P(k)=\binom{n}{k}\left(\frac{1}{2}\right)^k\left(\frac{1}{2}\right)^{n-k}$

$$P(5)=\binom{10}{5}\left(\frac{1}{2}\right)^5\left(\frac{1}{2}\right)^{10-5}$$
$$=\frac{10!}{5!5!}\left(\frac{1}{32}\right)\left(\frac{1}{2}\right)^5$$
$$=252\left(\frac{1}{32}\right)\left(\frac{1}{32}\right)$$
$$=\frac{252}{1024}$$
$$\approx 0.25\text{; Answers will vary.}$$

45. a. $P(\text{exactly } 3)$

$$= \binom{5}{3}(1-0.347)^{5-3}(0.347)^3$$

$$= 10(0.653)^2(0.347)^3 \approx 0.178\,;\, 17.8\%$$

b. $P(\text{at least } 3) = P(3) + P(4) + P(5)$

$$= \binom{5}{3}(0.653)^2(0.347)^3 + \binom{5}{4}(0.653)^1(0.347)^4$$

$$+ \binom{5}{5}(0.653)^0(0.347)^4$$

$$= 0.1782 + 0.0473 + 0.01449 \approx 0.230\,;$$
23.0%

47. a. $P(\text{exactly } 5)$

$$= \binom{8}{5}(1-0.94)^{8-5}(0.94)^5$$

$$= 56(0.06)^3(0.94)^5 \approx 0.0088;\, 0.88\%$$

b. $P(\text{exactly } 6)$

$$\binom{8}{6}(1-0.94)^{8-6}(0.94)^6$$

$$= 28(0.06)^2(0.94)^6 \approx 6.9\%$$

c. $P(\text{at least } 6) = P(6) + P(7) + P(8)$

$$= \binom{8}{6}(0.06)^2(0.94)^6 + \binom{8}{7}(0.06)(0.94)^7$$

$$+ \binom{8}{8}(0.06)^0(0.94)^8$$

$$= 0.0695 + 0.3113 + 0.6096$$
$$= 0.9904 \approx 0.99;\, 99\%$$

d. $P(\text{none}) = P(\text{all on time})$

$$= \binom{8}{8}(0.06)^0(0.94)^8 \approx 0.610;\, 61.0\%$$

49. (a) $\%\text{error} = \dfrac{\text{approximate value}}{\text{actual value}}$

$$= \dfrac{1.476}{(1.02)^{20}} = 0.9933 = 99.33\%$$

(b) $\%\text{error} = \dfrac{\text{approximate value}}{\text{actual value}}$

$$\dfrac{1.4}{(1.02)^{20}} = 0.94216 = 94.22\%$$

51. $\dbinom{n}{k} = \dbinom{n}{n-k}$

$$\binom{6}{6} = \binom{6}{0}$$

$$\dfrac{6!}{6!0!} = \dfrac{6!}{0!6!}$$

$$1 = 1;$$

$$\binom{6}{4} = \binom{6}{2}$$

$$\dfrac{6!}{4!2!} = \dfrac{6!}{2!4!}$$

$$15 = 15;$$

$$\binom{6}{5} = \binom{6}{1}$$

$$\dfrac{6!}{5!1!} = \dfrac{6!}{1!5!}$$

$$6 = 6$$

53. $f(x) = \begin{cases} x+2 & x \le 2 \\ (x-4)^2 & x > 2 \end{cases}$

$$f(3) = (3-4)^2 = (-1)^2 = 1$$

55. $g(x) = x^3 - x^2 - 6x$

$$x^3 - x^2 - 6x = 0$$
$$x(x^2 - x - 6) = 0$$
$$x(x-3)(x+2) = 0$$

$x = 0; \ x = 3; \ x = -2$

x-intercepts: $(0, 0), (3, 0), (-2, 0)$
y-intercept: $(0, 0)$
$g(x) \uparrow : (-\infty, -1) \cup (2, \infty)$
$g(x) \downarrow : (-1, 2)$
$g(x) > 0 : (-2, 0) \cup (3, \infty)$

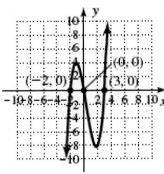

Chapter 10 Summary Exercises

1. $a_n = 5n - 4$

$a_1 = 5(1) - 4 = 5 - 4 = 1$;
$a_2 = 5(2) - 4 = 10 - 4 = 6$;
$a_3 = 5(3) - 4 = 15 - 4 = 11$;
$a_4 = 5(4) - 4 = 20 - 4 = 16$;
$a_{10} = 5(10) - 4 = 50 - 4 = 46$
$1, 6, 11, 16; a_{10} = 46$

3. $1, 16, 81, 256, \ldots$

$a_n = n^4$
$a_6 = 6^4 = 1296$

5. $\dfrac{1}{2}, \dfrac{1}{4}, \dfrac{1}{8}, \ldots$

$a_n = \dfrac{1}{2^n}$

$S_8 = a_1 + a_2 + a_3 + a_4 + a_5 + a_6 + a_7 + a_8$

$S_8 = \dfrac{1}{2} + \dfrac{1}{4} + \dfrac{1}{8} + \dfrac{1}{16} + \dfrac{1}{32} + \dfrac{1}{64} + \dfrac{1}{128} + \dfrac{1}{256}$

$S_8 = \dfrac{255}{256}$

7. $\displaystyle\sum_{n=1}^{7} n^2$

$$= 1^2 + 2^2 + 3^2 + 4^2 + 5^2 + 6^2 + 7^2$$
$$= 1 + 4 + 9 + 16 + 25 + 36 + 49$$
$$= 140$$

9. $a_n = \dfrac{n!}{(n-2)!}$

$a_2 = \dfrac{2!}{(2-2)!} = \dfrac{2!}{0!} = \dfrac{2 \cdot 1}{1} = 2$;

$a_3 = \dfrac{3!}{(3-2)!} = \dfrac{3!}{1!} = \dfrac{3 \cdot 2}{1} = 6$;

$a_4 = \dfrac{4!}{(4-2)!} = \dfrac{4 \cdot 3 \cdot 2!}{2!} = 12$;

$a_5 = \dfrac{5!}{(5-2)!} = \dfrac{5 \cdot 4 \cdot 3!}{3!} = 20$;

$a_6 = \dfrac{6!}{(6-2)!} = \dfrac{6 \cdot 5 \cdot 4!}{4!} = 30$

$2, 6, 12, 20, 30$

11. $\displaystyle\sum_{n=1}^{7} n^2 + \sum_{n=1}^{7} (3n - 2)$

$\displaystyle\sum_{n=1}^{7} (n^2 + 3n - 2)$

$$= (1^2 + 3(1) - 2) + (2^2 + 3(2) - 2)$$
$$+ (3^2 + 3(3) - 2) + (4^2 + 3(4) - 2)$$
$$+ (5^2 + 3(5) - 2) + (6^2 + 3(6) - 2)$$
$$+ (7^2 + 3(7) - 2)$$
$$= (1 + 3 - 2) + (4 + 6 - 2) + (9 + 9 - 2)$$
$$+ (16 + 12 - 2) + (25 + 15 - 2) + (36 + 18 - 2)$$
$$+ (49 + 21 - 2)$$
$$= 2 + 8 + 16 + 26 + 38 + 52 + 68$$
$$= 210$$

13. $3, 1, -1, -3 \ldots$; find a_{35}

$a_1 = 3; \ d = -2$
$a_n = a_1 + (n-1)d$
$a_n = 3 + (n-1)(-2)$
$a_{35} = 3 + (35 - 1)(-2) = 3 + 34(-2)$
$= 3 - 68 = -65$

15. $1 + 4 + 7 + 10 + \ldots + 88$

$a_n = a_1 + (n-1)d$

$88 = 1 + (n-1)(3)$

$88 = 1 + 3n - 3$

$88 = -2 + 3n$

$90 = 3n$

$30 = n$;

$a_1 = 1; \quad d = 3; \quad a_n = 88; \quad n = 30$;

$S_n = \dfrac{n(a_1 + a_n)}{2}$

$S_{30} = \dfrac{30(1+88)}{2} = \dfrac{30(89)}{2} = 1335$

17. $1 + \dfrac{3}{4} + \dfrac{1}{2} + \dfrac{1}{4} + \ldots; \; S_{15}$

$a_n = a_1 + (n-1)d$

$a_{15} = 1 + (15-1)\left(\dfrac{-1}{4}\right) = 1 + 14\left(\dfrac{-1}{4}\right) = \dfrac{-5}{2}$;

$a_1 = 1; \quad d = \dfrac{-1}{4}; \quad a_{15} = \dfrac{-5}{2}; \quad n = 15$

$S_n = \dfrac{n(a_1 + a_n)}{2}$

$S_{15} = \dfrac{15\left(1 - \dfrac{5}{2}\right)}{2} = \dfrac{15\left(\dfrac{-3}{2}\right)}{2} = -11.25$

19. $\displaystyle\sum_{n=1}^{40} (4n - 1)$

$a_1 = 4(1) - 1 = 4 - 1 = 3$;

$a_2 = 4(2) - 1 = 8 - 1 = 7$;

$a_{40} = 4(40) - 1 = 160 - 1 = 159$;

$a_1 = 3; \quad d = 4; \quad a_{40} = 159; \quad n = 40$;

$S_n = \dfrac{n(a_1 + a_n)}{2}$

$S_{40} = \dfrac{40(3 + 159)}{2} = \dfrac{40(162)}{2} = 3240$

21. $a_1 = 4, r = \sqrt{2}$; find a_7

$a_n = a_1 r^{n-1}$

$a_7 = 4\left(\sqrt{2}\right)^{7-1} = 4\left(\sqrt{2}\right)^6 = 4(8) = 32$

23. $16 - 8 + 4 - \ldots$ find S_7

$a_1 = 16; \quad r = \dfrac{-1}{2}$;

$S_n = \dfrac{a_1\left(1 - r^n\right)}{1 - r}$

$S_7 = \dfrac{16\left(1 - \left(\dfrac{-1}{2}\right)^7\right)}{1 - \left(\dfrac{-1}{2}\right)} = \dfrac{16\left(1 - \left(\dfrac{-1}{128}\right)\right)}{\dfrac{3}{2}}$

$= \dfrac{16\left(\dfrac{129}{128}\right)}{\dfrac{3}{2}} = 10.75$

25. $\dfrac{4}{5} + \dfrac{2}{5} + \dfrac{1}{5} + \dfrac{1}{10} + \ldots$ find S_{12}

$a_1 = \dfrac{4}{5}; \quad r = \dfrac{\dfrac{2}{5}}{\dfrac{4}{5}} = \dfrac{1}{2}$;

$S_n = \dfrac{a_1\left(1 - r^n\right)}{1 - r}$

$S_{12} = \dfrac{\dfrac{4}{5}\left(1 - \left(\dfrac{1}{2}\right)^{12}\right)}{1 - \dfrac{1}{2}} = \dfrac{\dfrac{4}{5}\left(1 - \dfrac{1}{4096}\right)}{\dfrac{1}{2}}$

$= \dfrac{\dfrac{4}{5}\left(\dfrac{4095}{4096}\right)}{\dfrac{1}{2}} = \dfrac{819}{512}$

27. $5 + 0.5 + 0.05 + 0.005 + \ldots$

$a_1 = 5; \quad r = \dfrac{0.5}{5} = \dfrac{1}{10}$;

$S_\infty = \dfrac{a_1}{1 - r}$

$S_\infty = \dfrac{5}{1 - \dfrac{1}{10}} = \dfrac{5}{\dfrac{9}{10}} = \dfrac{50}{9}$

29. $\displaystyle\sum_{n=1}^{8} 5\left(\frac{2}{3}\right)^n$

$a_1 = 5\left(\frac{2}{3}\right) = \frac{10}{3}$;

$a_2 = 5\left(\frac{2}{3}\right)^2 = 5\left(\frac{4}{9}\right) = \frac{20}{9}$;

$r = \dfrac{\frac{20}{9}}{\frac{10}{3}} = \frac{2}{3}$;

$a_1 = \frac{10}{3}$; $r = \frac{2}{3}$;

$S_n = \dfrac{a_1\left(1-r^n\right)}{1-r}$

$S_8 = \dfrac{\frac{10}{3}\left(1-\left(\frac{2}{3}\right)^8\right)}{1-\frac{2}{3}} = \dfrac{\frac{10}{3}\left(1-\frac{256}{6561}\right)}{\frac{1}{3}}$

$= \dfrac{\frac{10}{3}\left(\frac{6305}{6561}\right)}{\frac{1}{3}} = \dfrac{63050}{6561}$

31. $\displaystyle\sum_{n=1}^{\infty} 5\left(\frac{1}{2}\right)^n$

$a_1 = 5\left(\frac{1}{2}\right) = \frac{5}{2}$;

$a_2 = 5\left(\frac{1}{2}\right)^2 = 5\left(\frac{1}{4}\right) = \frac{5}{4}$;

$r = \dfrac{\frac{5}{4}}{\frac{5}{2}} = \frac{1}{2}$;

$a_1 = \frac{5}{2}$; $r = \frac{1}{2}$;

$S_\infty = \dfrac{a_1}{1-r}$

$S_\infty = \dfrac{\frac{5}{2}}{1-\frac{1}{2}} = \dfrac{\frac{5}{2}}{\frac{1}{2}} = 5$

33. $a_1 = 121500$; $r = \frac{2}{3}$

$a_n = a_1 r^n$

$a_7 = 121500\left(\frac{2}{3}\right)^7$

$= 121500\left(\frac{128}{2187}\right) \approx 7111.1 ft^3$

35. $1+2+3+4+5+\ldots+n$;

The needed components are:

$a_n = n$; $a_k = k$; $a_{k+1} = k+1$;

$S_n = \dfrac{n(n+1)}{2}$; $S_k = \dfrac{k(k+1)}{2}$;

$S_{k+1} = \dfrac{(k+1)(k+2)}{2}$

1. Show S_n is true for $n = 1$.

$S_1 = \dfrac{1(1+1)}{2} = \frac{2}{2} = 1$

Verified

2. Assume S_k is true:

$1+2+3+\ldots+k = \dfrac{k(k+1)}{2}$

and use it to show the truth of S_{k+1} follows. That is:

$1+2+3+\ldots+k+(k+1)$

$= \dfrac{(k+1)(k+1+1)}{2}$

$S_k + a_{k+1} = S_{k+1}$

Working with the left hand side:

$1+2+3+\ldots+k+(k+1)$

$= \dfrac{k(k+1)}{2}+k+1$

$= \dfrac{k(k+1)+2(k+1)}{2}$

$= \dfrac{(k+1)(k+2)}{2}$

$= S_{k+1}$

Since the truth of S_{k+1} follows from S_k, the formula is true for all n.

37. $S_n : 3^n \geq 2n+1$

$S_k : 3^k \geq 2k+1$

$S_{k+1} : 3^{k+1} \geq 2(k+1)+1$

$S_{k+1} : 3^{k+1} \geq 2k+3$

1. Show S_n is true for $n = 1$.

$S_1 :$

$3^1 \geq 2(1)+1$

$3 \geq 3$

2. Assume $S_k : 3^k \geq 2k+1$ is true and use it to show the truth of S_{k+1} follows.

That is: $3^{k+1} \geq 2k+3$.

Working with the left hand side:

$3^{k+1} = 3(3^k)$

$= 3(2k+1)$

$= 6k+3$

$\geq 2k+3$

Since k is a positive integer,

$6k+3 \geq 2k+3$ showing $3^{k+1} \geq 2k+3$.

Verified.

39. $3^n - 1$ is divisible by 2

1. Show S_n is true for $n = 1$.

$S_n : 3^n - 1 = 2m$

$S_1 :$

$3^1 - 1 = 2m$

$2 = 2m$

2. Assume $S_k : 3^k - 1 = 2m$ for $m \in Z$. and use it to show the truth of S_{k+1} follows.

That is: $S_{k+1} : 3^{k+1} - 1 = 2p$ for $p \in Z$.

Working with the left hand side:

$= 3^{k+1} - 1$

$= 3(3^k) - 1$

$= 3(2m+1) - 1$

$= 6m + 3 - 1$

$= 6m + 2$

$= 2(3m+1)$

$3^{k+1} - 1$ is divisible by 2, which is $= S_{k+1}$.

Verified

41. (a) $10 \cdot 9 \cdot 8 = 720$

(b) $10 \cdot 10 \cdot 10 = 1000$

43. $_{12}C_3 = \dfrac{_{12}P_3}{3!} = \dfrac{1320}{6} = 220$

45. a. $7! = 7 \cdot 6 \cdot 5 \cdot 4 \cdot 3 \cdot 2 \cdot 1 = 5040$

b. $_7P_4 = \dfrac{7!}{(7-4)!} = \dfrac{7 \cdot 6 \cdot 5 \cdot 4 \cdot 3!}{3!} = 840$

c. $_7C_4 = \dfrac{_7P_4}{4!} = \dfrac{840}{4 \cdot 3 \cdot 2 \cdot 1} = \dfrac{840}{24} = 35$

47. $\dfrac{_8P_8}{3!2} = \dfrac{8!}{3!2!} = \dfrac{8 \cdot 7 \cdot 6 \cdot 5 \cdot 4 \cdot 3!}{3!2 \cdot 1} = \dfrac{6720}{2} = 3360$

49. $P(\text{ten or face card}) = \dfrac{4}{52} + \dfrac{12}{52} = \dfrac{16}{52} = \dfrac{4}{13}$

51. $P(\sim 3) = 1 - P(3) = 1 - \dfrac{1}{6} = \dfrac{5}{6}$

53. $n(E) = {}_7C_4 \cdot {}_5C_3$; $n(S) = {}_{12}C_7$

$$P(E) = \frac{n(E)}{n(S)} = \frac{{}_7C_4 \cdot {}_5C_3}{{}_{12}C_7} = \frac{350}{792} = \frac{175}{396}$$

55. a. $\displaystyle \binom{7}{5} = \frac{7!}{5!2!} = \frac{7 \cdot 6 \cdot 5!}{5! \cdot 2 \cdot 1} = \frac{42}{2} = 21$

   b. $\displaystyle \binom{8}{3} = \frac{8!}{3!5!} = \frac{8 \cdot 7 \cdot 6 \cdot 5!}{3 \cdot 2 \cdot 1 \cdot 5!} = \frac{336}{6} = 56$

57. a. $\left(a + \sqrt{3}\right)^8$

$$\binom{8}{0}a^8 + \binom{8}{1}a^7\left(\sqrt{3}\right)$$
$$+ \binom{8}{2}a^6\left(\sqrt{3}\right)^2 + \binom{8}{3}a^5\left(\sqrt{3}\right)^3$$
$$= a^8 + 8a^7\sqrt{3} + \frac{8!}{2!6!}a^6(3)$$
$$+ \frac{8!}{3!5!}a^5\left(3\sqrt{3}\right)$$
$$= a^8 + 8\sqrt{3}a^7 + 28a^6(3) + 56a^5\left(3\sqrt{3}\right)$$
$$= a^8 + 8\sqrt{3}a^7 + 84a^6 + 168\sqrt{3}a^5$$

   b. $(5a + 2b)^7$

$$\binom{7}{0}(5a)^7 + \binom{7}{1}(5a)^6(2b)$$
$$+ \binom{7}{2}(5a)^5(2b)^2 + \binom{7}{3}(5a)^4(2b)^3$$
$$= 78125a^7 + 7\left(15625a^6\right)(2b)$$
$$+ \frac{7!}{2!5!}\left(3125a^5\right)\left(4b^2\right) + \frac{7!}{3!4!}\left(625a^4\right)\left(8b^3\right)$$
$$= 78125a^7 + 218750a^6b$$
$$+ 21\left(12500a^5b^2\right) + 35\left(5000a^4b^3\right)$$
$$= 78125a^7 + 218750a^6b$$
$$+ 262500a^5b^2 + 175000a^4b^3$$

Chapter 10 Mixed Review

1. a. $120, 163, 206, 249, \ldots$
   Arithmetic
   b. $4, 4, 4, 4, 4, 4, \ldots$
   $a_n = 4$
   c. $1, 2, 6, 24, 120, 720, 5040, \ldots$
   $a_n = n!$
   d. $2.00, 1.95, 1.90, 1.85, \ldots$
   Arithmetic
   e. $\dfrac{5}{8}, \dfrac{5}{64}, \dfrac{5}{512}, \dfrac{5}{4096}, \ldots$
   Geometric
   f. $-5.5, 6.05, -6.655, 7.3205, \ldots$
   Geometric
   g. $0.\overline{1}, 0.\overline{2}, 0.\overline{3}, 0.\overline{4}$
   Arithmetic
   h. $525, 551.25, 578.8125, \ldots$
   Geometric
   i. $\dfrac{1}{2}, \dfrac{1}{4}, \dfrac{1}{6}, \dfrac{1}{8}, \ldots$
   $a_n = \dfrac{1}{2n}$

3. $2 \cdot 25 \cdot 24 \cdot 23 = 27600$

5. $a_1 = 0.1, r = 5$
   $a_2 = 5(0.1) = 0.5$;
   $a_3 = 5(0.5) = 2.5$;
   $a_4 = 5(2.5) = 12.5$;
   $a_5 = 5(12.5) = 62.5$;
   $a_n = a_1 r^{n-1}$
   $a_{15} = 0.1(5)^{14} = 610,351,562.5$
   $0.1, 0.5, 2.5, 12.5, 62.5; a_{20} = 1907348632812$

7. $P(\sim \text{doubles}) = 1 - P(D) = 1 - \dfrac{6}{36} = \dfrac{30}{36} = \dfrac{5}{6}$

9. a. $\displaystyle\sum_{n=1}^{\infty}\left(\dfrac{2}{3}\right)^n$

$a_1 = \dfrac{2}{3}; \quad a_2 = \left(\dfrac{2}{3}\right)^2 = \dfrac{4}{9}; \quad r = \dfrac{\frac{4}{9}}{\frac{2}{3}} = \dfrac{2}{3}$

$S_\infty = \dfrac{a_1}{1-r}$

$S_\infty = \dfrac{\frac{2}{3}}{1-\frac{2}{3}} = \dfrac{\frac{2}{3}}{\frac{1}{3}} = 2$

b. $\displaystyle\sum_{n=1}^{10}(9+2n)$

$a_1 = 9 + 2(1) = 9 + 2 = 11;$

$a_{10} = 9 + 2(10) = 9 + 20 = 29;$

$S_n = \dfrac{n(a_1 + a_n)}{2}$

$S_{10} = \dfrac{10(11+29)}{2} = \dfrac{10(40)}{2} = 200$

c. $\displaystyle\sum_{n=1}^{5}12n + \sum_{n=1}^{5}(-5) + \sum_{n=1}^{5}n^2$

$\displaystyle\sum_{n=1}^{5}12n$

$a_1 = 12; \quad a_5 = 12(5) = 60;$

$S_5 = \dfrac{5(12+60)}{2} = \dfrac{5(72)}{2} = 180;$

$\displaystyle\sum_{n=1}^{5}(-5)$

$a_1 = -5; \quad a_5 = -5;$

$S_5 = \dfrac{5(-5-5)}{2} = \dfrac{5(-10)}{2} = -25;$

$\displaystyle\sum_{n=1}^{5}n^2 = 1 + 4 + 9 + 16 + 25 = 55;$

$\displaystyle\sum_{n=1}^{5}12n + \sum_{n=1}^{5}(-5) + \sum_{n=1}^{5}n^2$

$= 180 - 25 + 55 = 210$

11. $(a+b)^n$

a. first 3 terms for $n = 20$

$\dbinom{20}{0}a^{20} + \dbinom{20}{1}a^{19}b + \dbinom{20}{2}a^{18}b^2$

$= a^{20} + 20a^{19}b + \dfrac{20!}{2!\,18!}a^{18}b^2$

$= a^{20} + 20a^{19}b + 190a^{18}b^2$

b. last 3 terms for $n = 20$

$\dbinom{20}{18}a^2b^{18} + \dbinom{20}{19}ab^{19} + \dbinom{20}{20}b^{20}$

$= \dfrac{20!}{18!2!}a^2b^{18} + 20ab^{19} + b^{20}$

$= 190a^2b^{18} + 20ab^{19} + b^{20}$

c. fifth term where $n = 35$

$a = a; \quad b = b; \quad n = 35; \quad r = 4$

$\dbinom{n}{r}a^{n-r}b^r$

$\dbinom{35}{4}a^{35-4}b^4 = \dfrac{35!}{4!\,31!}a^{31}b^4$

$= 52360a^{31}b^4$

d. fifth term where $n = 35, p = 0.2, q = 0.8$

$\dbinom{n}{r}a^{n-r}b^r$

$\dbinom{35}{4}(0.2)^{35-4}(0.8)^4$

$= \dfrac{35!}{4!\,31!}(0.2)^{31}(0.8)^4$

$= 52360(0.2)^{31}(0.8)^4$

$= 4.6 \times 10^{-18}$

13. $3 + 6 + 9 + \ldots + 3n = \dfrac{3n(n+1)}{2}$

The needed components are:

$a_n = 3n$; $a_k = 3k$; $a_{k+1} = 3(k+1)$;

$S_n = \dfrac{3n(n+1)}{2}$; $S_k = \dfrac{3k(k+1)}{2}$;

$S_{k+1} = \dfrac{3(k+1)(k+2)}{2}$

1. Show S_n is true for $n = 1$.

$S_1 = \dfrac{3(1)(1+1)}{2} = \dfrac{3(2)}{2} = 3$

Verified

2. Assume S_k is true:

$3 + 6 + 9 + \ldots + 3k = \dfrac{3k(k+1)}{2}$

and use it to show the truth of S_{k+1}
follows. That is:

$3 + 6 + 9 + \ldots + 3k + 3(k+1)$

$= \dfrac{3(k+1)(k+1+1)}{2}$

$S_k + a_{k+1} = S_{k+1}$

Working with the left hand side:

$3 + 6 + 9 + \ldots + 3k + 3(k+1)$

$= \dfrac{3k(k+1)}{2} + 3(k+1)$

$= \dfrac{3k(k+1) + 6(k+1)}{2}$

$= \dfrac{(k+1)(3k+6)}{2}$

$= \dfrac{3(k+1)(k+2)}{2}$

$= S_{k+1}$

Since the truth of S_{k+1} follows from
S_k, the formula is true for all n.

15. $P(E_1 \text{ or } E_2) = \dfrac{15}{2000} + \dfrac{5}{550} \approx 0.01659$

17. $0.36 + 0.0036 + 0.00036 + 0.00000036 + \ldots$

$a_1 = 0.36$; $r = 0.01$

$S_\infty = \dfrac{a_1}{1-r}$

$S_\infty = \dfrac{0.36}{1-0.01} = \dfrac{0.36}{0.99} = \dfrac{4}{11}$

19. $\begin{cases} a_1 = 10 \\ a_{n+1} = a_n \left(\dfrac{1}{5} \right) \end{cases}$

$a_{1+1} = a_1 \left(\dfrac{1}{5} \right)$

$a_2 = 10 \left(\dfrac{1}{5} \right) = 2$;

$a_{2+1} = a_2 \left(\dfrac{1}{5} \right)$

$a_3 = 2 \left(\dfrac{1}{5} \right) = \dfrac{2}{5}$;

$a_{3+1} = a_3 \left(\dfrac{1}{5} \right)$

$a_4 = \dfrac{2}{5} \left(\dfrac{1}{5} \right) = \dfrac{2}{25}$;

$a_{4+1} = a_4 \left(\dfrac{1}{5} \right)$

$a_5 = \dfrac{2}{25} \left(\dfrac{1}{5} \right) = \dfrac{2}{125}$;

$10, 2, \dfrac{2}{5}, \dfrac{2}{25}, \dfrac{2}{125}$

Chapter 10 Practice Test

1. a. $a_n = \dfrac{2n}{n+3}$

$a_1 = \dfrac{2(1)}{1+3} = \dfrac{2}{4} = \dfrac{1}{2}$;

$a_2 = \dfrac{2(2)}{2+3} = \dfrac{4}{5}$;

$a_3 = \dfrac{2(3)}{3+3} = \dfrac{6}{6} = 1$;

$a_4 = \dfrac{2(4)}{4+3} = \dfrac{8}{7}$;

$a_8 = \dfrac{2(8)}{8+3} = \dfrac{16}{11}$;

$a_{12} = \dfrac{2(12)}{12+3} = \dfrac{24}{15} = \dfrac{8}{5}$

$\dfrac{1}{2}, \dfrac{4}{5}, 1, \dfrac{8}{7}$; $a_8 = \dfrac{16}{11}, a_{12} = \dfrac{8}{5}$

b. $a_n = \dfrac{(n+2)!}{n!}$

$a_1 = \dfrac{(1+2)!}{1!} = \dfrac{3!}{1!} = 3 \cdot 2 \cdot 1 = 6$;

$a_2 = \dfrac{(2+2)!}{2!} = \dfrac{4!}{2!} = \dfrac{4 \cdot 3 \cdot 2!}{2!} = 12$;

$a_3 = \dfrac{(3+2)!}{3!} = \dfrac{5!}{3!} = \dfrac{5 \cdot 4 \cdot 3!}{3!} = 20$;

$a_4 = \dfrac{(4+2)!}{4!} = \dfrac{6!}{4!} = \dfrac{6 \cdot 5 \cdot 4!}{4!} = 30$;

$a_8 = \dfrac{(8+2)!}{8!} = \dfrac{10!}{8!} = \dfrac{10 \cdot 9 \cdot 8!}{8!} = 90$;

$a_{12} = \dfrac{(12+2)!}{12!} = \dfrac{14!}{12!} = \dfrac{14 \cdot 13 \cdot 12!}{12!} = 182$;

$6, 12, 20, 30$; $a_8 = 90, a_{12} = 182$

c. $a_n = \begin{cases} a_1 = 3 \\ a_{n+1} = \sqrt{(a_n)^2 - 1} \end{cases}$

$a_{1+1} = \sqrt{(a_1)^2 - 1}$

$a_2 = \sqrt{(3)^2 - 1} = \sqrt{9-1} = \sqrt{8} = 2\sqrt{2}$;

$a_{2+1} = \sqrt{(a_2)^2 - 1}$

$a_3 = \sqrt{(2\sqrt{2})^2 - 1} = \sqrt{8-1} = \sqrt{7}$;

$a_{3+1} = \sqrt{(a_3)^2 - 1}$

$a_4 = \sqrt{(\sqrt{7})^2 - 1} = \sqrt{7-1} = \sqrt{6}$;

$a_{7+1} = \sqrt{(a_7)^2 - 1}$

$a_8 = \sqrt{(\sqrt{3})^2 - 1} = \sqrt{2}$;

$a_{11+1} = \sqrt{(a_{11})^2 - 1}$

$a_{12} = \sqrt{i^2 - 1} = \sqrt{-1-1} = \sqrt{-2} = i\sqrt{2}$;

$3, 2\sqrt{2}, \sqrt{7}, \sqrt{6}$; $a_8 = \sqrt{2}, a_{12} = i\sqrt{2}$

3. a. $7, 4, 1, -2, \ldots$

$a_1 = 7, d = -3, a_n = 10 - 3n$

b. $-8, -6, -4, -2, \ldots$

$a_1 = -8, d = 2, a_n = 2n - 10$

c. $4, -8, 16, -32, \ldots$

$a_1 = 4, r = -2, a_n = 4(-2)^{n-1}$

d. $10, 4, \dfrac{8}{5}, \dfrac{16}{25}, \ldots$

$a_1 = 10, r = \dfrac{2}{5}, a_n = 10\left(\dfrac{2}{5}\right)^{n-1}$

Chapter 10: Additional Topics in Algebra

5. a. $7+10+13+...+100$

$a_1 = 7; \quad a_n = 100; \quad d = 3; \quad n = 32$

$a_n = a_1 + (n-1)d$

$100 = 7 + (n-1)(3)$

$100 = 7 + 3n - 3$

$100 = 4 + 3n$

$96 = 3n$

$32 = n;$

$S_n = \dfrac{n(a_1 + a_n)}{2}$

$S_{32} = \dfrac{32(7+100)}{2} = \dfrac{32(107)}{2} = 1712$

b. $\displaystyle\sum_{k=1}^{37}(3k+2)$

$a_1 = 3(1) + 2 = 3 + 2 = 5;$

$a_2 = 3(2) + 2 = 6 + 2 = 8;$

$a_{37} = 3(37) + 2 = 111 + 2 = 113;$

$a_1 = 5; \quad a_{37} = 113; \quad d = 3; \quad n = 37;$

$S_n = \dfrac{n(a_1 + a_n)}{2}$

$S_{37} = \dfrac{37(5+113)}{2} = \dfrac{37(118)}{2} = 2183$

c. $4 - 12 + 36 - 108 + ...$ Find S_7

$a_1 = 4; \quad r = \dfrac{-12}{4} = -3;$

$S_n = \dfrac{a_1(1 - r^n)}{1 - r}$

$S_7 = \dfrac{4(1 - (-3)^7)}{1 - (-3)} = \dfrac{4(1 - (-2187))}{4}$

$= 2188$

d. $6 + 3 + \dfrac{3}{2} + \dfrac{3}{4} + ...$

$a_1 = 6; \quad r = \dfrac{1}{2}$

$S_\infty = \dfrac{a_1}{1 - r}$

$S_\infty = \dfrac{6}{1 - \dfrac{1}{2}} = \dfrac{6}{\dfrac{1}{2}} = 12$

7. $a_1 = 3000; \quad r = 1.07; \quad n = 12$

$a_n = a_1 r^n$

$a_{12} = 3000(1.07)^{12} = 6756.57$

\$6756.57

9. $a_n = 5n - 3; \quad a_k = 5k - 3;$
The needed components are:

$a_{k+1} = 5(k+1) - 3 = 5k + 2;$

$S_n = \dfrac{5n^2 - n}{2}; \quad S_k = \dfrac{5k^2 - k}{2};$

$S_{k+1} = \dfrac{5(k+1)^2 - (k+1)}{2}$

1. Show S_n is true for $n = 1$.

$a_1 = 5(1) - 3 = 5 - 3 = 2$

$S_1 = \dfrac{5(1)^2 - 1}{2} = \dfrac{5-1}{2} = \dfrac{4}{2} = 2$

Verified

2. Assume S_k is true:

$2 + 7 + 12 + ... + 5k - 3 = \dfrac{5k^2 - k}{2}$

and use it to show the truth of S_{k+1} follows. That is:

$2 + 7 + 12 + ... + 5k - 3 + (5(k+1) - 3)$

$= \dfrac{5(k+1)^2 - (k+1)}{2}$

$S_k + a_{k+1} = S_{k+1}$

Working with the left hand side:

$2 + 7 + 12 + ... + 5k - 3 + (5k + 2)$

$= \dfrac{5k^2 - k}{2} + 5k + 2$

$= \dfrac{5k^2 - k + 2(5k + 2)}{2}$

$= \dfrac{5k^2 + 10k - k + 4}{2}$

$= \dfrac{5(k^2 + 2k) - k + 4}{2}$

$= \dfrac{5(k+1)^2 - k + 4 - 5}{2}$

$= \dfrac{5(k+1)^2 - k - 1}{2}$

$= \dfrac{5(k+1)^2 - (k+1)}{2}$

Since the truth of S_{k+1} follows from S_k, the formula is true for all n.

11. a.

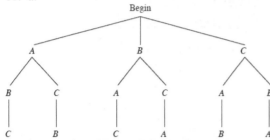

   b.   ABC, ACB, BAC, BCA, CAB, CBA

13. $_6C_6 +_6 C_5 +_6 C_4 +_6 C_3 +_6 C_2 +_6 C_1 +_6 C_0$

$$= \frac{6!}{6!0!} + \frac{6!}{5!1!} + \frac{6!}{4!2!} + \frac{6!}{3!3!} + \frac{6!}{2!4!} + \frac{6!}{1!5!} + \frac{6!}{0!6!}$$

$$= 1 + 6 + 15 + 20 + 15 + 6 + 1$$

$$= 64$$

15. $\dfrac{_{13}P_{13}}{2!3!4!4!} = \dfrac{6227020800}{6912} = 900900$

17. a. $(x - 2y)^4$

$$= x^4 - 4x^3(2y) + 6x^2(2y)^2$$
$$- 4x(2y)^3 + (2y)^4$$
$$= x^4 - 8x^3 y + 6x^2\left(4y^2\right)$$
$$- 4x\left(8y^3\right) + 16y^4$$
$$= x^4 - 8x^3 y + 24x^2 y^2 - 32xy^3 + 16y^4$$

   b. $(1+i)^4$

$$= 1^4 + 4(1)^3 i + 6(1)^2 i^2 + 4(1)i^3 + i^4$$
$$= 1 + 4i - 6 - 4i + 1$$
$$= -4$$

19. $1 - 0.011 = 0.989$

21. a. $0.05 + 0.03 = 0.08$
   b. $0.02 + 0.30 + 0.60 = 0.92$
   c. $0.02 + 0.30 + 0.60 + 0.05 + 0.03 = 1$
   d. 0
   e. $0.02 + 0.30 + 0.60 + 0.03 = 0.95$
   f. 0.03

23. a. P(woman or craftsman)
$$= \frac{50}{100} + \frac{18}{100} - \frac{9}{100} = \frac{59}{100}$$

   b. P(man or contractor)
$$= \frac{50}{100} + \frac{7}{100} - \frac{4}{100} = \frac{53}{100}$$

   c. P(man and technician) $= \dfrac{13}{100}$

   d. P(journeyman or apprentice)
$$= \frac{13}{100} + \frac{34}{100} = \frac{47}{100}$$

25. $1 + 2 + 3 + ... + n = \dfrac{n(n+1)}{2}$

$a_n = n$; $a_k = k$; $a_{k+1} = k+1$

$S_n = \dfrac{n(n+1)}{2}$; $S_k = \dfrac{k(k+1)}{2}$;

$S_{k+1} = \dfrac{(k+1)(k+2)}{2}$

1. Show S_n is true for $n = 1$.

$$S_1 = \frac{1(1+1)}{2} = \frac{2}{2} = 1$$

Verified

2. Assume S_k is true, show

$S_k + a_{k+1} = S_{k+1}$.

$1 + 2 + 3 + ... + k + (k+1)$

$$= \frac{k(k+1)}{2} + k + 1$$

$$= \frac{k(k+1) + 2(k+1)}{2}$$

$$= \frac{(k+1)(k+2)}{2}$$

$$= S_{k+1}$$

Verified

Calculator Exploration and Discovery

1. $a_1 = \dfrac{1}{3}$ and $r = \dfrac{1}{3}$

    Using a grapher: $\dfrac{1}{2}$

3. $a_n = \dfrac{1}{(n-1)!}$

    Using a grapher: e

Strengthening Core Skills

1. $\dfrac{{}_4C_1 \cdot {}_{13}C_5 - 40}{{}_{52}C_5} \approx 0.001970$

3. $\dfrac{4 \cdot {}_{13}C_4 \cdot {}_{39}C_1}{{}_{52}C_5} \approx 0.0429171669$

Chapters 1-10 Cumulative Review

1. a. $(9, 52)\ (11, 98)$

    $m = \dfrac{98 - 52}{11 - 9} = \dfrac{46}{2} = 23$

    23 cards are assembled each hour

    b. $23(8) = 184$

    184 cards

    c. $y - y_1 = m(x - x_1)$

    $y - 52 = 23(x - 9)$

    $y - 52 = 23x - 207$

    $y = 23x - 155$

    d. $\dfrac{52}{23} \approx 2.26$

    Approximately 2.25 hours before 9 am

    $\approx 6 : 45$ am

3. $y = \cos x$

x	y
0	1
$\dfrac{\pi}{6}$	$\dfrac{\sqrt{3}}{2}$
$\dfrac{\pi}{4}$	$\dfrac{\sqrt{2}}{2}$
$\dfrac{\pi}{3}$	$\dfrac{1}{2}$
$\dfrac{\pi}{2}$	0
$\dfrac{2\pi}{3}$	$-\dfrac{1}{2}$
$\dfrac{5\pi}{6}$	$-\dfrac{\sqrt{3}}{2}$
π	-1

5. $3x^2 + 5x - 7 = 0$

$$x = \frac{-5 \pm \sqrt{(5)^2 - 4(3)(-7)}}{2(3)}$$

$$x = \frac{-5 \pm \sqrt{25 + 84}}{6}$$

$$x = \frac{-5 \pm \sqrt{109}}{6}$$

$x \approx 0.91; x \approx -2.57$

7. a. $g(x) = 0$
   $x = 0$

   b. $g(x) < 0 : x \in (-1, 0)$

   c. $g(x) > 0 : x \in (-\infty, -1) \cup (0, \infty)$

   d. $g(x) \uparrow : x \in (-\infty, -1) \cup (-1, 1)$

   e. $g(x) \downarrow : (1, \infty)$

   f. Local Max
   $y = 3$ at $x = 1$

   g. Local Min
   None

   h. $g(x) = 2$
   $x \approx -2.3, x \approx 0.4, x \approx 2$

   i. $g(4) \approx \frac{1}{4}$

   j. $g(-1)$ undefined; does not exist

   k. as $x \to -1^+, g(x) \to -\infty$

   l. as $x \to \infty, g(x) \to 0$

   m. The domain of $g(x)$:
   $x \in (-\infty, -1) \cup (-1, \infty)$

9. $y = \begin{cases} -2 & -3 \le x \le -1 \\ x & -1 \le x \le 2 \\ x^2 & 2 \le x \le 3 \end{cases}$

Domain: $[-3, 3]$

Range: $y \in [-2, -2] \cup (1, 2) \cup [4, 9]$

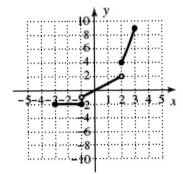

11. a. $f(x) = 2x^2 - 3x$

$$\frac{f(x+h) - f(x)}{h}$$

$$= \frac{\left(2(x+h)^2 - 3(x+h)\right) - \left(2x^2 - 3x\right)}{h}$$

$$= \frac{\left(2\left(x^2 + 2xh + h^2\right) - 3x - 3h\right) - \left(2x^2 - 3x\right)}{h}$$

$$= \frac{\left(2x^2 + 4xh + 2h^2 - 3x - 3h\right) - \left(2x^2 - 3x\right)}{h}$$

$$= \frac{2x^2 + 4xh + 2h^2 - 3x - 3h - 2x^2 + 3x}{h}$$

$$= \frac{4xh + 2h^2 - 3h}{h}$$

$$= \frac{h(4x + 2h - 3)}{h}$$

$$= 4x + 2h - 3$$

Chapter 10: Additional Topics in Algebra

b. $h(x) = \dfrac{1}{x-2}$

$$\dfrac{f(x+h) - f(x)}{h}$$

$$= \dfrac{\dfrac{1}{x+h-2} - \dfrac{1}{x-2}}{h}$$

$$= \dfrac{\dfrac{x-2-(x+h-2)}{(x-2)(x+h-2)}}{h}$$

$$= \dfrac{\dfrac{x-2-x-h+2}{(x-2)(x+h-2)}}{h}$$

$$= \dfrac{\dfrac{-h}{(x-2)(x+h-2)}}{h}$$

$$= \dfrac{-1}{(x+h-2)(x-2)}$$

13. $h(x) = \dfrac{2x^2 - 8}{x^2 - 1}$

$0 = 2x^2 - 8$

$0 = 2(x^2 - 4)$

$x^2 - 4 = 0$

$x^2 = 4$

$x = \pm 2$

x-intercepts: $(2, 0), (-2, 0)$

$h(0) = \dfrac{2(0)^2 - 8}{0^2 - 1} = 8$

y-intercept: $(0, 8)$

$0 = x^2 - 1$

$x = \pm 1$

Vertical asymptotes: $x = \pm 1$

Horizontal asymptote: $y = 2$

(deg num = deg den)

15. a. $3 = \log_x(125)$

$x^3 = 125$

b. $\ln(2x - 1) = 5$

$e^5 = 2x - 1$

17. a. $e^{2x-1} = 217$

$(2x - 1)\ln e = \ln 217$

$2x - 1 = \ln 217$

$2x = \ln 217 + 1$

$x = \dfrac{\ln 217 + 1}{2}$

$x \approx 3.19$

b. $\log(3x - 2) + 1 = 4$

$\log(3x - 2) = 3$

$10^3 = 3x - 2$

$1000 = 3x - 2$

$1002 = 3x$

$334 = x$

$x = 334$

19. $\begin{cases} 0.7x + 1.2y - 3.2z = -32.5 \\ 1.5x - 2.7y + 0.8z = -7.5 \\ 2.8x + 1.9y - 2.1z = 1.5 \end{cases}$

$A = \begin{bmatrix} 0.7 & 1.2 & -3.2 \\ 1.5 & -2.7 & 0.8 \\ 2.8 & 1.9 & -2.1 \end{bmatrix} \quad B = \begin{bmatrix} -32.5 \\ -7.5 \\ 1.5 \end{bmatrix}$

$X = A^{-1}B$

$X = \begin{bmatrix} 5 \\ 10 \\ 15 \end{bmatrix}$

$(5, 10, 15)$

21. $x^2 + 4y^2 - 24y + 6x + 29 = 0$

$$\left(x^2 + 6x\right) + 4\left(y^2 - 6y\right) = -29$$

$$(x+3)^2 + 4(y-3)^2 = -29 + 9 + 36$$

$$(x+3)^2 + 4(y-3)^2 = 16$$

$$\frac{(x+3)^2}{16} + \frac{(y-3)^2}{4} = 1$$

Center: $(-3, 3)$

Vertices: $(-7, 3), (1, 3)$

$a = 4, \ b = 2$

$c^2 = 16 - 4 = 12$

$c = 2\sqrt{3}$

Foci: $\left(-3 - 2\sqrt{3}, 3\right)\left(-3 + 2\sqrt{3}, 3\right)$

23. a) $\cos\left(\dfrac{\pi}{2} - \theta\right) = \sin\theta$

$$\cos\frac{\pi}{2}\cos\theta + \sin\frac{\pi}{2}\sin\theta = \sin\theta$$

$$0 \cdot \cos\theta + 1(\sin\theta) = \sin\theta$$

$$\sin\theta = \sin\theta$$

b) $\cos 15° = \cos\left(45° - 30°\right)$

$$= \cos 45°\cos 30° + \sin 45°\sin 30°$$

$$\frac{\sqrt{2}}{2}\frac{\sqrt{3}}{2} + \frac{\sqrt{2}}{2}\left(\frac{1}{2}\right) = \frac{\sqrt{6} + \sqrt{2}}{4}$$

25. $a_1 = 52; \ a_n = 10; \ d = -1$

$a_n = a_1 + (n-1)d$

$10 = 52 + (n-1)(-1)$

$10 = 52 - n + 1$

$10 = 53 - n$

$-43 = -n$

$43 = n;$

$$S_n = \frac{n(a_1 + a_n)}{2}$$

$$S_{43} = \frac{43(52 + 10)}{2} = \frac{43(62)}{2} = 1333$$

27. $P(\text{late}) = 0.04, P(\text{on time}) = 1 - 0.04 = 0.96$

(a) $P(10) = \dbinom{12}{10}(0.04)^2(0.96)^{10}$

$\approx 0.07 = 7\%$

(b) $P(x \geq 11) = \dbinom{12}{11}(0.04)^1(0.96)^{11}$

$+ \dbinom{12}{12}(0.04)^0(0.96)^{12} \approx 0.919 = 91.9\%$

(c) $P(x \geq 10) = \dbinom{12}{10}(0.04)^2(0.96)^{10}$

$+ \dbinom{12}{11}(0.04)^1(0.96)^{11}$

$+ \dbinom{12}{12}(0.04)^0(0.96)^{12} \approx 0.989 = 98.9\%$

(d) $P(x = 10)$

$= \dbinom{12}{0}(0.04)^{12}(0.96)^0 \approx 1.7 \times 10^{-17}$

virtually nil

29. $\cos(2\theta) = \begin{cases} \cos^2\theta - \sin^2\theta \\ 2\cos^2\theta - 1 \\ 1 - 2\sin^2\theta \end{cases}$

$$\cos(2\theta) = \frac{1}{2}$$

$$\frac{1}{2} = 1 - 2\sin^2\theta$$

$$-\frac{1}{2} = -2\sin^2\theta$$

$$\frac{1}{4} = \sin^2\theta$$

$$\sin\theta = \pm\sqrt{\frac{1}{4}}$$

$$\sin\theta = \pm\frac{1}{2}$$

$$\theta = \frac{\pi}{6}, \frac{5\pi}{6}, \frac{7\pi}{6}, \frac{13\pi}{6}$$

Chapter 10: Additional Topics in Algebra

Connections to Calculus

1. $f(x) = -x + 6$; $x \in [0, 4]$

   a) 4 rectangles; $W = 1$

   length $LW = f(x); x = i; i = 1, 2, 3, 4$

   $$\sum LW = \sum_{i=1}^{4} f(i)(1) = 1\sum_{i=1}^{4} f(i)$$
   $$= 1[f(1) + f(2) + f(3) + f(4)]$$
   $$= 1[(-1+6) + (-2+6) + (-3+6) + (-4+6)]$$
   $$= 1(5 + 4 + 3 + 2) = 14 \text{ units}^2$$

   b) 8 rectangles; $W = \dfrac{1}{2}$

   length $LW = f(x); x = \dfrac{1}{2}i; i = 1, 2, 3, 4, 5, 6, 7, 8$

   $$\sum LW = \sum_{i=1}^{8} f\left(\frac{1}{2}i\right)\left(\frac{1}{2}\right) = \frac{1}{2}\sum_{i=1}^{8} f\left(\frac{1}{2}i\right)$$
   $$= \frac{1}{2}\left[f\left(\frac{1}{2}\right) + f(1) + f\left(\frac{3}{2}\right) + f(2) + f\left(\frac{5}{2}\right) + f(3) + f\left(\frac{7}{2}\right)f(4)\right]$$
   $$= \frac{1}{2}\left[\frac{11}{2} + 5 + \frac{9}{2} + 4 + \frac{7}{2} + 3 + \frac{5}{2} + 2\right]$$
   $$= \frac{1}{2}\left[\frac{32}{2} + 14\right]$$
   $$= \frac{1}{2}[16 + 14] = \frac{1}{2}[30] = 15 \text{ units}^2$$

   Actual area $= \dfrac{1}{2}(6 + 2)4$

   $$= \frac{1}{2}(8)(4) = 16 \text{ units}^2$$

   More rectangles $\rightarrow$ better estimate

3. $f(x) = -x + 6$; $x \in [0, 4]$

   a) 32 rectangles

   Since the interval [0, 4] is 4 units wide and we're using 32 subintervals of equal length, the width of each interval (the width of each rectangle) will be $\dfrac{4}{32} = \dfrac{1}{8}$. The height of each rectangle is determined by a point of the graph of $f(x) = -x + 6$, so the length of the first rectangle is $f\left(\dfrac{1}{8}\right)$, the second length is $f\left(\dfrac{2}{8}\right)$ then is $f\left(\dfrac{3}{8}\right)$ and so on up to the 32nd rectangle.

Since A = LW, we multiply each length $= f\left(\dfrac{1}{8}i\right)$ by each width $= \dfrac{1}{8}$ and sum the areas of all such rectangles. Using i for a counter, this can be written as A =

$$\sum_{i=1}^{32} LW = \sum_{i=1}^{32} f\left(\frac{1}{8}i\right)\left(\frac{1}{8}\right).$$

Since all lengths are multiplied by $\dfrac{1}{8}$ (the counter i does not affect the constant $\dfrac{1}{8}$ in any case), we can factor out this term and evaluate

$f(x) = -x + 6$ at $x = \dfrac{1}{8}i$. The result is

$$\frac{1}{8}\sum_{i=1}^{32}\left(-\frac{1}{8}i + 6\right).$$

b) $f(x) = -x + 6$ at $x = \dfrac{1}{8}i$; $i = 1, 2, 3, ..., 32$

$$\sum_{i=1}^{32} f\left(\frac{1}{8}i\right)\left(\frac{1}{8}\right) = \frac{1}{8}\sum_{i=1}^{32} f\left(\frac{1}{8}i\right)$$
$$= \frac{1}{8}\sum_{i=1}^{32}\left(-\frac{1}{8}i + 6\right)$$
$$= \frac{1}{8}\left[\sum_{i=1}^{32}\left(-\frac{1}{8}i\right) + \sum_{i=1}^{32} 6\right]$$
$$= \frac{1}{8}\left[-\frac{1}{8}\sum_{i=1}^{32} i + \sum_{i=1}^{32} 6\right]$$
$$= \frac{1}{8}\left[-\frac{1}{8}\left(\frac{1056}{2}\right) + 6(32)\right]$$
$$= \frac{1}{8}[-66 + 192]$$
$$= \frac{1}{8}[126]$$
$$= 15.75 \text{ units}^2$$

11.1 Exercises

1. infinity

3. left–hand, right–hand, greater

5. Answers will vary.

7. $\lim\limits_{n\to\infty} V_n = \dfrac{4}{3}\pi r^3$

9. $\lim\limits_{t\to-\infty} e^{f(t)} = 0$

11. As x increases without bound

$\cos\left(\dfrac{1}{x}\right)$ approaches 0

$\lim\limits_{x\to\infty} \cos\left(\dfrac{1}{x}\right) = 0$

13. $A = nr^2 \tan\left(\dfrac{\pi}{n}\right).\ r = 10$ cm

$A = 10^2\pi = 100\pi \approx 314.16$

n	A
100	314.26
200	314.19
300	314.17
400	314.17
450	314.16
500	314.16

450 sides

15. $P = 2nr \tan\left(\dfrac{\pi}{n}\right)$, $r = 50$ mm

$C = 2\pi 50 = 100\pi \approx 314.16$

n	P
300	314.17
325	314.17
350	314.17
375	314.17
400	314.17
425	314.16
450	314.16

425 sides

17. $\lim\limits_{t\to 5} s_t = 5r$

19. $\lim\limits_{x\to a} \tan^{-1}\left[g(x)\right] = \dfrac{\pi}{3}$

21. $\lim\limits_{x\to-3} \dfrac{x+3}{x^2-9} = -\dfrac{1}{6}$

23. $p(x) = \cos x + \sin\dfrac{3x}{2}$; $x\to\pi$

x	$p(x)$
3.1116	−1.999
3.1216	−1.999
3.1316	−1.9998
3.1416	−2
3.1516	−1.9998
3.1616	−1.999
3.1716	−1.999

As x approaches π,
$p(x)$ approaches −2:
$\lim\limits_{x\to\pi} p(x) = -2$.

25. $v(x) = \dfrac{\cos\left(\dfrac{\pi}{x}\right)}{\sin\pi x}$; $x\to 2$

x	$v(x)$
1.997	.25038
1.998	.25025
1.999	.25013
2	Error
2.001	.24988
2.002	.24975
2.003	.24963

As x approaches 2,
$v(x)$ approaches $\dfrac{1}{4}$:

$\lim\limits_{x\to 2} v(x) = \dfrac{1}{4}$

27. $s(x) = \dfrac{2\cos x - 2}{x}; \ x \to 0$

x	$s(x)$
−.003	.003
−.002	.002
−.001	.001
0	Error
.001	−.001
.002	−.002
.003	−.003

As x approaches 0,
$s(x)$ approaches 0:

$\lim\limits_{x \to 0} s(x) = 0$

29. $r(x) = \dfrac{2x^2 - 7x + 6}{\sin(x - 2)}$

x	$r(x)$
1.999	.998
2	Error
2.001	1.002

$\lim\limits_{x \to 2} r(x) = 1$

$R(x) = \begin{cases} \dfrac{2x^2 - 7x + 6}{\sin(x - 2)} & x \neq 2 \\ 1 & x = 2 \end{cases}$

31. $f(x) = \dfrac{x^2 - 2x}{x^2 - 4}; \ x \to 2$

x	$f(x)$
1.997	0.49962
1.998	0.49975
1.999	0.49987
2	Error
2.001	0.50012
2.002	0.50025
2.003	0.50037

As x approaches 2,

$f(x)$ approaches $\dfrac{1}{2}$:

$\lim\limits_{x \to 2} f(x) = \dfrac{1}{2}$.

33. $g(x) = \dfrac{x^2 + 2x - 3}{x - 1}; \ x \to 1$

x	$v(x)$
0.997	3.997
0.998	3.998
0.999	3.999
1	Error
1.001	4.001
1.002	4.002
1.003	4.003

As x approaches 1,
$g(x)$ approaches 4:

$\lim\limits_{x \to 1} g(x) = 4$

35. $f(x) = 3x^2 - x - 2; x \to 1$

x	$f(x)$
0.997	−.015
0.998	−.01
0.999	−.005
1	0
1.001	.005
1.002	.01001
1.003	.01503

As x approaches 1,
$f(x)$ approaches 0;

$\lim\limits_{x \to 1} f(x) = 0$

37. $f(x) = 2x^2 - 3x$ as $x \to -3$
from the right.

x	$F(x)$
−3	27
−2.999	26.985
−2.998	26.97
−2.997	26.955

Ans: 27

39. $f(x) = \dfrac{x^3 + 8}{\frac{1}{2}x + 1}$ as $x \to -2 \backslash$
from the left.

x	$f(x)$
−2.003	24.036
−2.002	24.024
−2.001	24.012
−2	error

Ans: 24

41. $f(x) = \dfrac{\sin\sqrt{x}}{\sqrt{x}}$ as $x \to 0$

from the right.

x	$f(x)$
0	Error
0.001	.99983
0.002	.99967
0.003	.9995

Ans: 1

43. $\lim\limits_{x \to 3^-} I_x = 3\cos^2(R_1 + R_2)$

45. $\lim\limits_{x \to m^-} f = L$

47. Given $f(x) = \dfrac{\sin x}{|\sin x|}$, find:

a. $\lim\limits_{x \to \pi^-} f(x) = 1$

x	$f(x)$
3.1386	1
3.1396	1
3.1406	1
π	error

b. $\lim\limits_{x \to \pi^+} f(x) = -1$

x	$f(x)$
π	Error
3.1426	-1
3.1436	-1
3.1446	-1

49. Given $f(x) = \dfrac{x^2 - 10x + 24}{\sqrt{x^2 - 12x + 36}}$, find:

a. $\lim\limits_{x \to 6^-} f(x) = -2$

x	$f(x)$
5.997	-1.997
5.998	-1.998
5.999	-1.999
6	error

b. $\lim\limits_{x \to 6^+} f(x) = 2$

x	$f(x)$
6	Error
6.001	2.001
6.002	2.002
6.003	2.003

51. Given $f(x) = \dfrac{2}{x} - 3x^{\frac{1}{2}}$, find:

a. $\lim\limits_{x \to 4^-} f(x) = -\dfrac{11}{2}$

x	$f(x)$
3.997	-5.497
3.998	-5.498
3.999	-5.499
4	-5.5

b. $\lim\limits_{x \to 4^+} f(x) = -\dfrac{11}{2}$

x	$f(x)$
4	-5.5
4.001	-5.501
4.002	-5.502
4.003	-5.503

53. For $f(x) = \begin{cases} x^2 - 4 & x < 2 \\ \sin(x^2 - 4) & x \geq 2 \end{cases}$, find:

a. $\lim\limits_{x \to 2^-} x^2 - 4 = 0$

x	$f(x) = x^2 - 4$
1.997	-.012
1.998	-.008
1.999	-.004
2	0

b. $\lim\limits_{x \to 2^+} \sin(x^2 - 4) = 0$

x	$f(x) = \sin(x^2 - 4)$
2	0
2.001	.004
2.001	.008
2.003	.012

55. Given $f(x) = \begin{cases} 2x^2 - 7 & x < -5 \\ 3 - 2x & x \ge -5 \end{cases}$, find:

a. $\lim\limits_{x \to -5^-} 2x^2 - 7 = 43$

x	$f(x) = 2x^2 - 7$
-5.003	43.06
-5.002	43.04
-5.001	43.02
-5	43

b. $\lim\limits_{x \to -5^+} 3 - 2x = 13$

x	$f(x) = 3 - 2x$
-5	13
-4.999	12.998
-4.998	12.996
-4.997	12.994

c. $\lim\limits_{x \to 5} f(x)$ (dne)

because $\lim\limits_{x \to 5^-} f(x) \ne \lim\limits_{x \to -5^+} f(x)$

57. Given $f(x) = \dfrac{\sqrt{1 - \cos 2x}}{\sqrt{2} \sin x}$, find:

a. $\lim\limits_{x \to \pi^-} f(x) = 1$

x	$f(x)$
3.1386	1
3.1396	1
3.1406	1
π	error

b. $\lim\limits_{x \to \pi^+} f(x) = -1$

x	$f(x)$
π	Error
3.1426	-1
3.1436	-1
3.1446	-1

c. $\lim\limits_{x \to \pi} f(x)$ (dne)

because $\lim\limits_{x \to \pi^-} f(x) \ne \lim\limits_{x \to \pi^+} f(x)$

59. Given $f(x) = \begin{cases} \sin x & x < \dfrac{\pi}{4} \\ \tan x & x = \dfrac{\pi}{4} \\ \cos x & x > \dfrac{\pi}{4} \end{cases}$, find:

a. $\lim\limits_{x \to \frac{\pi}{4}^-} \sin x = \dfrac{\sqrt{2}}{2}$

x	$\sin x$
0.7824	0.70498
0.7834	0.70569
0.7844	0.7064
$\dfrac{\pi}{4}$	$\dfrac{\sqrt{2}}{2} \approx 0.70711$

b. $\lim\limits_{x \to \frac{\pi}{4}^+} \cos x = \dfrac{\sqrt{2}}{2}$

x	$\cos x$
$\dfrac{\pi}{4}$	$\dfrac{\sqrt{2}}{2} \approx 0.70711$
0.7864	0.7064
0.7874	0.70569
0.7884	0.70928

c. $\lim\limits_{x \to \frac{\pi}{4}} f(x) = \dfrac{\sqrt{2}}{2}$

because $\lim\limits_{x \to \frac{\pi}{4}^-} f(x) = \lim\limits_{x \to \frac{\pi}{4}^+} f(x)$

61. Given $f(x) = x^2 - \dfrac{3}{x+2}$,

find $\lim\limits_{x \to -2} f(x) = \left(\underset{\infty}{dne} \right)$

x	$f(x)$
-2.003	1004
-2.002	1504
-2.001	3004
-2	Error
-1.999	-2996
-1.998	-1496
-1.997	-996

Because $\lim\limits_{x \to -2^-} f(x) = \infty$

$\lim\limits_{x \to -2^+} f(x) = -\infty$

And $\lim\limits_{x \to -2^-} f(x) \ne \lim\limits_{x \to -2^+} f(x)$

11.1 Exercises

63. Given $f(x) = \dfrac{-4\cos\left(\dfrac{\pi}{4}x\right)}{x-2}$

find $\lim\limits_{x \to 2} f(x) = \pi$.

x	$f(x)$
1.997	3.1416
1.998	3.1416
1.999	3.1416
2	error
2.001	3.1416
2.002	3.1416
2.003	3.1416

65. Given $f(x) = \dfrac{2x}{x^3 + 216}$

find $\lim\limits_{x \to -6} f(x) = \left(\underset{-\infty}{dne}\right)$

x	$f(x)$
−6.003	37.037
−6.002	55.556
−6.001	111.11
−6	Error
−5.999	−111.1
−5.998	−55.56
−5.997	−37.04

because $\lim\limits_{x \to -6^-} f(x) \neq \lim\limits_{x \to -6^+} f(x)$

67. Given $f(x) = \sin\left(\dfrac{x+1}{x-1}\right)$

find $\lim\limits_{x \to 1} f(x) = \left(\underset{L}{dne}\right)$

x	$f(x)$
0.997	0.34381
0.998	0.02646
0.999	−0.8117
1	Error
1.001	0.1933
1.002	0.9
1.003	0.99694

because $\lim\limits_{x \to 1^-} f(x) \neq \lim\limits_{x \to 1^+} f(x)$

69. Given $f(x) = \tan(\cos x)$

find $\lim\limits_{x \to \frac{\pi}{2}} f(x) = 0$

x	$f(x)$
1.5678	.003
1.5688	.002
1.5698	.001
$\dfrac{\pi}{2}$	0
1.5718	−.001
1.5728	−.002
1.5738	−.003

because $\lim\limits_{x \to \frac{\pi}{2}^-} f(x) = \lim\limits_{x \to \frac{\pi}{2}^+} f(x) = 0$

71. $3x^4 - 19x^3 + 15x^2 + 27x - 10$

$\dfrac{\text{factors of } p}{\text{factors of } q} = \dfrac{\pm 1, \pm 2, \pm 5, \pm 10}{\pm 1, \pm 3}$

$= \left\{ \pm 1, \pm 2, \pm 5, \pm 10, \pm \dfrac{1}{3} \pm \dfrac{2}{3} \pm \dfrac{5}{3} \pm \dfrac{10}{3} \right\}$

$\underline{-1|}$ 3   −19   15   27   −10

      −3   22   −37   10

   3   −22   37   −10   0

$\underline{2|}$ 3   −22   37   −10

      6   −32   10

   3   −16   5   0

$(x+1)(x-2)(3x^2 - 16x + 5)$

$(x+1)(x-2)(3x-1)(x-5)$

$(x-5)(x-2)(x+1)(3x-1)$

73. $a = 5$ in.

$b = 8$ in.

$c = 9$ in.

Area: $a + b + c = 5 + 8 + 9 = 22$

Area: $= \dfrac{1}{4}\sqrt{22(22-2\cdot 5)(22-2\cdot 8)(22-2\cdot 9)}$

$= \dfrac{1}{4}\sqrt{22(12)(6)(4)}$

$= \dfrac{1}{4}\sqrt{6336}$

≈ 19.90 in^2

11.2 Exercises

1. sum; limits

3. root, nth;, $f(x) > 0$

5. Answers will vary.

7. $f(x) = x^2 + \dfrac{\sin x}{1000x}$

x	$f(x)$	x	$f(x)$
0.5	0.251	−0.5	0.251
0.4	0.161	−0.4	0.161
0.3	0.091	−0.3	0.091
0.2	0.041	−0.2	0.041
0.1	0.011	−0.1	0.011
0.01	0.0011	−0.01	0.0011
0.001	0.001001	−0.001	0.001001

a) limit appears to be 0

b)

x	$f(x)$
0.0001	0.001
0.00001	0.001
0.000001	0.001

limit is actually 0.001

9.

x	$y = x + \dfrac{\sqrt{x-2}}{10^x}$
2.7	≈2.7017
2.8	≈2.8014
2.9	≈2.9012
3	3.001
3.1	≈3.1008
3.2	≈3.2007
3.3	≈3.3006

11.

x	$y = x + \dfrac{\sqrt{x-2}}{10^x}$
2.99	≈2.991
2.999	≈ 3
2.9999	≈3.0009
3	3.001
3.0001	≈3.0011
3.001	≈3.002
3.01	≈3.011

13. $\lim\limits_{x \to 3}\left(x + \dfrac{\sqrt{x-2}}{10^x}\right)$

a) 3.001

b. $\lim\limits_{x \to 3}\left(x + \dfrac{\sqrt{x-2}}{10^x}\right) = 3.001$

15. $\lim\limits_{x \to -3}(3 - 2x) = 9$

x	$a(x)$
−3.1	9.2
−3.01	9.02
−3.001	9.002
−3	9
−2.999	8.998
−2.99	8.98
−2.9	8.8

17. $\lim\limits_{x \to -3}(1 - x^2) = -8$

x	$c(x)$
−3.1	−8.14
−3.01	−8.06
−3.001	−8.006
−3	−8
−2.999	−7.994
−2.99	−7.94
−2.9	−7.41

19. $\lim\limits_{x \to -3}\sqrt{x+7} = 2$

x	$f(x)$
−3.1	1.9748
−3.01	1.9975
−3.001	1.9997
−3	2
−2.999	2.0002
−2.99	2.0025
−2.9	2.0248

21. $\lim\limits_{x \to -3}h(x)$

$\lim\limits_{x \to -3}\sin(x+3) = 0$

x	$h(x)$
−3.1	−.0998
−3.01	−.01
−3.001	−.001
−3	0
−2.999	.001
−2.99	.01
−2.9	.0998

11.2 Exercises

23. $\lim\limits_{x\to -3} k(x)$

$\lim\limits_{x\to -3}\left|x^2-16\right|=7$

x	$k(x)$
-3.1	6.39
-3.01	6.9399
-3.001	6.994
-3	7
-2.999	7.006
-2.99	7.0599
-2.9	7.59

25. $\lim\limits_{x\to -3}\left[(3-2x)+\left(1-x^2\right)\right]=1$

x	$a(x)+c(x)$
-3.1	.59
-3.01	.9599
-3.001	.996
-3	1
-2.999	1.004
-2.99	1.0399
-2.9	1.39

$\lim\limits_{x\to -3} a(x)+\lim\limits_{x\to -3} c(x)=9+(-8)=1$

27. $\lim\limits_{x\to -3}\left[\sqrt{x+7}+\sin(x+3)+\left|x^2-16\right|\right]=9$

x	$f(x)+h(x)+k(x)$
-3.1	8.265
-3.01	8.9274
-3.001	8.9927
-3	9
-2.999	9.0072
-2.99	9.0724
-2.9	9.7147

$\lim\limits_{x\to -3} f(x)+\lim\limits_{x\to -3} h(x)+\lim\limits_{x\to -3} k(x)=$
$=2+0+7=9$

29. $\lim\limits_{x\to -3}\left[\sin(x+3)-\left(1-x^2\right)\right]=8$

x	$h(x)-c(x)$
-3.1	8.5102
-3.01	8.0501
-3.001	8.005
-3	8
-2.999	7.995
-2.99	7.9501
-2.9	7.5098

$\lim\limits_{x\to -3} h(x)-\lim\limits_{x\to -3} c(x)=0-(-8)=8$

31. $\lim\limits_{x\to -3}\left[\sqrt{x+7}+\left|x^2-16\right|-(3-2x)\right]=0$

x	$f(x)+k(x)-a(x)$
-3.1	-0.8352
-3.01	-0.0826
-3.001	-0.0083
-3	0
-2.999	0.00825
-2.99	0.0824
-2.9	0.81485

$\lim\limits_{x\to -3} f(x)+\lim\limits_{x\to -3} k(x)-\lim\limits_{x\to -3} a(x)=0$
$=2+7-9=0$

33. $\lim\limits_{x\to -3}\left[\left|x^2-16\right|+\left|x^2-16\right|\right]=14$

x	$k(x)+k(x)$
-3.1	12.78
-3.01	13.88
-3.001	13.988
-3	14
-2.999	14.012
-2.99	14.12
-2.9	15.18

$\lim\limits_{x\to -3} k(x)+\lim\limits_{x\to -3} k(x)=7+7=14$

35. $\lim\limits_{x\to -3}\left[\left(1-x^2\right)\left(\sqrt{x+7}\right)\right]=-16$

x	$c(x)\cdot f(x)$
-3.1	-17
-3.01	-16.1
-3.001	-16.01
-3	-16
-2.999	-15.99
-2.99	-15.9
-2.9	-15

$\left(\lim\limits_{x\to -3} c(x)\right)\left(\lim\limits_{x\to -3} f(x)\right)=-8\cdot 2=-16$

37. $\lim\limits_{x \to -3} \dfrac{\left|x^2 - 16\right|}{\sqrt{x+7}} = \dfrac{7}{2}$

x	$\dfrac{k(x)}{f(x)}$
−3.1	3.2357
−3.01	3.4743
−3.001	3.4974
−3	3.5
−2.999	3.5026
−2.99	3.5255
−2.9	3.7484

$\dfrac{\lim\limits_{x \to -3} k(x)}{\lim\limits_{x \to -3} f(x)} = \dfrac{7}{2}$

39. $\lim\limits_{x \to -3}\left[(3-2x)(\sin(x+3))\left(\left|x^2-16\right|\right)\right] = 0$

x	$a(x) \cdot h(x) \cdot k(x)$
−3.1	−5.869
−3.01	−0.626
−3.001	−0.063
−3	0
−2.999	0.06304
−2.99	0.63397
−2.9	6.6681

$\left(\lim\limits_{x \to -3} a(x)\right)\left(\lim\limits_{x \to -3} h(x)\right)\left(\lim\limits_{x \to -3} k(x)\right) = 9 \cdot 0 \cdot 7 = 0$

41. $\lim\limits_{x \to -3}\left[(1-x^2)(1-x^2)\right] = 64$

x	$c(x) \cdot c(x)$
−3.1	74.132
−3.01	64.965
−3.001	64.096
−3	64
−2.999	63.904
−2.99	63.045
−2.9	54.908

$\left(\lim\limits_{x \to -3} c(x)\right)\left(\lim\limits_{x \to -3} c(x)\right) = -8 \cdot -8 = 64$

43. $\lim\limits_{x \to -3} \dfrac{(3-2x)\cdot(1-x^2)}{\sqrt{x+7}} = -36$

x	$\dfrac{a(x) \cdot c(x)}{f(x)}$
−3.1	−40.11
−3.01	−36.4
−3.001	−36.04
−3	−36
−2.999	−35.96
−2.99	−35.61
−2.9	−32.2

$\dfrac{\left(\lim\limits_{x \to -3} a(x)\right)\left(\lim\limits_{x \to -3} c(x)\right)}{\lim\limits_{x \to -3} f(x)} = \dfrac{9 \cdot (-8)}{2} = -36$

45. $\lim\limits_{x \to -4}(x^3 - 5)$

$= \lim\limits_{x \to -4} x^3 - \lim\limits_{x \to -4} 5$

$= \left(\lim\limits_{x \to -4} x\right)^3 - \lim\limits_{x \to -4} 5$

$= (-4)^3 - 5$

$= -64 - 5 = -69$

47. $\lim\limits_{x \to -4}\left(2x^2 - x - 7\right)$

$= \lim\limits_{x \to -4} 2x^2 - \lim\limits_{x \to -4} x - \lim\limits_{x \to -4} 7$

$= 2\left(\lim\limits_{x \to -4} x\right)^2 - \lim\limits_{x \to -4} x - \lim\limits_{x \to -4} 7$

$= 2(-4)^2 - (-4) - 7$

$= 32 + 4 - 7$

$= 29$

49. $\lim\limits_{x \to 4}\left(x^2 - 5\sqrt{x} + 3\right)$

$= \lim\limits_{x \to 4} x^2 - \lim\limits_{x \to 4} 5\sqrt{x} + \lim\limits_{x \to 4} 3$

$= \left(\lim\limits_{x \to 4} x\right)^2 - 5\sqrt{\lim\limits_{x \to 4} x} + \lim\limits_{x \to 4} 3$

$= (4)^2 - 5\sqrt{4} + 3$

$= 16 - 5 \cdot 2 + 3$

$= 9$

11.2 Exercises

51. $\lim\limits_{x\to 2}\dfrac{2x^2-5}{x+3}$

$=\dfrac{\lim\limits_{x\to 2}\left(2x^2-5\right)}{\lim\limits_{x\to 2}\left(x+3\right)}$

$=\dfrac{2\left(\lim\limits_{x\to 2}x\right)^2-\lim\limits_{x\to 2}5}{\lim\limits_{x\to 2}x+\lim\limits_{x\to 2}3}$

$=\dfrac{2(2)^2-5}{2+3}$

$=\dfrac{8-5}{2+3}$

$=\dfrac{3}{5}$

53. $\lim\limits_{x\to 1}\dfrac{\dfrac{1}{x+1}-x^2}{\dfrac{1}{x}+2x^2-x}$

$=\dfrac{\lim\limits_{x\to 1}\left(\dfrac{1}{x+1}-x^2\right)}{\lim\limits_{x\to 1}\left(\dfrac{1}{x}+2x^2-x\right)}$

$=\dfrac{\lim\limits_{x\to 1}\dfrac{1}{x+1}-\lim\limits_{x\to 1}x^2}{\lim\limits_{x\to 1}\dfrac{1}{x}+\lim\limits_{x\to 1}2x^2-\lim\limits_{x\to 1}x}$

$=\dfrac{\dfrac{\lim\limits_{x\to 1}1}{\lim\limits_{x\to 1}x+\lim\limits_{x\to 1}1}-\left(\lim\limits_{x\to 1}x\right)^2}{\dfrac{\lim\limits_{x\to 1}1}{\lim\limits_{x\to 1}x}+2\left(\lim\limits_{x\to 1}x\right)^2-\lim\limits_{x\to 1}x}$

$=\dfrac{\dfrac{1}{1+1}-(1)^2}{\dfrac{1}{1}+2(1)^2-1}=\dfrac{\dfrac{1}{2}-1}{2}=\dfrac{-\dfrac{1}{2}}{2}=-\dfrac{1}{4}$

55. $\lim\limits_{x\to 3}\left(x^2-3\right)^3$

$=\left(\lim\limits_{x\to 3}x^2-\lim\limits_{x\to 3}3\right)^3$

$=\left(\left(\lim\limits_{x\to 3}x\right)^2-\lim\limits_{x\to 3}3\right)^3$

$=\left((3)^2-3\right)^3=(9-3)^3=6^3=216$

57. $\lim\limits_{x\to 10}5\sqrt[3]{2x+7}$

$=5\left(\sqrt[3]{\lim\limits_{x\to 10}\left(2x+7\right)}\right)$

$=5\left(\sqrt[3]{2\lim\limits_{x\to 10}x+\lim\limits_{x\to 10}7}\right)$

$=5\left(\sqrt[3]{2(10)+7}\right)$

$=5\left(\sqrt[3]{27}\right)=5\cdot 3=15$

59. $\lim\limits_{x\to -3}\left(\sqrt{x+7}-7x\right)$

$=\lim\limits_{x\to -3}\sqrt{x+7}-7\lim\limits_{x\to -3}x$

$=\sqrt{\lim\limits_{x\to -3}x+\lim\limits_{x\to -3}7}-7\lim\limits_{x\to -3}x$

$=\sqrt{-3+7}-7(-3)$

$=\sqrt{4}+21=2+21=23$

61. $\lim\limits_{x\to 2}\dfrac{x^3-2x-10}{2\sqrt[3]{5x^2+2x+3}}$

$=\dfrac{\lim\limits_{x\to 2}\left(x^3-2x-10\right)}{\lim\limits_{x\to 2}\left(2\sqrt[3]{5x^2+2x+3}\right)}$

$=\dfrac{\left(\lim\limits_{x\to 2}x\right)^3-2\left(\lim\limits_{x\to 2}x\right)-\lim\limits_{x\to 2}10}{2\sqrt[3]{5\left(\lim\limits_{x\to 2}x\right)^2+2\left(\lim\limits_{x\to 2}x\right)+\lim\limits_{x\to 2}3}}$

$=\dfrac{(2)^3-2(2)-10}{2\sqrt[3]{5(2)^2+2(2)+3}}$

$=\dfrac{8-4-10}{2\sqrt[3]{20+4+3}}$

$=\dfrac{-6}{2\sqrt[3]{27}}=\dfrac{-6}{2(3)}=\dfrac{-6}{6}=-1$

63. $\lim\limits_{x\to 3}\dfrac{\dfrac{2}{x+1}-3x}{\sqrt{x-2}+1}$

$=\dfrac{\lim\limits_{x\to 3}\left(\dfrac{2}{x+1}-3x\right)}{\lim\limits_{x\to 3}\left(\sqrt{x-2}+1\right)}$

$=\dfrac{\lim\limits_{x\to 3}\dfrac{2}{x+1}-3\left(\lim\limits_{x\to 3}x\right)}{\lim\limits_{x\to 3}\sqrt{x-2}+\lim\limits_{x\to 3}1}$

$=\dfrac{\dfrac{\lim\limits_{x\to 3}2}{\lim\limits_{x\to 3}x+\lim\limits_{x\to 3}1}-3\left(\lim\limits_{x\to 3}x\right)}{\sqrt{\lim\limits_{x\to 3}x-\lim\limits_{x\to 3}2}+\lim\limits_{x\to 3}1}$

$=\dfrac{\dfrac{2}{3+1}-3(3)}{\sqrt{3-2}+1}$

$=\dfrac{\dfrac{2}{4}-9}{\sqrt{1}+1}=\dfrac{\dfrac{1}{2}-9}{2}=\dfrac{-\dfrac{17}{2}}{2}=\dfrac{-17}{4}$

65. $\lim\limits_{x\to -2}\dfrac{3x^2-11x-4}{x^2-2x-8}$

$=\dfrac{\lim\limits_{x\to -2}\left(3x^2-11x-4\right)}{\lim\limits_{x\to -2}\left(x^2-2x-8\right)}$

$=\dfrac{3\left(\lim\limits_{x\to -2}x\right)^2-11\lim\limits_{x\to -2}x-\lim\limits_{x\to -2}4}{\left(\lim\limits_{x\to -2}x\right)^2-2\lim\limits_{x\to -2}x-\lim\limits_{x\to -2}8}$

$=\dfrac{3(-2)^2-11(-2)-4}{(-2)^2-2(-2)-8}$

$=\dfrac{12+22-4}{4+4-8}=\dfrac{30}{0}$

x	$\lim\limits_{x\to -2}\dfrac{3x^2-11x-4}{x^2-2x-8}$
−2.1	53
−2.01	503
−2.001	5003
−2	Error
−1.999	−4997
−1.99	−497
−1.9	−47

$\lim\limits_{x\to -2}\dfrac{3x^2-11x-4}{x^2-2x-8}=\underset{-\infty}{\left(dne\right)}.$

11.2 Exercises

67. $\lim\limits_{x \to 7} \dfrac{x^2 - 5x - 14}{x - 7}$

$= \dfrac{\lim\limits_{x \to 7}\left(x^2 - 5x - 14\right)}{\lim\limits_{x \to 7}(x - 7)}$

$= \dfrac{\left(\lim\limits_{x \to 7} x\right)^2 - 5\lim\limits_{x \to 7} x - \lim\limits_{x \to 7} 14}{\lim\limits_{x \to 7} x - \lim\limits_{x \to 7} 7}$

$= \dfrac{(7)^2 - 5(7) - 14}{7 - 7} = \dfrac{0}{0}$

x	$\lim\limits_{x \to 7} \dfrac{x^2 - 5x - 14}{x - 7}$
6.9	8.9
6.99	8.99
6.999	8.999
7	Error
7.001	9.001
7.01	9.01
7.1	9.1

$\lim\limits_{x \to 7} \dfrac{x^2 - 5x + 2}{x - 7} = 9.$

69. $\lim\limits_{x \to -2} \dfrac{x^2 + 5x + 6}{x^2 + 4x + 4}$

$= \dfrac{\lim\limits_{x \to -2}\left(x^2 + 5x + 6\right)}{\lim\limits_{x \to -2}\left(x^2 + 4x + 4\right)}$

$= \dfrac{\left(\lim\limits_{x \to -2} x\right)^2 + 5\lim\limits_{x \to -2} x + \lim\limits_{x \to -2} 6}{\left(\lim\limits_{x \to -2} x\right)^2 + 4\lim\limits_{x \to -2} x + \lim\limits_{x \to -2} 4}$

$= \dfrac{(-2)^2 + 5(-2) + 6}{(-2)^2 + 4(-2) + 4}$

$= \dfrac{4 - 10 + 6}{4 - 8 + 4} = \dfrac{0}{0}$

x	$\lim\limits_{x \to -2} \dfrac{x^2 + 5x + 6}{x^2 + 4x + 4}$
−2.1	−9
−2.01	−99
−2.001	−999
−2	Error
−1.999	1001
−1.99	101
−1.9	11

$\lim\limits_{x \to -2} \dfrac{x^2 + 5x + 6}{x^2 + 4x + 4} = \left(\underset{\infty}{dne}\right).$

71. $\lim\limits_{x \to -7} \dfrac{\sqrt{x^2+14x+49}}{x^2+8x+7}$

$= \dfrac{\lim\limits_{x \to -7}\left(\sqrt{x^2+14x+49}\right)}{\lim\limits_{x \to -7}\left(x^2+8x+7\right)}$

$= \dfrac{\sqrt{\lim\limits_{x \to -7}\left(x^2+14x+49\right)}}{\lim\limits_{x \to -7}\left(x^2+8x+7\right)}$

$= \dfrac{\sqrt{\left(\lim\limits_{x \to -7}x\right)^2 + 14\lim\limits_{x \to -7}x + \lim\limits_{x \to -7}49}}{\left(\lim\limits_{x \to -7}x\right)^2 + 8\lim\limits_{x \to -7}x + \lim\limits_{x \to -7}7}$

$= \dfrac{\sqrt{(-7)^2 + 14(-7) + 49}}{(-7)^2 + 8(-7) + 7}$

$= \dfrac{\sqrt{49-98+49}}{49-56+7} = \dfrac{0}{0}$

x	$\lim\limits_{x \to -7} \dfrac{\sqrt{x^2+14x+49}}{x^2+8x+7}$
-6.9	-0.16949
-6.99	-0.1669
-6.999	-0.1667
-7	Error
-7.001	0.16664
-7.01	0.16639
-7.1	0.16393

$\lim\limits_{x \to -7^-} f(x) = \dfrac{1}{6}$

$\lim\limits_{x \to -7^+} f(x) = -\dfrac{1}{6}$

$\lim\limits_{x \to -7} f(x) \quad \left(\underset{LH \ne RH}{dne}\right)$

73. $\lim\limits_{x \to 0} \dfrac{(x+3)^2-9}{x}$

$= \dfrac{\lim\limits_{x \to 0}\left((x+3)^2-9\right)}{\lim\limits_{x \to 0}x}$

$= \dfrac{\left(\lim\limits_{x \to 0}(x+3)\right)^2 - \lim\limits_{x \to 0}9}{\lim\limits_{x \to 0}x}$

$= \dfrac{\left(\lim\limits_{x \to 0}x + \lim\limits_{x \to 0}3\right)^2 - \lim\limits_{x \to 0}9}{\lim\limits_{x \to 0}x}$

$= \dfrac{(0+3)^2-9}{0} = \dfrac{9-9}{0} = \dfrac{0}{0}$

x	$\lim\limits_{x \to 0} \dfrac{(x+3)^2-9}{x}$
-0.1	5.9
-0.01	5.99
-0.001	5.999
0	Error
0.001	6.001
0.01	6.01
0.1	6.1

$\lim\limits_{x \to 0} \dfrac{(x+3)^2-9}{x} = 6$

75. $\lim\limits_{x \to 1} \dfrac{x^3 - x^2}{\sqrt{x^2 - 2x + 1}}$

$= \dfrac{\lim\limits_{x \to 1}\left(x^3 - x^2\right)}{\lim\limits_{x \to 1}\sqrt{x^2 - 2x + 1}}$

$= \dfrac{\left(\lim\limits_{x \to 1} x\right)^3 - \left(\lim\limits_{x \to 1} x\right)^2}{\sqrt{\lim\limits_{x \to 1}\left(x^2 - 2x + 1\right)}}$

$= \dfrac{\left(\lim\limits_{x \to 1} x\right)^3 - \left(\lim\limits_{x \to 1} x\right)^2}{\sqrt{\left(\lim\limits_{x \to 1} x\right)^2 - 2\lim\limits_{x \to 1} x + \lim\limits_{x \to 1} 1}}$

$= \dfrac{(1)^3 - (1)^2}{\sqrt{(1)^2 - 2(1) + 1}} = \dfrac{1 - 1}{\sqrt{1 - 2 + 1}} = \dfrac{0}{0}$

x	$\lim\limits_{x \to 1} \dfrac{x^3 - x^2}{\sqrt{x^2 - 2x + 1}}$
.9	−0.81
.99	−0.9801
.999	−0.998
1	Error
1.001	1.002
1.01	1.0201
1.1	1.21

$\lim\limits_{x \to 1^-} f(x) = -1$

$\lim\limits_{x \to 1^+} f(x) = 1$

$\lim\limits_{x \to 1} f(x) = \left(\underset{LH \neq RH}{dne}\right)$

77. $(2x - 3)(x + 1) \leq 0$

Zeroes: $2x - 3 = 0 \quad x + 1 = 0$
$\qquad\qquad 2x = 3 \qquad\quad x = -1$

Zeroes: $x = \dfrac{3}{2}, -1$

$x = -2 : (2(-2) - 3)(-2 + 1) = (-7)(-1) = 7 > 0$

$x = 0 : (2(0) - 3)(0 + 1) = (-3)(1) = -3 < 0$

$x = 2 : (2(2) - 3)(2 + 1) = (1)(3) = 3 > 0$

$p \leq 0 \;\to\; \left[-1, \dfrac{3}{2}\right]$

79. $\sec y - \cos y = \tan y \sin y$

$\dfrac{1}{\cos y} - \dfrac{\cos y}{1} = \tan y \sin y$

$\dfrac{1}{\cos y} - \dfrac{\cos y}{1} \cdot \dfrac{\cos y}{\cos y} = \tan y \sin y$

$\dfrac{1 - \cos^2 y}{\cos y} = \tan y \sin y$

$\dfrac{\sin^2 y}{\cos y} = \tan y \sin y$

$\dfrac{\sin y}{\cos y} \cdot \sin y = \tan y \sin y$

$\tan y \sin y = \tan y \sin y$

Mid-Chapter Check

1. $\lim\limits_{x \to -\infty} \dfrac{6x^2 - 3}{12x^3 - x - 1} = 0$

3. $A = P\left(1 + \dfrac{r}{n}\right)^{n \cdot t}$

   $A = Pe^{rt}$

   minimum number of compoundings per year
   needed to yield the same amount with the
   same $1000 principal.

x	y
1	1030.0000
2	1030.2250
3	1030.3010
4	1030.3392
5	1030.3622
6	1030.3775
7	1030.3885
8	1030.3967
9	1030.4031
10	1030.4083
15	1030.4237
20	1030.4314
25	1030.4360
30	1030.4391
35	1030.4413
40	1030.4429
41	1030.4432
42	1030.4435
43	1030.4438
44	1030.4440
45	1030.4442
46	1030.4445
47	1030.4447
48	1030.4449
49	1030.4451
50	1030.4453

49 per year (almost weekly)

5. $g(x) = \begin{cases} \dfrac{-x}{2x+5} & x < -5 \\ \cos(x+5) & x \geq -5 \end{cases}$

x	y
−5.1	−0.9808
−5.01	−09980
−5.001	−0.9998
−5	1
−4.999	1.00000
−4.99	0.99995
−4.9	0.99500

a. $\lim\limits_{x \to -5^-} g(x) = -1$

b. $\lim\limits_{x \to -5^+} g(x) = 1$

c. $\lim\limits_{x \to 5} g(x) = \left(\underset{LH \neq RH}{dne}\right)$

7. (a)

x	$y = \cos\left(\dfrac{\pi}{x}\right)$
0.1	1
0.01	1
0.001	1
0	Error
−0.001	1
−0.01	1
−0.1	1

(b) Answers may vary.

9. $\lim\limits_{x \to -3} \dfrac{2x+10}{2\sqrt[3]{11-x^2}}$

$= \dfrac{\lim\limits_{x \to -3}(2x+10)}{\lim\limits_{x \to -3} 2\sqrt[3]{11-x^2}}$

$= \dfrac{2\lim\limits_{x \to -3} x + \lim\limits_{x \to -3} 10}{2\lim\limits_{x \to -3}\left(11-x^2\right)^{\frac{1}{3}}}$

$= \dfrac{2\lim\limits_{x \to -3} x + \lim\limits_{x \to -3} 10}{2\left(\lim\limits_{x \to -3} 11 - \lim\limits_{x \to -3} x^2\right)^{\frac{1}{3}}}$

$= \dfrac{2\lim\limits_{x \to -3} x + \lim\limits_{x \to -3} 10}{2\left(\lim\limits_{x \to -3} 11 - \left(\lim\limits_{x \to -3} x\right)^2\right)^{\frac{1}{3}}}$

$= \dfrac{2(-3)+10}{2\left(11-(-3)^2\right)^{\frac{1}{3}}} = \dfrac{4}{2(2)^{\frac{1}{3}}}$

$= \dfrac{2}{\sqrt[3]{2}} \cdot \dfrac{\sqrt[3]{4}}{\sqrt[3]{4}} = \dfrac{2\sqrt[3]{4}}{2} = \sqrt[3]{4}$

11.3 Exercises

1. asymptotic, removable, jump

3. direct substitution

5. Answers will vary.

7. $f(x)$ at $x = -1$ is not continuous,; condition 1 is violated

9. $f(x)$ at $x = 1$ is continuous

11. $f(x)$ at $x = 0$ is not continuous condition 2 is violated

13. $g(x)$ at $x = -3$ is continuous

15. $g(x)$ at $x = -2$ is not continuous; condition 3 is violated

17. $\lim\limits_{x \to -3} 2x^2 - 5x + 3$

$= 2(-3)^2 - 5(-3) + 3$

$= 18 + 15 + 3 = 36$

19. $\lim\limits_{x \to 5} \sqrt{3x-19}$

direct substitution not possible because the domain is $\left[\dfrac{19}{3}, \infty\right)$ and 5 is not in the domain.

21. $\lim\limits_{x \to 2} \dfrac{x^2}{5x-2}$

$= \dfrac{(2)^2}{5(2)-2} = \dfrac{4}{10-2} = \dfrac{4}{8} = \dfrac{1}{2}$

23. $\lim\limits_{x \to -1} \dfrac{x+1}{x^2-1}$

direct substitution not possible; domain is $(-\infty, -1) \cup (-1, 1) \cup (1, \infty)$ and -1 is not in the domain

25. $\lim\limits_{x \to -5} \sqrt{x^2-6x}$

$= \sqrt{(-5)^2 - 6(-5)}$

$= \sqrt{25+30} = \sqrt{55}$

27. $\lim\limits_{x \to -1} \dfrac{x+1}{x^2-1}$

$= \lim\limits_{x \to -1} \dfrac{x+1}{(x+1)(x-1)}$

$= \lim\limits_{x \to -1} \dfrac{1}{x-1}$

$= \dfrac{1}{(-1)-1} = \dfrac{1}{-2}$

29. $\lim\limits_{x \to 4} \dfrac{x-4}{\sqrt{x}-2} = \left(dne \atop \infty\right)$ because 4 is not in the domain of the function.

31. $\lim\limits_{x \to 3} \dfrac{2x^2-3x-9}{x-3}$

$= \lim\limits_{x \to 3} \dfrac{(2x+3)(x-3)}{x-3}$

$= \lim\limits_{x \to 3} 2x+3 = 2(3)+3 = 9$

33. $\displaystyle\lim_{x\to-3}\frac{\sqrt{x+7}-2}{x+3}$

$\displaystyle=\lim_{x\to-3}\frac{\sqrt{x+7}-2}{x+3}\cdot\frac{\sqrt{x+7}+2}{\sqrt{x+7}+2}$

$\displaystyle=\lim_{x\to-3}\frac{x+7-4}{(x+3)\left(\sqrt{x+7}+2\right)}$

$\displaystyle=\lim_{x\to-3}\frac{x+3}{(x+3)\left(\sqrt{x+7}+2\right)}$

$\displaystyle=\frac{1}{\sqrt{-3+7}+2}=\frac{1}{\sqrt{4}+2}=\frac{1}{4}$

35. $\displaystyle\lim_{x\to-4}\frac{x^3+8x^2+16x}{x^2+7x+12}$

$\displaystyle=\lim_{x\to-4}\frac{x\left(x^2+8x+16x\right)}{(x+3)(x+4)}$

$\displaystyle=\lim_{x\to-4}\frac{x(x+4)(x+4)}{(x+3)(x+4)}$

$\displaystyle=\frac{-4(-4+4)}{-4+3}=\frac{-4(0)}{-1}=0$

37. $\displaystyle\lim_{h\to0}\frac{\left[2(x+h)^2-(x+h)\right]-\left(2x^2-x\right)}{h}$

$\displaystyle=\lim_{h\to0}\frac{\left[2\left(x^2+2xh+h^2\right)-(x+h)\right]-\left(2x^2-x\right)}{h}$

$\displaystyle=\lim_{h\to0}\frac{2x^2+4xh+2h^2-x-h-2x^2+x}{h}$

$\displaystyle=\lim_{h\to0}\frac{4xh+2h^2-h}{h}$

$\displaystyle=\lim_{h\to0}\frac{h(4x+2h-1)}{h}$

$\displaystyle=4x+2(0)-1=4x-1$

39. $\displaystyle\lim_{h\to0}\frac{\dfrac{3}{x+h+2}-\dfrac{3}{x+2}}{h}$

$\displaystyle=\lim_{h\to0}\frac{\dfrac{3(x+2)}{(x+h+2)(x+2)}-\dfrac{3(x+h+2)}{(x+2)(x+h+2)}}{h}$

$\displaystyle=\lim_{h\to0}\frac{3x+6-3x-3h-6}{h(x+h+2)(x+2)}$

$\displaystyle=\lim_{h\to0}\frac{-3h}{h(x+h+2)(x+2)}$

$\displaystyle=\lim_{h\to0}\frac{-3}{(x+h+2)(x+2)}$

$\displaystyle=\frac{-3}{(x+0+2)(x+2)}=\frac{-3}{(x+2)^2}$

41. $\displaystyle\lim_{h\to0}\frac{\sqrt{x+h+2}-\sqrt{x+2}}{h}$

$\displaystyle=\lim_{h\to0}\frac{\sqrt{x+h+2}-\sqrt{x+2}}{h}\cdot\frac{\left(\sqrt{x+h+2}+\sqrt{x+2}\right)}{\left(\sqrt{x+h+2}+\sqrt{x+2}\right)}$

$\displaystyle=\lim_{h\to0}\frac{\left(\sqrt{x+h+2}\right)^2-\left(\sqrt{x+2}\right)^2}{h\left(\sqrt{x+h+2}+\sqrt{x+2}\right)}$

$\displaystyle=\lim_{h\to0}\frac{(x+h+2)-(x+2)}{h\left(\sqrt{x+h+2}+\sqrt{x+2}\right)}$

$\displaystyle=\lim_{h\to0}\frac{x+h+2-x-2}{h\left(\sqrt{x+h+2}+\sqrt{x+2}\right)}$

$\displaystyle=\lim_{h\to0}\frac{1h}{h\left(\sqrt{x+h+2}+\sqrt{x+2}\right)}$

$\displaystyle=\lim_{h\to0}\frac{1}{\sqrt{x+h+2}+\sqrt{x+2}}$

$\displaystyle=\frac{1}{\sqrt{x+0+2}+\sqrt{x+2}}=\frac{1}{2\sqrt{x+2}}$

43. $\lim\limits_{h\to 0}\dfrac{(x+h+2)^3-(x+2)^3}{h}$

$=\lim\limits_{h\to 0}\dfrac{\left((x+2)+h\right)^3-(x+2)^3}{h}$

$=\lim\limits_{h\to 0}\dfrac{(x+2)^3+3(x+2)^2h+3(x+2)h^2+h^3-(x+2)^3}{h}$

$=\lim\limits_{h\to 0}\dfrac{3(x+2)^2h+3(x+2)h^2+h^3}{h}$

$=\lim\limits_{h\to 0}\dfrac{h\left(3(x+2)^2+3(x+2)h+h^2\right)}{h}$

$=\lim\limits_{h\to 0}3(x+2)^2+3(x+2)h+h^2$

$=3(x+2)^2+3(x+2)(0)+0^2$

$=3(x+2)^2$

45. $\lim\limits_{x\to\infty}\dfrac{5x^3+2}{10x^3-2x+1}=\dfrac{1}{2}$

x	y
50	0.50004
100	0.50001
150	0.50000
200	0.50000
1500	0.50000

47. $\lim\limits_{x\to\infty}\dfrac{6x^2-x+2}{2x^2+1}=3$

x	y
250	2.99799
500	2.99900
750	2.99933
1000	2.99950
1500	2.99967

49. $\lim\limits_{x\to\infty}\dfrac{7x^3}{5x^2+3x}=\left(dne\atop\infty\right)$

x	y
2500	3499.16
5000	6999.16
7500	10499.16
10000	13999.16
15000	20999.16

51. $\lim\limits_{x\to\infty}\dfrac{x^2+6x+9}{2x^3}=0$

x	y
2500	0.0002
5000	0.0001
7500	0.0001
10000	0.0001
15000	0.0000

53. $\lim\limits_{x\to-\infty}\dfrac{5x^3+2}{10x^3-2x+1}=\dfrac{1}{2}$

x	y
−50	0.50004
−100	0.50001
−150	0.50000
−200	0.50000
−1500	0.50000

55. $\lim\limits_{x\to-\infty}\dfrac{x^2+1}{2x-11}=\left(dne\atop-\infty\right)$

x	y
−5000	3499.16
−10000	6999.16
−15000	10499.16
−20000	13999.16
−25000	20999.16

57. $\lim\limits_{x\to-\infty}\dfrac{10-3x^2}{10-3x^3}=0$

x	y
−2500	−0.0002
−5000	−0.0001
−7500	−0.0001
−15000	−0.0000
−25000	−0.0000

59. Given $\lim\limits_{x\to\infty}\dfrac{1}{x}=0$, find the smallest

positive value of x such that $\dfrac{1}{x}\le 0.001$.

x	y
100	0.0100
200	0.0050
300	0.0033
400	0.0025
500	0.0020
600	0.0017
700	0.0014
800	0.0013
900	0.0011
1000	0.0010
1100	0.0009

$\dfrac{1}{x}\le 0.001$

$\dfrac{1}{x}\le\dfrac{1}{1000}$

$1000\le x$

$x=1000$

61. $\lim\limits_{x\to\infty}\dfrac{3x^2-2x+1}{8x^2+5}$

$=\lim\limits_{x\to\infty}\dfrac{\dfrac{3x^2}{x^2}-\dfrac{2x}{x^2}+\dfrac{1}{x^2}}{\dfrac{8x^2}{x^2}+\dfrac{5}{x^2}}$

$=\dfrac{\lim\limits_{x\to\infty}\left(3-\dfrac{2}{x}+\dfrac{1}{x^2}\right)}{\lim\limits_{x\to\infty}\left(8+\dfrac{5}{x^2}\right)}$

$=\dfrac{\lim\limits_{x\to\infty}3-\lim\limits_{x\to\infty}\dfrac{2}{x}+\lim\limits_{x\to\infty}\dfrac{1}{x^2}}{\lim\limits_{x\to\infty}8+\lim\limits_{x\to\infty}\dfrac{5}{x^2}}$

$=\dfrac{3-0+0}{8+0}=\dfrac{3}{8}$

63. $\lim\limits_{x\to-\infty}\dfrac{2x^2-1}{x^3+2x+12}$

$=\lim\limits_{x\to-\infty}\dfrac{\dfrac{2x^2}{x^3}-\dfrac{1}{x^3}}{\dfrac{x^3}{x^3}+\dfrac{2x}{x^3}+\dfrac{12}{x^3}}$

$=\dfrac{\lim\limits_{x\to-\infty}\left(\dfrac{2}{x}-\dfrac{1}{x^3}\right)}{\lim\limits_{x\to-\infty}\left(1+\dfrac{2}{x^2}+\dfrac{12}{x^3}\right)}$

$=\dfrac{\lim\limits_{x\to-\infty}\dfrac{2}{x}-\lim\limits_{x\to-\infty}\dfrac{1}{x^3}}{\lim\limits_{x\to-\infty}1+\lim\limits_{x\to-\infty}\dfrac{2}{x^2}+\lim\limits_{x\to-\infty}\dfrac{12}{x^3}}$

$=\dfrac{0-0}{1+0+0}=0$

65. $\lim\limits_{x\to\infty}\dfrac{\sqrt{36x^2-11}}{3x}$

$=\lim\limits_{x\to\infty}\dfrac{\sqrt{36x^2-11}}{\sqrt{9x^2}}$

$=\lim\limits_{x\to\infty}\sqrt{\dfrac{36x^2-11}{9x^2}}$

$=\sqrt{\lim\limits_{x\to\infty}\left(\dfrac{36x^2-11}{9x^2}\right)}$

$=\sqrt{\lim\limits_{x\to\infty}\left(\dfrac{\dfrac{36x^2}{x^2}-\dfrac{11}{x^2}}{\dfrac{9x^2}{x^2}}\right)}$

$=\sqrt{\dfrac{\lim\limits_{x\to\infty}\left(36-\dfrac{11}{x^2}\right)}{\lim\limits_{x\to\infty}9}}$

$=\sqrt{\dfrac{\lim\limits_{x\to\infty}36-\lim\limits_{x\to\infty}\dfrac{11}{x^2}}{\lim\limits_{x\to\infty}9}}$

$=\sqrt{\dfrac{36-0}{9}}=\sqrt{4}=2$

67. $\lim\limits_{x\to\infty} \dfrac{\sqrt{12x^2 - 6x + 1}}{7x}$

$= \lim\limits_{x\to\infty} \dfrac{\sqrt{12x^2 - 6x + 1}}{\sqrt{7x^2}}$

$= \lim\limits_{x\to\infty} \sqrt{\dfrac{12x^2 - 6x + 1}{49x^2}}$

$= \sqrt{\lim\limits_{x\to\infty} \left(\dfrac{\dfrac{12x^2}{x^2} - \dfrac{6x}{x^2} + \dfrac{1}{x^2}}{\dfrac{49x^2}{x^2}} \right)}$

$= \sqrt{\dfrac{\lim\limits_{x\to\infty}\left(12 - \dfrac{6}{x} + \dfrac{1}{x^2}\right)}{\lim\limits_{x\to\infty} 49}}$

$= \sqrt{\dfrac{\lim\limits_{x\to\infty} 12 - \lim\limits_{x\to\infty}\dfrac{6}{x} + \lim\limits_{x\to\infty}\dfrac{1}{x^2}}{\lim\limits_{x\to\infty} 49}}$

$= \sqrt{\dfrac{12 - 0 + 0}{49}} = \dfrac{2\sqrt{3}}{7}$

69. $\lim\limits_{x\to\infty} \dfrac{\sqrt[3]{216x^3 + 36x^2 - 6x + 1}}{2x}$

$= \lim\limits_{x\to\infty} \dfrac{\sqrt[3]{216x^3 + 36x^2 - 6x + 1}}{\sqrt[3]{(2x)^3}}$

$= \lim\limits_{x\to\infty} \dfrac{\sqrt[3]{216x^3 + 36x^2 - 6x + 1}}{\sqrt[3]{8x^3}}$

$= \sqrt[3]{\lim\limits_{x\to\infty} \dfrac{\left(\dfrac{216x^3}{x^3} + \dfrac{36x^2}{x^3} - \dfrac{6x}{x^3} + \dfrac{1}{x^3}\right)}{\left(\dfrac{8x^3}{x^3}\right)}}$

$= \sqrt[3]{\dfrac{\lim\limits_{x\to\infty}\left(216 + \dfrac{36}{x} - \dfrac{6}{x^2} + \dfrac{1}{x^3}\right)}{\lim\limits_{x\to\infty} 8}}$

$= \sqrt[3]{\dfrac{\lim\limits_{x\to\infty} 216 + \lim\limits_{x\to\infty}\dfrac{36}{x} - \lim\limits_{x\to\infty}\dfrac{6}{x^2} + \lim\limits_{x\to\infty}\dfrac{1}{x^3}}{\lim\limits_{x\to\infty} 8}}$

$= \sqrt[3]{\dfrac{216 + 0 - 0 + 0}{8}} = \dfrac{6}{2} = 3$

71. $f(0) + g(-1) = 1 + 2 = 3$

73. $g(4) - f(2)$
not possible since $f(2)$ is not defined

75. $g(0) + f(-4) = 2 + 1 = 3$

77. $\lim\limits_{x\to -4^-} f(x) - \lim\limits_{x\to 0^+} g(x) = 1 - 1 = 0$

79. $\lim\limits_{x\to -4^+} f(x) - \lim\limits_{x\to 0^-} g(x) = 3 - 2 = 1$

81. $\lim\limits_{x\to 5^-} f(x) + \lim\limits_{x\to -2^-} g(x)$
not possible, since $\lim\limits_{x\to -2^-} g(x)$ does not exist

83. $\lim\limits_{x\to 0}\left[f(x) - g(x)\right]$

$= \lim\limits_{x\to 0} f(x) - \lim\limits_{x\to 0} g(x)$

$\lim\limits_{x\to 0^+} g(x) = 1$

$\lim\limits_{x\to 0^-} g(x) = 2$

$= \lim\limits_{x\to 0} g(x) = \left(\underset{LH \neq RH}{dne}\right)$

not possible, since $\lim\limits_{x\to 0} g(x)$ does not exist

85. $\lim\limits_{x\to -3}\left[\dfrac{3f(x)}{g(x)}\right]$

$= \dfrac{3\lim\limits_{x\to -3} f(x)}{\lim\limits_{x\to -3} g(x)}$

$= \dfrac{3(2)}{-2} = -3$

87. $\lim\limits_{x \to -\infty} \left[f(x) - \sqrt[3]{g(x)} \right]$

$= \lim\limits_{x \to -\infty} f(x) - \lim\limits_{x \to -\infty} \sqrt[3]{g(x)}$

$= \lim\limits_{x \to -\infty} f(x) - \sqrt[3]{\lim\limits_{x \to -\infty} g(x)}$

$= 2 - \sqrt[3]{-1} = 2 - (-1) = 3$

89. a) $3x^2 + 4x - 12 = 0$

$a = 3 \quad b = 4 \quad c = -12$

$x = \dfrac{-b \pm \sqrt{b^2 - 4ac}}{2a}$

$x = \dfrac{-4 \pm \sqrt{(4)^2 - 4(3)(-12)}}{2(3)}$

$x = \dfrac{-4 \pm \sqrt{16 + 144}}{6}$

$x = \dfrac{-4 \pm \sqrt{160}}{6}$

$x = \dfrac{-4 \pm 4\sqrt{10}}{6} = \dfrac{-2 \pm 2\sqrt{10}}{3}$

b) $\sqrt{3x+1} - \sqrt{2x} = 1$

$\sqrt{3x+1} = 1 + \sqrt{2x}$

$\left(\sqrt{3x+1}\right)^2 = \left(1 + \sqrt{2x}\right)^2$

$3x + 1 = 1 + 2\sqrt{2x} + \left(\sqrt{2x}\right)^2$

$3x = 2\sqrt{2x} + 2x$

$x = 2\sqrt{2x}$

$(x)^2 = \left(2\sqrt{2x}\right)^2$

$(x)^2 = 4(2x)$

$x^2 = 8x$

$x(x - 8) = 0$

$x = 0 \quad \text{or} \quad x - 8 = 0$

$\qquad\qquad\qquad x = 8$

c. $\dfrac{1}{x+2} + \dfrac{3}{x^2 + 5x + 6} = \dfrac{2}{x+3}$

$\dfrac{1}{x+2} + \dfrac{3}{(x+2)(x+3)} = \dfrac{2}{x+3}$

$(x+2)(x+3)\left(\dfrac{1}{x+2} + \dfrac{3}{(x+2)(x+3)}\right) = \left(\dfrac{2}{x+3}\right)(x+2)(x+3)$

$x + 3 + 3 = 2(x+2); \ x \neq -2; \ x \neq -3$

$x + 6 = 2x + 4$

$x + 2 = 2x$

$2 = x$

$x = 2$

91. $A = 90° - 35° = 55°$

$C = 90°$

$\cos 35° = \dfrac{13.7}{c}$

$c(\cos 35°) = 13.7$

$c = \dfrac{13.7}{\cos 35°} \approx 16.7 \text{ cm}$

$\tan 35° = \dfrac{b}{13.7}$

$(13.7)(\tan 35°) = b$

$b = 9.6 \text{ cm}$

11.4 Exercises

1. difference

3. rectangles

5. Answers will vary.

7. instantaneous velocity at time t.

$f(t) = 500 + 88.2t - 4.9t^2$

$= \lim\limits_{h \to 0} \dfrac{\left[500 + 88.2(t+h) - 4.9(t+h)^2\right] - \left[500 + 88.2t - 4.9t^2\right]}{h}$

$= \lim\limits_{h \to 0} \dfrac{\left[500 + 88.2t + 88.2h - 4.9(t^2 + 2th + h^2)\right] - 500 - 88.2t + 4.9t^2}{h}$

$= \lim\limits_{h \to 0} \dfrac{500 + 88.2t + 88.2h - 4.9t^2 - 9.8th - 4.9h^2 - 500 - 88.2t + 4.9t^2}{h}$

$= \lim\limits_{h \to 0} \dfrac{88.2h - 9.8th - 4.9h^2}{h}$

$= \lim\limits_{h \to 0} \dfrac{h(88.2 - 9.8t - 4.9h)}{h}$

$= \lim\limits_{h \to 0} 88.2 - 9.8t - 4.9h$

$f(t) = 88.2 - 9.8t$

9. instantaneous velocity at time t.

$g(t) = 600 + 78.4t - 4.9t^2$

$= \lim\limits_{h \to 0} \dfrac{\left[600 + 78.4(t+h) - 4.9(t+h)^2\right] - \left[600 + 78.4t - 4.9t^2\right]}{h}$

$= \lim\limits_{h \to 0} \dfrac{\left[600 + 78.4t + 78.4h - 4.9(t^2 + 2th + h^2)\right] - 600 - 78.4t + 4.9t^2}{h}$

$= \lim\limits_{h \to 0} \dfrac{600 + 78.4t + 78.4h - 4.9t^2 - 9.8th - 4.9h^2 - 600 - 78.4t + 4.9t^2}{h}$

$= \lim\limits_{h \to 0} \dfrac{78.4h - 9.8th - 4.9h^2}{h}$

$= \lim\limits_{h \to 0} \dfrac{h(78.4 - 9.8t - 4.9h)}{h}$

$= \lim\limits_{h \to 0} 78.4 - 9.8t - 4.9h$

$g(t) = 78.4 - 9.8t$

11. maximum height of Frank's rocket occurs when $f(t) = 0$

$f(t) = 88.2 - 9.8t = 0$

$88.2 = 9.8t$

$9 = t$

$f(9) = 500 + 88.2(9) - 4.9(9)^2 = 896.9$ m

13. $d(t) = -4.9t^2 + 44.1$

$= \lim\limits_{h \to 0} \dfrac{\left[-4.9(t+h)^2 + 44.1\right] - \left[-4.9t^2 + 44.1\right]}{h}$

$= \lim\limits_{h \to 0} \dfrac{\left(-4.9(t^2 + 2th + h^2) + 44.1\right) + 4.9t^2 - 44.1}{h}$

$= \lim\limits_{h \to 0} \dfrac{-4.9t^2 - 9.8th - 4.9h^2 + 44.1 + 4.9t^2 - 44.1}{h}$

$= \lim\limits_{h \to 0} \dfrac{-9.8th - 4.9h^2}{h}$

$= \lim\limits_{h \to 0} \dfrac{h(-9.8t - 4.9h)}{h}$

$= \lim\limits_{h \to 0} -9.8t - 4.9h$

$d(t) = -9.8t$

15. $d(t) = -16t^2 + 256$

$= \lim\limits_{h \to 0} \dfrac{\left[-16(t+h)^2 + 256\right] - \left[-16t^2 + 256\right]}{h}$

$= \lim\limits_{h \to 0} \dfrac{\left(-16(t^2 + 2th + h^2) + 256\right) + 16t^2 - 256}{h}$

$= \lim\limits_{h \to 0} \dfrac{-16t^2 - 32th - 16h^2 + 256 + 16t^2 - 256}{h}$

$= \lim\limits_{h \to 0} \dfrac{-32th - 16h^2}{h}$

$= \lim\limits_{h \to 0} \dfrac{h(-32t - 16h)}{h}$

$= \lim\limits_{h \to 0} -32t - 16h$

$d(t) = -32t$

17. $f(x) = \dfrac{1}{2}x + 5$

$$\lim_{h \to 0} \frac{\left(\frac{1}{2}(x+h)+5\right)-\left(\frac{1}{2}(x)+5\right)}{h}$$

$$= \lim_{h \to 0} \frac{\frac{1}{2}x+\frac{1}{2}h+5-\frac{1}{2}x-5}{h}$$

$$= \lim_{h \to 0} \frac{\frac{1}{2}h}{h}$$

$$= \lim_{h \to 0} \frac{1}{2} = \frac{1}{2}$$

$$f'(x) = \frac{1}{2}$$

19. $f(x) = x^3$

$$\lim_{h \to 0} \frac{(x+h)^3 - x^3}{h}$$

$$= \lim_{h \to 0} \frac{x^3 + 3x^2h + 3xh^2 + h^3 - x^3}{h}$$

$$= \lim_{h \to 0} \frac{3x^2h + 3xh^2 + h^3}{h}$$

$$= \lim_{h \to 0} \frac{h\left(3x^2 + 3xh + h^2\right)}{h}$$

$$= \lim_{h \to 0} \left(3x^2 + 3xh + h^2\right)$$

$$f'(x) = 3x^2$$

23. $b(t) = 6 - \sqrt{t}$

$$= \lim_{h \to 0} \frac{\left(6-\sqrt{t+h}\right)-\left(6-\sqrt{t}\right)}{h}$$

$$= \lim_{h \to 0} \frac{6-\sqrt{t+h}-6+\sqrt{t}}{h}$$

$$= \lim_{h \to 0} \frac{\sqrt{t}-\sqrt{t+h}}{h} \cdot \frac{\sqrt{t}+\sqrt{t+h}}{\sqrt{t}+\sqrt{t+h}}$$

$$= \lim_{h \to 0} \frac{t-(t+h)}{h\left(\sqrt{t}+\sqrt{t+h}\right)}$$

$$= \lim_{h \to 0} \frac{t-t-h}{h\left(\sqrt{t}+\sqrt{t+h}\right)}$$

$$= \lim_{h \to 0} \frac{-h}{h\left(\sqrt{t}+\sqrt{t+h}\right)}$$

$$= \lim_{h \to 0} \frac{-1}{\sqrt{t}+\sqrt{t+h}}$$

$$b'(t) = \frac{-1}{\sqrt{t}+\sqrt{t}} = \frac{-1}{2\sqrt{t}}$$

21. $p(t) = 1.2\sqrt{t} + 40$

$$= \lim_{h \to 0} \frac{\left(1.2\sqrt{t+h}+40\right)-\left(1.2\sqrt{t}+40\right)}{h}$$

$$= \lim_{h \to 0} \frac{1.2\sqrt{t+h}+40-1.2\sqrt{t}-40}{h}$$

$$= 1.2\lim_{h \to 0} \frac{\sqrt{t+h}-\sqrt{t}}{h} \cdot \frac{\sqrt{t+h}+\sqrt{t}}{\sqrt{t+h}+\sqrt{t}}$$

$$= 1.2\lim_{h \to 0} \frac{t+h-t}{h\sqrt{t+h}+h\sqrt{t}}$$

$$= 1.2\lim_{h \to 0} \frac{h}{h\sqrt{t+h}+h\sqrt{t}}$$

$$= 1.2\lim_{h \to 0} \frac{1}{\sqrt{t+h}+\sqrt{t}}$$

$$p'(t) = 1.2\left(\frac{1}{\sqrt{t}+\sqrt{t}}\right) = \frac{1.2}{2\sqrt{t}} = \frac{0.6}{\sqrt{t}}$$

25. $f(x) = \dfrac{2}{x-1}$

$$\lim_{h \to 0} \dfrac{\dfrac{1}{(x+h)-1} - \dfrac{1}{x-1}}{h}$$

$$= \lim_{h \to 0} \left(\dfrac{2(x-1)}{(x-1)(x+h-1)} - \dfrac{2(x+h-1)}{(x-1)(x+h-1)} \right) \cdot \dfrac{1}{h}$$

$$= \lim_{h \to 0} \dfrac{2x-2-2x-2h+2}{(x-1)(x+h-1)} \cdot \dfrac{1}{h}$$

$$= \lim_{h \to 0} \dfrac{-2h}{h(x-h)(x+h-1)}$$

$$= \lim_{h \to 0} \dfrac{-2}{(x-1)(x+h-1)}$$

$$f(x) = \dfrac{-2}{(x-1)(x-1)} = \dfrac{-2}{(x-1)^2}$$

27. $h(x) = \dfrac{5x}{x+5}$

$$\lim_{h \to 0} \dfrac{\dfrac{5(x+h)}{(x+h)+5} - \dfrac{5x}{x+5}}{h}$$

$$= \lim_{h \to 0} \left(\dfrac{5(x+h)(x+5)}{(x+5)(x+h+5)} - \dfrac{5x(x+h+5)}{(x+5)(x+h+5)} \right) \cdot \dfrac{1}{h}$$

$$= \lim_{h \to 0} \dfrac{5(x^2+5x+hx+5h) - 5x^2 - 5xh - 25x}{h(x+5)(x+h+5)}$$

$$= \lim_{h \to 0} \dfrac{5x^2 + 25x + 5hx + 25h - 5x^2 - 5xh - 25x}{h(x+5)(x+h+5)}$$

$$= \lim_{h \to 0} \dfrac{25h}{h(x+5)(x+h+5)}$$

$$= \lim_{h \to 0} \dfrac{25}{(x+5)(x+h+5)}$$

$$h(x) = \dfrac{25}{(x+5)(x+5)} = \dfrac{25}{(x+5)^2}$$

29. $f(x)$ at $x = 3$

$$f(x) = \dfrac{-2}{(x-1)^2}$$

$$f(3) = \dfrac{-2}{(3-1)^2} = \dfrac{-2}{2^2} = \dfrac{-1}{2}$$

31. $h(x)$ at $x = 0$

$$h(x) = \dfrac{25}{(x+5)^2}$$

$$h(0) = \dfrac{25}{(0+5)^2} = \dfrac{25}{5^2} = 1$$

33. $q(x) = \dfrac{1}{2}x$

Area $= \dfrac{1}{2}bh$; $b = 6$, $h = 3$

$$= \dfrac{1}{2}(6)(3)$$

$$= \dfrac{1}{2} \cdot 18 = 9 \text{ units}^2$$

35. $t(x) = \dfrac{1}{2}x + 1$

Area $= \dfrac{1}{2}(b_1 + b_2)h$; $b_1 = 1$, $b_2 = 4$, $h = 6$

$$= \dfrac{1}{2}(1+4)(6)$$

$$= 15 \text{ units}^2$$

37. $\lim\limits_{n\to\infty}\sum\limits_{i=1}^{n}\left(\dfrac{1}{2}\cdot\dfrac{6}{n}i\right)\dfrac{6}{n}$

$\sum\limits_{i=1}^{n}\left(\dfrac{1}{2}\cdot\dfrac{6}{n}i\right)\dfrac{6}{n}$

$=\left(\dfrac{6}{n}\right)^{2}\left(\dfrac{1}{2}\right)\sum\limits_{i=1}^{n}(i)$

$=\dfrac{18}{n^{2}}\cdot\dfrac{n^{2}+n}{2}$

$=9\left(\dfrac{n^{2}}{n^{2}}+\dfrac{n}{n^{2}}\right)$

$=9\left(1+\dfrac{1}{n}\right)$

$\lim\limits_{n\to\infty}9+9\lim\limits_{n\to\infty}\dfrac{1}{n}=9\text{ units}^{2}$

39. $\lim\limits_{n\to\infty}\sum\limits_{i=1}^{n}\left(\dfrac{1}{2}\cdot\dfrac{6}{n}i+1\right)\dfrac{6}{n}$

$\dfrac{6}{n}\sum\limits_{i=1}^{n}\left(\dfrac{1}{2}\cdot\dfrac{6}{n}i+1\right)=\dfrac{6}{n}\left(\sum\limits_{i=1}^{n}\left(\dfrac{1}{2}\cdot\dfrac{6}{n}i\right)+\sum\limits_{i=1}^{n}1\right)$

$=\left(\dfrac{6}{n}\right)^{2}\cdot\dfrac{1}{2}\sum\limits_{i=1}^{n}i+\dfrac{6}{n}\sum\limits_{i=1}^{n}1$

$=\dfrac{36}{2n^{2}}\cdot\dfrac{n^{2}+n}{2}+\dfrac{6}{n}\cdot n$

$=\dfrac{9}{n^{2}}\left(n^{2}+n\right)+6$

$=9+\dfrac{9}{n}+6$

$\lim\limits_{n\to\infty}\left(9+\dfrac{9}{n}+6\right)=\lim\limits_{n\to\infty}15+\lim\limits_{n\to\infty}\left(\dfrac{9}{n}\right)$

$=15\text{ units}^{2}$

41. $\dfrac{4}{n}\sum\limits_{i=1}^{n}\left[\dfrac{1}{2}\left(\dfrac{4}{n}i\right)^{2}+3\right]$

$=\dfrac{4}{n}\left[\sum\limits_{i=1}^{n}\dfrac{1}{2}\left(\dfrac{4}{n}i\right)^{2}+\sum\limits_{i=1}^{n}3\right]$

summation properties (distribute)

$=\dfrac{4}{n}\left[\sum\limits_{i=1}^{n}\dfrac{1}{2}\dfrac{16}{n^{2}}i^{2}+\sum\limits_{i=1}^{n}3\right]$ simplify

$=\dfrac{4}{n}\left[\dfrac{16}{2n^{2}}\sum\limits_{i=1}^{n}i^{2}+\sum\limits_{i=1}^{n}3\right]$ factor $\dfrac{16}{n^{2}}$ from first

summation

$=\dfrac{4}{n}\left[\dfrac{8}{n^{2}}\left(\dfrac{2n^{3}+3n^{2}+n}{6}\right)+3n\right]$ apply summation formula

$=\left[\dfrac{32}{n^{3}}\left(\dfrac{2n^{3}+3n^{2}+n}{6}\right)+12\right]$ distribute $\dfrac{4}{n}$

$=\left[\dfrac{32}{6}\left(\dfrac{2n^{3}+3n^{2}+n}{n^{3}}\right)+12\right]$ rewrite denominators

$=\dfrac{16}{3}\left(2+\dfrac{3}{n}+\dfrac{1}{n^{2}}\right)+12$ rewrite rational expression

Applying the limit properties gives

$\dfrac{16}{3}\lim\limits_{n\to\infty}\left(2+\dfrac{3}{n}+\dfrac{1}{n^{2}}\right)+\lim\limits_{n\to\infty}12$, and the area

under the curve is $\left(\dfrac{16}{3}\right)(2)+12=\dfrac{68}{3}\text{ units}^{2}$.

The new employee has produced 22 complete parts.

43. $f(x) = -\dfrac{1}{2}x^2 + 4x$, $x \in [0,6]$

$A = \displaystyle\sum_{i=1}^{n} LW = \sum_{i=1}^{n} f\left(\dfrac{6}{n}i\right)\left(\dfrac{6}{n}\right)$ area formula,

rectangle method

$= \dfrac{6}{n}\displaystyle\sum_{i=1}^{n} f\left(\dfrac{6}{n}i\right)$ factor $\dfrac{6}{n}$

$= \dfrac{6}{n}\displaystyle\sum_{i=1}^{n}\left[-\dfrac{1}{2}\left(\dfrac{6}{n}i\right)^2 + 4\left(\dfrac{6}{n}i\right)\right]$ evaluate f at $\dfrac{6}{n}i$

$= \dfrac{6}{n}\left[-\dfrac{1}{2}\displaystyle\sum_{i=1}^{n}\left(\dfrac{6}{n}i\right)^2 + 4\sum_{i=1}^{n}\left(\dfrac{6}{n}i\right)\right]$ distribute

summation

$= \dfrac{6}{n}\left[-\dfrac{1}{2}\displaystyle\sum_{i=1}^{n}\dfrac{36}{n^2}i^2 + 4\sum_{i=1}^{n}\dfrac{6}{n}i\right]$ simplify

$= \dfrac{6}{n}\left[-\dfrac{36}{2n^2}i^2 + \dfrac{24}{n}\displaystyle\sum_{i=1}^{n}i\right]$ factor $\dfrac{36}{n^2}$ and $\dfrac{6}{n}$

$= \dfrac{6}{n}\left[-\dfrac{18}{n^2}\left(\dfrac{2n^3 + 3n^2 + n}{6}\right) + \dfrac{24}{n}\left(\dfrac{n^2 + n}{2}\right)\right]$ summation

formulas

$= -\dfrac{108}{n^3}\left(\dfrac{2n^3 + 3n^2 + n}{6}\right) + \dfrac{144}{n^2}\left(\dfrac{n^2 + n}{2}\right)$ distribute $\dfrac{6}{n}$

$= -\dfrac{108}{6}\left(\dfrac{2n^3 + 3n^2 + n}{n^3}\right) + \dfrac{144}{2}\left(\dfrac{n^2 + n}{n^2}\right)$ rewrite

denominators

$= -18\left(2 + \dfrac{3}{n} + \dfrac{1}{n^2}\right) + 72\left(1 + \dfrac{1}{n}\right)$ rewrite denominators

As $n \to \infty$, $\dfrac{3}{n}$, $\dfrac{1}{n^2} \to 0$, and the area is

$-(18)(2) + 72 = 36$ units^2

45. solve for x:

$-350 = 211e^{-0.025x} - 450$

$100 = 211e^{-0.025x}$

$\ln e^{-0.025x} = \ln\dfrac{100}{211}$

$-0.025x = \ln\dfrac{100}{211}$

$x = \dfrac{\ln\left(\dfrac{100}{211}\right)}{-0.025}$

$x \approx 29.87$

47. $x^3 - 5x^2 + 3x - 15 = 0$

$x^3 - 5x^2 + 3x - 15 = 0$

$(x - 5)(x^2 + 3) = 0$

$(x - 5) = 0$ or $(x^2 + 3) = 0$

$x = 5$ or $x^2 = -3$

$\sqrt{x^2} = \pm\sqrt{-3}$

$x = \pm i\sqrt{3}$

$x = 5,\ \pm\sqrt{3}i$

Summary and Concept Review

1. a. $\lim\limits_{x \to 2^-} f(x) = -3$

   b. $\lim\limits_{x \to 2^+} f(x) = 5$

   c. $\lim\limits_{x \to 2} f(x)$ $\left(\substack{dne \\ LH \neq RH}\right)$

3. $\lim\limits_{x \to \frac{3}{2}} \dfrac{\sin(2x-3)}{2x-3} = 1$

x	y
1.49	0.9999
1.499	1.0000
1.4999	1.0000
1.5	Error
1.5001	1.0000
1.501	1.0000
1.51	0.9999

   $\lim\limits_{x \to \frac{3}{2}^-} \dfrac{\sin(2x-3)}{2x-3} = 1$

   $\lim\limits_{x \to \frac{3}{2}^+} \dfrac{\sin(2x-3)}{2x-3} = 1$

5. $\lim\limits_{x \to -3} 2x^3 - 5x + 1$

   $= 2 \lim\limits_{x \to -3} x^3 - 5 \lim\limits_{x \to -3} x + \lim\limits_{x \to -3} 1$

   $= 2\left(\lim\limits_{x \to -3} x\right)^3 - 5 \lim\limits_{x \to -3} x + \lim\limits_{x \to -3} 1$

   $= 2(-3)^3 - 5(-3) + 1$

   $= 2(-27) + 15 + 1$

   $= -54 + 16 = -38$

7. $\lim\limits_{x \to 1} \dfrac{\sqrt{x}}{\left(x^2 - 12x + 9\right)^5}$

   $= \dfrac{\lim\limits_{x \to 1}(x)^{\frac{1}{2}}}{\lim\limits_{x \to 1}\left(x^2 - 12x + 9\right)^5}$

   $= \dfrac{\lim\limits_{x \to 1}(x)^{\frac{1}{2}}}{\left(\left(\lim\limits_{x \to 1} x\right)^2 - 12 \lim\limits_{x \to 1} x + \lim\limits_{x \to 1} 9\right)^5}$

   $= \dfrac{1^{\frac{1}{2}}}{\left((1)^2 - 12(1) + 9\right)^5}$

   $= \dfrac{1}{(-2)^5} = -\dfrac{1}{32}$

9. Name the x-values(s) where the graph has a:
   a. asymptotic discontinuity
      $x = -3$
   b. jump discontinuity
      $x = -2, \ -1, \ 3, \ 4$
   c. removable discontinuity
      $x = 1, \ 2$

11. $\lim\limits_{x \to 3^+} f(x) = -1$

13. $\lim\limits_{x \to -3^-} f(x) = \left(\substack{dne \\ -\infty}\right)$

15. $\lim\limits_{x \to 1} f(x) = -2$

Summary and Concept Review

17. $g(x) = \sqrt{2x-1}$

$$\lim_{x \to 0} \frac{\sqrt{2(x+h)-1} - \sqrt{2x-1}}{h} \cdot \frac{\sqrt{2(x+h)-1} + \sqrt{2x-1}}{\sqrt{2(x+h)-1} + \sqrt{2x-1}}$$

$$= \lim_{x \to 0} \frac{\left(\sqrt{2x+2h-1}\right)^2 - \sqrt{(2x-1)}^2}{h\left(\sqrt{2(x+h)-1} + \sqrt{2x-1}\right)}$$

$$= \lim_{x \to 0} \frac{2x+2h-1-2x+1}{h\left(\sqrt{2(x+h)-1} + \sqrt{2x-1}\right)}$$

$$= \lim_{x \to 0} \frac{2h}{h\left(\sqrt{2(x+h)-1} + \sqrt{2x-1}\right)}$$

$$= \lim_{x \to 0} \frac{2}{\sqrt{2(x+h)-1} + \sqrt{2x-1}}$$

$$= \frac{2}{\sqrt{2x+0-1} + \sqrt{2x-1}}$$

$$= \frac{2}{2\sqrt{2x-1}} = \frac{1}{\sqrt{2x-1}}$$

19. $f(x) = -x^2 + 3x$ at $x = 4$

$$\lim_{x \to 0} \frac{\left(-(x+h)^2 + 3(x+h)\right) - \left(-x^2 + 3x\right)}{h}$$

$$= \lim_{x \to 0} \frac{-\left(x^2 + 2xh + h^2\right) + 3x + 3h + x^2 - 3x}{h}$$

$$= \lim_{x \to 0} \frac{-x^2 - 2xh - h^2 + 3x + 3h + x^2 - 3x}{h}$$

$$= \lim_{x \to 0} \frac{-2xh - h^2 + 3h}{h}$$

$$= \lim_{x \to 0} \frac{h(-2x - h + 3)}{h}$$

$$= \lim_{x \to 0} (-2x - h + 3)$$

$f(x) = -2x - (0) + 3 = -2x + 3$

$f(4) = -2(4) + 3 = -5$

Slope at $x=4$; $m_{\tan} = -2x + 3 = -5$

Mixed Review

1. a. $\lim_{x \to a^+} f(x) = b$

   b. $\lim_{x \to \infty} f(x) = b$

   c. $\lim_{x \to a} f(x) = b$

   (d) $\lim_{x \to a^-} f(x) = \left(dne \atop \infty\right)$

3. $\lim_{x \to 0} \dfrac{2x^2 - 7x}{\sin(x)} = -7$

   $\lim_{x \to 0^-} \dfrac{2x^2 - 7x}{\sin(x)} = -7$

   $\lim_{x \to 0^+} \dfrac{2x^2 - 7x}{\sin(x)} = -7$

x	y
-0.1	-7.2120
-0.01	-7.0201
-0.001	-7.0020
0	Error
0.001	-6.9980
0.01	-6.9801
0.1	-6.8113

5. $f(x) = \sin|\csc x|$,

   $\lim_{x \to \pi} f(x) = \left(dne \atop \cancel{L}\right)$

x	y
3.1386	0.31932
3.1396	-0.4681
3.1406	0.82697
π	Error
3.1426	0.82697
3.1436	-0.4681
3.1446	0.31932
3.1456	-0.9704

7. $\lim_{x \to c} \sqrt{g(x) + \dfrac{6f(x)}{h(x)}}$

   $$= \sqrt{\lim_{x \to c} g(x) + \frac{6 \lim_{x \to c} f(x)}{\lim_{x \to c} h(x)}}$$

   $$= \sqrt{-13 + \frac{6(5)}{\frac{2}{3}}}$$

   $$= \sqrt{-13 + 30\left(\frac{3}{2}\right)}$$

   $$= \sqrt{-13 + 45} = \sqrt{32} = 4\sqrt{2}$$

638

9. $\lim\limits_{x \to 2} \dfrac{\dfrac{1}{3-x} - x^2}{5x^2 - x}$

$= \dfrac{\lim\limits_{x \to 2}\left(\dfrac{1}{3-x}\right) - \lim\limits_{x \to 2} x^2}{\lim\limits_{x \to 2} 5x^2 - \lim\limits_{x \to 2} x}$

$= \dfrac{\dfrac{\lim\limits_{x \to 2} 1}{\lim\limits_{x \to 2} 3 - \lim\limits_{x \to 2} x} - \left(\lim\limits_{x \to 2} x\right)^2}{5\left(\lim\limits_{x \to 2} x\right)^2 - \lim\limits_{x \to 2} x}$

$= \dfrac{\dfrac{1}{3-2} - (2)^2}{5(2)^2 - 2}$

$= \dfrac{\dfrac{1}{1} - 4^2}{20 - 2} = \dfrac{-3}{18} = \dfrac{-1}{6}$

11. $\lim\limits_{x \to -2} \dfrac{x^3 + 8}{x^2 - 4}$

$= \lim\limits_{x \to -2} \dfrac{(x+2)(x^2 - 2x + 4)}{(x+2)(x-2)}$

$= \lim\limits_{x \to -2} \dfrac{(x^2 - 2x + 4)}{(x-2)}$

$= \dfrac{\left(\lim\limits_{x \to -2} x\right)^2 - 2\lim\limits_{x \to -2} x + \lim\limits_{x \to -2} 4}{\lim\limits_{x \to -2} x - \lim\limits_{x \to -2} 2}$

$= \dfrac{(-2)^2 - 2(-2) + 4}{-2 - 2}$

$= \dfrac{4 + 4 + 4}{-4} = \dfrac{12}{-4} = -3$

13. $\lim\limits_{x \to \infty} \dfrac{-x^3 - 2x + 1}{-2x^3 + 5}$

$= \dfrac{\lim\limits_{x \to \infty} \dfrac{-x^3}{x^3} - \dfrac{2x}{x^3} + \dfrac{1}{x^3}}{\lim\limits_{x \to \infty} \dfrac{-2x^3}{x^3} + \dfrac{5}{x^3}}$

$= \dfrac{\lim\limits_{x \to \infty}(-1) - \lim\limits_{x \to \infty} \dfrac{2}{x^2} + \lim\limits_{x \to \infty} \dfrac{1}{x^3}}{\lim\limits_{x \to \infty}(-2) + \lim\limits_{x \to \infty} \dfrac{5}{x^3}}$

$= \dfrac{-1 - 0 + 0}{-2 + 0} = \dfrac{1}{2}$

15. $\lim\limits_{x \to 2}\left[\dfrac{f(x)}{\sqrt{g(x)}}\right]$

$= \dfrac{\lim\limits_{x \to 2} f(x)}{\sqrt{\lim\limits_{x \to 2} g(x)}}$

$= \dfrac{-1}{\sqrt{2}} = -\dfrac{\sqrt{2}}{2}$

17. rock's instantaneous velocity:
   a. at $t = 1$ second
      $d(1) = -32(1) - 10 = -42$ ft/sec
   b. at $t = 2$ seconds
      $d(2) = -32(2) - 10 = -74$ ft/sec
   c. when it was released ($t = 0$ sec)
      $d(0) = -32(0) - 10 = -10$ ft/sec

19. $P(t) = \dfrac{1}{2\sqrt{10t}}$ instantaneous rate of change
   in this percentage after:
   a. $\dfrac{2}{5}$ day

   $P\left(\dfrac{2}{5}\right) = \dfrac{1}{2\sqrt{10\left(\dfrac{2}{5}\right)}} = \dfrac{1}{2\sqrt{4}} = \dfrac{1}{2 \cdot 2} = \dfrac{1}{4}$

   $= 25\%$ of the staff per day

   b. 1.6 days

   $P(1.6) = \dfrac{1}{2\sqrt{10(1.6)}} = \dfrac{1}{2\sqrt{16}} = \dfrac{1}{2 \cdot 4} = \dfrac{1}{8}$

   $= 12.5\%$ of the staff per day

   c. 5 days

   $P(5) = \dfrac{1}{2\sqrt{10(5)}} = \dfrac{1}{2\sqrt{50}} = \dfrac{1}{2 \cdot 5\sqrt{2}} = \dfrac{1}{10\sqrt{2}}$

   $= \dfrac{\sqrt{2}}{20} \approx 7\%$ of the staff per day

Practice Test

1. The limit of $f(x)$ as x approaches 5 is 10.

3. False, a limit can exist even if c is not in the domain.

5. a. $\lim\limits_{x \to 1^-}\left(\sqrt{x-1}+2\right)=\left(\underset{\text{✗}}{\text{dne}}\right)$

   b. $\lim\limits_{x \to 1^+}\left(\sqrt{x-1}+2\right)$

   $= \sqrt{\lim\limits_{x \to 1^+} x - \lim\limits_{x \to 1^+} 1} + \lim\limits_{x \to 1^+} 2$

   $= \sqrt{1-1}+2 = 2$

   As the domain of g is $x \ge 1$,
   the limit in b. exists and the limit in a. does not.

7. a. $\lim\limits_{x \to -1^+} f(x) = 2$

   b. $\lim\limits_{x \to 3^+} f(x) = \left(\underset{\infty}{dne}\right)$

9. a. $f(0)-g(0)$

   $=1-(-2)=3$

   b. $g(-5)\square f(-4)$

   $=4\cdot 3 = 12$

11. a. $\lim\limits_{x \to -4}[f(x)+g(x)]$

   $= \lim\limits_{x \to -4} f(x) + \lim\limits_{x \to -4} g(x)$

   $= 3+1 = 4$

   b. $\lim\limits_{x \to 1^+} f(x)g(x)$

   $= \left[\lim\limits_{x \to 1^+} f(x)\right]\left[\lim\limits_{x \to 1^+} g(x)\right]$

   $= (2)(-2) = -4$

13. a. $\lim\limits_{x \to 0} \dfrac{\cos x - 1}{x} = 0$

   $\lim\limits_{x \to 0^+} \dfrac{\cos x - 1}{x} = 0$

   $\lim\limits_{x \to 0^-} \dfrac{\cos x - 1}{x} = 0$

x	y
-0.1	0.04996
-0.01	0.005
-0.001	0.0005
-0.0001	0.00005
0	Error
0.0001	-0.00005
0.001	-0.0005
0.01	-0.005
0.1	-0.05

   b. $\lim\limits_{x \to -1} \dfrac{x^3+1}{\sqrt{(x+1)^2}} = \left(\underset{LH \ne RH}{dne}\right)$

   $\lim\limits_{x \to -1^-} \dfrac{x^3+1}{\sqrt{(x+1)^2}} = -3$

   $\lim\limits_{x \to -1^+} \dfrac{x^3+1}{\sqrt{(x+1)^2}} = 3$

x	y
-1.1	-3.31
-1.01	-3.03
-1.001	-3.003
-1.0001	-3.0003
-1	Error
$-.9999$	2.9997
$-.999$	2.997
$-.99$	2.9701
$-.9$	2.71

c. $\lim\limits_{x \to 0} \dfrac{\sin x}{2x + \tan x} = \dfrac{1}{3}$

$\lim\limits_{x \to 0^-} \dfrac{\sin x}{2x + \tan x} = \dfrac{1}{3}$

$\lim\limits_{x \to 0^+} \dfrac{\sin x}{2x + \tan x} = \dfrac{1}{3}$

x	y
−0.1	0.33241
−0.01	0.33332
−0.001	0.33333
−0.0001	0.33333
0	Error
0.0001	0.33333
0.001	0.33333
0.01	0.33332
0.1	0.33241

d. $\lim\limits_{x \to 0} \dfrac{\sin(3x)}{3x} = 1$

$\lim\limits_{x \to 0^-} \dfrac{\sin(3x)}{3x} = 1$

$\lim\limits_{x \to 0^+} \dfrac{\sin(3x)}{3x} = 1$

x	y
−0.1	0.98507
−0.01	0.99985
−0.001	0.999999
0−.0001	0.99999999
0	Error
0.0001	0.99999999
.001	0.999999
0.01	0.99985
0.1	0.98507

15. $\lim\limits_{x \to -1} \dfrac{x^3 + 1}{x + 1}$

$= \lim\limits_{x \to -1} \dfrac{(x+1)(x^2 - x + 1)}{(x + 1)}$

$= \lim\limits_{x \to -1} (x^2 - x + 1)$

$= \lim\limits_{x \to -1} x^2 - \lim\limits_{x \to -1} x + \lim\limits_{x \to -1} 1$

$= 1 - (-1) + 1 = 3$

17. $\lim\limits_{x \to 25} \dfrac{\sqrt{x} - 5}{x - 25}$

$= \lim\limits_{x \to 25} \dfrac{\sqrt{x} - 5}{x - 25} \cdot \dfrac{\sqrt{x} + 5}{\sqrt{x} + 5}$

$= \lim\limits_{x \to 25} \dfrac{(x - 25)}{(x - 25)(\sqrt{x} + 5)}$

$= \lim\limits_{x \to 25} \dfrac{1}{(\sqrt{x} + 5)}$

$= \dfrac{\lim\limits_{x \to 25} 1}{\left(\sqrt{\lim\limits_{x \to 25} x} + \lim\limits_{x \to 25} 5\right)}$

$= \dfrac{1}{(\sqrt{25} + 5)}$

$= \dfrac{1}{5 + 5} = \dfrac{1}{10}$

19. $d(t) = -16t^2 + 224t$

a. $d(t) = -16t^2 + 224t$

$\lim\limits_{h \to 0} \dfrac{\left[-16(t+h)^2 + 224(t+h)\right] - \left[-16t^2 + 224t\right]}{h}$

$= \lim\limits_{h \to 0} \dfrac{-16(t^2 + 2th + h^2) + 224t + 224h + 16t^2 - 224t}{h}$

$= \lim\limits_{h \to 0} \dfrac{-16t^2 - 32th - 16h^2 + 224t + 224h + 16t^2 - 224t}{h}$

$= \lim\limits_{h \to 0} \dfrac{-32th - 16h^2 + 224h}{h}$

$= \lim\limits_{h \to 0} \dfrac{h(-32t - 16h + 224)}{h}$

$= \lim\limits_{h \to 0} (-32t - 16h + 224)$

$= \lim\limits_{h \to 0} (-32t) - \lim\limits_{h \to 0} 16h + \lim\limits_{h \to 0} 224$

$= -32t - 16(0) + 224$

$d'(t) = -32t + 224$

b. $d'(2) = -32(2) + 224 = 160$, the debris is rising at a velocity of 160 ft/sec;

$d'(6) = -32(6) + 224 = 32$, the upward velocity of the debris has slowed to 32 ft/sec;

$d'(7) = -32(0) + 224 = 0$, the debris has reached its maximum height (velocity is 0 ft/sec);

$d'(11) = -32(11) + 224 = -128$, the velocity of the debris is now in the downward direction ($v < 0$) at 128 ft/sec.

Cumulative Review Chapters 1–11

1. $f(x) = 2x^3 - 3x^2 - 9 + 6x$
$= x^2(2x-3) + 3(2x-3)$
$= (2x-3)(x^2+3)$

$(2x-3) = 0 \qquad (x^2+3) = 0$
$2x = 3 \qquad\quad x^2 = -3$
$x = \dfrac{3}{2} \qquad\quad x = \pm\sqrt{-3}$
$\qquad\qquad\qquad x = \pm i\sqrt{3}$

3. $\begin{cases} x + 3y - 2z = 6 \\ 2x + y + z = 2 \\ -3x + 4y - 2z = 3 \end{cases}$

$\begin{cases} (x + 3y - 2z = 6) & -2R1 \to R1 \\ 2x + y + z = 2 \end{cases}$

$\begin{cases} -2x - 6y + 4z = -12 \\ \underline{2x + y + z = 2} \end{cases}$
$\qquad -5y + 5z = -10$

$\begin{cases} x + 3y - 2z = 6 \\ -3x + 4y - 2 = 3 \end{cases} \quad 3R1 \to R1$

$\begin{cases} 3x + 9y - 6z = 18 \\ \underline{-3x + 4y - 2z = 3} \end{cases}$
$\qquad 13y - 8z = 21$

$\begin{cases} -5y + 5z = -10 & 8R1 \to R1 \\ 13y - 8z = 21 & 5R2 \to R2 \end{cases}$
$-40y + 40z = -80$
$\underline{65y - 40z = 105}$
$25y \qquad\quad = 25$
$\qquad y = 1;$

$-5(1) + 5z = -10$
$-5 + 5z = -10$
$5z = -5$
$z = -1;$
$x + 3(1) - 2(-1) = 6$
$x + 3 + 2 = 6$
$x + 5 = 6$
$x = 1$
$(1, 1, -1)$

5. $\tan 78° = \dfrac{x}{134}$
$134 \tan 78° = x$
$x \approx 630 \text{ ft}$

7. $a_1 = \dfrac{1}{3}, \quad r = \dfrac{1}{9} \div \dfrac{1}{3} = \dfrac{1}{9} \cdot \dfrac{3}{1} = \dfrac{1}{3}$

$S = \dfrac{a_1}{1-r} = \dfrac{\frac{1}{3}}{1 - \frac{1}{3}} = \dfrac{\frac{1}{3}}{\frac{2}{3}} = \dfrac{1}{3} \cdot \dfrac{3}{2} = \dfrac{1}{2}$

9. $f(x) = x^3 - 2x^2 - 5x + 6$
$\dfrac{\pm 1, \pm 2, \pm 3, \pm 6}{\pm 1} = \pm 1, \pm 2, \pm 3, \pm 6$

$\begin{array}{r|rrrr} 1 & 1 & -2 & -5 & 6 \\ & & 1 & -1 & -6 \\ \hline & 1 & -1 & -6 & \underline{|0} \end{array}$

$x^2 - x - 6 = 0$
$(x-3)(x+2) = 0$
$f(x) = (x-3)(x-1)(x+2)$

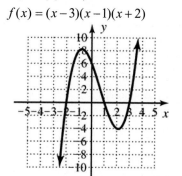

11. $z^6 = (2+2i)^6$, $\quad \theta = \dfrac{\pi}{4}$

$r = \sqrt{2^2 + 2^2} = \sqrt{8} = 2\sqrt{2}$

$z^6 = \left(2\sqrt{2}\right)^6 \left(\cos 6\left(\dfrac{\pi}{4}\right) + i \sin 6\left(\dfrac{\pi}{4}\right)\right)$

$= 512\left(\cos \dfrac{3\pi}{2} + i \sin \dfrac{3\pi}{2}\right)$

$= 512\left(0 + i(-1)\right)$

$= -512i$

13 a. $(0,19)$ $(30,9)$

$m = \dfrac{9-19}{30-0} = \dfrac{-10}{30} = \dfrac{-1}{3}$

$y - 19 = -\dfrac{1}{3}(x-0)$

$y = -\dfrac{1}{3}x + 19$

$L(x) = -\dfrac{1}{3}x + 19$

b. $L(15) = -\dfrac{1}{3}(15) + 19$

$= -\dfrac{1}{3}(15) + 19 = 14$ cm

c. $L(x) = 11$

$11 = -\dfrac{1}{3}(x) + 19$

$-8 = -\dfrac{1}{3}x$

$24 = x$

24 days

15. $f(x) = \begin{cases} (x+2)^2, & x \le 0 \\ x+2 & x > 0 \end{cases}$

$f(0)$ exists; $f(0) = (0+2)^2 = 4$

$\lim\limits_{x \to 0} f(x) = \left(\underset{LH \ne RH}{\text{dne}}\right)$

Because $\lim\limits_{x \to 0^-} f(x) \ne \lim\limits_{x \to 0^+} f(x)$

$f(x)$ is not continuous.

17. $-3e^{2x-1} = -28.08$

$\dfrac{-3e^{2x-1}}{-3} = \dfrac{-28.08}{-3}$

$e^{2x-1} = \dfrac{234}{25}$

$\ln e^{2x-1} = \ln \dfrac{234}{25}$

$(2x-1)\ln e = \ln \dfrac{234}{25}$

$2x - 1 = \ln \dfrac{234}{25}$

$2x = 1 + \ln \dfrac{234}{25}$

$x = \dfrac{1 + \ln \dfrac{234}{25}}{2}$

$x \approx 1.618$

19. $$\lim_{x \to -\infty} \frac{6x}{\sqrt{4x^2 + 5}}$$

$$= \frac{\lim_{x \to -\infty} 6x}{\lim_{x \to -\infty} \sqrt{4x^2 + 5}}$$

$$= -\frac{\lim_{x \to -\infty} \dfrac{6x}{x}}{\sqrt{\lim_{x \to -\infty} \dfrac{4x^2}{x^2} + \lim_{x \to -\infty} \dfrac{5}{x^2}}}$$

$$= -\frac{\lim_{x \to -\infty} 6}{\sqrt{\lim_{x \to -\infty} 4 + \lim_{x \to -\infty} \dfrac{5}{x^2}}}$$

$$= -\frac{6}{\sqrt{4 + 0}} = -\frac{6}{2} = -3$$

$$\lim_{x \to -\infty} f(x) = -3$$

21.

$$x^2 + x - 2 \overline{\smash{\big)}\, x^2 + x + 4} \quad \overset{\textstyle 1}{}$$

$$\underline{x^2 + x - 2}$$
$$ 6$$

$$\frac{6}{(x+2)(x-1)} = \frac{A}{x+2} + \frac{B}{x-1}$$

$$6 = A(x-1) + B(x+2)$$

$$6 = Ax - A + Bx + 2B$$

$$6 = Ax + Bx + (-A + 2B)$$

$$6 = (A+B)x + (-A + 2B)$$

$$(A+B) = 0 \qquad -A + 2B = 6$$

$$A + B = 0$$

$$-A + 2B = 6$$

$$3B = 6$$

$$B = 2$$

$$A + 2 = 0$$

$$A = -2$$

$$\frac{x^2 + x + 4}{x^2 + x - 2} = 1 + \frac{6}{(x+2)(x-1)}$$

$$1 + \frac{-2}{x+2} + \frac{2}{x-1}$$

23. $$3500 = 7000 \sin\left(4\theta - \frac{\pi}{6}\right) + 10$$

$$3490 = 7000 \sin\left(4\theta - \frac{\pi}{6}\right)$$

$$\frac{3490}{7000} = \sin\left(4\theta - \frac{\pi}{6}\right)$$

$$\frac{349}{700} = \sin\left(4\theta - \frac{\pi}{6}\right)$$

$$4\theta - \frac{\pi}{6} = \frac{\pi}{6}$$

$$4\theta = \frac{\pi}{3}$$

$$\theta = \frac{\pi}{12}$$

$$\theta = \frac{\pi}{12} + \frac{\pi}{2}k$$

25. Vertices at (−4, 1) and (6, 1);
Foci at (−3, 1) and (5, 1)

$$2a = \sqrt{(-4-6)^2 + (1-1)^2}$$

$$2a = \sqrt{(-10)^2 + 0^2} = 10$$

$$a = 5$$

$$2c = \sqrt{(-3-5)^2 + (1-1)^2}$$

$$2c = \sqrt{(-8)^2 + 0^2} = 8$$

$$c = 4$$

$$a^2 - b^2 = c^2$$

$$5^2 - b^2 = 4^2$$

$$25 - b^2 = 16$$

$$25 - 16 = b^2$$

$$b^2 = 9$$

$$b = 3$$

Center: $\left(\dfrac{-4+6}{2}, \dfrac{1+1}{2}\right)$

$$\left(\frac{2}{2}, \frac{2}{2}\right) = (1,1)$$

$$\frac{(x-1)^2}{25} + \frac{(y-1)^2}{9} = 1$$

27. a. $\lim\limits_{x \to -3^-} f(x) = 1$

   b. $\lim\limits_{x \to -3^+} f(x) = 0$

   c. $\lim\limits_{x \to 2} f(x) = -3$

   d. $\lim\limits_{x \to 5} f(x)$, 5 not in domain

29. $f(x) = -\dfrac{1}{4}x^2 + 9$ for $x \in [0,4]$

$$\square x = \frac{4-0}{n} = \frac{4}{n}$$

$$x_i = \frac{4}{n}i$$

$$\lim_{n \to \infty} \sum_{i=1}^{n}\left(-\frac{1}{4}\left(\frac{4}{n}i\right)^2 + 9\right)\left(\frac{4}{n}\right)$$

$$= \lim_{n \to \infty}\left[\frac{4}{n}\left(\sum_{i=1}^{n}\left(-\frac{1}{4}\right)\left(\frac{4}{n}\right)^2 i^2 + \sum_{i=1}^{n}9\right)\right]$$

$$= \lim_{n \to \infty}\left[\left(\frac{4}{n}\sum_{i=1}^{n}\left(-\frac{1}{4}\right)\left(\frac{4}{n}\right)^2 i^2\right) + \frac{4}{n}\sum_{i=1}^{n}9\right]$$

$$= \lim_{n \to \infty}\left[\left(\frac{4}{n}\right)\left(\frac{4}{n}\right)^2\left(-\frac{1}{4}\right)\sum_{i=1}^{n}i^2 + \left(\frac{4}{n}\right)9n\right]$$

$$= \lim_{n \to \infty}\left[\frac{4^3}{n^3}\left(-\frac{1}{4}\right)\left(\frac{2n^3 + 3n^2 + n}{6}\right) + 36\right]$$

$$= \lim_{n \to \infty}\left[\frac{64}{-24}\left(\frac{2n^3}{n^3} + \frac{3n^2}{n^3} + \frac{n}{n^3}\right) + 36\right]$$

$$= \lim_{n \to \infty}\left[\frac{64}{-24}\left(2 + \frac{3}{n} + \frac{1}{n^2}\right) + 36\right]$$

$$= \frac{-8}{3}\left(\lim_{n \to \infty}2 + \lim_{n \to \infty}\frac{3}{n} + \lim_{n \to \infty}\frac{1}{n^2}\right) + \lim_{n \to \infty}36$$

$$= \frac{-8}{3}(2 + 0 + 0) + 36 = 30.\overline{6} \text{ units}^2$$

Notes

Notes

Notes

Notes

Notes

Notes

Notes